Smart and Low Carbon Emission-Oriented Maritime Traffic Management and Controlling

Smart and Low Carbon Emission-Oriented Maritime Traffic Management and Controlling

Guest Editors

Xinqiang Chen
Salvatore Antonio Biancardo

MDPI

Basel • Beijing • Wuhan • Barcelona • Belgrade • Novi Sad • Cluj • Manchester

Guest Editors
Xinqiang Chen
Institute of Logistics Science and Engineering
Shanghai Maritime University
Shanghai
China

Salvatore Antonio Biancardo
Department of Civil, Construction and Environmental Engineering
University of Naples Federico II
Naples
Italy

Editorial Office
MDPI AG
Grosspeteranlage 5
4052 Basel, Switzerland

This is a reprint of the Special Issue, published open access by the journal *Journal of Marine Science and Engineering* (ISSN 2077-1312), freely accessible at: https://www.mdpi.com/journal/jmse/special_issues/786IZ97F0C.

For citation purposes, cite each article independently as indicated on the article page online and as indicated below:

Lastname, A.A.; Lastname, B.B. Article Title. *Journal Name* **Year**, *Volume Number*, Page Range.

ISBN 978-3-7258-5723-4 (Hbk)
ISBN 978-3-7258-5724-1 (PDF)
https://doi.org/10.3390/books978-3-7258-5724-1

Contents

Xinqiang Chen and Salvatore Antonio Biancardo
Smart and Low Carbon Emission-Oriented Maritime Traffic Management and Controlling
Reprinted from: *J. Mar. Sci. Eng.* **2025**, *13*, 1793, https://doi.org/10.3390/jmse13091793 . . . . . **1**

Chunchang Zhang and Ji Zeng
An Attention-Averaging-Based Compression Algorithm for Real-Time Transmission of Ship Data via Beidou Navigation System
Reprinted from: *J. Mar. Sci. Eng.* **2024**, *12*, 300, https://doi.org/10.3390/jmse12020300 . . . . . . **7**

Zhi Yang, Xinping Yan, Wu Ouyang, Hongfen Bai and Jinhua Xiao
Multi-Parameter Fuzzy-Based Neural Network Sensorless PMSM Iterative Learning Control Algorithm for Vibration Suppression of Ship Rim-Driven Thruster
Reprinted from: *J. Mar. Sci. Eng.* **2024**, *12*, 396, https://doi.org/10.3390/jmse12030396 . . . . . . **23**

Shuling Zhao and Sishuo Zhao
Ship Global Traveling Path Optimization via a Novel Non-Dominated Sorting Genetic Algorithm
Reprinted from: *J. Mar. Sci. Eng.* **2024**, *12*, 485, https://doi.org/10.3390/jmse12030485 . . . . . . **39**

Minghui Shao, Biao Wu, Yan Li and Xiaoli Jiang
Research on the Deployment of Professional Rescue Ships for Maritime Traffic Safety under Limited Conditions
Reprinted from: *J. Mar. Sci. Eng.* **2024**, *12*, 497, https://doi.org/10.3390/jmse12030497 . . . . . . **52**

Liang Chen and Jiayu Liu
Identification of Shipborne VHF Radio Based on Deep Learning with Feature Extraction
Reprinted from: *J. Mar. Sci. Eng.* **2024**, *12*, 810, https://doi.org/10.3390/jmse12050810 . . . . . . **67**

Xiaoqiong Bao, Zhi-Hua Hu and Yanling Huang
Routing a Fleet of Drones from a Base Station for Emission Detection of Moving Ships by Genetic Algorithm
Reprinted from: *J. Mar. Sci. Eng.* **2024**, *12*, 891, https://doi.org/10.3390/jmse12060891 . . . . . . **88**

Shunquan Huang and Yuyang Li
New Exploration of Emission Abatement Solution for Newbuilding Bulk Carriers
Reprinted from: *J. Mar. Sci. Eng.* **2024**, *12*, 973, https://doi.org/10.3390/jmse12060973 . . . . . . **109**

Yanguo Huang, Sishuo Zhao and Shuling Zhao
Ship Trajectory Planning and Optimization via Ensemble Hybrid A* and Multi-Target Point Artificial Potential Field Model
Reprinted from: *J. Mar. Sci. Eng.* **2024**, *12*, 1372, https://doi.org/10.3390/jmse12081372 . . . . . **130**

Xiaoyi Li
Ship Segmentation via Combined Attention Mechanism and Efficient Channel Attention High-Resolution Representation Network
Reprinted from: *J. Mar. Sci. Eng.* **2024**, *12*, 1411, https://doi.org/10.3390/jmse12081411 . . . . . **142**

Tao Liu, Yun Ye, Zhengling Lei, Yuchi Huo, Xiaocai Zhang, Fang Wang, et al.
A New Efficient Ship Detection Method Based on Remote Sensing Images by Device–Cloud Collaboration
Reprinted from: *J. Mar. Sci. Eng.* **2024**, *12*, 1422, https://doi.org/10.3390/jmse12081422 . . . . . **156**

Yuqing Wang, Xinqiang Chen, Yuzhen Wu, Jiansen Zhao, Octavian Postolache and Shuhao Liu
Visual Navigation Systems for Maritime Smart Ships: A Survey
Reprinted from: *J. Mar. Sci. Eng.* **2024**, *12*, 1781, https://doi.org/10.3390/jmse12101781 . . . . . **181**

Yaoming Wei, Huan Du, Qinyou Hu and Hu Wang
Optimizing Ship Draft Observation with Wave Energy Attenuation and PaddlePaddle-OCR in an Anti-Fluctuation Device
Reprinted from: *J. Mar. Sci. Eng.* **2024**, *12*, 1865, https://doi.org/10.3390/jmse12101865 . . . . . **205**

Tianrui Zhou, Jinggai Wang, Qinyou Hu and Zhihui Hu
A Novel Approach to Enhancing the Accuracy of Prediction in Ship Fuel Consumption
Reprinted from: *J. Mar. Sci. Eng.* **2024**, *12*, 1954, https://doi.org/10.3390/jmse12111954 . . . . . **225**

Fang Lu, Yubin Tian, Hongda Liu and Chuyuan Ling
Distributionally Robust Optimal Scheduling of Hybrid Ship Microgrids Considering Uncertain Wind and Wave Conditions
Reprinted from: *J. Mar. Sci. Eng.* **2024**, *12*, 2087, https://doi.org/10.3390/jmse12112087 . . . . . **245**

Xubiao Xu, Xingyu Liu, Lin Feng, Wei Yim Yap and Hongxiang Feng
Emission Estimation and Spatiotemporal Distribution of Passenger Ships Using Multi-Source Data: A Case from Zhoushan (China)
Reprinted from: *J. Mar. Sci. Eng.* **2025**, *13*, 168, https://doi.org/10.3390/jmse13010168 . . . . . . **263**

Rong Zhen, Meng Sun and Qionglin Fang
Optimization of Inbound and Outbound Vessel Scheduling in One-Way Channel Based on Reinforcement Learning
Reprinted from: *J. Mar. Sci. Eng.* **2024**, *13*, 237, https://doi.org/10.3390/jmse13020237 . . . . . . **286**

Guanhua Xu and Feilong Liu
Effects of High-Frequency Vibration on Residual Stress and Microstructure of Carbon Steel for Marine Structures: Comparative Analysis with Tempering
Reprinted from: *J. Mar. Sci. Eng.* **2025**, *13*, 408, https://doi.org/10.3390/ **306**

Yu Guo, Rui Yang, Zhiheng Zhang and Bing Han
ANFIS-Based Course Controller Using MMG Maneuvering Model
Reprinted from: *J. Mar. Sci. Eng.* **2025**, *13*, 490, https://doi.org/10.3390/jmse13030490 . . . . . . **326**

Gen Li, Haidong Zhang, Shibo Li and Chunchang Zhang
Risk Assessment of Hydrogen Fuel System Leakage in Ships Based on Noisy-OR Gate Model Bayesian Network
Reprinted from: *J. Mar. Sci. Eng.* **2025**, *13*, 523, https://doi.org/10.3390/jmse13030523 . . . . . . **350**

Haiying Yang, Feiyang Ren, Jingbo Yin, Siqi Wang and Rafi Ullah Khan
Tramp Ship Routing and Scheduling with Integrated Carbon Intensity Indicator (CII) Optimization
Reprinted from: *J. Mar. Sci. Eng.* **2025**, *13*, 752, https://doi.org/10.3390/jmse13040752 . . . . . . **369**

Akhahenda Whitney Khayenzeli, Woo-Ju Son, Dong-June Jo and Ik-Soon Cho
An AIS-Based Study to Estimate Ship Exhaust Emissions Using Spatio-Temporal Approach
Reprinted from: *J. Mar. Sci. Eng.* **2025**, *13*, 922, https://doi.org/10.3390/jmse13050922 . . . . . . **391**

Xinqiang Chen, Peishi Wu, Yajie Zhang, Xiaomeng Wang, Jiangfeng Xian and Han Zhang
Smart Maritime Transportation-Oriented Ship-Speed Prediction Modeling Using Generative Adversarial Networks and Long Short-Term Memory
Reprinted from: *J. Mar. Sci. Eng.* **2025**, *13*, 1045, https://doi.org/10.3390/jmse13061045 . . . . . **419**

Miaomiao Sun, Chengji Liang, Yu Wang and Salvatore Antonio Biancardo
Maritime Port Freight Flow Optimization with Underground Container Logistics Systems Under Demand Uncertainty
Reprinted from: *J. Mar. Sci. Eng.* **2025**, *13*, 1173, https://doi.org/10.3390/jmse13061173 . . . . . **449**

Journal of
*Marine Science
and Engineering*

MDPI

Editorial

Smart and Low Carbon Emission-Oriented Maritime Traffic Management and Controlling

Xinqiang Chen [1,*] and Salvatore Antonio Biancardo [2]

1 Institute of Logistics Science and Engineering, Shanghai Maritime University, Shanghai 201306, China
2 Department of Civil, Construction and Environmental Engineering, University of Naples Federico II, 80125 Naples, Italy; salvatoreantonio.biancardo@unina.it
* Correspondence: chenxinqiang@stu.shmtu.edu.cn

As the global shipping industry faces mounting pressure to reduce carbon emissions, AI-powered intelligent shipping systems are playing a pivotal role in optimizing vessel routes, fuel consumption, and port operations. AI-driven systems process real-time data from IoT sensors, weather forecasts, and Automatic Identification System (AIS) signals to recommend optimal fuel-efficient sailing routes, while circumventing congested waterways and adverse weather conditions. Machine learning algorithms bolster predictive maintenance capabilities, mitigating excessive emissions caused by engine inefficiency. Furthermore, AI-guided autonomous and remotely controlled vessels enable precise maneuvering, curtailing port idle time and ensuring just-in-time arrivals, which significantly reduces unnecessary fuel consumption [1]. Ports equipped with intelligent traffic management systems can dynamically allocate berths and optimize cargo handling processes, yielding further energy savings. AI-driven predictive maintenance helps to maintain engine efficiency, thereby further reducing carbon emissions [2]. The rise in autonomous and remotely controlled vessels also holds promise for more precise control, enabling vessels to achieve just-in-time arrivals, reduce idle time at ports, and minimize fuel waste. By leveraging AI-driven decision support systems, the shipping industry can transition to a low-carbon, high-efficiency future aligned with global sustainability goals, ensuring that shipping management systems become smarter and greener.

AI-driven intelligent shipping systems play a crucial role in optimizing vessel routes, fuel consumption, and port operations. Machine learning algorithms not only enhance predictive maintenance capabilities, reducing the excessive emissions caused by engine inefficiencies, but can also integrate circular economy principles to optimize resource flows through digital monitoring and reverse logistics [3]. The convergence of blockchain and AI technologies further strengthens the traceability and decision-making efficiency of maritime supply chains, ensuring that route optimization and emission monitoring comply with sustainability standards [4]. For example, dynamic path planning algorithms can efficiently generate low-carbon optimal routes by controlling the search space scale and restricting search directions, complementing the precise control of AI-guided autonomous and remotely operated vessels to reduce port idle time and ensure punctual arrivals, thereby minimizing fuel waste [5].

Additionally, ports equipped with intelligent traffic management systems can adopt centralized logistics and spatial planning strategies to dynamically allocate berths and optimize cargo handling processes, achieving more efficient energy utilization [6,7]. By integrating AI-driven decision support systems with predictive maintenance technologies, the shipping industry can transition toward a low-carbon, high-efficiency future aligned with global sustainable development goals, ensuring smarter and greener shipping management systems. The comprehensive application of this interdisciplinary approach not only improves maritime operational efficiency but also provides practical solutions for achieving environmentally friendly maritime traffic management.

Recent advances in maritime technology demonstrate significant progress across five interconnected domains: (1) vessel systems optimization, (2) intelligent navigation, (3) emission control, (4) port operations, and (5) safety management. The research trajectory begins with Zhang et al. (Contribution 1) addressing bandwidth limitations in BeiDou satellite communications through an innovative attention-averaging compression algorithm. This technical solution employs a three-stage processing pipeline (quantization → attention-averaging-based predictive coding → arithmetic coding) to achieve the efficient real-time transmission of ship status data while maintaining data integrity, providing a cost-effective monitoring solution for ocean-going vessels. Yang et al. (Contribution 2) enhanced propulsion system stability through neural network control, proposing a second-order sliding-mode observer integrating the Super-Twisting Algorithm (STA) to address speed estimation errors and torque vibration in permanent magnet synchronous motors (PMSMs). By dynamically adjusting control parameters via fuzzy neural networks and implementing online compensation through Iterative Learning Control (ILC), they significantly improved operational stability and the energy efficiency of propulsion systems. These foundational achievements enabled Zhao et al. (Contribution 3) to develop a multi-objective route optimization model. Their enhanced Non-dominated Sorting Genetic Algorithm II (NSGA-II) incorporates ship stability constraints and complex sea condition interference factors, establishing a dual-objective optimization model (energy consumption vs. time) that achieves an optimal balance between safety and economy in transoceanic route planning. Complementing these developments, Shao et al. (Contribution 4) and Chen et al. (Contribution 5) strengthened maritime safety through rescue vessel deployment optimization and VHF radio identification technology, respectively. Shao's team designed a multi-level coverage strategy to significantly improve emergency response efficiency, while Chen's group developed a hybrid deep learning model (CNN-BiLSTM) that achieves ship radio identification in non-cooperative conditions through CAM feature extraction and bidirectional temporal feature fusion, pioneering new approaches for maritime traffic awareness.

Building upon these achievements, Bao et al. (Contribution 6) developed an emission reduction decision prediction model, pioneering the application of the XGBoost algorithm to analyze shipowners' preferences between Very Low-Sulfur Fuel Oil (VLSFO) and scrubbers. Their predictive model incorporates key features such as deadweight tonnage and annual sailing distance, establishing a data-driven reference framework for post-pandemic emission reduction decisions. This work has technical complementarity with Huang et al.'s (Contribution 7) innovative drone monitoring system, which employs a genetic algorithm based on a Sequential Insertion Heuristic (SIH). The proposed DroneByDrone strategy significantly enhances emission detection efficiency through optimized UAV speed parameters, creating a new paradigm for mobile maritime surveillance. Navigation capabilities were strengthened through Huang et al.'s (Contribution 8) collision avoidance algorithm and Li et al.'s (Contribution 9) high-precision ship segmentation technology. By incorporating the COLREGs rules constraints and Collision Risk Index (CRI) evaluation, their approach generates kinematically feasible safe paths for both static obstacles and dynamic encounter scenarios. The CA^2HRNET model innovatively integrates channel attention, multi-scale spatial attention, and weight self-adjustment mechanisms, maintaining a precise segmentation capability for small target vessels even under strong noise interference through cross-scale attention fusion in feature pyramid networks. Liu et al. (Contribution 10) improved detection efficiency through cloud-edge collaboration, migrating computationally intensive tasks like semantic segmentation to the cloud while retaining only mask-based detection at the front end. Their optimization of coordinate attention modules and confluence algorithms significantly boosts detection speed without compromising accuracy. Wang et al. (Contribution 11) systematically mapped the technological landscape, conducting an in-depth analysis of key technical bottlenecks including SLAM and 3D reconstruction. They first proposed the "perception-decision-control" technical roadmap for visual navigation systems. Wei et al. (Contribution 12) achieved automation of critical draft measurement through innovative dynamically adjusted measurement tube positioning, addressing the pain points of traditional visual observation being susceptible to human factors.

Energy optimization emerged as another crucial pillar. Zhou et al. (Contribution 13) realized accurate fuel prediction by pioneering an interval prediction framework combining the Gaussian Process and Quantile Regression. By capturing fuel consumption volatility characteristics, their model generates probability-based prediction intervals, providing uncertainty quantification support for ship energy efficiency management and significantly improving bunkering planning reliability. Lu et al. (Contribution 14) ensured the robust management of hybrid power systems through a Distributionally Robust Optimization (DRO)-based bi-level scheduling model addressing dual uncertainties of wind–wave and PV conditions. By employing linear decision rules and column-and-constraint generation algorithms, their approach achieves the coordinated optimization of operational costs and carbon emissions while guaranteeing on-time arrival. Xu et al. (Contribution 15) provided critical emission insights through a bottom-up emission inventory model based on AIS trajectories and meteorological data. Zhen et al. (Contribution 16) optimized port traffic flow through intelligent scheduling, solving the NP-hard problem of one-way channel scheduling by designing state–action space encoding rules. Their real-time learning of vessel arrival dynamics and channel occupancy status enables the intelligent optimization of inbound/outbound sequences, effectively alleviating port congestion with improved computational efficiency compared to traditional algorithms. Xu et al.'s (Contribution 17) breakthroughs in materials science further enhanced vessel reliability through comparative studies of vibration stress relief and tempering methods for marine-grade steel components.

The refinement of the autonomous navigation ecosystem owes much to Guo et al.'s (Contribution 18) adaptive control system, which designed an ANFIS (Adaptive Neuro-Fuzzy Inference System) fuzzy neural network controller based on the MMG (Maneuvering Model Group) maneuvering model. By employing a backpropagation–least squares hybrid training strategy, the system achieved reduced response times under normal sea conditions and maintained rudder effectiveness stability in harsh sea conditions, thereby addressing the challenge of rigid parameter settings in traditional PID (Proportional–Integral–Derivative) control. Li et al.'s (Contribution 19) hydrogen safety framework integrated the Bow-tie model with fuzzy set theory to construct a Bayesian network incorporating Noisy–OR gates. Yang et al. (Contribution 20) incorporated carbon intensity indicators (CIIs) into route decision-making, establishing a profit–CII dual-objective model. They utilized the Gale–Shapley algorithm for the stable matching of cargo pallets and vessels, combined with a genetic algorithm for route optimization, achieving increased profits while meeting IMO (International Maritime Organization) rating requirements. The comprehensive emission analysis by Khayenzeli et al. (Contribution 21) and the advanced speed prediction model by Chen et al. (Contribution 22) provided support for these efforts. They pioneered a GAN-LSTM (Generative Adversarial Network–Long Short-Term Memory) hybrid architecture, leveraging Generative Adversarial Networks to reduce error accumulation in long-term LSTM predictions. Through spatiotemporal feature adversarial training, they offered forward-looking speed estimates for intelligent collision avoidance systems. Sun et al. (Contribution 23) integrated genetic algorithm-simulated annealing with deep Q-learning, driving innovation in port logistics through a hybrid intelligent algorithm-powered underground container system. This completed the construction of a comprehensive technological value chain from ship to port.

In the era of carbon neutrality and digital transformation, Smart and Low-Carbon-Emission-Oriented Maritime Traffic Management and Controlling has emerged as a pivotal direction for the sustainable development of the shipping industry. Leveraging the advancements in artificial intelligence (AI), big data analytics, and intelligent sensing, future maritime traffic systems aim to reduce carbon emissions while enhancing the safety, efficiency, and resilience of maritime operations. One of the core challenges lies in integrating green energy strategies with intelligent traffic control mechanisms, such as adaptive ship routing, energy-efficient voyage planning, and dynamic speed optimization. These solutions rely heavily on real-time data streams from AISs (Automatic Identification Systems), remote sensing, and onboard IoT devices, which enable intelligent decision-making through data-driven predictive models. Moreover, computer vision techniques applied to maritime surveillance videos allow for the real-time recognition of vessel behaviors, collision risk prediction, and automated traffic anomaly detection, thus supporting smart navigation in congested or low-visibility conditions. Outlier suppression in AIS data, multi-source data fusion, and the development of digital twins of ship traffic systems further contribute to the realization of intelligent emission-reducing strategies. As AI continues to evolve, its integration into maritime traffic situation awareness, autonomous ship control, and environmentally adaptive traffic scheduling is expected to play a transformative role in achieving low-carbon maritime logistics.

Funding: This work was jointly supported by the National Natural Science Foundation of China (52472347, 52331012).

Conflicts of Interest: The authors declare no conflicts of interest.

List of Contributions:

1. Zhang, C.; Zeng, J. An Attention-Averaging-Based Compression Algorithm for Real-Time Transmission of Ship Data via Beidou Navigation System. *J. Mar. Sci. Eng.* **2024**, *12*, 300. https://doi.org/10.3390/jmse12020300
2. Yang, Z.; Yan, X.; Ouyang, W.; Bai, H.; Xiao, J. Multi-Parameter Fuzzy-Based Neural Network Sensorless PMSM Iterative Learning Control Algorithm for Vibration Suppression of Ship Rim-Driven Thruster. *J. Mar. Sci. Eng.* **2024**, *12*, 396. https://doi.org/10.3390/jmse12030396
3. Zhao, S.; Zhao, S. Ship Global Traveling Path Optimization via a Novel Non-Dominated Sorting Genetic Algorithm. *J. Mar. Sci. Eng.* **2024**, *12*, 485. https://doi.org/10.3390/jmse12030485
4. Shao, M.; Wu, B.; Li, Y.; Jiang, X. Research on the Deployment of Professional Rescue Ships for Maritime Traffic Safety under Limited Conditions. *J. Mar. Sci. Eng.* **2024**, *12*, 497. https://doi.org/10.3390/jmse12030497
5. Chen, L.; Liu, J. Identification of Shipborne VHF Radio Based on Deep Learning with Feature Extraction. *J. Mar. Sci. Eng.* **2024**, *12*, 810. https://doi.org/10.3390/jmse12050810
6. Bao, X.; Hu, Z.-H.; Huang, Y. Routing a Fleet of Drones from a Base Station for Emission Detection of Moving Ships by Genetic Algorithm. *J. Mar. Sci. Eng.* **2024**, *12*, 891. https://doi.org/10.3390/jmse12060891
7. Huang, S.; Li, Y. New Exploration of Emission Abatement Solution for Newbuilding Bulk Carriers. *J. Mar. Sci. Eng.* **2024**, *12*, 973. https://doi.org/10.3390/jmse12060973
8. Huang, Y.; Zhao, S.; Zhao, S. Ship Trajectory Planning and Optimization via Ensemble Hybrid A* and Multi-Target Point Artificial Potential Field Model. *J. Mar. Sci. Eng.* **2024**, *12*, 1372. https://doi.org/10.3390/jmse12081372
9. Li, X. Ship Segmentation via Combined Attention Mechanism and Efficient Channel Attention High-Resolution Representation Network. *J. Mar. Sci. Eng.* **2024**, *12*, 1411. https://doi.org/10.3390/jmse12081411

10. Liu, T.; Ye, Y.; Lei, Z.; Huo, Y.; Zhang, X.; Wang, F.; Sha, M.; Wu, H. A New Efficient Ship Detection Method Based on Remote Sensing Images by Device–Cloud Collaboration. *J. Mar. Sci. Eng.* **2024**, *12*, 1422. https://doi.org/10.3390/jmse12081422
11. Wang, Y.; Chen, X.; Wu, Y.; Zhao, J.; Postolache, O.; Liu, S. Visual Navigation Systems for Maritime Smart Ships: A Survey. *J. Mar. Sci. Eng.* **2024**, *12*, 1781. https://doi.org/10.3390/jmse12101781
12. Wei, Y.; Du, H.; Hu, Q.; Wang, H. Optimizing Ship Draft Observation with Wave Energy Attenuation and PaddlePaddle-OCR in an Anti-Fluctuation Device. *J. Mar. Sci. Eng.* **2024**, *12*, 1865. https://doi.org/10.3390/jmse12101865
13. Zhou, T.; Wang, J.; Hu, Q.; Hu, Z. A Novel Approach to Enhancing the Accuracy of Prediction in Ship Fuel Consumption. *J. Mar. Sci. Eng.* **2024**, *12*, 1954. https://doi.org/10.3390/jmse12111954
14. Lu, F.; Tian, Y.; Liu, H.; Ling, C. Distributionally Robust Optimal Scheduling of Hybrid Ship Microgrids Considering Uncertain Wind and Wave Conditions. *J. Mar. Sci. Eng.* **2024**, *12*, 2087. https://doi.org/10.3390/jmse12112087
15. Xu, X.; Liu, X.; Feng, L.; Yap, W.Y.; Feng, H. Emission Estimation and Spatiotemporal Distribution of Passenger Ships Using Multi-Source Data: A Case from Zhoushan (China). *J. Mar. Sci. Eng.* **2025**, *13*, 168. https://doi.org/10.3390/jmse13010168
16. Zhen, R.; Sun, M.; Fang, Q. Optimization of Inbound and Outbound Vessel Scheduling in One-Way Channel Based on Reinforcement Learning. *J. Mar. Sci. Eng.* **2025**, *13*, 237. https://doi.org/10.3390/jmse13020237
17. Xu, G.; Liu, F. Effects of High-Frequency Vibration on Residual Stress and Microstructure of Carbon Steel for Marine Structures: Comparative Analysis with Tempering. *J. Mar. Sci. Eng.* **2025**, *13*, 408. https://doi.org/10.3390/jmse13030408
18. Guo, Y.; Yang, R.; Zhang, Z.; Han, B. ANFIS-Based Course Controller Using MMG Maneuvering Model. *J. Mar. Sci. Eng.* **2025**, *13*, 490. https://doi.org/10.3390/jmse13030490
19. Li, G.; Zhang, H.; Li, S.; Zhang, C. Risk Assessment of Hydrogen Fuel System Leakage in Ships Based on Noisy-OR Gate Model Bayesian Network. *J. Mar. Sci. Eng.* **2025**, *13*, 523. https://doi.org/10.3390/jmse13030523
20. Yang, H.; Ren, F.; Yin, J.; Wang, S.; Khan, R.U. Tramp Ship Routing and Scheduling with Integrated Carbon Intensity Indicator (CII) Optimization. *J. Mar. Sci. Eng.* **2025**, *13*, 752. https://doi.org/10.3390/jmse13040752
21. Khayenzeli, A.W.; Son, W.-J.; Jo, D.-J.; Cho, I.-S. An AIS-Based Study to Estimate Ship Exhaust Emissions Using Spatio-Temporal Approach. *J. Mar. Sci. Eng.* **2025**, *13*, 922. https://doi.org/10.3390/jmse13050922
22. Chen, X.; Wu, P.; Zhang, Y.; Wang, X.; Xian, J.; Zhang, H. Smart Maritime Transportation-Oriented Ship-Speed Prediction Modeling Using Generative Adversarial Networks and Long Short-Term Memory. *J. Mar. Sci. Eng.* **2025**, *13*, 1045. https://doi.org/10.3390/jmse13061045
23. Sun, M.; Liang, C.; Wang, Y.; Biancardo, S.A. Maritime Port Freight Flow Optimization with Underground Container Logistics Systems Under Demand Uncertainty. *J. Mar. Sci. Eng.* **2025**, *13*, 1173. https://doi.org/10.3390/jmse13061173

References

1. Wu, B.; Wang, Y.; Zhang, J.; Savan, E.E.; Yan, X. Effectiveness of maritime safety control in different navigation zones using a spatial sequential DEA model: Yangtze River case. *Accid. Anal. Prev.* **2015**, *81*, 232–242. [CrossRef] [PubMed]
2. Chen, X.; Ma, F.; Wu, Y.; Han, B.; Luo, L.; Biancardo, S.A. MFMDepth: MetaFormer-based monocular metric depth estimation for distance measurement in ports. *Comput. Ind. Eng.* **2025**, *207*, 111325. [CrossRef]
3. Sassanelli, C.; Terzi, S. Circular Economy and Sustainable Business Performance Management. *Sustainability* **2023**, *15*, 8619. [CrossRef]
4. Skalkos, D. Prospects, Challenges and Sustainability of the Agri-Food Supply Chain in the New Global Economy II. *Sustainability* **2023**, *15*, 12558. [CrossRef]
5. Peng, T.; Yang, X.; Xu, Z.; Liang, Y. Constructing an Environmental Friendly Low-Carbon-Emission Intelligent Transportation System Based on Big Data and Machine Learning Methods. *Sustainability* **2020**, *12*, 8118. [CrossRef]

6. Guimarães, P. Cities and Retail: Sustainable Transformation of Retail in Urban Environments. *Sustainability* **2023**, *15*, 12743. [CrossRef]
7. Xiao, G.; Wang, Y.; Wu, R.; Li, J.; Cai, Z. Sustainable maritime transport: A review of intelligent shipping technology and green port construction applications. *J. Mar. Sci. Eng.* **2024**, *12*, 1728. [CrossRef]

Journal of
Marine Science and Engineering

MDPI

Article

An Attention-Averaging-Based Compression Algorithm for Real-Time Transmission of Ship Data via Beidou Navigation System

Chunchang Zhang [1] and Ji Zeng [2,*]

1 Merchant Marine College, Shanghai Maritime University, Shanghai 201306, China; cczhang@shmtu.edu.cn
2 Zhiyuan College, Shanghai Jiao Tong University, Shanghai 200240, China
* Correspondence: zeng_ji@sjtu.edu.cn

Abstract: The real-time transmission of ship status data from vessels to shore is crucial for live status monitoring and guidance. Traditional reliance on expensive maritime satellite systems for this purpose is being reconsidered with the emergence of the global short message communication service offered by the BeiDou-3 navigation satellite system. While this system presents a more cost-effective solution, its bandwidth is notably insufficient for handling real-time ship status data. This inadequacy necessitates the compression of such data. Therefore, this paper introduces an algorithm tailored for real-time compression of sequential ship status data. The algorithm is engineered to ensure both accuracy and the preservation of valid data range integrity. Our methodology integrates quantization, predictive coding employing an attention-averaging-based predictor, and arithmetic coding. This combined approach facilitates the transmission of succinct messages through the BeiDou Navigation System, enabling the live monitoring of ocean-going vessels. Experimental trials conducted with authentic data obtained from ship monitoring systems validate the efficiency of our approach. The achieved compression rates closely approximate theoretical minimum values. Consequently, this method exhibits substantial promise for the real-time transmission of parameters across various systems.

Keywords: ship data compression; predictive coding; attention averaging; Convolutional Neural Networks (CNNs)

1. Introduction

The advent of concepts such as smart ships and unmanned vessels has highlighted the growing need for remote ship status monitoring. Ships generate a multitude of operational parameters, such as speed, engine torque, and fuel consumption. By analyzing these data, insights into the ship's operational status and its energy-saving impact can be derived. The real-time transmission of these data allows onshore management personnel to analyze the system's status based on these parameters and issue real-time instructions to enhance energy efficiency and safety.

With the help of real-time data, common risk factors, such as equipment failure, human operational errors, and navigation accidents, can be promptly identified and eliminated by the shore-based data analysis system [1,2]. The shore-based system can issue more efficient dispatch instructions, such as Virtual Arrival, to save carbon emissions by utilizing the ship's position and speed data [3].

Traditional monitoring systems typically rely on costly maritime satellite systems to transmit data. To address the cost issue, the global short message communication service offered by the BeiDou-3 navigation satellite system can be leveraged for the nearly cost-free transmission of ship monitoring data [4,5]. However, its bandwidth falls significantly short of the requirements for this scenario. Specifically, within the global short message communication service of the BeiDou-3 navigation satellite system, each message cannot exceed 560 bits in length, and a single civil communication card can only send one message per minute [4]. The success rate decreases as the message length increases, sometimes dropping as low as 50% [6–8].

In our dataset, for instance, there are 41 fields to transmit, which would require 1312 bits when encoded in float32. The monitoring system records a data point every 10 s, with no tolerance for data loss. Although existing methods, such as utilizing multiple communication cards to increase bandwidth [9] and employing channel coding to eliminate data loss [8,9], can be adopted, the capacity of Beidou falls significantly short of the number of data to be transmitted. Therefore, data compression is necessary to minimize its length. Furthermore, both the compression algorithm's allowable calculation time and the computing power of both the sender on the ship and the receiver in the database are relatively abundant. Hence, optimizing the compression rate becomes the primary objective in the compression problem for the real-time transmission of ship data via Beidou Navigation System.

In this paper, we introduce our QAAPA method, a novel method combining **Q**uantization, **A**ttention-**A**veraging-based **P**redictive coding, and **A**rithmetic coding and facilitating the real-time transmission of ship data via the Beidou Navigation System. Our contributions can be summarized as follows:

- To address the challenge of economically transmitting real-time ship status data via the Beidou system, we have developed the QAAPA Method. This method comprises a series of tailored steps designed to compress data in real-time while maintaining fixed precision.
- Unlike traditional compression methods that transmit entire datasets, our approach transmits only the residuals, resulting in a significant reduction in data size. This transmission of residuals is made possible through a critical step in our method: attention-averaging-based predictive coding. Here, we compute the difference between actual and predicted data, reconstructing it at the receiver's end. Our predictor is built on a foundation of linear regression models and Convolutional Neural Networks (CNNs), incorporating appropriate weights. This framework ensures accurate real-time predictions by leveraging historical data, enabling precise reconstruction of transmitted residuals.
- Experimental Validation: We conducted experiments using actual ship monitoring system data. Results showcase that QAAPA achieves compression rates approaching theoretical minimums estimated by commercial algorithms, surpassing traditional approaches for similar problems.

The remainder of this paper is structured as follows: Section 2 discusses existing compression solutions, highlighting the absence of suitable methods for real-time ship status data transmission via Beidou. Section 3 details the QAAPA method tailored for Beidou-based ship data transmission. Section 4 presents experiments validating QAAPA's performance. Finally, Section 5 concludes the paper and outlines future work.

2. Related Work

The BeiDou-3 navigation satellite system's global short message communication service is integral to numerous monitoring systems operating where ground-based communication systems are unavailable [10–13]. However, prevalent encoding formats within these systems often lack significant compression effects or prove unsuitable for our specific challenge. Traditional methods usually use related techniques of signal processing [11,13]. In recent years, many studies have used machine learning for data compression [14]. Compression algorithms combined with machine learning are mainly used for images, sound and video, and text [15–18], and machine learning is rarely used in the compression of instrument status data. Common compression algorithms applied to numerical data, such as vector quantization [19] and transformation coding [20], fail to ensure our fixed precision goals while maintaining high compression rates.

Predictive coding presents the potential to achieve both objectives. Crucial to ensuring fixed precision is the encoding of residuals in predictive coding. The approach proposed by Zhidan Yan et al. [21] involves reducing the value range of residuals, while another approach proposed by Fout Nathaniel and Ma Kwan-Liu [22] adopts a variable precision floating-point format for encoding residuals, reducing precision for larger absolute values. However, both of them neglected meeting fixed precision goals, while neither approach guarantees fixed precision, especially considering sharp changes in ship states during engine start/stop cycles, rendering these methods unsuitable.

The utilization of machine learning-based prediction in predictive coding holds promise for enhancing the precision of predicted values. Past research in the compression of text or genome data has demonstrated successful applications of predictive coding. For example, Goyal Mohit et al. [23] and Cox David [18] approached data as symbol sequences, employing prediction methodologies across the spectrum of potential data values. However, our scenario involving ship status data predominantly comprises floating-point data spanning an extensive range of potential values. This broad spectrum renders predicting every possible value's probability nearly unattainable, a challenge not present in the more confined scope of letters or genomes. Consequently, the application of their compression methodologies to our scenario is unfeasible. Chen Yong [13] proposed a data compression method specifically tailored for enabling communication through Beidou's short message system. This method combines the Spinning Door Transformation algorithm for initial lossy compression and the Pre-special byte method for subsequent lossless compression. Nevertheless, this compression methodology is optimized for scalar sequences and cannot be readily adapted and optimized for our specific ship-status data structure, which consists of vector sequential data.

The state-of-the-art data compression techniques available are primarily designed for general or specific scenarios that diverge significantly from the unique nature of our situation regarding real-time transmission of ship status data, particularly vector sequential data. Consequently, none of the existing methodologies can be directly applied to our context. This predicament underscores the necessity of devising a bespoke data compression methodology tailored explicitly to our specific scenario. This necessity has served as the driving force behind our research endeavors, which will be comprehensively expounded upon in the subsequent sections.

3. Methodology

3.1. Preliminaries

3.1.1. Formal Definition

The compression problem for the real-time transmission of ship data via the BeiDou Navigation System can be described as follows:

Periodically, the sender generates a set of ship status data, denoted as $\mathbf{x}_t \in \mathbb{R}^n$, representing variables such as the ship's speed, engine torque, and fuel consumption, at the current time t. Both the sender and receiver maintain a record of all historical data sequences, $\mathbf{x}_{<t}$, which have been previously transmitted. These historical sequences are integral to the compression algorithm.

The sender employs an encoding function, denoted as E, to encode $\mathbf{x}_t$ into a binary string $c = E(\mathbf{x}_t|\mathbf{x}_{<t})$, which is then transmitted via the BeiDou short message communication system. Upon receiving this message c, the receiver performs decoding, resulting in an approximation of the original message, $\mathbf{x}'_t = D(c|\mathbf{x}_{<t})$.

Additionally, predetermined upper and lower limits exist, denoted as $\mathbf{L}_u, \mathbf{L}_d \in \mathbb{R}^n$, as well as transmission decimal places, denoted as $\mathbf{a} \in \mathbb{Z}^n$. These limits and decimal places are determined by the monitoring system's administrator according to their specific requirements. Our compression algorithm must encompass all potential values of $\mathbf{x}_t$, ensuring that the decoded approximation, $\mathbf{x}'_t$, meets the required level of accuracy. Moreover, the algorithm must strive to minimize the average coding length of the message c.

Formally, the compression algorithm must guarantee that:

$$\forall \mathbf{x}_t \in \mathbb{R}^n, \mathbf{x}_{t,i} \in [\mathbf{L}_{d,i}, \mathbf{L}_{u,i}], c = E(\mathbf{x}_t|\mathbf{x}_{<t}), \mathbf{x}'_t = D(c|\mathbf{x}_{<t}), |\mathbf{x}'_{t,i} - \mathbf{x}_{t,i}| \leq 10^{-\mathbf{a}_i}, \quad 0 \leq i < n \tag{1}$$

and minimize the average coding length of the message c. When applying this problem in practical scenarios, we make the following key assumptions:

- The interpretation of each dimension within the data vector, denoted as $\mathbf{x}$, remains constant, and each dimension represents a parameter associated with the ship's internal state.
- Both the sender and receiver utilize computers, as opposed to circuit boards, to process the data. They possess sufficient computing power and time (several seconds) to execute intricate compression algorithms.
- With the help of appropriate channel coding [8,9], the communication process is reliable.

3.1.2. Quantization

Quantization is a fundamental process involving the approximation of continuous signal values or a vast array of discrete values into a finite and often reduced set of discrete values [24]. It is used to ensure the fixed accuracy goal required in the problem while eliminating the unnecessary precision of the data.

3.1.3. Predictive Coding

Predictive coding, utilized in data compression, minimizes the data needed to represent information by transmitting or storing only the information that cannot be accurately predicted. Initially, it identifies repetition and similarity within the data, generates predicted values using a model, and compares them with actual values to derive residuals. These residuals, ideally smaller in size than the original data, facilitate more efficient transmission or storage [25]. The accuracy of the chosen predictive model significantly influences the effectiveness of predictive coding. It particularly excels in scenarios with high correlation or predictability between successive data points.

3.1.4. Arithmetic Coding

Arithmetic coding is a technique used in data compression to encode a sequence of symbols or characters with variable-length codes. Unlike more common methods like Huffman coding, which assigns fixed-length codes to individual symbols, arithmetic coding encodes entire sequences of symbols into a single code. Its typical process involves the following steps:

- Symbol Probability Modeling: The process starts with a probability model that estimates the probability of each symbol in the input sequence.
- Defining the Range: Represent the entire range of possible values with a fractional range between 0 and 1. This range is divided proportionally to the probabilities of the symbols.
- Mapping Symbols to Subranges: Each symbol in the input sequence is then mapped to a subrange within the total range based on their probabilities.
- Iterative Encoding: As each symbol is encoded, the subrange corresponding to that symbol is selected and used to narrow down the overall range further.
- Outputting the Code: The final output is a binary fraction within the narrowed range, which represents the compressed form of the entire input sequence.

Arithmetic coding is advantageous in that it can achieve higher compression ratios compared to methods like Huffman coding because it utilizes the probabilities of entire sequences rather than individual symbols. However, it requires more computational resources and may be more complex to implement efficiently.

3.1.5. Linear Regression and Convolutional Neural Networks (CNNs)

In our ship status data compression process, linear regression and CNN models are used. Linear regression is a fundamental statistical method used to model the relationship between two or more variables. Its simplicity and interpretability make it a widely applied technique in various fields, including economics, social sciences, health sciences, and machine learning.

At its core, linear regression aims to establish a linear relationship between a dependent variable (the outcome of interest) and one or more independent variables (predictors or explanatory variables). The model assumes that the relationship between these variables can be approximated by a straight line.

Mathematically, the linear regression model can be represented as:

$$\hat{y} = \theta_0 + \theta_1 x_1 + \theta_2 x_2 + \ldots + \theta_n x_n \tag{2}$$

The predictive values $\hat{y}$ are calculated as the weighted sum of input features, with n representing the number of features, x_n representing the ith feature value, and θ_j representing the jth model parameter (particularly, θ_j is the intercept, indicating the value of y when all independent variables are zero) [26]. However, it should be noted that these linear regression models assume a linear relationship by default [27].

In recent years, it has become popular to use neural networks to simulate the learning process of the human brain to construct models and provide automated solutions within the field of Artificial Intelligence (AI). Neural networks consist of interconnected neurons with numerous parameters, and they exhibit layers connecting input and output [28]. Convolutional Neural Networks (CNNs) [29] represent a class of deep neural networks specifically designed to process visual data. They have revolutionized various fields, ranging from computer vision to natural language processing, due to their capability to automatically extract intricate features from input data. Originally inspired by the visual processing mechanism of the human brain, CNNs have demonstrated remarkable success in tasks such as image classification, object detection, and segmentation. The essence of CNNs lies in their architecture, which comprises convolutional layers, pooling layers, and fully connected layers. Convolutional layers apply learnable filters to input images, extracting features hierarchically by convolving across the image. Pooling layers subsequently downsample these features, reducing computational complexity while retaining essential information. Fully connected layers, situated at the end of the network, utilize these learned features for classification or regression tasks. If the two-dimensional convolution in its convolutional layer is changed to a one-dimensional convolution, it can also be used for sequence data. In our ship data compression scenario, we applied CNNs in the predictive coding step and achieved relatively higher prediction accuracy than other deep neural networks, such as RNNs (Recurrent Neural Networks).

3.2. Overview of Proposed Method

The real-time transmission of ship status data comprises three primary stages (see Figure 1). Initially, at the ship's end, the ship status data undergoes encoding to reduce the message length for transmission. This encoding process encompasses quantization, predictive coding, and arithmetic coding steps. Subsequently, the compressed message is transmitted via the Beidou system to the intended receiver. Finally, upon reception, the receiver decodes (recovers) the message, initiating a reverse arithmetic decoding step followed by another predictive coding step. It is noteworthy that all blocks, except for quantization, maintain a lossless nature.

For a formal definition of the input, steps, and output involved in the encoding and decoding processes, refer to Algorithms 1 and 2 for detailed descriptions. Subsequent sections will elaborate on each step sequentially, providing in-depth insights into the methodologies.

Algorithm 1 Encoding algorithm

Input:
$\mathbf{x}_t \in \mathbb{R}^n$: current data to be compressed
$\{\mathbf{y}_\tau\}_{\tau=t-m}^{t-1} \in \mathbb{Z}^{n \cdot m}$: synced recent historical data
$P : \mathbb{Z}^{n \cdot m} \rightarrow \mathbb{Z}^n$ predictor
$Q : \mathbb{R}^n \rightarrow \mathbb{Z}^n$ quantizer
$A_e : \mathbb{Z}^n \rightarrow$ binary string arithmetic encoder with pre-calculated frequency table
Output: c: encoded message
$\mathbf{y}_t \leftarrow Q(\mathbf{x}_t)$
$\mathbf{y}'_t \leftarrow P(\{\mathbf{y}_\tau\}_{\tau=t-m}^{t-1})$
$\mathbf{e}_{t,i} \leftarrow (\mathbf{y}_t - \mathbf{y}'_t)_i \bmod L_i$
$c \leftarrow A_e(\mathbf{e}_t)$

Algorithm 2 Decoding algorithm

Input:
c: encoded message
$\{\mathbf{y}_\tau\}_{\tau=t-m}^{t-1} \in \mathbb{Z}^{n \cdot m}$: synced recent historical data
$P : \mathbb{Z}^{n \cdot m} \rightarrow \mathbb{Z}^n$ pre-trained predictor
A_d : binary string $\rightarrow \mathbb{Z}^n$ arithmetic decoder with pre-calculated frequency table
Output: $\mathbf{x}'_t \in \mathbb{R}^n$ approximation of the original data
$\mathbf{e}_t \leftarrow A_d(c)$
$\mathbf{y}'_t \leftarrow P(\{\mathbf{y}_\tau\}_{\tau=t-m}^{t-1})$
$\mathbf{y}_{t,i} \leftarrow (\mathbf{y}'_t + \mathbf{e}_t)_i \bmod L_i$
$\mathbf{x}'_t \leftarrow \mathbf{y}_t$

3.3. Quantization

Quantization, as introduced in Section 3.1.2, serves as a pivotal process to approximate continuous signal values or extensive discrete datasets into a finite and often condensed set of discrete values. In the domain of ship sensor data, this principle holds significant relevance for several reasons. Firstly, sensor measurement accuracy faces inherent limitations owing to environmental noise, technological constraints, or sensor precision, making the transmission of sensor readings with absolute accuracy unnecessary for the monitoring system. Instead, the system can maintain data precision to a specific decimal level. Moreover, ensuring sensor readings adhere to a predefined range is crucial, preventing the inclusion of outliers or erroneous values due to sensor anomalies. Effectively mapping floating-point sensor readings to a finite set of integers capitalizes on these properties. Quantization, as the initial step in the compression process, offers the system the capability to establish accurate thresholds consistently for all $\mathbf{x}_t \in STH$—a critical factor in aligning the algorithm with precise ship monitoring requirements. In essence, quantization strikes a balance between data precision and efficiency. It reduces data volume while maintaining a level of precision that precisely aligns with monitoring needs. Hence, we employed quantization to reduce ship sensor data volume while preserving the necessary precision. Fulfilling the demands of data precision and efficiency, the quantization step necessitates a configuration table. This table elucidates the practical significance of ship sensor data and defines the effective range required by the Beidou system. Practically, we prepared and maintained the configuration table in JSON format. Formally, the quantization step is described as follows:

$$Q(\mathbf{x})_i = \text{round}(\frac{\mathbf{x}_i - \mathbf{L}_{d,i}}{10^{\mathbf{a}_i}}) \tag{3}$$

where round stands for the round function, which finds the closest integer of any real number.

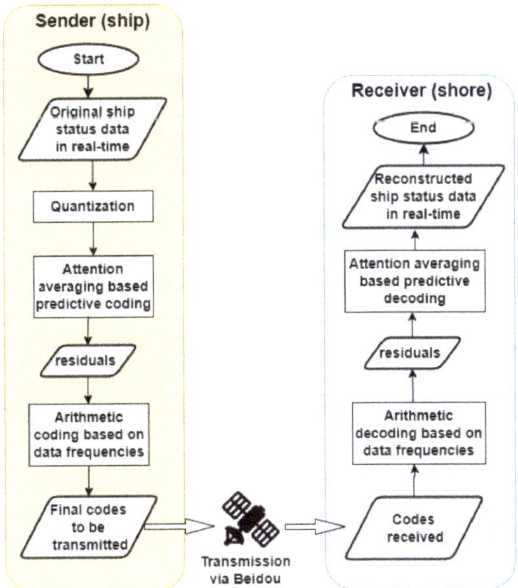

Figure 1. Flow chart of our proposed method. This figure is only used to show the global process of our data compression approach; therefore, it does not contain the technique details of each step. The corresponding specific structures or implementation details can be found in the subsequent sections.

The post-quantization data $\mathbf{y}_t \in \mathbb{Z}^n$ are defined as $\mathbf{y}_t = Q(\mathbf{x}_t)$. Each dimension of $\mathbf{y}_t$ is limited to L_i possible values, a constraint arising from finite accuracy goals and specific upper and lower limits. Notably, L_i exhibits considerable variability across our dataset, ranging from 100 to 100,000. This substantial variation poses a challenge during subsequent encoding procedures.

3.4. Attention-Averaging-Based Predictive Coding

Following the quantization step, a pivotal aspect of our proposed method to reduce data size involves the attention-averaging-machine-learning-based predictive coding step. In this process, a predictive model anticipates data values, comparing them with the actual values to derive errors or residuals. These residuals, representing the difference between predicted and actual data, are transmitted or stored in a reduced data size instead of the entire dataset.

The effectiveness of predictive coding hinges on the precision of the applied predictive model. Machine-learning-based prediction offers potential for generating more accurate predicted values, which are notably successful in compressing textual and genomic data [18,23]. However, text and genomic data's limited symbol variations enable the generation of probability values for each input symbol, facilitating their application in arithmetic encoding. Yet, directly applying existing predictive coding methods to ship status data presents a challenge. The extensive range and volume of ship status data values make generating probability values for each integer point impractical.

To address this, we designed a tailored predictive coding process for ship status data, enabling the transmission of residual errors between predicted and true values instead of transmitting the entire sensor data for arithmetic coding. This approach proves feasible for ship status data due to its periodic generation and transmission, maintaining continuous time correlations. Leveraging transmitted historical data at the receiver end, we can pre-train a predictor based on this historical ship sensor data. Calculating the residual error between predicted and true values of current real-time ship data enables us to transmit only these residuals to the next arithmetic coding step. These residuals can be reconstructed at the receiver end using the same predictor, ensuring the retrieval of the complete required data.

Figure 2 illustrates the specific process of our custom-designed predictive coding method. This method's key sub-step involves pre-training a predictor based on historical ship sensor data, utilized to generate predicted values for both ship and shore-side data. This pre-trained predictor enables the transmission of significantly smaller residual errors rather than the entire dataset to the subsequent arithmetic coding step. Formally, the predictor P forecasts the quantized current data $\mathbf{y}_t$ using the most recent m quantized historical data $\{\mathbf{y}_\tau\}_{\tau=t-m}^{t-1}$ [1].

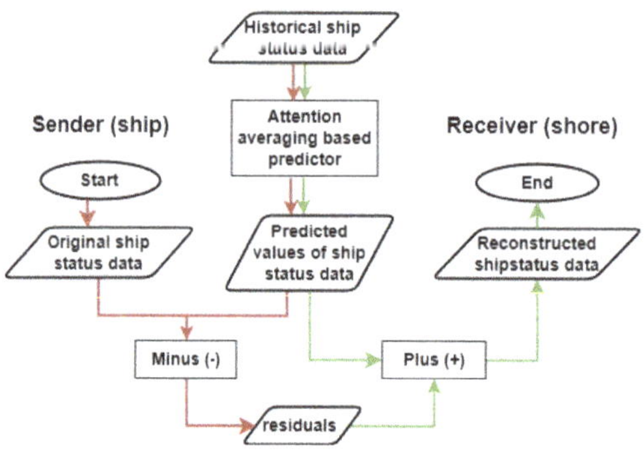

Figure 2. Flowchart of attention averaging based predictive coding.

Subsequently, only the predicted error $\mathbf{e}_{t,i} = (\mathbf{y}_t - \mathbf{y}'_t)_i \bmod L_i$ is transmitted, facilitating the reconstruction of $\mathbf{y}_t$ as $\mathbf{y}_{t,i} = (\mathbf{e}_t + \mathbf{y}'_t)_i \bmod L_i$. Both sender and receiver independently calculate $\mathbf{y}'_t$ using synchronized historical data $\{\mathbf{y}\tau\}_{\tau=t-m}^{t-1}$, ensuring identical results. Given that $\mathbf{e}_t$ tends to be close to zero most of the time, predictive coding substantially reduces message length, aided by arithmetic coding in Section 3.5.

3.4.1. Attention-Averaging-Based Predictor

In predictive coding, the accuracy of the predictor directly influences the length of transmitted information. Optimal predictor selection significantly impacts the efficacy of predictive coding. In our approach, we employed a predictor utilizing a blending concept akin to the attention-based averaging method proposed in [30]. This approach enables the blending of a linear model with a CNN, enhancing prediction accuracy. Our blending strategy stems from empirical observations of solely employing the linear or CNN models to predict ship status data. These observations revealed differing strengths of both models across different data fields. Consequently, to leverage their respective advantages, we designed an attention-averaging-based method, assigning different weights to the linear model and CNN based on distinct data properties.

Figure 3 illustrates the structure of our attention-averaging-based predictor. Initially, the predictor takes historical data as input for both the linear regression and CNN models. The linear regression model, unique in this paper, operates without normalization, differing from conventional linear prediction methods common in data compression. It processes 2-D historical data as input and produces a 1-D vector as a predicted value for all fields. The weights in the linear model are calculated using the SVD algorithm, optimizing their performance based on historical training data rather than being theoretically derived.

Concerning the CNN model, we adopted the WaveNet architecture [31], renowned for its success in audio processing. Within a WaveNet block, multi-layer convolutional kernels of size 2 and gradually increased diffusion rates are employed. The global CNN model structure includes normalization in the first layer, followed by a WaveNet block, a flattening layer, and two densely connected layers (as shown in Figure 4b). To optimize the model size, we applied valid padding.

Subsequently, the outputs from both the linear regression and CNN models are fed into our attention averaging network alongside the original historical data. This network, comprising three densely connected layers, determines the suitable weights for the linear regression and CNN models based on current data characteristics. Notably, the weights generated by the attention averaging network are vectors rather than scalars. This distinction allows field-specific variation in the weights of the linear model and CNN in the weighted average, enabling the model to adapt predictors to diverse field characteristics.

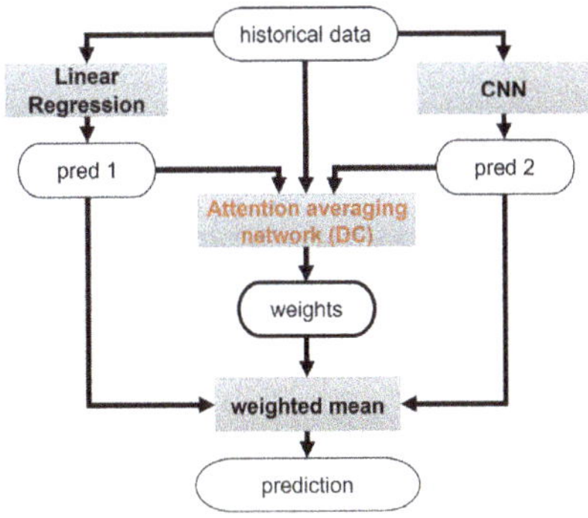

Figure 3. Structure of predictor.

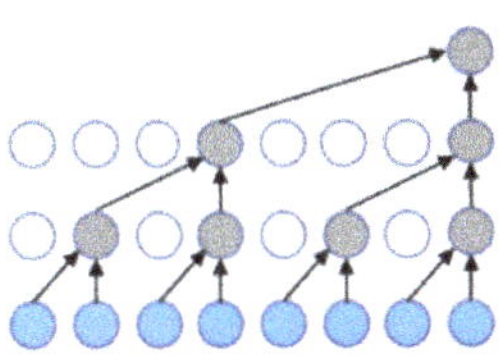

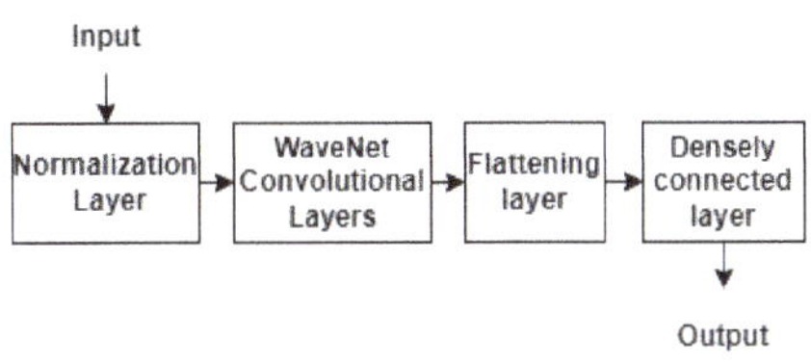

(**a**) (**b**)

Figure 4. CNN model used in the attention averaging based predictor. (**a**) WaveNet Convolutional Layers. (**b**) The global CNN architecture.

Moreover, the attention averaging model, by considering both the original inputs and outputs of the prediction models, dynamically adjusts weights based on different input conditions. Consequently, this approach amalgamates the strengths of linear models and neural networks. Additionally, to ensure the sum of output weights in each dimension equals 1, the Softmax function serves as the output layer activation function.

Once the attention averaging-based predictor determines suitable weights for the two models, it computes the final predicted values using these generated weights. The aim is to optimize prediction accuracy specifically tailored to ship data.

3.4.2. Residual Generation

Following the prediction by our attention averaging-based predictor, the values are quantized into integers. This quantization enables the calculation of residuals—derived by subtracting the quantized predicted values from the quantized actual values of the current ship data—to be fed into the arithmetic encoder, generating the final codes.

The process of computing residuals involves only integer operations, eliminating potential precision issues. Assuming the data to be sent is $\mathbf{x}$, the most recent historical data (typically the latest 50 data points in this context) is fed into the predictor to compute a predicted value $\mathbf{x}'$. Given that the receiver (typically the shore side) possesses an identical predictor and historical data, transmitting solely the predicted residual $\mathbf{x} - \mathbf{x}'$ facilitates complete reconstruction of the original data $\mathbf{x}$ at the receiver's end.

3.5. Data-Frequency-Based Arithmetic Coding

Our observations on ship status data reveal varying frequency distributions among values. For instance, in the "Boiler Heavy Oil Density" field, although values range from 825 to 845, a significant portion tends to hover around 835. Leveraging this insight, arithmetic coding algorithms, with the typical ideas assigning shorter encoding for frequently occurring values and longer encoding for occasional ones, are quite suitable to be applied for our ship status data. This approach compresses data further while minimizing information loss, functioning as an almost optimal entropy encoding method [32].

The challenge of applying arithmetic coding algorithm to our scenario lies in the fact that our maritime vessel state data can be viewed as a fixed-dimensional multidimensional vector, whereas arithmetic coding is typically utilized in symbol sequences of variable lengths. Given the diversity of fields in ship status data, each field necessitates an independent frequency table. Subsequently, the code vectors generated for all fields are combined into a single binary string—the final form transmitted to the receiver (typically the shore side). Notably, both sender (typically the ship side) and receiver share estimated data frequencies. During the encoding process, the frequency statistic tables remain static and are not dynamically updated or transmitted. Estimating frequencies for each value within a field is challenging directly, thus necessitating estimation from historical data. For every field in the dataset, the estimated frequency p_i of the i-th value is computed using the formula:

$$p_i = \frac{f_i + 1}{m + l} \tag{4}$$

Here, m represents the total historical data entries, l denotes the possible values in the field, and f_i signifies the occurrence count of the i-th value in the data.

Subsequently, the residuals obtained from the preceding predictive coding step are encoded using these frequency tables and transmitted to the receiver.

4. Experiments and Results

In this section, we introduce our experimental settings and discuss the experiment results we obtained.

4.1. Experimental Setting

4.1.1. Our Used Datasets

Our dataset comprises real sensor monitoring data collected from a ship's energy efficiency monitoring system over a two-month period. We have quantized the dataset according to the precision requirements and numerical limits of actual ship monitoring system, using quantization techniques discussed in Section 3.3. Our ship status data include 41 floating-point fields (Figure 5 shows a part of the data fields), which mainly describe the real-time fuel consumption related data fields, the rotate speed, the torque and the power of the main engine, the ship speed data fields, ship-drought-related data fields, and the real-time weather-related data fields. With data logged approximately every 10 s, this compilation spans 479,742 rows.

DGMGODensity	BlrMGODensity	DG1Power	DG2Power	DG3Power	MERpm	METorque	ShipSpdToWater	ShipSpd	ShipHeel	ShipTrim
279	84	899	727	89	136	586	174	1212	650	693
279	78	905	724	89	136	580	174	1212	667	716
279	82	885	706	89	136	578	173	1204	684	643
279	84	886	711	89	135	580	173	1204	689	641
279	82	871	688	89	136	585	173	1204	717	606
279	82	883	710	89	136	589	174	1190	667	631
279	85	879	698	89	136	585	173	1190	663	697
279	83	872	690	89	136	585	172	1190	646	712
279	84	881	707	89	135	592	171	1190	645	670
279	78	873	686	89	136	606	172	1190	657	691
279	82	871	687	89	136	594	172	1190	688	748
279	83	865	675	89	135	584	172	1190	682	610
279	83	873	695	89	135	592	173	1190	632	646
279	81	881	705	89	136	602	173	1190	668	752
279	82	894	709	89	136	597	173	1190	679	673

Figure 5. The screenshot of our quantized ship status dataset.

In applying our compression algorithm to the ship monitoring system, the number of fields for transmission in the practical application may not precisely match those in the experimental dataset. However, we assume similarity in the nature of the data between the application scenario and our experimental datasets.

To avoid overfitting, we chronologically divided the dataset into training, validation, and test sets, adhering to a 7:2:1 ratio. The training set serves for predictor training in our predictive coding step and for estimating data frequencies applied in arithmetic coding. The validation set aids in adjusting hyperparameters and method comparison. Lastly, the test set is reserved to validate the efficacy of our QAAPA approach, mitigating potential overfitting concerns.

4.1.2. Experimental Environments

Throughout our research, we conducted experiments using various compression methods for ship status data under the following configurations: The primary remote machine[2] boasted robust specifications with an Intel Core i9-10900 CPU[3], Nvidia GeForce RTX 3090 GPU[4], and 128GB of RAM; Operating on Ubuntu 20.04, we utilized Python 3.9, TensorFlow 2.5.0, tmux for prolonged program execution, and a Jupyter server hosted at a research institute under the Chinese Academy of Sciences. We accessed the remoted machine via SSH. After SSH connection, the Jupyter server on the remote machine linked to the local machine's 8888 port. We executed and debugged code using Microsoft Visual Studio Code[5]'s Jupyter Notebook and Python. The high specifications of the remote machine expedited training and model exploration, typically requiring mere minutes for neural network model training. For practical implementation, less resource-intensive servers or cloud services could be employed to curtail costs during model training.

4.1.3. Experimental Design

Our primary objective in the ship status data transmission scenario is to compress real-time data approaching its theoretical minimal length, enabling real-time transmission via the Beidou navigation system. Our optimization focus lies in achieving a long-term average compression rate for compression algorithms, reflected in the average code length post-compression. Given the steps involved in these algorithms—initially estimating symbol occurrence probabilities followed by arithmetic coding—the average code length closely aligns with the information entropy value calculated from these probabilities.

In our approach, we estimate the summed code length of encoding methods using the information entropy of estimated probabilities. This estimation enables us to gauge the compression rate of different algorithms. Sequential information entropy is estimated using commercial compression software, calculated as the compressed file size divided by the number of data points. Notably, the minimum achieved result across various configurations of compression software stands at 141 bit.

We devised a series of experiments to assess the effectiveness and reliability of our designed compression algorithm for real-time ship status data:

- **Experiments to estimate data frequency estimation effectiveness:** We evaluated the validity of the frequency estimation method employed in our study for future data. This analysis involved learning curves and performance comparison on training and validation sets. Estimation used the early part of the training data to gauge frequency and calculate information entropy.
- **Ablation Experiments on our compression algorithm (QAAPA):** We conducted experiments to analyze the key steps in our QAAPA method—Quantization, Attention-Averaging-based Predictive Coding, and Arithmetic Coding. We trained the attention averaging-based predictor by initially training the linear and CNN models separately. The subsequent training of the attention averaging model involved freezing linear model parameters for 20 epochs and using a constant learning rate of 5e-4. We implemented models in Keras with TensorFlow as the backend, trained on an Nvidia Quadro K620 GPU. We compared the summed coding length before and after applying

our QAAPA method across training, validation, and test datasets. Additionally, we assessed individual steps within QAAPA by comparing it with alternative methods:

- Original: only encoding data in *float64* format without data compression
- Q Method: Solely employing Quantization for data compression.
- QA method: Employing Quantization and Arithmetic Coding for data compression.

All of the above methods remove one or more steps from our QAAPA method. The following two methods replace the averaging-attention-based predictor of our QAAPA method:

- QLRPA: Replacing the attention-averaging-based predictor in QAAPA with linear-regression-based predictive coding and including Quantization, linear-regression-based predictive coding, and Arithmetic coding for data compression.
- QCNNPA: Replacing the attention-averaging-based predictor in QAAPA with CNN-based predictive coding and including Quantization, CNN-based Predictive coding, and Arithmetic coding for data compression.

- **Experiments on the reliability of our QAAPA method:** This experiment involved comparing compression results using our QAAPA method across training, validation, and test sets. The compression results are evaluated by the following descriptive statistics, including the mean value, standard deviation, minimum value, 25th percentile, median (50th percentile), 75th percentile, and maximum value. In addition, we analyzed the transmitted code length via Beidou over time to evaluate the reliability of our method.

4.2. Results and Discussion

4.2.1. Results on the Effectiveness of Data Frequency Estimation

Figure 6 shows our frequency estimation evaluation results. The information entropy calculated from the estimated frequency remains stable when the data constitute more than 15% of the total training dataset. This stability across varying data indicates that the frequency estimation method exhibits no significant overfitting tendencies. Conclusively, the frequency estimation derived from historical data demonstrates consistency, validating its suitability for future data and ensuring the reliability of the subsequent arithmetic coding step.

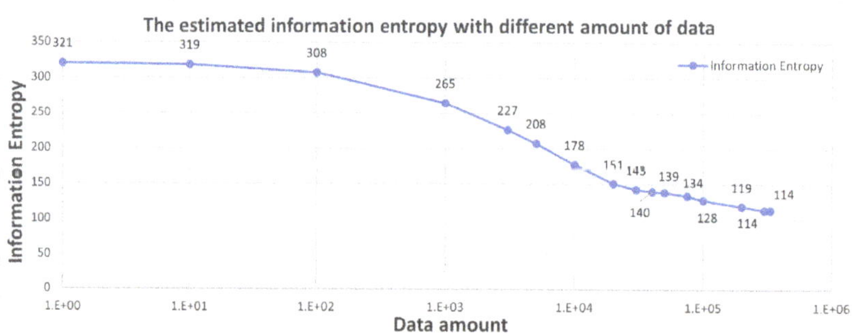

Figure 6. The estimated information entropy with different number of data.

4.2.2. Results on Ablation Experiments on our QAAPA Method

As introduced in Section 4.1.3, the coding lengths are obtained by summing the information entropy of all data fields. Figure 7 shows the separate information entropy of examples data fields. As we can see from it, with our tested dataset, the linear regression predictor is able to provide shorter coding lengths than the CNN predictor for most of the data fields, which indicates that a simple algorithm is sufficient to provide satisfactory compression effectiveness for most cases. However, there are still data fields that can achieve shorter coding lengths using more complicated CNN models, which proves the necessity to combine different algorithms. As expected, our attention-averaging-based algorithm can achieve the shortest coding lengths by assigning suitable weights for both algorithms for all data fields.

Based on the above-introduced summation method, we can obtain the average coding lengths of using different data compression methods shown in Table 1[6]. As we can see, Quantization alone reduced the average coding length to approximately 40% of the original data size. Incorporating arithmetic coding further halved the average coding length. Introducing predictive coding enhanced compression, with linear-regression-based prediction outperforming CNN-based prediction for most ship status data fields. The complete QAAPA method achieved the most substantial compression, reducing the original data size to less than 10%. These results collectively demonstrate the effectiveness of each step within the QAAPA approach and highlight the superior performance of the complete method in compressing real-time ship status data.

	Coding Lengths		
Data Fields	LR	CNN	Attention Averaging
DG1Power	5.01	8.56	5.01
DG2Power	4.4	8.69	4.4
DG3Power	4.7	8.52	4.7
MERpm	1.52	3.98	1.48
METorque	5.04	7.22	4.9
ShipSpdToWater	1.4	5.42	1.4
ShipSpd	3.39	9.29	3.39
ShipHeel	6.57	6.99	6.57
ShipTrim	6.79	8.39	6.79
WindSpd	4.99	8.26	4.99
_MEActCons	9.33	11.1	9.14
_DGActCons_1	7.77	9.91	7.76
_DGActCons_2	5.48	7.13	3.88
_ShipDraft_m2	6.81	9.56	6.79
_ShipDraught_L_R	6.69	9.54	6.68
_dlt_MEAccFOCons	5.94	3.81	2.36
_dlt_DGAccFOCons	5.34	3.68	1.42
_dlt_BlrAccFOCons	0.33	0.93	0.33
_dlt_DGAccMGOCons	2.69	0.46	0.33
_dlt_BlrAccMGOCons	0.00	0.00	0.00
_dlt_WindDir	4.11	4.05	4.10

Figure 7. The separate information entropy of data fields of the training dataset obtained using linear regression predictor (LR), CNN predictor (CNN), and our proposed attention-averaging-based predictor (Attention Averaging).

Table 1. Ablation study: average coding length of different data compression methods.

Methods	**Average Coding Lengths (bits)**	
	Training Dataset	**Validation Dataset**
Original	1312	1312
Q	504	504
QA	263	274
QLRA	122	120
QCNNPA	216	240
QAAPA	**110**	**108**

4.2.3. Experiments on the Reliability of Our QAAPA Method

We evaluated our proposed method in the test set and compared the performance on the test set with those on the train and valid set. The results of the test are shown in Table 2. As we can see from it, the compression effectiveness observed in the test set remained consistent with the training and validation sets, demonstrating stability and absence of overfitting. The graphical representation of generated code length versus time (Figure 8) illustrates the stability of compression effectiveness over time. The average coding length achieved by the proposed QAAPA method closely aligns with the sequential information entropy estimated earlier, indicating compression results nearing the theoretical minimum. These findings confirm the robustness and reliability of the QAAPA method for compressing real-time ship status data without significant overfitting issues.

Table 2. Average coding length of our QAAPA method.

Dataset	Statistics in Bits						
	Mean	Std	Min	25%	50%	75%	Max
train (70%)	109.96	18	68	96	108	120	360
valid (20%)	108.38	18	68	96	104	120	360
test (10%)	104.10	20	68	88	100	112	348

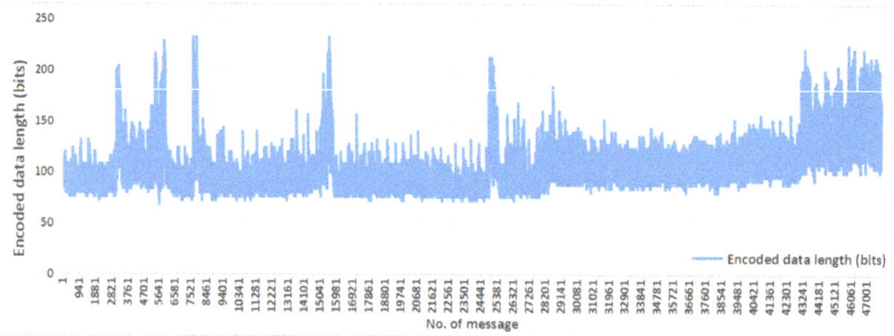

Figure 8. The generated code length versus time.

5. Conclusions and Future Work

This study presents QAAPA, a novel methodology facilitating real-time transmission of ship data via the Beidou Navigation System. The approach involves the real-time compression of sequential ship status data, achieving precision while adhering to predefined data ranges. QAAPA integrates quantization, attention-averaging-based predictive coding, and arithmetic coding algorithms to compress the data stream effectively. Experimental evaluation using real ship monitoring data showcases the efficacy of QAAPA. The results demonstrate its proximity to the theoretical minimum estimated through commercial compression algorithms, surpassing traditional approaches in similar contexts.

Future research avenues include expanding the method's utility in analogous scenarios, such as dynamically adjusting coding accuracy to limit message lengths and integrating incremental learning to eliminate the need for pre-training. Additionally, optimizing the structure of the predictive coding predictor stands as a promising area for enhanced efficiency and effectiveness of the QAAPA approach.

Author Contributions: Conceptualization, C.Z.; methodology, C.Z.; validation, J.Z.; formal analysis, J.Z.; investigation, J.Z.; resources, C.Z. and J.Z.; data curation, C.Z.; writing—original draft preparation, C.Z. and J.Z.; writing—review and editing, J.Z.; visualization, J.Z.; supervision, C.Z.; project administration, C.Z.; and funding acquisition, C.Z. All authors have read and agreed to the published version of the manuscript.

Funding: This research was funded by National Key R&D Program of China (2022YFB4301403).

Institutional Review Board Statement: Not applicable.

Informed Consent Statement: Not applicable.

Data Availability Statement: Data available on request due to restrictions: The data presented in this study are available on request from the corresponding author.

Conflicts of Interest: The authors declare no conflicts of interest.

Notes

1 m empirically set to 50.

2 sourced from Shanghai, China

3 Manufactured by Intel Corp., Santa Clara, CA, USA

4 Manufactured by Nvidia Corp., Santa Clara, CA, USA

5 Version 1.86.0

6 The bold numbers indicate that our method has the best performance compared to other methods.

References

1. Caban, J.; Brumerčík, F.; Vrábel, J.; Ignaciuk, P.; Misztal, W.; Marczuk, A. Safety of Maritime Transport in the Baltic Sea. *MATEC Web Conf.* **2017**, *134*, 00003. [CrossRef]
2. Yang, X.; Zhi, J.; Zhang, W.; Xu, S.; Meng, X. A Novel Data-Driven Prediction Framework for Ship Navigation Accidents in the Arctic Region. *J. Mar. Sci. Eng.* **2023**, *11*, 2300. [CrossRef]
3. Shao, T.; Du, W.; Ye, Y.; Li, H.; Dong, J.; Liu, G.; Zheng, P. A Novel Virtual Arrival Optimization Method for Traffic Organization Scenarios. *Sustainability* **2024**, *16*, 403. [CrossRef]
4. China Satellite Navigation Office. *Development of the BeiDou Navigation Satellite System (Version 4.0)*; China Satellite Navigation Office: Beijing, China, 2019.
5. China Satellite Navigation Office. *10 Application Scenarios of BDS in Africa*; China Satellite Navigation Office: Beijing, China 2021.
6. Li, G.; Guo, S.; Lv, J.; Zhao, K.; He, Z. Introduction to global short message communication service of BeiDou-3 navigation satellite system. *Adv. Space Res.* **2021**, *67*, 1701–1708. [CrossRef]
7. Nie, X.; Xie, J.; Liu, T.; Liu, C.; Zhao, K. Research on Key Performance of BeiDou Global Short Message Communication Service. In *Proceedings of the China Satellite Navigation Conference*; Springer: Singapore, 2020; pp. 435–442.
8. Xiaoxing, Y.; Ming, C.; Xiaojuan, G. Design of reliable Remote Communication system Based on Beidou Satellite. *Comput. Eng.* **2017**, *34*, 62–68.
9. Kai-wen, X.; Jing, F.; Qian-ru, W. Improvement Mechanism of Beidou Short Message Meteorological and Hydrological Data Transmission Efficiency. *Commun. Inf. Process.* **2021**, *40*, 5.
10. Lanhong, J.; Jinpeng, C.; Zengfeng, B. Application analysis of Beidou Satellite communication system flood exceeding designed level monitoring. *Water Resour. Informatiz.* **2021**, 68–71. [CrossRef]
11. Zhonghui, O.; Huijin, F.; Qinghua, C.; Daochang, H. Research on Compression Transmission Method of Special Vehicle Status Information Based on Beidou Short Message. *J. Ordnance Equip. Eng.* **2020**, *41*, 6.
12. Yue, Y.; Li-juan, Y.; Shou-qi, C.; Cheng-ming, C.; Bing-yi, Z. Design of River Water Quality Monitoring System Based on Beidou Communication. *Acta Metrol. Sin.* **2020**, *41*, 6.
13. Yong, C. Process Data Compression Method Based On Beidou's Short Message Communication. Ph.D. Thesis, Yan Shan University, Qinhuangdao, China, 2017.
14. Yang, Y.; Mandt, S.; Theis, L. An Introduction to Neural Data Compression. *Found. Trends Comput. Graph. Vis.* **2023**, *15*, 113–200. [CrossRef]
15. Li, M.; Zuo, W.; Gu, S.; Zhao, D.; Zhang, D. Learning convolutional networks for content-weighted image compression. In Proceedings of the IEEE Conference on Computer Vision and Pattern Recognition, Salt Lake City, UT, USA, 18–23 June 2018; pp. 3214–3223.
16. Possemiers, A.; Lee, I. Evaluating deep learned voice compression for use in video games. *Expert Syst. Appl.* **2021**, *181*, 115180. [CrossRef]
17. Lu, G.; Ouyang, W.; Xu, D.; Zhang, X.; Cai, C.; Gao, Z. Dvc: An end-to-end deep video compression framework. In Proceedings of the IEEE/CVF Conference on Computer Vision and Pattern Recognition, Long Beach, CA, USA, 15–20 June 2019; pp. 11006–11015.
18. Cox, D. Syntactically informed text compression with recurrent neural networks. *arXiv* **2016**, arXiv:1608.02893.
19. Makhoul, J.; Roucos, S.; Gish, H. Vector quantization in speech coding. *Proc. IEEE* **1985**, *73*, 1551–1588. [CrossRef]
20. Goyal, V.K. Theoretical foundations of transform coding. *IEEE Signal Process. Mag.* **2001**, *18*, 9–21. [CrossRef]
21. Yan, Z.; Wang, J.; Sheng, L.; Yang, Z. An effective compression algorithm for real-time transmission data using predictive coding with mixed models of LSTM and XGBoost. *Neurocomputing* **2021**, *462*, 247–259. [CrossRef]
22. Fout, N.; Ma, K.L. An adaptive prediction-based approach to lossless compression of floating-point volume data. *IEEE Trans. Vis. Comput. Graph.* **2012**, *18*, 2295–2304. [CrossRef] [PubMed]
23. Goyal, M.; Tatwawadi, K.; Chandak, S.; Ochoa, I. Deepzip: Lossless data compression using recurrent neural networks. *arXiv* **2018**, arXiv:1811.08162.
24. Gray, R.M.; Neuhoff, D.L. Quantization. *IEEE Trans. Inf. Theory* **1998**, *44*, 2325–2383. [CrossRef]
25. Sprevak, M. Predictive coding I: Introduction. *Philos Compass* **2023**, *19*, e12950. [CrossRef]
26. Géron, A. *Hands-on Machine Learning with Scikit-Learn, Keras, and TensorFlow*; O'Reilly Media, Inc.: Sebastopol, CA, USA, 2022.
27. Hackeling, G. *Mastering Machine Learning with Scikit-Learn*; Packt Publishing Ltd.: Birmingham, UK, 2017.

28. Taye, M.M. Understanding of Machine Learning with Deep Learning: Architectures, Workflow, Applications and Future Directions. *Computers* **2023**, *12*, 91. [CrossRef]
29. Chua, L.O.; Roska, T. The CNN paradigm. *IEEE Trans. Circuits Syst. I Fundam. Theory Appl.* **1993**, *40*, 147–156. [CrossRef]
30. Tulyakov, S.; Gehrig, D.; Georgoulis, S.; Erbach, J.; Gehrig, M.; Li, Y.; Scaramuzza, D. Time Lens: Event-based Video Frame Interpolation. In Proceedings of the 2021 IEEE/CVF Conference on Computer Vision and Pattern Recognition (CVPR), Nashville, TN, USA, 20–25 June 2021; pp. 16150–16159. [CrossRef]
31. Oord, A.v.d.; Dieleman, S.; Zen, H.; Simonyan, K.; Vinyals, O.; Graves, A.; Kalchbrenner, N.; Senior, A.; Kavukcuoglu, K. Wavenet: A generative model for raw audio. *arXiv* **2016**, arXiv:1609.03499.
32. Witten, I.H.; Neal, R.M.; Cleary, J.G. Arithmetic coding for data compression. *Commun. ACM* **1987**, *30*, 520–540. [CrossRef]

Journal of
Marine Science and Engineering

MDPI

Article

Multi-Parameter Fuzzy-Based Neural Network Sensorless PMSM Iterative Learning Control Algorithm for Vibration Suppression of Ship Rim-Driven Thruster

Zhi Yang [1,2], Xinping Yan [1,2,3], Wu Ouyang [1,2,3,*], Hongfen Bai [1,4] and Jinhua Xiao [1]

1 School of Transportation and Logistics Engineering, Wuhan University of Technology, Wuhan 430063, China; zhiyang@whut.edu.cn (Z.Y.); xpyan@whut.edu.cn (X.Y.); hfbai@shmtu.edu.cn (H.B.); xiaojinh@whut.edu.cn (J.X.)
2 Reliability Engineering Institute, National Engineering Research Center for Water Transport Safety, Wuhan 430063, China
3 State Key Laboratory of Maritime Technology and Safety, Wuhan University of Technology, Wuhan 430063, China
4 Key Laboratory of Transport Industry of Marine Technology and Control Engineering, Shanghai Maritime University, Shanghai 201306, China
* Correspondence: ouyangw@whut.edu.cn

Abstract: Aiming to reduce motor speed estimation and torque vibration present in the permanent magnet synchronous motors (PMSMs) of rim-driven thrusters (RDTs), a position-sensorless control algorithm using an adaptive second-order sliding mode observer (SMO) based on the super-twisting algorithm (STA) is proposed. In which the sliding mode coefficients can be adaptively tuned. Similarly, an iterative learning control (ILC) algorithm is presented to enhance the robustness of the velocity adjustment loop. By continuously learning and adjusting the difference between the actual speed and given speed of RDT motor through ILC algorithm, online compensation for the q-axis given current of RDT motor is achieved, thereby suppressing periodic speed fluctuations during motor running. Fuzzy neural network (FNN) training can be used to optimize the STA-SMO and ILC parameters of RDT control system, while improving speed tracking accuracy. Finally, simulation and experimental verifications have been conducted on the vector control system based on the conventional PI-STA and modified ILC-STA. The results show that the modified algorithm can effectively suppress the estimated speed and torque ripple of RDT motor, which greatly improves the speed tracking accuracy.

Keywords: rim-driven thruster (RDT); permanent magnet synchronous motor (PMSM); position-sensorless control; super-twisting algorithm (STA); iterative learning control (ILC); fuzzy neural network (FNN)

1. Introduction

Electric propulsion systems have undergone significant development in recent years due to their advanced control efficiency and performance. The rim-driven thruster (RDT) is a new type of low-carbon emission electric thruster, which integrates the propeller blade and the motor together by embedding the motor into the duct. This new type of electric thruster can eliminate the need for shaft transmission, mechanical seals, and oil circuit systems in traditional ship propulsion systems. It can save space for cabins on ships, improve motor efficiency and hydrodynamic efficiency, and reduce system installation complexity and manufacturing costs [1]. For the development of this new type of electric thruster, the selection of the propulsion motor RDT and its control algorithm are the key technologies.

At present, alternating current (AC) induction motors, brushless direct current (BLDC) motors, permanent magnet synchronous motors (PMSM), and high-temperature superconducting motors are included for the selection of RDT motors. Considering the work efficiency, manufacturing costs and processes performance of RDT, PMSM is currently considered the best choice for RDT propulsion motor. In order to accomplish the vector control of PMSM, it is necessary to obtain accurate rotor position information using position sensors. However, traditional mechanical position sensors (i.e., photoelectric encoders, resolvers, etc.) need to be mounted on the transmission shaft, which does not exist in RDT. They also have problems regarding large size, high costs, and low reliability, especially in specific engineering applications. RDT has a compact structure without the transmission shaft that can work in a harsh underwater environment, see Figure 1. Therefore, it is of great significance to develop anti-chattering position-sensorless control algorithms for RDT drive motors to realize high-performance control [2,3]. The PMSM position-sensorless control algorithms mainly include direct calculation methods, back-EMF integration methods, model reference adaptive methods, and various observer methods [4–8].

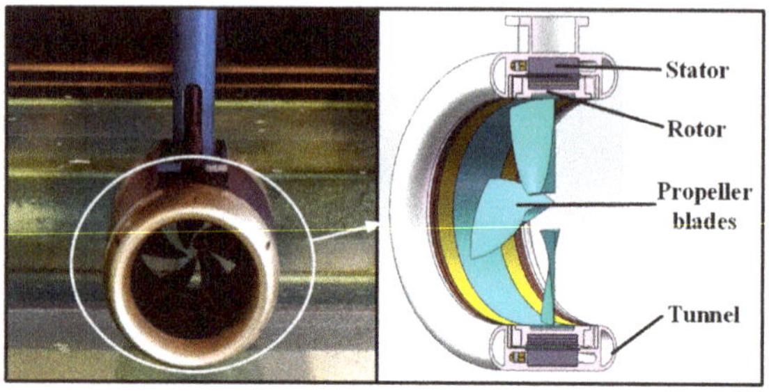

Figure 1. Physical diagram and geometric structure of an RDT.

Of these, the sliding mode observer (SMO) can reduce the dependence on motor parameters with advantages of having a fast response speed, strong robustness, and simple structure. Therefore, it is widely used in the position-sensorless control of PMSM [9]. The chattering will occur in the control system when using the SMO algorithm due to the switching of the sliding mode surface [10]. Especially in the vector control of PMSM at low speeds, it causes relatively large torque ripples. Therefore, it is necessary to eliminate chattering while maintaining control accuracy when using the SMO algorithm. The current research mainly focuses on the comprehensive improvement of switching function, sliding mode gain, estimation accuracy, etc. [11–14]. The first-order SMO always has chattering problems, which have a negative impact on the stability of the system. The second-order SMO can effectively eliminate chattering while maintaining the system robustness [15,16] and is well received by many scholars.

Due to the influences of motor design, inverter nonlinearities, and current detection errors, the PMSM control system suffers from torque vibration with current harmonics, which also reduces speed control accuracy. Usually, the vibration of motor torque is suppressed by optimizing motor design or improving the motor control strategy. However, it is difficult to modify the design after completing the original motor design [17]. With the development of control theory, many improved control strategies have been proposed to reduce torque vibration suppression, such as harmonic voltage injection, proportional-integral quasi-resonant compensator, and dead-zone feedforward compensation [18,19]. Iterative learning control (ILC) is an approach to improve the performance of a system that executes repetitively and periodically over a fixed time interval by learning from previous executions (trials, iterations, and passes) [20]. It does not rely on the mathematical model of systems and can be used in PMSM control to effectively improve tracking performance and compensate the control signal through continuous learning [21].

This paper presents a RDT position-sensorless control algorithm using an adaptive second-order SMO based on the super-twisting algorithm (STA). An adaptive algorithm is used to tune the sliding mode coefficients. Furthermore, a torque ripple suppression method based on ILC algorithm is proposed. A closed-loop PI-ILC algorithm with a forgetting factor is introduced to compensate for the q-axis given current through iterative learning, thereby reducing the real-time speed error. Considering the error and error rate of motor speed as input, and fuzzy neural network (FNN) training is used to optimize the ILC and STA-SMO parameters to improve the accuracy and stability of the control system. The proposed control algorithm can suppress the motor speed torque vibration and improve the speed tracking accuracy. The main innovations of this paper are as follows:

(1) A RDT control strategy using an adaptive second-order SMO based on the STA is designed.
(2) To reduce the torque ripple suppression of RDT motor, a closed-loop PI-ILC algorithm is introduced to compensate the speed real-time error to the q-axis given current.
(3) FNN training is used to optimize the ILC and STA-SMO parameters to improve the accuracy and stability of the current loop.

The rest of this paper is organized as follows: Section 2 introduces the materials and methods of RDT motor model and proposed control algorithm. Section 3 introduces the application of multi-parameter FNN for parameter tuning. Section 4 introduces the simulation results and discussions of a RDT motor sensorless control model on MATLAB R2017b/Simulink environment. Section 5 introduces the experimental results and the discussions of a RDT motor sensorless control system on experimental setup. The final section presents the conclusions and subsequent ideas of the topic covered in this paper.

2. Materials and Methods

The mathematic current model of the RDT motor in the $\alpha\beta$-axis reference system can be expressed as:

$$\begin{cases} \frac{di_\alpha}{dt} = -\frac{R_s}{L_s}i_\alpha + \frac{1}{L_s}u_\alpha - \frac{1}{L_s}e_\alpha \\ \frac{di_\beta}{dt} = -\frac{R_s}{L_s}i_\beta + \frac{1}{L_s}u_\beta - \frac{1}{L_s}e_\beta \end{cases} \tag{1}$$

where i_α and i_β and u_α and u_β represent the current and voltage in $\alpha\beta$-axis reference system, respectively. L_s and R_s represent the inductance and resistance of stator winding.

e_α and e_β represent the back EMFs in $\alpha\beta$-axis reference system and can be expressed as:

$$\begin{cases} e_\alpha = -\omega\psi_f \sin\theta \\ e_\beta = \omega\psi_f \cos\theta \end{cases} \tag{2}$$

where ψ_f is the flux linkages and ω and θ represent the rotor speed and position of the RDT motor.

2.1. Adaptive STA-SMO

The super-twisting algorithm (STA) was first proposed by A. Levant to effectively eliminate the chattering problem of first-order SMO. The basic form of STA with perturbation was designed as [22]:

$$\begin{cases} \frac{dx_1}{dt} = -k_1|\tilde{x}_1|^{1/2}\mathrm{sgn}(\tilde{x}_1) + x_2 + z_1(x_1,t) \\ \frac{dx_2}{dt} = -k_2\mathrm{sgn}(\tilde{x}_1) + z_2(x_2,t) \end{cases} \tag{3}$$

where x_1, x_2, and $\tilde{x}_1$ are the state variables, error between estimated values, and actual values of the state variables. k_1 and k_2 are designed to be the sliding-mode coefficients of STA. z_1 and z_2 are perturbation terms of system.

The stable conditions of STA have been proved in [23,24], demonstrating that the perturbation terms should be globally bounded by:

$$\begin{cases} |z_1| \le \delta_1|x_1|^{1/2} \\ |z_2| = 0 \end{cases} \tag{4}$$

And the sliding mode coefficients k_1 and k_2 should satisfy (5) to stabilize the system:

$$\begin{cases} k_1 > 2\delta_1 \\ k_2 > k_1 \frac{5\delta_1 k_1 + 4\delta_1^2}{2k_1 - 4\delta_1} \end{cases} \tag{5}$$

where $\delta_1 > 0$ and is a normal number.

In order to obtain the estimated value of the back EMF, the stator currents of RDT motor are estimated using STA based second-order SMO [15,16]:

$$\begin{cases} \frac{d\hat{i}_\alpha}{dt} = -\frac{R_s}{L_s}\hat{i}_\alpha + \frac{1}{L_s}u_\alpha + \frac{1}{L_s}k_1\left|\tilde{i}_\alpha\right|^{1/2}\mathrm{sgn}(\tilde{i}_\alpha) + \frac{1}{L_s}\int k_2\mathrm{sgn}(\tilde{i}_\alpha)dt \\ \frac{d\hat{i}_\beta}{dt} = -\frac{R_s}{L_s}\hat{i}_\beta + \frac{1}{L_s}u_\beta + \frac{1}{L_s}k_1\left|\tilde{i}_\beta\right|^{1/2}\mathrm{sgn}(\tilde{i}_\beta) + \frac{1}{L_s}\int k_2\mathrm{sgn}(\tilde{i}_\beta)dt \end{cases} \tag{6}$$

in which $\hat{i}_\alpha$ and $\hat{i}_\beta$ denote the estimated values of i_α and i_β. By comparing Equation (3) with Equation (6), it can be concluded that perturbation terms z_1 and z_2 are designed through:

$$\begin{cases} z_1(\hat{i}_\alpha, t) = -\frac{R_s}{L_s}\hat{i}_\alpha + \frac{1}{L_s}u_\alpha \\ z_1(\hat{i}_\beta, t) = -\frac{R_s}{L_s}\hat{i}_\beta + \frac{1}{L_s}u_\beta \end{cases}, z_2 = 0 \tag{7}$$

By substituting Equation (7) into Equation (4), we obtain:

$$\begin{cases} -\frac{R_s}{L_s}\hat{i}_\alpha + \frac{1}{L}u_\alpha - \delta_1\left|\hat{i}_\alpha\right|^{1/2} \leq 0 \\ -\frac{R_s}{L_s}\hat{i}_\beta + \frac{1}{L}u_\beta - \delta_1\left|\hat{i}_\beta\right|^{1/2} \leq 0 \end{cases} \tag{8}$$

For a sufficiently large δ_1, the validity of the above inequality can be met. The state equations of currents errors in $\alpha\beta$-axis reference system can be derived by subtracting Equation (1) from Equation (6):

$$\begin{cases} \frac{d\tilde{i}_\alpha}{dt} = -\frac{R_s}{L_s}\tilde{i}_\alpha + \frac{1}{L_s}(k_1\left|\tilde{i}_\alpha\right|^{1/2}\mathrm{sgn}(\tilde{i}_\alpha) + \int k_2\mathrm{sgn}(\tilde{i}_\alpha)dt + e_\alpha) \\ \frac{d\tilde{i}_\beta}{dt} = -\frac{R_s}{L_s}\tilde{i}_\beta + \frac{1}{L_s}(k_1\left|\tilde{i}_\beta\right|^{1/2}\mathrm{sgn}(\tilde{i}_\beta) + \int k_2\mathrm{sgn}(\tilde{i}_\beta)dt + e_\beta) \end{cases} \tag{9}$$

When the system reaches the sliding surface and the estimated error is close to 0 ($\tilde{i}_\alpha = 0, \tilde{i}_\beta = 0$). The equivalent back-EMFs e_α and e_β can be obtained by the equivalent control method, it can be concluded that:

$$\begin{cases} e_\alpha = -k_1\left|\tilde{i}_\alpha\right|^{1/2}\mathrm{sgn}(\tilde{i}_\alpha) - \int k_2\mathrm{sgn}(\tilde{i}_\alpha)dt \\ e_\beta = -k_1\left|\tilde{i}_\beta\right|^{1/2}\mathrm{sgn}(\tilde{i}_\beta) - \int k_2\mathrm{sgn}(\tilde{i}_\beta)dt \end{cases} \tag{10}$$

where k_1 and k_2 are the sliding mode coefficients. It is difficult to determine the actual values of k_1 and k_2 in practice. And different sets of k_1 and k_2 will influence the stability conditions of STA-SMO. Therefore, it is crucial to adaptively adjust the sliding mode coefficients. The adaptive algorithm is given as:

$$\begin{cases} k_1 = \sigma_1\omega \\ k_2 = \sigma_2\omega^2 \end{cases} \tag{11}$$

in which σ_1 and σ_2 are adaptive coefficients. The stability of the adaptive algorithm has been proved in [16].

The rotor position estimation method based on the arctangent function will directly introduce the sliding mode chattering into the division operation, resulting in significant estimation errors. Therefore, the PLL structure is used to extract the rotor position information of the RDT motor. It can suppress chattering and maintain good control performance. The structure block diagram of the adaptive STA-SMO is shown in Figure 2.

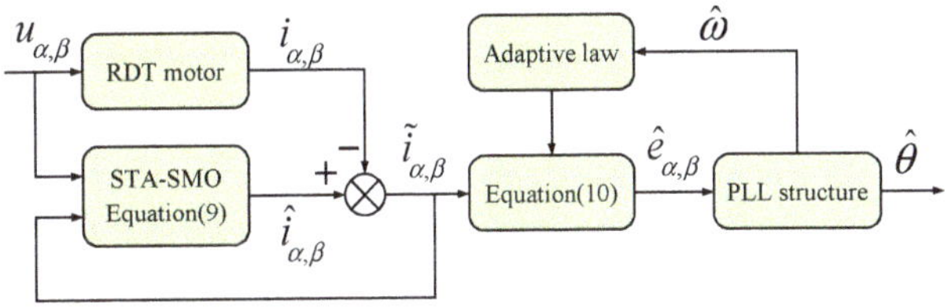

Figure 2. Structure block diagram of the adaptive STA-SMO.

2.2. Design of ILC Algorithm

By analyzing the previous control data, ILC can find appropriate control input through real-time online iteration, theoretically obtaining accurate tracking trajectories.

Traditional PI control achieves satisfactory results by selecting appropriate proportional and integral parameters but struggles to meet high-precision control requirements. By introducing the ILC algorithm to form an ILC controller, the q-axis given current of motor can be obtained with ILC controller and traditional PI controller.

The type of PI-ILC algorithm is adopted to acquire the learning law [19]:

$$i_{q,k+1}*(t) = (1-\alpha)i_{q,k}*(t) + \eta e_{k+1}(t) + \xi\int e_{k+1}(t)dt \tag{12}$$

where $i_{q,k+1}*(t)$ is the reference signal of the current cycle's q-axis current; $i_{q,k}*(t)$ is the reference signal of the q-axis current in the previous cycle called the "learning experience"; $e_{k+1}(t)$ is the current cycle speed deviation signal, which is mainly used to compensate for the "learning experience" loss caused by the forgetting factor; and η and ξ are the closed-loop learning coefficients. Formula (12) in the z-domain is shown as:

$$i_{q,k+1}*(z) = (1-\alpha)i_{q,k}*(z) + \eta e_{k+1}(z) + \xi\frac{z}{z-1}e_{k+1}(z) \tag{13}$$

The iterative process can be regarded as a delay link, which is shown as:

$$z^{-N}i_{q,k+1}*(z) = i_{q,k}*(z) \tag{14}$$

in the above equation, $N = f_{sample}/f_{signal}$, f_{sample} represents the sampling frequency of the digital control system, while f_{signal} is the system signal frequency.

When N takes a value of 1, the z-domain transfer function of the equation is shown as:

$$G(z) = \frac{i_{q,k+1}*(z)}{e_{k+1}(z)} = \frac{(\eta+\xi)z^2-\eta}{z^2-(2-\alpha)z+(1-\alpha)} \tag{15}$$

The structure block diagram of ILC process and closed-loop PI-ILC are shown in Figures 3 and 4.

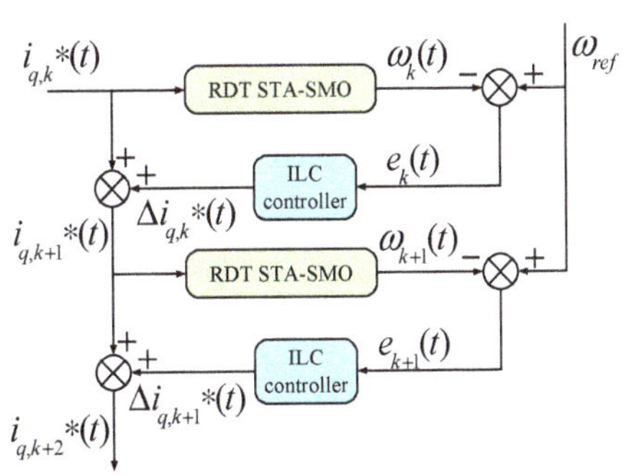

Figure 3. Structure block diagram of the ILC process.

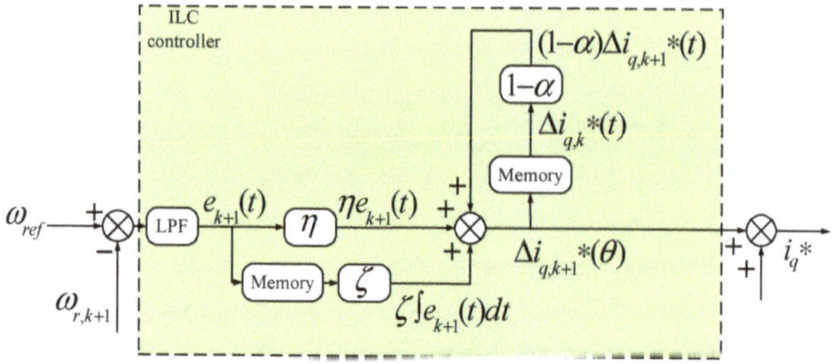

Figure 4. Structure block diagram of the closed-loop PI-ILC.

The design of the ILC controller should not only ensure its stability but also ensure that the iterative algorithm converges to its true value. The convergence conditions of the PI type iterative algorithm are determined as follows.

Combining Formulas (12) and (15) through $e_{k+1} = \omega_{ref} - \omega_{r,k+1}$, it can be seen that:

$$e_{k+1} = \omega_{ref} - \frac{n_p k_t}{J}\left\{\int (1-\alpha) i_{q,k} * + \int [\eta e_{k+1} + \xi \int e_{k+1} dt] dt\right\} \quad (16)$$

where J is the inertia moment, n_p is the pole pair of the motor, and k_t is the torque coefficient. Therefore, the following formulas are obtained:

$$\|e_{k+1}\|_\infty = \max_{0\le t\le T}\left|(1-\alpha)e_k + \alpha\omega_{ref} - \frac{n_p k_t}{J}\int_0^t [\eta e_{k+1} + \xi \int_0^t e_{k+1} dt] dt\right| \quad (17)$$

The definition of the infinite parametric number and the absolute value inequality leads to:

$$\left|1 - \frac{n_p k_t}{J}(\eta t + \frac{1}{2}\xi t^2)\right| \|e_{k+1}\|_\infty \le |1-\alpha| \|e_k\|_\infty + \alpha \|\omega_{ref}\|_\infty \quad (18)$$

When $\lim_{k\to\infty} e_k(t) = 0$, the algorithm converges by considering the convergence condition of the PI-ILC algorithm:

$$\left|\frac{J(1-\alpha)}{J - n_p k_t \eta T - 0.5 n_p k_t \xi t^2}\right| < 1 \quad (19)$$

where $t \in [0, T]$, T is the system tracking time.

3. Application of Multi-Parameter Fuzzy-Based Neural Network for Parameter Auto-Tuning

The principle of neural network is to analysis errors based on training and online results, modify weights and thresholds, and iteratively obtain models with the online results as outputs. It consists of input and output layer nodes, as well as one or more hidden layer nodes. The input information must be propagated forward to the hidden layer, and the activated information is passed to each output node after the activation function operation of the unit node to obtain the output results.

Fuzzy neural networks (FNNs) have good self-learning and adaptive abilities with strong robustness that only relies on fuzzy rules [25–29]. They can perform logical reasoning through established fuzzy rules which have good approximation abilities for certain complex nonlinear systems.

The proposed algorithm consists of an ILC algorithm, adaptive STA-SMO algorithm, and FNN algorithm. FNNs can generate fuzzy rules and adjust affiliation functions through strong self-learning and adaptive training. The control system transfers the speed deviation $e(t)$ and the deviation rate $ec(t)$ to the FNN, which can adjust the ILC parameters α, η, and

ξ and the STA-SMO parameters σ_1 and σ_2 to achieve better speed control performance after the fuzzification, fuzzy inference, and anti-fuzzification of the output.

The structure block diagram of FNN is shown as Figure 5. The four layers are input, fuzzification, fuzzy inference, and output layers in order. The calculation formula for input layer is shown as follows:

$$\begin{cases} x_1 = e(t) \\ x_2 = ec(t) \\ o_1(i) = [x_1, x_2] \end{cases} \tag{20}$$

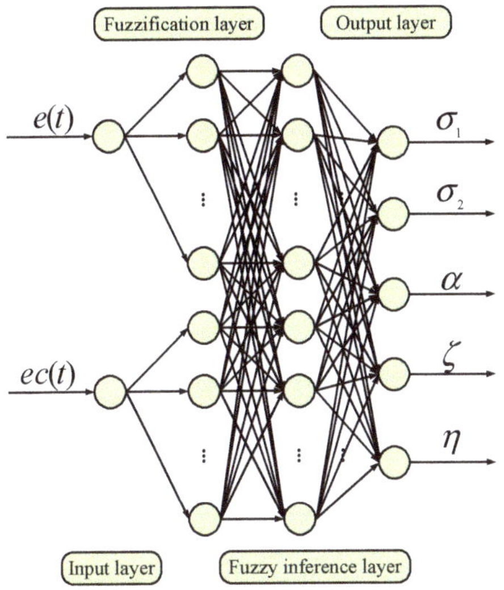

Figure 5. Structure block diagram of FNN.

The fuzzification layer considers a Gaussian membership function as the analysis method, which is shown as:

$$\begin{cases} o_2(i,j) = \exp\left(-[o_1(i) - c_{ij}]^2 / (b_{ij})^2\right) \\ i = 1,2; j = 1,2,\cdots,n \end{cases} \tag{21}$$

in which c_{ij} is the central value of membership function; b_{ij} represents the width of function curve; and n is the number of fuzzy subsets.

The calculation formula for the fuzzy inference layer is shown as:

$$\begin{cases} o_3(i) = o_2(1,k) \times o_2(2,k) \\ k_1 = 1,2,\cdots,5; k_2 = 1,2,\cdots,5 \\ i = 1,2,\cdots,25 \end{cases} \tag{22}$$

The calculation formula for the output layer is shown as follows:

$$\begin{cases} o_4(i) = \sum\limits_{j=1}^{25} o_3(j) \times \omega_{ij} \\ i = 1,2,3 \end{cases} \tag{23}$$

in which ω_{ij} is the weight coefficient.

The ILC and STA-SMO control parameters output by the output layer are shown as:

$$o_4(1) = \sigma_1, o_4(2) = \sigma_2, o_4(3) = \eta, o_4(4) = \xi, o_4(5) = \alpha \tag{24}$$

The parameters to be learned by the FNN are the affiliation function center value c_{ij}, the width value b_{ij}, and the weight value ω_{ij}, and the objective function is shown as follows:

$$E(k) = \frac{1}{2}[r(k) - y(k)]^2 \tag{25}$$

in the formula, $r(k)$, $y(k)$, and $r(k) - y(k)$ are the expected output, actual output of each network learning, and the control error.

The learning algorithm for the centroid, width, and weight of the affiliation function is shown as:

$$\begin{cases} c_{ij}(k+1) = c_{ij}(k) + \lambda \frac{\partial E(k)}{\partial c_{ij}(k)} + \mu \Delta c_{ij}(k) \\ b_{ij}(k+1) = b_{ij}(k) + \lambda \frac{\partial E(k)}{\partial b_{ij}(k)} + \mu \Delta b_{ij}(k) \\ \omega_{ij}(k+1) = \omega_{ij}(k) + \lambda \frac{\partial E(k)}{\partial \omega_{ij}(k)} + \mu \Delta \omega_{ij}(k) \end{cases} \tag{26}$$

where k is the iterative steps of network, λ represents the learning rate, and μ represents the inertia coefficient.

The block diagram of FNN and control system is presented in Figure 6:

(1) Determine neural network framework through number of layers and nodes in each network layer and select initial weights, learning rate, and momentum factor.
(2) Calculate the objective function obtained by running k times.
(3) Calculate the input and output values of each network layer.
(4) Calculate the output of the self-disturbance rejection controller.
(5) Perform backpropagation calculations to correct the weights of output and hidden layers.
(6) If successful, end iteration; if not, return to step (2).

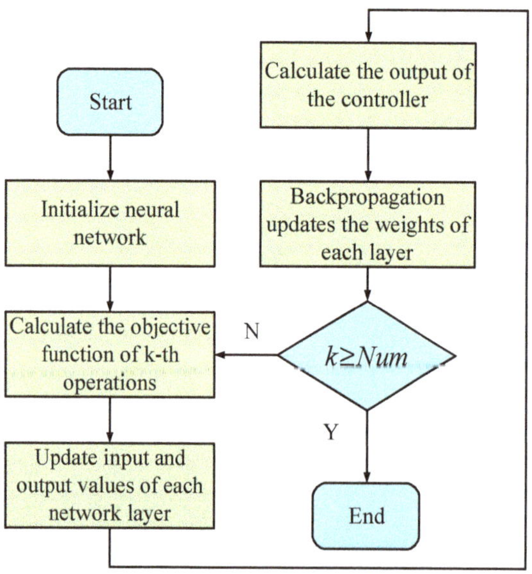

Figure 6. Block diagram of fuzzy neural network ILC control system.

4. Simulation Results and Discussions

The simulation model in this paper is based on MATLAB R2017b/Simulink environment. The structure block diagram of sensorless control model for RDT motor is shown in Figure 7. The main parameters of the RDT motor in simulation are summarized in Table 1.

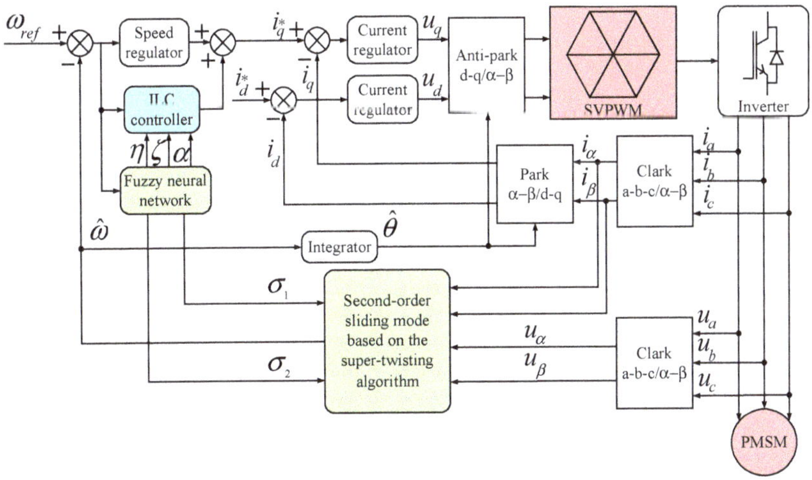

Figure 7. Structure block diagram of sensorless control for RDT motor.

Table 1. Main parameters of RDT motor.

Parameters	Value
Number of motor pole pairs (p)	10
Rotor flux (ψf)	0.0078 Wb
Rated voltage (U)	48 V
Inertia moment (J)	0.00106 kg·m^2
Rated power (Pe)	0.7 kW
Resistance (R)	0.061 Ω
Inductance (L)	0.12 mH

To verify the system control performance based on the proposed algorithm, comparisons between conventional PI-STA and modified ILC-STA have been carried out through simulations. The motor starts with the initial given speed of 1200 r/min, the load torque changes to 1.5 N·m at 0.4 s, the given speed changes to 800 r/min at 0.6 s, and the load torque changes back to 0 at 0.9 s.

The simulation results of the RDT motor using conventional PI-STA algorithm are shown in Figure 8. From top to bottom, the estimated speed and actual speed, the speed error, the given torque and electromagnetic torque, the estimated and actual position, and the position error and the three-phase stator currents are presented. It can be seen that the speed error is about −3 and 22 r/min under the 1200 r/min given speed and about −1 and 12 r/min under the 800 r/min given speed. The torque errors are about −1.2 and 1.1 Nm, −1.1 and 1.2 Nm, −1.3 and 1.2 Nm, and −0.9 and 0.8 Nm, respectively. The position error of the RDT motor is about 0.7 rad.

The simulation results of the RDT motor using modified ILC-STA are shown in Figure 9. It can be seen that the speed error is about −2 and 11 r/min under the 1200 r/min given speed and about −1 and 7 r/min under the 800 r/min given speed. The torque errors are about −0.4 and 0.4 Nm, −0.6 and 0.6 Nm, −0.5 and 0.5 Nm, and −0.5 and 0.5 Nm, respectively. The estimated position error of the RDT motor is about 0.7 rad.

From the simulation results, it can be seen that compared to the conventional PI-STA, the modified ILC-STA produces smaller speed and torque errors. The estimated position error is almost the same. The RDT motor speed, torque, and stator currents results display less chattering. This proves that the modified algorithm obviously maintains better control performance. Table 2 shows the performance comparison of different control algorithms.

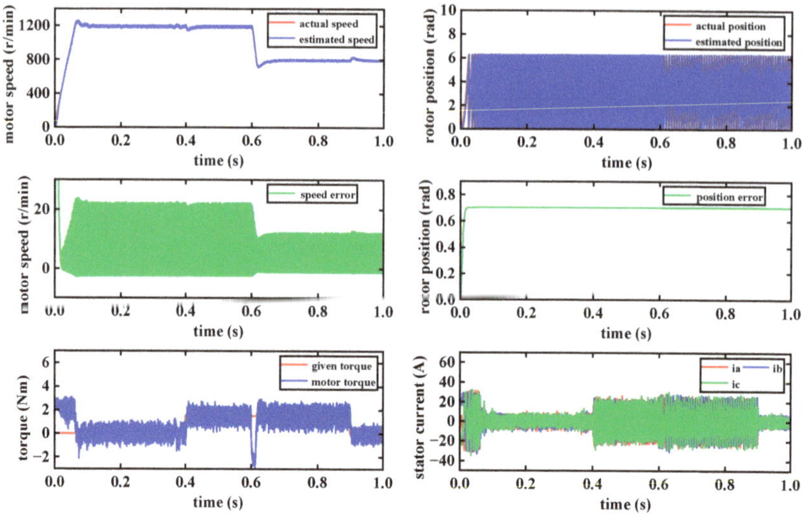

Figure 8. Simulation results using conventional PI-STA.

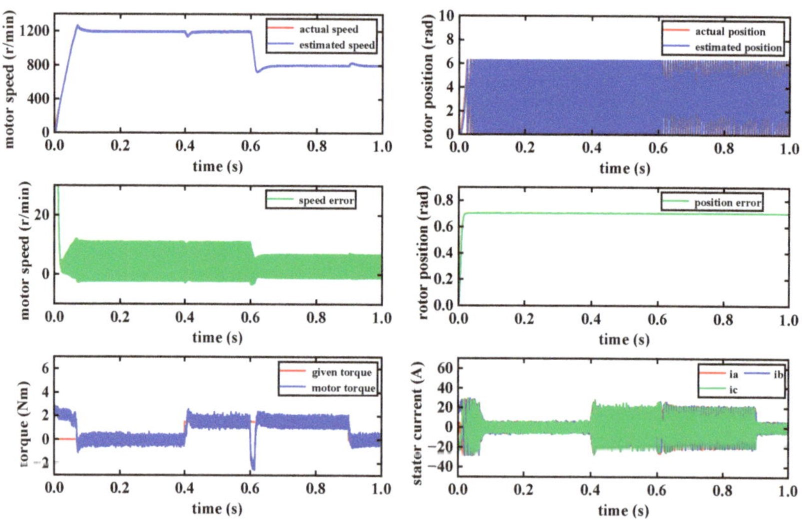

Figure 9. Simulation results using modified ILC-STA.

Table 2. Simulation performance comparison of different control algorithms.

Control Algorithm	Speed Error Ripple (r/min)	Torque Ripple (Nm)	Maximum Speed Error (r/min)
Conventional PI-STA	25/13	2.3/2.3/2.5/1.7	22/12
Modified ILC-STA	13/8	0.8/1.2/1/1	11/7

5. Experimental Results and Discussions

In order to verify the performance of the RDT motor sensorless control algorithm, a TMS320F28335 DSP-based experimental setup was built, as shown in Figure 10. The experimental setup consists of a magnetic powder brake, a torque sensor, and a test RDT motor, which are fixed together on a T-slotted platform and connected by elastic couplings.

The main parameters of the test RDT motor are the same as that in the simulation. And the experimental results under different conditions are shown in Figures 11–13.

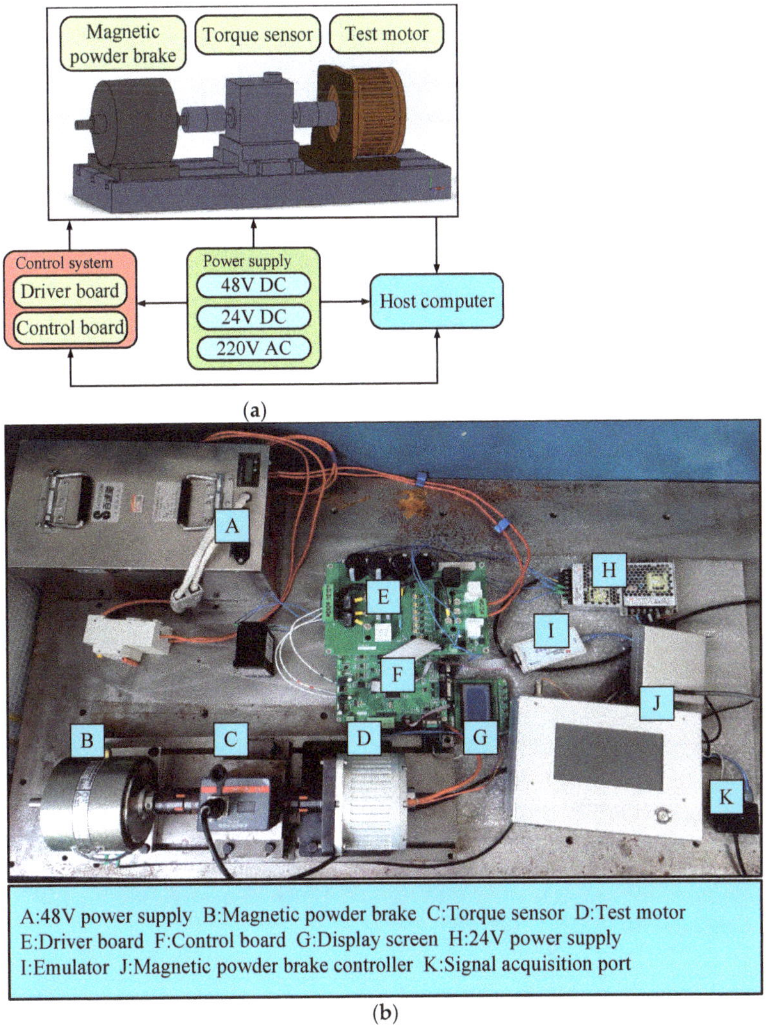

Figure 10. Experimental setup for the RDT motor. (**a**) Block diagram of experimental setup. (**b**) Diagram of physical experimental setup.

Condition 1: The tested given speed is set to 1200 r/min with sudden load torque changes from 0 to 1.5 Nm. Figure 11 shows the experimental results comparison between the conventional PI-STA and the modified ILC-STA in each case of step change in load torque. From top to bottom, the estimated speed and actual speed, the speed error, the given torque and electromagnetic torque, the estimated and actual position, the position error, and the three-phase stator currents are plotted. It can be observed that the stable error of the modified algorithm is smaller than that of the conventional algorithm. The motor speed estimation error ripple is about −2 and 22 r/min, respectively. The motor speed error ripple is about 24 r/min, and the electromagnetic torque ripples are about 2.2 Nm and 3 Nm, respectively. The rotor position error between the estimate and actual value of RDT motor is about 0.7 rad. When the modified ILC-STA is used, the motor speed estimation errors are about −2 and 11 r/min, respectively. The motor speed error ripple is about 13 r/min, and the electromagnetic torque ripples are about 0.9 Nm and 1 Nm, respectively. The rotor position error between the estimate and actual value of RDT motor is also about 0.7 rad. From Figure 11, it can be seen that the motor speed and electromagnetic torque waveforms are more stable, and the ripple is smaller than when using the conventional algorithm. The RDT motor speed error ripple deceases by 46% and the torque ripple deceases by 59% and 67%, respectively. The maximum speed error deceases by 11 r/min.

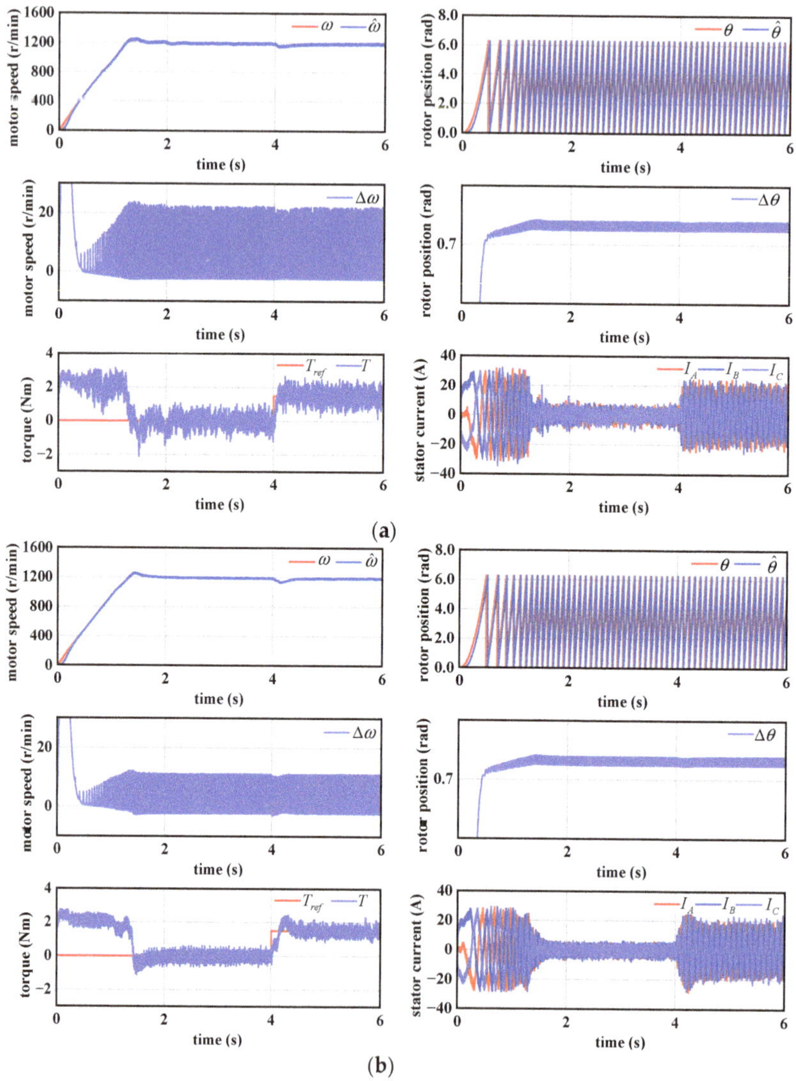

Figure 11. Experimental results under a step change of the load torque condition. (**a**) Conventional PI-STA. (**b**) Modified ILC-STA.

Condition 2: The tested given speed decreases from 1200 r/min to 800 r/min, with a continuous load torque of 1.5 Nm. Figure 12 shows the experimental results comparison between conventional PI-STA and modified ILC-STA in case of a step change in given speed. It is obvious that the stable error of the modified algorithm is smaller than that of the conventional algorithm. The motor speed estimation error ripples are about −1 and 12 r/min, respectively. The motor speed error ripple is about 13 r/min, and the electromagnetic torque ripple is about 2.5 Nm. When the modified ILC-STA algorithm is used, the motor speed estimation errors are about −1 and 7 r/min, respectively. The motor speed error ripple is about 8 r/min, and the electromagnetic torque ripple is about 1 Nm. From Figure 12, it can be seen that the motor actual speed and electromagnetic torque waveforms are more stable and the ripple is smaller than when using conventional algorithm. The RDT motor speed error ripple and torque ripple decease by 38% and 60%, respectively. The maximum speed error deceases by 5 r/min.

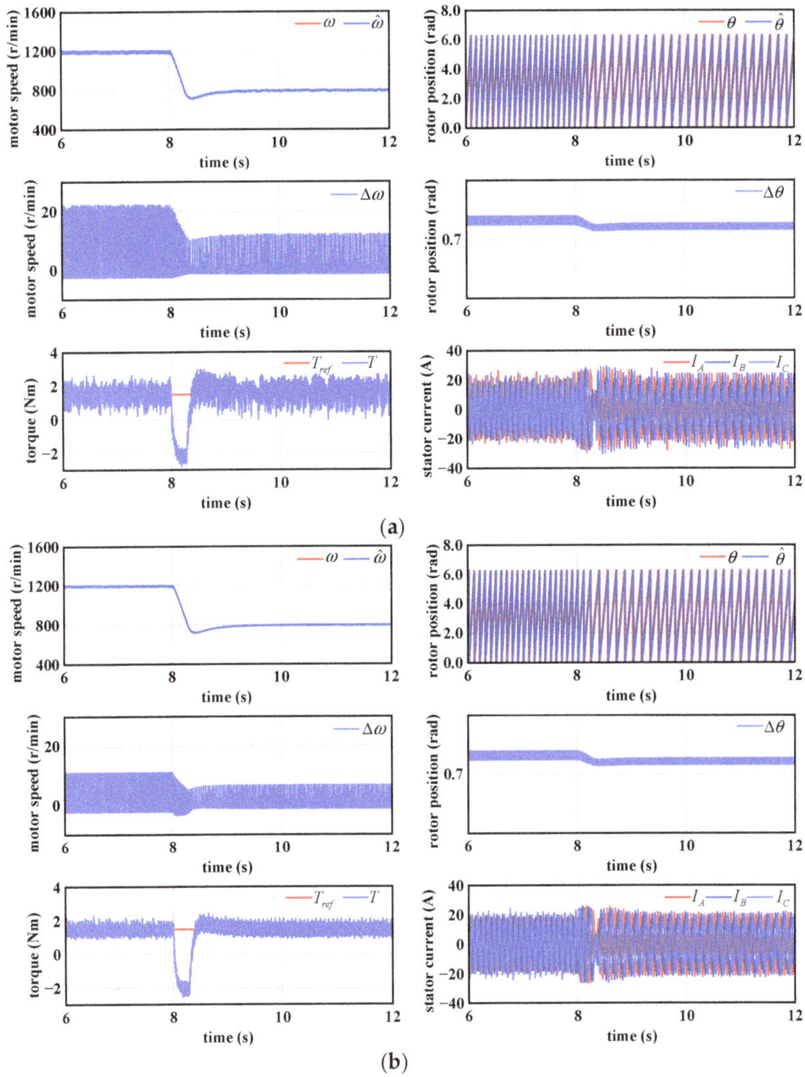

Figure 12. Experimental results under a step change of a given speed condition. (**a**) Conventional PI-STA. (**b**) Modified ILC-STA.

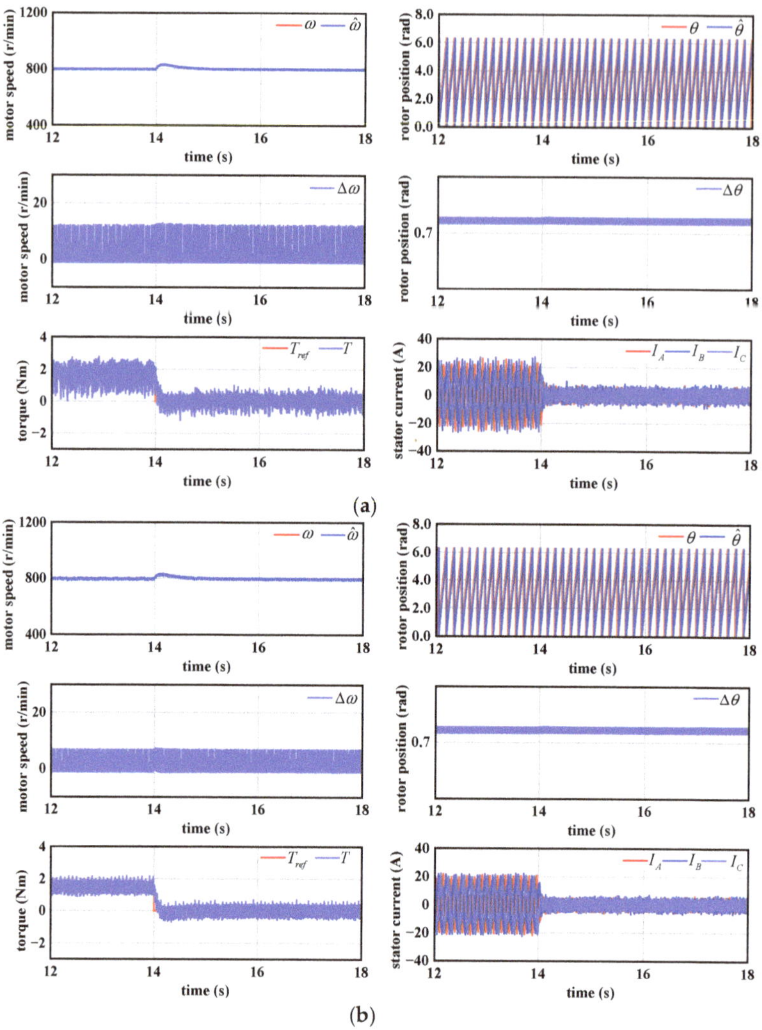

Figure 13. Experimental results under a step change of the load torque condition. (**a**) Conventional PI-STA. (**b**) Modified ILC-STA.

Condition 3: The tested given speed is set as the speed of 800 r/min with sudden load torque changes from 1.5 Nm to 0. Figure 13 shows the experimental results comparison between conventional PI-STA and modified ILC-STA in the case of step change in load torque. Similarly to the previous conditions, it is obvious that the stable error of the modified algorithm is smaller than that of the conventional algorithm. The motor speed estimation error ripples are about −1 and 12 r/min, respectively. The motor speed error ripple is about 13 r/min, and the electromagnetic torque ripple is about 1.7 Nm. When the modified ILC-STA algorithm is used, the motor speed estimation errors are about −1 and 7 r/min, respectively. The motor speed error ripple is about 8 r/min, and the electromagnetic torque ripple is about 1 Nm. Similarly to the previous conditions, it is obvious that the motor speed and electromagnetic torque waveform are more stable and the ripple is smaller than when using conventional method. The speed error ripple and torque ripple decease by 38% and 41%, respectively. The maximum speed error deceases by 5 r/min.

Similarly to the simulation results, it can be seen from the experimental results that compared to the conventional PI-STA, the modified ILC-STA produces lower speed errors and smaller electromagnetic torque errors. The motor speed, electromagnetic torque, and stator currents results also have less chattering. Table 3 summarizes the performance comparison of the different control algorithms.

Table 3. Experimental performance comparison of different control algorithms.

Control Algorithm		Speed Error Ripple (r/min)	Torque Ripple (Nm)	Maximum Speed Error (r/min)
Conventional PI-STA	Condition 1	24	2.2/3	22
	Condition 2	13	2.5	12
	Condition 3	13	1.7	12
Modified ILC-STA	Condition 1	13	0.9/1	11
	Condition 2	8	1	7
	Condition 3	8	1	7

6. Conclusions

In this paper, a RDT motor position-sensorless control algorithm that uses an adaptive second-order SMO based on STA is presented to reduce system chattering, and a closed-loop PI-ILC algorithm with the forgetting factor is used to suppress torque vibration. In addition, FNN training is used to optimize the ILC and STA-SMO parameters. The simulation and experimental analysis have been used to verify the proposed algorithm. The presented RDT sensorless control algorithm can be used to reduce the chattering generated in the RDT motor speed and torque ripple with improved speed tracking accuracy. The accuracy and complexity have been balanced and can be used to explore future research.

Our future work will apply the presented RDT sensorless control algorithm to a real ship, which will greatly contribute to the realization of smart and low carbon emission-oriented maritime traffic control technology.

Author Contributions: Conceptualization, Z.Y. and H.B.; methodology, Z.Y. and H.B.; software, Z.Y.; validation, Z.Y.; formal analysis, Z.Y.; investigation, Z.Y.; resources, Z.Y.; data curation, Z.Y.; writing—original draft preparation, Z.Y.; writing—review and editing, Z.Y., X.Y., W.O., H.B. and J.X.; visualization, Z.Y.; supervision, X.Y., W.O., H.B. and J.X.; project administration, X.Y., W.O. and H.B.; funding acquisition, X.Y., W.O. and H.B. All authors have read and agreed to the published version of the manuscript.

Funding: This research was partly funded by the National Key Research and Development Project, grant number 2022YFB4300802, and partly funded by the European Union's Horizon 2020 Research and Innovation Programme RISE, under grant agreement no. 823759, REMESH.

Institutional Review Board Statement: Not applicable.

Informed Consent Statement: Not applicable.

Data Availability Statement: Data are contained within the article.

Conflicts of Interest: The authors declare no conflicts of interest.

References

1. Yan, X.; Liang, X.; Ouyang, W.; Liu, Z.; Liu, B.; Lan, J. A review of progress and applications of ship shaft-less rim-driven thrusters. *Ocean Eng.* **2017**, *144*, 142–156. [CrossRef]
2. Yang, Z.; Yan, X.; Ouyang, W.; Bai, H. A Review of Electric Motor and Control Technology for Rim-Driven Thruster. *Trans. China Electrotech. Soc.* **2022**, *37*, 2949–2960. [CrossRef]
3. Wang, D.; Jin, S.B.; Wei, Y.S. Review on The Integrated Electric Propulsion System Configuration and its Applications. *Proc. CSEE* **2020**, *40*, 3654–3663.

4. Wang, G.L.; Valla, M.; Solsona, J. Position Sensorless Permanent Magnet Synchronous Machine Drives—A Review. *IEEE Trans. Ind. Electron.* **2020**, *67*, 5830–5842. [CrossRef]
5. Yang, H.; Yang, R.; Hu, W.; Huang, Z. FPGA-Based Sensorless Speed Control of PMSM Using Enhanced Performance Controller Based on the Reduced-Order EKF. *IEEE J. Emerg. Sel. Top. Power Electron.* **2021**, *9*, 289–301. [CrossRef]
6. Wang, G.; Hao, X.; Zhao, N.; Zhang, G.; Xu, D. Current Sensor Fault-Tolerant Control Strategy for Encoderless PMSM Drives Based on Single Sliding Mode Observer. *IEEE Trans. Transp. Electrif.* **2020**, *6*, 679–689. [CrossRef]
7. Bernard, P.; Praly, L. Estimation of Position and Resistance of a Sensorless PMSM: A Nonlinear Luenberger Approach for a Nonobservable System. *IEEE Trans. Autom. Control.* **2021**, *66*, 481–496. [CrossRef]
8. Abo Khalil, A.G.; Eltamaly, A.M.; Alsaud, M.S.; Sayed, K.; Alghamdi, A.S. Sensorless control for PMSM using model reference adaptive system. *Int. Trans. Electr. Energy Syst.* **2021**, *31*, e12733. [CrossRef]
9. Wu, L.; Liu, J.; Vazquez, S.; Mazumder, S.K. Sliding Mode Control in Power Converters and Drives: A Review. *IEEE-CAA J. Autom. Sin.* **2022**, *9*, 392–406. [CrossRef]
10. Ren, N.N.; Fan, L.; Zhang, Z. Sensorless PMSM Control with Sliding Mode Observer Based on Sigmoid Function. *J. Electr. Eng. Technol.* **2021**, *16*, 933–939. [CrossRef]
11. Kang, W.; Li, H. Improved sliding mode observer based sensorless control for PMSM. *IEICE Electron. Express* **2017**, *14*, 20170934. [CrossRef]
12. Ye, S.C.; Yao, X.X. An Enhanced SMO-Based Permanent-Magnet Synchronous Machine Sensorless Drive Scheme with Current Measurement Error Compensation. *IEEE J. Emerg. Sel. Top. Power Electron.* **2021**, *9*, 4407–4419. [CrossRef]
13. Yang, Z.B.; Ding, Q.F.; Sun, X.D.; Lu, C.L.; Zhu, H.M. Speed Sensorless Control of a Bearingless Induction Motor Based on Sliding Mode Observer and Phase-Locked Loop. *ISA Trans.* **2022**, *123*, 346–356. [CrossRef] [PubMed]
14. Gong, C.; Hu, Y.H.; Gao, J.Q.; Wang, Y.G.; Yan, L.M. An Improved Delay-Suppressed Sliding-Mode Observer for Sensorless Vector-Controlled PMSM. *IEEE Trans. Ind. Electron.* **2020**, *67*, 5913–5923. [CrossRef]
15. Liang, D.L.; Li, J.; Qu, R.H. Sensorless Control of Permanent Magnet Synchronous Machine Based on Second-order Sliding-mode Observer with Online Resistance Estimation. *IEEE Trans. Ind. Appl.* **2017**, *53*, 3672–3682. [CrossRef]
16. Liang, D.L.; Li, J.; Qu, R.H. Adaptive Second-order Sliding-mode Observer for PMSM Sensorless Control Considering VSI Nonlinearity. *IEEE Trans. Power Electron.* **2017**, *33*, 8994–9004. [CrossRef]
17. Nakao, N.; Akatsu, K. Suppressing pulsating torques: Torque ripple control for synchronous motors. *IEEE Ind. Appl. Mag.* **2014**, *20*, 33–44. [CrossRef]
18. Chen, Z.F.; Li, Z.X.; Ma, H.Z. A harmonic current injection method for electromagnetic torque ripple suppression in permanent-magnet synchronous machines. *Int. J. Appl. Electromagn. Mech.* **2017**, *53*, 327–336. [CrossRef]
19. Zhang, H.Y.; Xu, H.P.; Fang, C. Torque ripple suppression method of direct drive permanent magnet synchronous motor based on proportional-integral and quasi resonant controller. *Trans. China Electrotech. Soc.* **2017**, *32*, 41–51.
20. Liu, J.; Li, H.; Deng, Y. Torque ripple minimization of PMSM based on robust ILC via adaptive sliding mode control. *IEEE Trans. Power Electron.* **2018**, *33*, 3655–3671. [CrossRef]
21. Dang, X.J.; Miao, M.Y.; Jiang, H. Torque ripple suppression for switched reluctance motors based on current self-adaptive control. *J. Vib. Shock* **2018**, *37*, 66–71.
22. Levant, A. Sliding order and sliding accuracy in sliding mode control. *Int. J. Control* **1993**, *58*, 1247–1263. [CrossRef]
23. Moreno, J.A.; Osorio, M. Strict lyapunov functions for the super-twisting algorithm. *IEEE Trans. Autom. Control* **2012**, *57*, 1035–1040. [CrossRef]
24. Moreno, J.A.; Osorio, M. A lyapunov approach to second-order sliding mode controllers and observers. In Proceedings of the 47th IEEE Conference on Decision and Control, Cancun, Mexico, 9–11 December 2008.
25. Chen, X.; Liu, S.; Liu, R.W.; Wu, H.; Han, B.; Zhao, J. Quantifying Arctic oil spilling event risk by integrating an analytic network process and a fuzzy comprehensive evaluation model. *Ocean Coast Manag.* **2022**, *228*, 106326. [CrossRef]
26. Damien, F.; Ngac, K.N.; Patric, W. A self-learning solution for torque ripple reduction for non-sinusoidal permanent magnet motor drives based on artificial neural networks. *IEEE Trans. Ind. Electron.* **2014**, *61*, 655–666.
27. Wu, B.F.; Lin, C.H. Adaptive neural predictive control for permanent magnet synchronous motor systems with long delay time. *IEEE Access* **2019**, *7*, 108061–108069. [CrossRef]
28. Yang, S.Q.; Chen, C. Application of fuzzy neural network PID algorithm in oil pump control. In Proceedings of the 2019 International Conference on Computer Network, Electronic and Automation (ICCNEA), Xi'an, China, 27–29 September 2019.
29. Chen, X.; Wang, M.; Ling, J.; Wu, H.; Wu, B.; Li, C. Ship imaging trajectory extraction via an aggregated you only look once (YOLO) model. *Eng. Appl. Artif. Intell.* **2024**, *130*, 107742. [CrossRef]

Journal of
Marine Science and Engineering

MDPI

Article

Ship Global Traveling Path Optimization via a Novel Non-Dominated Sorting Genetic Algorithm

Shuling Zhao * and Sishuo Zhao

School of Electrical Engineering and Automation, Jiangxi University of Science and Technology, Ganzhou 341000, China; 6720221359@mail.jxust.edu.cn
* Correspondence: 9120010008@jxust.edu.cn

Abstract: Due to the intensification of economic globalization and the impact of global warming, the development of methods to reduce shipping costs and reduce carbon emissions has become crucial. In this study, a multi-objective optimization algorithm was designed to plan the optimal ship route for safe cross-ocean navigation under complex sea conditions. Based on the traditional non-dominated sorting genetic algorithm, considering ship stability and complex marine environment interference, a non-dominated sorting genetic algorithm model considering energy consumption was designed with the energy consumption and navigation time of the ship as the optimization objectives. The experimental results show that although the proposed method is 101.23 nautical miles more than the large ring route, and the voyage is increased by 10.1 h, the fuel consumption is reduced by 92.24 tons, saving 6.94%. Compared with the traditional genetic algorithm, the voyage distance and time are reduced by 216.93 nautical miles and 7.5 h, and the fuel consumption is reduced by 58.82 tons, which is almost 4.54%. Through experimental verification, the proposed model can obtain punctual routes, avoid areas with bad sea conditions, reduce fuel consumption, and is of great significance for improving the safety and economy of ship routes.

Keywords: route planning; energy consumption; weather forecast; ship stability; global optimization

1. Introduction

Maritime transportation is the main mode of transportation in trade, and the shipping industry accounts for more than 80% of global trade. However, with the continuous development of shipping, the direct result is an increase in global greenhouse gas (GHG) emissions. To this end, the International Maritime Organization (IMO) has established that by 2050, the global shipping industry will reduce carbon by at least 50% compared with 2008, and other shipping organizations have proposed a series of measures [1,2]. The acceleration of the global economic process has put forward higher requirements for maritime transportation. It has become an urgent task to improve shipping efficiency and reduce shipping costs [3]. However, due to the influence of the marine environment, the safety of ships has been greatly tested. Planning a route as economical and safe as possible has become a key factor in improving shipping efficiency.

The core of route optimization is to find an optimal route. Based on marine environment data, combined with ship performance and navigation tasks, a safe and economical route is selected. With the increasingly significant role of route optimization in maritime navigation, many methods for planning routes have been proposed. When navigating in complex weather conditions, the captain uses weather data to avoid potential dangers during the voyage and maximizes the safety of ship navigation. According to different theories, ship route planning methods can be divided into three categories: graph search method, meta-heuristic algorithm of dynamic path, and trajectory analysis method using big data [4].

In the early days, the graph search method usually set the route at a fixed speed, and then adjusted the speed according to the navigation loop. The navigation area is discretized into nodes and edges, and the route planning problem is transformed into directed graph search, such as Dijkstra algorithm and A-star algorithm [5], which are gradually applied to path planning to obtain the shortest time path and the shortest distance path. Although these algorithms can effectively avoid static obstacles, they are not suitable for dynamic and complex marine environments. In order to solve different problems, the improvement of these algorithms has been ongoing. Silveira et al. [6] and Wang et al. [7] made targeted improvements to the Dijkstra algorithm, although there is still more room for improvement. Liu et al. proposed a PE-A* algorithm, which uses the potential energy field to express the environmental field to achieve the effect of global planning and local collision avoidance [8]. Addressing this problem, Shin et al. proposed an improved A-star algorithm, which can avoid the limitations of the initial A-star algorithm [9]. Mannarini et al. proposed a new dynamic programming method (VISIR: Discovering Safe and Efficient Routes) [10], which evaluates the optimal path in time by solving the basic differential equation of the optimal path in dynamic wind and waves.

At present, various meta-heuristic algorithms are applied to the field of route planning, such as the genetic algorithm (GA) [11], simulated annealing algorithm, particle swarm optimization (PSO) [12], and ant colony algorithm. Zhou et al. proposed a hybrid genetic algorithm, which combines a simulated annealing algorithm with a genetic algorithm to avoid falling into a local optimum [13]. Zhang et al. combined the ant colony algorithm with the A-star algorithm [14], and introduced the Bessel curve method to smooth the path to obtain the optimal path.

The big data mining method is based on the statistical analysis of a large number of ship route data from the automatic identification system (AIS) to improve shipping efficiency [15–19]. This method usually considers historical ship trajectory data, using clustering methods for statistical analysis and establishing a route trajectory model to improve the safety of navigation.

Reasonable meteorological route arrangement can minimize the risks caused by total fuel consumption, sailing time, and adverse conditions, but in practice, it is difficult to pursue these three objectives at the same time. At present, the multi-objective optimization method is roughly divided into two kinds. One is to assign weights to the objective function and transform the multi-objective problem into a single-objective problem. This method simplifies the solution process, but the route involves conflicting objectives, such as reducing the total cost, emissions, and navigation time. The single-objective optimization method with heavy weight does not solve this conflict well. The latter finds the approximate optimal solution of the problem by finding the Pareto front and can complete the search in an acceptable time [20–24]. Du et al. proposed an improved fractional order particle swarm optimization (FOPSO) algorithm [25] that avoids problems such as falling into local optimization in solving multi-objective problems. Ma et al. used a non dominated sorting genetic algorithm to optimize the cost and time of navigation [26]. Considering carbon emissions and weather factors, Yuan et al. also proposed a weather uncertainty model [27], and added the weather probability model as a penalty factor to the non-dominated sorting genetic algorithm for route optimization. Polar navigation is also an important part of ship navigation. Szlapczynski et al. added the information of into the multi-objective evolutionary algorithm to find a route that is more in line with the decision makers' ideas [28], and considered the uncertainty of weather forecasts.

Addressing the problem of ship route optimization, this paper focuses on finding a safe and low-carbon route-optimization method, which designs a route for ocean-going navigation in bad sea conditions, takes the safety of ship navigation, sailing time, and fuel consumption as variables to optimize the route, and can update the weather and meteorology in time, avoid rough seas, and reach the destination safely. To summarize our views, the work we have done mainly includes the following aspects.

(1) A model for sea conditions using weather data is established. The complex sea area exceeding the threshold is marked as a dangerous area as a danger zone according to

the set safety threshold, and the calibrated dangerous area is updated by updating the environmental data.

(2) Considering the speed loss of ships in different marine environments, a speed-loss model of ships is established.

(3) On the basis of the above model, a route-optimization model considering both ship navigation time and fuel consumption is established by using a non-dominated sorting genetic algorithm (NSGA-II), and ship stability is added to the constraint factor to make the optimal path safer and more energy-saving.

The structure of the rest of this paper is as follows: In the second part, a complex marine environment model is proposed. By rasterizing the environmental data, a marine environment model is established to study the ship speed loss under different sea conditions. In the third part, the objective function of route optimization is designed, the constraints of navigation are proposed, and the NSGA-II is used to solve the route optimization problem. In the fourth part, the validity and reliability of the method are verified by comparing it to path optimization based on a genetic algorithm. Finally, the work done in this paper is summarized and discussed.

2. Materials and Methods

2.1. Complex Sea Environment Model

At present, meteorological data of weather forecasts can be obtained from the European Centre for Medium Weather Forecasting (ECMWF), National Centers for Environmental Prediction (NCEP), and National Oceanic and Atmospheric Administration (NOAA), etc. From these databases, weather and sea conditions that affect ship routes can be obtained.

Weather forecast data obtained from different agencies are slightly different in spatial and temporal resolution, but are usually expressed as a set of latitude–longitude grid data. At the weather forecast time-node t, the weather or sea-state element can be expressed as Equation (1):

$$W(\overline{x},\overline{y},\overline{t}) = f^{t}_{(p,v)} \tag{1}$$

where $\overline{x}$ and $\overline{y}$ are the latitude and longitude grid points, and W represents the weather forecast data of any position at any time t. The ship navigation environment information of any position $p(\varphi,\lambda)$ at any time $\overline{t}$ can be inferred from the weather forecast data set W, and for any position $p(\varphi,\lambda)$ at any time t, v represents the ship navigation speed at this time.

During the voyage, different ships can deal with different wind and wave conditions. According to the ship type, structural strength, operating performance, and loading capacity, the impact of wind and waves on the ship will be very different. Therefore, the actual situation of the ship should be taken into account in the specific definition of the strong wind and wave area. As long as the wind and wave environment in a certain sea area exceeds the bearing capacity of the ship, it can be defined as a strong wind and wave area. The environmental information is rasterized, and the rasterized environmental model is more conducive to the dangerous-area calibration of the sea conditions through which the route passes; the model is updated through real-time updated weather data, as shown in Figure 1. The route-optimization problem is transformed into an optimal path problem that can avoid obstacles.

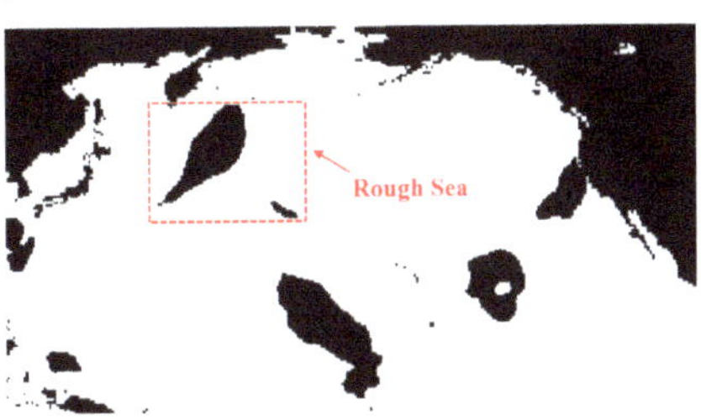

Figure 1. Sea state environment model.

2.2. Speed Loss Model of Ship

2.2.1. Definition of Route Model

In the route-optimized design, the whole route is defined as a combination of multiple route points, that is, the route is divided into multiple route segments, each including latitude and longitude coordinates (φ, λ) and navigation speed v. The starting position and the end position as well as the departure time and arrival time are set. Route optimization is used to find the shortest and safest route. The route consists of n waypoints, which are represented as $p = \{p_1, p_2, \cdots, p_n\}$, and the corresponding speed is represented by $v = \{v_1, v_2, \cdots, v_n\}$, as shown in Figure 2.

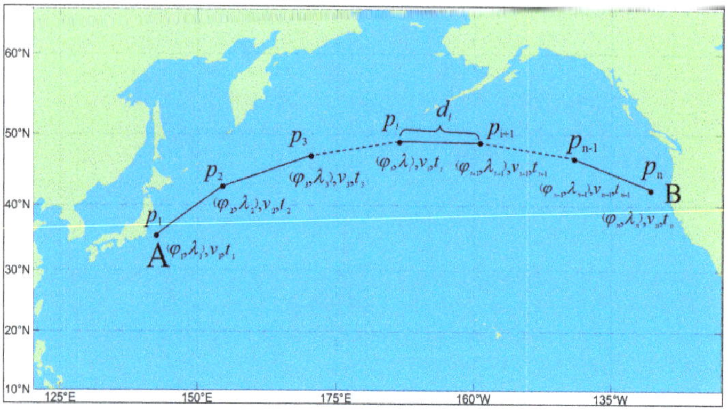

Figure 2. Route generation process.

2.2.2. Configuring Ship Speed

The external environment can have varying degrees of impact on the normal navigation of a ship in a complex and changing wind and wave environment. The ship will slow down if output power of the main engine remains constant, resulting in a lower actual speed than the still-water speed. This phenomenon is known as the natural speed loss of the ship. Of course, sometimes when the ship is sailing in bad sea conditions, the surrounding environment will pose a great threat to the safety of the ship route, and the captain will take the initiative to reduce the speed of the ship to ensure safe driving and avoid accidents. In ensuring safe running of the ship, reasonable planning of speed can make the ship sail at a suitable speed in different sea areas and achieve good operational benefits throughout the voyage. Under normal circumstances, the speed of a ship will be adjusted according to its own performance, navigation environment, and economic benefits, so as to keep it in a reasonable range, that is, the still-water speed v. To better study the influence of wind and waves on ship navigation, this paper assumes that the output power of the main engine of the ship is constant.

According to the empirical formula of Kwon [29], the speed loss of the ship can be shown as Equation (2):

$$\Delta v = \frac{v_i \times C_\beta C_U C_{Form}}{100} \tag{2}$$

Therefore, v_i represents the actual speed of the ship during navigation, and the ship water speed v_{wi} minus the speed loss Δv is obtained, as shown in Equation (3):

$$v_i = v_{wi} - \Delta v \tag{3}$$

However, the speed loss of ships during actual navigation is affected by factors such as the marine meteorological environment and ship performance. Specific factors are shown in Tables 1–3 [30]. Here, C_β represents the speed loss direction coefficient, which depends on the angle of direction of the weather and the value of the puff wind level, as shown in Table 1; C_U represents the deceleration coefficient, which varies with the blocking coefficient CB, the load condition, and the Froude number Fn, as shown in Table 2; C_{Form} represents the shape coefficient of the hull, which is related to the ship type, wind level, and ship displacement ∇ (m^3), as shown in Table 3.

Table 1. Speed loss direction coefficient C_β.

Wave (Wind) Direction	Direction Angle β	C_β
Top wave, headwind	0°	$2C_\beta = 2$
Oblique wave, facing wave	30–60°	$2C_\beta = 1.7 - 0.03(B_N - 4)^2$
Transverse wave, crosswinds	60–150°	$2C_\beta = 0.9 - 0.06(B_N - 6)^2$
Following seas, tailwind	150–180°	$2C_\beta = 0.4 - 0.03(B_N - 8)^2$

Table 2. Deceleration coefficient C_u.

Ship Blocking Coefficient CB	Ship Load Situation	C_u
0.55	Normal	$1.7 - 1.4Fn - 7.4(Fn)^2$
0.60	Normal	$2.2 - 2.5Fn - 9.7(Fn)^2$
0.65	Normal	$2.6 - 3.7Fn - 11.6(Fn)^2$
0.70	Normal	$3.1 - 5.3Fn - 12.4(Fn)^2$
0.75	Full load or normal	$2.4 - 10.6Fn - 9.5(Fn)^2$
0.80	Full load or normal	$2.6 - 13.1Fn - 15.1(Fn)^2$
0.85	Full load or normal	$3.1 - 18.7Fn - 28.0(Fn)^2$

Table 3. Hull shape coefficient C_{Form}.

Ship Form	C_{Form}
Full load conditions of all ship types (except cargo ships)	$0.5B_N + {B_N}^{6.5}/(2.7\nabla^{2/3})$
Preloading conditions of all ship types (except cargo ships)	$0.7B_N + {B_N}^{6.5}/(2.7\nabla^{2/3})$
Normal loading of cargo ships	$0.7B_N + {B_N}^{6.5}/(22.0\nabla^{2/3})$

3. Multi-Objective Route Optimization Model of the Ship

3.1. Objective Functions

In addition to the need to ensure the safety of navigation, a voyage must also consider the economic benefits of shipping, such as the amount of time it takes and how much fuel is used, in addition to ensuring the safety of the route. Therefore, ship route optimization is a complex problem involving multiple objectives, variables, and factors. The impact of sudden situations such as gusts is not taken into account by meteorological conditions to simplify the problem and facilitate the establishment and resolution of the model. The mathematical expression of the multi-objective route optimization model is as follows:

$$\min F(x) = [f_1(x), f_2(x)]^T \tag{4}$$

$$x = [p_1, p_2, \cdots, p_n, v_1, v_2, \cdots, v_n]^T \tag{5}$$

$$f_1(x) = \sum_{i=1}^{n} \frac{d_i}{v_{wi}} \tag{6}$$

$$f_2(x) = \sum_{i=1}^{n} \frac{d_i}{v_{wi}} \times \eta(v_i) \tag{7}$$

$$v_{\min} \le v_{wi} \le v_{\max}, i = 1, 2, \ldots, n \tag{8}$$

Here, x is the decision variable represented by the set of waypoints p and the corresponding actual speed v_i; n is the number of waypoints on the route. $F(\bullet)$ are the objective functions, where $f_1(\bullet)$ is the total time of n leglines by the ship along the route, and $f_2(\bullet)$ is the total fuel consumption of the ship during navigation, which is related to the speed of the ship in still water. d_i is the distance between waypoints i and $i+1$, v_{wi} is the still water speed of a ship sailing between waypoints i and $i+1$, $v_{\min}$ is the minimum speed, $v_{\max}$ is the maximum speed, and η is the fuel consumption rate determined by the performance of the main engine of the ship and the speed, in which the baseline fuel consumption of the ship can be obtained by using the Lagrange method based on a large number of voyage data of the ship.

During the voyage, ships will encounter shoals, reefs, and other obstacles, which greatly limit the scope of ship activities. Therefore, in the optimization process, it is necessary to stay as far away from them as possible. Equation (9) represents the geographical constraints of the ship in navigation:

$$D(p_i) > D_{draft} + D_{ukc}, i = 1, 2, \ldots, n-1 \tag{9}$$

where $D(p_i)$ is the water depth of the ship at the i-th waypoint, and D_{draft} and D_{ukc} respectively represent the average draft of the ship at this position and the gap under the keel.

During the voyage of the ship, under the combined action of external parameters such as wind, waves, and currents, the periodic change in the geometric shape of the immersed part of the hull and the shape of the waterline surface leads to the periodic change in the restoring moment of the ship, that is, the rolling phenomenon. This phenomenon can easily cause the ship to capsize and is a huge potential safety hazard. However, an important factor affecting rolling is the relationship between the encounter period and the natural period:

$$\mathrm{T}_n = \frac{2C \times B}{\sqrt{h_{GM}}} \tag{10}$$

$$T_e = \frac{2\pi}{\omega - \frac{\omega^2}{g} v \cos\chi} \tag{11}$$

In Equation (10), $C = 0.373 + 0.023 \times \frac{B}{d} - 0.00043L$; B is the width of the ship; L is the waterline length; d is the average water intake; and h_{GM} is the height of initial stable center.

In Equation (11), $\chi = \psi - \mu$ is the encounter angle; ψ is the heading angle; μ is the absolute wave direction angle; ω is the natural frequency of waves; v is the ship speed; and g is the acceleration due to gravity.

$$\left\{ \begin{array}{l} 1 - \tau_1 < \frac{T_e}{T_n} < 1 + \tau_1, \\ 1 - \tau_2 < \frac{2T_e}{T_n} < 1 + \tau_2, \\ 0.5Lpp \le \lambda \le 1.5Lpp \end{array} \right\} \tag{12}$$

Equation (12) addresses the stability constraint of the ship in navigation, where τ_1 and τ_2 are the thresholds, taken here to be 0.5; Lpp is the length of the ship; $\lambda = 1.56T^2$ is the wave length of the wave; and T is the wave period. Equation (13) represents the constraints of weather or sea condition that the ship can withstand:

$$\left\{ \begin{array}{l} H_{wave}(p_i) \le H_{\max}, \\ W_{wind}(p_i) \le W_{\max}, i = 1, 2, \ldots, n \end{array} \right\} \tag{13}$$

These conditions are usually simply set to deterministic thresholds, such as maximum allowable wave height and maximum allowable wind speed. H and W represent the wave height and wind speed encountered by the ship at the i-th waypoint, which do not exceed the maximum allowable wave height $H_{\max}$ and the maximum allowable wind speed $W_{\max}$.

3.2. Designing the Algorithm

By considering multiple objective functions and constraints at once, they can determine that a population approach is the most optimal solution in the feasible region.

The behavior of organisms during evolution is simulated using natural selection, elimination, and reproduction mechanisms. Selecting a suitable multi-objective evolutionary algorithm (MOEA) is crucial to finding the optimal solution for the multi-objective and multi-constraint route optimization problem.

Currently, the non-dominated sorting genetic algorithm is a mature MOEA. It achieves efficient search, uniform distribution, and diversity maintenance of solutions through fast non-dominated sorting, congestion comparison, and elite strategy, and has high convergence and robustness. Compared to other algorithms, the NSGA-II is more stable and efficient, so it is chosen to solve the multi-objective route optimization problem.

In the MOEA, the initial population is randomly generated in the solution space, and the characteristics of individuals are represented by chromosomes, and then the population is updated by crossover, mutation, and selection. Similarly, when solving the multi-objective route optimization problem, the individual ship route can be composed of variables determined by waypoints and speed information, as shown in Equation (14):

$$I_N = [p_1, p_2 \cdots, p_N, v_1, v_2, \cdots, v_N]^T, N = 1, 2, \cdots N \tag{14}$$

where p_1 and p_N are the departure position and destination position, respectively; and N is the number of individuals in the population.

In order to improve the calculation speed and accuracy of the algorithm, the initial population is generated randomly, that is, the routes around the great circle empirical routes are randomly generated as individual populations, and the navigation speed is also generated randomly. A new population Q_0 is generated by selecting, crossing, and mutating the initial population I_0 to ensure that the population sizes of I_0 and Q_0 are both N. I_t and Q_t are merged into C_t, and after the C_t is quickly non-dominated, a partial ordered set is established by calculating the crowding distance of all individuals in a certain self, and then individuals are selected in turn to enter I_{t+1} until the scale reaches N. Then, an iterative operation is performed to determine if the maximum number of iterations is satisfied. If not, the new generation of population Q_{t+1} is continued to be generated. If it is satisfied, the individual of I_{t+1} is output, that is, the non-dominated optimal solution set.

Route optimization is an optimization problem to find the minimum value of the objective function; the smaller the value, the better the individual's fitness and survival probability. In the process of population evolution, a penalty factor ζ is added to the objective function to evaluate the fitness of the individual. The individual increases the elimination rate of the next generation by increasing the value of the cost function, but the individual can evolve through operations such as cross mutation, which not only ensures the randomness of the individual, but also improves the efficiency of algorithm optimization. As shown in Equation (15):

$$F(I_N) = f_j(I_N) \times \zeta, j = 1, 2 \tag{15}$$

It is worth noting that the NSGA-II algorithm's discovery of the Pareto solution set is not the only optimal solution. Therefore, solving the route optimization problem by evolutionary algorithm is to find a set of solution sets that make the objective function more complementary, and then sort the solution sets according to the user's optimization criteria based on sorting method, and finally obtain the optimal route.

4. Experiment Analysis

4.1. Data Acquisition

Sea conditions and weather factors such as significant wave height (SWH), mean sea level pressure (MSLP), 10 m U-component of wind (U10), 10 m V-component of wind (V10), mean period of wind waves (MPWW), mean wave direction (MWD), and effective height of wind waves (SHWW) are obtained from the ECMWF. The meteorological data are stored in NetCDF format, with a grid resolution of 1 × 1.

Historical weather data from of the Pacific Ocean were selected as the input of the simulated environmental data. In order to make the experimental ship sail closer to the actual navigation environment, the environment data were updated every 1 h and re-entered into the model. Figures 3 and 4 show part of the data visualization.

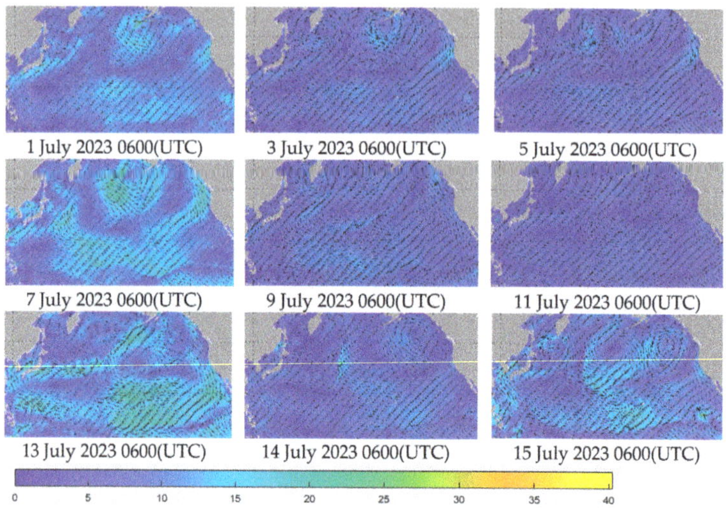

Figure 3. Wind field at different times (Pacific Ocean).

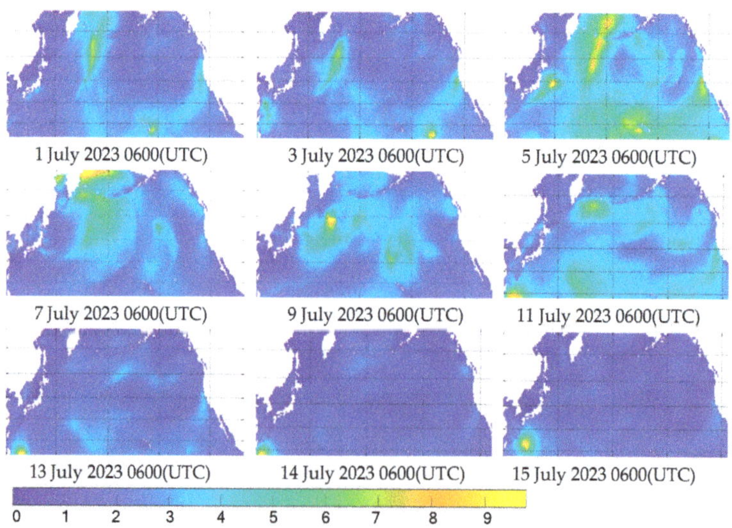

Figure 4. Effective wave heights at different times (Pacific Ocean).

4.2. Route Optimization Experiment

To verify the feasibility of the route optimization method proposed in this paper, bad weather conditions are simulated. In the experiment, the Pacific Ocean is selected as the navigation area, starting from Yokohama Port in Japan and ending at San Francisco Port in the United States, and the weather forecast data in July 2023 is selected as the navigation condition of the ship [31–33]. In addition, a typical container ship is simulated in the route optimization. The relevant parameters of the ship are shown in Table 4, and the speed of the ship is set to 12–20 knots. In this experiment, waves with heights higher than 6m are considered dangerous sea conditions. Based on the traditional speed fuel consumption table method this paper adopts the Lagrange interpolation polynomial fuel consumption baseline as shown in Figure 5.

Table 4. Relevant parameters of experimental ship.

Parameter	Numerical Value
Length/m	348
Ship width/m	51.2
Average draft/m	13.5
Displacement/m^3	169,700
Ship blocking coefficient	0.693
Initial metacentric height/m	4

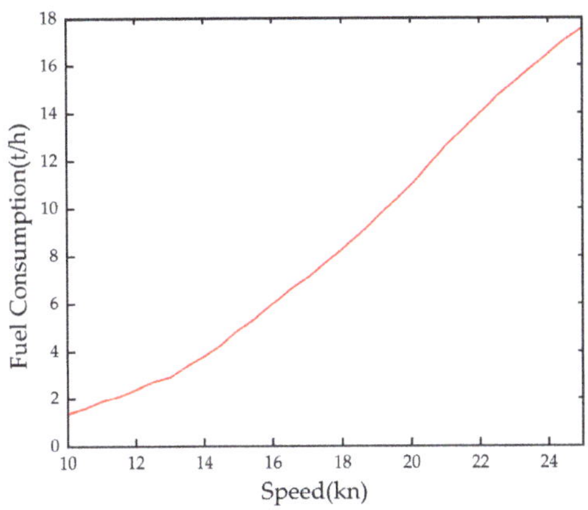

Figure 5. Fuel consumption benchmark curve.

In this experiment, in order to verify the feasibility of the proposed route optimization model in bad weather, this experiment adopts three methods for route planning and compares them under the same navigation conditions. The Great Circle Route is the shortest route between two points on the sphere, that is, in the ideal environment, the ship will take the Great Circle Route as the best choice; in reality, however, the ship is not easy to sail along the Great Circle Route, and considering the dangerous meteorological environment, the route will be adjusted. Therefore, three kinds of route planning are adopted to find the optimal route, including the Great Circle Route, the GA Optimization Method, and the NSGA-II Optimization Route, which are the classic ship routes. Figure 6a,b show the convergence rate of the genetic algorithm and Pareto solution set of the NSGA-II, respectively. Figure 6a uses a single-objective genetic algorithm with weights.

In the simulation experiment, the speed of the experimental ship is 15 knots, which is the speed of the circular route, that is, the speed of the empirical route. The speed of the other two optimized routes is calculated according to the speed loss calculation formula. Table 5 shows the parameter settings of the two methods. In the case of the same ship navigation environment, the routes obtained by the two optimization methods are compared with the classic routes. Table 6 shows the route information obtained by different methods. Figure 7 shows the optimal path optimized by the two algorithms.

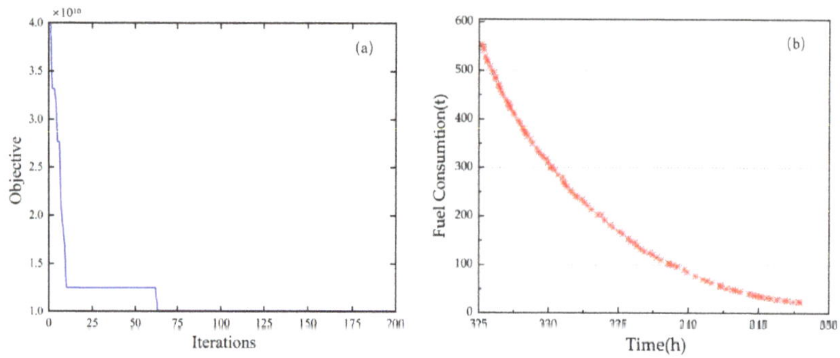

Figure 6. (**a**) Convergence curve of the genetic algorithm (GA). (**b**) Pareto front solution set of the NSGA-II algorithm.

Table 5. Relevant parameters of evolutionary algorithms.

Algorithm	Parameter	Value
Genetic algorithm	Population size	50
	Maximum number of iterations	200
	Penalty factor	1×10^{10}
	Cross probability	0.6
	Mutation probability	0.1
	Step factor	0.3
NSGA-II	Population size	50
	Iterations	200
	Penalty factor	1×10^{10}

Table 6. Comparison results of three routes.

Route	Distance/nm	Voyage Time/h	Fuel Consumption/t
Empirical route	4864.50	324.3	1329.78
GA route	5182.66	341.7	1296.36
NSGA route	4965.73	334.2	1237.54

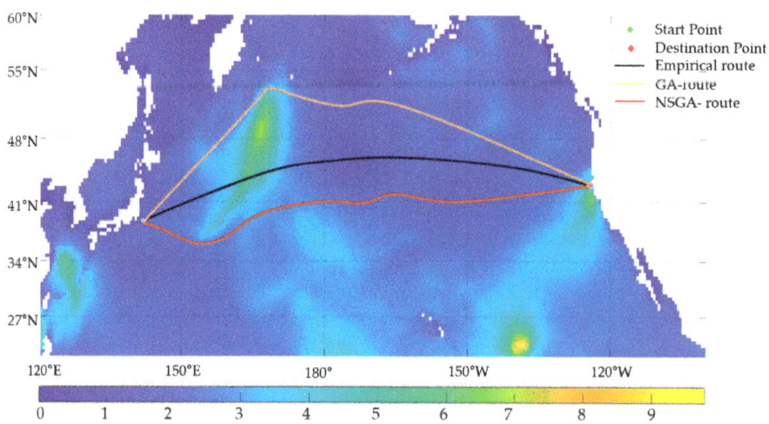

Figure 7. Route trajectories obtained by different methods.

As can be seen from Figure 7 and Table 6, the Great Circle Route obviously encounters waves higher than 6 m in the course of sailing, and the voyage is very risky. The route obtained by the multi-objective evolutionary method can avoid the dangerous sea state area, and a route with the lowest fuel consumption can be found within the given constraints. However, although the GA route avoids the dangerous area, it has a higher fuel consumption and takes longer. As for the route optimization method in complex weather, the voyage is 101.23 nautical miles longer than the voyage along the Great Circle Route, because it detours a certain distance to avoid bad weather. However, it can be clearly seen from the table that the fuel consumption of the route planned by the method proposed in this paper is 92.24 t (6.94%) less than that of the Great Circle Route, making it an economical and safe route.

Figure 8 shows the significant wave heights experienced by the three routes at different time and space positions; the red solid line represents the maximum allowable significant wave height (H = 6 m). From Figure 8, it can be found that during the period of segment 4–10, the wave height of the empirical route exceeds the allowable wave height limit, even exceeding 7 m. Therefore, there is a great risk in following the empirical route; in contrast, the other two optimized routes can avoid the high wind and wave area and rationally allocate the speed. Figure 9 shows the wind speed of the three routes at different time and space positions. It can be seen that the wind field of the two optimized routes is stable and safe compared with the Great Circle Route. Compared to the GA route, the NSGA route has the characteristics of short time consumption and low energy consumption.

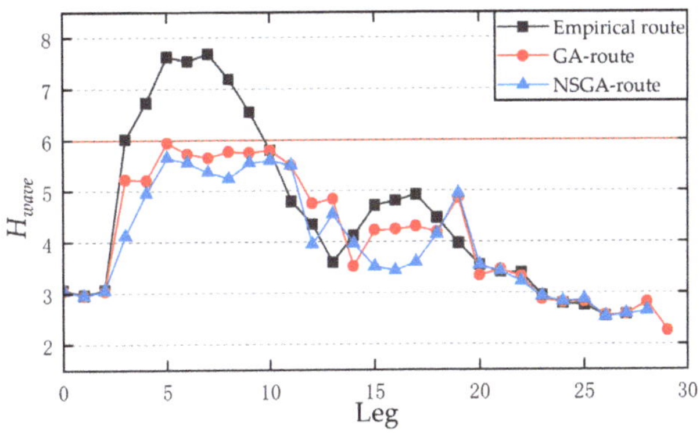

Figure 8. Effective wave heights encountered in different routes.

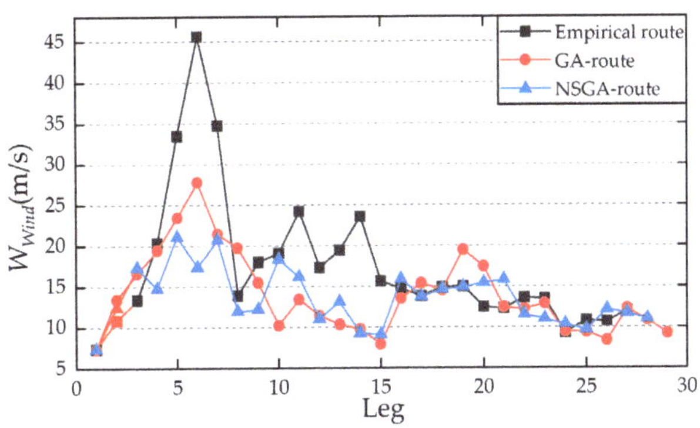

Figure 9. Wind speed encountered along different routes.

5. Conclusions

Ship navigation is forced to slow down or detour due to the influence of high wind and wave environment to ensure the safety of navigation, which will lead to increased shipping costs and increased risks. In order to find a safe and energy-saving green route within the specified time, this paper proposes a multi-objective route optimization method under a complex marine environment. Firstly, this paper rasterizes the marine environment information, establishes a complex marine environment model, and updates the model based on meteorological data. Taking the navigation position and speed as the decision variables, considering the influence of the marine environment on the ship's motion, the technical characteristics, and stability of the ship, the NSGA is used to find the optimal route. Through case analysis (Section 4), the proposed method is superior to the widely used genetic algorithm. It is worth noting that there are still some shortcomings in the route-optimization design method proposed in this paper, which need to be solved in future work. The main reason is that the environmental modeling in this paper only considers the influence of wind and waves on navigation, and there is a certain gap with the actual environment of maritime navigation. Therefore, it is necessary to consider comprehensive and more accurate weather data that affect the navigation status. In the future, a method of ship navigation safety assessment can be introduced to improve the accuracy of route optimization.

Author Contributions: Conceptualization, S.Z. (Shuling Zhao); methodology, S.Z. (Shuling Zhao) and S.Z. (Sishuo Zhao); writing—original draft preparation, S.Z. (Sishuo Zhao); writing—review and editing, S.Z. (Shuling Zhao) and S.Z. (Sishuo Zhao), funding acquisition, S.Z. (Shuling Zhao) All authors have read and agreed to the published version of the manuscript.

Funding: This work was jointly supported by the National Natural Science Foundation of China (72061016).

Institutional Review Board Statement: Not applicable.

Informed Consent Statement: Not applicable.

Data Availability Statement: The data can be accessed by sending an email to the corresponding author.

Conflicts of Interest: The authors declare no conflicts of interest.

References

1. Bullock, S.; Mason, J.; Broderick, J.; Larkin, A. Shipping and the Paris climate agreement: A focus on committed emissions. *BMC Energy* **2020**, *2*, 5. [CrossRef]
2. Müller-Casseres, E.; Edelenbosch, O.Y.; Szklo, A.; Schaeffer, R.; van Vuuren, D.P. Global futures of trade impacting the challenge to decarbonize the international shipping sector. *Energy* **2021**, *237*, 121547. [CrossRef]
3. Chen, X.; Liu, S.; Liu, R.W.; Wu, H.; Han, B.; Zhao, J. Quantifying Arctic oil spilling event risk by integrating an analytic network process and a fuzzy comprehensive evaluation model. *Ocean. Coast. Manag.* **2022**, *228*, 106326. [CrossRef]
4. Wu, G.; Atilla, I.; Tahsin, T.; Terziev, M.; Wang, L. Long-voyage route planning method based on multi-scale visibility graph for autonomous ships. *Ocean Eng.* **2021**, *219*, 108242. [CrossRef]
5. Seo, C.; Noh, Y.; Abebe, M. Ship collision avoidance route planning using CRI-based A∗ algorithm. *Int. J. Nav. Archit. Ocean. Eng.* **2023**, *15*, 100551. [CrossRef]
6. Silveira, P.; Teixeira, A.P.; Guedes-Soares, C. AIS based shipping routes using the Dijkstra algorithm. *TransNav Int. J. Mar. Navig. Saf. Sea Transp.* **2019**, *13*, 565–571. [CrossRef]
7. Wang, H.; Mao, W.; Eriksson, L. A Three-Dimensional Dijkstra's algorithm for multi-objective ship voyage optimization. *Ocean Eng.* **2019**, *186*, 106131. [CrossRef]
8. Liu, Y.; Wang, T.; Xu, H. PE-A* Algorithm for Ship Route Planning Based on Field Theory. *IEEE Access* **2022**, *10*, 36490–36504. [CrossRef]
9. Shin, Y.W.; Abebe, M.; Noh, Y.; Lee, S.; Lee, I.; Kim, D.; Bae, J.; Kim, K.C. Near-Optimal Weather Routing by Using Improved A* Algorithm. *Appl. Sci.* **2020**, *10*, 6010. [CrossRef]
10. Mannarini, G.; Subramani, D.N.; Lermusiaux, P.F.; Pinardi, N. Graph-search and differential equations for time-optimal vessel route planning in dynamic ocean waves. *IEEE Trans. Int. Trans. Syst.* **2019**, *21*, 3581–3593. [CrossRef]
11. Wang, L.; Zhang, Z.; Zhu, Q.; Ma, S. Ship route planning based on double-cycling genetic algorithm considering ship maneuverability constraint. *IEEE Access* **2020**, *8*, 190746–190759. [CrossRef]
12. Goez, G.D.; Velasquez, R.A.; Botero, J.S. UAV route planning optimization using PSO implemented on microcontrollers. *IEEE Lat. Am. Trans.* **2016**, *14*, 1705–1710. [CrossRef]

13. Zhou, P.; Zhou, Z.; Wang, Y.; Wang, H. Ship weather routing based on hybrid genetic algorithm under complicated sea conditions. *J. Ocean Uni. China* **2023**, *22*, 28–42. [CrossRef]
14. Zhang, G.; Wang, H.; Zhao, W.; Guan, Z.; Li, P. Application of improved multi-objective ant colony optimization algorithm in ship weather routing. *J. Ocean Uni. China* **2021**, *20*, 45–55. [CrossRef]
15. Svanberg, M.; Santén, V.; Hörteborn, A.; Holm, H.; Finnsgård, C. AIS in maritime research. *Mar. Policy* **2019**, *106*, 103520. [CrossRef]
16. Liu, D.; Rong, H.; Soares, C.G. Shipping route modelling of AIS maritime traffic data at the approach to ports. *Ocean Eng.* **2023**, *289*, 115868. [CrossRef]
17. Xiao, G.; Chen, L.; Chen, X.; Jiang, C.; Ni, A.; Zhang, C.; Zong, F. A hybrid visualization model for knowledge mapping: Scientometrics, SAOM, and SAO. *IEEE Trans. Int. Trans. Syst.* **2023**. [CrossRef]
18. Cai, J.; Chen, G.; Lützen, M.; Rytter, N.G.M. A practical AIS-based route library for voyage planning at the pre-fixture stage. *Ocean Eng.* **2021**, *236*, 109478. [CrossRef]
19. Gao, X.; Dong, Y.; Han, Y. An Optimized Path Planning Method for Container Ships in Bohai Bay Based on Improved Deep Q-learning. *IEEE Access* **2023**, *11*, 91275–91292. [CrossRef]
20. Comert, S.E.; Yazgan, H.R. A new approach based on hybrid ant colony optimization-artificial bee colony algorithm for multi-objective electric vehicle routing problems. *Eng. Appl. Artificial. Int.* **2023**, *123*, 106375. [CrossRef]
21. Chen, X.; Wang, M.; Ling, J.; Wu, H.; Wu, B.; Li, C. Ship imaging trajectory extraction via an aggregated you only look once (YOLO) model. *Eng. Appl. Art. Int.* **2024**, *130*, 107742. [CrossRef]
22. Srivastava, G.; Singh, A.; Mallipeddi, R. NSGA-II with objective-specific variation operators for multiobjective vehicle routing problem with time windows. *Expert Syst. Appl.* **2021**, *176*, 114779. [CrossRef]
23. Pingale, R.P.; Shinde, S.N. Multi-objective sunflower based grey wolf optimization algorithm for multipath routing in IoT network. *Wirel. Pers. Commun.* **2021**, *117*, 1909–1930. [CrossRef]
24. Zhang, L.; Bienkowski, A.; Macesker, M.; Pattipati, K.R.; Sidoti, D.; Hansen, J.A. Many-Objective Maritime Path Planning for Dynamic and Uncertain Environments. In Proceedings of the 2021 IEEE Aerospace Conference (50100), Big Sky, MT, USA, 6–13 March 2021; pp. 1–10.
25. Du, W.; Li, Y.; Zhang, G.; Wang, C.; Zhu, B.; Qiao, J. Ship weather routing optimization based on improved fractional order particle swarm optimization. *Ocean Eng.* **2022**, *248*, 110680. [CrossRef]
26. Ma, W.; Lu, T.; Ma, D.; Wang, D.; Qu, F. Ship route and speed multi-objective optimization considering weather conditions and emission control area regulations. *Mari Policy Manag.* **2021**, *48*, 1053–1068. [CrossRef]
27. Yuan, Q.; Wang, S.; Zhao, J. Uncertainty-informed ship voyage optimization approach for exploiting safety, energy saving and low carbon routes. *Ocean Eng.* **2022**, *266*, 112887. [CrossRef]
28. Szlapczynski, R.; Szlapczynska, J.; Vettor, R. Ship weather routing featuring w-MOEA/D and uncertainty handling. *Appl. Soft Computing.* **2023**, *138*, 110142. [CrossRef]
29. Kwon, Y.J. Speed loss due to added resistance in wind and waves. *Nav. Archit.* **2008**, *3*, 14–16.
30. Lu, R.; Turan, O.; Boulougouris, E.; Banks, C.; Incecik, A. A semi-empirical ship operational performance prediction model for voyage optimization towards energy efficient shipping. *Ocean Eng.* **2015**, *110*, 18–28. [CrossRef]
31. Uyanık, T.; Karatuğ, Ç.; Arslanoğlu, Y. Machine learning approach to ship fuel consumption: A case of container vessel. *Trans. Res. Part D Transport. Environ.* **2020**, *84*, 102389. [CrossRef]
32. Sang, Y.; Ding, Y.; Xu, J.; Sui, C. Ship voyage optimization based on fuel consumption under various operational conditions. *Fuel* **2023**, *352*, 129086. [CrossRef]
33. Cepowski, T.; Drozd, A. Measurement-based relationships between container ship operating parameters and fuel consumption. *Appl. Energy* **2023**, *347*, 121315. [CrossRef]

Journal of
Marine Science and Engineering

MDPI

Article

Research on the Deployment of Professional Rescue Ships for Maritime Traffic Safety under Limited Conditions

Minghui Shao [1,*], Biao Wu [2,*], Yan Li [1] and Xiaoli Jiang [1]

[1] College of Economics and Management, Heilongjiang Institute of Technology, Harbin 150050, China
[2] School of Automobile and Traffic Engineering, Heilongjiang Institute of Technology, Harbin 150050, China
* Correspondence: meteors_5@126.com (M.S.); wubiao0928@163.com (B.W.)

Abstract: This paper focuses on optimizing the deployment plan for standby points of professional rescue vessels based on the data of maritime incidents in the Beihai area of China. The primary objective is to achieve multi-level and multiple coverage of the jurisdictional waters of the Beihai Rescue Bureau. Models including the coverage quality of the jurisdictional waters, the coverage quality in high-risk areas, the maximum coverage of jurisdictional areas, and the maximum coverage of high-risk areas are constructed and solved using 0–1 integer programming. The optimal plan for eight standby points and their corresponding deployment plans for rescue vessels are obtained. A comparison with the current site selection plan of the Beihai Rescue Bureau validates the superiority of the proposed deployment plan for rescue vessel standby points in this paper.

Keywords: rescue vessel; standby point; 0–1 integer programming; site selection

1. Introduction

In recent years, there have been numerous maritime emergencies where rescue operations failed due to delays of rescue vessels in reaching the locations of incidents promptly, which can be attributed to the distant positioning of standby points for rescue vessels. How to reach the maritime scene for search and rescue (SAR) as soon as possible has been a hot research topic for experts and scholars. The scientific allocation and deployment of rescue vessels, as well as optimizing the selection of standby points, are key to solving this problem. Rational allocation of maritime rescue forces can improve the efficiency and success rate of rescue operations. The scientific selection of emergency standby points is particularly important as it enhances the capability of the maritime emergency response, reduces rescue distances, and minimizes rescue time. Currently, research on emergency location models mainly focuses on the selection of emergency standby points for land-based emergency operations. Studies on the selection of maritime rescue vessel locations are still primarily in the theoretical research stage, lacking quantitative vessel location models.

The research on emergency resource location problems originated with Hakimi's p-median problem proposed in 1964 [1,2]. This problem emphasizes the collective welfare of the final outcome, with the main objective being to select the locations of p facilities to achieve optimal overall or average performance [3]. Hakimi then proposed the p-center problem, which optimizes the chosen locations of p facilities for the worst-case scenario [4]. This approach has been widely applied in hospitals, the military, and service industries. In 1983, Shier and Dearing introduced the absolute center problem for the location of a single resource supply point [5]. The core of this type of location optimization problem is to find the absolute center point in the network, which is the point with the shortest distance to all other nodes. Ogryczak proposed a bi-objective model for the emergency service facility location problem [6]. This method simultaneously considers the median objective function and the center objective function, and can transform it into a single-objective model with parameter λ as the center point. Wang et al. established a mathematical model for optimizing the location of rescue centers [7], taking into account the probability of disaster occurrence, the spread function of the disaster, and the rescue function. Alinaghian et al. constructed a model for temporary rescue center location and a dynamic helicopter routing model for distributing essential supplies [8]. Compared to traditional static routing, dynamic routing resulted in less total allocation time. These algorithms are effective in solving the problems of finding the absolute centers and determining a single supply point. However, it is not suitable for solving problems involving multiple supply points, such as the deployment of standby points for search and rescue vessels.

Zero–one programming is applicable for solving problems involving multiple integer variables and where the variables have logical relationships of being either present or absent. It is also effective for optimizing the selection of multiple supply points. Chen et al. identified three key objectives in the multi-objective decision-making process for emergency system location selection [9]. Firstly, the distance or weighted distance between emergency rescue facility service points and demand points should be minimized. Secondly, emergency rescue facilities should have excess coverage, and the total weight value of the excess coverage area should be maximized. Lastly, the total weighted distance between emergency rescue facility service points and demand points should be minimized. Lai et al. constructed an optimization model for dual-objective emergency logistics center location, considering the emergency material demand at the disaster site, transportation costs, and time as discrete sets, and obtained optimal location solutions [10]. Moreno et al. explored the problem of heterogeneous fleets for road repair and rescue path planning, which had been overlooked in existing research [11]. Wu et al. proposed the concept of rescue zones for tourist attractions [12]. For each waypoint rescue zone within a tourist attraction, the effectiveness was calculated based on linear rewards. The rescue status and rescue areas of the tourist attraction rescue zones were determined based on the rewards of all rescue zones. Taking into account the reliability of the facility system, Nahavandi established an emergency logistics location-routing model to maximize the reliability of the facility system and ensure that the emergency logistics system can respond to demands even when certain facilities are unable to operate [13]. In the scenario where multiple suppliers provide goods, Khanchehzarrin considered the risk of the goods and developed a multi-objective bi-level model for the emergency logistics location-routing problem [14]. The study confirmed that when goods with higher risk and priority are provided, the efficiency of emergency logistics will be improved. Eliiyi established an emergency logistics location-routing model with the objective of minimizing response time and maximizing coverage [15].

There are several other models and algorithms designed to address resource allocation and site selection problems. Beiki developed a multi-objective mixed integer model for the location-path problem of medical relief during a disaster [16], with the objectives of minimizing response time, minimizing operational costs, and maximizing the reliability of the transportation network. The variable neighborhood search heuristic algorithm has been utilized to solve the central problem model for emergency warehouse location, taking into account constraints such as economic conditions, traffic conditions, and population distribution [17]. The parameters associated with the dynamic facility location vary over time. The tabu search algorithm is used to solve the site selection and path decision problems for disaster assessment [18]. The effectiveness of this method has been validated with earthquake cases. The ant colony algorithm is used to optimize the disaster vehicle delivery path, considering vehicle allocation and dynamic demand, including vehicle capacity, time to reach the distribution center, and time to reach the disaster area [19]. Scholars have also studied the problem of uncertain post-disaster emergency logistics routing, which can be handled using stochastic optimization and robust optimization methods. Sun et al. proposed a new emergency logistics location-path optimization model, which addresses uncertain demands using relative robust optimization and estimates equipment failure risks using genetic algorithms [20]. For post-disaster emergency blood supply, a multi-objective robust optimization model is used to solve the location selection of field hospitals and blood transportation path planning problems [21]. Search algorithms and ant colony algorithms are well suited for scenarios where there are relationships between variables and where search elements can be inputted into a search engine. However, they may not be the best choice for problems like the deployment of maritime rescue vessels involving multiple variables that are independent of each other. Robust algorithms are a less efficient choice compared with 0–1 programming.

In recent years, deep learning models [22] and geographic information systems (GISs) have also been used to predict the geographical coordinates of past merchant ships and alternative islands [23–25]. By employing the K-means clustering algorithm, they determined the positions of dynamic work points for establishing SAR bases at sea. A multi-objective plant growth simulation algorithm was designed to obtain optimal solutions that comprehensively consider the positions of dynamic SAR work points, the configurations of SAR vessels, and the locations of SAR bases. Wang et al. took China's Nansha Islands and reefs as an example and established a bi-objective optimization model [26]. By applying an improved non-dominated sorting genetic algorithm II (NSGA-II) with adaptive-scaling crowding distance and adaptive elitism preservation strategy, they obtained ship rescue base location solutions.

The Beihai area, with high maritime traffic density and complex sea conditions, faces severe maritime safety challenges. The scientific allocation and deployment of rescue vessels, as well as the selection of standby points, are crucial for improving search and rescue success rates and reducing casualties. This is the problem that this study aims to address.

2. Data

The deployment of rescue vessels in the Beihai area is managed by the Beihai Rescue Bureau. Currently, there is a total of 25 dynamic standby points of rescue vessels established in the jurisdictional waters [27]. The specific distribution of these standby points is shown in Table 1.

Table 1. The distribution of rescue vessel standby points of Beihai Rescue Bureau.

No.		Standby Position	Deployed Rescue Vessels	Remarks
1	Bohai Strait	Beihai No. 1 (38°35′ N/121°48′ E)	8000 kW and above	Key standby point
2		South Huangcheng Island		
3		Long Island	6000 kW and above	Key standby point
4		Longkou	Depending on the actual situation	Temporary standby point
5		Dongying	Depending on the actual situation	Key standby point
6	Qinhuangdao	Beihai No. 3 (39°10′ N/120°15′ E)	8000 kW and above	Key standby point
7		Jingtang Port		
8		Anchorage Outside The Harbor		
9		Base		
10	Dandong	Oceanic Island		
11		Anchorage Outside The Harbor		
12	Dalian	Dasanshan Island		
13		Dalian Port Anchorage	8000 kW and above	
14		Base	High-speed rescue ship	Mobile standby point

Table 1. *Cont.*

No.		Standby Position	Deployed Rescue Vessels	Remarks
15	Yantai	Yantai Port Anchorage	8000 kW and above	Key standby point
16	Tianjin	Beihai No. 2 (Caofeidian) (38°50′ N/118°25′ E)	8000 kW and above	Key standby point
17		Dagukou Port Anchorage		
18		Base	High-speed rescue ship	
19	Rongcheng	Beihai No. 4 (About 15 nautical miles southeast of Shidao)	8000 kW and above	Key standby point
20		Shidao Port	High-speed rescue ship	
21		Base	High-speed rescue ship	
22	Qingdao	Qingdao Port Anchorage	8000 kW and above	Key standby point
23		Chaolian Island		
24		Hui Island		
25		Rizhao	Depending on the actual situation	Temporary standby point

The Beihai Rescue Bureau currently has a total of 12 professional rescue vessels [28]. In coastal waters, shipwreck incidents are extremely uncommon. The more frequent emergency situations involve the rescue of injured crew members from fishing boats. Therefore, these rescue vessels are classified into three categories, large, medium, and small, based on their speed and endurance for the purpose of personnel rescue, as shown in Tables 2 and 3.

Table 2. Data of existing rescue vessels.

No.	Ship Name	Total Length (m)	Full Displacement (t)	Engine Power (kW)	Speed (Knots)	Wind Resistance	Endurance (n.m.)	Response Time (from Alarm to Departure)
1	101	117	6614	14,400	22	12	10,000	30 min (40 min in winter)
2	111	98	4891	9000	20	12	10,000	
3	112	98	4896	9000	20	12	10,000	
4	113	99	5143	9000	20	12	10,000	
5	115	99	5127	9000	20	12	10,000	
6	116	99	5198	9000	20	12	10,000	
7	117	98.5	5748	9000	17.3	12	10,000	
8	118	99	5748	9000	17.3	12	14,000	
9	119	99	5748	9000	17.3	12	14,000	
10	131	77	3211	6720	18	12	5000	
11	201	49.9	250	4480	32.5	6	500	20 min (30 min in winter)
12	203	49.9	278	5120	32.5	6	700	

Table 3. Classification of the professional rescue vessels.

Ship Type	Speed (Knots)	Endurance (n.m.)	Features
large	20	10,000	Advantages: high power, large towing force, strong wind resistance. Disadvantages: Long reaction time, poor maneuverability, and slow ship speed.
medium	17.3	5000 and 14,000	
small	32.5	500	Advantages: Fast, short reaction time. Disadvantages: Poor wind resistance, inability to sail in adverse weather, and poor endurance.

The typical response time for professional rescue vessels from receiving an alarm to departure, as managed by the Beihai Rescue Bureau, is usually 30 min (40 min during winter). However, due to their smaller size, Beihai Rescue 201 and 203 have a response time of 20 min (30 min during winter).

The North Sea area has been gridified, and the locations of incidents and existing rescue forces in the North Sea are marked on the nautical chart with icons of triangle and ship, respectively, as shown in Figure 1. According to the data provided by the Beihai Rescue Bureau, high-risk areas in the North Sea are depicted in Figure 2, with red grid areas representing high-risk zones.

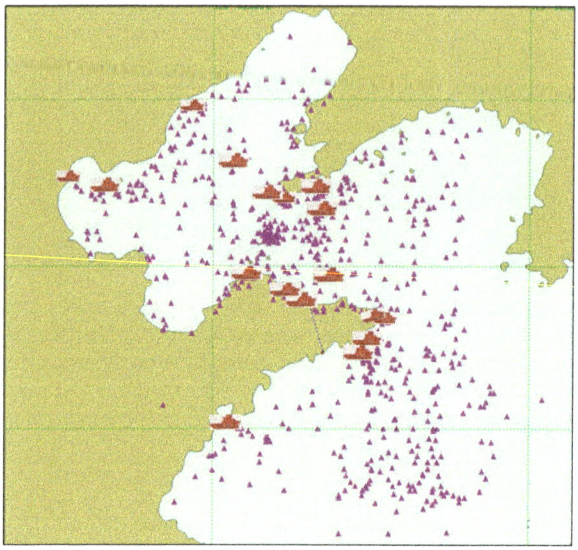

Figure 1. The distribution of the incidents points (triangles) and existing rescue forces (ships) in the Beihai area.

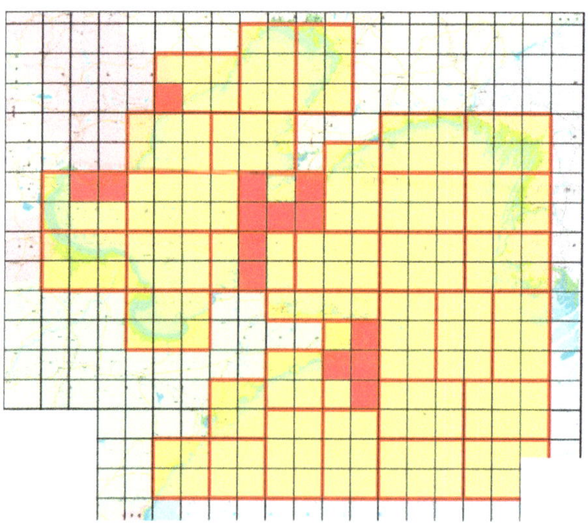

Figure 2. Positions of high-risk areas (red grids).

It can be found from Figure 1 that the incidents cover the entire North Sea area, mainly concentrated in four areas: Tianjin, Qinhuangdao, the Lüshun-to-Penglai line, and Rongcheng. The existing rescue forces of the Beihai Rescue Bureau are mainly deployed in these four areas.

The large rescue vessels have a cruising speed of 20 knots and a range of 10,000 nautical miles. The medium-sized rescue vessels, divided into two types, have a cruising speed of 17.3 knots and ranges of 5000 nautical miles and 14,000 nautical miles, respectively. The small rescue vessels have a cruising speed of 32.5 knots and a range of 500 nautical miles. Considering that it takes some time for the vessels to reach the rescue location and return, the coverage radius should not exceed one-third of the range. Here, it should be ensured that the vessels can arrive within 6 h, which includes the response time (from receiving the alert to departure, considering the longest response time in winter). The calculated coverage radii are shown in Table 4.

Table 4. Coverage radii of rescue vessels.

Ship Type	Speed (Knots)	Coverage Radius (Nautical Miles)	Coverage Radius (km)
Large rescue ship	20	106.67	203.72
Medium rescue ship	17.3	92.27	176.22
Small rescue ship	32.5	178.75	331.05

3. The Construction of the Deployment Model

Taking into account the advantages and disadvantages of the existing land site selection models, the ratio of the difference between the coverage radius of the rescue vessel and the distance to the incident location divided by the coverage radius of the rescue vessel is used as a measure of coverage quality. This approach ensures the maximum coverage quality for general areas while also maximizing coverage quality for high-risk areas. Based on the principles of "multi-source services" and "multi-layered coverage", the model emphasizes different service qualities of rescue forces at different levels and multiple quantity coverage at the same location. A multi-level and multi-layered coverage site selection model is proposed to improve the efficiency of emergency rescue.

3.1. Zero–One Construction of the Integer Programming Coverage Model

The location problem in this paper will be solved by using the 0–1 integer programming model. Considering different levels of rescue forces participating in accident response, the coverage situation is determined based on the distance between the emergency incident location and the standby points of the rescue forces. Each incident location can be served by different rescue standby points, providing varying levels of coverage. The expression is shown in Equation (1).

$$C_{ijp} = \begin{cases} 0, D_{ijp} > D_{pu} \\ \sum\limits_{j \in J, p \in P} \frac{D_{pu} - D_{ijp}}{D_{pu}}, D_{ijp} \leq D_{pu} \end{cases} \quad (1)$$

In this case, D_{pu} represents the maximum effective coverage radius of standby point P, and the coverage quality C_{ijp} is defined based on the linear distance (D_{ijp}) between incident point I and the rescue standby point (P_j). If $D_{ijp} < D_{pu}$, it indicates that incident point I is within the coverage capability of the multi-level rescue force. C_{ijp} is a partial-coverage degree function that decreases linearly with the increase in the distance between the rescue standby point and the incident point. $C_{ijp} = 0$ represents that the incident point is not covered by the standby point, and $0 < C_{ijp} < 1$ represents that the incident point is within the effective coverage capability. When selecting the location for maritime professional rescue vessels, the first priority is to ensure that the vessels can provide maximum coverage for the jurisdictional waters.

For accident point i, $i \in I$ (i belongs to the accident point set I, taking i = 1, 2, 3, ..., 143, a total of 143 accident points), introducing the decision variable g_i, as shown in Formulas (2) and (3):

$$g_i = \begin{cases} 1, \text{ the high-risk water area is covered.} \\ 0, \text{ not covered.} \end{cases} \quad (2)$$

that is

$$g_i = \begin{cases} 1, D_{ijp} \le D_{pu} \\ 0, D_{ijp} > D_{pu} \end{cases} \quad (3)$$

Then, the following set coverage problem can be established to ensure that the selected standby points can cover the accident water area in the jurisdiction to the maximum extent, with Formula (4):

$$\max \sum_{i \in I} g_i \quad (4)$$

z_{ijp}, x_{jp}, u_{jp}, d_p, and y_{ijp} are 0–1 decision variables in the maritime danger scenario. When the rescue resource reserve point (P_j) is selected, x_{jp} = 1; otherwise, it is 0. When accident point i is covered by rescue force P_j, u_{jp} = 1; otherwise, it is 0. If rescue force P_j participates in the rescue, then d_p = 1; otherwise, it is 0. When rescue force reserve point P_j has the ability to serve accident point i, y_{ijp} = 1 and z_{ijp} = 1; otherwise, they are 0. In the high-risk water area (m) within the jurisdiction (belonging to the accident point set, I), the rescue forces at all standby points should be able to cover the high-risk water area (m) to the maximum extent possible; D_{mjp} is the distance between the high-risk accident point (m) and the rescue reserve point, P_j. $D_{mjp} < D_{pu}$ indicates that the high-risk accident point (m) is within the coverage capacity of multi-level rescue forces; $C_{mjp} = 0$ indicates that the accident point (m) is not covered by the reserve point, and $0 < C_{mjp} < 1$ indicates that the accident point (m) is within the effective coverage capacity range. Formulas (5) and (6) introduce variable g_m.

$$g_m = \begin{cases} 1, D_{mjp} \le D_{pu} \\ 0, D_{mjp} > D_{pu} \end{cases} \quad (5)$$

$$\max \sum_{m \in I} g_m \quad (6)$$

The expression for the coverage quality (C_{mjp}) of high-risk water areas is shown in Formulas (7) and (8):

$$C_{mjp} = \begin{cases} 0, D_{mjp} > D_{pu} \\ \sum\limits_{j \in J, p \in P} \frac{D_{pu} - D_{mjp}}{D_{pu}}, D_{mjp} \le D_{pu} \end{cases} \quad (7)$$

$$\max \sum_{m \in I, j \in J, p \in P} Cmjp \quad (8)$$

The quality model of coverage in the jurisdiction is as follows:

$$\max \sum_{i \in I, j \in J, p \in P} d_p C_{ijp} z_{ijp} \quad (9)$$

$$\text{s.t.} \sum_{j \in J} x_{jp} = N_p; \forall p \in P \quad (10)$$

$$z_{ijp} \le x_{jp}; \forall i \in I, \forall j \in J, \forall p \in P \quad (11)$$

$$x_{jp} \le d_p; \forall j \in J, \forall p \in P \quad (12)$$

$$u_{jp} \le d_p; \forall i \in I, \forall p \in P \quad (13)$$

$$z_{ijp}, x_{jp}, u_{jp}, d_p, y_{ijp} \in \{0,1\}; \forall i \in I, \forall j \in J, \forall p \in P \tag{14}$$

Among them, the objective function (8) represents high-risk water areas, and the rescue force is the most effective coverage. The objective function (9) represents the most effective coverage of rescue force P_j in multiple rescue modes with cross-regional maritime hazard response levels. The constraint condition (10) indicates the number (N_p) of reserve points (P_j) participating in the response. In constraint (11), the premise for the accident site to receive rescue resource services is that the rescue reserve point is selected. The constraint conditions (12) and (13), respectively, indicate that only when the rescue force (P_j) is determined to participate in the rescue can P_j be selected to serve the accident point. The constraint condition (14) guarantees that z_{ijp}, x_{jp}, u_{jp}, d_p, and y_{ijp} are 0–1 integer decision variables.

3.2. Algorithm Design of 0–1 Integer Programming Coverage Model

There are multiple variables in this model, including a total of 25 standby points and three types of rescue vessels. After determining the standby points, the rescue vessels need to be allocated. It is evident that such a large number of variables cannot be solved directly. In the model, there is only one small rescue vessel with a speed of 32.5 knots, while the speed difference between the medium-sized rescue vessel (17.3 knots) and the large-sized rescue vessel (20 knots) is minimal. Therefore, we can initially assume that all rescue vessels are medium-sized rescue vessels with a speed of 17.3 knots, select the locations, and then adjust the types of vessels accordingly.

The most prioritized condition in the model is to maximize the coverage of the accident area and strive to eliminate blind spots. Therefore, the first step is to solve Equation (4) in order to achieve this objective.

Next, it is crucial to ensure maximum coverage of high-risk water areas and achieve the highest coverage quality. This involves solving Equations (6), (8), and (9) in order to obtain the optimal results.

In this paper, the optimal solution is obtained by using the **intlinprog** function in MATLAB, and the function prototype is show in Equation (15):

$$[\mathrm{x}, \mathrm{fval}, \mathrm{exitflag}] = \mathrm{intlinprog}(\mathrm{f}, \mathrm{intcon}, \mathrm{A}, \mathrm{b}, \mathrm{A_{eq}}, \mathrm{b_{eq}}, \mathrm{lb}, \mathrm{ub}) \tag{15}$$

where $\boldsymbol{A}$ and $\boldsymbol{b}$ are coefficient matrices, $\boldsymbol{A}_{\mathrm{eq}}$ and $\boldsymbol{b}_{\mathrm{eq}}$ are the matrix of coefficients and the column vector of the right-hand side values for those constraints, *lb* and *ub* are the lower and upper bounds of the variables, the return value $\boldsymbol{x}$ is the solution vector that minimizes or maximizes the objective function, and *fval* is the value of the objective function at the solution (x). Among them, *exitflag* is an integer indicating the termination condition of the algorithm, and ***intcon*** is a vector specifying the indices of the variables.

4. Case Analysis of Deployment in the North Sea Area (Results)

4.1. Existing Standby Positions of Rescue Vessels in the North Sea Area

The Beihai Rescue Bureau of China selected 15 locations out of the 25 candidate standby positions in Beihai based on actual operational conditions and the status of rescue vessels. The actual positions and their level of importance are listed in Table 3.

The Beihai Rescue Bureau of China plans to select eight standby points within its jurisdictional waters. Therefore, this article will utilize the 0–1 integer programming model constructed in the previous section to select eight points from the twenty-five candidate standby locations, and assign the corresponding rescue vessels to these eight points.

4.2. Numerical Simulation Analysis of Professional Rescue Ships in the North Sea Area

MAPGIS was used to convert the geographic coordinates of the risk points under the jurisdiction of the Beihai Rescue Bureau of China to the Cartesian coordinates, as shown in Figure 3.

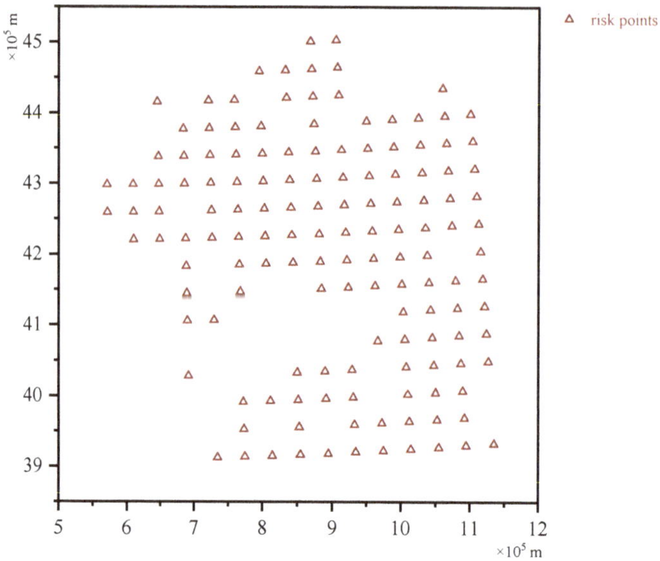

Figure 3. Risk points in the Cartesian coordinate system.

From Figure 3, it can be found that the risk points almost cover every unit position in the area under the jurisdiction of the Beihai Rescue Bureau. The blank area in Figure 3 is the land, involving the Bohai Sea Coast, Liaodong Peninsula, and Shandong Peninsula.

The geographic coordinates of high-risk points were also converted into the geodetic coordinates by using MAPGIS. The points with high risk are marked with red circles in Figure 4.

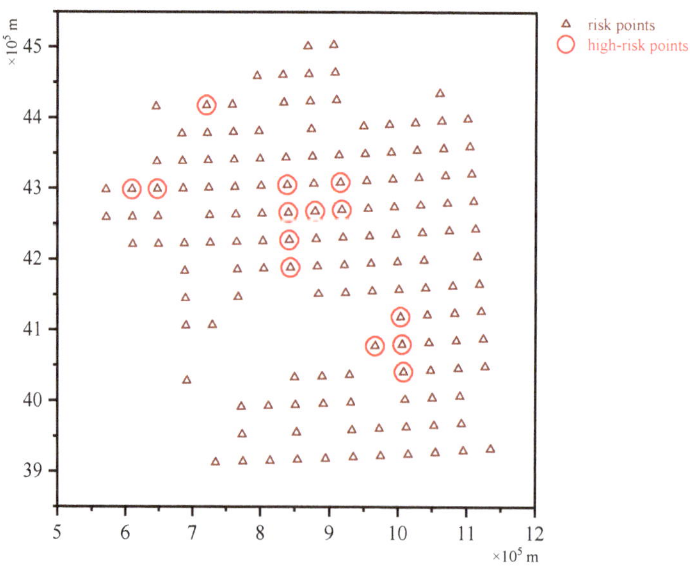

Figure 4. The units of high-risk points in the Cartesian coordinate system.

The geographic coordinates of 25 standby points for professional rescue vessels were converted into Cartesian coordinates. These standby points are marked in the map with stars, as shown in Figures 5–8.

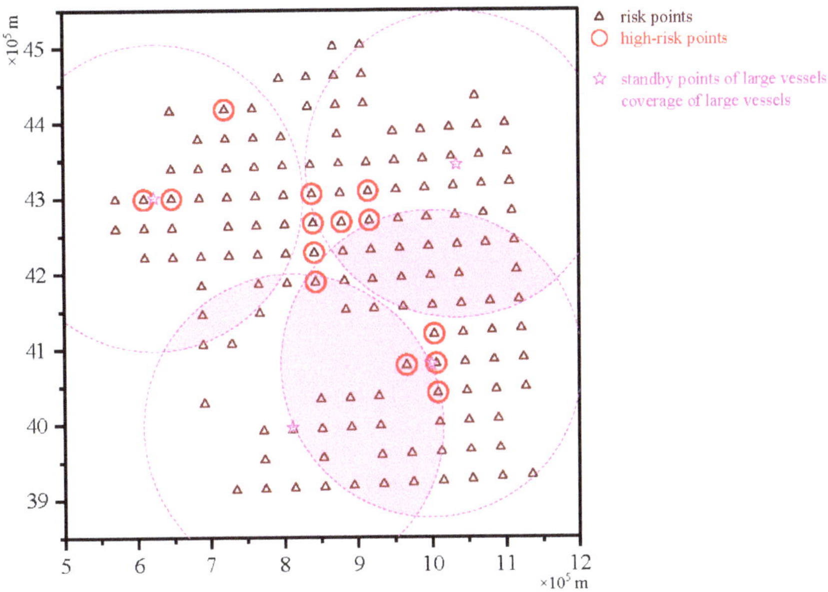

Figure 5. The coverage of the large rescue vessels in the high-risk area in the Cartesian coordinate system.

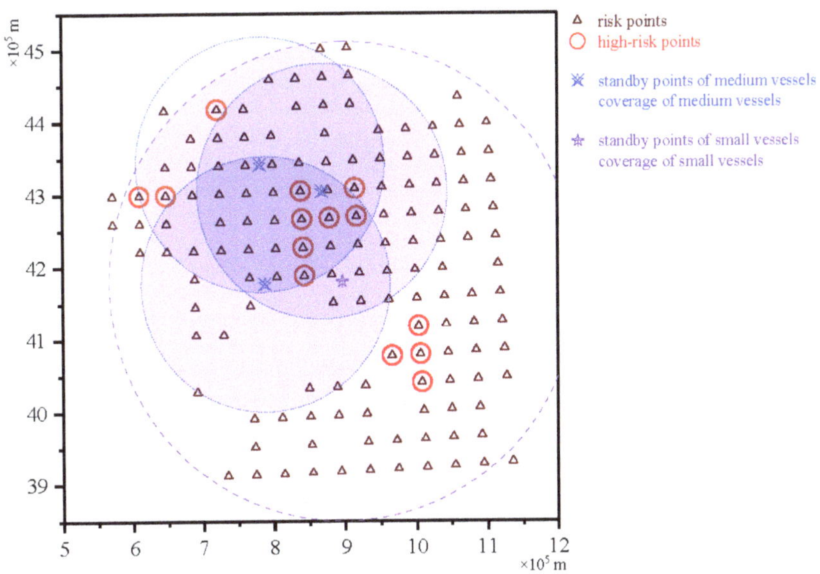

Figure 6. The coverage of the medium and small rescue vessels in the high-risk area in the Cartesian coordinate system.

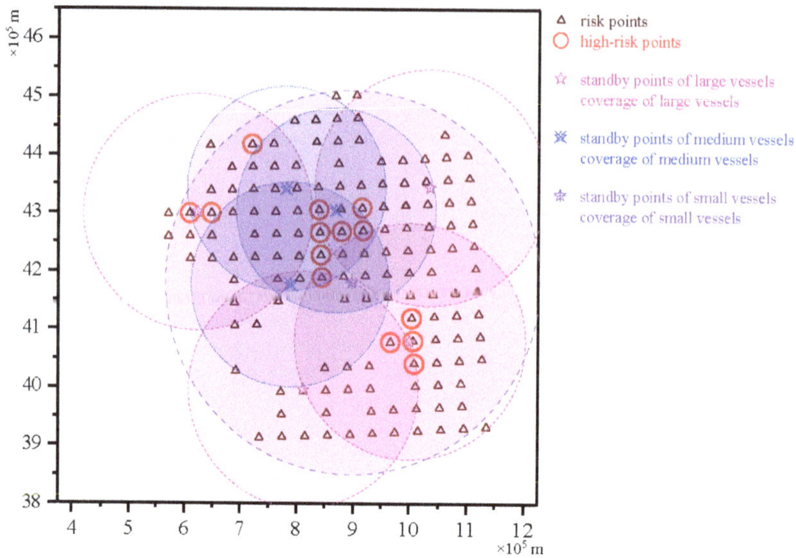

Figure 7. The coverage of the high-risk areas based on the results of the integer programming.

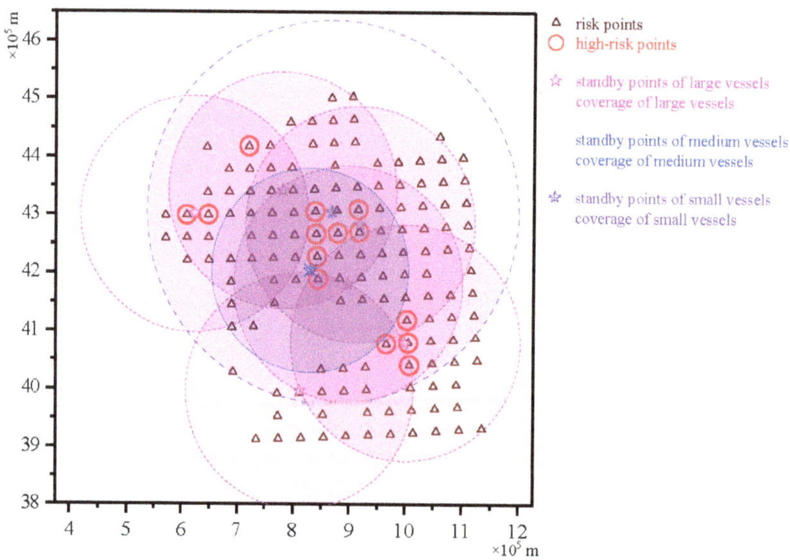

Figure 8. The coverage of the existing plan in the high-risk area of Beihai Rescue Bureau.

The first thing to ensure is that the standby point of the rescue vessels can cover all the waters within the jurisdiction. To satisfy this requirement, there must be at least one standby point in each of the five regions: Qinhuangdao, Tianjin, Rongcheng, Qingdao, and Dandong. The five standby points are shown in Table 5.

The remaining three standby points will be selected from the remaining twenty-one candidate points. The main consideration for site selection is based on the maximum coverage of jurisdictional areas and the maximum coverage of the high-risk area models.

Table 5. Positions of standby points with high risk level.

No.	Geographic Coordinates
10	Oceanic Island (39°3.93′ N/123°10.38′ E)
16	Beihai No. 2 (Caofeidian) (38°50′ N/118°25′ E)
19	Beihai No. 4 (About 15 nautical miles southeast of Shidao) (36°42.8′ N/122°35.3′ E)
22	Qingdao Port Anchorage (36°2.69′ N/120°27.44′ E)
6	Beihai No. 3 (39°10′ N/120°15′ E)

According to the previous assumption that all rescue vessels are medium rescue vessels with a speed of 17.3 knots, the optimal solutions for Formulas (7), (9) and (10) were obtained using the intlinprog function in MATLAB. Eight optimal points were selected, as shown in Table 6.

Table 6. Selection of standby points based on the calculation results.

No.	Standby Points	Cartesian Coordinates
4	Longkou (37°40.76′ N/120°15.89′ E)	(788,032.144, 4,176,939.38)
6	Beihai No. 3 (39°10′ N/120°15′ E)	(780,905.029, 4,342,043.581)
10	Oceanic Island (39°3.93′ N/123°10.38′ E)	(1,034,469.749, 4,343,967.765)
14	Base (38°48.05′ N/121°14.57′ E)	(868,637.853, 4,304,957.123)
15	Yantai Port Anchorage (37°40.31′ N/121°29.96′ E)	(897,055.492, 4,180,627.022)
16	Beihai No. 2 (Caofeidian) (38°50′ N/118°25′ E)	(623,010.888, 4,300,957.382)
19	Beihai No. 4 (About 15 nautical miles southeast of Shidao) (36°42.8′ N/122°35.3′ E)	(999,515.532, 4,079,289.45)
22	Qingdao Port Anchorage (36°2.69′ N/120°27.44′ E)	(811,598.577, 3,996,042.447)

According to the data from the Beihai Rescue Bureau [28], a total of twelve vessels are used for maritime search and rescue and material supply, including six large rescue ships, four medium-sized rescue ships, and two small rescue ships. Based on this ratio, four large rescue ships, three medium rescue ships, and two small rescue ship are selected and allocated to the eight standby points. Then, using the intlinprog function in MATLAB, the rescue force allocation is re-calculated based on Formulas (6), (8), and (9). The results are shown in Table 7. The coverage range of large rescue vessels is shown in Figure 5.

Table 7. Suggested deployment of the professional rescue vessels.

No.	Standby Points	Deployed Rescue Vessels
4	Longkou (37°40.76′ N/120°15.89′ E)	Medium
6	Beihai No. 3 (39°10′ N/120°15′ E)	Medium
10	Oceanic Island (39°3.93′ N/123°10.38′ E)	Large
14	Base (38°48.05′ N/121°14.57′ E)	Medium
15	Yantai Port Anchorage (37°40.31′ N/121°29.96′ E)	Small (Mobile standby point)
16	Beihai No. 2 (Caofeidian) (38°50′ N/118°25′ E)	Large
19	Beihai No. 4 (About 15 nautical miles southeast of Shidao) (36°42.8′ N/122°35.3′ E)	Large
22	Qingdao Port Anchorage (36°2.69′ N/120°27.44′ E)	Large

From Figure 5, it can be seen that large rescue vessels can cover almost all of the accident points within the water area, and they can arrive at the scene within 6 h (including alarm response time) for search and rescue operations.

Medium rescue ships primarily cover the Bohai Sea area and high-risk accident-prone areas. Small rescue ships are highly flexible, fast, and capable of reaching various water areas in their jurisdiction within 6 h, making them suitable as mobile standby points. However, in the event of severe weather conditions, small rescue ships are unable to sail due to their poor wind resistance.

According to the above rules of allocation, it can be ensured that rescue vessels will arrive at all accident water areas in their jurisdiction within 6 h (including the time from alarm to response). In high-risk water areas, different levels of rescue vessels can be deployed at different standby points to achieve multi-level and multiple coverage. The coverage is illustrated in Figure 7.

Based on this, the eight standby points divide the Beihai area into eight blocks: Bohai Strait, Qinhuangdao, Dandong, Dalian, Yantai, Tianjin, Rongcheng, and Qingdao. The rescue vessels can cover the entire Beihai area and achieve the fastest and most effective rescue for maritime emergencies. High-risk water areas can be covered by multiple standby points and vessel types. For emergencies near Tianjin Port, Beihai standby point No. 2 (Caofeidian) is primarily responsible for rescue operations, while the Ocean Island standby point is primarily responsible for emergencies near Dandong. For emergencies in the Yellow Sea, the Qingdao Port Anchorage and Rongcheng standby points are the main rescue points. Among them, Ocean Island, North Sea No. 2 (Caofeidian), North Sea No. 4, and Qingdao Port Anchorage are equipped with large rescue vessels, while Longkou, North Sea No. 3, and the base dock are equipped with medium rescue vessels. Although large and medium rescue vessels are lower, they have stronger endurance and wind resistance levels, allowing them to conduct maritime search and rescue operations for human lives, except in extremely adverse weather conditions. Small rescue vessels have poor wind resistance and cannot operate in severe weather conditions. Therefore, they are stationed at Yantai Port Anchorage as a mobile standby point, utilizing their advantages of covering all accident areas and high speed to achieve multi-level and multiple coverage.

4.3. Verification of the Superiority of the Deployment Plan

The Beihai Rescue Bureau has deployed twenty-five standby points, which are classified as seven important standby points, four temporary standby points, and one mobile standby point (small boats). Here, we take Beihai Rescue Bureau's deployment with seven important standby points and one mobile standby point, and make a comparison with the results obtained with the model in this paper. Rescue vessels with power above 8000 kW are classified as large rescue vessels, those above 6000 kW as medium rescue vessels, and high-speed rescue vessels as small rescue vessels. After coordinate conversion using MAPGIS, the eight standby points are listed in Table 8. The corresponding coverage areas are shown in Figure 8 on the Cartesian coordinate system.

Table 8. The existing standby points of Beihai Rescue Bureau and the deployment of rescue vessels.

No.	Standby Points	Cartesian Coordinates	Deployed Rescue Vessels
1	Beihai No. 1 (38°35′ N/121°48′ E)	(918,344.103, 4,283,194.123)	Large
3	Long Island	(829,653.656, 4,204,325.851)	Medium
6	Beihai No. 3 (39°10′ N/120°15′ E)	(780,905.029, 4,342,043.581)	Large
14	Base (38°48.05′ N/121°14.57′ E)	(868,637.853, 4,304,957.123)	Small (Mobile standby point)
15	Yantai Port Anchorage (37°40.31′ N/121°29.96′ E)	(897,055.492, 4,180,627.022)	Large
16	Beihai No. 2 (Caofeidian) (38°50′ N/118°25′ E)	(623,010.888, 4,300,957.382)	Large
19	Beihai No. 4 (About 15 nautical miles southeast of Shidao) (36°42.8′ N/122°35.3′ E)	(999,515.532, 4,079,289.45)	Large
22	Qingdao Port Anchorage (36°2.69′ N/120°27.44′ E)	(811,598.577, 3,996,042.447)	Large

The two deployment plans can be compared by solving models of the maximum coverage quality of the jurisdictional water areas and the maximum coverage quality in high-risk areas. The maximum coverage quality of the jurisdictional water areas in this deployment plan is 193.17, and the quality coverage of high-risk water areas is 25.53. The counterparts of Beihai Rescue Bureau's current plan are 169.88 and 23.15. Both results of this plan are higher. This means that the comprehensive distance between rescue points obtained in this work is smaller than that of the existing plan.

5. Conclusions

This study gathered data on maritime incidents within the jurisdictional water areas of the Beihai Rescue Bureau of China and developed a location model for selecting optimal standby points for professional rescue vessels. Models including the coverage quality of the jurisdictional waters, the coverage quality in high-risk areas, the maximum coverage of jurisdictional areas, and the maximum coverage of high-risk areas were constructed and solved using 0–1 integer programming. The resulting eight standby points are strategically located in the Bohai Strait, Qinhuangdao, Dandong, Dalian, Yantai, Tianjin, Rongcheng, and Qingdao regions. It ensures that rescue vessels can reach all accident areas within 6 h (including the time from receiving an alert to responding) and achieve multiple coverages of different levels in high-risk water areas. The standby points obtained from the model's solution theoretically improve both service effectiveness and efficiency. A comparison with the current site selection plan of the Beihai Rescue Bureau validates the superiority of the proposed deployment plan for rescue vessel standby points in this paper.

Author Contributions: Conceptualization, methodology, validation, formal analysis, and writing—original draft preparation, M.S.; data curation, M.S. and X.J.; writing—review and editing, M.S., B.W., X.J. and Y.L.; funding acquisition, M.S. and B.W. All authors have read and agreed to the published version of the manuscript.

Funding: This research was funded by "the Fundamental Research Funds for the Provincial Universities in Heilongjiang Province, 2023GJ01 and 2019BJ08" and the "Art and Science Planning Project of Heilongjiang, 2022B093".

Institutional Review Board Statement: Not applicable.

Informed Consent Statement: Not applicable.

Data Availability Statement: Data are contained within the article.

Acknowledgments: We sincerely appreciate the valuable comments provided by the editors and reviewers, which greatly contributed to the improvement of this work.

Conflicts of Interest: The authors declare no conflicts of interest.

References

1. Hakimi, S.L. Optimum locations of switching centers and the absolute centers and medians of a graph. *Oper. Res.* **1964**, *12*, 450–459. [CrossRef]
2. Hakimi, S.L. Optimum distribution of switching centers in a communication network and some related graph theoretic problems. *Oper. Res.* **1965**, *13*, 462–475. [CrossRef]
3. Kariv, O.; Hakimi, S.L. An algorithmic approach to network location problems. I: The p-centers. *SIAM J. Appl. Math.* **1979**, *37*, 513–538. [CrossRef]
4. Hakimi, S.L.; Schmeichel, E.F.; Pierce, J. On p-centers in networks. *Transp. Sci.* **1978**, *12*, 1–15. [CrossRef]
5. Shier, D.R.; Dearing, P.M. Optimal locations for a class of nonlinear, single-facility location problems on a network. *Oper. Res.* **1983**, *31*, 292–303. [CrossRef]
6. Ogryczak, W. On the distribution approach to location problems. *Comput. Ind. Eng.* **1999**, *37*, 595–612. [CrossRef]
7. Wang, D.-W.; Zhang, G.-X. Model and algorithm to optimize location of catastrophic rescue center. *J. Northeast. Univ.* **2005**, *26*, 953.
8. Alinaghian, M.; Aghaie, M.; Sabbagh, M.S. A mathematical model for location of temporary relief centers and dynamic routing of aerial rescue vehicles. *Comput. Ind. Eng.* **2019**, *131*, 227–241. [CrossRef]
9. Chen, Z.; You, J. A modeling approach to hierarchical location problem of urban disaster prevention and mitigation facilities. *J. Nat. Disaster* **2005**, *2*, 131–135.
10. Lai, Z.; Wang, Z.; Ge, D.; Chen, Y. A Robust Optimization Model for Multi objective Emergency Logistics Center Site Selection. *Oper. Res. Manag.* **2020**, *29*, 74–83.
11. Moreno, A.; Alem, D.; Gendreau, M.; Munari, P. The heterogeneous multicrew scheduling and routing problem in road restoration. *Transp. Res. Part B Methodol.* **2020**, *141*, 24–58. [CrossRef]
12. Wu, J.; Song, C.; Ma, J.; Wu, J.; Han, G. Reinforcement learning and particle swarm optimization supporting real-time rescue assignments for multiple autonomous underwater vehicles. *IEEE Trans. Intell. Transp. Syst.* **2021**, *23*, 6807–6820. [CrossRef]
13. Nahavandi, B.; Homayounfar, M.; Daneshvar, A.; Shokouhifar, M. Hierarchical structure modelling in uncertain emergency location-routing problem using combined genetic algorithm and simulated annealing. *Int. J. Comput. Appl. Technol.* **2022**, *68*, 150–163. [CrossRef]

14. Khanchehzarrin, S.; Panah, M.G.; Mahdavi-Amiri, N.; Shiripour, S. A bi-level multi-objective location-routing optimization model for disaster relief operations considering public donations. *Socio-Econ. Plan. Sci.* **2022**, *80*, 101165. [CrossRef]
15. Eliiyi, U. Artificial Intelligence for Smart Cities: Locational Planning and Dynamic Routing of Emergency Vehicles. In *The Impact of Artificial Intelligence on Governance, Economics and Finance*; Springer: Berlin/Heidelberg, Germany, 2022; Volume 2, pp. 41–63.
16. Beiki, H.; Seyedhosseini, S.M.; Mihardjo, L.W.; Seyedaliakbar, S.M. Multiobjective location-routing problem of relief commodities with reliability. *Environ. Sci. Pollut. Res.* **2021**, *28*, 1–10. [CrossRef]
17. Ye, F.; Zhao, Q.; Xi, M.; Dessouky, M. Chinese national emergency warehouse location research based on VNS algorithm. *Electron. Notes Discret. Math.* **2015**, *47*, 61–68. [CrossRef]
18. Balcik, B. Site selection and vehicle routing for post-disaster rapid needs assessment. *Transp. Res. Part E Logist. Transp. Rev.* **2017**, *101*, 30–58. [CrossRef]
19. Liu, J.; Xie, K. Emergency materials transportation model in disasters based on dynamic programming and ant colony optimization. *Kybernetes* **2017**, *46*, 656–671. [CrossRef]
20. Sun, H.; Xiang, M. Research on Robust Optimization of Emergency Location and Path for Uncertain Demand under Facility Failure Risk. *Chin. J. Manag. Sci.* **2020**, *28*, 199–207.
21. Razavi, N.; Gholizadeh, H.; Nayeri, S.; Ashrafi, T.A. A robust optimization model of the field hospitals in the sustainable blood supply chain in crisis logistics. *J. Oper. Res. Soc.* **2021**, *72*, 2804–2828. [CrossRef]
22. Chen, X.; Liu, S.; Zhao, J.; Wu, H.; Xian, J.; Montewka, J. Autonomous port management based AGV path planning and optimization via an ensemble reinforcement learning framework. *Ocean. Coast. Manag.* **2024**, *251*, 107087. [CrossRef]
23. Jin, Y.; Wang, N.; Song, Y.; Gao, Z. Optimization model and algorithm to locate rescue bases and allocate rescue vessels in remote oceans. *Soft Comput.* **2021**, *25*, 3317–3334. [CrossRef]
24. Xiao, G.; Chen, L.; Chen, X.; Jiang, C.; Ni, A.; Zhang, C.; Zong, F. A Hybrid Visualization Model for Knowledge Mapping: Scientometrics, SAOM, and SAO. *IEEE Trans. Intell. Transp. Syst.* **2023**, 1–14. [CrossRef]
25. Chen, X.; Lv, S.; Shang, W.-L.; Wu, H.; Xian, J.; Song, C. Ship energy consumption analysis and carbon emission exploitation via spatial-temporal maritime data. *Appl. Energy* **2024**, *360*, 122886. [CrossRef]
26. Wang, Y.; Wang, N.; Gao, Z.; Wu, D.; Lin, W. Optimization Model and Algorithm for Site Selection of Rescue Bases for Far Sea Islands. *Oper. Res. Manag.* **2022**, *31*, 45–50.
27. Deployment of Rescue Power of Beihai Rescue Bureau. Available online: www.bh-rescue.cn/jiuzhuzhinan/1580123544982265856.html (accessed on 2 November 2021).
28. Rescue Vessels in Beihai Rescue Bureau. Available online: www.bh-rescue.cn/product/6/ (accessed on 2 November 2021).

Journal of
Marine Science and Engineering

MDPI

Article

Identification of Shipborne VHF Radio Based on Deep Learning with Feature Extraction

Liang Chen * and Jiayu Liu

Merchant Marine College, Shanghai Maritime University, Shanghai 201306, China;
202330110040@stu.shmtu.edu.cn
* Correspondence: chenliang@shmtu.edu.cn

Abstract: In the feature identification of maritime VHF radio communication signals, shipborne VHF communication technology follows the same international technical standards formulated by IMO, uses analog communication technology and uses the same communication channel in the same area, and cannot effectively achieve signal feature identification by adding feature elements in the process of signal modulation. How to effectively identify the ship using VHF radio has always been a technical difficulty in the field of ship perception. In this paper, based on the convolutional neural network, combined with the feasibility of CAM feature extraction and BiLSTM feature extraction in non-cooperative signal recognition, a deep learning recognition model of shipborne VHF radio communication signals is established, and the deep learning approach is employed to discern the features of VHF signals, thereby accomplishing the identification and classification of transmitting VHF radio stations. Several experiments are designed according to the characteristics of ship communication scenes at sea. The experimental data show that the method proposed in this paper can provide a new feasible path for ship target perception in terms of radio signal characteristics and identification.

Keywords: ship target perception; VHF communication; signal recognition; deep learning

1. Introduction

The development of intelligent ships promotes innovation in the field of ship sensing technology, with maritime radio signal sensing emerging as a new research direction [1]. Ship navigation and maritime traffic management rely heavily on various radio signals, among which ship-borne Very High Frequency (VHF) Radios are most commonly used for navigation and traffic control purposes [2]. However, due to the utilization of analog communication technology in maritime VHF communication and the requirement for ships sailing in the same area to use a shared public channel for interactive communication, effectively identifying ships using VHF radios has always been a technical challenge within the realm of ship perception [3]. Ensuring accurate target identification during VHF radio communications is not only crucial for enabling autonomous driving and intelligent collision avoidance in intelligent ships but also essential for efficient water traffic management. The realization of this technology can significantly advance practical applications in intelligent ship development while simultaneously enhancing water traffic safety [4,5].

The identification of VHF radio falls under the domain of Specific Emitter Identification (SEI), which is an effective technique employed for identifying and locating sources of radio emissions in wireless communication networks. The SEI primarily emphasizes military scenarios, specifically the identification and classification of radar signal sources. The identification of the signal source is accomplished through statistical analysis of fundamental measurable parameters of the radar signal, such as radio frequency, amplitude, pulse width, or pulse repetition interval. Currently, hierarchical data clustering-based classification utilizing specific attributes serves as the primary technical approach for radar signal source identification [6,7]. However, there is still room for improvement in classifying identical types of radar emission sources [8].

The propagation objectives of wireless communication signals differ significantly from those of radar signals. Traditional radio signal classification primarily relies on the modulation characteristics of radio signals [9]. However, the likelihood-based (LB) and feature-based (FB) methods [10] require a prior design of feature symbols in the signal modulation process, which poses challenges for non-cooperative signals. The VHF voice communication at sea employs analog communication technology. However, in accordance with the international norms established by IMO, ships worldwide utilize shipboard VHF radio systems that adhere to identical technical standards and are verified by classification societies [11]. Unfortunately, these systems lack the capability to incorporate unique signal features during the modulation process for radio signal identification [12], resulting in a typical non-cooperative signal of the same type.

With the advancement of artificial intelligence-related technologies, deep learning methods are employed to devise learning and training models for radio signal characteristics in specific frequency bands [13], offering a novel technical pathway for perceiving and recognizing radio signals. The crux of utilizing deep learning techniques to identify radio signal features lies in establishing an effective mechanism for feature recognition [14]. Due to the operational mode of shipborne VHF radios, it is not feasible to construct a feature recognition model through supervised learning with prior information [15]. In practical maritime VHF voice communication, factors that can impact signal generation primarily stem from the radio frequency (RF) hardware structure of shipborne VHF radio stations on diverse vessels, the installation and operational environment within the equipment's location onboard ships, as well as the transmission environment from the feeder line to the antenna termination point [16]. These factors contribute to distinct individual characteristics even when VHF radio signals employ identical modulation modes.

The current radio signal feature recognition methods can be roughly divided into three categories. The first one is based on basic machine learning methods, which usually use classical machine learning algorithms such as support vector machine and decision tree to extract signal features and classify them. Niranjan, R. et al. developed a decision tree algorithm based on time-domain digital technology for real-time recognition and classification of various radar intrapulse modulation signals [17]. To solve the problem of misjudgment and confusion in the feature recognition of classified signals, a three-layer optimization algorithm is proposed by Wang S, Wang Z et al. [18]. S. Guo, S. Akhtar, and others proposed three kinds of RF fingerprints for radar model recognition using support vector machine, decision tree, random forest, gradient lifting, and the MDA algorithm [16]. The utilization of basic machine learning techniques enables the recognition and classification of diverse wireless signal types, exhibiting a certain degree of universality. However, these methods possess a broad focus on features and necessitate an extensive amount of labeled data for effective model training and learning. In the context of feature recognition within signals sharing the same modulation scheme, employing various machine learning algorithms can enhance efficiency in this process. Nevertheless, there exists a potential risk of overfitting identification outcomes.

The second is based on a deep learning method, which mainly uses a deep neural network to learn complex signal feature representation, such as the convolutional neural network (CNN), recurrent neural network (RNN), etc. This method shows superior performance in tasks of object detection, recognition, and classification. Hao C, Dang X et al. proposed a symbol detection and modulation classification detector (SDMCD) based on a deep neural network (DNN) to complete the mixed signal detection [19], but the parameter adjustment and optimization of the DNN model usually require some experience and expertise, and the maritime environment will also affect the performance and accuracy of the SDMCD model. Zhang J, Hu S proposed a digital signal modulation identification (DSMI) method suitable for orthogonal frequency division multiplexing (OFDM) under different multipath channels. This method can accurately detect the modulation characteristics rather than the channel characteristics to identify the modulation type, thereby reducing the amount of network training [20]. In the maritime VHF communication scenarios, the signal differences of the same modulation mode are mainly formed by the equipment environment and channel environment, so this kind of method lacks common attention to the signal and channel. Kim H, Kim Y-J, et al. proposed a deep signal recognition network based on multi-task learning [21]. The proposed modulation recognition model by Zha Y and Wang H aims to enhance the accuracy of recognizing high-order digital modulation signals in low SNR conditions, thereby significantly improving their performance [22]. Huynh-The, T. et al. focused on the application of classical neural networks (such as CNN, RNN, and FFNN) in deep learning for modulation recognition; Zhang, X, Wang, Y. et al.'s research focuses on the review of data sets, signal representation methods, and deep learning-based methods in the field of modulation recognition [23]. Generally speaking, in the scenario of VHF signal recognition at sea, it is necessary to focus on the feature recognition under the same signal modulation mode, and to describe the features from the interaction of the signal itself and environmental variables on the basis of deep learning, so as to improve the feature recognition of VHF signals.

The third is to combine the traditional machine learning method with the deep learning method, and make the concerns of the recognition model have a certain characteristic by fusing different algorithms, so as to improve the accuracy and robustness of signal recognition. F. Liu, Z. Zhang et al. proposed a modulation recognition method based on feature extraction and deep learning algorithms, which achieves high recognition rates even under low signal-to-noise ratios [24]. H. Xiang, B. Chen et al. proposed a feature-to-feature learning phase enhancement method based on 1D CNN and 2D CNN to improve the accuracy of DOA estimation [25], CNN can effectively extract the spatial features of signals from the marine environment, and can adapt to the complex changes of signals in various marine environments, but the large amount of computing resources and long training time in recognition enhancement may increase the limitations of this method in ship communication scenarios. Lu investigated a model that combines discrete wavelet transform and Extra Trees Classifier, and designed a model that combines continuous wavelet transform and CNN to perform individual identification, which requires a high computational cost and training time in terms of complexity and parameter tuning [26]. Tan, Y.; Jing, X. et al. adopted CNN to extract the features of the observed signals and built a new two-dimensional dataset of the received signals [27]. L. Zhang used the SCF-based dataset, combining the CNN denoising module and DNN-based classification design, H. Liu et al. can achieve better feature extraction and classification performance, thereby improving the signal recognition accuracy [28]. The combination of machine learning and deep learning can provide a more effective feature recognition model for signal recognition of specific targets, but it is also necessary to select appropriate methods for signal scene features to cope with computational complexity and training data demand.

In summary, numerous scholars have made significant advancements and progress in the field of signal modulation recognition. However, in the context of maritime VHF voice communication scenarios, it is crucial to fully consider the characteristics of maritime VHF signal recognition based on both signal modulation characteristics and the complex conditions present in ship communication scenarios, including factors such as waves, wind, and their impact on signal propagation. Therefore, this study initially aims to establish a fundamental feature recognition model for VHF voice communication signals in maritime scenarios using CNN methods. Subsequently, we attempt to incorporate Class Activation Mapping (CAM) and Bidirectional Long Short-Term Memory (BiLSTM) feature extraction algorithms. In the maritime scene specifically, CAM enables intuitive analysis of spatial dimension-based feature extraction by the model while BiLSTM effectively captures temporal evolution of signals as a timing model. This paper endeavors to combine CAM feature extraction with BiLSTM feature extraction to enhance the effectiveness of the basic model in identifying similar types and non-cooperative signal features. The contributions made by this paper primarily lie within the following aspects.

1. According to the modulation and demodulation characteristics of shipborne VHF signals, the effects of white noise and multipath propagation on the characteristics of VHF signals are studied.

2. Based on feature engineering and deep learning technology, a deep learning model for extracting the features of shipborne VHF radio signals is established by integrating the idea of a channel attention mechanism (ECANET), which can realize the recognition of different shipborne VHF radio signals.
3. Combined with the application scenario characteristics of VHF voice communication at sea, five VHF radios with significant similarity characteristics are selected, and the verification experiment of the correlation between communication voice and signal recognition is designed to verify the effectiveness and accuracy of the proposed method.

The paper is organized as follows. Section 1 analyzes the current main methods and technical paths in radio signal feature recognition, and analyzes the rationality of the selected method in this paper. Section 2 introduces the mathematical principles of the model and method designed in this paper. Section 3 verifies the model and method used in this paper by experiments. Section 4 analyzes and discusses the experimental results. Section 5 is the conclusion of this paper.

2. Materials and Methods

2.1. General Framework

The model consists of three core parts, which are VHF signal analysis, model optimization, and output, as shown in Figure 1. In the initial stage of signal processing, the model measures and identifies the propagation characteristics of VHF signals in different channel environments and the unique features of their modulation modes. Subsequently, it is input into the model optimization module, based on the CNN architecture. The model combines CAM and BiLSTM feature extraction methods and ECANET technology to deeply learn the time and frequency characteristics of signals, extract signal features, and achieve the accurate identification of various complex signals through training and optimized network parameters. Finally, the model outputs the recognition result of the signal to complete the whole process and achieve the accurate recognition of the input radio signal.

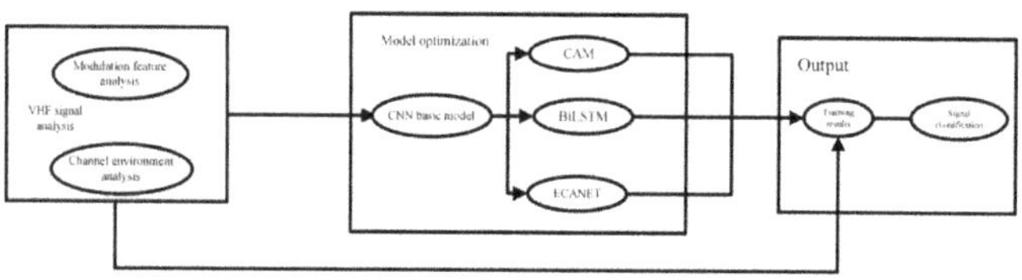

Figure 1. Overall framework.

2.2. VHF Signal Analysis

Signal identification for marine VHF is achieved by channel detection of marine VHF frequency band, which is able to collect and analyze the signals of the whole channel of marine VHF, and on this basis, extract the individual radio frequency fingerprint information of each module through feature engineering and deep learning; that is, by collecting and analyzing the characteristics of radio frequency signals. These characteristic parameters may include amplitude, phase, frequency, etc. of the signal. Due to the differences in the RF hardware and the influence of the communication environment, the radio signals received at the receiving end still have typical characteristics in the radio signals generated by the same modulation mode and RF hardware structure. By analyzing and comparing these characteristic parameters, the identity information can be determined. The RF fingerprint bound with the ship identity is constructed to distinguish the hardware differences of different shipborne VHF RF transmitting modules, and the identification and tracking of the shipborne VHF radio signal fingerprint are realized.

For communication emitters in practical application scenarios, the emitted RF signals need to be transmitted through the channel and demodulated before they can become available data sets. The modulation and demodulation part of the signal is to perform a certain operation on the original time domain signal to generate a signal of another frequency suitable for transmission in the channel; that is, to convert it into a form suitable for transmission, while the subsequent signal identification algorithm is to analyze the received signal. If there is an error or distortion in the modulation and demodulation process, the subsequent signal identification algorithm may be affected. As a result, the signal cannot be correctly identified or decoded, thus affecting the accuracy and reliability of identification. Therefore, after the acquisition equipment is used to complete the VHF signal acquisition, the signal demodulation processing should be carried out, and the analysis results should be stored in the database. The VHF signal received by any radio can be expressed as

$$S(n) = Acos[w_c n + \varphi(n)] \tag{1}$$

where A is the signal amplitude, w_c is the carrier angular frequency, and $\varphi(n)$ is the instantaneous phase:

$$S(n) = Acos\left(w_c n + K_f \Sigma f(n) + \varphi_0\right) \tag{2}$$

Orthogonal decomposition of the signal, the in-phase branch, is multiplied by $cos(w_c n)$ and the quadrature branch is multiplied by $\sin(w_c n)$, and decimated and low-pass filtered to obtain the in-phase and quadrature components.

Performing an arctangent operation on the quadrature component and the in-phase component:

$$\varphi(n) = arctan[\frac{X_0(n)}{X_1(n)}] = K_f \sum f(n) + \varphi_0 \tag{3}$$

The modulated signal $f(n)$ is then obtained by differencing the obtained phases and scaling them appropriately.

$$f(n) = \varphi(n) - \varphi(n-1) \tag{4}$$

That is, the required original signal. Thus, the demodulation of the VHF signal is completed.

The quality of maritime VHF communication is affected by many factors, such as the harsh environment of the sea, the transmission loss of sea reflection, and the transmission loss of free space. Therefore, considering the change of signal-to-noise ratio and the fading characteristics of the channel, it is necessary to preprocess the received shipborne VHF signals.

Firstly, different levels of Gaussian white noise are introduced, which can simulate the noise interference in the real environment and enhance the robustness and reliability of the signal data. Then, the Rayleigh channel function operation is employed, which can optimize the transmission characteristics of the signal and make it more suitable for subsequent analysis, identification, or transmission tasks.

(1) White noise is a random signal with a constant power spectral density. Ideal white noise has infinite bandwidth, so its energy is infinite, which is impossible in the real world. In fact, if the spectral width of a noise process is much larger than the bandwidth of the system on which it acts, and its power spectral density is essentially constant in this bandwidth, then it can be treated as white noise.

The amplitude of Gaussian white noise is subject to Gaussian distribution, and the power spectral density is uniformly distributed. Because Gaussian white noise can reflect the noise situation in the actual communication channel, it can better simulate the unknown real noise [29], and can be expressed by a specific mathematical expression, which is suitable for analyzing and calculating the anti-noise performance of the system.

Usually, the signal we obtain is discrete, and the power of the signal can be calculated directly.

$$P_{signal} = \frac{1}{n} \sum_{k=1}^{n} s_k^2 \tag{5}$$

With the signal power, if you want the signal-to-noise ratio (SNR) of the sample after adding the noise to be SNR, you can calculate the power of the noise from these two:

$$P_{noise} = \frac{P_{signal}}{10^{\frac{SNR}{10}}} \quad (6)$$

(2) Different from the wired channel whose transmission carrier is physical wire, the propagation environment of a shipborne VHF signal is more complex and changeable in the process of propagation on the radio channel. There are a lot of factors that affect the propagation of electromagnetic wave signals in the radio propagation environment, mainly the scatterers in the radio propagation environment, such as clouds, water, and so on. These factors restrict the transmission quality of electromagnetic wave signals. When the transmitted signal arrives at the receiver through different paths, the multipath propagation will lead to different time and direction, and the random amplitude and phase of different multipath components will cause the change of signal strength, resulting in small-scale fading and signal distortion, affecting the quality of communication. The propagation mode of electromagnetic waves in the radio propagation environment is shown in Figure 2:

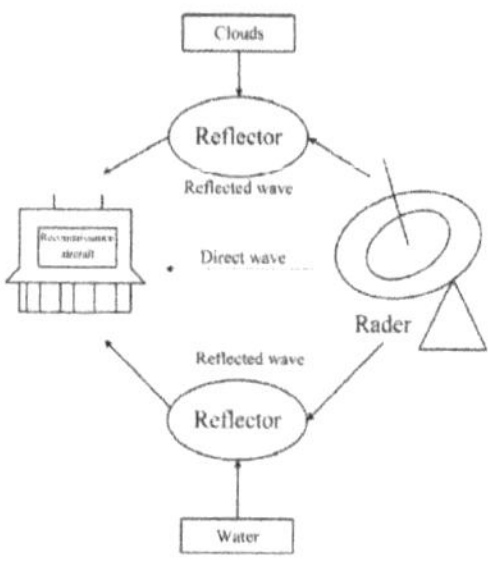

Figure 2. Signal Propagation Diagram.

Therefore, in the research of radio communication, the primary task is to model the channel. Considering the influence of Doppler frequency shift caused by the relative movement of the transmitter and receiver on the channel in the actual situation of maritime communication, statistical modeling is more suitable for outdoor channels. The Clarke model is a commonly used radio channel model, which is suitable for the scenario of slow mobile speed. The amplitude distribution and phase distribution are random and meet the Gaussian distribution. Due to the wide applicability of the Clarke model, the Clarke model can better simulate the channel characteristics in the complex maritime communication environment with high computational efficiency. The Clarke model can be used as a statistical modeling of the Rayleigh channel.

Assuming that the multipath scattering components of the signal are uniformly distributed and the average power of the signal in each path is the same, the in-phase and quadrature components of the channel are generated by two branches, respectively. Firstly, a complex Gaussian noise signal is generated on each branch of the model, and the two complex Gaussian noise signals are guaranteed to be independent of each other. These random signals are then shaped in the frequency domain by using a Doppler shaping filter to adapt to the variable environment of signal propagation. Subsequently, the two filtered signals are subjected to inverse Fourier transform as the in-phase component and the quadrature component of the received signal, respectively. Considering that the output of the Fourier transform is a real signal, the time domain signal of one of the branches needs to be rotated by 90 degrees to meet the requirement of orthogonal components. Finally, the two signals are superposed to obtain the output fading signal. The maximum Doppler frequency shift of the output signal is the bandwidth of the shaping filter, and the expression is

$$h(n) = h_I(n) + jh_Q(n) = \lim_{N \to \infty} \sum_{n=1}^{N} c_n exp(2\pi f_d n + \theta_n) \quad (7)$$

where N is the number of propagation paths; c_n is the attenuation coefficient; f_d is the maximum Doppler shift; θ_n is the Doppler phase, which follows the uniform distribution of $[-\pi, \pi]$; as N approaches infinity, $h(n)$ approaches a zero-mean complex Gaussian process, whose amplitude envelope follows the Rayleigh distribution according to the central limit theorem.

2.3. Foundation Model

CNN is a feed-forward neural network [30]. The application of CNN in signal recognition usually takes the original signal as the input, such as IQ signal, which is essentially a mapping from input to output, and can learn the mapping relationship between input and output [31], without precise mathematical expression, and automatically extract features and classify them through the training process. The training process includes two phases: the first phase is the forward propagation phase, in which the input is gradually transformed into a higher-level feature representation through convolution, activation function, pooling, and other operations, as shown in the following formula. The purpose of forward propagation is to calculate the prediction result of the network and compare it with the true label to calculate the loss function, which is used to measure the gap between the prediction result and the true result.

Convolve to convolutional layers:

$$z_i^{l+1}\sum_s a_s^l * w_{i,s}^{l+1} + b_i^{l+1} \tag{8}$$

$$a_i^{l+1}= sigmoid\left(z_i^{l+1}\right) \tag{9}$$

Convolutional layer to pooling layer:

$$ap_i^l = pooling\left(a_i^l\right) \tag{10}$$

Pooling layer to fully connected layer:

$$z_s^{l+1} = \sum_i ap_i^l w_{s,i}^{l+1} + b_s^{l+1} \tag{11}$$

The second stage is the backpropagation stage, where the gradient of the parameters in each layer is calculated by the chain rule and the gradient descent algorithm is used to update the parameter values. In this way, the network fine-tunes the parameters according to the gradient of the loss function in each training iteration to gradually optimize the predictive power of the model.

Fully connected layer error:

$$\delta^L \equiv \frac{\partial C}{\partial a^L}\sigma'(z^L) \tag{12}$$

Error propagation from fully connected layer to pooling layer:

$$\frac{\partial C}{\partial ap_i^l} = \sum_s w_{i,s}^{l+1}\delta_s^L \tag{13}$$

Error propagation from pooling layer to convolution layer:

$$\delta_i^l = upsample(\frac{\partial C}{\partial ap_i^l})\,\sigma'\left(z^L\right) \tag{14}$$

Error propagation between convolutional layer:

$$\delta_s^l = resize\left(\delta\right)_i^{l+1} * rot180^\circ\left(w\right)_{i,s}^{l+1}\sigma'\left(z_s^l\right) \tag{15}$$

Gradient of model parameters:

$$\frac{\partial C}{\partial w_{i,s}^{l+1}} = a_s^l * interpolation_d(\delta^{l+1})_i \tag{16}$$

When the signal is processed by the CNN, the convolutional layer automatically extracts features from the data. Through different filters, the convolution operation can capture the spatial correlation and characteristics of the signal, so that the input signal can be represented more abstractly after the convolution layer. These extracted features are then passed to the fully connected layer, which is responsible for learning a higher-level abstract representation of the signal and mapping it to the corresponding output category. Ultimately, the CNN outputs the recognition result to determine which category the input signal belongs to. This process realizes the end-to-end automatic recognition from the original signal to the final classification result, and realizes the automation and efficiency of signal recognition through machine learning. The model architecture of the CNN requires very little data preprocessing, so it is considered to be a robust deep learning method for the successful application of a truly multi-level network architecture [32]. This model can achieve or exceed the traditional classification method which relies on experts to extract features, and at the same time, because it does not need to extract features manually, the model has greater flexibility in expanding new signals.

2.4. Model Optimization

In this paper, an improved CNN model is used to recognize VHF signals, which combines CNN, CAM, and BiLSTM feature extraction technology, and incorporates the idea of a channel attention mechanism (ECANET) to enhance the adaptive learning ability of the network to the importance of channels. The depth feature extraction and fusion of VHF signal data are realized, and the basic structure of the model is shown in Figure 3.

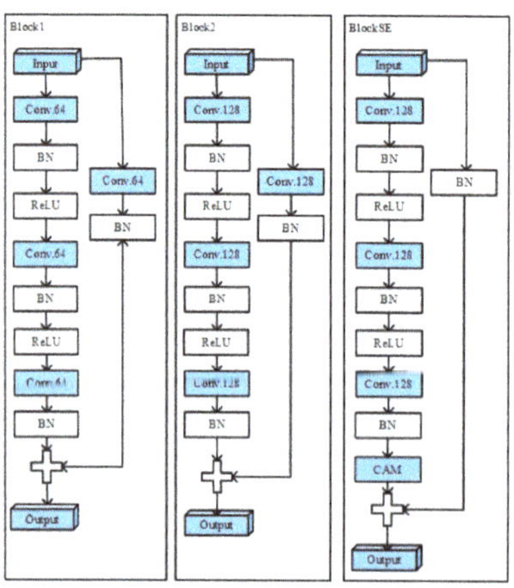

Figure 3. Improved CNN Model Structure.

The diagram is divided into three main sections, each of which represents a different block, and each of which has a different hierarchy and operation. Each block starts with "Input" and then passes through multiple layers, ending up with "Output". First, the Block1 part consists of three sequentially identical structures, each of which contains a convolution layer (CN), followed by a ReLU activation function and a batch normalization (BN) layer. After these three sets of operations, the output of Block1 is obtained. Block2 is similar to Block1, but with an increased number of convolution kernels per convolution layer. The input data in the BlockSE module first pass through several convolution layers (Conv), normalization layers (BN), and activation layers (ReLU), and finally through a channel attention mechanism layer (CAM), which can correlate the feature maps in the neural network with the classification results, thus helping to understand the classification basis of the network. The feature map is formed by multiplying and adding the parameters (weight matrix W) of the last fully connected layer and the final output feature map set; that is, the display model is based on which feature maps for decision classification. The extraction of CAM generally occurs in the convolution layer, especially in the last convolution layer.

CAM uses the principle of superposition of feature map weights to obtain the thermodynamic map, and the formula is as follows:

$$L_{CAM}^{c} = \sum_{k} w_{k}^{c} A^{k} \tag{17}$$

where A represents the size of the output of the last convolutional layer of the network, w represents the weight of the fully connected layer, and C is the category of classification.

Channel attention mechanism is a method used to enhance the neural network to learn feature representation between different channels, which can improve the effectiveness of feature representation by assigning different weights to different channels [33]. This mechanism has been widely used in Squeeze-and-Excitation Networks (SENet) and has been proved to significantly improve the performance of various computer vision tasks. In the traditional channel attention mechanism, the feature representation of different channels is usually adjusted by calculating the weight of each channel. Although this method can improve the network performance, it still has some limitations; for example, it may ignore the correlation between channels, resulting in information loss. This feedback connection allows the network to transfer information between different layers, so as to make better use of the correlation between channels. This feedback connection can help the network to better capture the dependence between specific channels, and improve the representation ability and generalization performance of the model. The improved CAM model is applied to the improved CNN model, as shown in Figure 4, which further improves the performance of the neural network in feature learning and representation.

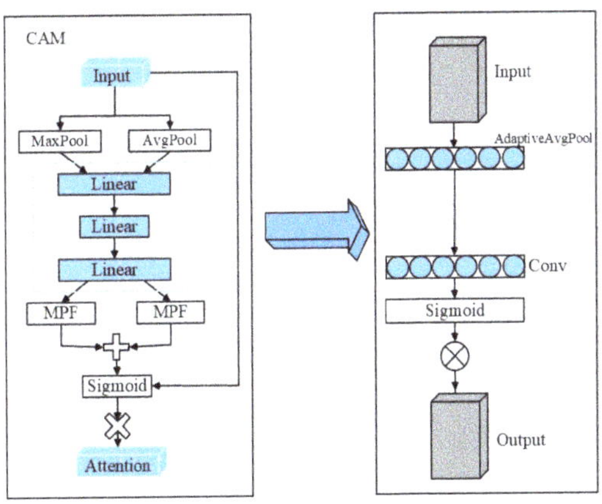

Figure 4. Improvement of Channel Attention Mechanism.

BiLSTM is a long and short-term memory network (LSTM) with forward and backward connections [34]. BiLSTM processes the input through two independent LSTM layers, one in chronological order and the other in reverse chronological order, capturing the characteristics of the input sequence from the forward and reverse directions, respectively. Specifically, the forward LSTM processes the input sequence from left to right in time steps, and the hidden state h_t and the cell state c_t at each time step are calculated by the following formulas:

$$i_t = \sigma(W_{ix}x_t + W_{ih}h_{t-1} + b_i) \tag{18}$$

$$f_t = \sigma\left(W_{fx}x_t + W_{fh}h_{t-1} + b_f\right) \tag{19}$$

$$o_t = \sigma(W_{ox}x_t + W_{oh}h_{t-1} + b_o) \tag{20}$$

$$\widetilde{c}_t = tanh(W_{cx}x_t + W_{ch}h_{t-1} + b_c) \tag{21}$$

$$c_t = f_t \odot c_{t-1} + i_t \odot \widetilde{c}_t \tag{22}$$

$$h_t = o_t \odot tanh(c_t) \tag{23}$$

where i_t, f_t, o_t and $\widetilde{c}_t$ are the input gate, forget gate, output gate, and candidate state representing the current information, respectively, and σ and tanh are the Sigmoid and tanh functions, respectively. Similarly, the reverse LSTM processes the input sequence from right to left in time steps, and the hidden state h'_t and the unit state c'_t for each time step can be calculated by similar formulas. Eventually, the output of BiLSTM is stitched from the hidden states in both directions: $y_t = [h_t; h'_t]$.

Through CAM feature extraction, the model can associate the feature map with the classification results, which enhances the classification ability of VHF signals, while BiLSTM feature extraction helps to capture the temporal information in VHF signal data [35]. Ultimately, deep feature fusion enables the improved CNN model to understand VHF signal data more comprehensively, thus improving the accuracy and robustness of recognition and classification.

In the original model structure, the convolution kernel size of a fixed scale is used, which may lead to the model unable to effectively capture the features of different scales in the input data. In order to improve the capability of feature extraction and target recognition in different scales, the multi-scale convolution module is introduced to make the model consider the information of different scales at the same time, so that the model can better adapt to targets of different scales. A common multi-scale convolution method is to use convolution kernels of different sizes for parallel convolution operations, and then splice the resulting feature maps in the channel dimension. As shown in Figure 5, two convolution kernels can be used to convolute the input image at the same time, and then the resulting feature maps can be spliced in the channel dimension. Multi-scale convolution can help the model to extract features at different scales, thus improving the recognition ability of the model for targets at different scales. Therefore, EBlock1 and EBlock2 are used instead of Block1 and Block2 in the improved CNN model structure.

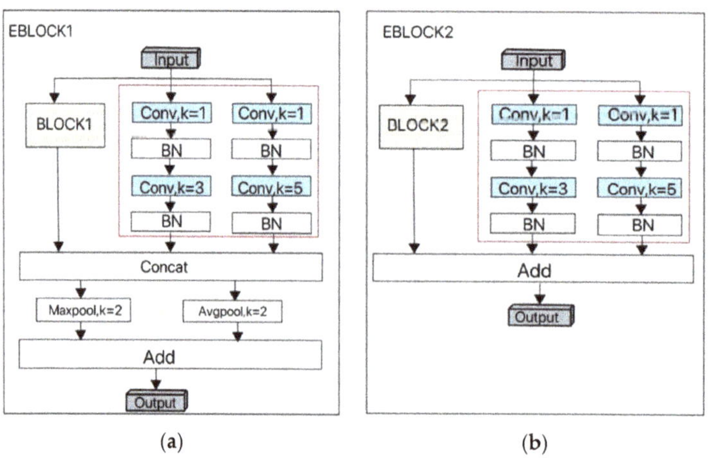

Figure 5. Model Improvement: (**a**) improved EBLOCK1 module; (**b**) improved EBLOCK2 module.

3. Results

3.1. Experimental Design

3.1.1. Experimental Scenario

In this experiment, four groups of experiments are carried out to collect the VHF signals with the same audio and different frequencies from five VHF interphones in the indoor experimental point X and outdoor experimental point Y, and the signals are recognized by using the improved CNN model. Very high frequency (VHF) radio is used most frequently when the ship is sailing in the channel. The characteristics of a VHF signal propagation environment in the sea channel mainly include the shelter of ship superstructure, water environment, and different ship radio installation and application environments. The greater the difference of these environments, the greater the difference of signal characteristics. In order to better verify the feasibility of the method in this paper, in the design of the experimental scene, based on the simulation of the scene of ship communication in the channel, the impact of environment and equipment on signal characteristics is minimized. In the experimental scene of this paper, five hand-held VHF walkie-talkies of the same model and old and new degree are selected as VHF communication equipment. The straight-line distance between the transmitting point and the experimental point Y in the experimental scene is about 100 m, and there are some buildings and lakes in the transmission path. At the same time, the weather of the outdoor experimental point Y is overcast and rainy. The experimental scene environment can be matched with the environment of ship communication in the channel. The following analogy can be made in order to make experiments performed indoors reliable for use in the marine environment.

1. Signal propagation characteristic:

In the indoor environment, signal propagation is affected by buildings, walls, and other obstacles, which may lead to signal attenuation, multipath effect, and other problems. It can be analogized that in the marine environment, signal propagation will be affected by seawater, ships, and other objects, which may lead to problems such as seawater reflection and multipath propagation. At the same time, the multipath propagation model is introduced, and the signal propagation characteristics are studied in depth to better understand the signal propagation mechanism in the marine environment

2. Interference sources:

There may be interference signals from electrical equipment, radio networks, etc. In the indoor environment, this affects the quality of the received signal. In the marine environment, ships, radar, and other equipment may produce interference signals, which interfere with the identification of radio reception signals.

3. Weather conditions:

In the indoor environment, the weather conditions are stable and are not affected by meteorological factors. In the marine environment, weather conditions may change, such as wind and waves, rain, and snow, which affect signal transmission and reception. Therefore, an experiment is designed for comparison in the outdoor rainy environment of experimental point Y. The straight-line distance between point Y and the receiving equipment is about 100 m, and there are also problems such as multipath propagation and interference.

4. Equipment:

The experiment is to collect VHF signals of the same type and batch of new walkie-talkies in the same environment for identification, and the difference between equipment and environment is very small. Considering the complexity and uncertainty of the marine environment, it is expected that in the marine environment, with different batches of equipment and VHF signals in different environments, the recognition rate of the experimental method may be further improved, which further verifies the feasibility of the experiment.

The purpose of the experiment is to verify the feasibility of the proposed method of feature identification and VHF radio classification. According to the analysis, the signal differences of ship-borne VHF radio are mainly reflected in the frequency domain (modulation) and time domain (propagation environment), in which the environmental impact and ship motion can increase the difference of signal characteristics. The purpose of the experimental scenario design in this paper is to simulate the lowest characteristic difference scenario in the ship communication scenario of the channel, including using the same type of walkie-talkie to obtain the lowest difference between devices. In the experimental scenario, there are buildings and water surfaces in the transceiver path to simulate the characteristics of the ship communication scenario, which can better verify the feasibility of this method.

3.1.2. Experimental Process

Firstly, respectively, data were acquired of the same audio frequency and different audio frequency signals sent by an interphone in an indoor environment and an outdoor environment by utilizing a signal acquisition equipment; secondly, the acquired audio frequency data were preprocessed, possibly comprising the steps of denoising, feature extraction and the like, modulating signals, and converting into an IQ data set. The size of 2×128 (I and Q) is used in the experiment to form a sample as the input layer of the CNN. Samples are used to train the model for audio signal recognition tasks, and then the accuracy of the model is verified to evaluate its performance in the recognition of the same audio and different audio signals, as shown in Figure 6. The models trained in the indoor environment are applied to the recognition tasks of the same audio and different audio signals in the outdoor environment to evaluate their generalization ability. Through experiments in indoor and outdoor environments, the performance of the model in different environments is compared, which can evaluate the robustness and generalization ability of the model more comprehensively, and provide more reliable reference and support for the application of the model in the marine environment in the future.

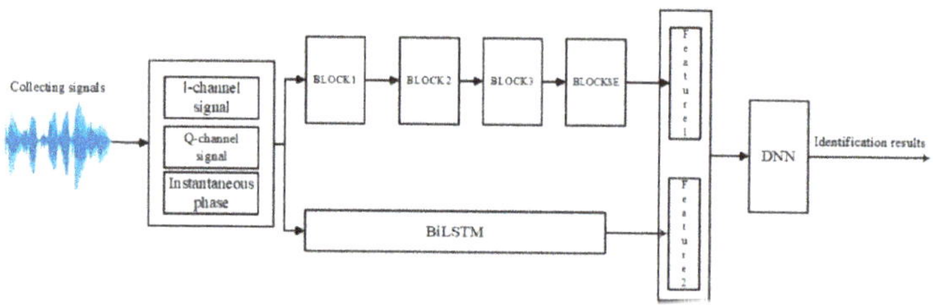

Figure 6. Experimental flow chart.

3.1.3. Parameter Selection

In the experiment, the above improved CNN model is concretized, and the structure of the model will directly affect the number and complexity of the parameters of the model. At the same time, by choosing reasonable parameters, the performance, generalization ability, and efficiency of the model can be optimized, so that the model can better adapt to specific tasks and data sets. In addition, the CNN extracts local features through a convolution kernel, and the model updates the convolution kernel parameter matrix through training, so that the convolution kernel can better extract features according to the purpose of the model. The number of convolution kernels determines the number of output feature maps of the layer. By adjusting the scale and number of convolution kernels, the dimension of feature maps and the expressive power of the network can be controlled. Therefore, the convolution kernel size of the Block1 convolution layer in the model structure is set to 68, and 68 convolution kernels are selected to effectively capture the characteristics of the input data without causing overfitting. In general, a smaller number of convolution kernels means that the model is simpler, easier to train, and may have better generalization ability. In Block2, the increase to 128 convolution kernels is to increase the representation capability of the network. As the network layer deepens, it is usually necessary to increase the number of convolution kernels to extract more abstract and complex features, based on further optimization of network performance and the need to capture more complex data patterns. In the improved multi-scale convolution module, different sizes of convolution kernels are used to increase the perception range of the neural network to the input data. In this paper, two sizes of convolution kernels, 3×3 and 5×5, are used. The size 3×3 is a smaller convolution kernel, which is mainly used to capture local details, while 5×5 is a larger convolution kernel, which is able to cover a wider area and thus extract more global feature information. At the same time, in practice, these two convolution kernels are widely used in deep learning models, and have achieved remarkable results. They show good performance and generalization ability in many visual tasks, providing a reliable basis for the optimization and application of the model.

3.1.4. Experimental Steps

Here are the specific steps of the experiment:

① Use the IC-M73 VHF walkie-talkie as shown in Figure 7a to transmit the signal carrying audio on channel 10.

② Use the sampling equipment as shown in Figure 7b for sampling. The center frequency of the sampling equipment is set as 156.5 MHz, and the bandwidth is set as 100 KHz; that is, the equipment collects signals from 156.45 MHz to 156.55 MHz of the VHF band, and can capture various VHF signals within this range. The sampling frequency is set to 100×10^3 and the sampling time is 5 s. The same and different audio signals of five IC-M73 walkie-talkies A\B\C\D\E at location X and location Y (location Y is farther from the receiving equipment than location X) are collected in turn as training samples.

③ The samples for training are manually labeled as the signal samples sent by which walkie-talkie, and then the improved CNN model is selected for training, with 100 generations of training each time and automatic testing every three generations. Check the accuracy, loss, and test results of the model after training.

④ The same and different audio signals of five IC-M73 walkie-talkies at X and Y are collected in turn as test samples.

⑤ Select the test samples without manual labelling and use the trained improved CNN model to identify the signals.

⑥ Check the final identification result.

(a) (b)

Figure 7. Experimental equipment: (**a**) model IC-M73 Interphone; (**b**) radio signal receiving equipment.

3.2. Experiment 1: Compare the Recognition Rate of the Same Audio and Different Audio Samples at Point X

At point X, the same audio samples used for training and testing in the experiment are the digital voice broadcast samples of "7408450651917459937611 5120" sent by five walkie-talkies, and the different audio samples for testing are the voice broadcast of different numbers.

The model is first trained using the samples, and the experimental results of the training are shown in Figures 8 and 9, with fold line 1 showing the accuracy and loss of training the model with the same audio samples at point X, and fold line 2 showing the accuracy and loss of training the model with different audio samples at point X.

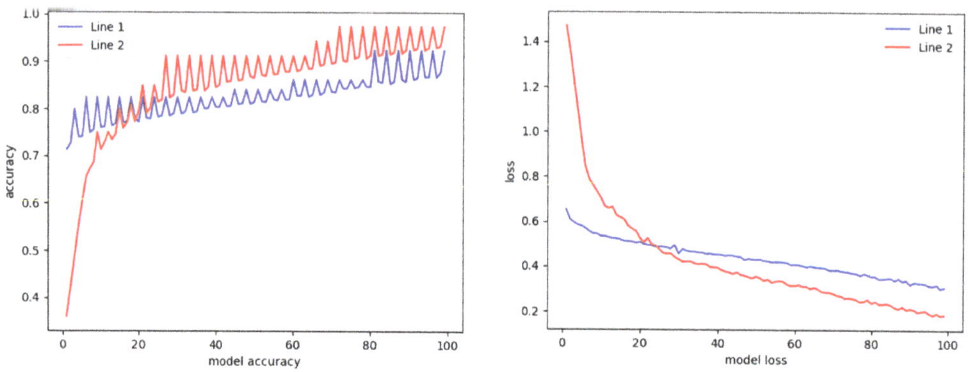

(**a**) (**b**)

Figure 8. Recognition Result of Experiment 1: (**a**) model accuracy; (**b**) model loss.

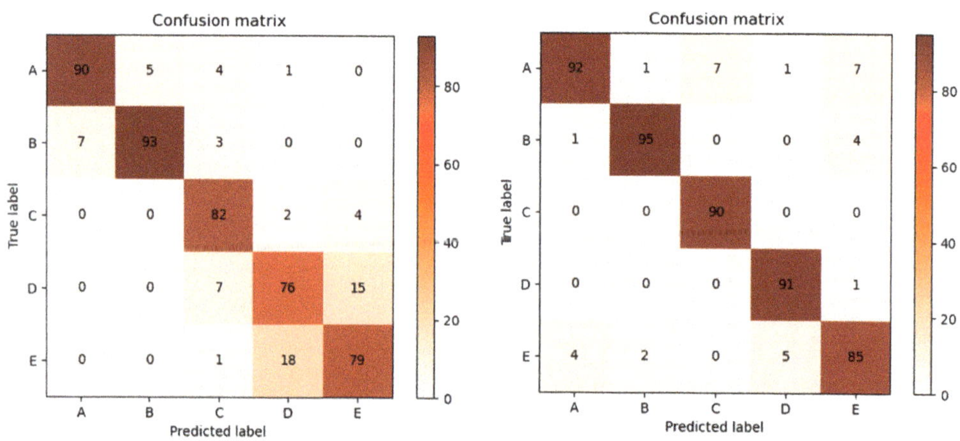

(**a**) (**b**)

Figure 9. Confusion matrix of Experiment 1: (**a**) same audio recognition result; (**b**) different audio recognition result.

From the image of model accuracy, the accuracy of "Line 1" rises rapidly at the beginning of training, then reaches an inflection point at about 20 iterations, and the accuracy begins to rise slowly and fluctuates around 0.9. "Line 2" is less accurate than "Line 1" from the beginning, but there is also an inflection point at about the same iteration of 20 times, after which the accuracy rises rapidly and reaches a plateau near 0.8. The reason for the inflection point may be that in the early stage of training, the model can often quickly learn some useful patterns from the data, resulting in a rapid increase in accuracy. As the training continues, the model may approach the upper limit of its ability, and it is difficult to further improve the accuracy significantly, resulting in the emergence of inflection points. At the same time, because "Line 2" is trained with different audio samples, it may cause the model to need more time to generalize different data features, resulting in fluctuations or delays in the process of increasing accuracy; in the loss diagram of the model, the model shown by "Line 1" declines faster in the early stage, which may mean that it learns faster. At a later stage, "Line 2" decreases slowly, but its loss value decreases to a similar level as "Line 1", indicating that the two methods may eventually have similar performance. The confusion matrix of the test results shows that the test effect of different audio is better and the recognition degree is higher.

After 10 test sample recognition experiments, the average accuracy of the comparison test is close to the accuracy of the training, which shows that the model has good performance and generalization ability, and can reliably predict or classify tasks in practical applications. The accuracy of recognition results is shown in Figure 10.

3.3. Experiment 2: Compare the Recognition Rates of the Same Audio Samples at Points X and Y

The same audio samples used for training and testing in the experiments at points X and Y are the digital voice broadcast samples of "74084506519174599376115120" emitted by five walkie-talkies.

First, the model is trained using samples. The experimental results of the training are shown in Figure 11. The broken line 1 is the accuracy and loss of training the same audio sample model at point X, and the broken line 2 is the accuracy and loss of training the same audio sample model at point Y.

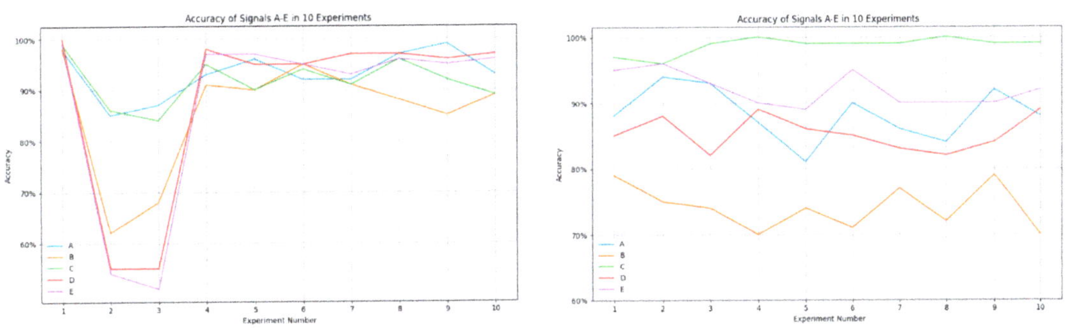

(a) (b)

Figure 10. Multiple Recognition Results of Experiment 1: (**a**) same audio recognition result; (**b**) different audio recognition result.

From the fluctuation of the fold line, we can see that the learning of the model is fluctuating throughout the training process, which is due to the fluctuation brought by the gradient update, or the model is trying to learn more complex patterns in the training set. It can be seen from the figure that when the same audio is trained, the accuracy of the model is high when it is close to the receiving point. It is possible that the signal may experience more attenuation or interference because point Y is farther away from the receiving device, resulting in a decrease in the quality of the input data, thus affecting the effect of model training. However, the overall difference is not significant, and the final "Line 2" loss is small.

The accuracy results of the recognition experiment for 10 test samples are shown in Figure 12.

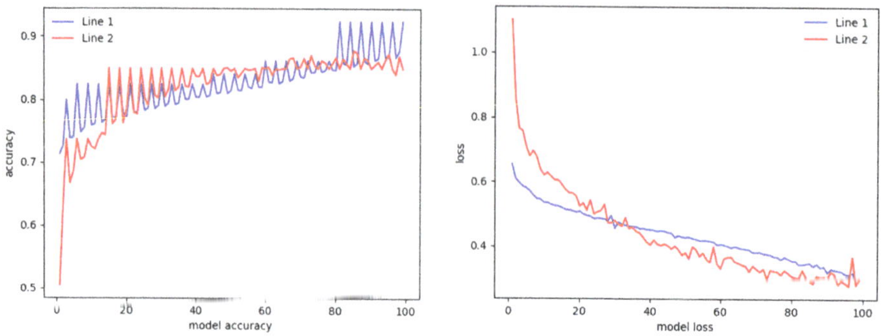

(**a**) (**b**)

Figure 11. Recognition Result of Experiment 2: (**a**) model accuracy; (**b**) model loss.

3.4. Experiment 3: Compare the Recognition Rates of Different Audio Samples at Points X and Y

In the experiment of point X and Y, different audio samples are used to train and test the voice broadcast of different numbers from five walkie-talkies.

First, use samples to train the model. The experimental results of the training are shown in Figure 13. Fold line 1 is the accuracy and loss of training different audio sample models at point X, and fold line 2 is the accuracy and loss of training different audio sample models at point Y.

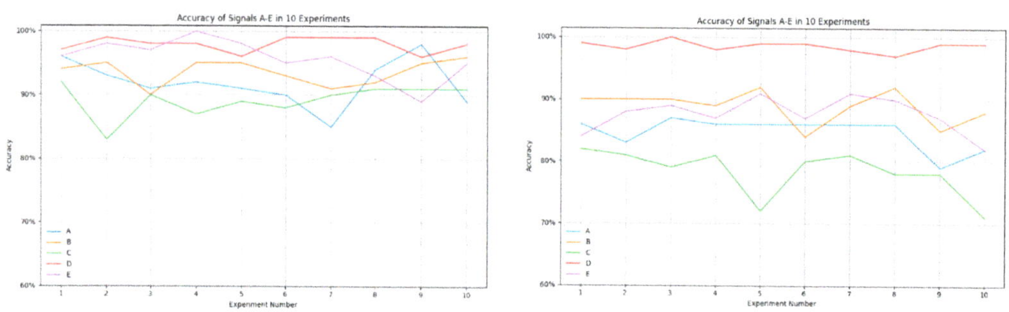

(**a**) (**b**)

Figure 12. Multiple Recognition Results of Experiment 2: (**a**) The same audio recognition result at point X; (**b**) the same audio recognition result at point Y.

The figure shows that both "Line 1" and "Line 2" increase sharply with the increase of model accuracy evaluation, and then tend to be stable, but the accuracy of Line 1 is higher than that of Line 2, and shows some fluctuations in the stable period. The inflection point occurs at the moment when the accuracy of the model is low, which may be due to the rapid learning of the basic patterns in the data at the beginning of the model training when the learning rate is high, resulting in a rapid increase in accuracy. As the model continues to train, the improvement in accuracy slows down, indicating that the model is beginning to reach some saturation point in its performance, at which point further tuning of parameters or more complex model structures may be needed to continue to improve accuracy. Meanwhile, the loss of Line 1 is also lower than that of Line 2, which indicates that compared with the signal close to the receiving device, the signal far away may cause more noise or distortion, making it difficult for the model to learn and predict accurately.

The accuracy results of the recognition experiment for 10 test samples are shown in Figure 14.

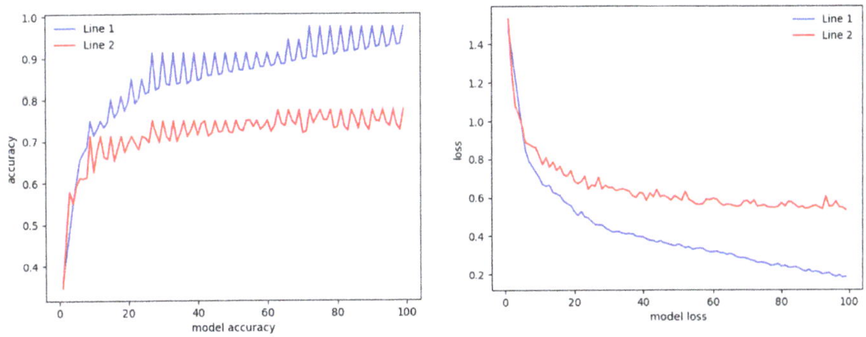

(**a**) (**b**)

Figure 13. Recognition Result of Experiment 3: (**a**) model accuracy; (**b**) model loss.

3.5. Experiment 4: Compare the Recognition Rate of the Same Audio and Different Audio Samples at Point Y

At point Y, the same audio samples used for training and testing in the experiment are the digital voice broadcast samples of "7408450651917459937611512O" sent by five walkie-talkies, and the different audio samples used for testing are the voice broadcast of different numbers.

First, the model is trained by using samples. The experimental results of the training are shown in Figures 15 and 16. Fold line 1 is the accuracy and loss of training the same audio sample model at point Y, and fold line 2 is the accuracy and loss of training different audio sample models at point Y.

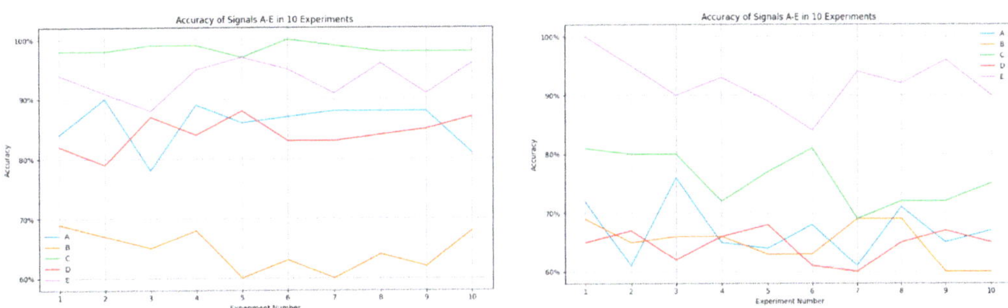

(**a**) (**b**)

Figure 14. Multiple Recognition Results of Experiment 3: (**a**) different audio recognition results at point X; (**b**) different audio recognition results at point Y.

Line 1 rises rapidly in the early iteration, then gradually stabilizes, and finally stabilizes at a higher accuracy level, which shows that the learning effect of the model on the training set is gradually saturated. Line 2 also shows an upward trend, but the accuracy rate increases more slowly, and the final accuracy rate is lower than that of Line 1. Line 2 means that different audio samples are used for training. In the case of complex experimental conditions (such as interference conditions), the accuracy of the model is naturally lower, reflecting the generalization ability of diverse data.

The experiment of receiving radio signals at sea will face many challenges, such as signal fading, noise interference, multipath effect, and so on. This may explain the low accuracy of Line 2: because the model needs to recognize signals in more diverse and complex environmental conditions, it puts forward higher requirements for the generalization capability of machine learning models.

The accuracy results of the recognition experiment of 10 test samples are shown in Figure 17, which shows that radio signals can be recognized even at a long distance.

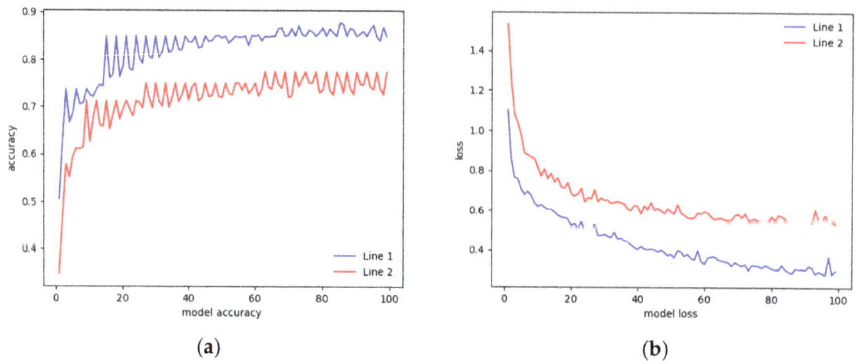

Figure 15. Recognition Result of Experiment 4: (**a**) model accuracy; (**b**) model loss.

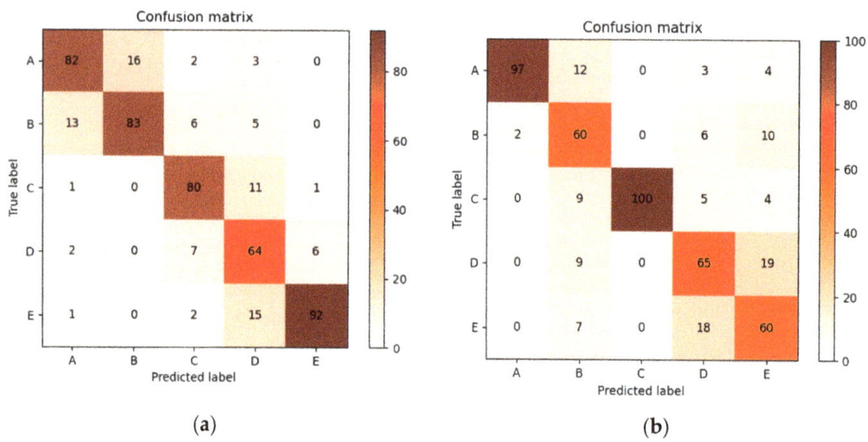

Figure 16. Confusion matrix of Experiment 4: (**a**) same audio recognition result; (**b**) different audio recognition result.

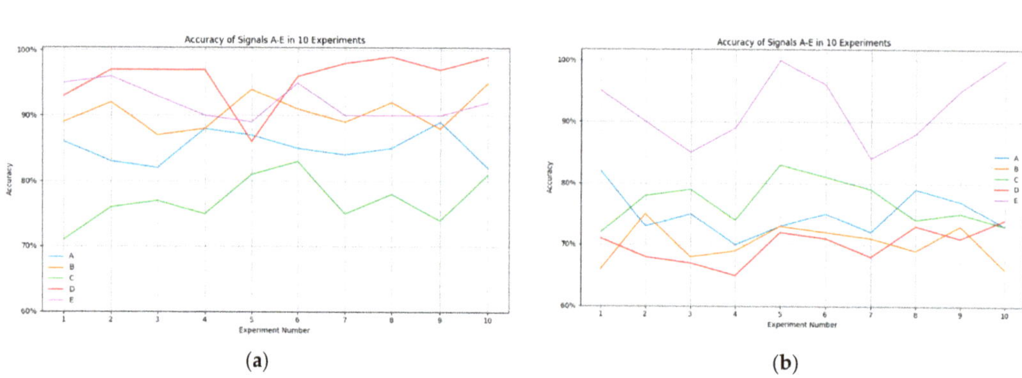

Figure 17. Multiple Recognition Results of Experiment 4: (**a**) the same audio recognition result; (**b**) different audio recognition result.

4. Discussion

Combining all the experimental results, it can be found that the model shows the characteristics of rapid learning in the early stage of training, and quickly learns the basic pattern from the data. With the progress of training, the accuracy of the model eventually tends to be stable and fluctuates at a relatively fixed level, indicating that the model can adapt to different data samples and environmental conditions to a certain extent and achieve certain recognition performance.

However, the accuracy of the model has a certain degree of fluctuation in the training process, and the amplitude and stability of this fluctuation are affected by the experimental conditions. Experiments 1 and 2 involve the same audio samples and different audio samples, and the model trained by the same audio samples has higher accuracy at the beginning of training and reaches a stable state faster, while different audio samples may need more training and adjustment to achieve similar accuracy levels. Experiment 3 and Experiment 4 involve different environmental conditions. It is found that the model in Experiment 4 needs more training and adjustment to improve its performance in the face of complex environments.

To sum up, the model shows some similarities and differences in the face of different data samples and environmental conditions, as shown in the following Table 1.

Table 1. Experimental information summary.

Experiment	Average Accuracy	Average Loss	Recognition Result
Experiment 1	82.60%	0.434	86.48%
Experiment 2	85.29%	0.418	80.34%
Experiment 3	82.14%	0.434	80.13%
Experiment 4	81.61%	0.451	90.64%

Through four groups of the same and different audio recognition experiments in different environments, and by comparing the average accuracy, average loss, and average accuracy of the final recognition results of the model in different experiments, it is found that although there are differences in the performance of the model in different experiments, it can still achieve certain recognition performance through reasonable training and adjustment, which indicates that the model has certain applicability and feasibility in the task of maritime radio signal feature recognition. Effective signal identification and communication can be realized.

5. Conclusions

In this study, a deep learning recognition model for shipborne VHF wireless signal features is established by integrating a convolutional neural network with CAM feature extraction and BiLSTM feature extraction. To validate the effectiveness of the proposed method in non-cooperative signal feature recognition and signal source classification for the same type of signals, an experimental scenario simulating marine ship communication is designed using five VHF radios of identical model and parameters. These experiments provide a novel technical approach towards ship target perception.

Due to the inherent randomness and discreteness of ship communication scenes, obtaining a large-scale dataset for training VHF communication signals is challenging. Therefore, this paper selects hand-held VHF radios to acquire a substantial amount of training data. Additionally, the experimental scene focuses on a limited dataset consisting of five targets with low computational complexity. Future research will delve deeper into perceiving actual ship VHF signals, collecting a wider range and larger volume of ship VHF signal data, integrating AIS data and VHF voice data for annotation purposes, and establishing dynamic training models to validate their feasibility in diverse maritime business scenarios.

Author Contributions: Conceptualization, L.C.; methodology, L.C. and J.L.; software, J.L.; validation, J.L.; formal analysis, L.C. and J.L.; resources, L.C.; data curation, L.C. and J.L.; writing—original draft preparation, J.L.; writing—review and editing, L.C.; visualization, L.C. and J.L.; supervision, L.C.; project administration, L.C.; funding acquisition, L.C. All authors have read and agreed to the published version of the manuscript.

Funding: This work was supported by the National Natural Science Foundation of China (Grant No. 52372316, 51909156).

Institutional Review Board Statement: Not applicable.

Informed Consent Statement: Not applicable.

Data Availability Statement: The data presented in this study are available on request from the corresponding author as the data require the use of the signal acquisition device which was used in the experiment.

Conflicts of Interest: The authors declare no conflict of interest.

References

1. Forti, N.; D'Afflisio, E.; Braca, P.; Millefiori, L.M.; Willett, P. Next-Gen Intelligent Situational Awareness Systems for Maritime Surveillance and Autonomous Navigation Point of View. *Proc. IEEE* **2022**, *110*, 1532–1537. [CrossRef]
2. Ahmad, Z.; Acarer, T.; Kim, W. Optimization of Maritime Communication Workflow Execution with a Task-Oriented Scheduling Framework in Cloud Computing. *J. Mar. Sci. Eng.* **2023**, *11*, 2133. [CrossRef]
3. Bisio, I.; Garibotto, C.; Lavagetto, F.; Sciarrone, A. Performance Analysis of VCA-Based Target Detection System for Maritime Surveillance. *IEEE Trans. Veh. Technol.* **2023**, *72*, 5010–5020. [CrossRef]
4. Djenouri, Y.; Belhadi, A.; Djenouri, D.; Srivastava, G.; Lin, J.C.W. Intelligent Deep Fusion Network for Anomaly Identification in Maritime Transportation Systems. *IEEE Trans. Intell. Transp. Syst.* **2023**, *24*, 2392–2400. [CrossRef]
5. Felski, A.; Zwolak, K. The Ocean-Going Autonomous Ship-Challenges and Threats. *J. Mar. Sci. Eng.* **2020**, *8*, 41. [CrossRef]
6. Dudczyk, J. Radar Emission Sources Identification Based on Hierarchical Agglomerative Clustering for Large Data Sets. *J. Sens.* **2016**, *2016*, 1879327. [CrossRef]
7. Dudczyk, J.; Kawalec, A. Specific emitter identification based on graphical representation of the distribution of radar signal parameters. *Bull. Pol. Acad. Sci.-Tech.* **2015**, *63*, 391–396. [CrossRef]
8. Dudczyk, J.; Rybak, L. Application of Data Particle Geometrical Divide Algorithms in the Process of Radar Signal Recognition. *Sensors* **2023**, *23*, 8183. [CrossRef]
9. Chen, J.; Guo, J.; Shan, X.; Kong, D. Signal Modulation Identification Based on Deep Learning in FBMC/OQAM Systems. *Mob. Inf. Syst.* **2021**, *2021*, 4809699. [CrossRef]
10. Peng, S.; Sun, S.; Yao, Y.-D. A Survey of Modulation Classification Using Deep Learning: Signal Representation and Data Preprocessing. *IEEE Trans. Neural Netw. Learn. Syst.* **2022**, *33*, 7020–7038. [CrossRef]
11. Demirci, S.M.E.; Cicek, K. Intelligent ship inspection analytics: Ship deficiency data mining for port state control. *Ocean Eng.* **2023**, *278*, 18. [CrossRef]
12. Rong, B. Reinforcement Learning for Maritime Communications. *IEEE Wirel. Commun.* **2023**, *30*, 12. [CrossRef]
13. Xiao, W.S.; Luo, Z.Q.; Hu, Q. A Review of Research on Signal Modulation Recognition Based on Deep Learning. *Electronics* **2022**, *11*, 2764. [CrossRef]
14. Salem, M.H.; Li, Y.J.; Liu, Z.Y.; AbdelTawab, A.M. A Transfer Learning and Optimized CNN Based Maritime Vessel Classification System. *Appl. Sci.* **2023**, *13*, 1912. [CrossRef]
15. Rawson, A.; Brito, M.; Sabeur, Z.; Tran-Thanh, L. A machine learning approach for monitoring ship safety in extreme weather events. *Saf. Sci.* **2021**, *141*, 11. [CrossRef]
16. Guo, S.; Akhtar, S.; Mella, A. A Method for Radar Model Identification Using Time-Domain Transient Signals. *IEEE Trans. Aerosp. Electron. Syst.* **2021**, *57*, 3132–3149. [CrossRef]
17. Niranjan, R.K.; Rao, C.B.R.; Singh, A.K. FPGA based Identification of Frequency and Phase Modulated Signals by Time Domain Digital Techniques for ELINT Systems. *Def. Sci. J.* **2021**, *71*, 79–86. [CrossRef]
18. Wang, S.; Wang, Z.; Zhang, T. Modulation signal identification and classification based on the OA algorithm model. *Int. J. Electron.* **2023**, *111*, 1012–1032. [CrossRef]
19. Hao, C.; Dang, X.; Li, S.; Wang, C. Deep Learning Based Low Complexity Symbol Detection and Modulation Classification Detector. *IEICE Trans. Commun.* **2022**, *E105B*, 923–930. [CrossRef]
20. Zhang, J.; Hu, S.; Du, Z.; Wu, W.; Gao, Y.; Cao, J. Deep learning-based digital signal modulation identification under different multipath channels. *IET Commun.* **2021**, *15*, 1950–1962. [CrossRef]
21. Kim, H.; Kim, Y.-J.; Kim, W.-T. Multitask Learning-Based Deep Signal Identification for Advanced Spectrum Sensing. *Sensors* **2023**, *23*, 9806. [CrossRef] [PubMed]

22. Zha, Y.; Wang, H.; Shen, Z.; Shi, Y.; Shu, F. Intelligent identification technology for high-order digital modulation signals under low signal-to-noise ratio conditions. *IET Signal Process.* **2023**, *17*, e12189. [CrossRef]
23. Huynh-The, T.; Pham, Q.V.; Nguyen, T.V.; Nguyen, T.T.; Ruby, R.; Zeng, M.; Kim, D.S. Automatic Modulation Classification: A Deep Architecture Survey. *IEEE Access* **2021**, *9*, 142950–142971. [CrossRef]
24. Liu, F.; Zhang, Z.; Zhou, R. Automatic modulation recognition based on CNN and GRU. *Tsinghua Sci. Technol.* **2022**, *27*, 422–431. [CrossRef]
25. Xiang, H.H.; Chen, B.X.; Yang, T.; Liu, D. Improved De-Multipath Neural Network Models with Self-Paced Feature-to-Feature Learning for DOA Estimation in Multipath Environment. *IEEE Trans. Veh. Technol.* **2020**, *69*, 5068–5078. [CrossRef]
26. Lu, L.J.; Mao, J.N.; Wang, W.Q.; Ding, G.X.; Zhang, Z.W. A Study of Personal Recognition Method Based on EMG Signal. *IEEE Trans. Biomed. Circuits Syst.* **2020**, *14*, 681–691. [CrossRef] [PubMed]
27. Tan, Y.H.; Jing, X.J. Cooperative Spectrum Sensing Based on Convolutional Neural Networks. *Appl. Sci.* **2021**, *11*, 4440. [CrossRef]
28. Zhang, L.; Liu, H.; Yang, X.L.; Jiang, Y.; Wu, Z.Q. Intelligent Denoising-Aided Deep Learning Modulation Recognition with Cyclic Spectrum Features for Higher Accuracy. *IEEE Trans. Aerosp. Electron. Syst.* **2021**, *57*, 3749–3757. [CrossRef]
29. Pereira, M.I.; Claro, R.M.; Leite, P.N.; Pinto, A.M. Advancing Autonomous Surface Vehicles: A 3D Perception System for the Recognition and Assessment of Docking-Based Structures. *IEEE Access* **2021**, *9*, 53030–53045. [CrossRef]
30. Yang, Y.; Shao, Z.P.; Hu, Y.; Mei, Q.; Pan, J.C.; Song, R.X.; Wang, P. Geographical spatial analysis and risk prediction based on machine learning for maritime traffic accidents: A case study of Fujian sea area. *Ocean Eng.* **2022**, *266*, 113106. [CrossRef]
31. Byeon, Y.-H.; Kwak, K.-C. Individual Identification by Late Information Fusion of EmgCNN and EmgLSTM from Electromyogram Signals. *Sensors* **2022**, *22*, 6770. [CrossRef] [PubMed]
32. Liu, R.W.; Yuan, W.Q.; Chen, X.Q.; Lu, Y.X. An enhanced CNN-enabled learning method for promoting ship detection in maritime surveillance system. *Ocean Eng.* **2021**, *235*, 109435. [CrossRef]
33. Chen, X.Q.; Liu, S.H.; Liu, R.W.; Wu, H.F.; Han, B.; Zhao, J.S. Quantifying Arctic oil spilling event risk by integrating an analytic network process and a fuzzy comprehensive evaluation model. *Ocean. Coast Manag.* **2022**, *228*, 106326. [CrossRef]
34. Park, J.; Jeong, J.; Park, Y. Ship Trajectory Prediction Based on Bi-LSTM Using Spectral-Clustered AIS Data. *J. Mar. Sci. Eng.* **2021**, *9*, 1037. [CrossRef]
35. Zhou, X.Y.; Liu, Z.J.; Wang, F.W.; Xie, Y.J.; Zhang, X.X. Using Deep Learning to Forecast Maritime Vessel Flows. *Sensors* **2020**, *20*, 1761. [CrossRef]

Journal of
Marine Science and Engineering

MDPI

Article

Routing a Fleet of Drones from a Base Station for Emission Detection of Moving Ships by Genetic Algorithm

Xiaoqiong Bao, Zhi-Hua Hu * and Yanling Huang

Logistics Research Center, Shanghai Maritime University, Shanghai 201306, China; xqbao@shmtu.edu.cn (X.B.); 202240510016@stu.shmtu.edu.cn (Y.H.)
* Correspondence: zhhu@shmtu.edu.cn

Abstract: A fleet of drones is considered in the routing problems with an offshore drone base station, considering the simultaneous movements of drones and ships. A model, entitled meeting model, between a drone and a moving ship is devised, and an extended model is developed based on the vehicle routing problem model. A genetic algorithm based on a sequential insert heuristic (SIH) is designed to solve the model as a holistic framework with two strategies to determine the sequential assignments of ships to drones, namely, the DroneByDrone, and ShipByShip strategies. The proposed models and solution algorithms are demonstrated and verified by experiments. Numerical studies show that the DroneByDrone strategy can overperform the ShipByShip strategy regarding traveling distances. In addition, when considering the simultaneous movement of the ship and drone, improving the drone flying speeds can reduce the flying time of drones rather than optimizing the ship's moving speed. The managerial implications and possible extensions are discussed based on modeling and experimental studies.

Keywords: ship emission estimation; emission control area; vehicle routing problem; drone routing problem; sequential insert heuristic model

1. Introduction

As an essential component of logistics transportation, maritime transportation plays a crucial role in connecting global markets and promoting international trade. Global maritime cargo trade is expected to nearly triple between 2020 and 2050 [1]. As a result, the growth in shipping demand is causing shipping pollution concerns. According to the findings of the International Maritime Organization, the shipping industry's carbon dioxide emissions constitute approximately 2.2% of worldwide total emissions. Ship fuel combustion also produces nitrogen oxides, sulfur oxides, and other pollutants. The oxides and pollutants negatively impact the environment and human health [2]. The pollutants emitted from ships at high altitudes can spread to an area with air currents, causing more severe impacts on the ground and atmosphere, especially in Port City. Therefore, the International Maritime Organization (IMO) institutes Emission Control Areas (ECAs) globally to regulate maritime emissions.

ECA aims to significantly reduce ship pollutant emissions and promote a green and sustainable shipping industry. Within ECAs, maritime regulations mandate the utilization of fuels with a sulfur content that does not exceed a specified threshold. Nonetheless, the absence of a cost–benefit for low sulfur oil compared to heavy oil leads numerous shipping enterprises and vessel owners to favor heavy oil for marine fuel. Consequently, the monitoring of ship emissions in ECAs becomes critically essential. The traditional detection method is to take samples on board, send them to a third party for inspection, or use a fast detector for detection. However, the operation has low detection efficiency and is complex. Hence, innovative emission detection and regulation methods are among the most essential ways to control shipping pollution. Among them, onboard drone-based sniffing technology can realize fast and remote online monitoring and can be used as a feasible and reliable means of pollution monitoring. However, drones' initial acquisition costs and ongoing maintenance are relatively high. As a result, reasonable scheduling and optimizing the routes for drones should be developed to ensure cost-effectiveness.

This study contributes to the related studies in the following three aspects. First, we consider the simultaneous movements of drones and ships according to their respective moving directions and speeds and propose a meeting model to optimize the catch-up time during drone ship emissions. Second, since a fleet of drones' detection scenario for ships is much like the classical VRP, we formulated the routing problem based on VRP models. However, the customer points visited by the vehicles in the VRP are stationary while the ships are moving. The general VRP models cannot be fully adapted to this problem. We extend the VRP models by proposing an extended model considering "moving customers". Finally, we develop a sequential insert heuristic (SIH) and SIH-based bi-stage solution algorithm to compute the drones' and ships' meeting times and positions and then create an SIH-based genetic algorithm to calculate the visiting order of ships by the drone for emissions detection.

The rest of the sections are organized as follows. Section 2 reviews the related studies on ECA, ship emission detection, and drone routing problems. Section 3 introduces the problem and investigates two routing modes, the TSP-based and VRP-based modes. Section 4 formulates the mathematical models. Section 5 presents a genetic algorithm based on a sequential insert heuristic algorithm. Then, we conduct a series of numerical experiments to examine the proposed models and algorithms and discuss the managerial implications in Section 6. We conclude this study in Section 7.

2. Related Studies

This study featured the drone routing problem inspired by drone applications in emission detection in emission control areas (ECAs).

2.1. Emission Control Area

An ECA regulates the release of harmful gases from ships, aiming to diminish the emissions of sulfur dioxide, nitrogen oxides, and particulate matter regarding the effects of ship emissions on human health and the environment [3]. Data from the IMO indicate that the implementation of ECAs in Europe has resulted in a 60% decrease in sulfur dioxide emissions from ships [4]. Implementing ECA has driven the development of new technologies, such as exhaust gas scrubbers and liquefied natural gas (LNG), which can effectively control ship emissions.

Table 1 summarizes the research on ECA in terms of the research questions, methods, and corresponding regions or ports. ECA is a universal emission control solution; so, most governments and ports have taken it as a policy-making and administration tool. The research in Table 1 investigated various features, including ECA formation, impact assessment, scheduling optimization based on ECA, and stakeholder behaviors. Four methods are used in the research: data-driven [5], mathematical analysis [6], optimization [7], and evaluation methods [8]. In Table 1, emission control is significant, and various technologies were applied. Drones should emerge as new devices for ECA management, challenging drone scheduling, and routing problems.

Table 1. Pioneering studies on ECA.

Study	Research Problem	Method	Region/Port
[9]	Design China's domestic ECA	BLP	China
[6]	The accomplishment of the ECA program development	A	China
[10]	Assess noxious gases of ship operations in a potential ECA	D	Incheon Port
[11]	The spatial relationship between the net benefit and the size of ECA	Simulation	China

Table 1. *Cont.*

Study	Research Problem	Method	Region/Port
[7]	Locate ECA to minimize the impacts of Sulfur emissions on human health	MILP	China, Africa
[12]	Dual environmental effects of ECA and reduced speed zones	MILP	Pacific Ocean
[5]	Investigate the heterogeneity of the ECA policy effectiveness	D	Yangtze River Delta
[8]	Examine whether ECA regulations impact the efficiency of ports	DEA	European Union, North America
[13]	Impact of Sulfur ECA on port state control's inspection outcome	A	-
[4]	Influence of sulfur emission control areas on particulate matter emission	A	China
[14]	Optimize ship route and speed with ECA simultaneously	A	-
[15]	Global impacts of ECA	A	Asia, Europe
[16]	Decision on liner shipping considering ECA regulations	MINP	North America
[17]	Analyze the ship's response strategy to ECA	MINP	North America

Note: A = Mathematical analytical model; BLP = Bi-level program; D = Statistics or other data-driven analysis; DEA = Data envelopment analysis; H = Heuristics; MILP = Mixed-integer linear program; MINP = Mixed-integer nonlinear program.

2.2. Ship Emission Detection

Ship emission detection technologies play a critical role in reducing the impact of ship emissions on the environment and public health [3]. Ship emission detection technologies are still developing, and future research should focus on improving the technology's accuracy, efficiency, and cost-effectiveness [18]. Drones equipped with sniffing technology have been tested and applied to detect atmospheric pollution on ships. Using drones for ship emission detection can achieve remote, online, and real-time detection. It is more efficient, flexible, and convenient than conventional detection methods.

Table 2 summarizes some groundbreaking research on ship emission detection in three areas: the research problem, research method, and detection method. The research problem includes ship emission detection methods, detection accuracy optimization, and detection effectiveness evaluation. Research methods related to ship detection issues include neural networks [18], simulation analysis [19], machine learning [20], optimization [21], and utilizing data such as Automatic Identification System (AIS) data [22], satellite images, and videos. Detection methods involve remote sensing, onboard measurement [23], and imaging pattern recognition [24]. This study applied a fleet of drones to ship emission detection and investigated the scheduling and routing problem of drones equipped with specialized emission detection devices.

Table 2. Pioneering studies on Ship emission detection.

Study	Research Problem	Detection	Solution
[18]	Individual ship pollution or ship plume evaluation	Onboard	NN
[20]	A big data analytics framework for ship performance monitoring	Onboard	ML
[25]	Apply an energy consumption approach to estimate container ship emissions	Onboard	EA
[22]	A method evaluating emissions in detailed	Onboard	EA
[23]	Precise identification of carbon emission hotspot areas and ship types in ports	Onboard	D
[19]	Monitor the SO_2 concentration in ship exhausts	Imaging	S
[26]	Estimate SO_2, CO_2, and NO concentrations of exhaust gas by imaging methods	Imaging	NN
[27]	Investigating the impacts of satellite fire observation accuracy on the top-down nitrogen oxide emission estimation in northeastern Asia.	RS	D
[21]	Develop a marine diesel SO_X online detection system	RS	H
[28]	Remotely detect sulfur contents in fuel oil used by ships in ECA	RS	S
[24]	Detect NO2 pollution plumes from ships with the TROPOMI/S5P satellite sensors.	RS	D

Note: RS = Remote sensing; D = Statistics or other data-driven methods; DFT = Data field theory; DL = Deep learning; EA = Empirical analysis; H = Heuristics; ML = Machine learning; NN = Neural network; S = Simulation.

2.3. Drone Routing Problem

Several complex stages are involved during ship emission detection, including law enforcement personnel boarding, document verification, fuel sampling, and actual detection [29]. Drones offer numerous benefits, including their high mobility, precise information gathering, and prompt data transmission, making them widely applicable in various fields [30]. Drones can detect and monitor ship emissions in the emission control area, constrained by energy consumption and cost [31]. Most importantly, drones used for ship emission detection must handle the simultaneous movements of ships, which is a distinct feature in the literature and challenges the existing routing models and solution algorithms.

Most of the literature formulates the drone routing problem as a mixed-integer linear programming (MILP) to determine the task assignment and the visited orders for each drone. Table 3 lists the research on drone routing problems under various scenarios. The VRPs are NP-hard; so, general MILP solvers can only solve small-scale tasks. Hence, to solve medium- and large-scale tasks, researchers usually develop heuristics [32] (e.g., greedy, neighborhood search), a metaphor approach [33,34] (e.g., genetic algorithm, Ant colony optimization), and a decomposition-based algorithm [35] (e.g., Branch-and-Cut, Branch-and-Price).

In terms of application, the delivery and distribution (mainly including emergency [36], last-mile delivery [33], and monitoring [37]) can inspire the model and algorithm developments. We can observe distinct features of the drone scheduling problems: endurance and collaborative delivery with trucks [38]. This study formulated the drone routing problem based on VRP modes and their concepts. In ship emission detection, the tasks performed by a fleet of drones and a drone may visit more than one ship.

Applying drones to ship emission detection has the characteristics of a basic VRP. Specifically, a drone flies from a drone base station to the ship, whose position can be determined by the AIS receivers connected with the station [22]. Namely, the ship to be detected corresponds to the customer in a typical VRP. As a difference, the ship is not a stationary object but moves with time. These distinct features challenge the existing models and solution algorithms studied in Table 3. To model the proposed problem, we review the related VRP with drones to provide a fruitful methodological reference.

Table 3. Pioneering studies on drone routing problems.

Study	Application Scenario	TSP/VRP Variance	Model	Algorithm
[39]	Last-mile package delivery	VRP with drones	MILP	VNS + TS
[32]	Truck–drone collaborative delivery	Parallel TSP	MILP	H
[31]	Truck–drone collaborative delivery	Multi-trip VRP with time windows	MILP	B&C
[40]	Truck–drone collaborative delivery	Multi-visit VRP	Analytical	H
[41]	Mothership-drone coordination	VRP with drones	MISOCP	MH
[35]	Last-mile package delivery	TSP with drones	MILP	B&P
[42]	Last-mile package delivery	VRP with drones	MILP	B&C
[36]	Emergency delivery	VRP with drones	Analytical	-
[43]	Delivery	VRP with drones	MILP	ACO
[44]	Island delivery	VRP with drones	RO	-
[45]	Truck–multi-drone team delivery	TSP with multiple flying sidekicks	Reasoning	ML
[33]	Last-mile package delivery	VRP with backhauls	MILP	SA + VNS
[38]	Truck–drone collaborative delivery	VRP with time-dependent road travel time	MILP	H
[46]	Surveillance	Nested VRP	MIQCP	NS

Note: ACO = Ant colony optimization; B&B = Branch-and-Bound; B&C = Branch-and-Cut; B&P = Branch-and-Price; H = Heuristic; MH = Math-heuristic; MILP = Mixed-integer linear program; MIQCP = Mixed-integer quadratically constrained programming; MISOCP = Mixed-integer second-order cone programming; ML = Machine learning; NS = Neighborhood search; RO = Robust optimization; SA = Simulated annealing; TS = Tabu search; VNS = Variable neighborhood search.

3. Problem Statement

In this section, we consider a routing problem with a single drone base station and a fleet of drones for ship emission detection (Section 2.1) in an ECA (Section 2.2). As studied in Section 2, the drone routing problem for ship emission detection is featured by the dynamic movements of drones and ships in an ECA. Therefore, it is a complicated extension of the vehicle routing problem, and it challenges the modeling and algorithm development due to the new feature.

The ships' sailing direction and speed are known, and the drones' speed is significantly higher than the ships' sailing speed. The drones fly toward the ships' direction of travel and meet them on the routes. The objective is to minimize the total flying time of the drones for efficient, speedy, and cost-effective ship emission detection. We analyzed a VRP-based mode for the routing problem of a fleet of drones.

The VRP-based mode employs a fleet of drones, and each drone visits a sub-set of ships, as depicted in Figure 1. Based on factors such as the number of ships, their locations, the movement directions, and speeds, it decides the routes for each drone to detect the ships. In the VRP-based mode in Figure 1, two drones start from the base station, and ship 4 is detected before ship 5 and the detection task is completed shortly after their journey begins.

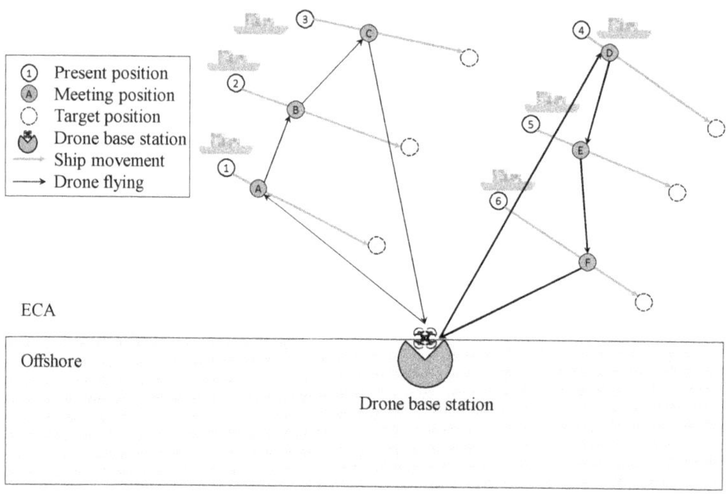

Figure 1. A conceptual diagram of the VRP-based mode.

Many factors are involved in ship pollution emission detection by drones: drone cost, endurance distance, drones, fly distance, detection efficiency, ships, efficiency, fly times, and emergency capability. The number of drones in the fleet will impact these factors. Changing the number of drones in the base station significantly affects the waiting time and detection sequence of ships in ECA. For example, more drones in the fleet may increase detection efficiency while incurring higher drone investment costs.

4. Modeling

Using drones for ship emission detection is an innovative and effective attempt to improve traditional emission detection. To better leverage the utility of drones, we propose a drone routing problem for ship emission detection. Considering the synchronized motion of drones and ships when ships arrive in the ECA, we propose a meeting mode between drones and ships and integrate the motion constraints of drones and ships into the classic VRP, constructing an extended model for this problem.

4.1. Meeting Model between a Drone and a Moving Ship

As studied above, when the drone navigates to the ship's predetermined "present" position, the ship will have advanced to a new location along its trajectory, as defined by its current and intended positions, as demonstrated in Figure 2. Therefore, it is advantageous to consider the movements of both the drone and ship simultaneously. We calculate the rendezvous point, considering their respective trajectories, and then direct the drone to this estimated position.

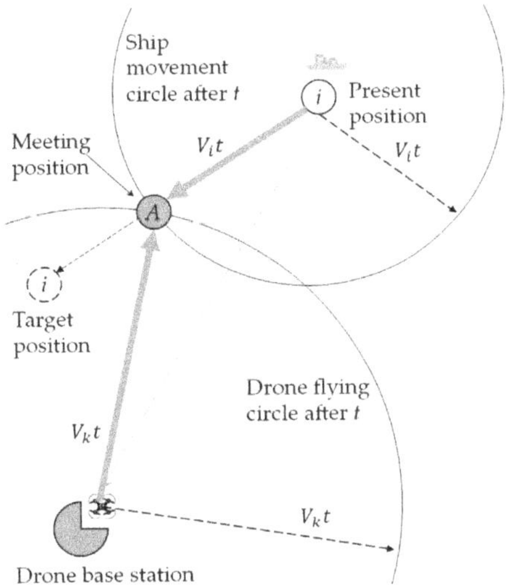

Figure 2. A meeting model for a drone and directional mobile ship.

For a ship i situated at coordinates (X_i, Y_i), its movement is directed toward a target location (X_i, Y_i) at a speed V_i. Meanwhile, a drone base station k is positioned at $(\overline{X}_k, \overline{Y}_k)$. If station k deploys a drone at speed $\overline{V}_k$ to approach ship i, the drone must travel for a duration t to intersect with the moving ship at its target trajectory.

In Figure 2, the point of intersection between the drone and the ship is depicted. The ship moves from point $a(X_i, Y_i)$ to point $b(X_i^t, Y_i^t)$. Two circles can be drawn: one centered at $(\overline{X}_k, \overline{Y}_k)$ for the drone with a radius of $\overline{V}_k t$, and another for the ship centered at (X_i, Y_i) with a radius of $V_i t$. The meeting point is where these two circles intersect on the line ab. This intersection is calculated using model (1), where V_i^X, V_i^Y are specified in Equations (2) and (3). The [Mmeet] model can be solved using Equations (4)–(10). A feasible meeting time is then ascertained by Equation (11), disregarding any negative values. The meeting coordinates (x_i, y_i) are determined by the meeting time t, as calculated in (1).

$$\begin{cases} x_i = X_i + t \cdot V_i^X \\ y_i = Y_i + t \cdot V_i^Y \\ \sqrt{(x_i - \overline{X}_k)^2 + (y_i - \overline{Y}_k)^2} = t \cdot \overline{V}_k \end{cases} \tag{1}$$

$$V_i^X = V_i \frac{X_i^t - X_i}{\sqrt{(X_i - X_i^t)^2 + (Y_i - Y_i^t)^2}} \tag{2}$$

$$V_i^Y = V_i \frac{Y_i^t - Y_i}{\sqrt{(X_i - X_i^t)^2 + (Y_i - Y_i^t)^2}} \tag{3}$$

$$(x_i - \overline{X}_k)^2 + (y_i - \overline{Y}_k)^2 = t^2 \cdot \overline{V}_k^2 \tag{4}$$

$$\left(X_i + t \cdot V_i^X - \overline{X}_k\right)^2 + \left(Y_i + t \cdot V_i^Y - \overline{Y}_k\right)^2 = t^2 \cdot \overline{V}_k^2 \tag{5}$$

$$\left(t \cdot V_i^X + X_i - \overline{X}_k\right)^2 + \left(t \cdot V_i^Y + Y_i - \overline{Y}_k\right)^2 = t^2 \cdot \overline{V}_k^2 \tag{6}$$

$$\left(\left(V_i^X\right)^2 + \left(V_i^Y\right)^2 - \overline{V}_k^2\right)t^2 + 2\left(V_i^X\left(X_i - \overline{X}_k\right) + V_i^Y\left(Y_i - \overline{Y}_k\right)\right)t + \left(\left(X_i - \overline{X}_k\right)^2 + \left(Y_i - \overline{Y}_k\right)^2\right) = 0 \tag{7}$$

$$at^2 + bt + c = 0 \tag{8}$$

$$\begin{cases} a = \left(V_i^X\right)^2 + \left(V_i^Y\right)^2 - \overline{V}_k^2 \\ b = 2\left(V_i^X\left(X_i - \overline{X}_k\right) + V_i^Y\left(Y_i - \overline{Y}_k\right)\right) \\ c = \left(X_i - \overline{X}_k\right)^2 + \left(Y_i - \overline{Y}_k\right)^2 \end{cases} \tag{9}$$

$$t = \frac{-b \pm \sqrt{b^2 - 4ac}}{2a} \tag{10}$$

The meeting position (x, y) and flying time t^{meet} (the valid value in (10)) can be computed for a drone at the base station k and a ship i by the above meeting model, denoted as follows:

$$t^{meet}, x_i, y_i \leftarrow f^{[\text{meet}]}\left(\overline{X}_k, \overline{Y}_k, X_i, Y_i, X_i^t, Y_i^t, \overline{V}_k, V_i\right).$$

4.2. Baseline VRP Model

The VRP represents a crucial and fundamental problem in combinatorial optimization, gathering extensive attention from numerous researchers [47,48]. In a VRP, the objective is to optimize the routes for a fleet of vehicles to minimize the total travel costs, considering a set of designated customer locations. Using a drone fleet for ship emission detection parallels a VRP scenario, where each drone embarks from a base station and sequentially visits a series of ships, ensuring a single visit per ship. Therefore, we will refer to the VRP models and establish the baseline model.

The baseline model encompasses three distinct sets: a set of ships denoted as $I = \{1, 2, \cdots\}$, indexed by i; a set of drones denoted as $K = \{1, 2, \cdots\}$, indexed by k; and a combined set of ships and drones denoted as $I^+ = \{1, 2, \cdots\}$, also indexed by i. The variable C_{ij} represents the time taken for a drone to fly from ship i to ship j. D denotes the total number of drones available at the station. The binary decision variable $x_{ijk} \in \{0, 1\}$ is set to 1 if drone k is assigned to visit ship j immediately after ship i, and 0 otherwise. Similarly, the binary decision variable $y_{ik} \in \{0, 1\}$, is 1 if drone k is scheduled to visit ship i, and 0 otherwise. Additionally, the binary decision variable u_{ik} signifies the sequential order in which ship i is monitored by drone k. Utilizing these data and variables, the baseline model is designed to optimize the allocation of ships to drones.

$$\min \sum\nolimits_{i,j,k} C_{ij} x_{ijk} \tag{11}$$

Subject to

$$\sum\nolimits_{k \in K} y_{ik} = 1, \ \forall i \in N \tag{12}$$

$$\sum\nolimits_{j \in I^+} x_{ijk} = y_{ik}, \ \forall i \in N, k \in K \tag{13}$$

$$\sum\nolimits_{j \in I^+} x_{ijk} = \sum\nolimits_{j \in I^+} x_{jik}, \ \forall i \in I^+, k \in K \tag{14}$$

$$\sum_{i \in I} y_{ik} \leq D, \forall k \in K \tag{15}$$

$$u_{ik} + 1 \leq u_{jk} + M\left(1 - x_{ijk}\right), \ \forall i \in I^+, j \in I, k \in K \tag{16}$$

$$x_{ijk}, y_{ik} \in \{0, 1\}, \ \forall i, j \in I^+, k \in K \tag{17}$$

$$u_{ik} > 0, i \in I^+, k \in K \tag{18}$$

In the baseline model, the objective function (11) represents the minimum flight distance of drones; Constraint (12) ensures that each ship must be assigned to an AGV for detection. In Constraint (13), the assignment variable (y_{ki}) is presented by the sequencing variable (x_{kij}). Namely, ship $i$is detected by drone k if ship i is assigned to drone k. Constraint (14) ensures that the in-degree and out-degree of visiting the ship are either one or zero. In Constraint (15), the number of ships detected by each drone is no more than the specific value. Constraint (16) sequences the drone operations, namely, the orders of ships the drones visited. The constraint avoids loops within a route by the order variable of a drone (k) visiting two ships (i, j). Here, $M = |I|$. When $x_{ijk} = 1$, the constraint is $u_{ik} + 1 \leq u_{jk}$, namely, the orders of visiting the ships i, j. Constraint (17) and Constraint (18) constrain the ranges of the variables.

4.3. Extended Model

In the baseline model, C_{ij} represents the traveling time (distance or other costs and metrics) from a position i to another position j. In other words, C_{ij} is constant because the positions of i, j are determined and not changed in the model. However, in the ship emission detection scenario, i, j are two moving ships, and the value of C_{ij} depends on the time of the drone visiting the ships. As studied in Section 3, in the ship emission detection scenario, C_{ij} represents the drone's flying time from a ship i to another ship j. A drone visits j just after visiting i. A route from the drone base station k can be denoted by $R_k =< k, i_1, i_2, \cdots, i_n, k >$, where k represents the drone and a virtual depot, and $i_1, i_2, \cdots, i_n$ represents the visiting sequence of ships. Apparently, $C_{i_a j_{a+1}}$ depends on the sequence $< k, i_1, \cdots, i_a, i_{a+1}]$. Formally, we introduce a function $f^C(R_k, i, j)$ to present the dependence,

$$C_{ij}, (x_j, y_j) \leftarrow f^C(R_k, i, j).$$

Here, $R_k = \left\{ x_{ijk} = 1 \middle| \forall i, j \right\}$. Thus,

$$C_{ij}, (x_j, y_j) \leftarrow f^C\left(\left\{ x_{ijk} = 1 \middle| \forall i, j \right\}, i, j \right).$$

So, we can update the baseline model and obtain an extended model.

$$\min \sum_{i,j,k} C_{ij} x_{ijk}$$

Subject to

$$((1)\text{–}(18))$$

$$C_{ij} = f^C\left(\left\{ x_{ijk} = 1 \middle| \forall i, j \right\}, i, j \right). \tag{19}$$

In the extended model, $f^C(\cdot)$ is a complicated operator that computes the traveling time dependent on the visiting sequence of ships. So, the objective function $C_{ij} x_{ijk}$ is nonlinear. Although we formulate the one-station drone VRP based on a linearized one-depot VRP, the extended model is a nonlinear mathematical program. It cannot be solved by one-the-shelf MILP solvers, e.g., Cplex (www.ibm.com/products/ilog-cplex-optimization-studio/cplex-optimizer, accessed on 18 April 2024) and Gurobi (www.gurobi.com, accessed on 18 April 2024).

5. Solution Methods

We develop an SIH-based genetic algorithm to solve this problem. A sequential insert heuristics algorithm is devised to compute the meeting positions and times for a sequence of ships to be visited. The genetic algorithm is used for the iterative solution.

5.1. Sequential Insert Heuristics (SIH)

Based on the meeting model, a sequential insert heuristics algorithm is devised to compute the meeting positions and times for a sequence of ships to be visited. First, initialize the meeting times t and the meeting positions (x, y). The following information is available: the present and target positions and speed of the ship; the position of the drone base station and the drone's speed; and a ship sequence to be visited by the drone. Then, by setting the iteration variable i, Set $i = i_1$ to compute the first meeting time t^{meet} and position(x_i, y_i), update $t = t + t^{meet}$, $x = x + |x_i|, y = y + |y_i|$. If $i = k$ return t, x, y.

The detailed process is shown in Algorithm 1. The algorithm takes a sequence of ships as inputs to generate the times and positions meeting with the drone k. In the given six steps, $f^{[\text{meet}]}(\cdot)$ is invoked to compute the meeting time and position for the ship in the given ship sequence. Then, calculate the meeting time and position with the next ship in this sequence.

Algorithm 1 Sequential insert heuristics (SIH)

Input	$X_i, Y_i, X_i^t, Y_i^t, V_i, \forall i$: the present and target positions and speed of the ship i $\overline{X}_k, \overline{Y}_k, \overline{V}_k$: the position of the drone base station and the drone's speed $R_k = < k, i_1, i_2, \cdots, i_n, k >$: a ship sequence to be visited by the drone from k
Output	t, x, y: The meeting times and positions
Steps	
Step 1	Initialize outputs Set $t = [], x = [], y = []$.
Step 2	Set the iteration variable i Set $i = i_1$;
Step 3	Compute the first meeting time and position. $t^{meet}, x_i, y_i \leftarrow f^{[\text{meet}]}(\overline{X}_k, \overline{Y}_k, X_i, Y_i, X_i^t, Y_i^t, \overline{V}_k, V_i)$. Update $t = t + [t^{meet}], x = x + [x_i], y = y + [y_i]$.
Step 4	Termination condition If $i = k$ Then, Return t, x, y
Step 5	Compute the next meeting time and position Set $j = next(R_k, i)$ $t^{meet}, x_j, y_j \leftarrow f^{[\text{meet}]}\left(x_i, y_i, X_j, Y_j, X_j^t, Y_j^t, \overline{V}_k, V_j\right)$. Update $t = t + [t^{meet}], x = x + \left[x_j\right], y = y + \left[y_j\right]$.
Step 6	Update the iteration variable i $i \leftarrow j$. Go to Step 4

Algorithm 1 can be denoted by $f^{\text{SIH}}(\cdot)$, as follows.

$$t, x, y \leftarrow f^{\text{SIH}}\left(\left[X_i, Y_i, X_i^t, Y_i^t, V_i \middle| \forall i\right], \overline{X}_k, \overline{Y}_k, \overline{V}_k, R_k\right)$$

By $f^{\text{SIH}}(\cdot)$, the meeting times t and positions (x, y) between a drone and a sequence of ships can be computed. This is a prerequisite for computing the drone's and target ship's total detection time in the following sections. The time complexity of Algorithm 1 is $O(n)$, where n represents the scale of the number of ships.

5.2. SIH-Based Bi-Stage Solution Algorithm

About the VRP, we have formulated the baseline model to determine the sequence in which a drone visits ships, assuming that the ships are stationary and awaiting detection. This model is then expanded, considering the simultaneous movements of both drones and ships. Subsequently, a two-stage algorithm is developed for this purpose. The first stage involves deriving an initial solution by solving the baseline model, which does not consider the concurrent movements of drones and ships. The second stage employs Sequential Insert Heuristics—Algorithm 1 (SIH)—to translate the ship sequences from the initial solution into actual drone routes, this time factoring in the simultaneous movements of the drones and ships.

The SIH-based bi-stage solution algorithm (Algorithm 2) first solves the VRP to obtain the visit sequence R of drones to ships (Step 1). Then, for each drone visit sequence in turn, the algorithm applies the SIH algorithm—$f^{\text{SIH}}(\cdot)$—to determine the meeting times and positions of each drone visiting the ships (Step 2).

Algorithm 2 Bi-stage algorithm based on One Station VRP and sequential insert decoding (MP + SIH)

Input	$X_i, Y_i, X_i^t, Y_i^t, V_i, \forall i$: the present and target positions and speed of the ship i
	$\overline{X}_k, \overline{Y}_k, \overline{V}_k$: the position of the drone base station and the drone's speed
	$R = \{R_k =< k, i_1, i_2, \cdots, i_n, k > \vert \forall k\}$: a ship sequence to be visited by the drone k, obtained by solving [VRP].
Output	t, x, y: The meeting times and positions
Steps	
Step 1	Solve [VRP].
	$t, R \leftarrow f^{[\text{VRP}]}\left(I, K, C_{ij}, Q_k\right)$.
Step 2	For each R_k in R
Step 2.1	$t^k, x^k, y^k \leftarrow f^{\text{SIH}}\left(\left[X_i, Y_i, X_i^t, Y_i^t, V_i \middle\vert \forall i\right], \overline{X}_k, \overline{Y}_k, \overline{V}_k, R_k\right)$.
Step 3	Return $\sum_k last\left(t^k\right), \left[t^k, x^k, y^k \middle\vert \forall k\right]$

In summary, the first stage (Step 1) solves the baseline model—a mathematical program—to obtain an initial solution, thus denoted by "MP"; in the second stage (Step 2), Algorithm 1 (Sequential Insert Heuristics) is used to convert this initial solution into a solution meeting the constraints in the extended model. So, Algorithm 2 is denoted by $f^{\text{MP+SIH}}$, as follows,

$$t, x, y \leftarrow f^{\text{MP+SIH}}\left(\left[X_i, Y_i, X_i^t, Y_i^t, V_i \middle\vert \forall i\right], \left\{\overline{X}_k, \overline{Y}_k, \overline{V}_k, R_k \middle\vert \forall k\right\}\right).$$

In Algorithm 2, the second stage can improve the first stage's solution, while its optimality is constrained by the first stage, which does not consider the simultaneous movements of drones and ships.

5.3. SIH-Based Genetic Algorithm

As a population-based optimization algorithm, the genetic algorithm (GA) is widely used in routing and scheduling problems [49]. During the solving process, the GA first encodes a chromosome string, imitating the biological evolution process, and performs selection, crossover, and mutation to generate chromosomes that are different from and superior to their parents. The new chromosomes are then subjected to the same operations of selection, crossover, and mutation, and this iteration continues until the optimal solution is found in the population. This study encoded ships as a sequence in the chromosome genes, representing their numeric order. The decoding process utilizes a sequential insertion algorithm (Algorithm 1) to translate these sequences into drone detection routes, thereby establishing the meeting positions and times.

The characteristics of the problem are integrated into the GA encoding and decoding techniques. As outlined in Algorithm 2, the solution is encoded in the form of a sequence of ships. The decoder then utilizes this sequence to calculate the specific time and position for the drone's detection of each ship. The GA devised in this study can be denoted by f^{GA} with three sets of algorithmic hyperparameters ($Hparams$) and two problem-related parameters ($Ships, Drones$). Executing f^{GA} can produce the meeting time and position for each ship, the assigned drone, and the drones' visiting sequences of ships.

$$\begin{gathered}\left[R_k, t^k, x^k, y^k \middle\vert \forall k\right] \leftarrow f^{\text{GA}}(Ships, Drones, Hparams) \\ Ships = \left[X_i, Y_i, X_i^t, Y_i^t, V_i \middle\vert \forall i\right] \\ Drones = \left\{\overline{X}_k, \overline{Y}_k, \overline{V}_k \middle\vert \forall k\right\} \\ Hparams = \left[P^m, P^x, P^s, P^g, \cdots\right]\end{gathered}$$

5.3.1. The Chromosome Presentation

In this study, the drones used are homogenous, and investment-related decisions predetermine the number. The chromosome consists of a sequence of ship identifiers. For example, a chromosome encoded as integers for 12 ships is as follows: $9 \rightarrow 8 \rightarrow 12 \rightarrow 10 \rightarrow 1 \rightarrow 4 \rightarrow 6 \rightarrow 5 \rightarrow 2 \rightarrow 3 \rightarrow 11$. The sequence is a permutation of the ships' identifiers, $1, 2, \cdots, 12$. The sequence indicates the priorities of ships to be considered for the assignments of ships to drones and the ships' emission detection orders. The decoding algorithm (Section 5.3.3) will determine the assignments of ships to drones and each drone's visiting orders to the ships.

5.3.2. The Decoding Algorithm

The decoding algorithm transforms genetic information generated by GAs into solutions for the given problem. This study uses a real-number encoding scheme. Given a chromosome (a sequence of ships), the ships to be inspected are divided into subsequences for drones using either the drone-by-drone sequence division strategy (DroneByDrone) or the ship-by-ship sequence division strategy (ShipByShip). Then, Algorithm 2 (SIH) can decode each subsequence into a feasible solution with determined meeting times and positions for drones and ships.

In summary, the decoding scheme consists of two stages. First, the ships are assigned to drones based on the priorities determined by the chromosome. This stage is entitled "sequence to subsequences" in Section 5.3.3. Second, the ships assigned to a drone constitute a subsequence of ships, which can be used to compute the meeting times and positions by Algorithm 2. In Algorithm 3, the two steps are formulated. The first stage (Step 2 of Algorithm 3) elucidated above is denoted by $divide(x, K)$. In the second stage (Step 3 of Algorithm 3), the meeting times and positions are computed by $f^{SIH}(\cdot)$ for each route generated by the first stage.

Algorithm 3 Decoding algorithm

Input	x: A chromosome of a sequence of ships $divide(x, K)$: an operator to divide x into $\lvert K \rvert$ sub-sequences for the stations $X_i, Y_i, X_i^t, Y_i^t, V_i, \forall i$: the present and target positions and speed of ship i $\overline{X}_k, \overline{Y}_k, \overline{V}_k$: the position of the drone base station and the drone's speed
Output	t: the total flying time $\left[t^k, x^k, y^k \middle| \forall k\right]$: The meeting times and positions of the drones
Steps	
Step 1.	Initialize the total flying time. $t = 0$.
Step 2	Divide the ship sequence into sub-sequences (see Algorithms 4 and 5) $R = \{R_k \mid k \in K\} \leftarrow divide(x, K)$.
Step 3	For each R_k in R: $t^k, x^k, y^k \leftarrow f^{SIH}\left(\left[X_i, Y_i, X_i^t, Y_i^t, V_i \middle| \forall i\right], \overline{X}_k, \overline{Y}_k, \overline{V}_k, R_k\right)$. $t = t + t^k$;
Step 4	Return $t, R, \left[t^k, x^k, y^k \middle| \forall k\right]$

5.3.3. Sequence-to-Subsequence Division Strategies

Sequence-to-subsequence division strategies divide a sequence into subsets under specific principles and rules. Section 5.3.3 introduces two methods, namely, DroneByDrone and ShipByShip.

The DroneByDrone strategy involves computing the average number of ships that each drone needs to inspect, given a known sequence of drones performing ship pollution inspections and the number of drones available. This average value, denoted by n, is used to group about n ships as a batch and assign them to the same drone. Thus, the algorithmic hyperparameters are given as follows: $Hparams = [P^m, P^x, P^s, P^g, decode = \text{DroneByDrone}]$. Algorithm 4 gives an implementation of the DroneByDrone strategy.

Algorithm 4 Drone-by-drone sequence division strategy (DroneByDrone)	
Input	x: A chromosome of a sequence of ships K: the set of drones
Output	$[R_k =< k, i_1, i_2, \cdots, i_n, k >, \forall k]$: the sequences for the drones
Steps	
Step 1.	Compute the maximum number of ships visited by a drone. $n = \lfloor |x|/|K| \rfloor$.
Step 2.	Set the sub-sequence for each drone base station. $R_k =< k, x_{n(k-1)+1}, \cdots, x_{nk}, k >, \forall k \in K \backslash \{|K|\}$. $R_k =< k, x_{n(k-2)+1}, \cdots, x_{|x|}, k >, k = |K|$.
Step 3.	Return $\{R_k | \forall k\}$

The ShipByShip strategy is an alternative approach to the sequence-to-subsequence division strategies. Under the ShipByShip strategy, ships are sequentially assigned to their corresponding drones based on a regular arithmetic progression of the detected ship order. Algorithm 5 implements the ShipByShip strategy with $Hparams = [P^m, P^x, P^s, P^g, decode = \text{ShipByShip}]$.

Algorithm 5 Ship-by-ship sequence division strategy (ShipByShip)	
Input	x: A chromosome of a sequence of ships K: the drone base stations
Output	$[R_k =< k, i_1, i_2, \cdots, i_n, k >, \forall k]$: the sequences for the drones
Steps	
Step 1.	Compute the maximum number of ships visited by a drone. $n = \lceil |x|/|K| \rceil$.
Step 2.	Set the sub-sequence for each drone base station. $R_k =< x_k, \cdots, x_{nk} | \forall i \in \{1, 2, \cdots, n\}, \exists x_{nk} >, \forall k$. $R_k =< k, R_k, k >, \forall k$.
Step 3.	Return $\{R_k | \forall k\}$

6. Numerical Experiments

This study analyzes the optimization problem of routing a fleet of drones for ship pollution emission detection in ECAs. To validate the effectiveness of the devised models (Section 4) and the GA and its invoking sub-algorithms (Section 5), experimental studies were conducted using Python programming language on a computer with 11th Gen Intel(R) Core (TM) i5-11300H configuration (Intel Corporation, Santa Clara, CA, USA).

6.1. Dataset Generation

In this study, the drone's speed is fixed at $V_0 = 25$m per second. Ships within the ECA navigate at speeds uniformly distributed between 10 and 20 knots, roughly equivalent to $V_i \sim Uniform[10, 20]$ meters per second. Here, $Uniform[\text{a}, \text{b}]$ represents a uniform distribution with the lower and upper limits a, b, respectively. The ECA is considered to be located at the Yangtze River estuary, spanning an area of 20 km by 10 km. To abstract from specific real-world locations, this virtual area represents a generic ECA, emphasizing methodological generalizability. The drone's endurance range varies depending on battery technology, with different types having different capabilities. It is assumed that the drone's endurance is sufficient for this study, which may influence the setup of ships and the virtual area.

Datasets are generated with the following specifications: The drone base station is positioned at coordinates (0, 0). For each ship, the present and target positions are generated within the 20 × 10 (km × km) range. The position farther from the drone base station is designated as the present position, while the closest is the target. The pair of present and target positions determine the ship's direction of movement. The dataset is named in the format "KkNnVvXxYY", where K indicates the number of drones at the base station, N the number of ships, V is the drone's flying speed, X is the horizontal range, and Y the vertical range. For instance, "K2N5V25X20Y10" denotes an ECA spanning 20 × 10 km^2

with two drones at the base station tasked to detect emissions from five ships, and the drones traveling at a speed of 25 m/s.

6.2. Algorithmic Parameter Tuning

The hyperparameters of GA include the number of population iterations (G), the population size (ρ), the crossover rate (α), and the mutation probability (β). The parameter-tuning experiments used the dataset "K3N30X20Y10V25". The numerical values of the algorithm parameters are determined through parameter-tuning experiments, where each parameter is adjusted and, others are fixed, and the values that produce the best results are selected.

Figure 3 depicts the optimal values of the GA for various combinations of crossover and mutation probabilities. These two parameters jointly influence the computational efficiency and performance of the GA. The results reveal that a crossover rate of 0.8 and a mutation probability of 0.7 produce the best optimization outcomes. The gray area in Figure 4 shows the parameter combination with the highest objective values.

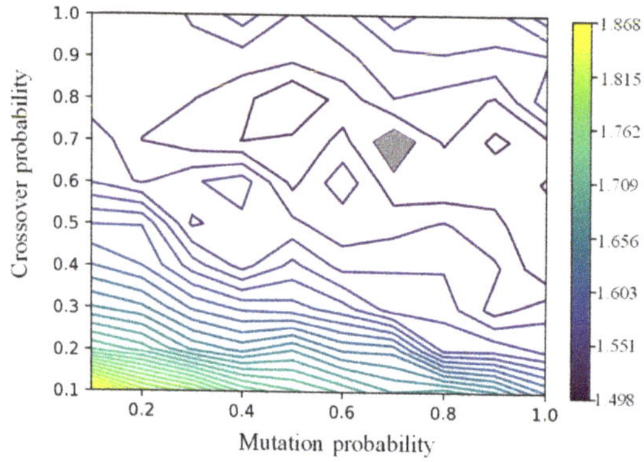

Figure 3. Objectives of various mutation and crossover probabilities.

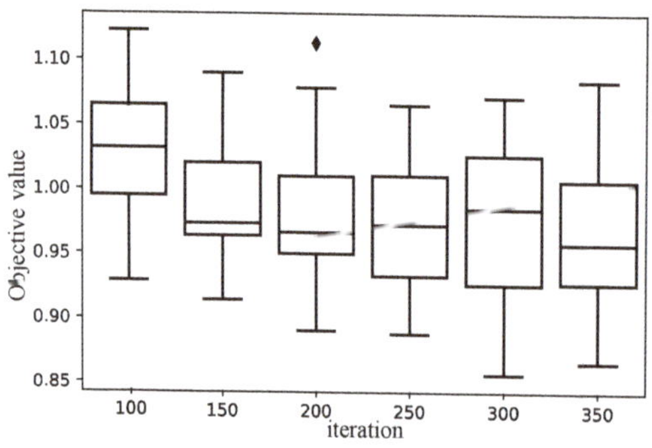

Figure 4. Distributions of objectives for different iterations.

As shown in Figure 4, the mean and distribution of the detection time for multiple drones and ships were investigated for the iterations of the GA. As indicated by the results, we can choose the settings of 250 iterations.

Table 4 shows the impact of population size on the optimal value of the GA. The analysis indicates that the smallest objective function value of the optimal solution is obtained when the population size is 45. As the population size increases, the objective function value remains almost constant while the computation time increases.

Table 4. Average objective values and computing times for various population sizes.

Population Size (G)	Best Objective	Computation Time
35	1.354	4.332
40	1.335	4.953
45	1.319	5.515
50	1.330	6.142
55	1.329	6.731
60	1.317	7.329

These results demonstrate the vital role of parameter tuning and iteration adjustment in achieving optimal performance for the GA. Based on the results, the following settings are obtained: $(G, \rho, \alpha, \beta) = (250, 45, 0.7, 0.8)$.

6.3. Demonstration of the Models and Algorithms

The dataset "K2N10V10X20Y10" demonstrates the effectiveness of the proposed models and the GAs by using four solutions, as presented in Figure 5. For the demonstration purpose, the present positions of the ships are all set to $(0, y_i)$ and y_i is distributed from 5 to 10 uniformly. Moreover, the target positions of the ships share the same values of y_i. Notably, the devised model and algorithms do not require the ship movement tracks to be paralleled horizontally. So, the ships should move horizontally.

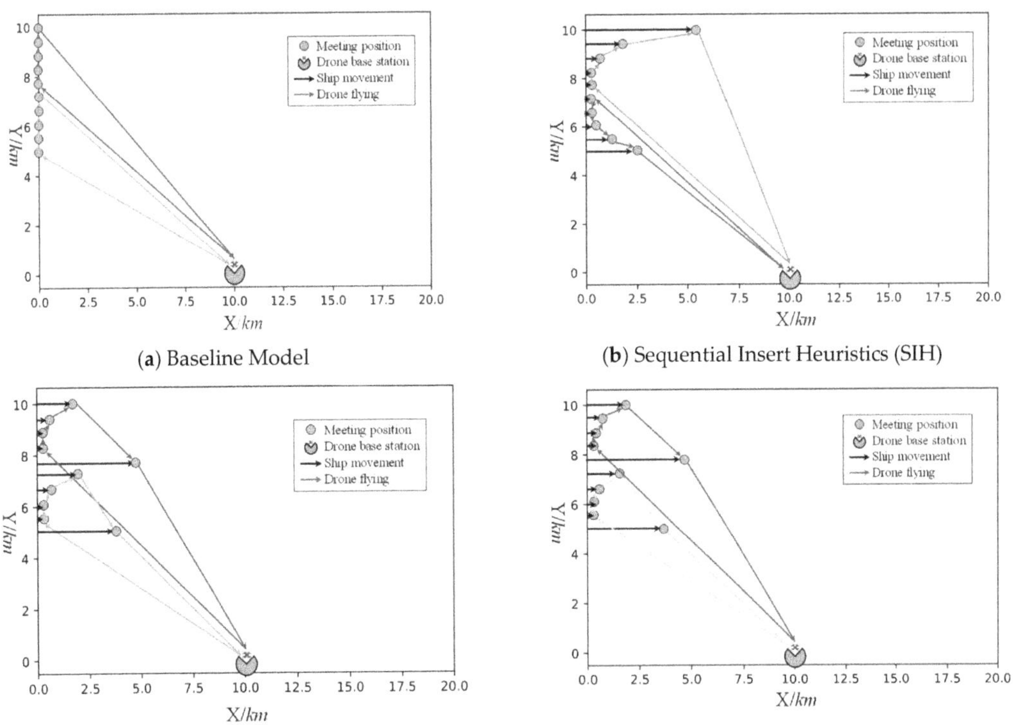

(a) Baseline Model

(b) Sequential Insert Heuristics (SIH)

(c) GA + ShipByShip

(d) GA + DroneByDrone

Figure 5. Demonstrations of four solution methods.

In Figure 5a, the ships are presumed to be stationary, not moving, and just waiting for emission detection. Therefore, the meeting positions are still located at $(0, y_i)$. In Figure 6b, the bi-stage SIH is used. In the first stage, the ships are presumed to be stationary, and the baseline model is used to obtain the assignments of ships to the two drones and the sequence of ships for each drone. So, in Figure 5b, the assignments and sequences of ships to each drone are the same as in Figure 5a. However, the second stage algorithm [SIH] changes the meeting positions and times. In Figure 5b, the movements of ships can be observed. In Figure 5c,d, the ShipByShip and DroneByDrone-based GAs are executed to generate demonstrative results. In these two figures, we can observe no differences. However, when the algorithms are repeated 50 times, the average objectives will incur minor differences, as presented in Table 5. The two decoding strategies will also incur discrepancies in convergence speeds and quality, as shown in Figure 6. The DroneByDrone strategy can achieve better solutions while lowering convergency performance more than the ShipByShipstrategy.

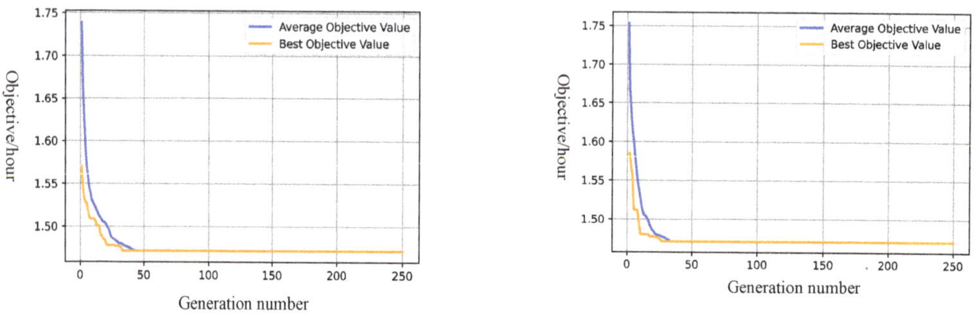

(**a**) DroneByDrone (**b**) ShipByShip

Figure 6. Trace plots of the GAs with DroneByDrone and ShipByShip.

Table 5. The objective values for four different solution approaches.

Solution Method	Objective	Using Meeting Model?
Baseline Model	0.7607	No
Sequential Insert Heuristics (SIH)	0.7515	Yes
GA		
DroneByDrone	0.7429	Yes
ShipByShip	0.7445	Yes

6.4. Comparisons of the Division Strategies

The impacts of dataset sizes on the division strategies are investigated in Figure 7. The solution methods, [GA + DroneByDrone] and [GA + ShipByShip], are executed for the datasets with different numbers of drones (K) and ships ($|I|$), $K = 2, 3$, and $|I| = 10, 15, \cdots, 50$. With the same number of ships to be detected, the total time for detecting ship pollutants is significantly shorter when using three drones compared to that when using two drones. However, under the same drone quantity condition, adopting the DroneByDrone strategy results in shorter drone flight times than when using the ShipByShip strategy. As described in Section 5.3.3, the DroneByDrone strategy prioritizes scheduling drone visits to ships based on the number of ships a drone can access, resulting in a sequence of nearby visits. In contrast, the ShipByShip strategy assigns each ship to a corresponding drone. Compared with the DroneByDrone strategy, the ShipByShip strategy may require the detection of ships that are further apart from each other, and as ships and drones move synchronously, the flight time during the drone's visit to a ship increases when the ship being visited is farther away.

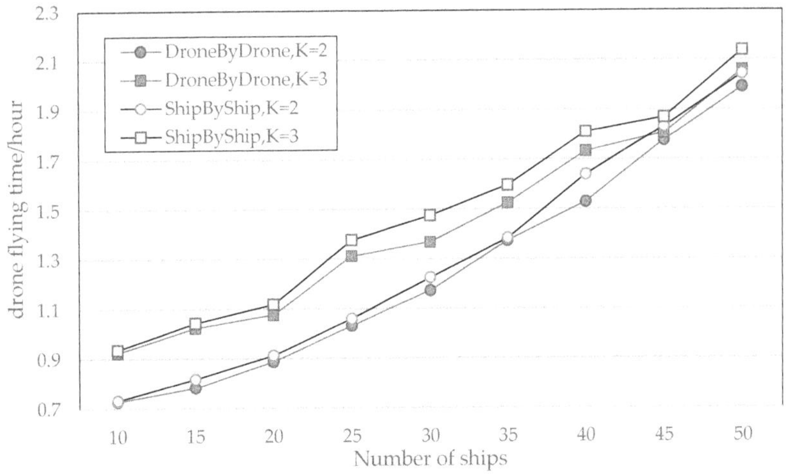

Figure 7. Comparisons of the DroneByDrone strategy and ShipByShip strategy.

6.5. Sensitivities of Ships and Drones' Speeds

The speeds of both drones V_0 and ships V_i are contingent upon technological advancements and investments. Increasing the speeds of ships and drones results in higher energy expenditures and augmented emissions. Additionally, technological limitations can restrict the maximum achievable speeds of drones. Accelerating drone speeds might necessitate innovation and further investment. In the subsequent analysis, we explore how variations in the speeds of drones and ships affect the total operational time.

Table 6 presents the sensitivity of the ships' speeds (V_i) by varying them −5% and 5% and the sensitivity of the drones' speeds (V_i) by varying them −5% and 5%. From the information in Table 6, we can see that the increase or decrease in the shipping speed does not show an apparent correlation regularity on the effect of the objective function, while the rise in the drone flight speed can effectively decrease the value of the objective function; the decline in the drone flight speed will make the value of the objective function grow larger. As presented in Figure 7, improving the flight speed of drones is probably more effective for optimizing the detection time of multi-drone ship pollution emissions.

Table 6. Sensitivity of the ships' speeds (V_i) and drones' speeds (V_k).

			V_i			V_k	
Stations	**Ships**	**(100 − 5)%**	**100%/h**	**(100 + 5)%**	**(100 − 5)%**	**100%/h**	**(100 + 5)%**
2	10	−0.04	0.76	−0.20	4.70	0.76	−2.86
2	15	−0.07	0.83	−3.48	3.82	0.83	−7.11
2	20	−2.80	0.93	−2.14	2.75	0.93	−3.66
2	25	1.97	1.04	1.54	5.58	1.04	2.44
2	30	−1.24	1.22	−8.40	4.32	1.22	−15.50
2	35	4.69	1.28	6.75	9.48	1.28	−10.90
2	40	−1.14	1.51	−2.24	2.98	1.51	−11.74
2	45	5.01	1.63	−2.54	6.58	1.63	3.31
2	50	−3.27	2.00	−2.54	−2.42	2.00	−6.83
3	10	−0.13	0.95	0.24	5.24	0.95	−5.04
3	15	2.87	1.08	4.27	6.63	1.08	−5.29
3	20	0.04	1.12	1.25	11.79	1.12	0.69
3	25	2.32	1.32	5.72	10.56	1.32	−3.78
3	30	−2.78	1.49	1.73	1.86	1.49	−7.58
3	35	−1.31	1.68	−2.77	−2.18	1.68	−14.29
3	40	2.48	1.75	4.64	7.15	1.75	−10.20
3	45	−2.31	2.02	−9.46	−0.62	2.02	−7.87
3	50	2.79	2.25	−6.38	−0.60	2.25	−8.05

As presented in Figure 8, it is evident that the drones' speed significantly impacts the flight time required to complete ship-pollution-monitoring tasks. The effect of drone speed on flight time was analyzed for two and three drones, with task numbers N ranging from 10 to 50. The results show that increasing drone speed leads to a noticeable decrease in flight time, whereas decreasing drone speed leads to a significant increase in flight time. Figure 9 reveals that under the same conditions of drone quantity, $V_i(1-5\%)$, V_i, and $V_i(1+5\%)$ almost overlap. This indicates that changes in ship speed do not significantly impact the flight time required for drones to complete ship-pollution-monitoring tasks.

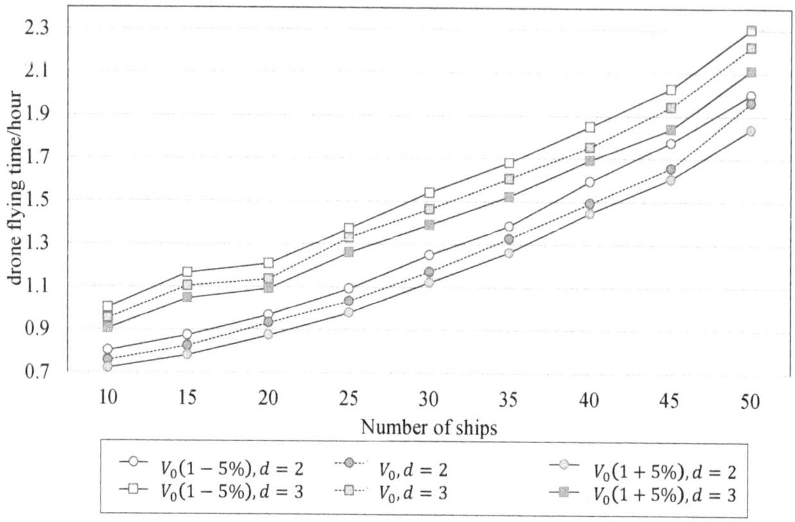

Figure 8. Sensitivity analysis of drones' speed.

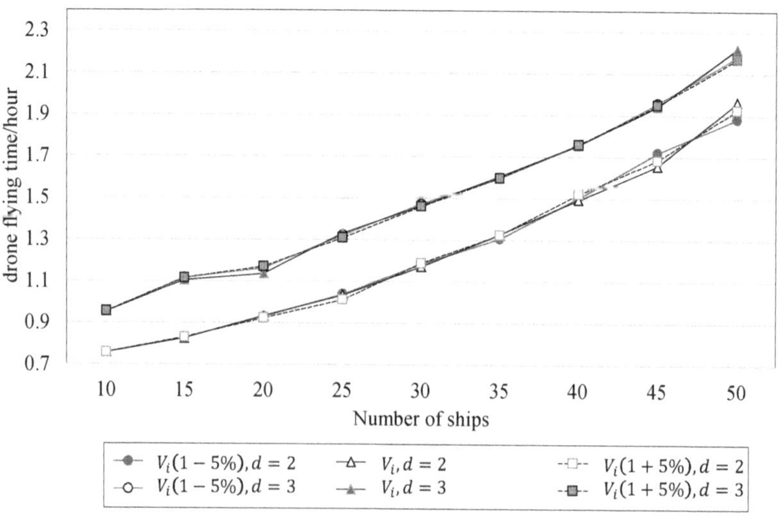

Figure 9. A diagram of ships' speed sensitivity analysis.

6.6. Discussions and Managerial Implications

Using drones for ship pollution monitoring in ECA is an innovative way to monitor ship emissions at sea, and it offers higher monitoring efficiency than traditional monitoring methods. The establishment of ECA has set clear targets and regulations for ship emission reduction in specific areas, but practical implementation of these policies requires efficient regulatory methods and execution capabilities. To achieve effective regulation, it is necessary to optimize the deployment of drones for ship monitoring tasks. In this study, we formulate the problem of efficient and reasonable ship emission detection as a multi-drone routing optimization problem.

We draw the following conclusions as managerial implications based on the experimental results.

(1) The optimization of drone flight routing for ship emission detection is a dynamic problem involving the simultaneous movement of drones and ships. The proposed meeting model can appropriately characterize the dynamic problem. The model is an extension and theoretical increment to the static VRP problem. In addition, the meeting model complements the study of operations research theory on dynamic problems. In practice, compared to single drones, multi-drone and ship meeting models based on the VRP model can overcome the difficulties of low detection efficiency and long flight distances. Experiments have verified their advantages in applications.

(2) Regarding detection efficiency, the DroneByDrone strategy has an advantage over the ShipByShip strategy in terms of performance. The reason is that with drones traveling at much higher speeds than ships, the DroneByDrone strategy prioritizes the drone detection sequence, thereby reducing the detection time. In addition, in the meeting problem, fast-moving objects should be used as the entry point for algorithm design. Therefore, the DroneByDrone strategy should be prioritized to handle the drone routing problem. However, the meeting model is nonlinear, and it is not easy to find an exact solution. Thus, linearization methods need further research to save costs.

(3) Our research focuses on the scheduling problem of drones at the operational level, with the understanding that coordination mechanisms and information systems are vital determinants of operational performance. First, an effective drone scheduling system must dynamically track the positions of ships using data acquired from AIS receivers. Second, drone base stations must exchange scheduling information and integrate their schedules into a centralized, top-level scheduling system. This integration facilitates the global deployment of drones. Third, local ECA authorities should collaborate in sharing the schedules for ship emission detection and the resulting data, ensuring comprehensive monitoring and management.

7. Conclusions

The adverse effects of ship emissions on the ecological environment and human health in coastal areas have prompted a need for innovative ship emission detection methods. In this regard, we investigated the optimization problem of drone routing in ship emission detection. Firstly, effective detection methods are necessary for the ECA to achieve good ship emission monitoring and control effects, thereby reducing the impact of ship emissions on people's health in the coastal areas. Traditional detection methods suffer from low efficiency and cumbersome procedures. Using drones for ship emission detection is an innovative and effective attempt to improve traditional emission detection. To better leverage the utility of drones, we propose a multi-drone routing optimization problem for ship emission detection. Secondly, considering the synchronized motion of drones and ships when ships arrive in the ECA, we propose a meeting mode between drones and ships and integrate the motion constraints of drones and ships into the classic VRP, constructing an extended model for this problem. Thirdly, we develop an SIH-based genetic algorithm to solve this problem. The sequence insertion algorithm determines the meeting time and location of drones and ships, and the genetic algorithm is used for the iterative solution. Finally, taking the Yangtze River Delta ECA as an example, we develop a dataset generation method and conduct experiments to demonstrate and study the proposed model, algorithm, and sensitivity analysis.

Our study proposes an innovative and practical approach to ship emission detection using drones and develops a multi-drone routing optimization model and an SIH-based genetic algorithm for solving it. The experimental results demonstrate the effectiveness and feasibility of the proposed method, which can be applied to other ECA or environmental monitoring tasks. The baseline Model (Section 4.2) can obtain routes while not considering dynamic simultaneous movements of ships and drones; it, coupled with SIH, can constitute the bi-stage solution algorithm. Further, a GA can use SIH as a decoding algorithm with two strategies to allocate ships to drone routes, DroneByDrone, and ShipByShip. The bi-stage and the GAs (with DroneByDrone or ShipByShip) can solve the drone routing problem, while the bi-stage is not competitive compared with the latter two methods.

In future research, this study can be expanded in several ways.

(1) The base station for drone flights is currently fixed and singular. Future research could explore the possibility of having multiple base stations in an ECA and optimizing the selection of base station locations in conjunction with drone detection routing to minimize drone detection routing and achieve better management and economic value. Additionally, the genetic algorithm based on sequence insertion used in this study is effective but cannot guarantee optimality. Future research could develop more efficient algorithms to solve this problem.

(2) Implementing machine learning and artificial intelligence techniques could be explored to improve the accuracy and efficiency of the drone detection system. This could involve training models to identify patterns in drone flight routing and predict potential breaches of the ECA. Additionally, the use of advanced sensors and imaging technologies could be investigated to enhance the system's detection capabilities.

(3) The potential for integrating the drone detection system with other security systems could be explored. For example, the system could be integrated with perimeter intrusion detection or video surveillance systems to provide a comprehensive security solution. This would require careful consideration of data integration and sharing protocols to ensure seamless integration and optimal performance.

Author Contributions: Conceptualization, Z.-H.H.; methodology, Z.-H.H.; software, X.B.; validation, X.B. and Y.H.; formal analysis, X.B.; investigation, X.B.; resources, Y.H.; data curation, Y.H.; writing—original draft preparation, Z.-H.H. and X.B.; writing—review and editing, Z.-H.H.; visualization, Y.H.; supervision, Z.-H.H.; project administration, Z.-H.H.; funding acquisition, Z.-H.H. All authors have read and agreed to the published version of the manuscript.

Funding: This research was funded by the Shanghai Municipal Commission of Science and Technology, grant number 23ZR1426500.

Institutional Review Board Statement: Not applicable.

Informed Consent Statement: Informed consent was obtained from all subjects involved in the study. Written informed consent has been obtained from the patient(s) to publish this paper.

Data Availability Statement: The model data and its generator can be obtained by contacting the author.

Conflicts of Interest: The authors declare no conflicts of interest.

References

1. Liu, J.; Law, A.W.-K.; Duru, O. Abatement of atmospheric pollutant emissions with autonomous shipping in maritime transportation using Bayesian probabilistic forecasting. *Atmos. Environ.* **2021**, *261*, 118593. [CrossRef]
2. Tokuslu, A. Estimating greenhouse gas emissions from ships on four ports of Georgia from 2010 to 2018. *Environ. Monit. Assess.* **2021**, *193*, 385. [CrossRef] [PubMed]
3. Wan, Z.; Zhang, Q.; Xu, Z.; Chen, J.; Wang, Q. Impact of emission control areas on atmospheric pollutant emissions from major ocean-going ships entering the Shanghai Port, China. *Mar. Pollut. Bull.* **2019**, *142*, 525–532. [CrossRef] [PubMed]
4. Wan, Z.; Zhou, X.; Zhang, Q.; Chen, J. Influence of sulfur emission control areas on particulate matter emission: A difference-in-differences analysis. *Mar. Policy* **2021**, *130*, 104584. [CrossRef]
5. Zhang, Q.; Liu, H.; Wan, Z. Evaluation on the effectiveness of ship emission control area policy: Heterogeneity detection with the regression discontinuity method. *Environ. Impact Assess. Rev.* **2022**, *94*, 106747. [CrossRef]

6. Jiang, B.; Wang, X.; Xue, H.; Li, J.; Gong, Y. An evolutionary game model analysis on emission control areas in China. *Mar. Policy* **2020**, *118*, 104010. [CrossRef]
7. Sun, Y.; Yang, L.; Zheng, J. Emission control areas: More or fewer? *Transp. Res. Part D Transp. Environ.* **2020**, *84*, 102349. [CrossRef]
8. Chang, Y.-T.; Park, H.; Lee, S.; Kim, E. Have Emission Control Areas (ECAs) harmed port efficiency in Europe? *Transp. Res. Part D Transp. Environ.* **2018**, *58*, 39–53. [CrossRef]
9. Tian, X.; Yan, R.; Qi, J.; Zhuge, D.; Wang, H. A Bi-Level Programming Model for China's Marine Domestic Emission Control Area Design. *Sustainability* **2022**, *14*, 3562. [CrossRef]
10. Chang, Y.-T.; Roh, Y.; Park, H. Assessing noxious gases of vessel operations in a potential Emission Control Area. *Transp. Res. Part D Transp. Environ.* **2014**, *28*, 91–97. [CrossRef]
11. Okada, A. Benefit, cost, and size of an emission control area: A simulation approach for spatial relationships. *Marit. Policy Manag.* **2019**, *46*, 565–584. [CrossRef]
12. Dong, G.; Tae-Woo Lee, P. Environmental effects of emission control areas and reduced speed zones on container ship operation. *J. Clean. Prod.* **2020**, *274*, 122582. [CrossRef]
13. Fan, L.; Yu, Y.; Yin, J. Impact of Sulphur Emission Control Areas on port state control's inspection outcome. *Marit. Policy Manag.* **2022**, *50*, 1–16. [CrossRef]
14. Ma, D.; Ma, W.; Jin, S.; Ma, X. Method for simultaneously optimizing ship route and speed with emission control areas. *Ocean Eng.* **2020**, *202*, 107170. [CrossRef]
15. Chen, L.; Yip, T.L.; Mou, J. Provision of Emission Control Area and the impact on shipping route choice and ship emissions. *Transp. Res. Part D Transp. Environ.* **2018**, *58*, 280–291. [CrossRef]
16. Ma, W.; Hao, S.; Ma, D.; Wang, D.; Jin, S.; Qu, F. Scheduling decision model of liner shipping considering emission control areas regulations. *Appl. Ocean Res.* **2021**, *106*, 102416. [CrossRef]
17. Li, L.; Gao, S.; Yang, W.; Xiong, X. Ship's response strategy to emission control areas: From the perspective of sailing pattern optimization and evasion strategy selection. *Transp. Res. Part E Logist. Transp. Rev.* **2020**, *133*, 101835. [CrossRef]
18. Šilas, G.; Rapalis, P.; Lebedevas, S. Particulate Matter (PM1, 2.5, 10) Concentration Prediction in Ship Exhaust Gas Plume through an Artificial Neural Network. *J. Mar. Sci. Eng.* **2023**, *11*, 150. [CrossRef]
19. Zhang, Z.; Zheng, W.; Cao, K.; Li, Y.; Xie, M. Simulation Analysis on the Optimal Imaging Detection Wavelength of SO2 Concentration in Ship Exhaust. *Atmosphere* **2020**, *11*, 1119. [CrossRef]
20. Bui, K.Q.; Perera, L.P. Advanced data analytics for ship performance monitoring under localized operational conditions. *Ocean Eng.* **2021**, *235*, 109392. [CrossRef]
21. Guo, J.; Guo, L. Research on SOX online detection system of marine diesel engine based on LabVIEW. *J. Mar. Sci. Technol.* **2022**, *27*, 1179–1191. [CrossRef]
22. Rapalis, P.; Šilas, G.; Žaglinskis, J. Ship Air Pollution Estimation by AIS Data: Case Port of Klaipeda. *J. Mar. Sci. Eng.* **2022**, *10*, 1950. [CrossRef]
23. Zhou, C.; Huang, H.; Liu, Z.; Ding, Y.; Xiao, J.; Shu, Y. Identification and analysis of ship carbon emission hotspots based on data field theory: A case study in Wuhan Port. *Ocean Coast. Manag.* **2023**, *235*, 106479. [CrossRef]
24. Georgoulias, A.K.; Boersma, K.F.; van Vliet, J.; Zhang, X.; Zanis, P.; de Laat, J. Detection of NO2 pollution plumes from individual ships with the TROPOMI/S5P satellite sensor. *Environ. Res. Lett.* **2020**, *15*, 124037. [CrossRef]
25. Czermański, E.; Cirella, G.T.; Oniszczuk-Jastrząbek, A.; Pawłowska, B.; Notteboom, T. An Energy Consumption Approach to Estimate Air Emission Reductions in Container Shipping. *Energies* **2021**, *14*, 278. [CrossRef]
26. Zhang, Z.; Zheng, W.; Cao, K.; Li, Y.; Xie, M. An improved method for optimizing detection bands of marine exhaust SO2 concentration in ultraviolet dual-band measurements based on signal-to-noise ratio. *Atmos. Pollut. Res.* **2022**, *13*, 101479. [CrossRef]
27. Beirle, S. Estimate of nitrogen oxide emissions from shipping by satellite remote sensing. *Geophys. Res. Lett.* **2004**, *31*, L18102. [CrossRef]
28. Peng, X.; Huang, L.; Wu, L.; Zhou, C.; Wen, Y.; Chen, H.; Xiao, C. Remote detection sulfur content in fuel oil used by ships in emission control areas: A case study of the Yantian model in Shenzhen. *Ocean Eng.* **2021**, *237*, 109652. [CrossRef]
29. Kuzniar, M.; Pawlak, M.; Orkisz, M. Comparison of Pollutants Emission for Hybrid Aircraft with Traditional and Multi-Propeller Distributed Propulsion. *Sustainability* **2022**, *14*, 15076. [CrossRef]
30. Anand, A.; Wei, P.; Gali, N.K.; Sun, L.; Yang, F.H.; Westerdahl, D.; Zhang, Q.; Deng, Z.Q.; Wang, Y.; Liu, D.G.; et al. Protocol development for real-time ship fuel sulfur content determination using drone based plume sniffing microsensor system. *Sci. Total Environ.* **2020**, *744*, 140885. [CrossRef]
31. Paradiso, R.; Roberti, R.; Laganá, D.; Dullaert, W. An Exact Solution Framework for Multitrip Vehicle-Routing Problems with Time Windows. *Oper. Res.* **2020**, *68*, 180–198. [CrossRef]
32. Dell'Amico, M.; Montemanni, R.; Novellani, S. Matheuristic algorithms for the parallel drone scheduling traveling salesman problem. *Ann. Oper. Res.* **2020**, *289*, 211–226. [CrossRef]
33. Salama, M.R.; Srinivas, S. Collaborative truck multi-drone routing and scheduling problem: Package delivery with flexible launch and recovery sites. *Transp. Res. Part E-Logist. Transp. Rev.* **2022**, *164*, 102788. [CrossRef]
34. Xu, J.; Liu, C.; Shao, J.; Xue, Y.; Li, Y. Collaborative orchard pesticide spraying routing problem with multi-vehicles supported multi-UAVs. *J. Clean. Prod.* **2024**, *458*, 142429. [CrossRef]

35. Roberti, R.; Ruthmair, M. Exact Methods for the Traveling Salesman Problem with Drone. *Transp. Sci.* **2021**, *55*, 315–335. [CrossRef]
36. Gentili, M.; Mirchandani, P.B.; Agnetis, A.; Ghelichi, Z. Locating platforms and scheduling a fleet of drones for emergency delivery of perishable items. *Comput. Ind. Eng.* **2022**, *168*, 108057. [CrossRef]
37. Mukhamediev, R.I.; Yakunin, K.; Aubakirov, M.; Assanov, I.; Kuchin, Y.; Symagulov, A.; Levashenko, V.; Zaitseva, E.; Sokolov, D.; Amirgaliyev, Y. Coverage Path Planning Optimization of Heterogeneous UAVs Group for Precision Agriculture. *IEEE Access* **2023**, *11*, 5789–5803. [CrossRef]
38. Wang, Y.; Wang, Z.; Hu, X.; Xue, G.; Guan, X. Truck-drone hybrid routing problem with time-dependent road travel time. *Transp. Res. Part C-Emerg. Technol.* **2022**, *144*, 103901. [CrossRef]
39. Schermer, D.; Moeini, M.; Wendt, O. A hybrid VNS/Tabu search algorithm for solving the vehicle routing problem with drones and en route operations. *Comput. Oper. Res.* **2019**, *109*, 134–158. [CrossRef]
40. Poikonen, S.; Golden, B. Multi-visit drone routing problem. *Comput. Oper. Res.* **2020**, *113*, 104802. [CrossRef]
41. Amorosi, L.; Puerto, J.; Valverde, C. Coordinating drones with mothership vehicles: The mothership and drone routing problem with graphs. *Comput. Oper. Res.* **2021**, *136*, 105445. [CrossRef]
42. Tamke, F.; Buscher, U. A branch-and-cut algorithm for the vehicle routing problem with drones. *Transp. Res. Part B-Methodol.* **2021**, *144*, 174–203. [CrossRef]
43. Huang, S.-H.; Huang, Y.-H.; Blazquez, C.A.; Chen, C.-Y. Solving the vehicle routing problem with drone for delivery services using an ant colony optimization algorithm. *Adv. Eng. Inform.* **2022**, *51*, 101536. [CrossRef]
44. Jung, H.; Kim, J. Drone scheduling model for delivering small parcels to remote islands considering wind direction and speed. *Comput. Ind. Eng.* **2022**, *163*, 107784. [CrossRef]
45. Miguel Leon-Blanco, J.; Gonzalez, P.L.; Andrade-Pineda, J.L.; Canca, D.; Calle, M. A multi-agent approach to the truck multi-drone routing problem. *Expert Syst. Appl.* **2022**, *195*, 116604. [CrossRef]
46. Zeng, F.; Chen, Z.; Clarke, J.-P.; Goldsman, D. Nested vehicle routing problem: Optimizing drone-truck surveillance operations. *Transp. Res. Part C-Emerg. Technol.* **2022**, *139*, 103645. [CrossRef]
47. Braekers, K.; Ramaekers, K.; Van Nieuwenhuyse, I. The vehicle routing problem: State of the art classification and review. *Comput. Ind. Eng.* **2016**, *99*, 300–313. [CrossRef]
48. Pillac, V.; Gendreau, M.; Gueret, C.; Medaglia, A.L. A review of dynamic vehicle routing problems. *Eur. J. Oper. Res.* **2013**, *225*, 1–11. [CrossRef]
49. Potvin, J.Y. State-of-the Art Review Evolutionary Algorithms for Vehicle Routing. *Inf. J. Comput.* **2009**, *21*, 518–548. [CrossRef]

Journal of
Marine Science and Engineering

MDPI

Article

New Exploration of Emission Abatement Solution for Newbuilding Bulk Carriers

Shunquan Huang and Yuyang Li *

College of Economics and Management, Shanghai Maritime University, Shanghai 201306, China; sqhuang@shmtu.edu.cn
* Correspondence: 202230710058@stu.shmtu.edu.cn

Abstract: With the implementation of the International Maritime Organization's (IMO) sulfur cap 2020, shipowners have had to choose suitable sulfur oxide emission abatement solutions to respond to this policy. The use of Very Low Sulfur Fuel Oil (VLSFO) and the installation of scrubbers are the main response solutions for bulk carriers today. In recent years, the epidemic has gradually improved, and the options facing shipowners may change. Based on the Clarkson Shipping Intelligence Network, this paper collects data related to newbuilding bulk carriers after the implementation of this policy, considers several factors affecting shipowners' decision, and adopts a machine learning approach for the first time to build a model and make predictions on emission abatement solutions to provide some reference for shipowners to choose a more suitable solution. The results of the study show that the Extreme Gradient Boosting (XGBoost) model is more suitable for the problem studied in this paper, and the highest prediction accuracy of about 84.25% with an Area Under the Curve (AUC) value of 0.9019 is achieved using this model with hyperparameter adjustment based on a stratified sampling divided data set. The model makes good predictions for newbuilding bulk carriers. In addition, the deadweight tonnage and annual distance traveled of a ship have a greater degree of influence on the choice of its option, which can be given priority in the decision making. In contrast to traditional cost–benefit analyses, this study incorporates economic and non-economic factors and uses machine learning methods for effective classification, which have the advantage of being fast, comparable, and highly accurate.

Keywords: newbuilding bulk carrier; emission abatement solution; imbalanced dataset; XGBoost model

1. Introduction

The World Health Organization (WHO) declared in May 2023 that the COVID-19 pandemic no longer constituted a "public health emergency of international concern". International trade is gradually returning to business as usual and the global supply chain is beginning to restart, prompting a gradual return to normalcy in the shipping industry.

As an indispensable part of the international shipping industry, bulk carriers mainly transport grain, coal, ore, and other cargoes, and they are mainly divided into four categories according to their tonnage: Capesize, Panamax, Handymax, and Handysize. As a complex and volatile market, the dry bulk shipping market is susceptible to the influence of the commodity market and fuel market [1], and shipowners' investment in ships is closely related to the market, operation, corporate strategy, and industry cluster factors [2]. In recent years, with the changes in the external economic and trade environment, it has become increasingly difficult for shipowners to make decisions [3].

Emissions of sulfur oxides are only behind nitrogen oxides, and are also an environmental issues worthy of consideration [4]. As the International Maritime Organization (IMO) requires the implementation of the "sulfur cap" from 1 January 2020, shipowners' decisions on sulfur oxide emission abatement solutions have become a new topic of discussion, requiring a global reduction in the maximum sulfur content of ship fuels to 0.5% for environmental improvements and human health [5]. In response, the shipping industry is always looking for more economical ways to reduce sulfur emissions, and there tend to be different decision options for different ship types and operating models [6]. In the case of newbuilding ships, retrofitting is generally not carried out in the short term after the determination of the scheme, so it is more necessary to combine the initial investment and long-term operating costs to choose a suitable emission abatement solution [7], which is more conducive to maintaining long-term competitiveness.

Currently, the three most popular approaches are using Very Low Sulfur Fuel Oil (VLSFO), continuing to use High Sulfur Fuel Oil (HSFO) but installing sulfur oxide scrubbers to reduce the exhaust gases, and using alternative fuels. VLSFO refers to fuel oil with a sulfur content of no more than 0.5%. The scrubber, which filters the exhaust gas to reduce the sulfur content, is a high investment in the initial construction. The initial investment in choosing to use VLSFO is relatively low compared to installing a scrubber tower, making it more cost-effective in the short term, as scrubbers require a significant up-front capital outlay and regular maintenance and monitoring [8]. Karatuğ et al. concluded that although SOx emissions were effectively reduced due to the installation of scrubbers, other types of emissions increased with the use of HSFO and a higher fuel consumption (including CO_2), thus contradicting the decarbonization strategy assigned by IMO, which is to reduce the carbon intensity of all ships by 40% by 2030, compared to the 2008 baseline [9]. The most typical representative of alternative fuels is Liquefied Natural Gas (LNG), and ammonia has also been gradually developed. Even before the policy was implemented, LNG was seen as a promising alternative fuel that could effectively reduce pollution, but also had certain safety concerns [10]. It is mainly characterized by its susceptibility to leaks that can cause serious accidents [11], especially in coastal areas where there is a lack of complete infrastructure for fuel refueling [12]. Although some government subsidies are widely used [13], LNG prices may change abruptly at any time [14]. It is more difficult to grasp the cost, causing many ship owners to remain in the wait-and-see stage or take the "LNG Ready" option, which is to design for future conversion to LNG. It has also been suggested that improving energy consumption efficiency will help maritime policy makers to provide more reasonable regulations for improving the energy consumption of ships [15]. Through the survey and analysis of existing bulk carriers in service after 2020 and their new ship orders, the use of VLSFO and the installation of scrubbers are the two most common emission abatement solutions currently used by bulk carriers. Therefore, this paper will also make a decision among these two emission abatement solutions.

The data in this article come from the Clarkson Maritime Intelligence Network, which manually collected data on about 680 newbuilding bulk carriers. Although the amount of data is small, each sample is highly representative. The purpose of this paper is to find out the factors affecting newbuilding bulk carriers regarding the selection of sulfur oxide emission abatement solutions through big data, and to establish a classification prediction model, including Logistic Regression (LR), K-Nearest Neighbor (KNN), Support Vector Machine (SVM), Random Forest (RF), Adaptive Boosting (AdaBoost), and Extreme Gradient Boosting (XGBoost), to compare their prediction accuracy and select the best model to provide some reference and suggestions for shipowners' decisions on sulfur abatement solutions. Our analysis, at the theoretical level, suggests that the XGBoost model is more suitable for solving this research problem.

The remainder of the paper is structured as follows. Section 2 reviews the existing literature, provides an overview of the existing research results, and highlights the innovation of the paper. Section 3 details the data processing and machine learning methods used in this paper. Section 4 describes the data used in this paper and the pre-processing procedure and provides a descriptive analysis of the data. Section 5 conducts an empirical analysis where multiple machine learning models are used to make predictions and the predictive effects of different models are compared through hyperparameter tuning. In addition, the importance of each influencing factor is analyzed and forecasts are made for newbuilding bulk carriers. Section 6 summarizes the full paper and provides an outlook for future research.

2. Related Work

2.1. Machine Learning

Machine learning is one of the fastest growing technology fields today, comprising supervised and unsupervised types, and has been widely used in various fields [16]. Supervised types mainly include two directions: classification and regression, which can effectively and automatically learn from data, better handle high-dimensional and high-volume data, and have a higher prediction accuracy compared to traditional models. In recent years, it has also been gradually used in the field of green shipping. Kim et al. used deep learning for the route optimization of ships to reduce fuel consumption, which also helps improve maritime safety and efficiency [17]. Chen et al. proposed a path optimization model for port environments based on an artificial potential field and a twin delay depth deterministic policy gradient framework, which can facilitate the efficient operation and management of ports [18]. Xiao et al. conducted a systematic literature study on the application areas of AI technology in the shipping industry using bibliometric methods to help researchers understand the current status and development trends of AI technology in shipping applications [19]. This paper handles a dichotomous problem for emission abatement solutions, considering multiple factors and, therefore, using a machine learning model has a higher prediction accuracy, while the degree of influence of the variables can be ranked more effectively.

XGBoost is an excellent integrated learning algorithm that has achieved remarkable results in many fields and, in recent years, has been applied to the shipping industry as well [20]. Its advantages include the following aspects. First, by parallel computing, it greatly improves the model training speed and prediction accuracy [21]. Second, performing second-order Taylor expansion on the objective function and using second-order derivatives during training accelerate the convergence of the model. Third, by setting class weights and using an area Under the Curve (*AUC*) as the evaluation criterion, it can handle imbalanced training data well [22]. Fourth, it can output the factors' importance magnitude. The reasons for preferring this algorithm for prediction in this paper are mainly due to the inclusion of the regularization term to prevent overfitting to a certain extent, the fact that the problem studied in this paper is a sample imbalance problem, and the need to understand the magnitude of the influence of each factor. In summary, we believe that the XGBoost model will be more beneficial to solve this problem.

2.2. Existing Studies on Emission Abatement Solution

Our extensive literature survey of the research on emission abatement solutions divides them into three main directions. The minimization of transportation costs is often considered as one of the optimization goals [23], which is mainly based on a cost–benefit approach to economic analysis. Studies on the choice of fuel oil have been available since before IMO proposed the "sulfur cap" in 2016, and Lindstad et al. mentioned that there is no single answer for the best emission abatement solution, which is influenced by a function of each ship's engine size, annual fuel consumption in the Emission Control Area (ECA), and the foreseen future fuel prices, and the choice of emission abatement solutions is made by evaluating the cost and fuel consumption [24]. Zis et al. presented a cost–benefit methodology to assess emission abatement investments from shipowners, and they discovered that the drop in fuel prices can significantly delay the payback period for ships with scrubbers [25]. Fan et al. used a cost–benefit framework to analyze the ships' compliance choices for fuel switching and blending scrubbers in the ECA region and applied it to routes in the China ECA [26]. The results indicated that the proportion of ships sailing in the ECA, the difference between low and high sulfur fuels, and the cost of scrubbers would be important influencing factors in the decision. Wu et al. amortized the scrubber lifetime and calculated the total incremental cost of the two solutions, taking into account the loss of cargo space, fuel prices, etc., and concluded that the incremental cost of the VLSFO is higher than the scrubber for the first 4.14 years after the solution choice [8]. Studies on cost-effectiveness effectively analyze the feasibility of each scenario, but in practical scenarios, there are often deviations from existing studies, while non-economic factors may be ignored, so new research methods have also emerged.

Growing interest in the development and impact of maritime sulfur emission regulations, new optimization problems, and usage and route selection models are being applied to the field [27]. The second category of research directions incorporates more non-economic factors than the first category and uses mathematical methods that are more adaptive to changes in the environment. Zis et al. considered the speed differences inside and outside the ECAs region and the operating hours of the ships, estimated the economic and environmental impacts of compliance with sulfur limits for various representative ship types, and suggested the creation of a new decision support system that considers a wider range of stakeholders [28]. They confirm that scrubber investments are more profitable when fuel prices are rising and for vessels that spend relatively more time sailing, in addition to the fact that the potential for speed differentiation inside and outside the ECAs has diminished. Zhao et al. combined three sulfur reduction technologies, fuel switching, scrubbers, and an LNG dual-fuel engine, and developed a robust optimization model based on two-stage stochastic linear programming to effectively cope with future uncertainties, which can reduce a certain amount of sulfur emissions [29]. Lagemann et al. formulated the choice of alternative fuels and the corresponding ship power system as an objective integer optimization problem, and the model was applied to a very large dry bulk carrier, where the uncertainty of external factors can be explained by a stochastic model, with the advantage of providing the corresponding parameters into the model, which can lead to a reliable solution [30].

The third category of research solutions is to build statistical models for analysis and prediction by collecting big data on ships in service. This category of methods presupposes the collection of a sufficient amount of data and is a method that has emerged in recent years. It can look for patterns in the emission abatement solutions of active ships and use big data for effective predictions. Li et al. used a Multinomial Logistic Model (MNL) to determine the decision factors of ships in response to the "sulfur cap", and concluded that different ship ages and ship types have their corresponding preferences; it is through this statistical method that the significance of the influence of each factor and the magnitude of the influence can be effectively verified [31]. Therefore, Bao et al. studied the selection factors of three emission abatement solutions in the cruise industry using VLSFO, installing scrubbers, and using LNG with MNL. They found that the options depended on the size, age, and flag of the cruise ship, but not on the sailing area, and that fluctuations in fuel prices did not play a decisive role. All the above studies were completed through MNL, and although the key factors of shipowners' decision preferences were found, the models were not predicted so that prediction accuracy could be discussed [32]. In terms of research on new ships, Zhang et al. still used MNL to model the emission abatement solutions for a wide range of ship types to study their key determinants, and they obtained a large correlation between the decisions affecting the ship and the type of ship, the oil price difference, and the nationality of the ship owner. They made predictions with the model, but did not consider the sample imbalance [33]. Bai et al. then used a logit model to determine the key factors affecting different ship types, and tested the prediction effect of the model after considering the sample imbalance. The prediction accuracy for bulk carriers was 76.59%, which provided some reference for shipowners' selection, but there was still the problem of a low prediction accuracy [34].

The innovation of this paper is to summarize the features of the existing studies, integrating economic and non-economic factors, and further refining the experience under the cost scenarios by taking the data of the ships put into service after the implementation of the policy to be able to predict more accurate scenarios more quickly for the shipowners. Shipowners can also input multiple combinations to obtain the appropriate output solution.

For the first time, the imbalance data processing technique in the XGBoost model is applied to the selection of sulfur oxide emission abatement solutions. The optimal hyperparameters are determined by the grid search method, which further improves the prediction accuracy of the model, and at the same time, the importance of each influencing factor is analyzed. This study provides a reference for shipowners to find a suitable emission abatement solution.

3. Methodology

3.1. Modeling Process

We modeled the bulk carrier data according to the method in Section 2, and the process is shown in Figure 1. Since the sample ratio for the two categories was 1.559:1, we used three methods to calculate the classification effect separately: no balancing process, manual synthesis of the data using Borderline Synthetic Minority Oversampling Technique (SMOTE) to achieve balance, and stratified sampling of the data. Based on these three methods, we built six machine learning models, LR, KNN, SVM, RF, AdaBoost, and XGBoost, to predict the emission abatement solution for newbuilding bulk carriers, and compared the prediction effects of different models to find the more optimal balance method and prediction model. Then, we adjust the hyperparameters of these models to further find the model with the best accuracy and output the importance of each feature. The first three of these methods are single classifier algorithms, while the last three are integrated learning algorithms. We implemented each balanced method and machine learning algorithm using the package of R 4.2.2.

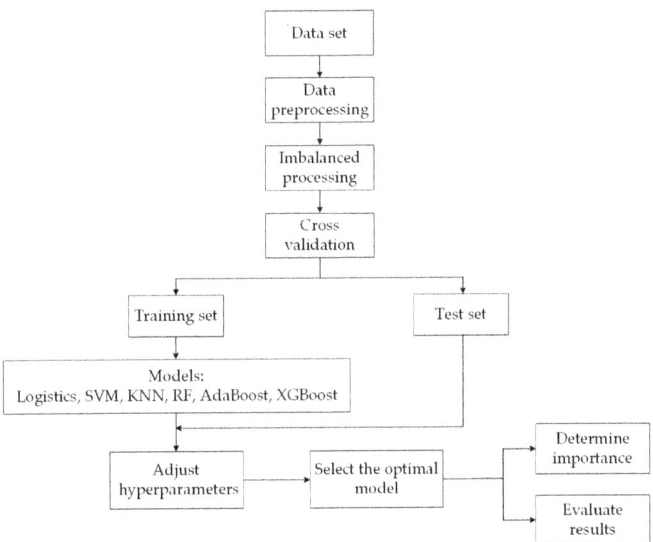

Figure 1. The process of modeling.

3.2. Data Imbalance Processing

The emission abatement solutions are mainly divided into two categories, VLSFO and scrubbers, and thus can be classified as a binary classification problem. Through statistics, the sample size of the VLSFO sample is slightly higher than the scrubber sample, which is about 1.559:1 and belongs to the imbalanced dataset, which may lead to the prediction results favoring the class with higher sample size. Therefore, the following three types of treatments were considered when dividing the data set.

3.2.1. No Processing

The sample ratio of the original data set is maintained. At this time, the VLSFO-to-scrubber sample ratio is 1.559:1, and random sampling is taken when dividing the training set and test set in the later section, without considering the sample imbalance problem.

3.2.2. Artificial Synthetic Data Technology

This is carried out by adding a few categories of samples to achieve a balance between the two types of sample data. Since the traditional oversampling technique is by simply replicating the samples, this may lead to overfitting the results. Rather than simply copying samples, SMOTE selects minority class samples that are close together and uses them to generate synthetic new samples [35]. This algorithm is widely used because it can learn from more new samples that are more favorable for minority class classification. The Borderline SMOTE artificial data synthesis technique was proposed based on SMOTE [36], which focuses more on the instances located on the decision boundary and learns the boundary of each class accurately, overcoming the drawback that SMOTE can only be generated within the region, and the process of artificially synthesizing the data to reach the balance is shown in Figure 2a.

(a) SMOTE (b) Stratified Sampling

Figure 2. Artificially synthesized data to balance and stratified sampling.

3.2.3. Stratified Sampling

Through the stratified sampling in statistics, it makes the ratio of the two types of samples in the training set and the test set consistent with the original data when drawing the samples, i.e., the ratio of both types of samples in the training set and the test set is 1.559:1, as shown in Figure 2b. This method ensures that the sample set and the original data set have similar distribution.

3.3. Methodology

Based on the above three different sample imbalance processing methods, six more machine learning models are used, respectively, to classify and study the emission abatement solution problem of bulk carriers, including LR, KNN, SVM, RF, AdaBoost, and XGBoost. The XGBoost model is the method we mainly consider, and the following is the introduction of each method.

3.3.1. XGBoost

XGBoost is an integrated learning model based on decision trees, which is a parallelized boosting algorithm. By constructing and combining multiple weak learners to accomplish the learning task, many data science problems can be solved quickly and accurately. The basic idea is to continuously generate new trees, and each tree is learned based on the difference between the previous lesson tree and the target value, thus reducing the bias of the model. The XGBoost model also uses regularization to avoid overfitting and improve the accuracy of the model, and also greatly improves the speed of computing, which is the method focused on in this paper, and the specific implementation steps are as follows:

XGBoost is an additive model consisting of t base models (decision trees):

$$\hat{y}_i = \sum_{k-1}^{t} f_k(x_i) \tag{1}$$

The objective function of the XGBoost model has a model loss function and a canonical term Ω that suppresses the complexity of the model:

$$Obj = \sum_{i=1}^{n} l(\hat{y}_i, y_i) + \sum_{k=1}^{t} \Omega(f_k) \tag{2}$$

The boosting model is forward additive, and using the model at step t, as an example, the model predicts the i-th sample x_i as follows:

$$\hat{y}_i^{(t)} = \hat{y}_i^{(t-1)} + f_t(x_i) \tag{3}$$

where $\hat{y}_i^{t-1}$ is the predicted value given by the model at step $t-1$. After adding the predicted value of the new model, the objective function can be written as:

$$Obj^{(t)} = \sum_{i=1}^{n} l(y_i, \hat{y}_i^{(t-1)} + f_t(x_i)) + \sum_{i=1}^{t} \Omega(f_i) \tag{4}$$

The t-th round of regularization is obtained from the sum of the regularization terms of all trees and consists of the difference between the true and predicted values and the regularization function, which is used to prevent overfitting during the training of the XGBoost model:

$$\Omega(f_t) = \gamma T + \frac{1}{2}\lambda \sum_{j=1}^{T} \omega_j^2 \tag{5}$$

where T is the number of leaf nodes, γ and λ are penalty factors, and ω_j is the j-th leaf node score.

The second-order Taylor formula for Obj is expanded as:

$$Obj^{(t)} = \sum_{i=1}^{n} [l(y_i, \hat{y}_i^{(t-1)}) + g_i f_t(x_i) + \frac{1}{2} h_i f_t^2(x_i)] \tag{6}$$

where g_i and h_i are the first- and second-order partial derivatives of the loss function for the i-th sample.

When the model is trained, $G_j = \sum g_i$ and $H_j = \sum h_j$, its objective function is:

$$Obj^{(t)} \approx \sum_{j=1}^{\mathrm{T}} [G_j \omega_j + \frac{1}{2}(H_j + \lambda)\omega_j^2] + \gamma T \tag{7}$$

where g_i and h_i are the results obtained from step $t-1$, whose values are known to be considered as constants and only the leaf node ω_j of the last tree is uncertain. We then find the first-order derivative of the objective function with respect to ω_j and make it equal to 0, and the weights corresponding to the leaf node j can be found as follows:

$$\omega_j = -\frac{G_j}{H_j + \lambda} \tag{8}$$

The final objective function can be reduced to:

$$Obj^{(t)} = -\frac{1}{2} \sum_{j=1}^{\mathrm{T}} \frac{G_j^2}{H_j + \lambda} + \gamma T \tag{9}$$

In addition, the XGBoost model can output the importance ranking of each feature. In the XGBoost model, the importance of a feature indicates the sum of the number of times it appears in all trees. In other words, the more a variable is used to build decision trees in the model, the higher its importance is, and finally, normalization is performed to obtain the importance ratio of each feature.

3.3.2. Other Models

The LR algorithm outputs a predicted probability between 0 and 1 by linearly weighting the input variables with a Sigmoid function, which is a typical binary linear model. It can obtain the degree of influence of each factor through regression coefficients and determine the significance of each factor.

The KNN algorithm is an instance-based learning method that determines the class of a predicted sample by calculating the distance between samples and selecting the K nearest neighbors and voting on the classification labels of these K neighbors.

SVM is a binary classification method whose basic model is a linear classifier defined on the feature space with maximum interval, but at the same time, it can also handle the nonlinear case by projecting the samples to a high-dimensional space using kernel functions.

The RF algorithm is one of the most common integration algorithms, consisting of multiple decision tree classifiers, where each dataset is randomly selected with putback, while some features are randomly selected as input. A strong learner algorithm developed based on the Bagging integration method is widely used in classification and regression. When used for classification, different decision trees will make their own decisions and take more than one class as the prediction result.

The AdaBoost algorithm is part of the Boosting model, which is different from the bagging model (random forest). Among the multiple classifiers that AdaBoost will construct, each classifier determines its own weight according to its own accuracy and then merges. At the same time, AdaBoost will adjust the data weights according to the previous classification effect.

3.4. Model Evaluation

3.4.1. Cross-Validation

For the division of the dataset, this paper uses the K-fold cross-validation method to complete the principle. That is, the original dataset is divided into K copies, K-1 copies as the training set, 1 copy as the test set, and finally, into K models. Finally, the average of K model classification indicators is used as the final evaluation indicator of the model, which can effectively avoid overfitting and underfitting and make the prediction more accurately [37]. In this paper, we use the ten-fold cross-validation method for partitioning, i.e., the ratio of training set to test set of samples is 9:1. The advantage of this method is that the randomly generated subsamples are applied repeatedly for training and validation at the same time, and the results are validated once each time. We also use this method for hyperparameter adjustment.

3.4.2. Accuracy and Error Rate

By constructing the above six models, the confusion matrix of prediction can be obtained, see Table 1, and then the Accuracy (*ACC*) of the respective models and the error rate (*Error*) of the two types of samples can be measured to judge the effectiveness of the classification of the models.

Table 1. Confusion matrix.

Solution		Actual Values	
		VLSFO	**Scrubber**
Predicted Values	VLSFO	True Positive (TP)	False Positive (FP)
	Scrubber	False Negative (FN)	True Negative (TN)

ACC indicates the proportion of correctly classified samples to the total sample, and the error rate of two categories of samples indicates that the true result is the proportion of prediction errors in the results of the corresponding categories, and it can avoid the increase in the overall accuracy rate due to the high accuracy of one of the categories, calculated as follows:

$$ACC = \frac{TP + TN}{TP + FP + FN + TN} \tag{10}$$

$$Error_{\text{VLSFO}} = \frac{FN}{TP + FN} \tag{11}$$

$$Error_{Scrubber} = \frac{FP}{FP + TN} \tag{12}$$

where $Error_{VLSFO}$ is the error rate of the VLSFO and $Error_{Scrubber}$ is the error rate of the scrubber.

3.4.3. ROC Curves and AUC Value

The ROC curve, also called receiver operating characteristic curve, can be used to measure the predictive performance of a binary classification model [38]. In general, the horizontal coordinate of the ROC curve is the False Positive Rate (*FPR*) and the vertical coordinate is the True Positive Rate (*TPR*). When different thresholds are chosen, the ROC curve can reflect the corresponding trend as the ratio of *TPR* and *FPR* as follows:

$$TPR = \frac{TP}{TP + FN}, \tag{13}$$

$$FPR = \frac{FP}{TN + FP} \tag{14}$$

The *AUC* value is the area under curve, its size is proportional to the quality of the model, and it takes values between 0.5 and 1. When the value of *AUC* is equal to 0.5, the model is a random guess and the model has no predictive ability, and the higher the *AUC* value is, the better the classification effect of the model is. Therefore, the *AUC* value is chosen as other indicators to evaluate the classification effect of the model.

3.4.4. Grid Search

Grid search, originally proposed in 1998, is a commonly used method for hyperparameter tuning by discretizing the range of values of different hyperparameters into a discrete number of points, forming a multidimensional grid of these points, with each grid point corresponding to a hyperparameter combination, training the model with different hyperparameter combinations in turn and evaluating it on the validation set, and finally, selecting the hyperparameter combination that displays the best performance on the validation set. The combination of hyperparameters with the best performance on the validation set is selected as the hyperparameter of the final model. In the field of machine learning, grid search is widely used in the training and tuning process of the model, which can effectively improve the accuracy and generalization ability of the model and find the global optimal solution, combining with ten-fold cross-validation to jointly complete the fine-tuning of hyperparameters [39].

4. Data Description and Preprocessing

4.1. Data Sources

This paper uses data from the Clarkson Shipping Intelligence Network. The data set collected includes 683 bulk carriers in service between January 2020 and October 2023, covering three types of bulk carriers: Capesize, Panamax, and Handymax. To avoid the duplication of statistics, only one of the sister vessels ordered by the same company was taken as a sample. Meanwhile, the Handysize bulker almost tended to use VLSFO in the emission abatement solution, so they are not included in the research sample of this paper.

Among active bulk carriers and new ship orders, ships using alternative fuels as an emission abatement solution occupy a very small proportion, with only eight active bulk carriers using alternative fuels between 2020 and 2023, and only 38 ships using alternative fuels, accounting for only 4.26% of all new bulk carrier orders, according to Clarkson's new ship orders. We understand that the emission abatement solution for bulk carriers is mainly based on VLSFO and the installation of scrubbers, so in the choice of a solution, this paper focuses on these two.

A total of nine representative characteristics were collected in this paper, which are the Age of the Ship (Age), Deadweight Tonnage (DWT), Main Engine Power (Power), Annual Average Speed (Speed), Annual Distance Traveled (Distance), Price Difference between Low and High Sulfur Fuel (PD), Baltic Dry Index (BDI), Ship Docking Time in Port (DT), and the Proportion of Sailing Time in the Sulfur Emission Control Area (ECA).

The age of the ship is measured in years, and the collected ships were put into service from 2020 to 2023, so the value is taken to be between 1 and 4. The DWT reflects the actual weight of the ship, and the unit of Power is kW. The data suggest that there may be a strong correlation between the DWT and Power.

In the process of obtaining ship features data, the DWT, Age, and Power can be directly obtained. Speed is the average speed of the ship in a service year in knots. The Distance is the average distance traveled by the ship in a service year in nautical miles. For the distance traveled by the ship in 2023, it can be derived from the distance traveled and the months traveled. The fuel price difference is the price difference between HSFO and Marine Gas Oil (MGO). Here, since VLSFO has been available since 2020, MGO is used instead of VLSFO in this paper, and the fuel price is the current month's fuel price when the shipbuilding contract is signed by the shipowner. Singapore is the world's largest bunker oil trading market, so the current month's bunker price in Singapore is chosen as a feature variable in this paper. BDI is also derived from the corresponding value in the month when the shipbuilding contract is signed. In addition, this paper also introduces the sailing time weighting and ship docking time in port in the ECAs, which includes four regions, the Baltic Sea, the North Sea, North America, and the U.S. Caribbean, where the sulfur content of ship fuel is required to be no more than 0.1%. We also take the Hainan and Yangtze River of China into consideration.

4.2. Data Preprocessing

Data preprocessing is an essential preparation before data analysis. We mainly deal with missing values and outliers, and then further standardize them.

In dealing with the missing data, some ships are missing important variables, such as the distance traveled and the time when the shipbuilding contract was signed. Since the missing samples are small and have no major impact on the construction of the model, the ships with missing data are directly excluded from this paper.

For the outliers in the individual features of the ships, mainly for the outliers of the Speed and Distance, we considered the data other than the mean plus or minus three times the standard deviation as outliers and deleted the corresponding ships. In addition, we also fully considered whether the ships had been repaired or modified during service, and corrected the sailing distances for such samples. Finally, we obtained 683 valid samples, which are detailed in Table 2.

Table 2. Number of three types of ships.

Type	2020	2021	2022	2023	Total
Capesize	48	36	26	21	131
Panamax	93	65	72	66	296
Handymax	96	49	57	54	256
Total	237	150	155	141	683

To reduce the variation in the magnitude of the data, we standardized the data by the Z-score standardization method, which is based on the mean and standard deviation of the original data. The formula used is as follows:

$$X' = \frac{X - \overline{X}}{Std} = \frac{X - \overline{X}}{\sqrt{\frac{1}{N-1}\sum_{i=1}^{n}(X - \overline{X})^2}} \quad (15)$$

where X' is the normalized data, X is the original value of the sample, and $\overline{X}$ is the mean value of the corresponding feature of the sample.

4.3. Descriptive Analysis

Table 3 calculates the mean, median, and standard deviation of each feature of the two emission abatement solutions for newbuilding bulk carriers, respectively. Figure 3 shows the box plots of each feature separately, where the distribution can be seen further. Based on the descriptive statistical analyses and box plots, the data can be better understood and certain conclusions can be drawn to provide some references for further empirical analyses.

Table 3. Descriptive statistics.

Feature	VLSFO			Scrubber		
	Mean	Median	Std. Dev.	Mean	Median	Std. Dev.
Age/a	1.84	2.00	0.83	2.38	3.00	0.77
DWT	77,347.89	75,771.00	35,443.22	126,994.60	82,264.00	84,943.96
Power/kW	9022.86	8302.00	2728.09	11,958.6	9801.00	5029.91
Speed/knots	11.68	11.80	0.68	11.90	11.92	0.46
Distance/nm	54,857.48	56,172.00	13,007.77	66,525.92	66,255.50	9887.81
PD/$	173.16	175.13	58.23	190.06	192.25	44.91
BDI	1350.71	1293.11	588.53	1367.24	1314.11	449.69
ECA/%	9.90	8.38	8.25	7.46	4.38	9.08
DT/d	4.64	4.49	1.76	4.43	4.02	1.89

According to the data in Figure 3a, we can find that the proportion of bulk carriers using VLSFO has been increasing in recent years. In addition, Figure 3b,c shows that as the DWT and Power increase, the more likely the shipowner is to use scrubbers. Figure 3d,e represent that those ships using scrubbers tend to be faster and travel longer distances, which may have some relationship with their fuel costs. Figure 3f,g indicate that the higher the PD and the higher the BDI, the more likely the shipowner is to use scrubbers, but the effect is not significant. Figure 3h shows that the longer the time sailing in ECA is, the more likely the ship owner is to use VLSFO. According to the data in Figure 3i, the ships using VLSFO have a shorter DT. However, this result may require further validation. The above conclusions can also be obtained from the descriptive statistics charts in Table 3.

Finally, we calculated the correlation coefficients among the indicators, and Figure 4 shows the correlation coefficients between VLSFO and the scrubbers. We found that there is a highly positive correlation between the DWT and Power among the ships using scrubbers in Figure 4b, and the ships with a larger tonnage also traveled relatively more distance, so we need to consider the multicollinearity problem in the modeling process later. Secondly, there are also correlations between the DWT and the Power, Distance, and PD.

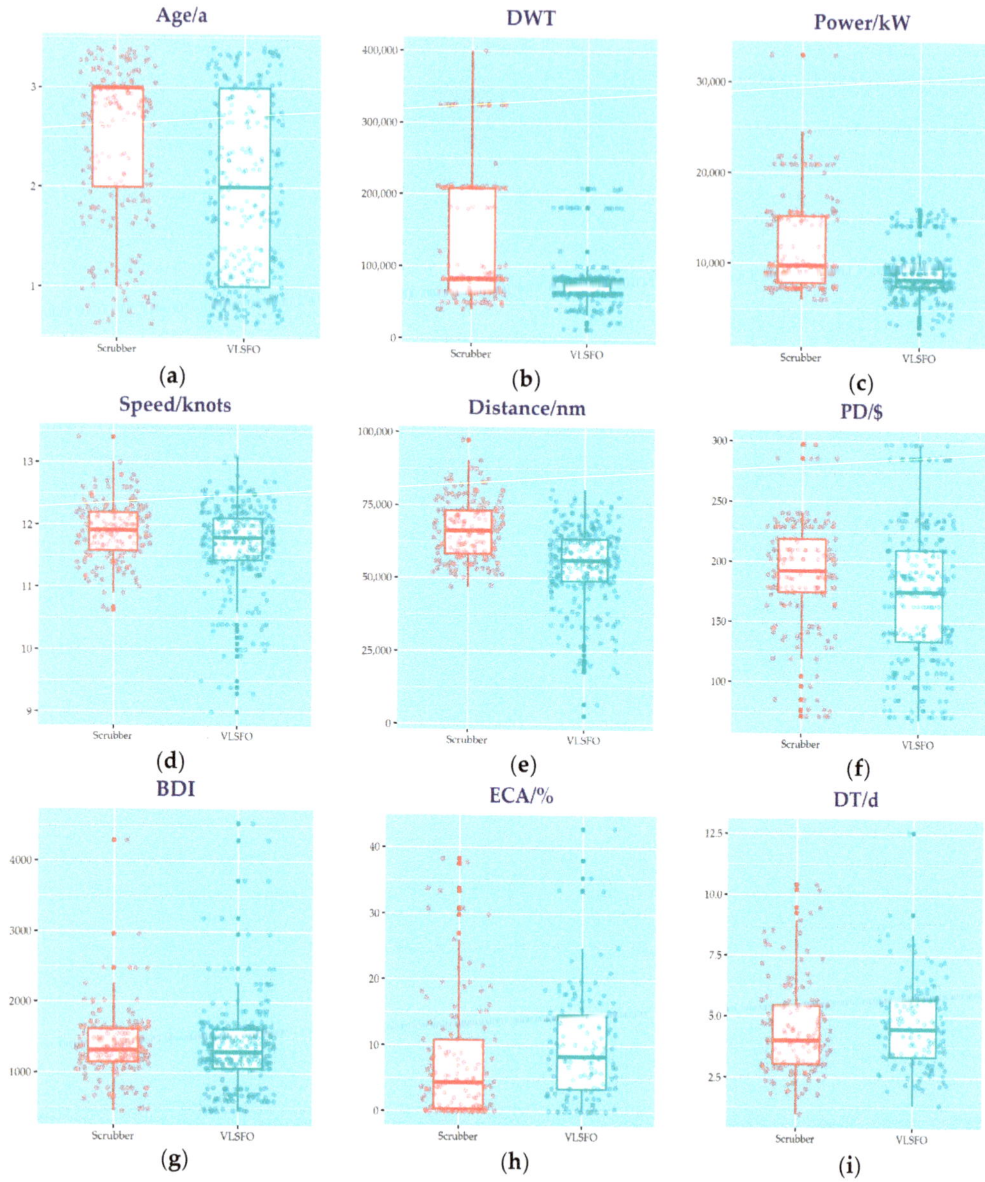

Figure 3. Box plots of each feature.

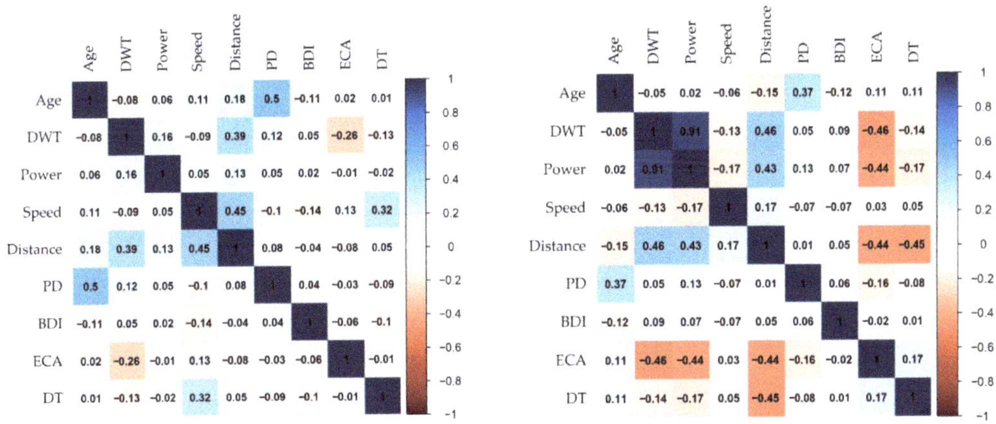

(**a**) VLSFO (**b**) Scrubbers

Figure 4. Correlation coefficient chart.

5. Results and Analysis

In this section, we analyze the predictions of the model and Table 4 shows the list of acronyms for the features and evaluation metrics involved.

Table 4. List of acronyms.

Abbreviations	Connotation
Age	Age of ship
DWT	Deadweight tonnage
Power	Main machine power
Speed	Annual average speed of navigation
Distance	Annual distance traveled
PD	Price difference between the two types of fuel oil at contract time
BDI	Baltic dry index
ACC	Accuracy of the model
Error	False judgement rate for a category
AUC	The size of the area under the ROC curve

5.1. Model Predictions

By screening different features for iterative modeling comparisons, we finally selected seven features as the input variables for the prediction model, namely Age, DWT, Power, Speed, Distance, PD, and BDI. We excluded the two features ECA and DT because their introduction does not significantly improve the prediction and may even lead to a decrease in the prediction results, like the results of Zis et al. [28]. In addition, we mentioned in Section 3 that there is a strong correlation between the DWT and Power, so when using LR, we excluded Power to avoid the covariance problem. In the machine learning model, there is no need to consider the multicollinearity problem.

After the standardization of the samples, we first tried to balance the samples using three different methods and constructed the model using the default hyperparameters to compare the effects of the three methods. The results are shown in Table 5, where we can see that the method of using Borderline SMOTE to synthesize the data is less effective. Therefore, our next work will be based on the two methods of using raw data and stratified sampling to divide the dataset to adjust the hyperparameters of the model.

Table 5. Prediction results of the six models.

Model	Imbalanced Processing	ACC	$Error_{VLSFO}$	$Error_{Scrubber}$	AUC
LR	No Processing	73.36%	17.82%	40.29%	0.8066
	Artificial Synthetic Data	69.17%	36.30%	22.25%	0.8062
	Stratified Sampling	73.52%	16.71%	41.55%	0.8015
KNN	No Processing	70.02%	23.26%	40.44%	0.5482
	Artificial Synthetic Data	66.47%	38.02%	26.57%	0.5526
	Stratified Sampling	69.78%	23.26%	41.02%	0.5538
SVM	No Processing	75.44%	**9.60%**	47.78%	0.7607
	Artificial Synthetic Data	70.21%	35.30%	**21.26%**	0.7839
	Stratified Sampling	74.56%	10.56%	48.48%	0.7701
RF	No Processing	79.60%	10.60%	35.09%	0.8745
	Artificial Synthetic Data	77.93%	19.51%	26.08%	0.8629
	Stratified Sampling	78.97%	12.30%	34.64%	0.8717
AdaBoost	No Processing	80.64%	14.70%	26.61%	0.8950
	Artificial Synthetic Data	80.01%	16.05%	26.08%	0.8843
	Stratified Sampling	80.43%	15.04%	26.55%	**0.8968**
XGBoost	No Processing	**81.04%** [1]	13.01%	28.25%	0.8852
	Artificial Synthetic Data	78.34%	19.17%	25.56%	0.8731
	Stratified Sampling	79.99%	14.39%	28.83%	0.8781

[1] The optimal results for each metric are indicated in bold.

After the experimental comparison, we found that the highest ACC rate of 81.04% can be obtained using the XGBoost model without the data balancing process. Additionally, the SVM model has an advantage in terms of the individual error rate. In addition, we also found that the error rate of the VLSFO samples is lower compared to the scrubber samples. This may be due to the higher similarity of the features of some of the ships installed with scrubbers and those using VLSFO. Therefore, in future research work, we need to introduce relevant features with more differentiation. Overall, the integrated learning algorithm outperforms single classifiers by combining the predictions of multiple classifiers, reducing the risk of the overfitting of individual classifiers and improving generalization capabilities, so we have focused on further tuning the hyperparameters of these three integrated learning models.

5.2. Adjustment of Hyperparameters

We have listed their main hyperparameters and their adjusted ranges for the three models, and the details of the hyperparameters and their adjusted ranges are shown in Table 6. For the unmentioned hyperparameters, we took the default values because they do not positively affect the models during the adjustment process. By using a grid search combined with ten-fold cross-validation, we obtained the results of the optimal hyperparameters with the goal of finding the highest ACC. In Table 7, we display the obtained optimal hyperparameters and the ACC for each model on the test set under the two imbalance treatments.

Table 6. Range of hyperparameter adjustment.

Model	Hyperparameter	Explanation	Adjustment Range
RF	n_estimators	Number of decision trees	[50, 1000] step = 50
	max_depth	Maximum depth of the tree	[2, 10] step = 1
	max_features	Number of splitting features	[1, 5] step = 1
AdaBoost	n_estimators	Number of classifiers	[50, 1000] step = 50
	learning_rate	Learning rate	[0, 1] step = 0.1
	max_depth	Maximum depth of the tree	[2, 10] step = 1
XGBoost	n_estimators	Number of classifiers	[50, 1000] step = 50
	learning_rate	Learning rate	[0, 1] step = 0.1
	max_depth	Maximum depth of the tree	[2, 10] step = 1

Table 7. Results of optimal hyperparameters.

Method	Model	Hyperparameter	Value	*ACC*
No Processing	RF	n_estimators max_features	300 5	81.26%
	AdaBoost	n_estimators learning_rate max_depth	200 0.3 5	82.50%
	XGBoost	n_estimators learning_rate max_depth	850 0.4 8	83.34%
Stratified Sampling	RF	n_estimators max_features	200 5	80.84%
	AdaBoost	n_estimators learning_rate max_depth	500 0.3 6	81.89%
	XGBoost	n_estimators learning_rate max_depth	800 0.4 5	**84.25%**

By comparing the results, we have found that the highest *ACC* of 84.25% is obtained after adjusting the three hyperparameters of *'n_estimators'*, *'learning_rate'*, *and 'max_depth'* under the method of stratified sampling to divide the dataset. *The 'n_estimators'* is used to control the number of trees. The *'learning_rate'* and *'max_depth'* adjustments have a large fluctuation effect on the *ACC*. The *'learning_rate'* controls the weight reduction factor of each weak learner. The smaller the value is, the faster the model converges to the optimal solution, but it may also increase the risk of overfitting. The *'max_depth'* is used to control the maximum depth of the tree; a larger value allows the model to predict the results more accurately, but it also leads to a longer training time. Therefore, Figure 5 shows the effect of the variation of these two hyperparameters on the *ACC* when *'n_estimators'* is taken as 800, with the horizontal and vertical coordinates of *'learning_rate'* and *'max_depth'*, and the vertical coordinates indicating the *ACC* after tenfold cross-validation on the test set, from which it can be clearly seen that most of the *ACC* is distributed around 0.82 to 0.83. The highest *ACC* of 84.25% is obtained when the values of *'learning_rate'* and *'max_depth'* are 0.4 and 5, respectively.

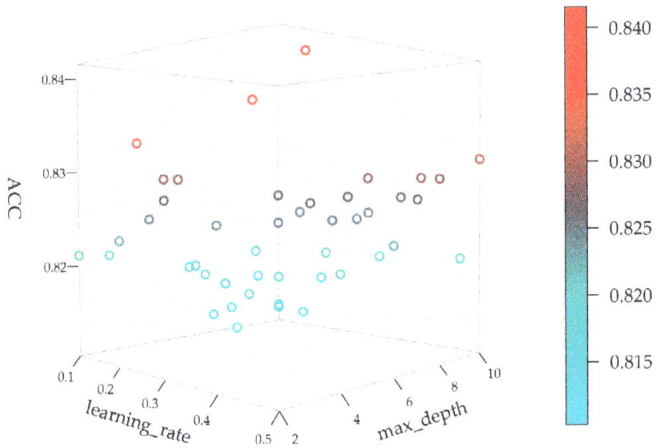

Figure 5. ACC of different hyperparameters.

The remaining hyperparameters are default values, mainly *'Alpha'* is the L1 regularization parameter in the range [0, +∞) with a default of 0. The larger the parameter, the less likely it is to overfit. The '*Lambda*' is used to control the regularization part of XGBoost, i.e., it is used to control the complexity of the model, with a default of 0, which means that the default regularization term is used. When the value of the '*Lambda*' parameter is greater than 0, it means that a stricter regularization term is used, which can reduce the risk of overfitting. The '*Gamma*' is used to control the splitting strategy of the tree, i.e., the splitting operation is initiated when the loss function of a node drops below a set threshold. When its value is taken as large values, the model will be more conservative and not split easily, which leads to more stable leaf nodes of the tree. When its value is small, the model will be more aggressive and tends to split the child nodes earlier to obtain higher gains with the highest *ACC* when it is taken as 0.

We introduced in Section 2 that the superior performance of the XGBoost model may be more appropriate for this study, and the results are consistent with the expectations. The RF model is integrated by the averaging of votes and uses random subsets of features and the randomness of decision trees to improve model diversity, compared to the XGBoost model which uses more complex weak classifiers (e.g., decision trees with variable depths) and provides a better fit and prediction performance through effective techniques such as gradient calculations and feature subsampling to update the model parameters more accurately in each iteration. Thus, it improves the model performance providing a better fitting ability and predictive performance. The AdaBoost model uses loss functions, but focuses only on misclassified samples in each iteration, while XGBoost updates the model parameters more accurately in each iteration. The XGBoost model is designed with parallel computing and efficiency optimization in mind to improve the training speed and efficiency, which may cause the AdaBoost model to perform less well than the XGBoost model in some cases.

Figure 6 presents the effect of the number of iterations of the XGBoost model on the loss value of the training set for the optimal hyperparameters found above. '*logloss*' refers to the loss value of the training set calculated using log loss as an evaluation metric during the training process. The training process of the XGBoost model aims to minimize the '*Train_logloss*' in order to improve the model's ability to fit the training set. When the '*Train_logloss*' is smaller, it indicates that the model's predictions on the training set fit the true labels better. However, in order to avoid overfitting, it is also necessary to pay attention to the performance of the model on the test set. It is clear from the Figure 6 that the decreasing trend of the model logloss value slows down when the number of iterations is greater than 200, and finally, when the number of iterations reaches 800, the value of '*Train_logloss*' is about 0.0039 and the value of '*Test_logloss*' is about 0.0550, which reaches a low level and continues to decrease, indicating that the model has reached some kind of steady state and there is no overfitting. We have also searched through the grid to obtain the highest *ACC* on the test set at a number of iterations of 800.

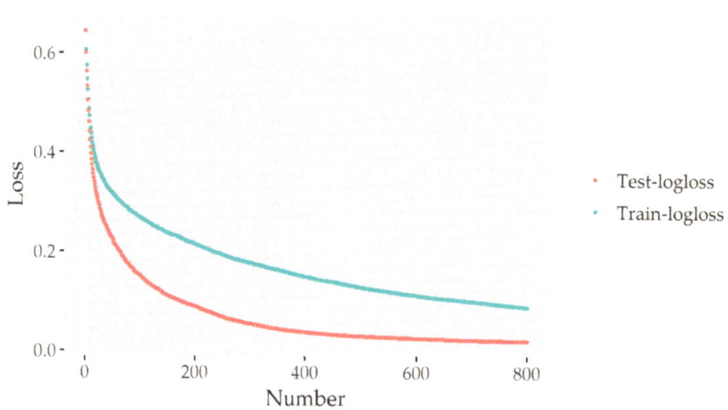

Figure 6. Variation of log loss with the number of iterations.

In Table 8, we compare the changes in the relevant metrics before and after hyperparameter tuning, and find that the *ACC*, the false positive rate of both types of samples, and the *AUC* value are all optimized to some extent. Specifically, our *ACC* significantly improved from 79.99% to 84.25%, and the false positive rate also decreased, while its *AUC* value improved from 0.8781 to 0.9019. These results further indicate that the model quality is significantly improved after hyperparameter adjustment.

Table 8. Changes in indicators before and after adjustment.

Indicators	Before Adjustment	After Adjustment
ACC	79.99%	84.25%
$Error_{VLSFO}$	14.39%	11.66%
$Error_{Scrubber}$	28.83%	21.34%
AUC	0.8781	0.9019

5.3. The Importance of Output Variables

We calculated the important share ranking of each feature in the decision making of the ship abatement scheme by the XGBoost model based on stratified sampling, and the results show that the DWT, Distance, and Power are the three most important features in the decision-making process, and their importance shares are 0.2757, 0.2461, and 0.1277, respectively. In contrast, Age, BDI, and PD have less influence on decision making, and their importance ratios are relatively small. The ranking of the importance of these features can provide a direct reference basis for shipowners' decision making and help them to choose the optimal emission abatement solution more accurately. The results of the feature importance ranking are shown in Figure 7.

Figure 7. Ranking the importance of each feature.

DWT and Distance are two important features of bulk carriers, accounting for more than half of the total. The descriptive statistical analysis shows that these two features have a strong differentiation. On large bulk carriers, they are more inclined to use scrubbers. Therefore, when shipowners need to make a decision, they can first consider the feature of deadweight tonnage. In addition, travel distance is another very important feature, and there is a large difference in the travel distance between the two different emission abatement solutions. Bulk carriers using the scrubber solution tend to travel longer distances, which is also related to the vessel's travel cost. Referring to the results of Bai et al. [34], the larger the ship tonnage is and the longer the distance traveled, the more likely the scrubber solution will be used. Therefore, when considering emission abatement solutions, it is recommended to give priority to these two features. Shipowners can refer to the importance of each feature when making decisions, so as to choose a more appropriate decision solution.

5.4. Certification of Newbuilding Bulk Carriers

To verify the practical applicability of the model on the newest ships, we again collected 51 ships that were built and commissioned in January 2024 and introduced their characteristics into the model to make the predictions. New ships built at this time have relatively smooth characteristics and are also representative of the new trend.

Due to the time delay of data collection, it was not possible to obtain an accurate annual distance traveled by the ship, so the distance traveled in the last 30 days was used for the estimation, and the data were collected in the year 2024 in May.

The 51 sample vessels were brought into the optimal model in the previous section for prediction, and their *ACC*, *Error Rate*, and *AUC* values were obtained as 86.27%, 15.79%, 15.38%, and 0.9017, respectively. The prediction results might have been affected by the errors in their sailing distances and speeds. However, overall, this can indicate a better prediction for new bulk carriers. Some of the prediction samples are shown in Table 9. In addition, the input characteristics of these samples are shown in Appendix A.

Table 9. Forecast results for newbuilding bulk carriers.

Name	Solution	Prediction	Result
Cape Pleasure	VLSFO	VLSFO	True
Mount Bandeira	Scrubber	Scrubber	True
Mount Heng	VLSFO	VLSFO	True
SG Ocean	VLSFO	Scrubber	False
Mineral Luxembourg	VLSFO	VLSFO	True
Green Rio Grande	VLSFO	VLSFO	True
Kerynia	VLSFO	Scrubber	False
Duchess Emerald	Scrubber	Scrubber	True
Star Voyager	Scrubber	Scrubber	True
Shine Jade	VLSFO	Scrubber	False
Glbs Hero	VLSFO	VLSFO	True
NS Ningbo	VLSFO	VLSFO	True
Northern Venture	Scrubber	Scrubber	True
Brighton	VLSFO	VLSFO	True
Emerald Liuheng	Scrubber	Scrubber	True
Chang Hang Wei hai	VLSFO	VLSFO	True
BBC Ceres	VLSFO	VLSFO	True
Seacon Bangkok	VLSFO	VLSFO	True
Amalia	VLSFO	VLSFO	True
Polsteam Dabie	VLSFO	VLSFO	True

6. Conclusions

Taking the selection of newbuilding emission abatement options as a binary classification problem, this study uses four years of newbuilding bulk carrier data for preprocessing and constructs a classification model based on three methods of balancing samples and six machine learning models to bring into the test set for prediction, and adjusts the hyperparameters of the model by ten-fold cross-validation. The results show that the XGBoost model based on stratified sampling to divide the dataset has the highest *ACC* of 84.25% with an *AUC* value of 0.9019, which is also consistent with our theoretical rationale for choosing the XGBoost model. Forecasts for newbuilding bulk carriers in 2024 have also been more successful. In addition, this study outputs the importance level of each feature to be considered for decision making. We found that the deadweight tonnage of the ship and the sailing distance can be considered as priority factors. Combined with the results of descriptive statistics, we found that the larger the tonnage of the ship is and the longer the annual distance sailing is, the more inclined the shipowner is to install scrubbers. The above results are a guide for shipowners when making decisions on emission abatement solutions. Shipowners can focus on the deadweight tonnage of the ship and the future route to be sailed. The XGBoost model can be constructed by calculating the approximate sailing distance and ship speed, combining the current fuel price, BDI, and other factors, and bringing in the expected feature variables for prediction, in order to come up with a suitable solution. At the same time, it is possible to flexibly adapt these features, especially those related to the ship itself, for comparison between solutions.

This paper has the following shortcomings, which can be further improved in future work: 1. VLSFO and scrubbers are the most important emission abatement solutions for bulk carriers at present. In the future, as the proportion of alternative fuel ships is expanding, shipowners may face new choices. Therefore, when the sample of alternative fuel is increased, it will be more reasonable to build a decision model for shipowners. Furthermore, the data used are from three years after the implementation of the "sulfur cap" policy. 2. Consider adding other relevant economic and non-economic factors, such as national policies, the impact of other major pollutants, antitrust exemptions, and shipping alliances [40], to consider the factors of shipowner's choice of emission abatement solutions from a more comprehensive perspective, and also whether the COVID-19 outbreak had a certain impact on carbon emission reduction in shipping [41] and on sulfur emission abatement. How will the factors affecting shipowners' decision making change in the post-epidemic era? 3. Applying the ideas of this study to other types of ships and comparing the differences in the models and results between different types of ships will better demonstrate the models and prediction methods adapted to different types of ships. 4. Explore and apply emerging machine learning models to improve the accuracy and efficiency of predictive models, making research more meaningful and providing stronger support for ship owners to make the right emission abatement solutions.

Author Contributions: Conceptualization, S.H.; methodology, Y.L.; software, Y.L.; writing—original draft preparation, Y.L.; writing—review and editing, Y.L.; supervision, S.H.; project administration, S.H.; funding acquisition, S.H. All authors have read and agreed to the published version of the manuscript.

Funding: This research was funded by National Natural Science Foundation of China, grant number 52002243.

Institutional Review Board Statement: Not applicable.

Informed Consent Statement: Not applicable.

Data Availability Statement: The data presented in this study are available on request from the corresponding author.

Conflicts of Interest: The authors declare no conflicts of interest.

Appendix A

The input characteristics regarding the January 2024 newbuilding bulk carriers in 5.4 are shown in the table below.

Table A1. Characteristics of newbuilding bulk carriers.

Name	Age	DWT	Power	Speed	Distance	PD	BDI
Cape Pleasure	0.42	182,096	13,360	11.3	93,624	296	1835
Mount Bandeira	0.42	208,479	15,840	11.4	88,104	177	2932
Mount Heng	0.42	208,954	15,840	11.7	57,816	173	3188
SG Ocean	0.42	210,933	13,900	12.2	72,288	415	2220
Mineral Luxembourg	0.42	210,197	14,400	11.3	75,372	148	3720
Green Rio Grande	0.42	77,000	10,600	13.1	22,020	296	1835
Kerynia	0.42	82,114	7760	10.9	90,600	256	1761
Duchess Emerald	0.42	82,464	9000	11.5	96,480	296	1835
Star Voyager	0.42	82,459	9000	11.7	57,648	296	1835
Shine Jade	0.42	82,377	9000	12.1	85,932	424	2464

Table A1. *Cont.*

Name	Age	DWT	Power	Speed	Distance	PD	BDI
Glbs Hero	0.42	63,742	6670	11.1	75,768	657	1814
NS Ningbo	0.42	64,129	6949	11.9	50,304	192	4820
Northern Venture	0.42	64,636	7100	11.7	97,800	177	2932
Brighton	0.42	64,701	7100	12.7	44,580	136	4288
Emerald Liuheng	0.42	63,875	6949	12.8	44,232	713	2389
Chang Hang Wei hai	0.42	36,469	4200	9.8	24,792	230	2780
BBC Ceres	0.42	40,550	6480	11.5	59,852	296	1835
Seacon Bangkok	0.42	40,540	5150	12	35,525	576	1487
Amalia	0.42	14,309	2999	9.9	27,840	415	2220
Polsteam Dabie	0.42	37,592	8900	11.6	32,587	256	1761

References

1. Kavussanos, M.; Tsouknidis, D.; Visvikis, I. *Freight Derivatives and Risk Management in Shipping*; Routledge: Oxford, UK, 2021.
2. Akgul, E.; Cetin, I. An investigation on behavioral biases in ship investments of small-sized shipping companies. *Marit. Bus. Rev.* **2021**, *6*, 377–391. [CrossRef]
3. Poulsen, R.; Viktorelius, M.; Varvne, H.; Rasmussen, H.; von Knorring, H. Energy efficiency in ship operations-exploring voyage decisions and decision-makers. *Transp. Res. Part D Transp. Environ.* **2022**, *102*, 103120. [CrossRef]
4. Xiao, G.; Xiao, Y.; Shu, Y.; Ni, A.; Jiang, Z. Technical and economic analysis of battery electric buses with different charging rates. *Transp. Res. Part D Transp. Environ.* **2024**, *132*, 104254. [CrossRef]
5. Winebrake, J.; Corbett, J.; Green, E.; Lauer, A.; Eyring, V. Mitigating the health impacts of pollution from oceangoing shipping: An assessment of low-sulfur fuel mandates. *Environ. Sci. Technol.* **2009**, *43*, 4776–4782. [CrossRef]
6. Lindstad, H.; Rehn, C.; Eskeland, G. Sulphur abatement globally in maritime shipping. *Transp. Res. Part D Transp. Environ.* **2017**, *57*, 303–313. [CrossRef]
7. Fan, L.; Gu, B. Impacts of the increasingly strict sulfur limit on compliance option choices: The case study of Chinese SECA. *Sustainability* **2019**, *12*, 165. [CrossRef]
8. Wu, P.; Lin, C. Cost-benefit evaluation on promising strategies in compliance with low sulfur policy of IMO. *J. Mar. Sci. Eng.* **2020**, *9*, 3. [CrossRef]
9. Karatuğ, Ç.; Arslanoğlu, Y.; Guedes Soares, C. Feasibility analysis of the effects of scrubber installation on ships. *J. Mar. Sci. Eng.* **2022**, *10*, 1838. [CrossRef]
10. Iannaccone, T.; Landucci, G.; Scarponi, G.E.; Bonvicini, S.; Cozzani, V. Inherent safety assessment of alternative technologies for LNG ships bunkering. *Ocean Eng.* **2019**, *185*, 100–114. [CrossRef]
11. Xie, C.; Huang, L.; Wang, R.; Deng, J.; Shu, Y.; Jiang, D. Research on quantitative risk assessment of fuel leak of LNG-fuelled ship during lock transition process. *Reliab. Eng. Syst. Saf.* **2022**, *221*, 108368. [CrossRef]
12. Qi, J.; Wang, S. LNG bunkering station deployment problem—A case study of a Chinese container shipping network. *Mathematics* **2023**, *11*, 813. [CrossRef]
13. Qi, J.; Wang, H.; Zheng, J. Promoting liquefied natural gas (LNG) bunkering for maritime transportation: Should ports or ships be subsidized? *Sustainability* **2022**, *14*, 6647. [CrossRef]
14. Kim, K.; Lim, S.; Lee, C.; Lee, W.; Jeon, H.; Jung, J.; Jung, D. Forecasting liquefied natural gas bunker prices using artificial neural network for procurement management. *J. Mar. Sci. Eng.* **2022**, *10*, 1814. [CrossRef]
15. Chen, X.; Lv, S.; Shang, W.; Wu, H.; Xian, J.; Song, C. Ship energy consumption analysis and carbon emission exploitation via spatial-temporal maritime data. *Appl. Energy* **2024**, *360*, 122886. [CrossRef]
16. Jordan, M.I.; Mitchell, T.M. Machine learning: Trends, perspectives, and prospects. *Science* **2015**, *349*, 255–260. [CrossRef] [PubMed]
17. Kim, M.; Kim, J.; Yang, H. Optimal route generation and route-following control for autonomous vessel. *J. Mar. Sci. Eng.* **2023**, *11*, 970. [CrossRef]
18. Chen, X.; Liu, S.; Zhao, J.; Wu, H.; Xian, J.; Montewka, J. Autonomous port management based AGV path planning and optimization via an ensemble reinforcement learning framework. *Ocean Coast. Manag.* **2024**, *251*, 107087. [CrossRef]
19. Xiao, G.; Yang, D.; Xu, L.; Li, J.; Jiang, Z. The application of artificial intelligence technology in shipping: A bibliometric review. *J. Mar. Sci. Eng.* **2024**, *12*, 624. [CrossRef]
20. Chen, T.; Guestrin, C. Xgboost: A scalable tree boosting system. In Proceedings of the 22nd ACM SIGKDD International Conference on Knowledge Discovery and Data Mining, San Francisco, CA, USA, 13–17 August 2016; pp. 785–794.
21. Zhang, D.; Chen, H.; Zulfiqar, H.; Yuan, S.; Huang, Q.; Zhang, Z.; Deng, K. iBLP: An XGBoost-Based Predictor for Identifying Bioluminescent Proteins. *Comput. Math. Methods Med.* **2021**, *2021*, 6664362. [CrossRef]
22. Ma, M.; Zhao, G.; He, B.; Li, Q.; Dong, H.; Wang, S.; Wang, Z. XGBoost-based method for flash flood risk assessment. *J. Hydrol.* **2021**, *598*, 126382. [CrossRef]

23. Lu, J.; Wu, X.; Wu, Y. The construction and application of dual-objective optimal speed model of liners in a changing climate: Taking Yang Ming route as an example. *J. Mar. Sci. Eng.* **2023**, *11*, 157. [CrossRef]
24. Lindstad, H.; Sandaas, I.; Strømman, A. Assessment of cost as a function of abatement options in maritime emission control areas. *Transp. Res. Part D Transp. Environ.* **2015**, *38*, 41–48. [CrossRef]
25. Zis, T.; Angeloudis, P.; Bell, M.; Psaraftis, H. Payback period for emissions abatement alternatives: Role of regulation and fuel prices. *Transp. Res. Rec.* **2016**, *2549*, 37–44. [CrossRef]
26. Fan, L.; Gu, B.; Luo, M. A cost-benefit analysis of fuel-switching vs. hybrid scrubber installation: A container route through the Chinese SECA case. *Transp. Policy* **2020**, *99*, 336–344. [CrossRef]
27. Zis, T.; Cullinane, K. The desulphurisation of shipping: Past, present and the future under a global cap. *Transp. Res. Part D Transp. Environ.* **2020**, *82*, 102316. [CrossRef]
28. Zis, T.; Cullinane, K.; Ricci, S. Economic and environmental impacts of scrubbers investments in shipping: A multi-sectoral analysis. *Marit. Policy Manag.* **2021**, *49*, 1097–1115. [CrossRef]
29. Zhao, Y.; Ye, J.; Zhou, J. Container fleet renewal considering multiple sulfur reduction technologies and uncertain markets amidst COVID-19. *J. Clean. Prod.* **2021**, *317*, 128361. [CrossRef]
30. Lagemann, B.; Lindstad, E.; Fagerholt, K.; Rialland, A.; Ove Erikstad, S. Optimal ship lifetime fuel and power system selection. *Transp. Res. Part D Transp. Environ.* **2022**, *102*, 103145. [CrossRef]
31. Li, K.; Wu, M.; Gu, X.; Yuen, K.; Xiao, Y. Determinants of ship operators' options for compliance with IMO 2020. *Transp. Res. Part D Transp. Environ.* **2020**, *86*, 102459. [CrossRef]
32. Bao, Z.; Zhang, X.; Fu, G. Factors influencing decision to sulphur oxide emission abatement for cruise shipping companies. *Int. J. Logist. Res. Appl.* **2022**, 1–20. [CrossRef]
33. Zhang, X.; Bao, Z.; Ge, Y. Investigating the determinants of shipowners' emission abatement solutions for newbuilding vessels. *Transp. Res. Part D Transp. Environ.* **2021**, *99*, 102987. [CrossRef]
34. Bai, X.; Hou, Y.; Yang, D. Choose clean energy or green technology? Empirical evidence from global ships. *Transp. Res. Part E Logist. Transp. Rev.* **2021**, *151*, 102364. [CrossRef]
35. Lee, H.T.; Lee, J.S.; Son, W.J.; Cho, I.S. Development of machine learning strategy for predicting the risk range of ship's berthing velocity. *J. Mar. Sci. Eng.* **2020**, *8*, 376. [CrossRef]
36. Li, Y.; Lai, X.; Wang, M.; Zhang, X. C-saso: A clustering-based size-adaptive safer oversampling technique for imbalanced sar ship classification. *IEEE Trans. Geosci. Remote Sens.* **2022**, *60*, 5231112. [CrossRef]
37. Bassam, A.M.; Phillips, A.B.; Turnock, S.R.; Wilson, P.A. Ship speed prediction based on machine learning for efficient shipping operation. *Ocean Eng.* **2022**, *245*, 110449. [CrossRef]
38. Wu, S.; Chen, X.; Shi, C.; Fu, J.; Yan, Y.; Wang, S. Ship detention prediction via feature selection scheme and support vector machine (SVM). *Marit. Policy Manag.* **2022**, *49*, 140–153. [CrossRef]
39. Kim, J.; Lee, H.; Oh, J. Study on prediction of ship's power using light GBM and XGBoost. *J. Korean Soc. Mar. Eng.* **2020**, *44*, 174–180. [CrossRef]
40. Xiao, G.; Wang, T.; Shang, W.; Shu, Y.; Biancardo, S.A.; Jiang, Z. Exploring the factors affecting the performance of shipping companies based on a panel data model: A perspective of antitrust exemption and shipping alliances. *Ocean Coast. Manag.* **2024**, *253*, 107162. [CrossRef]
41. Xu, L.; Yang, Z.; Chen, J.; Zou, Z. Impacts of the COVID-19 epidemic on carbon emissions from international shipping. *Mar. Pollut. Bull.* **2023**, *189*, 114730. [CrossRef]

Journal of
Marine Science and Engineering

MDPI

Article

Ship Trajectory Planning and Optimization via Ensemble Hybrid A* and Multi-Target Point Artificial Potential Field Model

Yanguo Huang, Sishuo Zhao and Shuling Zhao *

School of Electrical Engineering and Automation, Jiangxi University of Science and Technology, Ganzhou 341000, China; yghuang@jxust.edu.cn (Y.H.); 6720221359@mail.jxust.edu.cn (S.Z.)
* Correspondence: 9120010008@jxust.edu.cn

Abstract: Ship path planning is the core problem of autonomous driving of smart ships and the basis for avoiding obstacles and other ships reasonably. To achieve this goal, this study improved the traditional A* algorithm to propose a new method for ship collision avoidance path planning by combining the multi-target point artificial potential field algorithm (MPAPF). The global planning path was smoothed and segmented into multi-target sequence points with the help of an improved A* algorithm and fewer turning nodes. The improved APF algorithm was used to plan the path of multi-target points locally, and the ship motion constraints were considered to generate a path that was more in line with the ship kinematics. In addition, this method also considered the collision avoidance situation when ships meet, carried out collision avoidance operations according to the International Regulations for Preventing Collisions at Sea (COLREGs), and introduced the collision risk index (CRI) to evaluate the collision risk and obtain a safe and reliable path. Through the simulation of a static environment and ship encounter, the experimental results show that the proposed method not only has good performance in a static environment but can also generate a safe path to avoid collision in more complex encounter scenarios.

Keywords: ship trajectory planning; improved A* model; multi-target point artificial potential field model; smart shipping; maritime traffic safety

1. Introduction

Big data and artificial intelligence have driven the development of smart autonomous ships, which are expected to dominate future maritime transportation. The safety of ship navigation has become the primary concern. Ship navigation accidents can result in significant loss of life and property of both the ship and crew [1]. According to a survey, over 75% of accidents are caused by human decision-making, and the use of autonomous navigation on intelligent ships has partially alleviated losses. One of the primary research directions is to avoid ship collisions, which is one of the urgent problems for intelligent ships. There are three steps in the process of ship collision avoidance (CA): obstacle detection, CA decision-making, and CA action. The detection of obstacles by radar, AIS, and other navigation aids is crucial for ships to make collision avoidance decisions [2]. International regulations for preventing collisions at sea (COLREGs) contain precise advice on how to avoid ship collisions in various encounter scenarios. It is a challenge for ships to avoid collision when considering COLREGs and ship kinematics, particularly in complex encounter scenes.

Many scholars have proposed a variety of path planning algorithms for avoiding ship collisions [3,4]. The effects of these methods vary depending on the driving environment. Meta-heuristic optimization algorithms, such as genetic algorithms, have been proposed by various studies to optimize the shortest time or shortest distance path [5]. Machmudah, A. improved path efficiency by optimizing the UAV turning mechanism; this provided a great reference for ship heading optimization [6]. The simulated annealing algorithm and the multi-objective optimization algorithm have also been applied to ship path planning [7,8].

Graph search is a dynamic path-planning method that is based on environment modeling. This method plans the path by rasterizing the navigation environment image and combining it with the corresponding algorithm. The Dijkstra algorithm and A* algorithm are widely used. The Dijkstra algorithm [9] combines the concepts of breadth search algorithms (BFS) and greedy strategies to calculate the lowest cost path from the source node to the destination during operation. Padhy used the Dijkstra algorithm to solve the route optimization problem and achieved success; the application of this algorithm in route optimization has been the main focus of researchers since then [10]. The A* algorithm [11–13] combined with the heuristic function of the Dijkstra algorithm results in the optimal path by calculating the cost of the route to reach the target. Lan proposed a ship path-planning algorithm fusing the A-star and ant colony algorithms, which solved the problems of slow convergence speeds and long search times [14,15]. Singh et al. proposed a new A* algorithm, which takes into collision risk and driving rules, to find an optimal and safest path [16]. He et al. proposed a collision avoidance algorithm that is dynamic and can still be effective in complex multi-ship encounter scenes [17]. Donggyun Kim proposed a Samsung Autonomous Ship (SAS) collision avoidance system that can visualize the best route searched and reflect the relevant parameters of consumers in real time [18]. Hyeong-Tak Lee et al. proposed a new method to obtain the shortest navigation distance in the navigable area by taking automatic identification system data into consideration [19–21]. One of its disadvantages is that it perhaps lapses into local minimization so that it cannot obtain the correct results. In addition, there are heuristic algorithms for swarm intelligence, including the ant colony algorithm, particle swarm algorithm (PSO), and grey wolf algorithm. The optimal path is achieved by genetic algorithms by solving the optimal solution using the phenomenon of chromosome crossover and mutation. PSO simulates the foraging behavior of birds in nature and obtains the optimal result through the evolution process of the whole particle swarm from disorder to order in the solving space [22,23]. The PSO algorithm has fast convergence and simple convergence. Despite the application of these methods in ship route planning, there have always been problems with search speed and local optimization. Therefore, various algorithms have been improved and combinations of different algorithms have been developed to solve the defects of a single algorithm.

The artificial potential field algorithm (APF) is a popular algorithm in path planning because of its small amount of calculation, ease of understanding, and flexibility. Lyu and Yin proposed a vector environment information model to obtain a safe shortest path by using the repulsion from different obstacles during navigation [24]. However, this method does not solve the defect whereby the traditional APF method falls into the local optimum. Lazarowska proposed the discrete APF to optimize a ship's risk-free trajectory, which greatly improves the safety of ship navigation [25]. Farhad proposed a route optimization method based on the electrostatic potential field (EPF), which uses the high potential value generated by the electrostatic field to prevent the vehicle from falling into the local optimum [26]. Therefore, some scholars have proposed a hybrid algorithm combining APF with other algorithms [27], fully utilizing the advantages and disadvantages of different algorithms. Artificial potential field theory was introduced to collision avoidance systems by Mousazadeh and others, resulting in a solution to the optimization problem concerning collision prevention [28]. There has been rapid development of random trees (RRT) [29–31] and deep reinforcement learning algorithms [32–35]. They are also used to obtain better ship trajectories.

In previous studies, rasterization or binarization thresholds are usually used to process electronic charts and environmental images and then intelligent algorithms are used to plan paths on the unitized map. In order to obtain the best results, the previous methods analyze the optimized results repeatedly, which makes it easy to fall into the local cycle and increases the calculation time cost.

In view of the above shortcomings, this paper makes the following contributions:

(1) In this paper, a new A*path planning algorithm is proposed, which considers the constraints of ship motion, improves the traditional A*algorithm, eliminates redundant nodes, and smoothes the path.
(2) Considering the phenomenon that the APF algorithm can easily fall into local loops, an artificial potential field multi-target points (MPAPF) approach is proposed. The global path derived by the A* algorithm is divided into multi-target points, and MPAPF is used to optimize each target point locally to obtain an optimal path.
(3) In route planning, the collision risk factor (CRI) is introduced, and, under the rules proposed by the International Maritime Organization (IMO), the CRI value of each segment is continuously calculated to ensure that a safe, economical, and reliable route is found.

The structure of the rest of this paper is as follows: In the second part, the ship motion planning model is established and ship motion planning is considered. Then, in Section 2.2, the ship path-planning algorithm is introduced, namely, the improved A* algorithm and the multi-objective artificial market algorithm. In the third part, the simulation experiments of ship collision avoidance path planning are carried out in two scenarios of static sea area and ship encounter and compared with the traditional A* algorithm, and the effectiveness and reliability of the method are verified. Finally, the work conducted in this study is summarized and prospects are discussed.

2. Methods

2.1. Ship Motion Model

The motion model of a ship is a very complex, nonlinear problem. Due to the diverse influence of hydrodynamic forces and environments, a ship's power system is considerably different from conventional linear systems like vehicles and ground robots. It is not possible to disregard the motion characteristics of ships for collision avoidance. We established a ship motion model as shown in Figure 1, with O-XYZ as the inertial coordinate system and o-xyz as the appendage coordinate system (ship motion coordinate system). When describing the motion of ships on water, a six-degree-of-freedom motion is commonly used to describe it, which includes sway, surge, heave, pitch, roll, and yaw. The efficiency of ship trajectory planning was considered in this study by considering three degrees of freedom motions: pitch, roll, and yaw. The definition symbols are shown in Table 1.

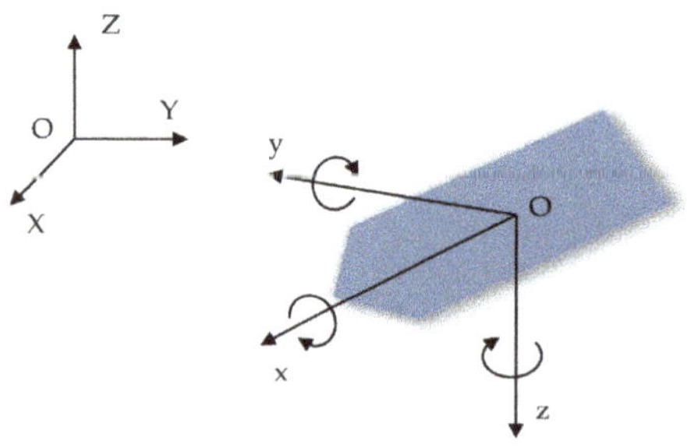

Figure 1. Coordinate of ship motion.

Table 1. Symbol definitions.

Freedom	Force and Momentum	Linearity/Angular Velocity	Position/Attitude Angle
Surge	X	u	x
Sway	Y	v	y
Yaw	N	r	ψ

In this study, the unified hydrodynamic model was used to establish the motion model of the vehicle, which could be expressed by Equations (1) and (2):

$$\dot{\eta} = \mathrm{J}(\psi) \cdot v \tag{1}$$

$$M\dot{v} + C(v)v + D(v)v = \tau + \tau_e \tag{2}$$

where $v = [u, v, r]^T$ represents the vector matrix of heave, pitch, and roll motions of the ship and $\eta = [x, y, \psi]^T$ is the position (x, y) and attitude matrix of the ship. The driving force matrix $\tau = [\tau_u, \tau_v, \tau_r]^T$ and external disturbance $\tau_e = [\tau_{eu}, \tau_{ev}, \tau_{er}]^T$. $\mathrm{J}(\psi)$ is the transformation matrix between the inertial coordinate system and the body coordinate system, M is the inertia matrix, C is a skew-symmetric matrix, and D is a linear damping matrix, and its parameters are defined as in Equations (3)–(6):

$$\mathrm{J}(\psi) = \begin{bmatrix} \cos\psi & -\sin\psi & 0 \\ \sin\psi & \cos\psi & 0 \\ 0 & 0 & 1 \end{bmatrix} \tag{3}$$

$$M = \begin{bmatrix} m - X_{\dot{u}} & 0 & 0 \\ 0 & m - Y_{\dot{v}} & mX_g - Y_{\dot{r}} \\ 0 & mX_g - Y_{\dot{r}} & I_z - N_{\dot{r}} \end{bmatrix} \tag{4}$$

$$C = \begin{bmatrix} 0 & 0 & -(m - Y_{\dot{v}})v - (mX_g - Y_{\dot{r}})r \\ 0 & 0 & (m - X_{\dot{u}})u \\ (m - Y_{\dot{v}})v + (mX_g - Y_{\dot{r}})r & -(m - X_{\dot{u}})u & 0 \end{bmatrix} \tag{5}$$

$$D = \begin{bmatrix} X_u & 0 & 0 \\ 0 & -Y_v & -Y_r \\ 0 & -N_V & -N_r \end{bmatrix} \tag{6}$$

In addition, considering the flexibility of ship movement and navigation efficiency, certain constraints were imposed on the ship's maneuverability, namely, Equation (7):

$$u \le u_{\max}, |r| \le r_{\max} \tag{7}$$

where $u_{\max}$ represents the maximum surge velocity and $r_{\max}$ represents the maximum yaw speed of the ship.

2.2. Ship Path Planning Algorithm

The path-planning algorithm proposed in this paper is mainly composed of two parts: using the improved A* algorithm to plan the global path and generate an optimal path. To plan the local path, the multi-target point artificial potential field method (MPAPF) was utilized and the maneuvering motion model was considered. Finally, a safe and efficient path was obtained.

2.2.1. Improved A* Global Path Optimization Algorithm

As a classic and popular heuristic search algorithm, the A* algorithm is widely used to find the shortest path in a unit graph. Before the A* algorithm is implemented, the ENC is converted into a bitmap of cells. The attribute of each cell determines whether there are obstacles, with a = 0 indicating that there are no obstacles and a = 1 indicating that there are obstacles. When the starting and ending points are identified, the A* algorithm will locate the most efficient route by utilizing the heuristic method of finding adjacent zero-value cells and Euclidean distance. Specifically, A* algorithm chooses the path that minimizes these by calculating the cost value F, as shown in Equation (8):

$$F(n) = g(n) + h(n) \tag{8}$$

where $F(n)$ is the cost value of the n th cell, $g(n)$ represents the path length from the starting point to the n-th cell, and $h(n) = \sqrt{(n_i - p_i)^2 + (n_j - p_j)^2}$ is the heuristic distance from the n-th cell to the destination point p.

Although the traditional A* algorithm can generate the shortest global path, it does not guarantee vehicle safety or take into account turning costs and path redundancies. Therefore, an improvement was made.

(a) Adding the turning cost ρ to the cost function could reduce the number of turns of the ship, thereby improving the efficiency and safety of the ship, as shown in Equation (9):

$$F(n) = g(n) + h(n) + \rho \tag{9}$$

(b) By constraining the search distance and path length, the search range could be narrowed and the search time could be reduced, as shown in Equation (10):

$$\begin{cases} F(n) \leq \lambda \cdot l_{\max} \\ h(n) \leq k \cdot d_{\max} \end{cases} \tag{10}$$

where λ and k are scale factors.

2.2.2. Hybrid Improved A* and Multi-Target Point Artificial Potential Field Algorithm

The APF has the characteristics of strong computational power, ease of understanding, and strong robustness in solving the local path-planning problem of ships. This paper proposes a multi-target point artificial potential field algorithm (MPAPF) that is combined with the improved A* algorithm to deal with the phenomenon of A* algorithm's tendency to fall into the regional air circulation. In addition, the A* algorithm can address the problem of the APF algorithm's weak global search capability.

When the traditional APF carries out path planning, when the target point and the obstacle are in a straight line, the repulsion force is greater than the target gravity, which will cause the ship to be unable to reach the target point through the obstacle. When the ship sails to a certain area, when the repulsion and gravity are equal, the resultant force is zero, which may cause the algorithm to fall into the local optimal value and cause the ship to stop traveling.

Considering the shortcomings of traditional APF algorithms, the following improvements were made:

(a) The traditional vector calculation method was modified to the potential energy calculation method and the effect boundary value R of the gravitational field function was set. When the relative distance between the ship and the target position $G \leq R$, the value of the gravitational field was proportional to the k-th power of G, when $G > R$, the value of the gravitational field was proportional to G, thus reducing the influence range of the gravitational potential field [36]. The specific equation is shown in Equations (11)–(13):

$$\mathrm{U}_t = \mathrm{U}_a - \mathrm{U}_r \tag{11}$$

$$\mathrm{U}_a = \begin{cases} \frac{1}{2}\alpha \cdot G^k, G \leq R \\ R \cdot \alpha \cdot G, G > R \end{cases} \tag{12}$$

$$\mathrm{U}_r = r \cdot \left(\frac{1}{O}\right)^k \tag{13}$$

where U_a is the gravitational field, U_r is the repulsive field, and U_t is the total field. G is the distance from the current point to the goal point, O is the distance from the current position to the obstacle, and α and r are the proportional factors of the gravitational potential field and the repulsive potential field, respectively, both of which are greater than zero, and $k = 2.5$ is the index of the calculated field.

(b) The route achieved by the A* algorithm was divided into multi-target points, which were sub-target points of the APF algorithm, and the ship's motion model was taken into account. The ship's speed was nearly zero when it fell into a local minimum. At this time, the sub-target point sequence found the next point as the current target point to continue to run; this ensured that when the program fell into the local optimum, the current point was skipped to find the optimal solution, greatly improving the fluency and rapidity of the algorithm. A flow chart of the ship path-planning method proposed in this paper is shown in Figure 2.

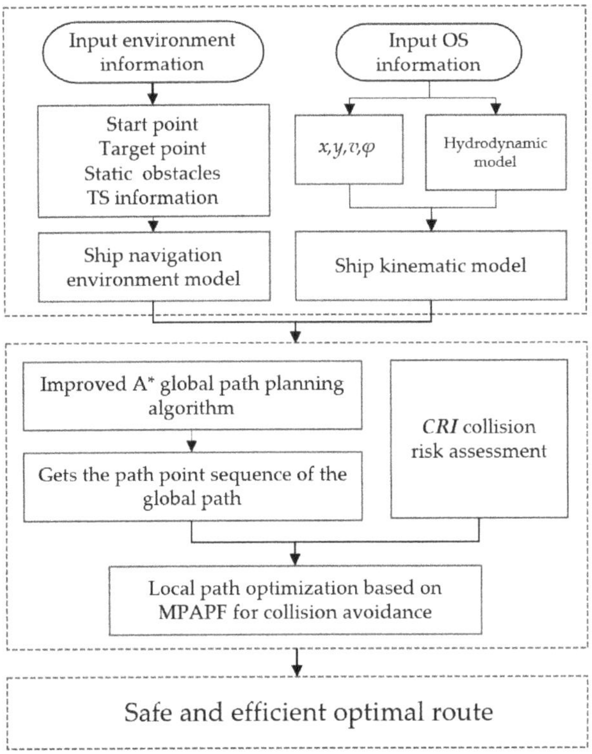

Figure 2. Process of the proposed ship path-planning method.

3. Experiments

3.1. Static Sea Environment Path Planning

To verify the effectiveness of the method, this study selected the Bohai Bay and Aegean coral islands areas as experimental planning areas for simulation, respectively. First, the improved A* algorithm was used to obtain the global planning route and then the MPAPF algorithm was used to plan the local route, taking into account the ship motion constraint model, and, finally, a safe and efficient route was obtained.

To grid the satellite image, it was converted into a greyscale image, and each pixel represented a grid. Figures 3 and 4 show the satellite map (a), grey image (b), and image (c) transformed into the unit grid of the offshore waters of China and the Aegean islands area, respectively. Figure 3c has 687 × 630 pixels, the length of each pixel is about 1710 m, and the width is about 1500 m. Figure 4c has 687 × 715 pixels, each pixel is about 144 m in length and 196 m in width. The range of each unit grid was determined by the scale of the satellite map and the number of pixels. Usually, in order to ensure the safety of the planned route, the path algorithm considered more distance between the ship and the shoreline, that is, a path with fewer pixels was selected.

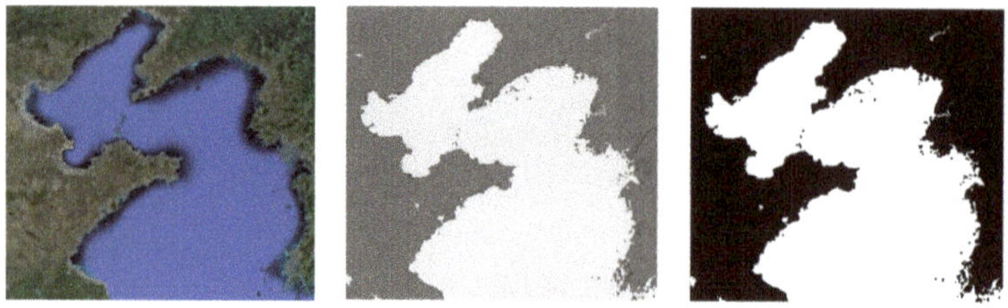

(**a**) (**b**) (**c**)

Figure 3. China offshore area. (**a**) Satellite maps. (**b**) Greyscale image. (**c**) Grid cell images.

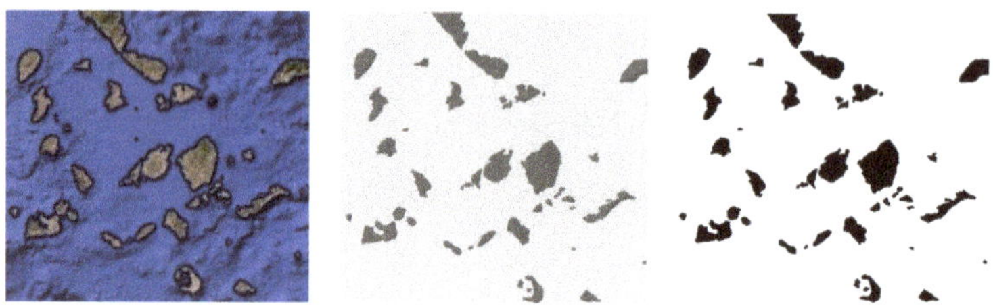

(**a**) (**b**) (**c**)

Figure 4. Aegean islands area. (**a**) Satellite maps. (**b**) Greyscale image. (**c**) Grid cell images.

In addition, when planning the path, the black part represented the land (island) and the white part represented the navigation area. It is worth noting that the cell image may have some errors due to the color and resolution of the satellite image, which can impact the final path planning. Before verifying the proposed method, the definition of obstacles was assumed, which meant that the land in the cell image was complete. After completing the preliminary work, the path planning simulation was carried out for the two environments, and the routes are shown in Figures 5 and 6. Among them, the green route and the blue route represent the routes reached by the traditional A* algorithm and the improved A* algorithm, respectively; the red route represents the route optimized by the method in this paper. The route information is shown in Tables 2 and 3. It is worth noting that IMPAPF * is our proposed method. It can be seen that by improving the A* algorithm, the steering nodes of the route were significantly reduced and the efficiency of navigation was effectively improved.

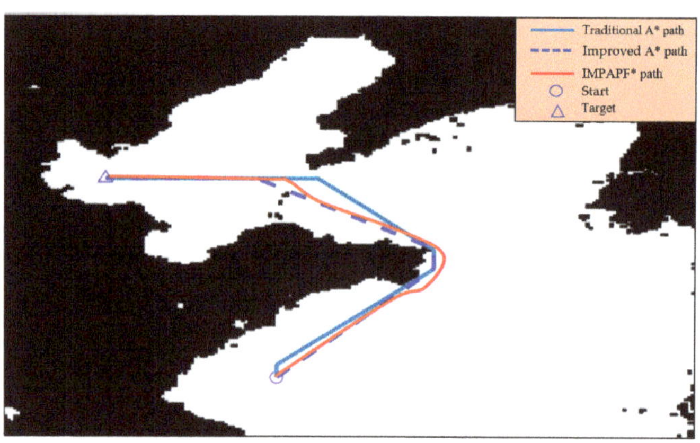

Figure 5. Route planning of China offshore area.

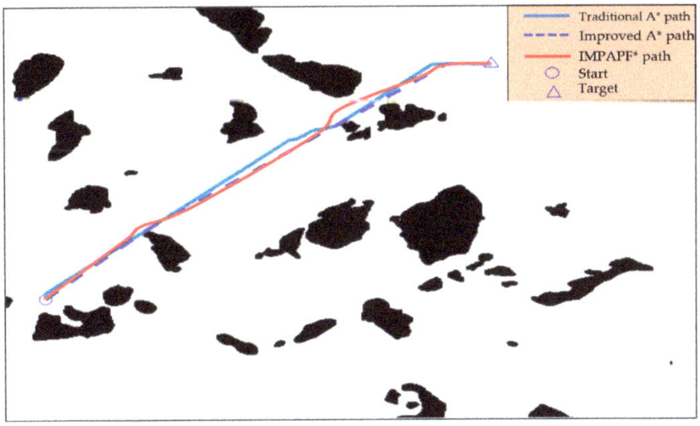

Figure 6. Route planning of Aegean islands area.

Table 2. Route information of China offshore waters.

Method	Calculation Time/s	Steering Node	Path Length/n Mile
Traditional A*	427.61	153	596.4
Improved A*	349.23	11	490.6
IMPAPF*	385.97	17	512.8

Table 3. Route information of Aegean islands area.

Method	Calculation Time/s	Steering Node	Path Length/n Mile
Traditional A*	235.55	267	73.9
Improved A*	133.86	7	59.4
IMPAPF*	167.16	12	63.5

3.2. Ship Collision Avoidance Decision

According to the actual situation encountered in the process of ship running, the encounter between two ships can be divided into three situations: head-on, crossing, and overtaking. In the process of collision avoidance, the ship conducts collision avoidance operations in accordance with the international maritime collision avoidance rules. Before that, the risk of collision needs to be assessed. In this study, CRI was used to evaluate the risk of vehicle collision, as shown in Equation (14):

$$CRI = W \cdot U = \begin{pmatrix} 0.4 \\ 0.367 \\ 0.133 \\ 0.067 \\ 0.033 \end{pmatrix}^T \times \begin{bmatrix} u_{DCPA} \\ u_{TCPA} \\ u_{D_R} \\ u_{\theta_T} \\ u_{\varepsilon} \end{bmatrix} \tag{14}$$

W is the weight vector, a constant matrix. U is the membership matrix, the speed ratio of the target ship and the ship. $\varepsilon = V_T/V_O$ is the jet velocity ratio of the target ship to our ship, and other factors are defined in reference [37]. The value of collision risk was between 0 and 1, and the closer the value was to 1, the higher the collision risk was. Usually, when $CRI > 0.7$, it is considered that the risk of collision is higher, and collision avoidance operations should be carried out at this time. When searching the path, the algorithm proposed in this paper masked the nodes with $CRI > 0.7$ to prevent them from becoming the next node. This made route planning safer, more stable, and more efficient than general methods.

In order to verify the effectiveness of the proposed method, the ship encounter situation was simulated and verified, and three situations, namely, head-on encounter, cross encounter, and overtaking, were considered. The main parameters of the simulation ship are shown in Table 4. Figure 7 shows the collision avoidance operation when a single ship meets. OS stands for the ship and TS stands for the target ship. Without any steering operation by TS, when two ships approached each other, the collision risk rose at this time, and the algorithm performed collision avoidance operations in an emergency. According to the international collision rules, the ship changed its course to starboard to avoid the obstacle ship and drive safely.

Table 4. The main parameters of the simulation ship.

Names and Symbols	Value	Unit
Length	116	m
Width	18	m
Draft	5.4	m
Max rudder angle	30	°
Speed	10	knot
Block coefficient	0.56	-it
Maneuverability index (K)	0.31	s^{-1}
Maneuverability index (T)	62.4	s

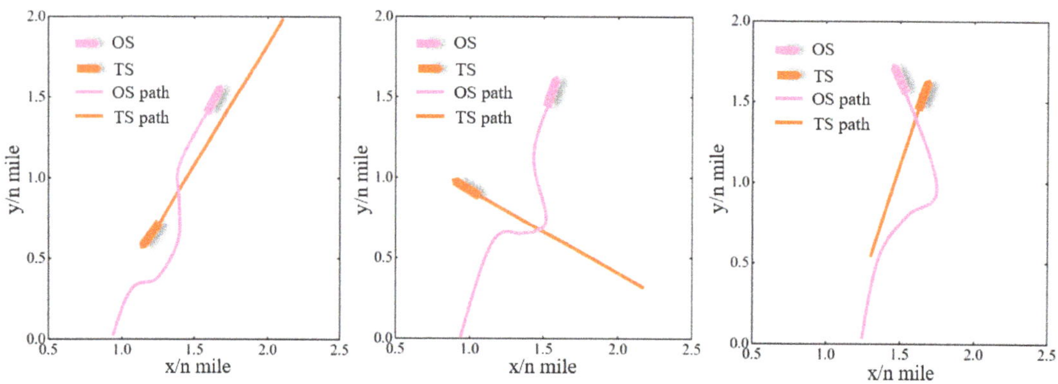

(a) Head on situation (b) Crossing situation (c) Overtaking situation

Figure 7. Encounter situations of two ships.

Figure 8a shows the safe path planned by this algorithm when multiple ships meet. It can be seen that when the ship met other ships, the ship followed the planned path. When the ship detected the target ship and the CRI with the target ship was high, the ship avoided other ships by changing direction. In addition, after collision avoidance, the ship returned to the planned route. Figure 8b shows the course of our ship. When the ship was running normally, the ship kept a certain course. When meeting with the target ship, the ship's course changed greatly, and, after collision avoidance, the ship's course returned to its original direction. Figure 8c shows the rudder angle change of our ship; Figure 9 shows the distance between the ship and the target ship. It can be seen that the relative distance between the ship and the target ship was greater than 1 n mile, which is a safe distance. Therefore, the route planned by the method proposed in this paper was reliable and safe.

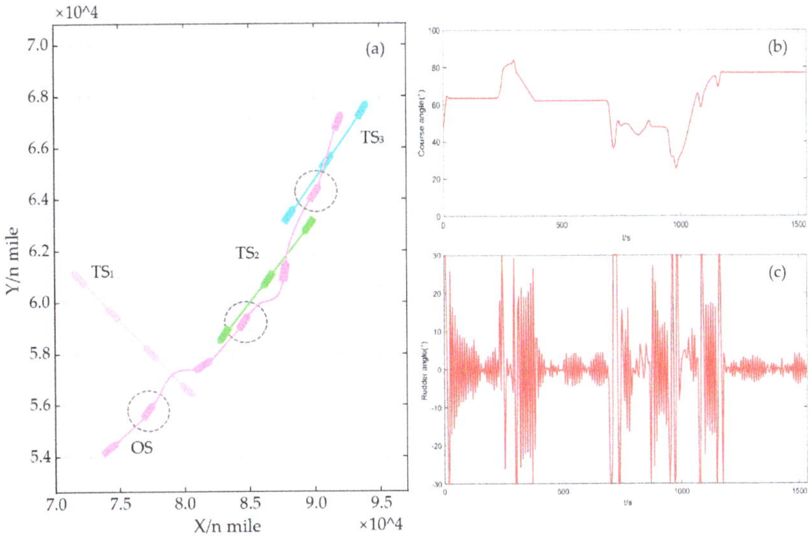

Figure 8. Route planning results of multi-ship meeting area. (**a**) The trajectory of the target ship and the collision avoidance path of our ship when multiple ships meet. (**b**) The course angle of our ship. (**c**) The rudder angle change of our ship.

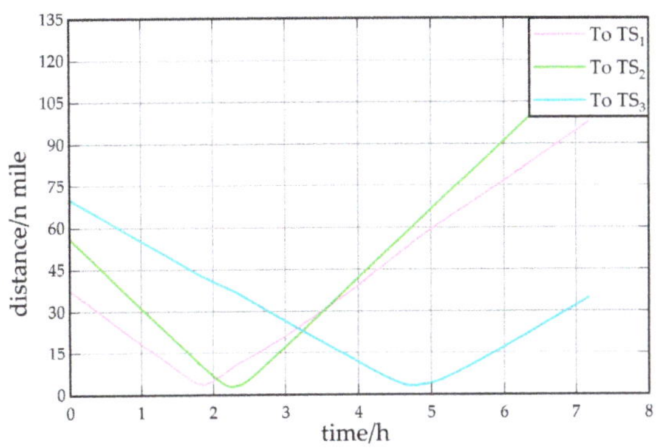

Figure 9. Distance between the ship and the target ship.

4. Conclusions

In this paper, a ship collision avoidance path-planning method IMPAPF * algorithm based on an improved A* algorithm and multi-target point artificial potential field algorithm (MPAPF) was proposed, and the static environment and ship encounter were analyzed and verified. The improved A*algorithm was used to obtain the optimal path and then the multi-target point-target point artificial potential field algorithm was used for local path planning; a ship motion model was introduced to ensure the reliability and safety of the path. In addition, when ships met, in order to prevent the occurrence of collision risk, CRI was used to evaluate the collision risk and avoid collision according to international maritime rules. At the end of this study, the proposed method was simulated, and the results show that the path obtained by the proposed method was safe and effective.

It is worth noting that although the method proposed in this paper can plan ship paths well in a static sea environment and ship encounter scene, this study did not take into account the situation of a ship emergency collision. In future work, how to solve the emergency collision avoidance problem will be incorporated into the planning algorithm. In addition, weather conditions such as ocean currents, wind waves, and tides will have an impact on ship trajectory. Therefore, in follow-up work, the impact of external weather factors on ship motion planning will also be considered.

Author Contributions: Conceptualization, Y.H. and S.Z. (Sishuo Zhao); methodology, Y.H., S.Z. (Sishuo Zhao), and S.Z. (Shuling Zhao); writing—original draft preparation, S.Z. (Sishuo Zhao) and S.Z. (Shuling Zhao); writing—review and editing, S.Z. (Sishuo Zhao) and Y.H.; funding acquisition, Y.H. All authors have read and agreed to the published version of the manuscript.

Funding: The work was jointly supported by the National Natural Science Foundation of China (72061016). Jiangxi Province Key Laboratory of Multidimensional Intelligent Perception and Control, China (No. 2024SSY03161).

Institutional Review Board Statement: Not applicable.

Informed Consent Statement: Not applicable.

Data Availability Statement: The data can be accessed by sending an email to the corresponding author.

Conflicts of Interest: The authors declare no conflicts of interest.

References

1. Yan, X.P.; Wang, S.-W.; Ma, F.; Liu, Y.C.; Wang, J. A novel path planning approach for smart cargo ships based on anisotropic fast marching. *Expert Syst. Appl.* **2020**, *159*, 113558. [CrossRef]
2. Sawada, R.; Sato, K.; Majima, T. Automatic ship collision avoidance using deep reinforcement learning with LSTM in continuous action spaces. *J. Mar. Sci. Technol.* **2021**, *26*, 509–524. [CrossRef]
3. Öztürk, Ü.; Akdağ, M.; Ayabakan, T. A review of path planning algorithms in maritime autonomous surface ships: Navigation safety perspective. *Ocean Eng.* **2022**, *251*, 111010. [CrossRef]
4. Sukumar, N.; Srivastava, A. Exact imposition of boundary conditions with distance functions in physics-informed deep neural networks. *Comput. Methods Appl. Mech. Eng.* **2022**, *389*, 114333. [CrossRef]
5. Wang, L.; Zhang, Z.; Zhu, Q.; Ma, S. Ship route planning based on double-cycling genetic algorithm considering ship maneuverability constraint. *IEEE Access* **2020**, *8*, 190746–190759. [CrossRef]
6. Machmudah, A.; Shanmugavel, M.; Parman, S.; Manan TS, A.; Dutykh, D.; Beddu, S.; Rajabi, A. Flight trajectories optimization of fixed-wing UAV by bank-turn mechanism. *Drones* **2022**, *6*, 69. [CrossRef]
7. Yuan, Q.; Wang, S.; Zhao, J.; Hsieh, T.H.; Sun, Z.; Liu, B. Uncertainty-informed ship voyage optimization approach for exploiting safety, energy saving and low carbon routes. *Ocean Eng.* **2022**, *266*, 112887. [CrossRef]
8. Lee, S.M.; Roh, M.I.; Kim, K.S.; Jung, H.; Park, J.J. Method for a simultaneous determination of the path and the speed for ship route planning problems. *Ocean Eng.* **2018**, *157*, 301–312. [CrossRef]
9. Sedeño-Noda, A.; Colebrook, M. A biobjective Dijkstra algorithm. *Eur. J. Oper. Res.* **2019**, *276*, 106–118. [CrossRef]
10. Padhy, C.P.; Sen, D.; Bhaskaran, P.K. Application of wave model for weather routing of ships in the North Indian Ocean. *Nat. Hazards* **2008**, *44*, 373–385. [CrossRef]
11. Oyekanlu, E.A.; Smith, A.C.; Thomas, W.P.; Mulroy, G.; Hitesh, D.; Ramsey, M.; Sun, D. A review of recent advances in automated guided vehicle technologies: Integration challenges and research areas for 5G-based smart manufacturing applications. *IEEE Access* **2020**, *8*, 202312–202353. [CrossRef]
12. Grifoll, M.; Borén, C.; Castells-Sanabra, M. A comprehensive ship weather routing system using CMEMS products and A* algorithm. *Ocean Eng.* **2022**, *255*, 111427. [CrossRef]
13. Miao, T.; El Amam, E.; Slaets, P.; Pissoort, D. An improved real-time collision-avoidance algorithm based on Hybrid A* in a multi-object-encountering scenario for autonomous surface vessels. *Ocean Eng.* **2022**, *255*, 111406. [CrossRef]
14. Lan, X.; Lv, X.; Liu, W.; He, Y.; Zhang, X. Research on robot global path planning based on improved A-star ant colony algorithm. In Proceedings of the 2021 IEEE 5th Advanced Information Technology, Electronic and Automation Control Conference (IAEAC), Chongqing, China, 12–14 March 2021; Volume 5, pp. 613–617. [CrossRef]
15. Zhang, M.; Ren, H.; Zhou, Y. Research on global ship path planning method based on improved ant colony algorithm. *IEEE Open J. Intell. Transp. Syst.* **2023**, *4*, 143–152. [CrossRef]
16. Singh, Y.; Sharma, S.; Sutton, R.; Hatton, D.; Khan, A. A constrained A* approach towards optimal path planning for an unmanned surface vehicle in a maritime environment containing dynamic obstacles and ocean currents. *Ocean Eng.* **2018**, *169*, 187–201. [CrossRef]
17. He, Z.; Liu, C.; Chu, X.; Negenborn, R.R.; Wu, Q. Dynamic anti-collision A-star algorithm for multi-ship encounter situations. *Appl. Ocean. Res.* **2022**, *118*, 102995. [CrossRef]

18. Kim, D.; Kim, J.S.; Kim, J.H.; Im, N.K. Development of ship collision avoidance system and sea trial test for autonomous ship. *Ocean. Eng.* **2022**, *266*, 113120. [CrossRef]
19. Lee, H.T.; Choi, H.M.; Lee, J.S.; Yang, H.; Cho, I.S. Generation of Ship's passage plan using data-driven shortest path algorithms. *IEEE Access* **2022**, *10*, 126217–126231. [CrossRef]
20. Chen, X.; Wei, C.; Yang, Y.; Luo, L.; Biancardo, S.A.; Mei, X. Personnel trajectory extraction from port-like videos under varied rainy interferences. *IEEE Trans. Intell. Transp. Syst.* **2024**, *25*, 6567–6579. [CrossRef]
21. Asl, A.N.; Menhaj, M.B.; Sajedin, A. Control of leader–follower formation and path planning of mobile robots using Asexual Reproduction Optimization (ARO). *Appl. Soft Comput.* **2014**, *14*, 563–576. [CrossRef]
22. Jovanović, I.; Perčić, M.; BahooToroody, A.; Fan, A.; Vladimir, N. Review of research progress of autonomous and unmanned shipping and identification of future research directions. *J. Mar. Eng. Technol.* **2024**, *23*, 82–97. [CrossRef]
23. Wang, X.; Feng, K.; Wang, G.; Wang, Q. Local path optimization method for unmanned ship based on particle swarm acceleration calculation and dynamic optimal control. *Appl. Ocean. Res.* **2021**, *110*, 102588. [CrossRef]
24. Lyu, H.; Yin, Y. COLREGS-constrained real-time path planning for autonomous ships using modified artificial potential fields. *J. Navig.* **2019**, *72*, 588–608. [CrossRef]
25. Lazarowska, A. A discrete artificial potential field for ship trajectory planning. *J. Navig.* **2020**, *73*, 233–251. [CrossRef]
26. Bayat, F.; Najafinia, S.; Aliyari, M. Mobile robots path planning: Electrostatic potential field approach. *Expert Syst. Appl.* **2018**, *100*, 68–78. [CrossRef]
27. Gao, J.; Xu, X.; Pu, Q.; Petrovic, P.B.; Rodić, A.; Wang, Z. A Hybrid Path Planning Method Based on Improved A* and CSA-APF Algorithms. *IEEE Access* **2024**, *12*, 39139–39151. [CrossRef]
28. Mousazadeh, H.; Jafarbiglu, H.; Abdolmaleki, H.; Omrani, E.; Monhaseri, F.; Abdollahzadeh, M.R.; Makhsoos, A. Developing a navigation, guidance and obstacle avoidance algorithm for an Unmanned Surface Vehicle (USV) by algorithms fusion. *Ocean Eng.* **2018**, *159*, 56–65. [CrossRef]
29. Zaccone, R. COLREG-compliant optimal path planning for real-time guidance and control of autonomous ships. *J. Mar. Sci. Eng.* **2021**, *9*, 405. [CrossRef]
30. Liu, Z.; Cui, J.; Meng, F.; Xie, H.; Dan, Y.; Li, B. Research on Intelligent Ship Route Planning Based on the Adaptive Step Size Informed-RRT* Algorithm. *J. Marine. Sci. Appl.* **2024**. [CrossRef]
31. Zhao, W.; Wang, H.; Geng, J.; Hu, W.; Zhang, Z.; Zhang, G. Multi-objective weather routing algorithm for ships based on hybrid particle swarm optimization. *J. Ocean. Univ. China* **2022**, *21*, 28–38. [CrossRef]
32. Almasan, P.; Suárez-Varela, J.; Rusek, K.; Barlet-Ros, P.; Cabellos-Aparicio, A. Deep reinforcement learning meets graph neural networks: Exploring a routing optimization use case. *Comput. Commun.* **2022**, *196*, 184–194. [CrossRef]
33. Chen, X.; Dou, S.; Song, T.; Wu, H.; Sun, Y.; Xian, J. Spatial-Temporal Ship Pollution Distribution Exploitation and Harbor Environmental Impact Analysis via Large-Scale AIS Data. *J. Mar. Sci. Eng.* **2024**, *12*, 960. [CrossRef]
34. Sang, H.; You, Y.; Sun, X.; Zhou, Y.; Liu, F. The hybrid path planning algorithm based on improved A* and artificial potential field for unmanned surface vehicle formations. *Ocean. Eng.* **2021**, *223*, 108709. [CrossRef]
35. Liang, C.; Zhang, X.; Watanabe, Y.; Deng, Y. Autonomous collision avoidance of unmanned surface vehicles based on improved A star and minimum course alteration algorithms. *Appl. Ocean. Res.* **2021**, *113*, 102755. [CrossRef]
36. Moradi, M.H.; Brutsche, M.; Wenig, M.; Wagner, U.; Koch, T. Marine route optimization using reinforcement learning approach to reduce fuel consumption and consequently minimize CO_2 emissions. *Ocean Eng.* **2022**, *259*, 111882. [CrossRef]
37. Seo, C.; Noh, Y.; Abebe, M.; Kang, Y.J.; Park, S.; Kwon, C. Ship collision avoidance route planning using CRI-based A$*$ algorithm. *Int. J. Nav. Arch. Ocean Eng.* **2022**, *15*, 100551. [CrossRef]

Journal of
Marine Science and Engineering

MDPI

Article

Ship Segmentation via Combined Attention Mechanism and Efficient Channel Attention High-Resolution Representation Network

Xiaoyi Li

Merchant Marine College, Shanghai Maritime University, Shanghai 201306, China; lixiaoyi@stu.shmtu.edu.cn

Abstract: Ship segmentation with small imaging size, which challenges ship detection and visual navigation model performance due to imaging noise interference, has attracted significant attention in the field. To address the issues, this study proposed a novel combined attention mechanism and efficient channel attention high-resolution representation network (CA2HRNET). More specially, the proposed model fulfills accurate ship segmentation by introducing a channel attention mechanism, a multi-scale spatial attention mechanism, and a weight self-adjusted attention mechanism. Overall, the proposed CA2HRNET model enhances attention mechanism performance by focusing on the trivial yet important features and pixels of a ship against background-interference pixels. The proposed ship segmentation model can accurately focus on ship features by implementing both channel and spatial fusion attention mechanisms at each scale feature layer. Moreover, the channel attention mechanism helps the proposed framework allocate higher weights to ship-feature-related pixels. The experimental results show that the proposed CA2HRNET model outperforms its counterparts in terms of accuracy (Acc_s), precision (P_c), F1-score ($F1_s$), intersection over union (IoU), and frequency-weighted IoU (F_{IoU}). The average Acc_s, P_c, $F1_s$, IoU, and F_{IoU} for the proposed CA2HRNET model were 99.77%, 97.55%, 97%, 96.97%, and 99.55%, respectively. The research findings can promote intelligent ship visual navigation and maritime traffic management in the smart shipping era.

Keywords: ship segmentation; attention mechanism; smart shipping; visual navigation

1. Introduction

It is a big challenge to detect ships from visual sensory data sources when the ship imaging size is small due to a small signal-to-noise ratio and low signal intensity. The background in maritime images imposes additional ship detection and segmentation challenges. However, it is quite important to detect ships far away from your own ship in order for collision avoidance measures to be implemented. More specifically, the ship's course can be adjusted to avoid a potential maritime traffic collision by changing the ship's heading direction, lowering/increasing the ship's speed, etc. Ship identification far from the distance to closest point of approach (DCPA) can help a ship's crew take action to prevent ship collisions. From the perspective of ship visual navigation, ship identification can be roughly divided into ship detection from image sequences and ship recognition in terms of pixel-wise ship segmentation. Ship detection plays a crucial role in ship navigation and maritime-traffic surveillance-related tasks. More specifically, the on-board ship crew can take action to control a ship's speed and direction with the help of ship detection results (i.e., a potential ship collision can be identified via ship detection results and measures can be taken to avoid a maritime accident). In addition, maritime traffic regulators can determine illegal ship activities from ship detection results. Currently, the ship detection task is primarily implemented by finding the maximum similarity between input ship images and pre-trained models [1]. Ship detection models show satisfactory performance by labeling to-be-detected ships with bounding boxes in maritime images. Deep-learning-based models (such as the faster region-convolution neural network (R-CNN), you only look once, etc. [2,3]) show better performance compared to traditional ship detection models (contour-based ship detection logic [4]). Ren et al. proposed a novel saliency-guided feature ship detection network to tackle distant ship detection under complex background interference [5].

A ship with a small imaging size can also be successfully detected by the saliency-guided detection model. Lee et al. utilized a deep learning model to detect ships from Korea Multipurpose Satellite (KOMPSAT) image sequences, which aimed to efficiently detect various ships sailing on the ocean and inland waterways [6]. Yasir et al. systematically analyzed a ship detection model constructed using synthetic aperture radar (SAR) data, and they conducted a comprehensive literature review by focusing on the merits and demerits of existing ship detection models to identify future directions in ship detection [7]. Jiang et al. developed a lightweight real-time ship detection model with the help of the YOLOv7-tiny model as the baseline model, and a partial convolution mechanism was employed to reduce parameters in the proposed framework [8]. The experimental results suggested that ship detection accuracy was larger than 97%, whilst the frame per second rate was 119.84 (i.e., the model can be implemented in real time). Wang et al. proposed an efficient convolutional-neural-network-based ship detection model by introducing an attention mechanism to enhance small ship detection performance, and a dilated convolution layer was further introduced to a traditional CNN to increase the CNN's receptive perception field. The proposed framework promoted ship detection in remote imagery data. However, ship detection in a complex maritime environment (e.g., large traffic volume, ships moving at high speed, ship occlusion) challenges ship detection model performance [9]. Similar studies can also be found elsewhere [10–12].

Much attention is also paid to recognizing ships from maritime images via traditional computer vision models and deep-learning-supported models. It is found that traditional ship-detection-related models may fail to design a distinct feature extraction mechanism, and thus, ship recognition and detection accuracy may sharply vary due to input image quality. CNN-based deep learning models have demonstrated powerful performance in ship-recognition-related tasks compared to traditional ship segmentation models. Ship semantic segmentation models can achieve pixel-wise ship identification and recognition, whilst background interference can be minimized as background and ship pixels can be successfully and accurately identified by the semantic models. In addition, semantic-segmentation-based models can clearly retain object boundary pixels in complex maritime traffic situations (i.e., when ships and background-related objects occupy the majority of maritime images).

A large number of studies have been conducted to accomplish ship segmentation via a multiscale attention ship instance segmentation mechanism, transformer, lightweight multi-layer perceptron, etc. [11,13,14]. George et al. proposed a novel framework to determine optimal ship speed under a fixed course by segmenting the overall path into small corridors, and the interdependency for each sub-trajectory optimization is also considered [15]. David et al. developed context information exploitation with the help of an automatic identification system (AIS) and a ship's trajectory is further identified and segmented by segmentation-related models [16]. Sun et al. suggested that instance segmentation for identifying small objects from maritime images plays quite an important role in the maritime visual navigation task, and small imaging size, low image coverage rate and blurry ship visual features challenge model performance. Sun et al. proposed a dual-branch activation network for small imaging size ship detection and segmentation [17]. Overall, previous studies mainly focus on ship detection and tracking by identifying ships with a bounding box (i.e., a bounding box indicating a ship). However, background-related pixels may be wrongly considered as ship pixels in previous studies. In addition, the semantic models try to identify ship-contour-related pixels, and trivial ship features and pixels may be discarded. To address the issue, this study proposes a novel ship segmentation framework via a combined attention mechanism and efficient channel attention high-resolution representation network (CA2HRNET). The remaining sections are organized as follows.

Section 2 describes the method used in the study, Section 3 describes the experimental results, and finally, conclusions are drawn in Section 4.

2. Methodology

2.1. Basic Ship Segmentation Structure

It should be noted that traditional convolutional neural network models implement ship feature extraction in a top-down manner. A traditional CNN employs down-sampling and max-pooling to compress the width and height of an input image. The feature extraction procedure can obtain different shapes with the help of feature layers. A traditional CNN implements top-down feature extraction, while a high-resolution representation network (HRNet) employs both high-resolution and low-resolution convolution operations (in a parallel manner) to extract object trivial and distinct features from images. It is found that both down-sampling and up-sampling mechanisms help the deep learning framework extract and merge ship shape features for the purpose of exploiting distinctive features. The HRNet model was first employed to detect human posture, via semantic segmentation, from video clips and showed quite powerful performance [18]. An HRNet-based model can efficiently learn global context and spatial information from input images, which mitigates the disadvantages of recovering low-resolution feature maps. The object shape and boundary features may be easily discarded during feature transfer from a lower encoder layer to a higher encoder layer. HRNet-network-generated prediction maps can obtain higher accuracy for achieving the object recognition task.

Overall, the basic HRNet network consists of a backbone layer, a feature integration layer, and a prediction head layer. The backbone layer uses four stages to perform multi-resolution feature fusion tasks via feature exchange in the layer. A schematic overview of the HRNet structure is shown in Figure 1. As can be seen, the HRNet network model consists of two key characteristics. First, the high-resolution and low-resolution layers are connected in a parallel manner. Second, feature map information is constantly exchanged between high- and low-resolution maps. The high-resolution feature map helps the model obtain accurate ship spatial information, while the low-resolution map ensures that boundary-related pixels can be clearly identified.

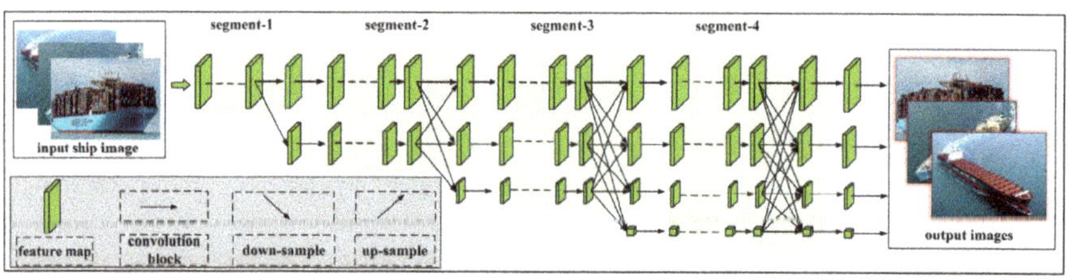

Figure 1. Overview of the HRNet Network.

2.2. CA2HRNET Framework

Traditional semantic segmentation models may lead to spatial and global contextual information loss due to resolution degradation during the feature extraction procedure. While HRNet-based models can alleviate this disadvantage with the help of high-resolution layer-based feature extraction, the information loss problem cannot be completely addressed. Our study proposes a novel CA2HRNET framework by tackling the information loss challenge with the help of an attention module. The attention module is introduced and embedded at the end of the high-resolution subnetwork convolutional layer. The proposed framework implements consistently high-resolution feature extraction by following the rule used in the HRNet network, and the feature resolution is simultaneously analyzed, which differs from traditional semantic segmentation models. The proposed network starts with a high-resolution sub-network as the first stage, and high-resolution layers and corresponding feature maps are added into low-resolution layers and networks to further implement the feature extraction procedure. Moreover, multi-scale feature fusion is repetitively implemented to enrich feature maps.

2.2.1. Channel Attention Mechanism

Deep-learning-based models employ an attention mechanism to focus on a region of interest, and this attention mechanism can be adjusted to fulfill different visual tasks by changing channel features or spatial structures. The convolutional block attention mechanism (CBAM) consists of a channel attention mechanism and a spatial attention mechanism, which outperforms the conventional attention mechanism. Usually, channel attention (CA) is used to learn the importance of each channel, while spatial attention is used to learn the importance of each spatial location for an input video clip. This study enhances CBAM attention mechanism performance by introducing a correlation attention module (CAM) attention mechanism, which is abbreviated as C2BA2M2 in our study. The CAM mechanism helps the ship segmentation framework better identify ship trivial features while ship features are extracted by varied channel layers. More specifically, the proposed C2BA2M2 integrates channel attention, multi-scale spatial attention, and adaptive weight fusion, combines multiple modules, and enhances the ability of attention modeling from different perspectives. An overview of the proposed C2BA2M2 module is shown in Figure 2. The C2BA2M2 module better focuses on the important regions in the input image due to its ability to capture spatial information at different scales by introducing multiple attention units with different convolutional kernel sizes in the spatial attention module. In other words, the model receptive field can be significantly improved and target imaging scale variation does not impose an obviously negative impact on the C2BA2M2 module. Moreover, a weight fusion module is introduced to help the model adaptively determine the importance of channel attention and spatial attention based on the input features and dynamically adjust the weight setups. This mechanism makes attention allocation more flexible and efficient.

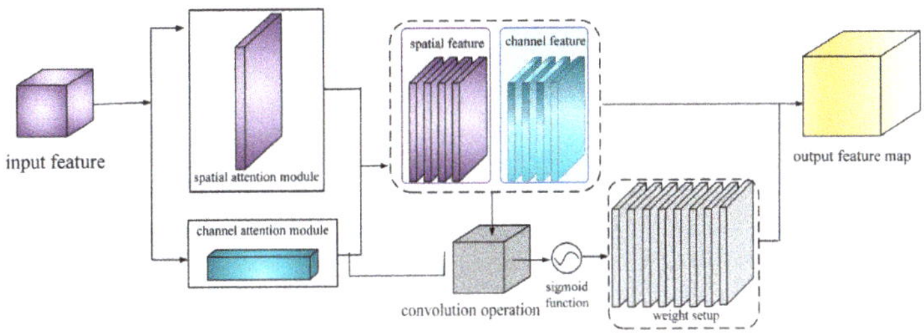

Figure 2. Schematic overview of the proposed C2BA2M2 module.

The spatial attention module introduces multiple attention units with different convolutional kernel sizes based on the original CBAM network. It then performs a weighted fusion of features obtained from different convolutions to obtain the spatial information of the multi-scale feature map. A schematic diagram of the spatial attention module is shown in Figure 3. Note that the multi-scale convolutional feature extraction is computed with Equations (1) and (2). The multi-scale fusion operation can output the ship features based on Equation (3). The fused features are subjected to global max-pooling and average pooling operations from the perspective of channel dimensions, and the convolution layer is implemented to reduce model dimensionality. Moreover, Sigmoid operations are used to generate ship spatial attention features.

$$F_i = Conv(K_i \times K_i)\ X_i,\ i = 0,\ 1,\ 2\ldots,\ S-1 \quad (1)$$

$$K_i = 2 \times (i+1) + 1 \quad (2)$$

$$F = Con(F_0,\ F_1,\ \ldots F_{s-1}) \quad (3)$$

where F_i is the feature extracted by convolution at different scales, K_i is the convolution kernel size and X_i is the input feature. The Con symbol represents the concatenate function.

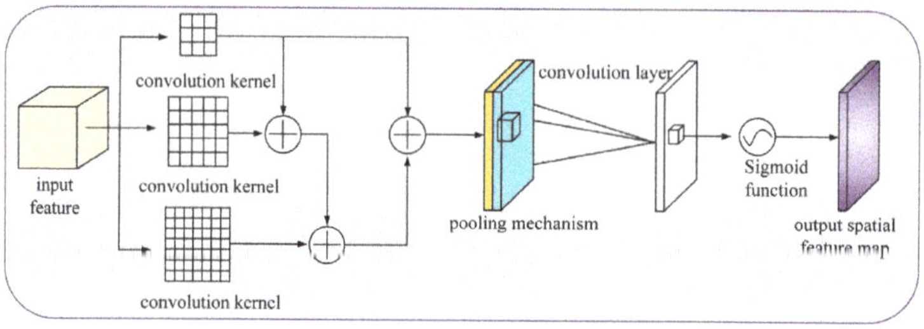

Figure 3. Spatial attention module used in the study.

The channel attention mechanism performs max-pooling and average pooling operations on the input feature maps in the width and height dimensions, respectively. The pooling layer outputs are then transmitted with a fully connected layer for weight parameter sharing with the help of weighting operation on the parameter-shared features. In addition, the sigmoid activation operation is applied to generate the final channel attention features. A schematic representation of the channel attention module is shown in Figure 4.

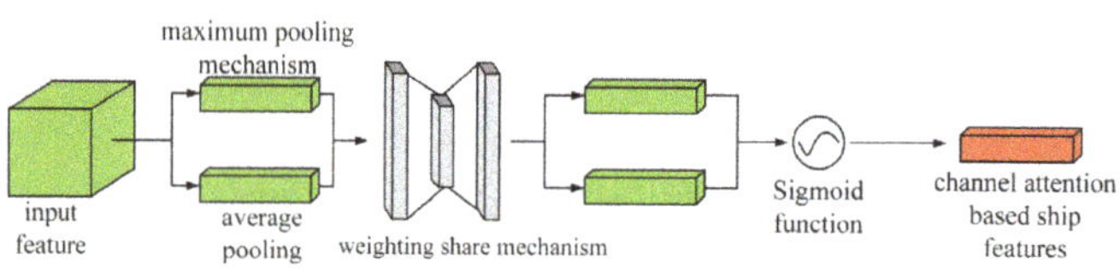

Figure 4. Channel Attention Module used in the study.

2.2.2. Details about CA2HRNET Framework

Utilization of the CAM attention mechanism allows the model to give more attention to the image spatial and channel scales in the feature extraction stage. Ship features will be further enhanced with the help of an efficient channel attention mechanism (ECA) after the up-sampling procedure. More important ship features will be assigned larger weights with channel-related information and mechanism. A schematic overview of the proposed CA2HRNET framework is shown in Figure 5. The entire CA2HRNET network consists of a backbone section, a feature integration section, and a prediction header section. In the backbone, the CA2HRNET network performs both down-sampling and up-sampling during the feature extraction process. In that way, different shapes of feature maps are obtained and feature fusion is performed. In the feature integration section, the network performs feature fusion of all the acquired feature maps by up-sampling the feature maps with smaller widths and heights. We can obtain a global feature map via this feature map fusion procedure, which primarily involves convolution, normalization, and activation operations. In the prediction header section, the network adjusts the number of channels to the number of categories with convolution operation (e.g., typical convolution kernel is 1×1). The potential width and height for the output layer are adjusted based on the input image size.

The proposed CA2HRNET network backbone is divided into four segments, segment-1, segment-2, segment-3, and segment-4. Segment-1 is the first part and performs the initial feature extraction. For a given input ship image, the network employs two convolution operations with step size (2, 2), convolution kernel size (3, 3) and 64 channels for height and width compression and feature extraction. Then, four bottleneck-block residual convolution operations are performed to merge ship features. The bottleneck-block layer involves two convolution operators, a normalization operator, and an activation function, alongside one convolution and normalization in the trunk section. Moreover, the residual edge section either bypasses any processing or undergoes minimal processing before directly connecting to the output. The size of the feature map is obtained by implementing four bottleneck-block operations, while the feature map size is supposed to be (120, 120, 256).

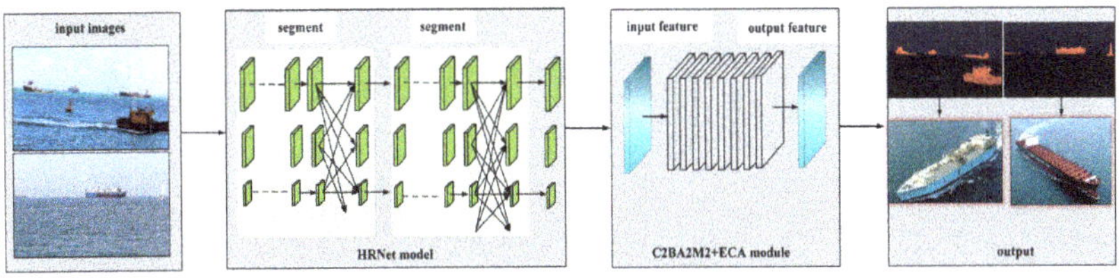

Figure 5. The proposed CA2HRNET framework overview.

Segment-2 is the second part and performs the feature extraction procedure. The feature layer input into segment-2 is processed using a convolution with a step size of (1, 1) and 32 channels and a convolution with a step size of (2, 2) and 64 channels to obtain a feature layer of (120, 120, 32) and a feature layer of (60, 60, 64). After that, four basic-block residual convolution operations are processed for each of the two feature layers. In the basic-block module, the backbone sections consist of a convolution, normalization, activation function and a convolution, normalization; The residual edge section either bypasses any processing or undergoes minimal processing before directly connecting to the output. After implementing four basic-block operations, a (120, 120, 32) feature layer and a (60, 60, 64) feature layer are obtained. The feature layer of (120, 120, 32) is then down-sampled and then added to the feature layer of (60, 60, 64); the feature layer of (60, 60, 64) is up-sampled and then added to the feature layer of (120, 120, 32). The potential output feature map sizes obtained are (120, 120, 32) and (60, 60, 64).

Segment-3 is responsible for further feature extraction. For the feature layer input into segment-3, convolution with a step size of (2, 2) and 128 channels is applied to process the (60, 60, 64) feature layer, and the output is a (30, 30, 128) feature layer. Therefore, there are three feature layers. The four basic-block residual convolution operations are performed on these three feature layers, and these three feature layers are utilized for up-sampling and down-sampling to establish a connection between the different feature maps, which can help achieve the high-level of feature fusion.

Segment-4 is responsible for the final feature fusion. For the feature layer input into segment-4, convolution with a step size of (2, 2) and 256 channels is applied to process the (30, 30, 128) feature layer, resulting in a (15, 15, 256) feature layer. Therefore, the four feature layers are processed with four basic-block residual convolution operations, which are used to establish connections with the up-sampling and down-sampling operations. Segment-4 outputs distinct yet discriminative features with the feature fusion procedure.

By constructing the above-mentioned segments, we can obtain four effective feature layers in segment-4, which are described as (128, 128, 32), (60, 60, 64), (30, 30, 128), and (15, 15, 256). These four feature layers enable network focus on the crucial yet non-trivial features in the image space and channel (i.e., the advantages of an attention mechanism), respectively. Then, feature fusion is performed on these four effective feature layers. The first three effective feature layers are first up-sampled, the height and width of the feature layers are adjusted to 128 $\times$ 128, and then, the adjusted feature layers are stacked, before, finally, the feature integration of these four adjusted effective feature layers is performed using a convolution, normalization, and activation function. The features of the input image can be obtained through the above steps. The extracted features are allowed to pass through an ECA attention mechanism that allows the network to assign higher weights to the important channel information. Then, a 1 $\times$ 1 convolution is utilized to adjust the channels to the number of categories for the final feature layer, and up-sampling is carried out by utilizing a resize operation to make the width and height of the final output layer consistent with that of the input image.

3. Experimental Section

3.1. Experimental Data

For the ship recognition study, three maritime video clips of typical maritime traffic situations were collected to validate the results of the semantic segmentation of ships. We used a similar dataset and dataset collection rule in our previous study [11]. Scenario 1 is a scene with lower visibility and the presence of occlusion by two large ships. Scenario 2 is a scene with good weather conditions and the presence of occlusion by two large ships. Scenario 3 is a scene with good weather conditions and the presence of a fast-moving near target. Part of the dataset is shown in Figure 6 and detailed maritime video parameters are shown in Table 1. For the purpose of model training, we have made a set of training image sequences with the help of annotation segmentation software labelMe (Version 5.0.1). The raw images in Figure 6a,c,e demonstrate typical ship image samples, while Figure 6b,d,f are the corresponding labeling images. To mitigate the image overfitting phenomenon, data augmentation was applied to the input training set in the dataset (e.g., image rotation, specifically in the form of left-right flipping and up-down flipping) to improve the stability and robustness of the model.

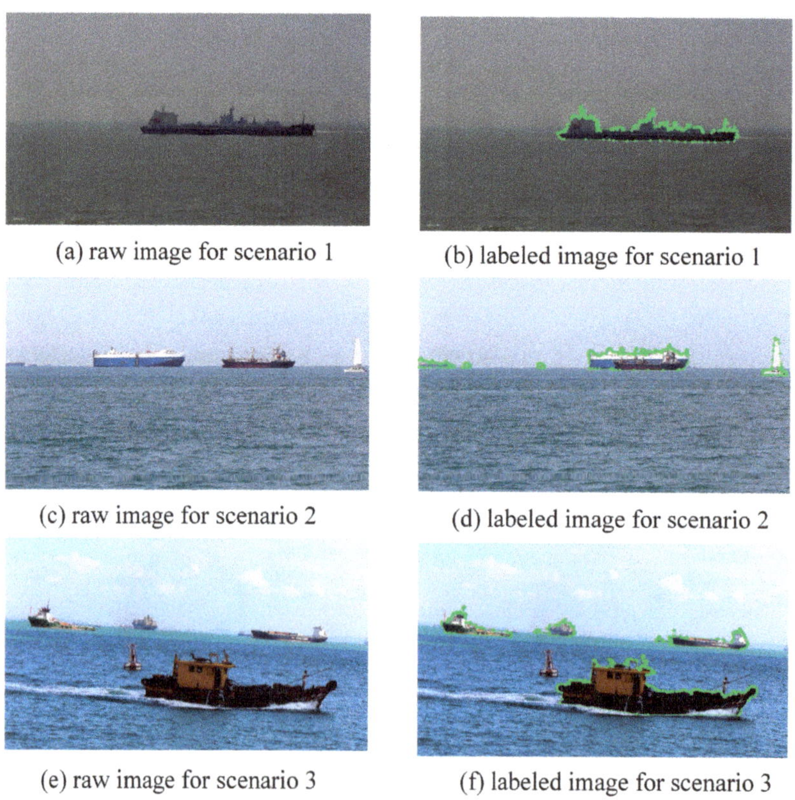

Figure 6. The input raw maritime images and training samples.

Table 1. Descriptions of the Maritime Video clips.

Video Number	Frame Rate	Image Resolution	Video Duration	Environmental Feature
scenario 1	30	1920 × 1080	15	slightly cloudy, fewer ships, no waves, low visibility
scenario 2	30	1920 × 1080	18	slightly cloudy, fewer ships, no waves, high visibility
scenario 3	30	1920 × 1080	9	cloudy, many ships, waves, high visibility

3.2. Experimental Setups

The experiments are implemented on the Ubuntu operation system (the corresponding version is Ubuntu 23.10), and the TensorFlow deep learning framework is used as a baseline for the experiment. The experimental platform has 16 GB memory, a 4 GHz CPU, and an RTXA4000 GPU. Python 3.7 version is used to implement the various models used in the study. During the experiment, CuDNN, Cuda, OpenCV, and other third-party dependent libraries are used, and corresponding software packages are installed according to the requirements for each model to ensure the implementation of the network. The CuDNN version is 8.8.0, and Cuda toolkit is 12.2.2, and opencv version is 3.4.16.

3.3. Evaluation Index

Ship segmentation can be regarded as a type of binary classification of pixels. Overall, the to-be-detected ship target pixels are considered the foreground and the rest of the pixels are considered the background. To accurately evaluate the segmentation results, the true positions of the ship target pixel are manually marked in the experiment. The effectiveness of the algorithm is evaluated by calculating the difference between the detected and ground-truth values in the test set. The study employs accuracy (Acc_s), precision (P_c), F1-score ($F1_s$), intersection over union (IoU), and frequency-weighted IoU ($\mathrm{F_{IoU}}$) indicators to quantify model performance. The formulas for calculating Acc_s, P_c, $F1_s$, IoU, and $\mathrm{F_{IoU}}$ can be found in Equations (4)–(8), respectively.

$$Acc_s = \frac{D_B + D_H}{D_B + D_H + P_B + P_H} \tag{4}$$

$$P_c = \frac{D_H}{D_H + P_H} \tag{5}$$

$$F1_s = 2 \times \frac{P_c \times R_e}{P_c + R_e} \tag{6}$$

$$\mathrm{IoU} = \frac{D_H}{D_H + P_H + P_B} \tag{7}$$

$$\mathrm{F_{IoU}} = \frac{D_H + P_B}{D_H + D_B + P_H + P_B} \times \frac{D_H}{D_H + P_H + P_B} \tag{8}$$

where D_B represents the number of pixels correctly classified as background; D_H represents the number of pixels correctly classified as a ship target; P_B represents the number of background pixels misclassified as ship targets; and P_H represents the number of ship target pixels misclassified as background.

3.4. Results and Analysis

For the purpose of verifying the ship segmentation accuracy of the proposed CA2HRNET network, we have also implemented other segmentation models that use an attention mechanism as well as other typical semantic segmentation networks. More specifically, the conventional HRNet [19] is implemented, and we have enhanced the performance of the basic HRNet model by adding a squeeze and excitation mechanism (abbreviated as HRNet-SE). In addition, the basic HRNet model is also improved by incorporating a coordinate attention mechanism (abbreviated as HRNet-CA). Moreover, the PSP Net is also implemented to verify the segmentation performance of the proposed framework [20]. The UNet segmentation model [21] and DeeplabV3+ [22] are also implemented for the purpose of model performance comparison.

3.4.1. Scenario 1 Experimental Results

There are two large ships overlapping each other in scenario 1. The imaging size of the ships is quite large and the traffic volume is low. Since the ships are in the distance (i.e., far away from the camera), the contours of the ships are blurred, which challenges model segmentation accuracy. Despite this, our proposed framework achieves an excellent performance. The segmentation results of the different models are shown in Figure 7. It can be seen from the figure that each network has identified the ship contours; in particular, the HRNet with added attention mechanism exhibited better effect than the unimproved HRNet. Especially in the bow, CA2HRNET better expresses the details of the foremast of the forecastle and the superstructure and antenna of the stern are also well expressed. Compared with other networks, the UNet network performs well, and the corresponding contour of the ship is well expressed. DeepLabv3+ network performance is not good, from the enlarged detail map, it can be seen that the details of the bow part are not well expressed, with instances of misidentification of hulls and backgrounds. The PSPNet network displays the worst ship segmentation accuracy, with no corresponding contour of the hull at all, and the foremast being completely unrecognized in the zoomed-in detail image.

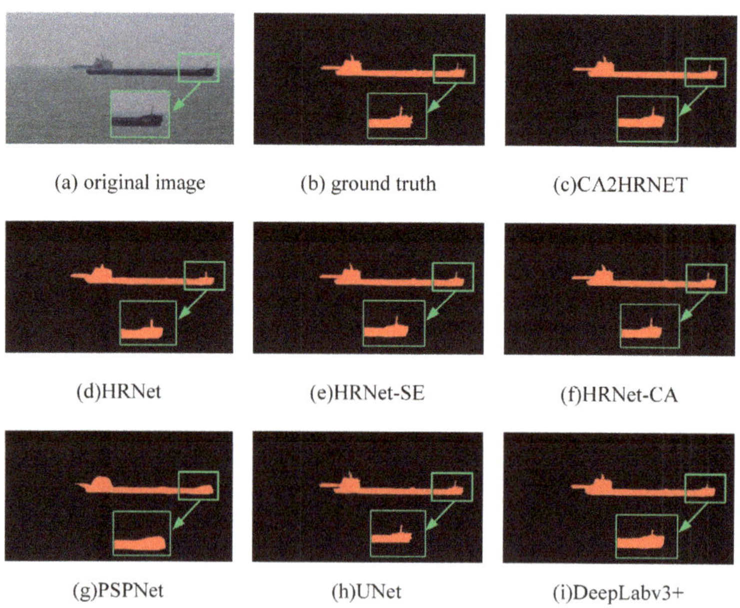

Figure 7. Ship Semantic segmentation of scenario 1.

The segmentation results of the different models are shown in Table 2. According to the results of the five evaluation indexes, the varied ship segmentation models have shown satisfactory results, as the accuracy values are larger than 99% for scenario 1. Compared with the original segmentation network, the networks with different attention mechanisms achieved better segmentation results. The proposed CA2HRNET model outperformed its counterparts with a segmentation accuracy of 97.81%. More specifically, the majority of ship-related pixels can be correctly identified by the proposed CA2HRNET model. Compared with the HRNet-SE network and HRNET-CA network, the accuracy of the CA2HRNET network was 0.02% and 0.28% higher, respectively. This may be because the newly added attention mechanism results in a significant improvement.

Table 2. Ship segmentation performance comparison for scenario 1.

MODEL	Acc_s	P_c	$F1_s$	IoU	F_{IoU}
CA2HRNET	99.81%	97.81%	97.56%	97.53%	99.63%
HRNet-SE	99.81%	97.79%	97.53%	97.49%	99.63%
HRNet-CA	99.81%	97.53%	97.52%	97.50%	99.62%
HRNet	99.67%	95.31%	95.78%	95.78%	99.36%
PSPNet	99.41%	93.03%	92.34%	92.58%	98.87%
UNet	99.77%	98.47%	96.97%	96.94%	99.54%
DeepLabv3+	99.68%	94.22%	95.89%	95.88%	99.37%

The proposed framework achieved better ship segmentation performance because the spatial attention mechanism and channel attention mechanism helped the model obtain ship trivial visual features, and thus, the ship's bow-like pixels can be more accurately identified and segmented by the model. The segmentation accuracy and precision may be slightly or significantly improved based on the background interference severity in the maritime images.

3.4.2. Experimental Results in Scenario 2

The traffic volume in scenario 2 is obviously larger than that in scenario 1, and ships in scenario 2 can be clearly identified from the maritime images (i.e., ship imaging sizes in scenario 2 were significantly larger than those in scenario 1). For ships with large imaging areas, the network can better learn the hull features due to the presence of many semantic pixels of ships, which can significantly help the model recognize ship contours from images. However, for ships with small imaging areas, it is difficult for the network to classify the pixels of the ship from those of the background as there are fewer pixels for the network to learn. Therefore, it is more challenging for the network to classify ship and background pixels accurately and ship identification with small imaging sizes is more important.

We have also selected a series of image segmentation results to demonstrate varied model performance for the second scenario. The segmentation results of the different models are shown in Figure 8. It can be observed that for ships with larger imaging areas, there is more overlapping area between general cargo ships and bulk carriers. The CA2HRNET network better expresses the detailed contours of the superstructure of the overlapping part. The HRNet-SE network performed better than the other non-CA2HRNET networks, as only a few contour details were missed. For small target ships with small imaging areas, the various HRNet series networks successfully identified the small ships near the skyline. It can be inferred that the HRNet network retains the high-resolution feature map throughout the whole process, and the spatial information of the image is successfully captured by the segmentation model. The important information of the spatial domain and channel domain is extracted effectively by the introduced attention mechanism. The multi-level up-sampling mechanism of feature maps used by the network implemented better fusion and expression of feature maps with different resolutions.

Table 3 shows the segmentation performance of the different models according to five statistical indicators. It is found that the CA2HRNET network segmentation reaches 97.14% in terms of accuracy, which is 0.14% and 0.19% larger than the HRNet-SE network and HRNet-CA network, respectively. The CA2HRNET model outperformed the PSPNet, UNet, and DeepLabv3+ networks by 3.26%, 0.14%, and 1.69%, respectively, in terms of segmentation accuracy. It can be seen that the values of the statistical indicators in scenario 2 obtained by the various models are relatively smaller than those in scenario 1. After carefully checking the ship segmentation results, it was found that the false-detection rate in scenario 2 was higher than that in scenario 1 due to interference caused by small and medium-sized ship targets in the maritime images (i.e., ship boundaries were also too blurred to be detected). Based on the values of five evaluation indexes, the CA2HRNET network achieved excellent results, and the accuracy of the HRNet network with attention mechanism reached 99.80%, which is 0.04% higher than the second-best UNet network.

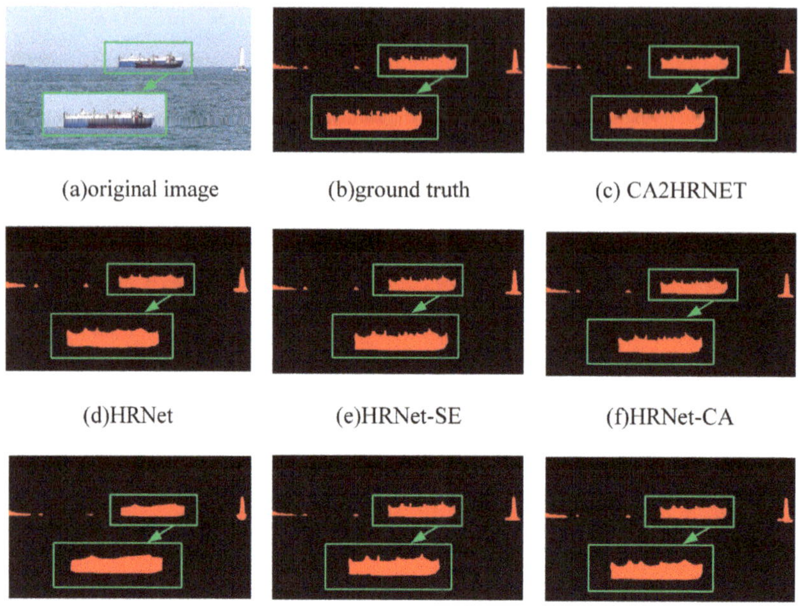

(a)original image (b)ground truth (c) CA2HRNET

(d)HRNet (e)HRNet-SE (f)HRNet-CA

(g)PSPNet (h)UNet (i)DeepLabv3+

Figure 8. Ship Semantic segmentation of scene 2.

Table 3. Comparison of segmentation accuracy for Scenario 2.

MODEL	Accuracy	Precision	F1-Score	mIoU	FWIoU
CA2HRNET	99.80%	97.14%	96.59%	96.60%	99.61%
HRNet-SE	99.80%	97.00%	96.55%	96.57%	99.61%
HRNet-CA	99.80%	96.95%	96.53%	96.55%	99.60%
HRNet	99.63%	92.65%	93.61%	93.80%	99.28%
PSPNet	99.57%	93.88%	92.45%	92.76%	99.16%
UNet	99.76%	97.00%	95.74%	95.79%	99.52%
DeepLabv3+	99.65%	95.45%	94.00%	94.16%	99.33%

3.4.3. Experiment Results in Scenario 3

The traffic volume in scenario 3 is larger than that of both scenario 2 and scenario 1, and ship speeds in the third scenario were on average higher than in the previous two scenarios (it is noted that the fishing boat in the third video moves quite fast). Due to the fast speed of the fishing boat, the position feature of the fishing boat pixels showed a rapid change. This imposes a significant challenge for the learning of the network. However, at the same time, there are more abundant image pixels of ships near the skyline, which helps the network to learn and recognize ship contour details.

The segmentation results of the different models for the scenario 3 are illustrated in Figure 9. It can clearly be seen that each segmentation model can well separate ship pixels from those of the skyline. Similar to the segmentation performance in scenario 1 and scenario 2, the CA2HRNET network has a good performance in the segmentation of the detailed contours of the prospective ships. Indeed, the proposed CA2HRNET network showed the best segmentation results and the overlap between the detail map and the real label is quite large. In addition, the HRNet networks that integrate other attention mechanism networks (i.e., HRNet-CA and HRNet-SE) also show better segmentation than the original HRNet network.

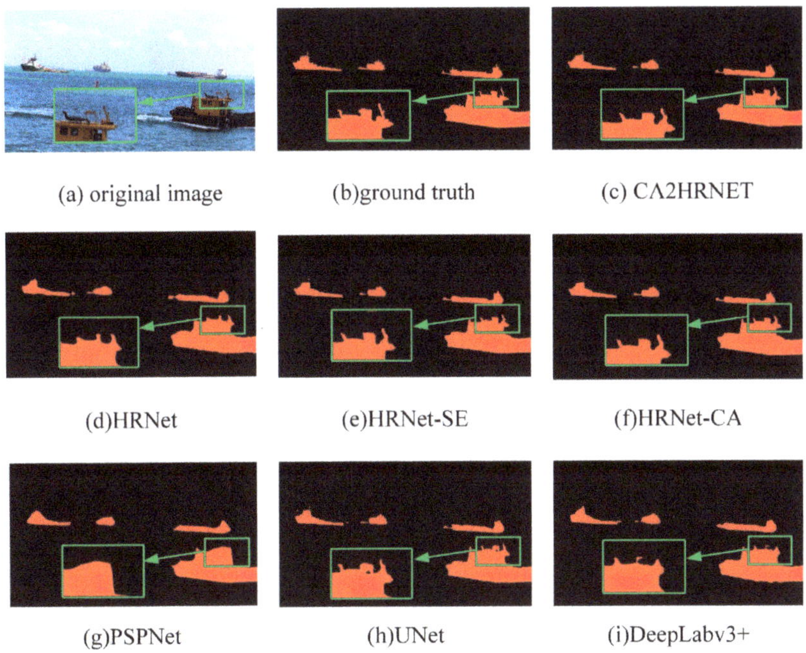

Figure 9. Semantic segmentation of scene 3.

Statistical evaluators for the third scenario are shown in Table 4. It can be seen that the accuracy of the CA2HRNET network reaches 97.69%, which outperforms the HRNet-CA network and HRNet network by 0.09% and 4.11%, respectively, in terms of the precision-related indicator. Moreover, the CA2HRNET network outperformed the PSPNet, UNet, and DeepLabv3+ networks by 3.69%, 0.33%, and 4.62%, respectively. Based on the above-mentioned analysis, the CA2HRNET network achieved better performance due to the advantages provided by the introduced hybrid attention mechanisms. The accuracy of the HRNet network with the introduced attention mechanism reaches about 99.680%, which is 0.02% higher than that of the UNet network.

Table 4. Comparison of segmentation accuracy for Scenario 3.

MODEL	Accuracy	Precision	F1-Score	mIoU	FWIoU
CA2HRNET	99.69%	97.69%	96.85%	96.79%	99.40%
HRNet-SE	99.68%	97.69%	96.72%	96.66%	99.37%
HRNet-CA	99.68%	97.60%	96.76%	96.70%	99.38%
HRNet	99.43%	93.58%	94.26%	94.27%	98.91%
PSPNet	99.33%	93.73%	93.06%	93.16%	98.70%
UNet	99.66%	97.36%	96.51%	96.45%	99.33%
DeepLabv3+	99.42%	93.43%	94.14%	94.16%	98.88%

4. Conclusions

To address small to medium-sized ship identification and recognition from maritime images, the study proposes a novel CA2HRNET model with the help of spatial attention and channel attention mechanisms. The proposed CA2HRNET model extracts effective feature maps in terms of channels and spatial dimensions from input maritime image sequences. The proposed ship semantic segmentation model generates and connects feature maps with high-resolution to low-resolution in the encoder layer. Overall, the CA2HRNET network retained the advantages of HRNet by generating four feature maps with different resolutions in a parallel connection manner rather than a cascaded manner. In this way, ship features with high resolution for each layer can be successfully retained, and thus, ship segmentation accuracy is significantly enhanced. The proposed model has shown good performance for maritime images overwhelmed with blurred ship contours. The CAM attention mechanism is further proposed to enrich the diversity of receptive fields due to the limited number of the four parallel layers. Channel attention, multi-scale spatial attention, and adaptive weight fusion are also integrated to improve ship segmentation model performance, which is robust against ship scale variation and visual shape variation challenges. Experimental results show that the proposed ship segmentation model showed accurate ship segmentation model performance (i.e., the average ship segmentation accuracy was larger than 97%). The following improvements can be explored to expand our study. First, we can further enhance our model performance utilizing a lightweight model, and model computational cost can also be evaluated. Second, we can collect maritime videos under additional adverse weather conditions (e.g., storm, low-visibility, night-time weather) for the purpose of improving model usage in real-world maritime activities. Third, it was found that ship detection performance for the various models was quite similar. The primary reasons can be ascribed to ship detection and segmentation model robustness and noise-free input maritime images. In future work, we may also verify ship detection and segmentation model performance when using low-quality maritime image clips.

Funding: This work was jointly supported by National Natural Science Foundation of China (52331012, 52102397).

Institutional Review Board Statement: Not applicable.

Informed Consent Statement: Not applicable.

Data Availability Statement: The data can be accessed by sending an email to the author.

Conflicts of Interest: The author declares no conflicts of interest.

References

1. Cheng, S.; Zhu, Y.; Wu, S. Deep learning based efficient ship detection from drone-captured images for maritime surveillance. *Ocean Eng.* **2023**, *285*, 115440. [CrossRef]
2. Pinault, L.J.; Yano, H.; Okudaira, K.; Crawford, I.A. YOLO-ET: A Machine Learning model for detecting, localising and classifying anthropogenic contaminants and extraterrestrial microparticles optimised for mobile processing systems. *Astron. Comput.* **2024**, *47*, 100828. [CrossRef]
3. Gladis, K.A.; Madavarapu, J.B.; Kumar, R.R.; Sugashini, T. In-out YOLO glass: Indoor-outdoor object detection using adaptive spatial pooling squeeze and attention YOLO network. *Biomed. Signal Process. Control* **2024**, *91*, 105925.
4. Chang, H.; Fu, X.; Lu, J.; Guo, K.; Dong, J.; Zhao, C.; Feng, C.; Li, Z.; Zhang, Y. SPANet: A Self-Balancing Position Attention Network for Anchor-Free SAR Ship Detection. *IEEE J. Sel. Top. Appl. Earth Obs. Remote Sens.* **2023**, *16*, 8363–8378. [CrossRef]
5. Ren, Z.; Tang, Y.; Yang, Y.; Zhang, W. SASOD: Saliency-Aware Ship Object Detection in High-Resolution Optical Images. *IEEE Trans. Geosci. Remote Sens.* **2024**, *62*, 1–15. [CrossRef]
6. Lee, K.-J.; Lee, S.-J.; Chang, J.-Y. A Study on Ship Detection and Classification Using KOMPSAT Optical and SAR Images. *Ocean Sci. J.* **2024**, *59*, 10. [CrossRef]
7. Yasir, M.; Jianhua, W.; Mingming, X.; Hui, S.; Zhe, Z.; Shanwei, L.; Colak, A.T.I.; Hossain, M.S. Ship detection based on deep learning using SAR imagery: A systematic literature review. *Soft Comput.* **2023**, *27*, 63–84. [CrossRef]
8. Jiang, X.; Cai, J.; Wang, B. YOLOSeaShip: A lightweight model for real-time ship detection. *Eur. J. Remote Sens.* **2024**, *57*, 2307613. [CrossRef]

9. Wang, W.; Fu, Y.; Dong, F.; Li, F. Semantic segmentation of remote sensing ship image via a convolutional neural networks model. *IET Image Process.* **2019**, *13*, 1016–1022. [CrossRef]
10. Manar, A.; Kim, S. IR/EO Ship Detection and Tracking Using SiamMask. In Proceedings of the 2022 22nd International Conference on Control, Automation and Systems (ICCAS), Jeju, Republic of Korea, 27 November–1 December 2022; pp. 1604–1606.
11. Chen, X.; Wang, M.; Ling, J.; Wu, H.; Wu, B.; Li, C. Ship imaging trajectory extraction via an aggregated you only look once (YOLO) model. *Eng. Appl. Artif. Intell.* **2024**, *130*, 107742. [CrossRef]
12. Chen, X.; Dou, S.; Song, T.; Wu, H.; Sun, Y.; Xian, J. Spatial-Temporal Ship Pollution Distribution Exploitation and Harbor Environmental Impact Analysis via Large-Scale AIS Data. *J. Mar. Sci. Eng.* **2024**, *12*, 960. [CrossRef]
13. Sharma, R.; Saqib, M.; Lin, C.T.; Blumenstein, M. MASSNet: Multiscale Attention for Single-Stage Ship Instance Segmentation. *Neurocomputing* **2024**, *594*, 127830. [CrossRef]
14. Zhang, Y.; Li, C.; Shang, S.; Chen, X. SwinSeg: Swin transformer and MLP hybrid network for ship segmentation in maritime surveillance system. *Ocean Eng.* **2023**, *281*, 114885. [CrossRef]
15. Tzortzis, G.; Sakalis, G. A dynamic ship speed optimization method with time horizon segmentation. *Ocean Eng.* **2021**, *226*, 108840. [CrossRef]
16. Sánchez Pedroche, D.; García Herrero, J.; Molina López, J.M. Context learning from a ship trajectory cluster for anomaly detection. *Neurocomputing* **2024**, *563*, 126920. [CrossRef]
17. Sun, Y.; Su, L.; Yuan, S.; Meng, H. DANet: Dual-Branch Activation Network for Small Object Instance Segmentation of Ship Images. *IEEE Trans. Circuits Syst. Video Technol.* **2023**, *33*, 6708–6720. [CrossRef]
18. Sun, K.; Xiao, B.; Liu, D.; Wang, J. Deep High-Resolution Representation Learning for Human Pose Estimation. In Proceedings of the 2019 IEEE/CVF Conference on Computer Vision and Pattern Recognition (CVPR), Long Beach, CA, USA, 15–20 June 2019; pp. 5686–5696.
19. Wang, J.; Sun, K.; Cheng, T.; Jiang, B.; Deng, C.; Zhao, Y.; Liu, D.; Mu, Y.; Tan, M.; Wang, X.; et al. Deep High-Resolution Representation Learning for Visual Recognition. *IEEE Trans. Pattern Anal. Mach. Intell.* **2021**, *43*, 3349–3364. [CrossRef] [PubMed]
20. Liu, Q.; Wang, S.; Dai, Y.; Zhang, J.; Wang, Y.; Zhou, R. Improved PSP-Net Segmentation Network for Automatic Detection of Neovascularization in Color Fundus Images. In Proceedings of the 2022 IEEE International Conference on Visual Communications and Image Processing (VCIP), Suzhou, China, 13–16 December 2022; pp. 1–5.
21. Ortega-Ruíz, M.A.; Karabağ, C.; Roman-Rangel, E.; Reyes-Aldasoro, C.C. DRD-UNet, a UNet-Like Architecture for Multi-Class Breast Cancer Semantic Segmentation. *IEEE Access* **2024**, *12*, 40412–40424. [CrossRef]
22. Anilkumar, P.; Venugopal, P.; Maddikunta, P.K.R.; Gadekallu, T.R.; Al-Rasheed, A.; Abbas, M.; Soufiene, B.O. An Adaptive DeepLabv3+ for Semantic Segmentation of Aerial Images Using Improved Golden Eagle Optimization Algorithm. *IEEE Access* **2023**, *11*, 106688–106705. [CrossRef]

Journal of
Marine Science and Engineering

MDPI

Article

A New Efficient Ship Detection Method Based on Remote Sensing Images by Device–Cloud Collaboration

Tao Liu [1], Yun Ye [1], Zhengling Lei [2,*], Yuchi Huo [3], Xiaocai Zhang [4], Fang Wang [2], Mei Sha [1] and Huafeng Wu [5]

1 College of Transport & Communications, Shanghai Maritime University, Shanghai 201306, China; dlmult@hotmail.com (T.L.)
2 College of Engineering Science and Technology, Shanghai Ocean University, Shanghai 201306, China
3 State Key Lab of CAD&CG, Zhejiang University, Hangzhou 310058, China
4 Faculty of Engineering and Information Technology, University of Melbourne, Parkville 3010, Australia
5 Merchant Marine College, Shanghai Maritime University, Shanghai 201306, China
* Correspondence: zllei@shou.edu.cn

Abstract: Fast and accurate detection of ship objects in remote sensing images must overcome two critical problems: the complex content of remote sensing images and the large number of small objects reduce ship detection efficiency. In addition, most existing deep learning-based object detection models require vast amounts of computation for training and prediction, making them difficult to deploy on mobile devices. This paper focuses on an efficient and lightweight ship detection model. A new efficient ship detection model based on device–cloud collaboration is proposed, which achieves joint optimization by fusing the semantic segmentation module and the object detection module. We migrate model training, image storage, and semantic segmentation, which require a lot of computational power, to the cloud. For the front end, we design a mask-based detection module that ignores the computation of nonwater regions and reduces the generation and postprocessing time of candidate bounding boxes. In addition, the coordinate attention module and confluence algorithm are introduced to better adapt to the environment with dense small objects and substantial occlusion. Experimental results show that our device–cloud collaborative approach reduces the computational effort while improving the detection speed by 42.6% and also outperforms other methods in terms of detection accuracy and number of parameters.

Keywords: ship detection; remote sensing image; device-cloud collaboration; semantic segmentation

1. Introduction

Ship detection plays a key role in maritime surveillance and protection of offshore installations [1–4]. The main purpose of ship detection is to identify the ship object, ship location, and ship type. Ship detection not only plays an important role in the automatic navigation, collision avoidance, and efficient communication of intelligent ships but also plays a key role in traffic management, safety supervision, environment protection, and emergency response in traffic supervision. The application of intelligent ship detection has greatly improved the efficiency and safety of maritime management and promoted the development of smart ships. With the development of deep learning methods and aerospace measurements, deep learning-based object detection models can quickly and accurately detect targets in remote sensing images and have been widely used in the military, navigation, rescue, and other fields [5,6]. Despite significant progress in deep learning-based object detection algorithms, they still face the challenge of high computation and memory resource consumption. First, we analyze the classification of existing deep learning-based object detection algorithms and their limitations. Existing deep learning-based object detection algorithms can be categorized into anchor-free, anchor-based, and Transformer-based methods [7–9]. The anchor-based method can directly classify the target and regress the bounding box, improving small object detection ability. However, it generates many redundant anchor boxes, increasing the computation and memory consumption. Although the anchor-free method predicts the target directly based on the feature map instead of a preset anchor box, it is more susceptible to noise interference, especially for remote sensing images. Remote sensing images have a complex structure and contain a large number of small objects. Fast and accurate ship detection in remote sensing images must overcome these factors' interference. The detection model based on the Transformer architecture relies on the global attention mechanism, which needs to compute the relationship between all the positions [10]. This computation becomes particularly complex and time-consuming with large-size inputs (e.g., remote sensing images). The model contains a large number of parameters, so it suffers from the problems of slow convergence speed and insufficient detection performance of small targets. Therefore, designing an efficient and accurate remote sensing image ship detection model is an important research direction [11].

In addition, semantic segmentation of remote sensing images is an important research area closely related to object detection. Semantic segmentation aims to extract regions of interest in an image and determine the class attribution of each image pixel. Deep learning-based semantic segmentation algorithms are widely used for land classification of remote sensing images. Since ships involved in maritime safety do not appear in areas outside the water, ship detection calculations in remote sensing images do not need to consider areas outside the water. This provides a vital idea for improving the efficiency of ship detection computation in remote sensing images. The semantic segmentation and object detection models based on deep learning have a large number of parameters and require significant computing power support for the model training. This puts high requirements on ordinary mobile devices.

In this paper, we propose a novel and efficient ship detection framework based on device–cloud collaboration, which integrates joint optimization of semantic segmentation and object detection and migrates some tasks that consume a lot of computing and memory resources to the cloud. This distributed device–cloud collaborative computing approach enables our system to efficiently manage computing resources and computational loads while maintaining high-precision detection. By excluding the nonwater regions, the computation and postprocessing time for these areas can be saved, thereby improving detection accuracy and efficiency. The main contributions of this paper are as follows:

1. A novel ship detection framework based on device–cloud collaboration is proposed, which integrates semantic segmentation and object detection functions. By migrating the segmentation module to the cloud, the semantic segmentation results are stored in the cloud for the device-end object detection exploitation to optimize the allocation and utilization of computational resources.
2. A Mask module and Anchor Head module are designed to reduce the computation and postprocessing time of the network.
3. The performance of semantic segmentation and ship detection is improved by joint training of semantic segmentation and object detection.
4. The postprocessing step and loss function of the original YOLO model are improved. Additionally, a Coordinate Attention module is introduced to better adapt to environments with dense small objects and heavy occlusion.
5. In order to verify the effectiveness of the proposed method, we have conducted extensive ablation and comparison experiments. The experimental results show that our device–cloud collaborative method significantly improves detection efficiency while ensuring detection accuracy.

The following sections are divided as follows: Section 2 describes previous work closely related to this research, including semantic segmentation, detection method, and device–cloud collaboration architecture. Section 3 describes our definition of ship detection for remote sensing images. Section 4 describes the method proposed in this paper, including the device–cloud collaboration framework, the Backbone structure of SD-Net, the Neck and Head structure, and some computation details. Section 5 introduces the experiments and analysis. Finally, Section 6 is the conclusion of this paper.

2. Related Work

In our framework, we use the semantic segmentation model to find the attention area and then the object detection model to obtain the final results. We also propose a new efficient computation framework based on device–cloud collaboration to integrate semantic segmentation and object detection. Therefore, this section mainly reviews the work on semantic segmentation, object detection, and device–cloud collaboration.

2.1. Semantic Segmentation in Remote Sensing Images

Currently, semantic segmentation methods based on deep learning can be divided into two main categories: one is based on the convolutional neural network, and the other is based on the Transformer network structure. Among these methods, convolutional neural networks are widely used to analyze land classification and land cover information in remote sensing images. FCN [12] is the first effective knot method to solve the semantic segmentation problem in an end-to-end way. However, it suffers from issues such as the loss of semantic information and the lack of correlation between pixels. To address these problems, researchers mainly explore three strategies: combination with traditional algorithms, multiscale methods, and data fusion. For the method combination, Shen and Gao [13] designed a semantic segmentation network based on DeeplabV3+ to enhance the accuracy of pixel classification of natural landform edges in remote sensing images by fusing the edge detection module. For the multiscale strategy, Li et al. [14] proposed a multi-field-of-view deep adaptive fusion network (MFVNet) to obtain different scales of views from multiple FOVs by pyramid sampling and select a scale-specific model to make the best prediction. Cai et al. [15] found that different classes of objects at different scales have their preferred resizing scales for more accurate semantic segmentation, leading to the proposal of a stack-based semantic segmentation (SBSS) framework that improves segmentation results. For the data fusion, Saralioglu and Gungor [16] used a 3D-2D convolutional neural network model that combines spectral and spatial information to improve the classification accuracy of high-resolution satellite images.

Recently, many works have applied the Transformer network architecture to semantic segmentation tasks, originally used to process and analyze sequence data. Wu et al. [17] proposed a semantic segmentation network fusing the convolutional neural network with a multiscale Transformer and constructed a Transformer decoder to perform semantic segmentation by processing the output features of the CNN encoder. Xiao et al. [18] proposed a novel hybrid architecture for HRRS image segmentation, combining the advantages of convolutional network and Transformer to enhance multiscale representation learning. He et al. [19] proposed a novel framework for semantic segmentation of remote sensing images by embedding Swin Transformer into the classical CNN-based UNet.

2.2. Ship Detection in Remote Sensing Images

Detecting ships in remote sensing images is a challenging task. Deep learning-based methods have made significant progress. We only list the work that is more relevant to our research. For the simplification of detection, researchers have proposed ideas similar to our method. They use different strategies to reduce the computational effort of detection, including step-by-step filtering and information fusion. Li et al. [20] first distinguished suspicious regions from images based on the Hue, Saturation, and Value (HSV) differences between the ship and the background and extracted regions of interest from the input image. Then, they designed an improved YOLO algorithm based on a high-resolution image network algorithm to match candidate ships more efficiently. To expel false candidate targets in nontarget regions, You et al. [21] improved the detection accuracy and reduced false alarms by combining a near-shore scene mask with the detection framework. In addition, Tian et al. [22] proposed a small two-stage detection method.

Candidate region extraction is first performed to localize potential ship locations, and then detection calculation is conducted. Liu et al. [23] proposed an unsupervised deep learning ship depth estimation method for sensing images. Thus, ship detection can fuse the depth information. However, these methods lack careful consideration for the deployment of lightweight front ends.

For lightweight ship detection models, researchers have also conducted extensive work based on the YOLO model for lightweight ship detection. The authors in Xu et al. [24] proposed a lightweight shipboard SAR ship detector based on the YOLOv5 algorithm, reducing the computation and improving detection performance through network pruning and multimodule integration. The WGC-PANet proposed in [25] reduces the model parameter and computational complexity to ensure the light weight of the network. Peng et al. [26] offered a lightweight YOLO (AMFLW-YOLO), which reduces the model parameter by employing depth-separable convolution and introducing Coordinate Attention (CA) modules and bidirectional feature pyramids to enhance feature extraction. Liu et al. [27] proposed a lightweight ship detection model based on joint optimization of defogging and detection, which has improved the YOLOv5s model.

2.3. Cloud Computing and Device–Cloud Collaboration

With the widespread application of deep learning models, the computational power of the front end makes it difficult to meet the growing computation demand of deep learning, leading to a decrease in model training accuracy and prediction efficiency. However, cloud computing technology has been widely used and has demonstrated flexible and powerful computing capabilities. To improve the front end's computational efficiency and storage capacity, researchers have begun to utilize cloud computing to design device–cloud collaboration frameworks to solve problems with extensive computation and storage demands.

Regarding computing power optimization, current research focuses on computing power architecture design, computing offloading, and model decomposition. To alleviate the problem of limited computational capacity of edge servers, Liu et al. [28] constructed a hybrid computing task offloading architecture for cloud–edge–terminal collaboration and proposed a multiobjective offloading algorithm SPMOO based on the improved Strength Pareto Evolutionary Algorithm SPEA2, which aims to optimize the average latency and energy consumption of complex dependent tasks. Liu et al. [29] investigated joint task offloading and resource allocation for cloud–edge–terminal collaboration by formulating a new optimization problem to minimize the task processing delay. Zhuang et al. [30] proposed a novel module-level model decomposition design that decomposes the cloud model into multiple composable submodules, enabling it to generate compact, task-specific submodels for different edge devices. In terms of application, researchers have also conducted many valuable explorations. Based on the concept of cloud–edge–terminal collaboration, Wang et al. [31] proposed a four-layer networked architecture that breaks through traditional battery management systems' computational power and storage space limitations. Zhang et al. [32] proposed a device–cloud collaboration-based fault detection model with a stacked auto-encoder deep neural network (SAES-DNN). The model is trained in the cloud and deployed on the front end through a transfer learning strategy. Tan et al. [33] designed user experience metrics to measure object detection and energy consumption comprehensively, then proposed an adaptive scheduling algorithm to schedule object detection and tracking tasks through network bandwidth and transmission delay prediction.

3. Definition of the Problem

In port and waterway areas where ship management is required, front-end equipment obtains real-time images to detect ships. However, ship targets only exist in water areas, so nonwater information in images is redundant. This redundant information increases the amount of calculation required for ship detection, which in turn increases the computing power consumption of front-end equipment. Therefore, this paper mainly focuses on the high complexity and high consumption of front-end detection calculations, especially for ship monitoring of UAV applications. Existing ship detection models are large, consume a lot of power, and run on high-performance computers. However, when front-end equipment such as UAV performs ship detection, it faces limitations in terms of chip computing efficiency, power consumption, storage capacity, and network capacity. Specifically, the limitations can be described as follows:

1. Chip computing efficiency affects the speed of ship detection.
2. Power consumption affects the working time of front-end equipment. For example, the maximum flight time of the DJI Mavic 3E is 45 min.
3. Storage capacity affects the detection capacity and range of front-end equipment. For example, the storage capacity of DJI Mavic 3E is usually 512 GB.
4. Network capacity affects the speed at which information can be transmitted.

To deal with these limitations, we redefine the remote sensing image detection problem under the device–cloud collaborative framework, refine the calculation process, improve the accuracy of each subprocess, reduce the front-end calculation amount, and achieve fast and accurate ship detection.

Definition 1. *Detection model* $Q(x)$.

$$Q(x) = F(x) \otimes H_1(x) \otimes H_2(x). \tag{1}$$

Since ships involved in maritime safety do not appear in regions other than water, focusing our attention on the water region can effectively reduce unnecessary computations and thus improve the efficiency and accuracy of the overall detection process. Therefore, ship detection $Q(x)$ does not need to consider regions other than water, and ship detection can be categorized into semantic segmentation $F(x)$, object detection in water region $H_1(x)$, and object detection in the nonwater region $H_2(x)$.

Definition 2. *Semantic segmentation* $F(x)$.

Semantic segmentation determines the category to which each pixel of the image belongs. We first categorize the pixels of remote sensing images by semantic segmentation $F(x)$ to distinguish between water and nonwater areas.

Definition 3. *Object detection in water* $H_1(x)$.

In the water region of the image, we perform ship detection.

Definition 4. *Object detection in nonwater* $H_2(x)$.

In the nonwater region of the image, ship detection is not performed by utilizing semantic features and mask images, thus reducing the amount of computation.

4. Methodology

To efficiently process remote sensing images and improve the accuracy of ship detection, we design a novel joint optimization network for semantic segmentation and object detection (SD-Net, shown in Figure 1). The network is capable of joint training and can be deployed separately to achieve low coupling between modules. We introduce residual connectivity between the segmentation and detection modules and use the α-IoU loss as the regression loss for the bounding box to ensure that the network converges stably. Considering the dense targets and severe occlusion in remote sensing images, we use the Confluence algorithm as a postprocessing step for ship detection. In the head layer of the detection network, we integrate the CA (Coordinate Attention) module to improve the model's sensitivity to the target location and the detection accuracy. In addition, we design the Mask module and Anchor Head module to reduce the computation of network convolution operation and the number of candidate bounding boxes based on the mask feature, thus reducing the overall computation and postprocessing time.

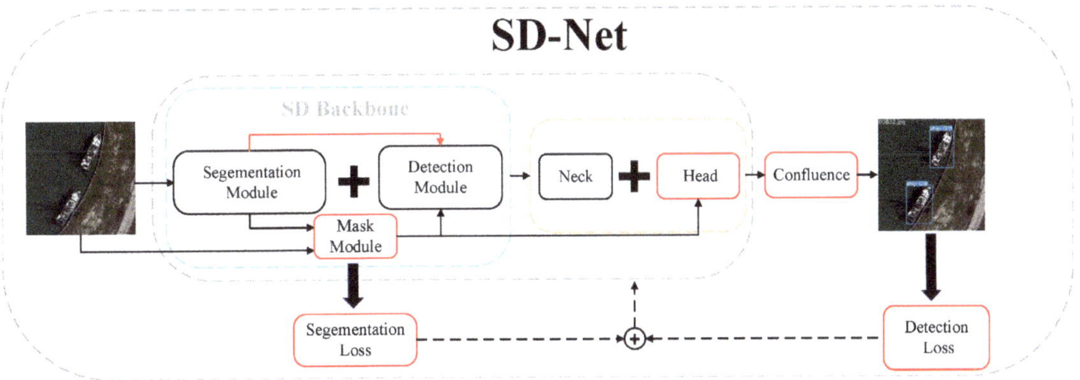

Figure 1. Schematic diagram of SD-Net network structure. Solid arrows indicate forward processes and dashed arrows indicate backward processes. The red boxes are the main modules designed in this paper. The red arrow denotes the residual connection we designed.

In this paper, the device–cloud collaboration computation framework is used to place each subprocess of the detection model $Q(x)$ either in the cloud or in the front end, depending on its computation characteristics. We first describe the device–cloud collaboration framework and then explain the structure of our proposed SD-Net network and the improvements to the YOLO model.

4.1. Device–Cloud Collaboration Framework

With the rapid development of cloud computing, massive computing programs can be split into smaller subtasks. This advancement provides new opportunities for our ship detection research, which is very important for the front end in the 5G era. First, most front ends have relatively low computational power and small storage capacity. However, model training and model prediction of the deep learning-based semantic segmentation and object detection tasks require high computing power and large storage capacity. In order to reduce the computing consumption of the front end, some tasks that consume a lot of computing and memory resources should be migrated to the cloud. Second, the computing property of each task is different. Semantic segmentation can be defined as a "one-time calculation". That is, after the input image has undergone a semantic segmentation operation, the optimized results can be retained for subsequent operations. The ship detection calculation can use the obtained results to identify ship objects in the water area. However, object detection in this paper is defined as "repetitive calculation" because this calculation will be repeated for each input image.

Based on the above analysis, we designed a device–cloud collaboration framework for ship detection, which migrates operations that require lots of computing and memory resources to the cloud. Specifically, we assign the "one-time calculation" and the corresponding output to the cloud side and assign the "repetitive calculation" that can be run directly on the front end and the corresponding output to the front end.

In the cloud, we train SD-Net by cloud computing, and then we divide SD-Net into two modules: one module is used for semantic segmentation, and the other is used for ship detection. Since the range of the detection area is usually administratively fixed, the cloud-first performs semantic segmentation on the remote sensing images of the detection area to distinguish between water parts and nonwater parts and stores the obtained mask images in the cloud. Figure 2 shows our device–cloud collaboration scenario. Under good weather conditions, normal 5G network, and normal operation of the device–cloud devices, when the front-end device (UAV) reaches a certain location in the surveillance area, it uploads the current location information to the cloud. After receiving the location information, the cloud sends mask images of the corresponding area to the front end.

By applying these mask images to the real-time images, the front-end device is able to reduce the convolution operations in nonwater areas and reduce the number of candidate bounding boxes generated in the nonwater part of the network, thus reducing the overall computation and shortening the postprocessing time. Based on this distributed computing approach, our framework can effectively manage the computational load while maintaining high detection accuracy.

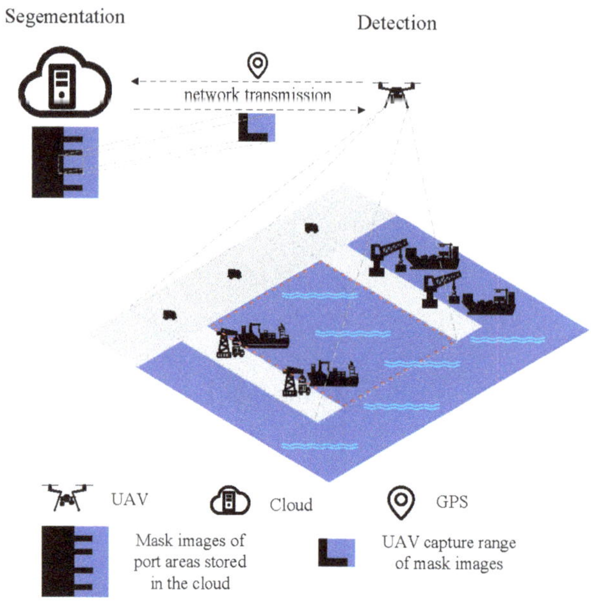

Figure 2. Schematic diagram of the device–cloud collaboration scenario. The area surrounded by the gray dashed line is the detection area of the UAV, and the area surrounded by the red dashed line is the water area.

4.2. Backbone Structure

4.2.1. Basic Module

In this paper, we design our SD Backbone with reference to the ideas of DeepLab [34] and YOLO [35] models. As shown in Figure 3, the Backbone structure fuses multiscale information to improve the accuracy of segmentation based on an Atrous Spatial Pyramid Pooling Integrated (ASPP) module [36]. By applying multiple parallel null convolutions with different sampling rates, the ASPP module can increase the receptive field while maintaining information integrity, which is especially important for distinguishing targets of various sizes. The features extracted at each sampling rate are further processed in separate branches to obtain multiscale object information. The Efficient Layer Aggregation Networks (ELAN) module is a critical component of our network, which enhances the network's learning capability and robustness by controlling the shortest and longest gradient paths. In addition, we also use the Spatial Pyramid Pooling and Fully Connected Spatial Pyramid Convolution (SPPCSPC) module, which reduces the computation and improves the processing speed and accuracy.

To avoid the vanishing and exploding gradient problem, we design residual connections between the segmentation module and the detection module (the red solid lines shown in Figure 3), which help to optimize the stability of the network training.

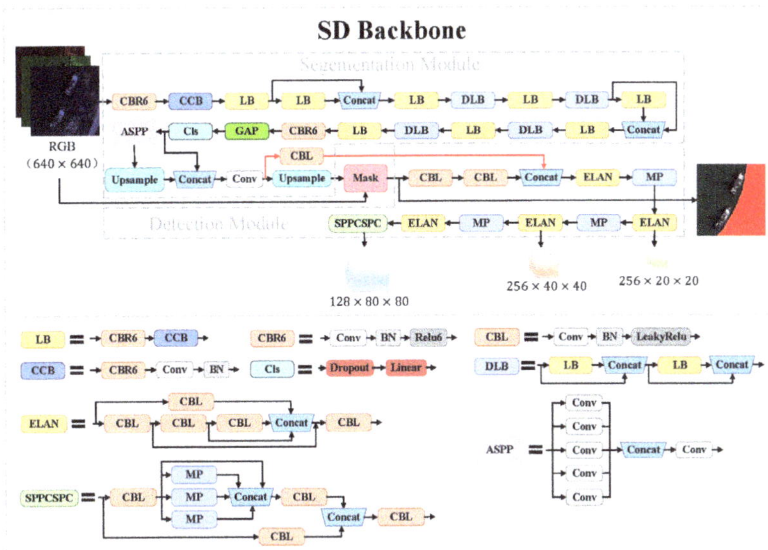

Figure 3. Diagram of the SD Backbone Network. The land area where the mask is added in the remote sensing image is marked in red.

4.2.2. Mask Module

We design a Mask module to filter the input image based on the semantic segmentation results and fuse the semantic feature. By analyzing the elements on the feature map and their corresponding positions on the original map, we decide whether to perform subsequent convolution operations or not. For regions that are not within the watershed, we do not process the convolution operation, significantly reducing the computational effort of the network by this method. It can optimize not only the detection calculation but also the segmentation results. The Mask module can be represented by Equation (2).

$$I' = \begin{cases} (0,0,0), & \text{if } M(x_i, y_j) = 0; \\ I(x_i, y_j), & \text{if } M(x_i, y_j) \neq 0. \end{cases} \quad (2)$$

where I denotes the input image, $I(x_i, y_j)$ denotes the pixel value of the image at (i, j), M is the mask image, $M(x_i, y_j)$ is the pixel value of the mask image at (i, j), I' is the output image after adding mask, and $I'(x_i, y_j)$ denotes the pixel value of the image at (i, j). The use of the Mask module is described in Section 4.3.2.

4.3. Neck and Head Structure

4.3.1. Basic Module

To fully utilize the feature information extracted from the SD Backbone and the masked images, we design a network structure to integrate the aspects of information, as shown in Figure 4. It can integrate three-scale feature information output from Backbone to achieve more effective object detection. Specifically, our strategy is to perform object detection on three different scale feature maps, 80×80 feature map for small object detection, 40×40 feature map for medium-sized object detection, and 20×20 feature map for large object detection. This multiscale detection approach allows the model to recognize and distinguish between targets of different sizes more accurately. Last, we use feature fusion and multiscale processing methods to improve the detection efficiency for the complex content of remote sensing images.

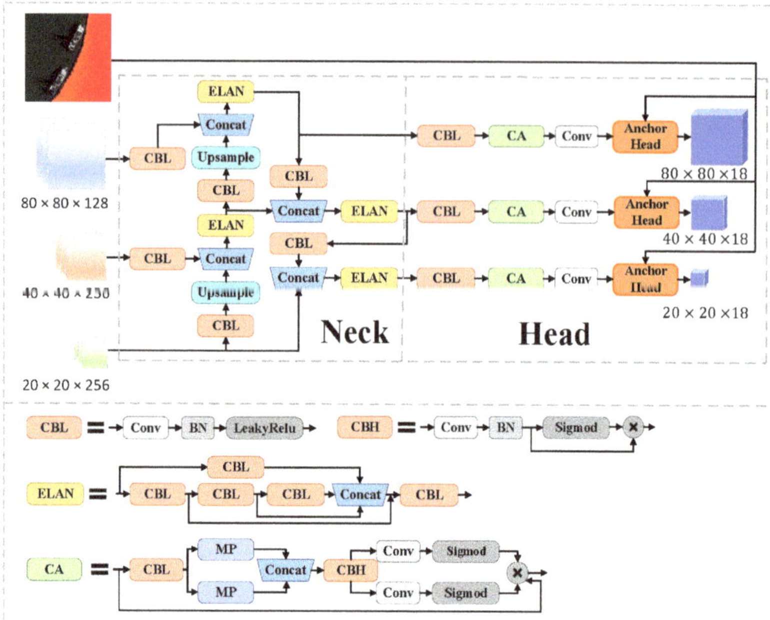

Figure 4. Diagram of Neck and Head network structure. The land area where the mask is added in the remote sensing image is marked in red.

4.3.2. Anchor Head Module

The postprocessing step usually requires a lot of computing resources. Optimization of this step is crucial in the object detection process. To address this challenge, we design a novel Anchor Head module in the detection head section for fusing semantic features and generating candidate bounding boxes only in the water portion, thus reducing the computational effort. This process can be described by Equations (3) and (4).

$$B_k = \begin{cases} \text{anchor}, & \text{if } S_k + \sum\limits_{(i,j)\in \text{Neighbord}_k} S'_{ij} > T; \\ \text{no anchor}, & \text{otherwise.} \end{cases} \tag{3}$$

$$S_k = \begin{cases} 1, & \text{if } (x_k, y_k) \in L; \\ 0, & \text{if } (x_k, y_k) \notin L. \end{cases} \tag{4}$$

where (x_k, y_k) is the coordinate position of the centroid of the kth grid, L denotes the pixel of the water portion, S_k is an indicator function of whether the centroid of the kth grid is in the water or not, and T is a threshold for determining whether or not to generate a candidate bounding box, we set $T = 4$ in this paper. We not only consider whether or not the current grid's centroid is in the water but also consider the states of the neighboring locations. This provides a more nuanced and dynamic way of determining whether to generate a candidate bounding box in the current region. The Anchor Head module determines whether to generate a candidate bounding box and thus decides whether to conduct a prediction.

4.4. Coordinate Attention Module

The attention mechanism focuses on localization information, which can suppress useless information and amplify useful information. Therefore, incorporating the attention mechanism can effectively enhance the extraction of localization features from the region of interest. Because the ship targets in remote sensing images are often small and dense, we add an efficient CA module [37] in the Head layer, which embeds the position information into the channel attention so that the model can more accurately locate and recognize the ship in the water area with little computation load. The processing of the CA module is as follows:

$$f = \delta(F_1([z^h, z^w])). \tag{5}$$

The input x is first subjected to a global pooling operation in the horizontal and vertical directions to produce the outputs Z^h and Z^w, respectively. These two features are subsequently spliced along the spatial dimension and fed into a convolution layer. The output of the convolution layer is processed through a sigmoid activation function, and the generated features are then divided into two parts along the spatial dimension: f^h and f^w.

$$g^h = \sigma(F_h(f^h)), \tag{6}$$

$$g^w = \sigma(F_w(f^w)), \tag{7}$$

$$y_c(i,j) = x_c(i,j) \times g_c^h(i) \times g_c^w(j). \tag{8}$$

By feeding f^h and f^w into a convolution layer, the attention weights g^h and g^w will be obtained followed by a sigmoid function. Each element in the attentional feature maps g^h and g^w reflects whether the ship object of interest is present in the corresponding row and column. This encoding process allows our coordinate attention to more accurately localize the exact location of the ship, which in turn helps the overall model to identify the ship better. g^h and g^w will be multiplied with the original input x to generate coordination features.

The CA module decomposes channel attention into two 1D feature encoding processes that aggregate features along different directions. Long-range dependencies are captured along one direction, and precise location information is retained along the other spatial direction. The output feature maps are then encoded separately to form a pair of direction-aware and location-sensitive feature maps, which can be applied complementarily to the input feature map to enhance the representation of the interest target.

4.5. Postprocessing Design

The YOLO model uses the NMS algorithm to retain the bounding box. However, NMS [38] relies heavily on the classification confidence score to determine the optimal box, which is not always an effective description of localization accuracy. Ship targets in remote sensing images are usually small, densely distributed, and often occluded. The NMS algorithm leads to significant problems of missed and false detections. In order to deal with these small and dense targets more effectively, we adopt the Confluence algorithm [39] as the postprocessing step of the network.

The Confluence algorithm reduces unnecessary bounding boxes while preserving key detection results. The confidence-weighted Manhattan distance is first used to assess the similarity between different candidate bounding boxes. Then, dense clusters are obtained based on normalized proximity metrics. An optimal box is selected where the gradient converges to zero, and other bounding boxes fused with the optimal box are removed. Rather than simply suppressing a large number of detection results, the Confluence algorithm determines the optimal box by identifying the box that intersects the most with other bounding boxes. As a result, this algorithm reduces the likelihood of false detections and thus improves the overall accuracy. It is more robust when dealing with densely distributed and small targets.

The Manhattan distance is the sum of the vertical and horizontal distances between two points, and the Manhattan distance $MD_{(u,v)}$ between point $u(x_1, y_1)$ and point $v(x_2, y_2)$ is calculated according to Equation (9).

$$MD_{(u,v)} = |x_2 - x_1| + |y_2 - y_1|. \tag{9}$$

The proximity measure between two candidate bounding boxes can be expressed as the sum of the Manhattan distances of the upper-left and lower-right points (as shown in Figure 5). It is calculated according to Equation (10).

$$P_{(u,v,m,n)} = MD_{(u,m)} + MD_{(v,n)}, \quad (10)$$

where $P_{(u,v,m,n)}$ is the proximity measure between the two candidate bounding boxes, $MD_{(u,m)}$ is the Manhattan distance between the upper left corner points of the two boxes, and $MD_{(v,n)}$ is the Manhattan distance between the lower right corner points of the two boxes. P denotes the degree of intersection of two frames. A larger P indicates that two frames are less likely to represent the same object. For frames within a cluster, we take the frame with the smallest intra-cluster P-value as the best frame. Considering that different frame sizes may result in particular sensitivity to the threshold parameter, the P-value between any two intersection frames will be less than two by normalizing all coordinates to a range from 0 to 1. Thus, if the P-value between two frames is less than 2, they can be considered to belong to the same cluster. Once the clusters are identified, the optimal intracluster boxes can be determined. Next, a threshold is set to eliminate false detections by removing all boxes that have a P-value less than this threshold with the optimal box. This process is repeated for all candidate boxes to optimize the overall detection.

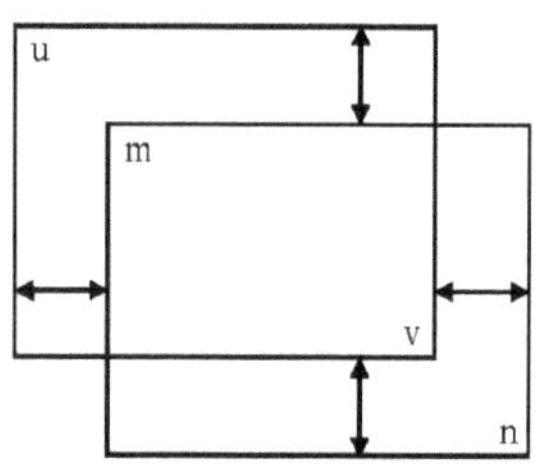

Figure 5. Schematic diagram of the Manhattan distance between two bounding boxes.

4.6. IoU Loss Design

Remote sensing images contain both small and large ship targets. In order to enable the model to adapt to different types of targets, we use α-IoU loss [40] to enhance the sensitivity of the loss function to bounding boxes of different sizes and aspect ratios. This does not add additional training parameters and does not increase the computation burden. The Loss is defined according to Equation (11).

$$L_{\alpha-\mathrm{IoU}} = \frac{1-\mathrm{IoU}^{\alpha}}{\alpha}, \quad \alpha > 0, \quad (11)$$

where α is a positive parameter. Most of the existing loss functions can be derived by adjusting the value of α. The α-IoU loss in the two cases of $\alpha \to 0$ and $\alpha \nrightarrow 0$ is calculated according to Equation (12).

$$L_{\alpha-\mathrm{IoU}} = \begin{cases} -\log(\mathrm{IoU}), & \alpha \to 0; \\ 1-\mathrm{IoU}^{\alpha}, & \alpha \nrightarrow 0. \end{cases} \quad (12)$$

For the case of $\alpha \geq 1$, the α-IoU loss under the condition of $\alpha \nrightarrow 0$ can be extended to a more general form by introducing a power regularization term α_1 (as shown in Equation (13)).

$$L_{\alpha-\mathrm{IoU}} = 1 - \mathrm{IoU}^{\alpha_1} + \rho^{\alpha_2}(B, B^{gt}), \quad (13)$$

where $\alpha_1, \alpha_2 > 0$, B and B^{gt} are the predicted box and the ground truth, respectively, and $\rho^{\alpha_2}(B, B^{gt})$ is a penalty term calculated according to B and B^{gt}.

When $\alpha_1 > 1$, the loss function will focus more on high IoU objects by increasing the weight of their relative loss. Thus, it helps the object detector to learn objects with a high degree of IoU faster. At the same time, the absolute amount of loss increased by α_1 is more conducive to optimizing objects at all levels.

4.7. Joint Optimization

Object detection and semantic segmentation are two separate remote sensing image processing tasks. However, these two tasks are closely related: accurate segmentation can help object detection to localize objects more accurately, while effective object detection can provide essential clues about object classes in the segmentation task. By integrating object detection and image segmentation in a unified framework, joint optimization of the two tasks can be achieved with the help of information flow between them.

We use a joint loss function to guide this joint optimization process. The joint loss function is shown in Equation (14).

$$L_{MLT} = \lambda_{seg} L_{seg} + (1 - \lambda_{seg}) L_{de}, \tag{14}$$

where L_{seg} is the segmentation loss, L_{de} is the detection loss, and λ_{seg} is the weight parameter. At the beginning of training, we prioritize accurate segmentation since it is essential for subsequent ship detection. We set the initial value of λ_{seg} to a high value of 0.9. As the training progresses, this weight is gradually reduced to 0.5. This strategy first ensures high-quality segmentation results and then gradually enhances ship detection, thereby achieving an effective balance between the two tasks.

The segmentation loss L_{seg} includes CE Loss and Dice Loss. The former calculates the loss of each pixel equally, and the latter can alleviate the negative impact of foreground and background imbalance in the dataset. The two losses are calculated according to Equation (15) and Equation (16), respectively.

$$\text{CE Loss} = -\sum_{n=1}^{N} y_n \times \log(\hat{y}_n), \tag{15}$$

$$\text{Dice Loss} = 1 - \frac{2\sum_{n=1}^{N} y_n \hat{y}_n}{\sum_{n=1}^{N} y_n + \sum_{n=1}^{N} \hat{y}_n}, \tag{16}$$

where y_n is the label, $\hat{y}_n$ is the prediction result, and N is the number of pixels. The detection module L_{det} uses the α-IoU loss (as in Equation (11) to accurately assess the degree of overlap between the predicted bounding box and the ground truth and uses BCE Loss to assess the probability of the presence of the detected target. The BCE Loss is calculated according to Equation (17).

$$\text{BCE Loss} = -\frac{1}{N}\sum_{m=1}^{M} \left[y_m \times \log(\hat{y}_m) \quad + (1 - y_m) \times \log(1 - \hat{y}_m) \right], \tag{17}$$

where y_m is the label, $\hat{y}_m$ is the prediction result, and M is the number of targets. Based on this joint optimization, the information flow between these two tasks forms a positive feedback loop, making the two tasks more accurate and effective.

5. Experiments and Analysis

5.1. Implementation Details

5.1.1. Dataset

The first challenge of this research is that few datasets are dedicated to both ship detection and semantic segmentation. To address this issue, we construct a new dataset based on publicly available datasets, including DIOR [41], DOTA [42], and LEVIR [43]. We select remote sensing images labeled with "ship" and exclude all nonship remote sensing images. Then, we label these remote sensing images with water and nonwater areas. The newly constructed dataset contains 2891 images. We divide these data into training and validation sets in the ratio of 80/20.

Figure 6 shows the visual analysis of our dataset. Figure 6a shows that there are more than 40,000 ship instances in our dataset. Figure 6b shows the sizes and numbers of all our ground truths. Figure 6c is the distribution of the centroids of ground truths with respect to the position of the whole image, and it can be seen that most of the instances are concentrated in the central region of the image. Figure 6d shows the distribution of the number of different instances, and it can be seen that the majority of instances of the dataset are small ships. Figure 6e shows the distribution of water and nonwater pixels in the dataset. The water portion can account for about one-third of the dataset.

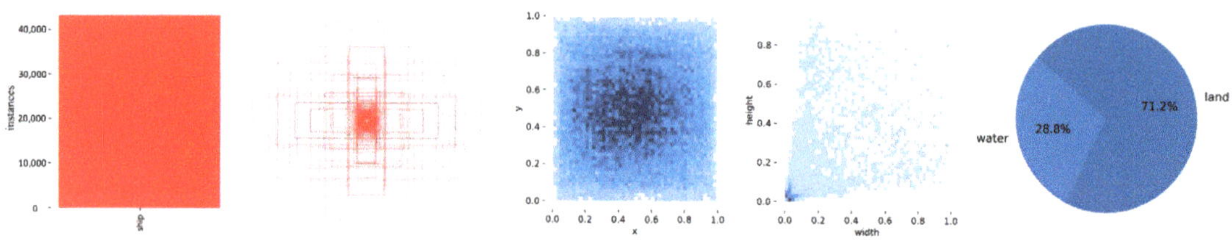

(a) (b) (c) (d) (e)

Figure 6. Dataset visualization: (**a**) Number of ship instances, (**b**) the distribution of ground truths, (**c**) the distribution of instance center points, (**d**) the distribution of instance sizes, and (**e**) the proportion of water and land areas. The darker the color of the point, the more the number of instances in (**c**,**d**).

5.1.2. Model Training

During the training process, we first resize the images to a uniform size of 640 × 640. We employ various data enhancement methods, including image flipping, affine transformation, and Gaussian blurring, to further enhance the model's generalization ability and robustness. We use Stochastic Gradient Descent (SGD) as the optimizer and train the model for 200 epochs with a starting learning rate of 0.01, weight decay of 0.0005, batch size of 32, and cosine annealing used to adjust the learning rate dynamically. All experiments are conducted based on the Pytorch framework and run on NVIDIA 4090 GPU. We use the constructed dataset to train our model and the validation set to validate the results of each epoch during model training. Figure 7 shows the graphs of Box loss (Box), Objectness loss (Objectness), Classification loss (Classification), Precision, Recall, mAP@0.5, and mAP@0.5:0.95 during the training of our model. Since we are testing for only one category, ships, the category loss is not recorded. As shown in Figure 7, the val box loss and val Objectness loss drop rapidly in the first hundred epochs and then stabilize. From Figure 7, mAP@0.5 and mAP@0.5:0.95, we can see that our model can obtain good detection results, and there is no overfitting problem.

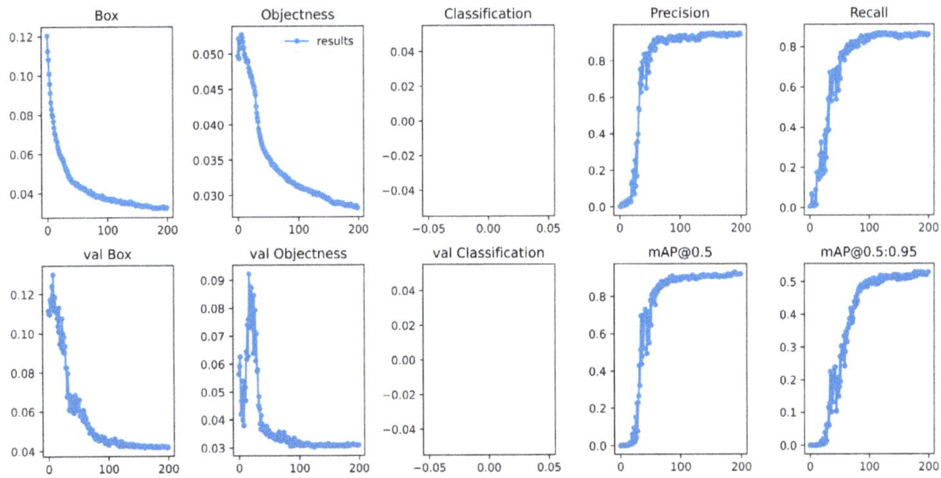

Figure 7. Performance indicators of model training, including bounding box regression loss (Box), objectness loss (Objectness), classification loss (Classification), Precision, Recall, and mean average precision (mAP@0.5: IoU = 0.5, mAP@0.5:0.95: IoU $\in [0.5, 0.95]$). val denotes the corresponding result on the validation set.

5.2. Evaluation Indicators

In this paper, we choose the inference time of the model at $batch = 1$ as a metric for detection speed, use mAP to evaluate the detection performance of the model, and adopt Mean Intersection over Union (MIoU), Mean Pixel Accuracy (MPA), and Accuracy to evaluate the segmentation performance. The metrics are defined as follows:

$$\text{mAP} = \frac{1}{C}\sum_{k=1}^{N^{th}} P(k)\Delta R(k), \tag{18}$$

$$\text{MIoU} = \frac{1}{T+1}\sum_{i=0}^{T}\frac{TP_i}{TP_i + FN_i + FP_i}, \tag{19}$$

$$\text{MPA} = \frac{1}{T+1}\sum_{i=0}^{T}\frac{TP_i}{TP_i + FP_i}, \tag{20}$$

$$\text{Accuracy} = \frac{TP + TN}{TP + FP + FN + TN}, \tag{21}$$

where C represents the number of target categories ($C = 1$ in this paper), N^{th} denotes the number of IoU thresholds, k denotes the IoU threshold, $P(k)$ denotes the precision under threshold, $R(k)$ denotes the recall rate under threshold; TP denotes the number of pixels correctly predicted as water; FP refers to the number of pixels incorrectly predicted as nonwater; FN denotes the number of pixels incorrectly predicted as water; TN denotes the number of pixels correctly predicted as nonwater; and $T + 1$ denotes the number of categories of pixels ($T = 1$ in this paper).

5.3. Parameter Analysis

In the cloud, we optimize semantic segmentation and object detection by joint training with the loss function defined in Equation (14). λ_{seg} is a parameter that controls the weight of the segmentation loss. To explore the effect of this parameter on segmentation and detection, we perform a parameter sensitivity analysis on λ_{seg}. The specific results of the experiment are shown in Table 1. The effect of λ_{seg} on the evaluation metrics of segmentation and detection is not significant. In this paper, we choose to set the initial value of the segmentation loss weight λ_{seg} to a high value of 0.9 at the beginning of the training, and gradually reduce this weight to 0.5 as the training progresses. Such a strategy ensures high-quality segmentation results while ensuring that the detection achieves a high mAP.

Table 1. Sensitivity analysis of the λ_{seg} parameter in the loss function.

λ_{seg}	Segmentation			Detection	
	MIoU (%)	MPA (%)	Accuracy (%)	mAP@0.5 (%)	mAP@0.5: 0.95 (%)
Constant at 0.5	93.0	96.2	97.0	92.3	52.9
Increasing from 0.5 to 0.9	92.8	96.1	97.1	92.2	52.9
Decreasing from 0.9 to 0.5	93.0	96.2	97.1	92.4	52.9

5.4. Analysis of the Calculation Amount

In this paper, we perform object detection based on the corresponding mask constructed in the cloud. Before the detection task, the mask image is transmitted from the cloud to the front end, and the masked part enters the object detection convolutional network without generating any anchor frames or performing the corresponding subsequent computations.

Assuming that the detection network has a fixed amount of computation for each region of the image, it is assumed that the reduced amount of computation is related to the proportion of mask in the original image. It can be expressed by Equation (22).

$$W' = \tau(1 - k)\omega, \tag{22}$$

where W' is the calculation amount required for network postprocessing after adding the Mask module and Anchor Head module, k is the percentage of the mask in the original image, and ω is the calculation amount required for all grids in the network. Let $W = \tau\omega$ be the calculation amount required for network postprocessing of the original image.

The above scenario is ideal without considering the network latency between the front end and the cloud. In practical applications, network latency is an unavoidable problem. The network latency will affect the effect of the calculation reduction of our framework. Therefore, we update the Equation (22) to Equation (23).

$$W' = \tau(1 - k)\omega + \lambda\phi, \tag{23}$$

where λ is the average network latency time in milliseconds, and ϕ is the average MFLOPS (Million Floating-point Operations per Second) of the CPU/GPU. $\lambda\phi$ is usually a very small value. These variables need to be obtained through statistics of practical applications.

As shown in Figure 8, we take three remote sensing images in the dataset as an example. After these images go through the detection network, each image is divided into an 80×80 grid. After adding the mask, the area where our network has to perform computation is significantly reduced, and no anchor frame is generated within the mask. Compared with the detection without masked features, our solution reduces the computational effort for the generation of anchor frames and the subsequent computation of the masked portion.

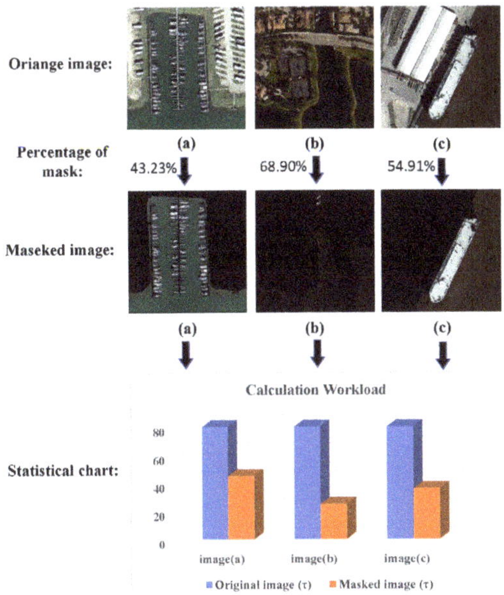

Figure 8. Schematic diagram of mask operation and calculation reduction statistics. (**a**–**c**) are three remote sensing images in the dataset. The computation is reduced by adding a mask to the original image and not computing on the nonwater part.

5.5. Comparison with Other Detection Models

To analyze the ship detection performance of our proposed model, we compare the proposed object detection model with YOLOv3-tiny [44], YOLOv4s-tiny [45], YOLOv5s, YOLOv6s [46], YOLOv7-tiny [35], YOLOV8s, RT-DETR-R18 [47], and YOLOv5-ODConvNeXt [48]. We use mAP, Parameters, Weight file size, and Giga Floating-point Operations Per Second (GFLOPs) as evaluation metrics. The comparison results are shown in Table 2. By jointly optimizing semantic segmentation and object detection, our detection model improves ship detection accuracy and performs well in terms of the number of parameters, weight file size, GFLOPs, and Speed. Our detection model achieves 92.4% accuracy. Compared with YOLOv3-tiny, YOLOv4-tiny, YOLOv5s, YOLOv6s, YOLOv7-tiny , YOLOv8s, RT-DETR-R18, and YOLOv5-ODConvNeXt, our model's accuracy is 21.7%, 26.2%, 3.7%, and 13.3% higher, respectively, 2.9%, 2.3%, 14.7% and 0.4%. In terms of the number of parameters, our detection model has fewer parameters than other models: 30.2% fewer than YOLOv3-tiny, 1.6% fewer than YOLOv4-tiny, 14.3% fewer than YOLOv5s, 67.6% fewer than YOLOv6s, 45.9% fewer than YOLOv8s, 70% fewer than RT-DETR-R18, and 14.3% fewer than YOLOv5-ODConvNeXt. In terms of weight file size, the weight file size of our detection model is 30.9%, 1.1%, and 15.5% smaller than YOLOv3-tiny, YOLOv4-tiny, YOLOv5s, YOLOv6s, YOLOv7tiny, YOLOv8s, RT-DETR-R18, and YOLOv5-ODConvNeXt, respectively, 70.3%, 1.3%, 64.2%, 70.3%, and 15.2%. In terms of GFLOPs, our detection model has 24.8%, 41.2%, 38.6%, 78.8%, 25.4%, 65.8%, 83.8%, and 34.5% lower GFLOPs compared to YOLOv3-tiny, YOLOv4-tiny, YOLOv5s, YOLOv6s, YOLOv7-tiny, YOLOv8s, RT-DETR-R18, and YOLOv5-ODConvNeXt, respectively. The experiments show that our object detection model achieves lightweight while maintaining high accuracy.

Table 2. Performance metrics of our object detection model. Our proposed method is compared with YOLOv3-tiny, YOLOv4-tiny, YOLOv5s, YOLOv6s, YOLOv7-tiny, YOLOv8s, RT-DETR-R18, YOLOv5-ODConvNeXt. The bold indicates the best result in each column, and the underline indicates the second-best result in each column.

Models	mAP (%)	Params (M)	Weight File (MB)	GFLOPs
YOLOv3-tiny	75.9	8.6	66.3	<u>12.9</u>
YOLOv4-tiny	73.2	<u>6.1</u>	<u>46.3</u>	16.5
YOLOv5s	89.1	7.0	54.2	15.8
YOLOv6s	81.5	18.5	154.3	45.7
YOLOv7-tiny	89.8	**6.0**	46.4	13.0
YOLOv8s	90.3	11.1	128.0	28.4
RT-DETR-R18	80.5	20.0	154.3	60.0
YOLOv5-ODConvNeXt	<u>92.0</u>	7.0	54.0	14.8
Ours (Detection)	**92.4**	**6.0**	**45.8**	**9.7**

5.6. Ablation Experiment

In this section, we perform the ablation experiments, which include postprocessing, semantic segmentation, and detection module.

5.6.1. Postprocessing

We compare different postprocessing methods, including NMS, Soft NMS [49], DIoU NMS [50], and the Confluence algorithm. Confluence algorithm determines the optimal box by identifying the box that intersects the most with other bounding boxes. It is more robust when dealing with densely distributed and small targets. Due to this feature, it shows better performance than other postprocessing methods. The comparison of detection results is shown in Table 3.

Table 3. The ablation experiment results for different postprocessing methods. Bold indicates the best result in each column, and underline indicates the second-best result in each column.

Methods	mAP@0.5 (%)	mAP@0.95 (%)
NMS	<u>90.3</u>	<u>51.9</u>
Soft NMS	86.6	50.6
DIoU NMS	88.6	51.5
Ours (Confluence)	**92.2**	**52.9**

5.6.2. Segmentation Module

We conduct ablation experiments to explore the effect of joint optimization on the performance of semantic segmentation. According to Table 4, it can be seen that joint optimization will improve the performance of semantic segmentation. MIoU, MPA, and Accuracy of our model are 1%, 0.7%, and 0.5% higher than that of the segmentation-only model, respectively. Some segmentation results on the validation dataset are shown in Figure 9 to visualize the improvement. Our model can achieve more accurate segmentation results, especially preserving the ship target information.

Table 4. The ablation experiment results for joint optimization of segmentation module. Bold indicates the best result in each column.

Methods	MIoU (%)	MPA (%)	Accuracy (%)
Segmentation only	92.0	95.5	96.6
Segmentation (joint optimization)	**93.0**	**96.2**	**97.1**

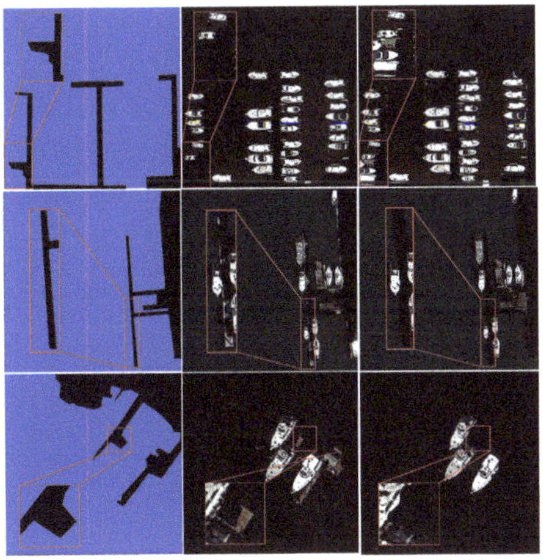

Figure 9. Segmentation results of different models. The first column shows the segmentation results of the segmentation-only model, the second column shows the results of the joint optimization model, and the third column shows the ground truths. The red rectangle shows a partially enlarged detail of the image.

5.6.3. Detection Module

To quickly and accurately detect ship objects in remote sensing images, we design a segmentation and detection joint optimization network and use the Confluence algorithm, CA module, α-IoU loss function, Mask module, and Anchor Head module to improve the YOLO model. A series of ablation experiments are conducted on the validation dataset to analyze which improvement plays a more critical role in improving detection performance. The results are shown in Table 5. When the joint optimization of semantic segmentation and object detection is used, mAP@0.5 and mAP@0.95 of the detection model increase by 0.9% and 1.6%, respectively, compared with the object-detection-only model. After introducing the Confluence algorithm, α-IoU loss, CA module, Mask module, and Anchor Head module, mAP@0.5 and mAP@0.95 of the detection model reach up to 92.4% and 52.9%, which increase by 2.7% and 6.9%, respectively, compared with the object-detection-only model.

Table 5. The ablation experiment results for the detection module. Bold indicates the best result in each column, and underline indicates the second-best result in each column.

Methods	Confluence	CA	α-IoU ($\alpha = 2$)	α-IoU ($\alpha = 3$)	Mask	Anchor Head	mAP@0.5 (%)	mAP@0.95 (%)
Detection-only	×	×	×	×	×	×	89.8	49.5
Detection (Joint optimization)	×	×	×	×	×	×	90.6	50.3
	✓	×	×	×	×	×	91.0	51.8
	✓	✓	×	×	×	×	91.6	52.4
	✓	✓	✓	×	×	×	<u>92.2</u>	**52.9**
	✓	✓	×	✓	×	×	91.9	<u>52.7</u>
	✓	✓	✓	×	✓	×	**92.4**	**52.9**
	✓	✓	✓	×	✓	✓	**92.4**	**52.9**

In this paper, we use the α-IoU loss function defined in Equation (11). α is a hyperparameter. To further illustrate the detection effect of different values of α, we visualize some detection cases in Figure 10. When $\alpha = 2$, both large and small ships can be accurately detected. When $\alpha = 3$, there will be bounding box overlap (first and second rows in Figure 10) and missed detection (third row in Figure 10) problems for small ships. The increase in α causes the detection model to be overly sensitive to small targets, resulting in bounding box overlaps for small ships.

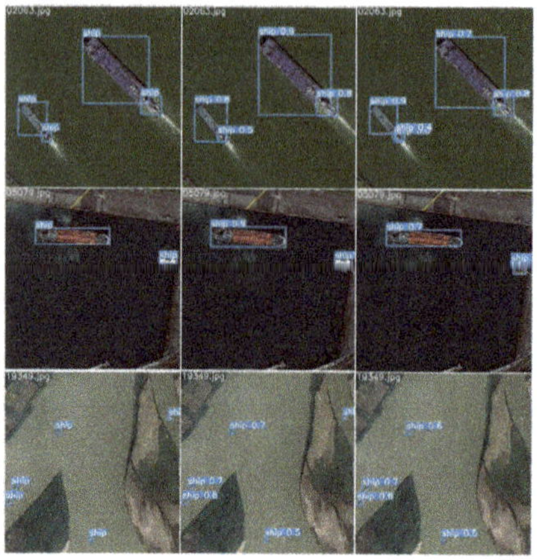

Figure 10. Sensitivity analysis of α in the loss function. The first column shows the detection results when $\alpha = 2$, the second column shows the detection results when $\alpha = 3$, and the third column shows the ground truths.

In the device–cloud collaboration architecture, the role of the cloud is mainly to store the semantic mask images and transmit the corresponding mask images to the front end based on the geolocation information sent by the front end. In this process, the computational burden of the cloud is almost negligible. Although the complete image data still need to be input into the network at the front end, we design the Mask module and Anchor Head module to use the mask images received from the cloud to exclude the nonwater portion and reduce the number of candidate bounding boxes, thus reducing the computational burden on the front end. To validate the effectiveness of our approach, we show three visualization examples of candidate bounding boxes output by the model before postprocessing in Figure 11. For these cases, the number of candidate bounding boxes is reduced by 18.2%, 12.9%, and 54.2%, respectively. We further analyze the memory and time consumption of the postprocessing procedure. As shown in Table 6, by applying mask images to the network, our network reduces the processing time by 0.1ms without increasing the peak memory and total memory footprint, which is essential when real-time ship detection is implemented on a front-end device with limited computation resources.

Table 6. Comparison results of postprocessing time consumption, peak memory, and average memory usage.

Methods	Postprocessing Time Consumption (ms)	Memory Peak (bytes)	Memory Used (bytes)
YOLO Model	0.9	1099.0	195.0
Ours	0.8	1099.0	195.0

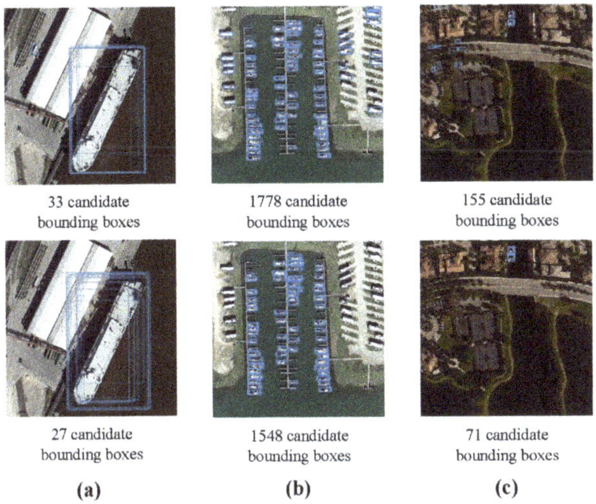

Figure 11. Candidate bounding box visualization generated by YOLO model before NMS (first row) and candidate bounding box visualization generated by our model before NMS (second row). (**a–c**) are three remote sensing images in the dataset.

5.7. Efficiency Analysis

As shown in Table 7, we analyze the efficiency of semantic segmentation, network transmission, and detection network in the device–cloud collaborative framework, where Cloud denotes the time required for semantic segmentation in the cloud, Transmission denotes the time required to transmit the mask image from the cloud to the front end, and End denotes the time required for ship detection in the front end. First, we verify the effect of the improvement on the detection speed by ablation experiments and then compare the performance of other models in terms of ship detection speed.

Table 7. Device–Cloud Collaboration Efficiency Analysis Experiment (ablation of the model and comparison with other models). Bold indicates the best result in each column, and underline indicates the second-best result in each column.

Model	Methods						Cloud (ms)	Transmission (ms)	End (ms)
	Confluence	CA	α-IoU (α = 2)	α-IoU (α = 3)	Mask	Anchor Head			
Detection-only	×	×	×	×	×	×	×	×	5.4
Detection (Joint optimization)	×	×	×	×	×	×	2.1	0.16	5.4
	✓	×	×	×	×	×	2.1	0.16	5.4
	✓	✓	×	×	×	×	2.1	0.16	5.4
	✓	✓	✓	×	×	×	2.1	0.16	5.4
	✓	✓	×	✓	×	×	2.1	0.16	5.4
	✓	✓	✓	×	✓	×	2.1	0.16	<u>3.1</u>
	✓	✓	✓	×	✓	✓	2.1	0.16	<u>3.1</u>
YOLOv3-tiny	×	×	×	×	×	×	×	×	**1.8**
YOLOv4-tiny	×	×	×	×	×	×	×	×	3.6
YOLOv5s	×	×	×	×	×	×	×	×	5.5
YOLOv6s	×	×	×	×	×	×	×	×	6.0
YOLOv7-tiny	×	×	×	×	×	×	×	×	5.4
YOLOv8s	×	×	×	×	×	×	×	×	5.3
RT-DETR-R18	×	×	×	×	×	×	×	×	5.8
YOLOv5-ODConvNeXt	×	×	×	×	×	×	×	×	5.4

The time required for semantic segmentation in the cloud is about 2.1 ms. It can be assumed that the size of the segmented mask image is 10 KB and the 5G bandwidth is 500 Mbit/s. The transmission time of a mask image is about 0.16 ms without considering the network latency and other interferences. As shown in Table 7, the addition of the MASK module reduces the calculation of nonwater areas and reduces the average ship detection speed from 5.4 ms to 3.1 ms, which is a 42.6% improvement in detection speed. Compared with other models, the inference speed of YOLOv3-tiny is the fastest, but there is a large sacrifice in accuracy, which makes it difficult to meet the requirements of ship detection tasks. Compared with YOLOv4-tiny, YOLOV5s, YOLOv6s, YOLOv7-tiny, YOLOv8s, RT-DETR-R18, and YOLOv5-ODConvNeXt, our method has faster inference speed.

For our device–cloud collaborative approach, the ship detection process does not directly perform the inference of the semantic segmentation network. The cloud is responsible for receiving the position information sent by the front-end device and returning the corresponding mask image stored in the cloud to the front end. Therefore, the cloud's processing time can be negligible. Through this device–cloud collaborative approach, not only the speed of ship detection is improved but also the accuracy of ship detection is ensured.

5.8. Attention Visualization Analysis

In order to visualize the specific effect of the CA attention mechanism on the detection process, we select three images in the dataset and perform a heatmap visualization of the first detection layer, as shown in Figure 12. Specifically, the attention area of the original model is not enough, and the correlation between the thermal distribution and the ship's geometric distribution is weak. After adding the CA module, our model significantly improves this shortcoming with a more concentrated heat distribution and a stronger correlation between the thermal distribution and the ship's geometric distribution. This indicates that the CA attention mechanism can effectively enhance the model's attention to the ship region and improve detection accuracy.

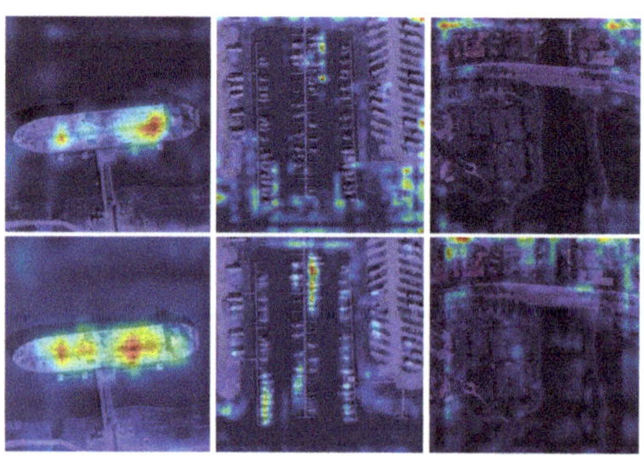

Figure 12. Visualization of attention. The first row shows the heatmap visualization results without adding the CA attention mechanism, and the second row shows the heatmap visualization results after adding the CA attention mechanism. The blue area represents the part where the model pays less attention, while the yellow and red areas represent the parts where the model pays more attention. Specifically, the red area shows higher attention relative to the yellow area. This color difference reflects the attention level and priority of the model when processing the visual input.

5.9. Visualization of Results

5.9.1. Semantic Segmentation

We deploy the segmentation module in the cloud and utilize the weight file obtained through joint optimization to perform the segmentation task. After completing this task, we store the semantic segmentation results in the cloud for subsequent calculations. Some segmentation results and their corresponding ground truths are shown in Figure 13. It can be seen that our segmentation model can effectively segment water and nonwater areas.

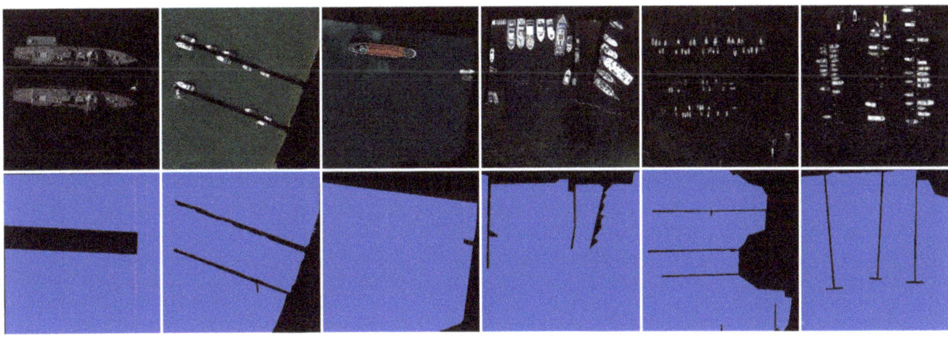

Figure 13. Segmentation results (first row) and ground truths (second row) visualization.

5.9.2. Object Detection

We deploy the object detection module in the front end and utilize the weight file obtained through joint optimization to perform the ship detection task. The corresponding semantic segmentation results in the cloud will be used for detection. Some ship detection results and their corresponding ground truths are shown in Figure 14. It can be seen that our model can accurately detect ships in remote sensing images.

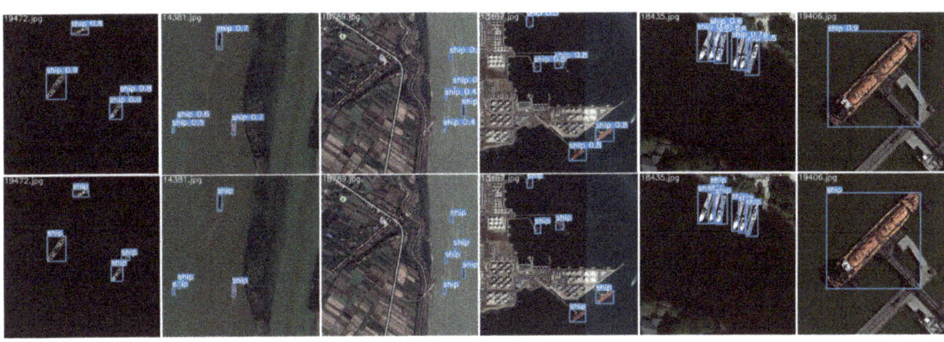

Figure 14. Ship detection results (first row) and ground truths (second row) visualization.

6. Conclusions

In order to achieve fast and accurate detection of ship targets in remote sensing images under the environment of limited front-end computing power, this paper proposes an efficient ship detection model based on device–cloud collaboration. The model improves the overall performance of the system by jointly optimizing semantic segmentation and object detection. The ship targets in remote sensing images are small, and the image background is complex. We introduce the coordinate attention mechanism, α-IoU loss, and Confluence algorithm to enhance the detection accuracy of small targets. The constructed SD-Net outperforms the current detection methods in terms of mAP, Params, and GFLOPs. By utilizing the mask image sent by the cloud, the calculation required for detection in nonwater areas is reduced, reducing the detection time from 5.4 ms to 3.1 ms (improving the detection speed by 42.6%). This is particularly suitable for front-end devices (UAV) with limited computing and power resources. It not only maintains high detection accuracy but also reduces the consumption of computing resources, thereby increasing the endurance of UAVs.

The device–cloud collaboration method is limited by network bandwidth to a certain extent. To reduce the impact of network bandwidth, we will consider using spike neural networks to further reduce the amount of calculations and combine knowledge distillation techniques to further improve the reasoning speed. With the development of information transmission technology, we will also explore the use of edge computing to optimize the model computing architecture. For the changeable climate environment at sea, we will also explore the ship detection methods under different weather conditions in the future.

Author Contributions: Conceptualization, T.L. and Y.Y.; methodology, T.L., Y.Y. and Z.L.; software, T.L. and Y.Y.; validation, T.L., Y.Y. and Z.L.; formal analysis, Y.H. and X.Z.; investigation, T.L. and Y.Y.; resources, Z.L.; data curation, Y.Y.; writing—original draft preparation, T.L. and Y.Y.; writing—review and editing, F.W., M.S. and H.W.; visualization, T.L. and Y.Y.; supervision, M.S. and H.W.; project administration, T.L. and Z.L.; funding acquisition, T.L. and Z.L. All authors have read and agreed to the published version of the manuscript.

Funding: This work was supported in part by the National Natural Science Foundation of China under Grant 52071201, Grant 52301420, and Grant 52331012, in part by the Science and Technology Commission of Shanghai Municipality (STCSM) Capacity Building Project of Local Universities under Grant 23010502200, in part by the Special Funding for the Development of Science and Technology of Shanghai Ocean University under Grant A2-2006-21-200207, in part by the Open Fund of Key Laboratory of High-Performance Ship Technology (Wuhan University of Technology), Ministry of Education under Grant GXNC23052801, in part by the Open Project Program of the State Key Laboratory of CAD&CG (Zhejiang University) under Grant A2326, and in part by Fund from State Key Laboratory of Maritime Technology and Safety. *(Corresponding author: Zhengling Lei).*

Institutional Review Board Statement: Not applicable.

Informed Consent Statement: Not applicable.

Data Availability Statement: The data that support the findings of this study are available from the corresponding author, [Z.L. Lei], upon reasonable request.

Conflicts of Interest: The authors declare no conflicts of interest.

References

1. Zhenbo, B.; Shiyou, Z.; Hua, Y.; Yuanhong, W. Survey of Ship Detection in Video Surveillance Based on Shallow Machine Learning. *J. Syst. Simul.* **2021**, *33*, 16.
2. Dudczyk, J.; Rybak, Ł. Application of data particle geometrical divide algorithms in the process of radar signal recognition. *Sensors* **2023**, *23*, 8183. [CrossRef] [PubMed]
3. Dudczyk, J.; Kawalec, A. Specific emitter identification based on graphical representation of the distribution of radar signal parameters. *Bull. Pol. Acad. Sci. Tech. Sci.* **2015**, *63*, 391–396. [CrossRef]
4. Dudczyk, J. Radar emission sources identification based on hierarchical agglomerative clustering for large data sets. *J. Sens.* **2016**, *2016*, 1879327. [CrossRef]
5. Chen, X.; Wu, H.; Han, B.; Liu, W.; Montewka, J.; Liu, R.W. Orientation-aware ship detection via a rotation feature decoupling supported deep learning approach. *Eng. Appl. Artif. Intell.* **2023**, *125*, 106686. [CrossRef]
6. Chen, X.; Wei, C.; Yang, Y.; Luo, L.; Biancardo, S.A.; Mei, X. Personnel trajectory extraction from port-like videos under varied rainy interferences. *IEEE Trans. Intell. Transp. Syst.* **2024**, *25*, 6567–6579. [CrossRef]
7. Carion, N.; Massa, F.; Synnaeve, G.; Usunier, N.; Kirillov, A.; Zagoruyko, S. End-to-end object detection with transformers. In *Proceedings of the European Conference on Computer Vision*; Springer: Berlin/Heidelberg, Germany , 2020; pp. 213–229.
8. Dai, X.; Chen, Y.; Yang, J.; Zhang, P.; Yuan, L.; Zhang, L. Dynamic detr: End-to-end object detection with dynamic attention. In Proceedings of the IEEE/CVF International Conference on Computer Vision, Montreal, BC, Canada, 11–17 October 2021; pp. 2988–2997.
9. Li, J.; Tian, P.; Song, R.; Xu, H.; Li, Y.; Du, Q. PCViT: A Pyramid Convolutional Vision Transformer Detector for Object Detection in Remote Sensing Imagery. *IEEE Trans. Geosci. Remote Sens.* **2024**, *62*, 5608115. [CrossRef]
10. Yan, R.; Yan, L.; Geng, G.; Cao, Y.; Zhou, P.; Meng, Y. ASNet: Adaptive Semantic Network Based on Transformer-CNN for Salient Object Detection in Optical Remote Sensing Images. *IEEE Trans. Geosci. Remote Sens.* **2024**, *62*, 5608716. [CrossRef]
11. Zhao, T.; Wang, Y.; Li, Z.; Gao, Y.; Chen, C.; Feng, H.; Zhao, Z. Ship Detection with Deep Learning in Optical Remote-Sensing Images: A Survey of Challenges and Advances. *Remote Sens.* **2024**, *16*, 1145. [CrossRef]
12. Long, J.; Shelhamer, E.; Darrell, T. Fully convolutional networks for semantic segmentation. In Proceedings of the IEEE Conference on Computer Vision and Pattern Recognition, Boston, MA, USA, 7–12 June 2015; pp. 3431–3440.

13. Shen, Q.; Gao, C. Research on Semantic Segmentation of Natural Landform Based on Edge Detection Module. *J. Syst. Simul.* **2022**, *34*, 293–302.
14. Li, Y.; Chen, W.; Huang, X.; Gao, Z.; Li, S.; He, T.; Zhang, Y. MFVNet: A deep adaptive fusion network with multiple field-of-views for remote sensing image semantic segmentation. *Sci. China Inf. Sci.* **2023**, *66*, 140305. [CrossRef]
15. Cai, Y.; Fan, L.; Fang, Y. SBSS: Stacking-Based Semantic Segmentation Framework for Very High-Resolution Remote Sensing Image. *IEEE Trans. Geosci. Remote Sens.* **2023**, *61*, 1–14. [CrossRef]
16. Saralioglu, E.; Gungor, O. Semantic segmentation of land cover from high resolution multispectral satellite images by spectral-spatial convolutional neural network. *Geocarto Int.* **2022**, *37*, 657–677. [CrossRef]
17. Wu, H.; Huang, P.; Zhang, M.; Tang, W.; Yu, X. CMTFNet: CNN and Multiscale Transformer Fusion Network for Remote Sensing Image Semantic Segmentation. *IEEE Trans. Geosci. Remote Sens.* **2023**, *61*, 2004612. [CrossRef]
18. Xiao, T.; Liu, Y.; Huang, Y.; Li, M.; Yang, G. Enhancing Multiscale Representations with Transformer for Remote Sensing Image Semantic Segmentation. *IEEE Trans. Geosci. Remote Sens.* **2023**, *61*, 1–16. [CrossRef]
19. He, X.; Zhou, Y.; Zhao, J.; Zhang, D.; Yao, R.; Xue, Y. Swin transformer embedding UNet for remote sensing image semantic segmentation. *IEEE Trans. Geosci. Remote Sens.* **2022**, *60*, 1–15. [CrossRef]
20. Li, L.; Jiang, L.; Zhang, J.; Wang, S.; Chen, F. A complete YOLO-based ship detection method for thermal infrared remote sensing images under complex backgrounds. *Remote Sens.* **2022**, *14*, 1534. [CrossRef]
21. You, Y.; Cao, J.; Zhang, Y.; Liu, F.; Zhou, W. Nearshore ship detection on high-resolution remote sensing image via scene-mask R-CNN. *IEEE Access* **2019**, *7*, 128431–128444. [CrossRef]
22. Tian, Y.; Liu, J.; Zhu, S.; Xu, F.; Bai, G.; Liu, C. Ship Detection in Visible Remote Sensing Image Based on Saliency Extraction and Modified Channel Features. *Remote Sens.* **2022**, *14*, 3347. [CrossRef]
23. Liu, T.; Jia, Z.; Lei, Z.; Zhang, X.; Huo, Y. Unsupervised depth estimation for ship target based on single view UAV image. *Int. J. Remote Sens.* **2022**, *43*, 3216–3235. [CrossRef]
24. Xu, X.; Zhang, X.; Zhang, T. Lite-yolov5: A lightweight deep learning detector for on-board ship detection in large-scene sentinel-1 sar images. *Remote Sens.* **2022**, *14*, 1018. [CrossRef]
25. Tian, Y.; Wang, X.; Zhu, S.; Xu, F.; Liu, J. LMSD-Net: A Lightweight and High-Performance Ship Detection Network for Optical Remote Sensing Images. *Remote Sens.* **2023**, *15*, 4358. [CrossRef]
26. Peng, G.; Yang, Z.; Wang, S.; Zhou, Y. AMFLW-YOLO: A Lightweight Network for Remote Sensing Image Detection Based on Attention Mechanism and Multi-scale Feature Fusion. *IEEE Trans. Geosci. Remote Sens.* **2023**, *61*, 4600916. [CrossRef]
27. Liu, T.; Zhang, Z.; Lei, Z.; Huo, Y.; Wang, S.; Zhao, J.; Zhang, J.; Jin, X.; Zhang, X. An approach to ship target detection based on combined optimization model of dehazing and detection. *Eng. Appl. Artif. Intell.* **2024**, *127*, 107332. [CrossRef]
28. Liu, L.; Chen, H.; Xu, Z. SPMOO: A Multi-Objective Offloading Algorithm for Dependent Tasks in IoT Cloud-Edge-End Collaboration. *Information* **2022**, *13*, 75. [CrossRef]
29. Liu, F.; Huang, J.; Wang, X. Joint Task Offloading and Resource Allocation for Device-Edge-Cloud Collaboration with Subtask Dependencies. *IEEE Trans. Cloud Comput.* **2023**, *11*, 3027–3039. [CrossRef]
30. Zhuang, Y.; Zheng, Z.; Shao, Y.; Li, B.; Wu, F.; Chen, G. ECLM: Efficient Edge-Cloud Collaborative Learning with Continuous Environment Adaptation. *arXiv* **2023**, arXiv:2311.11083.
31. Wang, Y.; Xu, R.; Zhou, C.; Kang, X.; Chen, Z. Digital twin and cloud-side-end collaboration for intelligent battery management system. *J. Manuf. Syst.* **2022**, *62*, 124–134. [CrossRef]
32. Zhang, K.; Huang, W.; Hou, X.; Xu, J.; Su, R.; Xu, H. A fault diagnosis and visualization method for high-speed train based on edge and cloud collaboration. *Appl. Sci.* **2021**, *11*, 1251. [CrossRef]
33. Tan, G.; Li, C.; Zhan, Z. Adaptive Scheduling Algorithm for Object Detection and Tracking Based on Device-Cloud Collaboration. *J. South China Univ. Technol. Nat. Sci. Ed.* **2021**, *49*.
34. Chen, L.C.; Zhu, Y.; Papandreou, G.; Schroff, F.; Adam, H. Encoder-decoder with atrous separable convolution for semantic image segmentation. In Proceedings of the European Conference on Computer Vision (ECCV), Munich, Germany, 8–14 September 2018; pp. 801–818.
35. Wang, C.Y.; Bochkovskiy, A.; Liao, H.Y.M. YOLOv7: Trainable bag-of-freebies sets new state-of-the-art for real-time object detectors. In Proceedings of the IEEE/CVF Conference on Computer Vision and Pattern Recognition, Vancouver, BC, Canada, 17–24 June 2023; pp. 7464–7475.
36. Chen, L.C.; Papandreou, G.; Schroff, F.; Adam, H. Rethinking atrous convolution for semantic image segmentation. *arXiv* **2017**, arXiv:1706.05587.
37. Hou, Q.; Zhou, D.; Feng, J. Coordinate attention for efficient mobile network design. In Proceedings of the IEEE/CVF Conference on Computer Vision and Pattern Recognition, Nashville, TN, USA, 19–25 June 2021; pp. 13713–13722.
38. Neubeck, A.; Van Gool, L. Efficient non-maximum suppression. In Proceedings of the 18th International Conference on Pattern Recognition (ICPR'06), Hong Kong, China, 20–24 August 2006; Volume 3, pp. 850–855.
39. Shepley, A.J.; Falzon, G.; Kwan, P.; Brankovic, L. Confluence: A robust non-IoU alternative to non-maxima suppression in object detection. *IEEE Trans. Pattern Anal. Mach. Intell.* **2023**, *45*, 11561–11574. [CrossRef]
40. He, J.; Erfani, S.; Ma, X.; Bailey, J.; Chi, Y.; Hua, X.S. α-IoU: A family of power intersection over union losses for bounding box regression. *Adv. Neural Inf. Process. Syst.* **2021**, *34*, 20230–20242.

41. Li, K.; Wan, G.; Cheng, G.; Meng, L.; Han, J. Object detection in optical remote sensing images: A survey and a new benchmark. *ISPRS J. Photogramm. Remote Sens.* **2020**, *159*, 296–307. [CrossRef]
42. Xia, G.S.; Bai, X.; Ding, J.; Zhu, Z.; Belongie, S.; Luo, J.; Datcu, M.; Pelillo, M.; Zhang, L. DOTA: A large-scale dataset for object detection in aerial images. In Proceedings of the IEEE Conference on Computer Vision and Pattern Recognition, Salt Lake City, UT, USA, 18–23 June 2018; pp. 3974–3983.
43. Zou, Z.; Shi, Z. Random access memories: A new paradigm for target detection in high resolution aerial remote sensing images. *IEEE Trans. Image Process.* **2017**, *27*, 1100–1111. [CrossRef]
44. Redmon, J.; Farhadi, A. Yolov3: An incremental improvement. *arXiv* **2018**, arXiv:1804.02767.
45. Bochkovskiy, A.; Wang, C.Y.; Liao, H.Y.M. Yolov4: Optimal speed and accuracy of object detection. *arXiv* **2020**, arXiv:2004.10934.
46. Li, C.; Li, L.; Jiang, H.; Weng, K.; Geng, Y.; Li, L.; Ke, Z.; Li, Q.; Cheng, M.; Nie, W.; et al. YOLOv6: A single-stage object detection framework for industrial applications. *arXiv* **2022**, arXiv:2209.02976.
47. Lv, W.; Xu, S.; Zhao, Y.; Wang, G.; Wei, J.; Cui, C.; Du, Y.; Dang, Q.; Liu, Y. Detrs beat yolos on real-time object detection. *arXiv* **2023**, arXiv:2304.08069.
48. Cheng, S.; Zhu, Y.; Wu, S. Deep learning based efficient ship detection from drone-captured images for maritime surveillance. *Ocean Eng.* **2023**, *285*, 115440. [CrossRef]
49. Bodla, N.; Singh, B.; Chellappa, R.; Davis, L.S. Soft-NMS–improving object detection with one line of code. In Proceedings of the IEEE International Conference on Computer Vision, Venice, Italy, 22–29 October 2017; pp. 5561–5569.
50. Zheng, Z.; Wang, P.; Liu, W.; Li, J.; Ye, R.; Ren, D. Distance-IoU loss: Faster and better learning for bounding box regression. In Proceedings of the AAAI Conference on Artificial Intelligence, New York, NY, USA, 7–12 February 2020; Volume 34, pp. 12993–13000.

Journal of
Marine Science and Engineering

MDPI

Review

Visual Navigation Systems for Maritime Smart Ships: A Survey

Yuqing Wang [1], Xinqiang Chen [2,*], Yuzhen Wu [3], Jiansen Zhao [4], Octavian Postolache [5] and Shuhao Liu [4]

1. School of Economics & Management, Shanghai Maritime University, Shanghai 201306, China; 202210711370@stu.shmtu.edu.cn
2. Institute of Logistics Science and Engineering, Shanghai Maritime University, Shanghai 201306, China
3. Shandong Port Group, Co., Ltd., Qingdao 266000, China; wuyz@qdport.com
4. Merchant Marine College, Shanghai Maritime University, Shanghai 201306, China; jszhao@shmtu.edu.cn (J.Z.); liushuhao0058@stu.shmtu.edu.cn (S.L.)
5. ISCTE—Instituto Universitário de Lisboa, Lisbon University Institute, 1649-004 Lisbon, Portugal; octavian.adrian.postolache@iscte-iul.pt

* Correspondence: chenxinqiang@stu.shmtu.edu.cn

Abstract: The rapid development of artificial intelligence has greatly ensured maritime safety and made outstanding contributions to the protection of the marine environment. However, improving maritime safety still faces many challenges. In this paper, the development background and industry needs of smart ships are first studied. Then, it analyzes the development of smart ships for navigation from various fields such as the technology industry and regulation. Then, the importance of navigation technology is analyzed, and the current status of key technologies of navigation systems is deeply analyzed. Meanwhile, this paper also focuses on single perception technology and integrated perception technology based on single perception technology. As the development of artificial intelligence means that intelligent shipping is inevitably the trend for future shipping, this paper analyzes the future development trend of smart ships and visual navigation systems, providing a clear perspective on the future direction of visual navigation technology for smart ships.

Keywords: smart ship; visual navigation technology; situation awareness; maritime safety

1. Introduction

Research on navigation safety has a long history. With the continuous development of science and technology, the design of safer and smarter navigation systems has become mainstream. Maritime transportation has always been considered a high-risk job, and as the size and number of ships continue to increase, factors affecting ship safety need to be given high priority. The government expects maritime security organizations to ensure the smooth running of maritime operations and at the same time handle maritime accidents efficiently in order to secure the country's economic interests. The majority of maritime accidents during the period 2002–2016 were of a navigational nature, including groundings, collisions, close quarters and contact, with a rate of 52.8 percent [1]. Figure 1 shows the components of maritime accidents, and it is clear that the majority of accidents in 2014–2019 are caused by human factors [2]. On the other hand, maritime accidents not only bring about loss of economic benefits, but also lead to air and water pollution, which has a bad impact on the environment.

Shipping is one of the most internationalized and dangerous of all large industries in the world. The history of shipping has been one of increasing safety of navigation and efficiency of transportation. Human factors such as fatigue, work stress and environmental factors are contributing factors to shipping accidents [3]. The development of smart ships effectively replaces pilots to a certain extent. The advent of smart ships and drones will reduce the number of people at risk at sea, and even if autopilot sailing does not reduce the number of accidents, this means that the safety of maritime navigation will increase. It is expected that the number of accidents will decrease with the introduction of smart ships due to the high impact of the human factor. However, quantifying what percentage of accidents can be prevented by smart ships is difficult since accidents are usually not caused by pure human error. de Vos J et al. (2021) evaluated the percentage reduction in loss of life for different scenarios where autonomous shipping is applied, assuming that the types of events affected are only related to navigation. The percentage reductions for the scenarios were 47.4% for small cargo ships being unmanned, 69.5% for all cargo ships being unmanned and 100% for all ships being unmanned. As can be seen from the data, the number of maritime shipping accidents is likely to be reduced with unmanned ships, and the development of smart ships will greatly enhance the safety of maritime navigation [4]. In previous studies on the safety of ship navigation, researchers have tended to focus mainly on how to quantify maritime risks and to focus on the relationship between humans and maritime safety and on the impact of human situation awareness on ship safety. Most of the studies only mention the development related to smart ships and the improvement of autonomous collision avoidance systems. In this paper, the development of collision avoidance technology and the importance of improving situation awareness are also studied relevantly. However, unlike the traditional approach of previous studies, this paper outlines the IMO-related guidelines and industry specifications for smart ships, which provides a background for subsequent studies on smart ships. At the same time, this paper also outlines the development and progress of smart ships, not only studying the progress of collision avoidance technology, which has mainly appeared since the development of smart ships, but also focusing on different kinds of visual perception technology, integrating different technologies into one system. By studying the organic combination of visual perception technology and collision avoidance technology, we expect to further enhance the safety of maritime navigation. In addition, this paper does not discuss a single visual perception technology, but analyzes and integrates the main visual perception technologies, which effectively deepens the depth of the research on visual perception. Through computer assistance, a visual navigation system visualizes the decisions of a visual navigation unit in order to reduce the cognitive workload of the staff and further enhance the situation awareness of the pilots. With the development of a new generation of artificial intelligence technology, autonomous systems have been widely used in the fields of unmanned vehicles, underwater vehicles and unmanned aircraft. Various technologies and advanced research are applied in navigation systems, including perception technology, motion control technology, collision avoidance technology and communication technology. Through the research on these technologies, the application of visual perception navigation technology to intelligent ships can greatly reduce the possibility of maritime accidents caused by human factors.

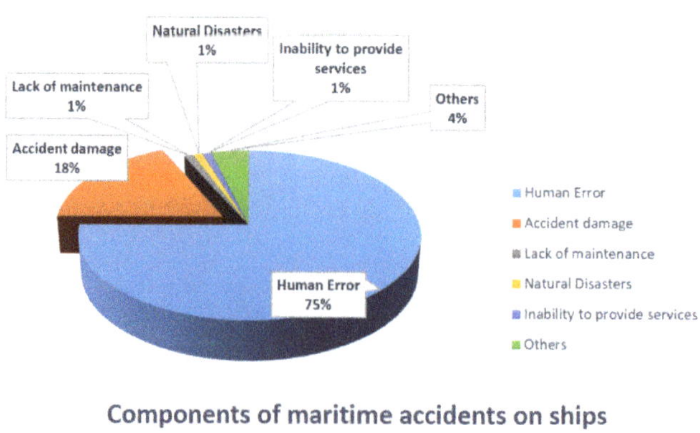

Figure 1. The components of maritime accidents.

This article analyzes the current opportunities and challenges of navigation technology for smart ships, providing the following contributions:

- A summary of the IMO guidelines and industry codes for smart ships;
- A review of the development of navigation technology and an analysis of its advancement;
- A characterization of the combination and application of different visual navigation techniques;
- An overview of the current status and future development of visual navigation technology and smart ships.

2. Smart Ships and Navigation Technology

2.1. Smart Ships

With the development and application of smart ships, the operational pressure on staff has been reduced, the safety of maritime navigation has been greatly improved and pollution of the oceans has been drastically reduced. Section 2.1 describes and analyzes the IMO's guidance on smart ships and the industry requirements for smart ships, as well as the progress and development of smart ships.

2.1.1. IMO Guidance and Industry Requirements for Smart Ships

The concept of an smart ship, first introduced in the 19th century, refers to a ship that is capable of carrying out a series of well-defined operations without or with a small number of crew members supervising it [5]. In order to give sustainable impetus to the development of the shipping industry in Europe, Norway was the first to develop technology for smart ships. There are three categories of more leading smart ship projects that examine the feasibility of developing viable business models in the short to medium term, reducing accident fatalities and analyzing the feasibility of smart ship scenarios [6]. Smart ships have been designed and developed for a number of reasons: to provide a better working environment for crews; to mitigate possible future crew shortages; to reduce risks in transportation; to be environmentally friendly; and to provide greater safety of navigation. The introduction of smart ships will bring about a new shift in cost efficiency, accident prevention and human resources. While smart ships are expected to be safer than conventional ships, the emergence of new technologies also brings new risks, such as the emergence of cybersecurity threats and misinformation. It is therefore important to define standards for shore-based operators of smart ships and provide relevant training. With regard to legal issues, the difference between the time it takes for technology to mature and the time it takes for regulations to be developed and put into practice may have a negative impact on the adoption of advanced technologies, and it is therefore important to recalibrate the relevant regulatory approaches. Smart ships may also raise ethical issues in emergency situations, such as problems of untimely human–machine and machine-to-machine communication, as well as when searching for and rescuing ships, crews or passengers in urgent need of help. A more comprehensive, international and harmonized approach to the development of a new regulatory framework for smart ships is important before they become fully operational. By understanding the impacts on different sectors and their associated linkages, it will be possible to prevent maritime accidents, protect the environment and maximize commercial benefits in the context of ongoing developments [7].

The International Maritime Organization (IMO) is primarily concerned with the development of international standards for the enhancement of safety and security and the protection of the marine environment [8]. The IMO therefore has a responsibility to consider the safety of international shipping for smart ships. Since the Maritime Safety Committee (MSC) placed the issue of smart ships on its agenda in January 2017, the MSC has been working to guide countries in the development and testing of autonomous navigation technologies. The development and application of smart ship technologies from 2017 to 2020 has been driven by the MSC to achieve real progress. "Autonomous" ships are not all similar, and the International Maritime Organization has classified MASSs into four categories [9], including ships with a crew and some automated systems (Class 1), ships with a crew that can be remotely controlled (Class 2), remotely controlled ships without a crew (Class 3), and ships capable of autonomous navigation without a crew and without human remote control (Class 4). The current standard for navigation is Class 1 vessels, whereas Class 2 vessels are supervised by on-board personnel who are ready to take over the operation of the system, so it is Class 3 and 4 vessels that could potentially pose a threat to the safety of shipping. The IMO, through its survey, and in order to safeguard the safe and environmentally friendly navigation of smart ships, decided in 2017 to carry out a Regulatory Scoping Exercise (RSE) aimed at assessing the extent to which the existing regulatory framework has been impacted in order to identify problems and gaps [10]. Four IMO committees—the Maritime Safety Committee (MSC), the Legal Committee (LEG), the Facilitation Committee (FAL) and the Marine Environment Protection Committee—are involved in this task, which affects many areas. At the same time, the introduction of technologically advanced Information and Communications Technology (ICT) is expected to lead to steady growth of the maritime industry [7]. The MSC has taken note of the test standards for real ship experiments on smart ship development submitted by relevant countries, provided advice on the development of standards, and contributed to the safety of smart ships and the prevention of pollution of the environment. The MSC held its 101st session in 2019, and approved interim guidelines on MASS testing, which include guidelines related to the safety, security and protection of the marine environment and list the mandatory instruments of compliance. These guidelines also cover the staffing and qualifications of participants in MASS experiments, human factors and infrastructure to ensure the safe conduction of experiments, management of cyber risks, etc.

2.1.2. Progress and Development of Smart Ships

Autonomous driving has long been utilized in many areas, and the question of autonomous driving in the shipping industry is no longer if it will happen, but when it will happen. The degree of autonomy has been categorized into three phases: In the first phase of autonomy, the crew, supported by the system, makes decisions from information or data collected during the operation of the ship. In the second stage, under the supervision and authorization of the crew, the ship can make decisions and initiate actions. And in the last stage, a fully smart ship can be realized [11]. Smart ships also have an impact on many aspects. The global maritime industry is characterized by rapid scientific and technological developments, and while manned and smart ships have some aspects in common, regulatory provisions should also take into account the characteristics of smart ships and develop specific regulatory policies. In addition, the most important thing to consider for smart ships is safety, and a range of technologies should be researched to control autonomous functions in order to improve safety. In terms of the industry, which has previously relied on the technology of the crew, autonomous technology will now re-invent the field of shipping by way of smart ships. More ship design and port facilities need to be considered to make smart ships efficient and reliable [7].

Intelligent ship autonomous technology is the integration of a variety of technologies, including autonomous navigation technology (navigation situational awareness, behavioral decision making and motion control), intelligent cabin operation and maintenance, communication technology, intelligent hulls, integrated experiments, etc. [12]. Intellectualization, greening and automation will become mainstream in global ship development. With the continuous improvement of ship intelligence, the development of unmanned cabin maintenance, assisted piloting technology and self-diagnosis of faults will all gradually reduce the labor demand of ships. In the future, the number of crew members with rich experience is likely to decrease, which will increase our concern for ship safety. The accuracy of driving-behavior decision making is directly related to the safety of water transportation, so the research on related decision-making algorithms is of great significance. The long-term navigation phase of smart ships requires comprehensive consideration of theoretical technology levels, unmanned ship navigation rules, human–machine cooperation, etc. Therefore, the study of human–machine interaction in smart ships is also a key area for the development of smart ships in the future. Human beings have advantages in understanding complex situations and reasoning, while machines have the advantages of high computational power and reliability in specific environments. Promoting the study of human–machine cooperation theories and technologies for ships can give full play to the advantages of human–machine cooperation and promote the prosperous development of smart ships. One aspect of the digital transformation of the shipping industry is the increasing digitization of shipboard systems, which is achieved by combining information technology with operating systems, and this high degree of interconnectivity and interdependence increases the likelihood of cyberattacks and cyber risks to a ship's digital infrastructure. Therefore, although the risk of cyberattacks can hardly be reduced to zero, it is necessary to study the related cyber decision making and the design of security architectures for ships. The most common propulsion systems used in modern ships are two-stroke diesel engines, steam and gas turbines, or combined marine propulsion systems, and these require the use of different fuels, lubricants, and so on. These complex systems require adequate personnel on board to monitor and maintain them. And as ships move towards autonomy, the number of crew members will be reduced and agencies will need to consider specialized training and education for personnel involved in special functions to further develop smart ships.

Smart ships have a positive impact in terms of capital savings, lower fuel costs, energy savings, improved safety at sea and environmental protection. However, there are challenges in terms of port operations, port capacity and insurance. Putting smart ships into practical use reduces human error, which in turn reduces the costs associated with accidents and subsequent insurance and optimizes operations and processes [13]. Increased cargo handling capacity and up to a 90% reduction in labor costs will result as smart ships achieve targeted efficiencies without the need for personnel [14]. Figure 2 shows the overall percentages for the annual cost of cargo transportation. Smart ships can also increase productivity and reduce fuel consumption, with each smart ship saving more than USD 7 million over 25 years in fuel consumption, crew wages and supplies [15]. In addition, the use of smart ships can ensure more effective surveillance activities and facilitate the integration and visualization of global trade supply chains. Through more advanced and efficient automated energy management systems and improved navigation systems, smart ships can also increase the effectiveness and efficiency of other shipping activities, such as loading and unloading and lock access.

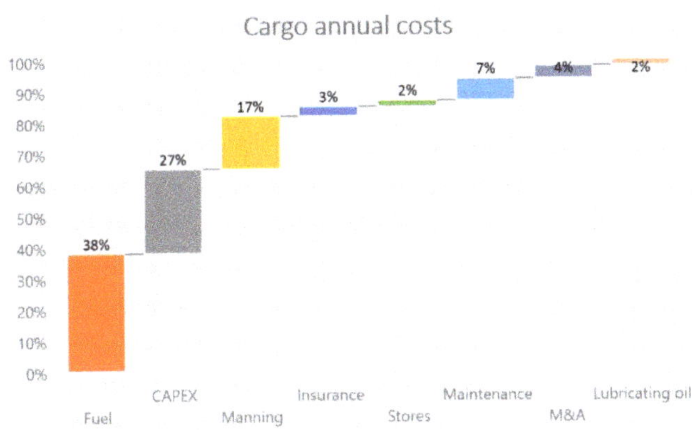

Figure 2. Cargo annual costs. Source: Rolls-Royce data.

The development of smart ships cannot be separated from the progress of technology, and the development of smart ship technology is a gradual evolutionary process. Perception technology is the basis for the development of smart ships, and it has undergone an evolution from single sensors to multi-sensor fusion. Initially, perception technology relies on basic sonar systems, which can only provide limited environmental information. As technology progressed, perception technology began to integrate multiple sensors, such as radar, infrared and vision systems, to collect richer data. These data include not only static obstacle information, but also dynamic currents and weather conditions. With the continuous optimization of algorithms, perception technology has become more accurate and efficient, capable of processing a large amount of data in real time to provide strong support for ship navigation and decision making. The development of communication technology is closely linked to the progress of perception technology. Initially, ship communication mainly relied on radio waves, and although this communication method can realize basic information transmission, it is limited by its signal coverage and transmission speed. With the emergence of satellite communication and broadband network technology, the communication capability of ships has been greatly enhanced. Modern smart ships are able to transmit a large amount of information in real time through high-speed data links, including the ship's position, status and environmental data, thus realizing more effective ship management and maritime traffic coordination. Motion control technology is the core of the autonomous navigation capability of smart ships. In the early days, the motion control of ships mainly relied on manual operations, and the crew needed to adjust a ship's heading and speed based on experience. With the development of automation technology, motion control systems began to integrate advanced algorithms and sensor data to realize automatic navigation and path planning. The motion control system of modern smart ships can automatically optimize routes, reduce fuel consumption, and maintain the stability and safety of ships under complex sea conditions. Collision avoidance technology is the key to ensuring the safe operation of smart ships. With a deeper understanding of ship dynamics and the marine environment, collision avoidance technology has evolved from relying on manual judgment and operations to advanced systems that can automatically detect potential collision risks and take avoidance measures. The development of these technologies has significantly improved the safety of ships in complex sea conditions. With the development of single perception technologies, smart ships are beginning to explore multi-sensor integrated perception technologies. Such fusion technologies are able to integrate data from different sensors to provide a more comprehensive and accurate perception of the environment. The combined application of technologies such as AIS, radar, infrared and vision systems can provide more comprehensive environmental sensing capabilities. This integrated perception not only improves the accuracy of the perception, but also enhances the robustness of a system, which maintains an effective perception capability even when some sensors fail. Through the gradual development and mutual integration of these technological fields, smart ships are moving towards a higher level of automation and intelligence and the innovation and development of the maritime transportation industry continues to be promoted.

2.2. Navigation Technology

Navigation technology is the top priority of the whole smart ship system, and appropriate navigation technology can improve the safety of navigation, reduce operational costs and improve the efficiency of navigation at the same time. Determining collision avoidance strategies and paths by sensing and learning sea information in intelligent ships during navigation is a problem that needs to be solved through further research. Effective collision avoidance and safe navigation of smart ships on the sea rely heavily on efficient and useful smart navigation systems. Therefore, the research on smart ship navigation technology in Section 2.2 is mainly categorized into perception technology, communication technology, motion control technology and collision avoidance technology. Their relationships can be represented in Figure 3:

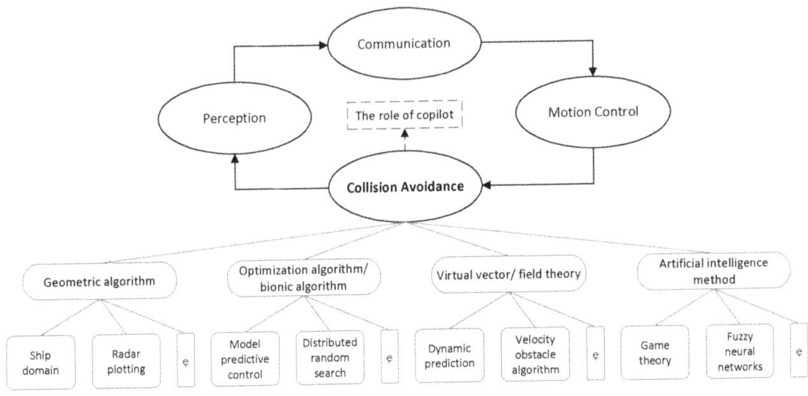

Figure 3. The relationships between navigation technologies.

2.2.1. Perception Technology

The sensing technology of ship navigation needs to use radar, LIDAR, ship automatic identification system (AIS), probe, camera and other sensors to collect environmental data when ships are sailing and build intelligent algorithms such as perception enhancement, data fusion, target classification, decision recommendation, etc., to process and analyze the sensed data automatically, distinguish potential dangers and abnormalities, and at the same time formulate contingency measures, such as adjusting the ship's sailing routes and speed, so as to improve the safety of navigation [16]. For smart ships, smart perception technology may have been developed to a more advanced stage. Ship identification, static obstacle perception, visibility impact, speed perception, distance perception, observation angle and cost are all issues that still need to be addressed [12]. Currently, radar is commonly used to detect targets, but the echo of radar cannot feedback the shape and appearance of the target, while the long distance and the motion of the ship itself will affect ship perception, so Liu et al. (2019) proposed a novel ship recognition and tracking system based on a deep learning framework, using a deep residual network and cross-layer jump connection strategy to extract advanced ship features to improve the recognition classification accuracy [17]. Leng et al. (2019) proposed a new method for ship detection based on single-channel synthetic aperture radar (SAR) image complex signal klick (CSK), which first detects the potential location of a ship based on the area proposal and then obtains the final ship target based on the target identification, which is free from the influence of false alarms caused by radio-frequency interferences (RFIs) and avoids the leakage of the detection situation under dense targets [18]. During navigation, the position of the ship to be sensed is often uncertain, and the problem of recognizing multiple types of ships is also faced, all of which undoubtedly add to the difficulty of perception. Wang et al. (2022) proposed a real-time ship collision risk perception model based on two domain proximity parameters, DDV (Degree of Domain Violation) and TDV (Time to Domain Violation), which adopts the inverse computation method and incorporates the information of the home route in the parameter computation and effectively solves the problem caused by the uncertainty of ship position prediction [19]. In order to achieve the detection and recognition of multiple types of ships, Zhou et al. (2021) improved the YOLOv5s algorithm by optimizing the loss function, regrouping the K-mean of the target initial frame at the data input side and enlarging the sensory field of view at the output frame, which greatly improved the precision and accuracy of ship image detection [20]. The variability of the maritime environment poses many challenges for the development of perception techniques; tracking a moving vessel in the context of moving ocean dynamics is one of them, and researchers overcame the challenge of tracking and recognizing moving vessels at sea by using background subtraction with real-time approximation of curve evolution based on level sets to delineate the contours of a moving vessel in the ocean [21].

At present, the research on navigation systems is still inseparable from environment perception in rain, snow, haze, low-light and other bad weather conditions; it is difficult for navigation systems to realize fast and accurate environment perception and meet the navigation and safety requirements of ships [22]. To ensure that ship detection remains accurate in low visibility, Guo et al. (2022) proposed a low-visibility enhancement network (called LVENet) based on the Retinex theory combined with depth-separable convolution and further proposed synthetic degraded image generation and hybrid loss functions to enhance visibility under low-light imaging conditions [23]. Liu et al. (2023) proposed an integrated low-visibility enhancement network (called AiOENet) that uses a unified encoder–decoder network architecture to flexibly recover low-visibility images, reducing the impact of harsh conditions on navigational safety [24]. Wang et al. (2023) constructed a foggy navigation database using a differentiated deep learning architecture and an EfficientNet neural network and a new method for foggy images and perceived visibility, which by replacing the SE module and combining the convergent block-attention module and the focal point loss function identified the best enhancement algorithm depending on the fog [25]. Li et al. (2024) proposed an inshore ship real-time detection transformer (ISRt-detr) that combines multiple coordinated attention and contrast learning to enhance the ability to perceive ships at night [26]. Detection of small targets at sea is an important area of ship navigation perception. AIS and radar systems may have the problem of missed detection; therefore, Wang et al. (2024) introduced an SSIM-based region of interest search method and an ellipse-weighted fusion method, proposed the ship panoramic vision stitching algorithm SSIM-EW, and proposed a perceptual model (YOLOv8-SGW) for sea-surface target detection, which improves the intelligence level of ships [16]. Yan et al. (2022) established relevant datasets of different complex sea states and introduced the improved Faster R-CNN for target detection in small sample datasets, which solved the problem of multiple complex sea-state data being difficult to collect simultaneously [27]. Traditional multi-feature detectors can only handle three or fewer features. Xu et al. (2020) proposed a multi-feature detector based on the Isolated Forest (iForest) algorithm in order to detect small floating targets in sea clutter, which breaks through the limitation of the number of features and reduces the cost of computation [28].

The flow of ship perception is shown in Figure 4:

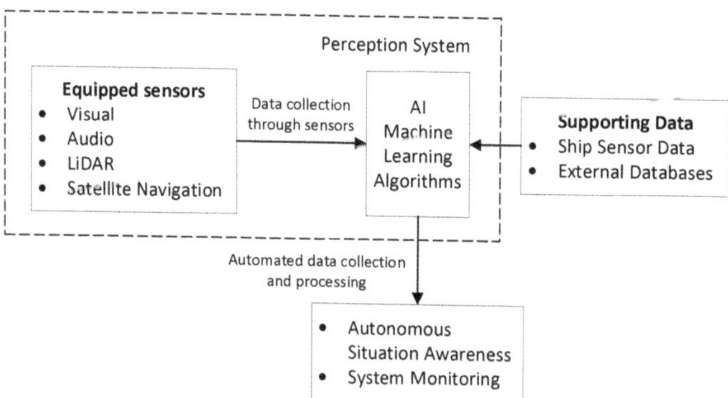

Figure 4. The flow of ship perception.

2.2.2. Communication Technology

Maritime radio first appeared on board ships at the beginning of the 20th century and in the early days was mainly used for transmitting and receiving telegrams from passengers. Subsequently, the International Convention for the Safety of Life at Sea (the SOLAS Convention) was introduced, and maritime radio communications were developed in the fields of both applied technology and communication procedures, providing an effective technical means for the exchange of information and communication between ship and shore and between ships. In the 1970s, telex, telephone, and facsimile communications were gradually applied to ship communications. Narrowband direct printing telegraphy (NBDP) and radiotelephone (RT) technology were applied to terrestrial communications systems in the 1970s and 1980s. Satellite communication systems have also been used occasionally. By the end of the 1980s, satellite systems became more widely used and had a larger share of applications in ship-to-shore communications [12]. Amendments to the GMDSS on radio-communications were adopted in 1998 and fully implemented in 1992 to improve the safety of ships in the field of radio-communications [29]. At present, the information exchange between ship and shore includes not only the ship's report and some instruction telegrams, but also image, voice and various other data about the ship. The information is digitized and transmitted, laying a certain foundation for the development of intelligent navigation in smart ships. Interference in communication networks between ships may lead to data distortion. Xie et al. (2019) proposed a reliability modeling method for ship communication networks based on the Apriori algorithm, constructed a transmission link model for ship communication networks and adjusted the reliability of the forwarding nodes of the links in ship communication networks through a fuzzy PID neural network control model, which improved the accuracy of data transmission and greatly improved the quality of communication [30]. Hoole et al. (2013) used the parameters of a communication system for ship navigation to a predetermined port terminal, utilizing electromagnetic beams to keep the ship within a marked trajectory boundary line, improving the accuracy of ship navigation [31]. Bin et al. (2024) proposed the Stackelberg Q-learning-based multi-hop cooperative routing algorithm (SQMCR) that balances packet forwarding benefits and energy costs. They also formulated collaborative communication strategies to ensure the reliability and efficiency of communication and further improve the performance of the routing algorithm [32].

The communication architecture of a smart ship is shown in Figure 5:

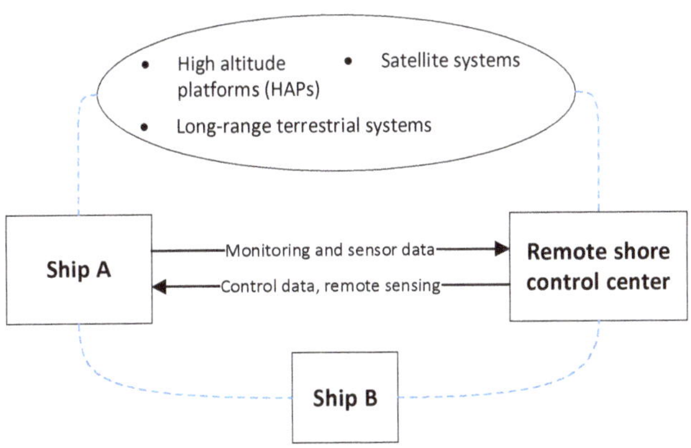

Figure 5. The communication architecture of a smart ship.

2.2.3. Motion Control Technology

Uncertainties in ship dynamics, random environmental perturbations and inaccuracies in measurement information are the main problems facing ship motion control [33]. The types of ship motion control are shown in Figure 6. The last three categories concern the control of heading, path, speed and stability, with the focus on which control strategy to choose. Different types of ship motion control correspond to different control algorithms. The basic algorithms used for ship heading control are proportional–integral–differential (PID) control [34], fuzzy logic control [35], predictive control [36], sliding mode control [37], autoresistant control (ADRC) [38] and artificial intelligence (AI) [39] algorithms, among others. On this basis, other algorithms are used to improve the effectiveness of heading control, such as optimizing PID controllers using Particle Swarm Optimization (PSO) [40], applying Genetic Algorithms (GA) to the optimization of the parameters of fuzzy controllers, and so on. For the study of ship stability control, the basic algorithms used are PID control [41], fuzzy control [42], predictive control [43], sliding mode control [44], ADRC [45], etc. Optimization algorithms are combined with basic algorithms on this basis, such as the combination of predictive algorithms with fuzzy control [46] and PID algorithms [47], to ensure the stability of ship navigation. For path tracking, the basic algorithms used include PID control [48], fuzzy logic control [49], predictive control [50], sliding mode control [51], ADRC [52], etc. The basic algorithms can be used in combination, with adaptivity being progressively improved as artificial intelligence is applied to path tracking. For ship trajectory tracking, the basic algorithms used include PID control [53], ADRC [54] and artificial intelligence algorithms [55]. Optimization algorithms, such as fuzzy logic control combined with PID control [56], PSO combined with ADRC [57], and neural networks combined with predictive control [58], are combined with basic algorithms, which can be used to achieve trajectory tracking and ship collision avoidance in different situations.

Types	(1)Ship speed control	(2)Course control
Control Object	Speed	Course
Types	(3)Stabilization	(4)Path-following
Control Object	Course,speed and path	Path
Types	(5)Trajectory tracking	(6)Path planning, obstacle avoidance and guidance
Control Object	Trajectory	Path
Types	(7)Automatic docking	(8)Multi-ships formation cooperative control
Control Object	Course, speed and path	Course, speed and path

Figure 6. Types of ship motion control.

The structure of ship motion control is shown in Figure 7:

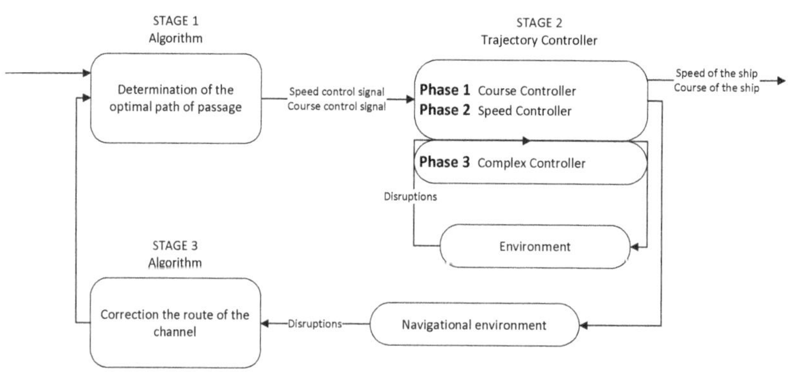

Figure 7. The structure of ship motion control.

2.2.4. Collision Avoidance Technology

Early on, since humans were at the center of collision prevention, researchers developed navigational aids to enhance the situation awareness of human operators. More recently, with the development of autonomous driving, the focus of research has become solving collisions through machines [59]. The techniques involved in collision avoidance are called collision prevention techniques and involve three basic processes, namely, motion prediction, conflict detection and conflict resolution.

Motion prediction is the basic module of collision avoidance for ships and involves the process of predicting the trajectories of operating systems and obstacles, which involves the estimation of the future state of a ship's motion. This is usually performed based on the ship's current position, speed, heading and possible influence of external factors, such as wind currents. A five- to ten-second prediction of the ship's motion can give an operator enough time to avoid a serious collision. Ding et al. (2022) proposed a new algorithm for ship motion prediction based on the modified covariance (MCOV) method and a neural network, which firstly utilizes the MCOV method for spectral analysis of a ship's motion and then utilizes the main spectra of the ship's motion to build a ship motion model through the neural network (NN), which advances the ship motion prediction time to ensure the consistency of the accuracy [60]. Path planning for smart ships has been a difficult problem in research. He et al. (2021) proposed an easy-to-implement multi-ship-encounter collision-avoidance path-planning method for smart ships, which combines a ship motion model and a PID controller to predict a ship's automatic heading change process, providing an effective method for collision avoidance [61]. Kim et al. (2022) proposed an algorithm that can predict a ship's state and plan the optimal route based on the uncertainty of the number of ships, using the Untraceable Kalman Filter (UKF) to derive the Time to the Closest Point of Approach (TCPA) and the Distance to the Closest Point of Approach (DCPA) from the geometric data of the home ship and the target ship [62]. To reduce the limitations associated with the low accuracy and high complexity of previous methods, Abebe et al. (2022) investigated a hybrid Autoregressive Integrated Moving Average (ARIMA)–Long Short-Term Memory (LSTM) model to accurately estimate near-term trajectories and assess collision risk [63].

Conflict detection refers to whether and when a ship should take evasive action and is based on motion prediction to determine whether two or more ships are at risk of collision in the future. Conflict detection in practice consists of three steps: recognizing a potential collision and alerting the operator; triggering the ship's self-help system to find ways to avoid the collision; and evaluating the risks posed by alternative solutions or avoidance actions [59]. The first two focus on risk-informed scenarios, and the last step focuses on decision making in response to risk. Ship collision risk modeling and probabilistic and collision risk assessment can all provide effective strategies to reduce the risk of ship collisions. Liu et al. (2021) developed a dynamic ship domain model to propose a systematic approach to detect possible collision scenarios and identify the distribution of collision risk hotspots in a given area, providing support for identifying areas of high collision risk and the distribution of risk in time and space [64]. Szlapczynski et al. (2021) proposed a new ship collision risk model utilizing the ship domain concept and domain-based collision risk parameters for collision risk assessment using many auxiliary parameters derived from the ship domain concept, such as Time to Domain Violation (TDV), Time to Domain Exit (TDE), etc. [65]. Liu et al. (2022) further modeled and visualized ship collision prediction analysis by introducing the Quadratic Ship Domain (QSD) into the Vessel Conflict Ranking Operator (VCRO), visualizing risk using the Kernel Density Estimation (KDE) model and finally predicting future risk using the Convolutional Long and Short-Term Memory Network (ConvLSTM) model [66]. In terms of probability and collision risk estimation, many scholars have utilized AIS data for analysis. Liu et al. (2020) established a unified collision risk assessment framework based on an autoencoder (AE) to clean up historical ship movement data and estimate the frequency of ship collisions [67]. Xin et al. (2021), on the other hand, quantified the uncertainty distribution of trajectories using AIS data, and a two-stage Monte Carlo simulation algorithm was used to ensure the accuracy of the estimates [68]. Shi et al. (2022) established a framework for collision risk assessment at sea considering trajectory-to-trajectory collision based on a multi-factor Douglas–Peucker algorithm adapted to ship trajectory compression considering speed and turning constraints [69].

At the heart of ship collision avoidance is conflict resolution, which is the process of taking appropriate measures to avoid a collision after the risk of collision has been detected. Many technologies for ship collision avoidance have emerged today, which can be broadly categorized into six main groups. The first one is rule-based collision avoidance. Perera et al. (2011) formulated and implemented a new fuzzy inference system based on an if–then-rule-based decision-making process and integration using the MATLAB software platform; this fuzzy logic-based intelligent decision-making system greatly improves the safety of ships through collision avoidance [70]. The second one is to determine a ship's motion using a virtual vector field. Park et al. (2021) used a Bayesian theory-based Relevance Vector Machine (RVM) to estimate a ship's collision risk, and it improves the accuracy and efficiency of ship parameter prediction [71]. The third one is to discretize the possible scenarios and thus select the optimal one. Chen et al. (2020) proposed an improved Time-Discretized Non-Linear Velocity Obstacle (TD-NLVO) algorithm and determined hazardous scenarios based on predefined criteria, which improves the accuracy of analyzing the geometrical probability of the risk of collision of ships [72]. Tengesdal et al. (2021), on the other hand, introduced a Cross-Entropy (CE)-based estimation of ship collision probability, which is able to obtain a low variance estimate of the probability of small collisions [73]. The fourth is to represent collisions as constraints, and Cheng et al. (2024) used the Velocity Obstacle (VO) algorithm to support collision avoidance with a trajectory non-linearly predictable target vessel and reflect the effect of the remaining collision time on the outcome by being sensitive to the effect of the remaining collision time, which can provide early warning of dangerous encounters and buy more valuable time for the ship to perform collision avoidance maneuvers [74]. In order to minimize collision avoidance operation times, Li et al. (2019) proposed a multi-vessel rolling optimization method for predicting and calculating the collision risk between vessels [75]. The fifth is the replanning method, which converts the ship collision avoidance problem into a path planning problem. Zhu et al. (2023) proposed the improved route-plan-guided Artificial Potential Field (APF) method, the Fast Local Path Planning (FLPP) method and a dynamic goal-guided (APF) method, which provide technical support for ships' autonomous navigation and collision avoidance [76]. Ren et al. (2021) proposed an autonomous collision avoidance algorithm for ships based on the improved speed barrier method that includes a path replanning algorithm by fusing the dynamics model of unmanned ships, the motion model of the encountering ship and the international maritime collision avoidance rules [77]. The last one consists of hybrid algorithms, such as the hybrid algorithm for smart ship path planning, which improves the typical Artificial Potential Field (APF) algorithm and the Velocity Obstacle (VO) method while combining them with the international rules for collision avoidance at sea [78], and the intelligent hybrid collision avoidance algorithm based on deep reinforcement learning, which improves the original sampling mechanism of DDPG, among others [79].

A comparison of the collision avoidance decision processes for manned ships and smart ships is shown in Figure 8:

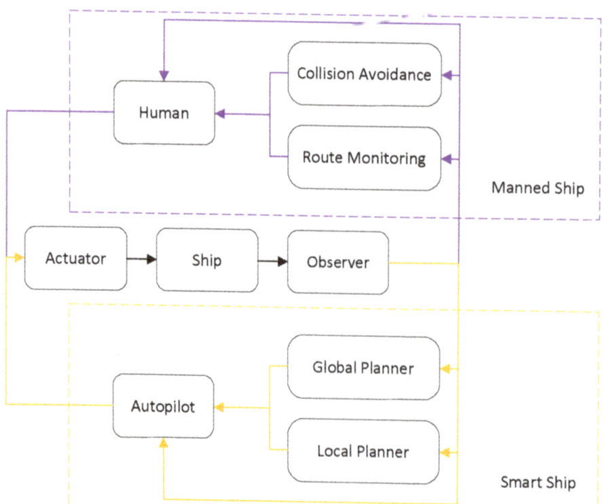

Figure 8. A comparison of collision avoidance decision processes.

3. Visual Perception Navigation Technology

Visual aid navigation is an important function of autonomous navigation in smart ships. In order to reduce the workload of smart ship pilots, a visual assistance system visualizes and processes the decisions of intelligent navigation devices for computer-aided visual navigation, which improves the situation awareness of smart ship pilots. The navigation information of the existing visual navigation system mainly comes from the traditional equipment on the ship's bridge, such as AISs, radar, infrared sensors, electronic nautical charts, etc. Through augmented reality technology, the above navigation information will be matched to the video imagery of the ship's video camera so as to realize the overlap of the virtual information and the real imagery and adjust the routes and sailing speeds according to the current situation of the ship, as well as make avoidance decisions so as to improve the safety of the ship [80]. In a navigation system, it is crucial to accurately detect the navigation information of the target ship. The environmental perception of ship navigation also means using AIS, radar, depth sounder, camera and other sensors to collect environmental data for ship navigation and constructing algorithms, such as data integration and perception enhancement algorithms, to automatically process and analyze perceived data, identify potential dangers, implement collision avoidance measures and safeguard the safety of navigation. The visual perception technology of ships cannot be separated from the environmental data perceived by the sensors; however, the use of any one of these methods for perception alone may have the problems of inconsistent information and insufficient data accuracy. At this stage, in addition to single-sensor perception, the main source of environmental data is multi-sensor data fusion. Generally, when designing a multi-sensor data fusion architecture, centralized fusion, distributed fusion and hybrid fusion are involved. We first analyze the technologies related to single perception, followed by a further study of integrated perception based on single perception.

3.1. Single Perception

3.1.1. AISs

AIS is the acronym for a universal shipborne automatic identification system; it is a kind of maritime mobile-band VHF broadcasting system, which is mainly used in the field of ship navigation and monitoring, realizing the data exchange between ship and ship and between ship and shore through the VHF radio band. The core function of an AIS system is to send and receive the dynamic and static information about a ship. Dynamic information includes the ship's position, speed, heading, destination and expected arrival time, while static information involves the ship's name, call sign, type, size, deadweight tonnage and other basic attributes [81]. The real-time updating and exchange of this information enables the ship to understand the surrounding environment and the dynamics of other ships in a timely manner so as to make more accurate navigational decisions. AISs track ship movements through the electronic exchange of navigational data between ships via shipboard transceivers and terrestrial or satellite base stations and uses the information collected for maritime security [82]. In order to detect multi-vessel encounters, Zhu et al. (2022) proposed a model to mine multi-vessel encounter situation awareness, analyze the spatial process of encounters and make collision avoidance decisions from automatic identification system (AIS) data, which overcomes the problem that the traditional collision risk assessment methods are only applicable to the differences between two ships and ship awareness [83]. Meanwhile, Rong et al. (2022) proposed a new method based on an improved sliding window algorithm to automatically identify ship collision avoidance behavior from ship trajectories, which can be used to accurately detect ship collision avoidance behavior from AIS trajectory data [84]. With continuous research and discoveries, researchers started to analyze AIS data using visual analysis tools. Öztürk et al. (2021) proposed a tangible visual analysis tool to analyze the changes in maritime traffic using the spatio-temporal basis of AISs—a novel way to advance the development of judging the safety of navigation of vessels and waters [85]. He et al. (2021) focused more on the visual model and human–computer interaction and designed and applied a visual analytics (VA) method called AIS Data Quality Visualization (ADQvis) to assess and explore the data quality of AISs [86]. Navigation technologies for smart ships utilize collected information to generate computer images, but these images also need to be converted into usable navigation data through human interpretation, increasing the likelihood of human error. To address this problem, Carter et al. (2023) designed an AIS-based visually enhanced simulation system for ship navigation that superimposes computer-generated navigation information onto real-world scenarios in real time, improving the safety and usability of the navigation system [87].

3.1.2. Radar

Radar, radio detection and ranging technology comprise an electronic system that uses electromagnetic waves for target detection and localization. It realizes precise measurement of the position, speed and shape of a target by transmitting electromagnetic waves and receiving reflected signals. The working principle of radar is based on the propagation characteristics of electromagnetic waves, and target information is obtained by analyzing the time delay, frequency change and phase difference of the echo signal. Radar technology is also being developed to be gradually applied in visual perception navigation systems. Xia et al. (2022) innovatively proposed a visual transformer framework based on contextual joint representation learning, referred to as CRTransSar, for the synthetic aperture radar (SAR) target detection task, combining the global contextual information perception capability of a transformer with the local feature representation capability of a convolutional neural network (CNN), which improves detection accuracy [88]. Chen et al. (2023) proposed a novel robust CFAR procedure for background clutter fitting, which provides a new interdisciplinary perspective for SAR image segmentation and achieves excellent detection performance [89]. Mou et al. (2019) introduced a maritime target detection method based on improved Faster R-CNN for navigation radar PPI (Plane Position Indicator) images, which improved the target detection performance of radar [90]. In addition to ordinary sea navigation, ice navigation is likewise an important research problem in path planning. Hsieh et al. (2021) used nautical radar imaging to reconstruct ice navigation scenarios, establish a sea-ice warning visualization function and select the optimal path planning scheme [91]. Naus et al. (2021), on the other hand, evaluated the accuracy of ship positioning based on the observation of navigational radar echoes from electronic nautical charts, providing the prerequisites for an autonomous navigation system that replaces the Global Navigation Satellite System (GNSS) [92]. Chen et al. (2022) proposed a maritime target detection method based on Marine-Faster R-CNN algorithm in complex environments, which utilizes convolutional neural networks (CNNs) for feature extraction and target identification of PPI images generated by radar echoes with enhanced generalization capability [93]. In order to solve the problems of the low target resolution of traditional radar and insufficient intuition of ship navigation, Li et al. (2022) proposed a multi-camera ship "video radar" enhanced navigation system and a deep learning-based feature point detection method (SRSuperpoint) to improve the navigation interface stitching effect and present a more comprehensive and intuitive navigational environment [94]. Xu et al. (2023) used radar sequence images to estimate ship dynamic features, which provides a reliable method for ship collision risk assessment in ship traffic service centers and improves operator interactivity [95].

3.1.3. Infrared

Infrared technology (IRT), or IRT for short, is an advanced navigation aid technology that detects and recognizes objects based on their thermal radiation properties. The technology uses infrared sensors to capture the thermal radiation signals emitted by the target object, and by analyzing the wavelength, intensity and distribution of these signals, it realizes the precise measurement of the target's position, speed and shape. Its high-precision detection capability enables ships to navigate the sea more safely and efficiently.

Infrared technology is applied in the form of infrared thermometers, infrared spectrometers, infrared communication devices, etc., and infrared cameras are major tools in the application of infrared technology for sensing and thus assisting navigation systems, so we mainly analyze the research on the technology related to infrared cameras. Zhou et al (2021) proposed a robust foreground detection method based on background modeling combined with multiple features in the Fourier domain (BMMFF) for ship target detection in dynamic sea-surface backgrounds, which outperforms the relevant comparable algorithms in different challenging sea-surface scenes [96]. When the weather at sea is clear, the maritime target is located exactly between the sunlight and the infrared camera, and the gray value of the background of the infrared image is larger than that of the target. Dong et al. (2019) designed an infrared maritime target detection algorithm based on the Visual Attention Model (VAM), which improves the accuracy and efficiency of the search for maritime targets in backlit environments [97]. In order to improve the target tracking efficiency of ships, Liu et al. (2022) proposed an efficient method of SiamRPN++ based on cross-connected and spatially transformed networks of AlexNet, which effectively improved the method to 63.9 FPS [98]. Infrared thermal imaging technology has been widely used in the field of target detection. It can not only adapt to different light-intensity environments; at the same time, it has the characteristics of strong concealment, strong detection ability, long detection distance and high detection sensitivity. Cao et al. (2023) proposed an improved Canny segmentation algorithm based on the maximum interclass variance method on the basis of infrared thermal imaging, which stabilized the flow field of background light to achieve the effective segmentation of a ship's image and obtained a high-definition operational target [99]. In addition to the above advantages, infrared technology also has an inherent advantage in small-target detection. Gao et al. (2023) took the lightweight model of infrared small-ship detection as the research object, constructed an infrared small-ship dataset and preprocessed the images using gamma transformation, and the model parameters were reduced by 83% compared with the YOLOv5m model [100]. To compensate for the limitations of radar and AISs, such as blind spots and limited detection ranges, Park et al. (2024) utilized an infrared radiometric camera to supplement conventional perception sensors which is capable of accurately detecting obstacles at sea and estimating their dynamic motions based on enhanced tracking results [101].

3.1.4. Visual Technology

Visual Technology, as a high-tech means of simulating human visual perception, gives machines the ability to "see" through image capture, processing and analysis. This technology captures images of the surrounding environment through high-resolution cameras and utilizes advanced image processing and pattern recognition algorithms to provide real-time visual information to ships. It is capable of recognizing channel markers, obstacles and other vessels, thus assisting navigation decisions. The application of vision technology not only enhances the ability of ships to navigate in complex sea conditions, but also provides additional security for maritime safety. Li et al. (2019) designed an overlapping-sector laser-propagation navigation system that utilizes red and green lasers to form a sector-shaped safety zone for safe navigation, which is easy to operate and highly sensitive [102]. Pan et al. (2020) proposed a fine-grained classification RMA (ResNet–Multi-Scale–Attention) model based on deep learning for analyzing subtle and local differences between different types of beacons for beacon recognition [103]. Electro-optical (EO) sensors such as video cameras are used to complement marine radar for accurate detection of objects on the sea surface, and Shao et al. (2022) proposed an enhanced convolutional neural network, called VarifocalNet*, that improves the detection of objects in harsh marine environments [104]. There is still a lack of low-cost and reliable sensing devices. Bi et al. (2023) designed a vision-based method to recognize ships and their microscopic features for navigational planning of ship collision avoidance, which can effectively and efficiently identify the navigation signals of the target ship [105]. There are many challenges in maritime navigation, such as narrow visibility for ship navigators, the limited view length angle of a single camera, complex maritime environments, etc. Wang et al. (2024) proposed panoramic visual perception-assisted navigation technology, introduced an SSIM-based region-of-interest search method and an ellipsoid-weighted fusion method, and put forward the ship panoramic visual stitching algorithm SSIM-EW. This technology obviously improves the ability to detect small targets at sea, which can expand the mariner's field of view, identify the targets missed by AIS and radar systems, guarantee navigation safety, and improve the level of ship intelligence [16].

3.2. Integrated Perception

With the continuous development and advancement of technology, integrated perception technology plays an increasingly important role in the field of visual perception. It realizes comprehensive understanding and accurate mapping of complex environments by integrating data from different sensors and utilizing image processing and machine learning algorithms. The core advantage of integrated perception is its ability to overcome the limitations of a single perception method and significantly improve the accuracy and reliability of target identification through the synergistic effect of multi-source data. For the fusion of AIS and vision technologies, AIS data can provide the vessel of interest with a unique Maritime Mobile Service Identity (MMSI), position coordinates, ground heading and ground speed, while cameras can directly display the visual appearance of the vessel. Qu et al. (2023) fused AIS and vision data to improve maritime traffic surveillance and were able to obtain more accurate ship tracking results and motion characteristics, taking full advantage of multiple sources of data [106]. To address the lack of detailed information important for real-world applications in deep learning object detection methods, Gülsoylu et al. (2024) developed a technique for fusing automatic identification system (AIS) data with ships detected in images to create a dataset enriched with ship images [107]. Ding et al. (2024) also combined AIS data with visual data and showed that this technique outperformed the pure AIS data technique during the daytime and outperformed the pure visual data technique at night, providing more reliable data for collision avoidance [108].

For the fusion of visual data and radar, both electronic charts and radar are essential equipment in a ship's navigation system, and the fusion of the two ensures that the ship is able to clearly display the dangers on the sea surface. Guo et al. (2020) proposed a deep learning-based data fusion algorithm which extracts robust features from radar images and merges marine radar and electronic chart data to provide more comprehensive information [109]. Zhang et al. (2021) proposed a radar image denoising algorithm based on the concept of the Generative Adversarial Network (GAN) with Wasserstein distance to reduce the noise interference brought by external factors and hardware and utilize the sparse theory to parallelize the high-frequency and low-frequency sub-band coefficients of detected images obtained by Fast Fourier Transformation to realize the fusion of images [110]. For the combination of visual and infrared data, Gao et al. (2022) used refined feature fusion and Sobel loss for the fusion of marine infrared and visible images [111], while Jeon et al. (2023) introduced an integrated approach to visual and infrared detection and ranging, combining deep learning-based computer vision techniques with real-time physics-based EO/IR data processing algorithms [112], all of which enhance the ability of ships to navigate autonomously in a variety of marine environments.

For the fusion of radar and AIS systems, in the maritime environment, radar signals are often blocked by islands and large ships, resulting in degradation of radar target detection and performance. Sun et al. (2023) utilized a priori environmental knowledge from electronic nautical charts (ENCs) and information from automatic identification systems (AISs) to propose an AIS-Assisted Integrated Probabilistic Data Association (A-IPDA) algorithm, which is well able to counteract long-time target occlusion in complex environments at sea and maintain tracking continuity and tracking accuracy [113]. To address the fact that false echoes in radar images and packet loss in AIS data transmission can lead to uncertainty in ship speed and heading estimation, Xu et al. (2023) proposed a ship speed and heading estimation method utilizing the fusion of radar sequential images and AIS data to quantitatively characterize the state of ship navigation [114]. In addition to the two-by-two combination of sensing technologies, some scholars have also studied the combination of multiple sensing data to enhance the navigation safety of smart ships. The use of any one ship motion sensing method alone may have problems such as inconsistent information and inaccurate data. Therefore, Wu et al. (2022) proposed a multi-sensor integrated perception system for motion detection of ships and constructed a multi-sensor integrated ship motion perception system hardware platform consisting of radar, AIS, camera and other attachments, which significantly improved the consistency of the information and the accuracy of the data for monitoring ship motion [115].

4. Future Trends

4.1. Trends in Intelligence and Automation

The relevant data for our study can be found in [4,14,16,115] as well as Rolls-Royce data, which provide us with a solid foundation for the study of smart ship navigation technology. In particular, the Rolls-Royce data provide valuable references for our research by providing information about the composition of ship cargo costs, which provides us with the advantages brought about by the development of smart ships. Through these data, we were able to gain an in-depth understanding of the pulse of the development of smart ship navigation technologies, as well as the performance and potential of each technology in practical applications. The accumulation of these research results adds persuasive power to our thesis and provides clear guidelines for our future research direction. By comprehensively reviewing the development history of smart ships, industry guidelines and current navigation technologies, this paper provides a clear perspective on the future direction of visual navigation technology for smart ships. Future ship visual navigation systems will realize the integrated processing of AIS, radar, infrared and visual data through the fusion of multi-modal perception technologies, thus improving the robustness and adaptability of the systems. The development of smart ships will benefit from the convergence of cross-domain technologies, especially the Internet of Things, big data analytics and cloud computing. The integration of these technologies will provide smart ships with powerful data processing and analysis capabilities, enabling real-time remote monitoring and decision support.

With the rapid development of artificial intelligence technology, the intelligent transformation of the future shipping industry has become an irreversible trend. The core of smart ship technology lies in the intelligence and automation of the ship's visual navigation system. The future direction of this system will rely on the in-depth application of deep learning and machine learning algorithms to achieve accurate identification and analysis of dynamic changes in the marine environment. Through these advanced algorithms, the system will be able to provide comprehensive information support to the crew, enhance their understanding of the surrounding environment and improve their ability to anticipate potential risks, thereby significantly improving their situation awareness.

4.2. Future Challenges and Opportunities for Smart Ship Visual Navigation Systems

Smart ships are becoming more autonomous, but the challenges that come with this cannot be ignored. Human–machine collaboration still plays a crucial role in ensuring safe ship operation. Future smart ship designs will need to focus more on optimizing human–machine interaction, providing intuitive user interfaces and decision support tools that will enable crew members to effectively supervise and control ship automation systems. In addition, with the development of technology, how to ensure data security and privacy protection against cyberattacks is also a major challenge that smart ship visual navigation systems need to face. The core of ship visual navigation is security and risk management. With the advancement of data analysis technology, intelligent systems will be able to assess navigation risks more effectively, issue timely warnings to crews and help them take preventive measures, thus improving the safety of navigation. However, this also requires intelligent systems to have a high degree of reliability and stability, and any technical failure or miscalculation may lead to serious consequences. In future developments, the field of ship navigation will focus more on environmental adaptability and sustainability, adopting more environmentally friendly technologies, optimizing energy consumption and reducing the impact on the marine environment. At the same time, the application fields of ship visual navigation technology will continue to expand, extending from the traditional shipping field to a wider range of fields, such as marine scientific research, marine operations and environmental monitoring, providing more diversified technical support for the development and protection of marine resources and promoting the sustainable development of the marine economy.

The future of visual navigation systems for smart ships at sea is full of challenges and opportunities. Continuing advances in technology will drive smart ships towards higher levels of autonomy and safety. However, this will also bring new requirements, and the smart ship industry needs to work closely with governments, research institutes and international organizations to address the challenges and ensure the healthy development and widespread application of smart ship technology [116]. This includes the development of new regulations and standards to accommodate the special needs of smart ships, as well as the development of new education and training programs to ensure that crews are able to adapt to future technological changes. It is only through these combined efforts that smart ship visual navigation systems can realize their full potential and revolutionize the shipping industry.

5. Conclusions

As the scale of maritime transportation continues to expand, maritime safety issues are becoming more and more prominent, and the research and development of visual navigation systems for smart ships as a key technology to enhance navigation safety is of great significance. By comprehensively reviewing the development history of smart ships, the industry guidelines of the International Maritime Organization (IMO) and current navigation technologies, this paper provides a clear perspective on the future direction of the development of visual navigation technology for smart ships.

This paper first summarizes the IMO guidelines and industry codes for smart ships and reviews the development of smart ships. The emergence of smart ships has not only reduced the risk of maritime navigation; it has also reduced the impact on the environment and improved operational efficiency. Through the integrated application of smart ship technologies, the autonomy of ships has been significantly improved and operators' situation awareness has been enhanced. Then, this paper discusses the latest progress in the key technology areas of perception technology, communication technology, motion control technology and collision avoidance technology, and analyzes the combination and application characteristics of different visual navigation technologies. Together, these technologies contribute to the performance enhancement of visual navigation systems for smart ships by improving ships' environmental sensing capabilities, communication efficiency and motion control accuracy and the effectiveness of collision avoidance strategies. This paper also emphasizes the importance of integrated perception techniques. By integrating data from different sensors, such as AIS, radar, infrared and vision systems, it is possible to provide a more comprehensive and accurate understanding of the environment, significantly improving the accuracy and reliability of target recognition. Finally, this paper looks forward to the future development trend of visual navigation systems for smart ships. With the integration of technologies such as artificial intelligence, the Internet of Things, big data analysis and cloud computing, future smart ships will realize higher levels of autonomy and safety. At the same time, human–machine collaboration will continue to be a key factor in ensuring the safe operation of ships, and the design of future smart ships will pay more attention to the optimization of human–machine interaction.

The research on visual navigation technology for smart ships is not only of great significance for enhancing maritime navigation safety, but also has a profound impact in terms of promoting the sustainable development of the marine economy. Future research should continue to focus on technological innovation, cross-domain technology integration, optimization of human–machine collaboration and sustainability.

Author Contributions: Conceptualization, Y.W. (Yuqing Wang), X.C. and Y.W. (Yuzhen Wu) methodology, Y.W. (Yuqing Wang), X.C. and J.Z.; validation, O.P. and S.L.; formal analysis, Y.W. (Yuqing Wang) and X.C. writing—original draft preparation, Y.W. (Yuqing Wang), X.C., O.P. and S.L.; writing—review and editing, J.Z.; visualization, X.C.; supervision, X.C.; project administration, X.C.; funding acquisition, X.C. All authors have read and agreed to the published version of the manuscript.

Funding: This work was jointly supported by the National Natural Science Foundation of China (52331012, 52102397, 52071200 and 52201403, 52472347) and the Shanghai Committee of Science and Technology, China (23010502000).

Informed Consent Statement: Not applicable.

Data Availability Statement: The research does not contain data, and thus there is no need to publish any.

Conflicts of Interest: Author Yuzhen Wu was employed by the company Shandong Port Group, Co., Ltd. The remaining authors declare that the research was conducted in the absence of any commercial or financial relationships that could be construed as a potential conflict of interest.

List of Abbreviations

ADRC	Active Disturbance Rejection Control
AE	Autoencoder
AI	Artificial Intelligence
AIS	Automatic Identification System
ADQvis	AIS Data Quality Visualization
A-IPDA	AIS-Assisted Integrated Probabilistic Data Association
APF	Artificial Potential Field
ARIMA	Autoregressive Integrated Moving Average
BMMFF	Background Modeling Combined with Multiple Features in the Fourier Domain
CE	Cross-Entropy
CFAR	Constant False Alarm Rate
CNN	Convolutional Neural Network
ConvLSTM	Convolutional Long and Short-Term Memory Network
CSK	Complex Signal Klick
DCPA	Distance to the Closest Point of Approach
DDPG	Deep Deterministic Policy Gradient
DDV	Degree of Domain Violation
ENCs	Electronic Nautical Charts
EO	Electro-Optical
FAL	The Facilitation Committee
FLPP	Fast Local Path Planning
GA	Genetic Algorithms
GAN	Generative Adversarial Network
GMDSS	Global Maritime Distress and Safety System
GNSS	Global Navigation Satellite System
ICT	Information and Communications Technology
IMO	International Maritime Organization
IR	Infrared
IRT	Infrared Technology
ISRt-detr	Inshore Ship Real-Time Detection Transformer
KDE	Kernel Density Estimation
LEG	The Legal Committee

LIDAR	Light Detection and Ranging
LSTM	Long Short-Term Memory
LVENet	Low-Visibility Enhancement Network
MASS	Maritime Autonomous Surface Ship
MCOV	Modified Covariance
MMSI	Maritime Mobile Service Identity
MSC	Maritime Safety Committee
NBDP	Narrowband Direct Printing Telegraphy
NN	Neural Network
PID	Proportional–Integral–Differential
PPI	Plane Position Indicator
PSO	Particle Swarm Optimization
QSD	Quadratic Ship Domain
R-CNN	Region Convolutional Neural Network
RFIs	Radio-Frequency Interferences
RMA	ResNet–Multi-Scale–Attention
RSE	Regulatory Scoping Exercise
RT	Radiotelephone
RVM	Relevance Vector Machine
SAR	Synthetic Aperture Radar
SE	Squeeze and Excitation
SGW	Serving Gate Way
SiamRPN++	Siamese Region Proposal Network Plus Plus
SOLAS	The Safety of Life at Sea
SQMCR	Stackelberg Q-learning-Based Multi-Hop Cooperative Routing Algorithm
SSIM-EW	Structural Similarity Index Measure–Elliptical Weighted Algorithm Stitcher
TCPA	Time to the Closest Point of Approach
TDE	Time to Domain Exit
TD-NLVO	Time-Discretized Non-Linear Velocity Obstacle
TDV	Time to Domain Violation
UKF	Untraceable Kalman Filter
VA	Visual Analytics
VAM	Visual Attention Model
VCRO	Vessel Conflict Ranking Operator
VHF	Very High Frequency
VO	Velocity Obstacle

References

1. Sakar, C.; Toz, A.C.; Buber, M.; Koseoglu, B. Risk analysis of grounding accidents by mapping a fault tree into a Bayesian network. *Appl. Ocean Res.* **2021**, *113*, 102764. [CrossRef]
2. Zalewski, P.; Posacka, K. Analysis of ship accidents based on European statistical surveys. *Zesz. Nauk. Akad. Morskiej Szczecinie* **2021**, *68*, 17–25.
3. Fan, S.; Yang, Z. Accident data-driven human fatigue analysis in maritime transport using machine learning. *Reliab. Eng. Syst. Saf.* **2024**, *241*, 109675. [CrossRef]
4. de Vos, J.; Hekkenberg, R.G.; Banda, O.A.V. The impact of autonomous ships on safety at sea—A statistical analysis. *Reliab. Eng. Syst. Saf.* **2021**, *210*, 107558. [CrossRef]
5. Zhang, W.; Zhang, Y. Navigation Risk Assessment of Autonomous Ships Based on Entropy–TOPSIS–Coupling Coordination Model. *J. Mar. Sci. Eng.* **2023**, *11*, 422. [CrossRef]
6. Moon, K.D.; Jeong, C.Y.; Kim, M.S.; Park, Y.K.; Lee, K. Develop and evaluate of intelligent autonomous-ship framework. *IOP Conf. Ser. Mater. Sci. Eng.* **2020**, *929*, 012006. [CrossRef]
7. Kim, M.; Joung, T.-H.; Jeong, B.; Park, H.-S. Autonomous shipping and its impact on regulations, technologies, and industries. *J. Int. Marit. Saf. Environ. Aff. Shipp.* **2020**, *4*, 17–25. [CrossRef]
8. Álvarez, P.S. From maritime salvage to IMO 2020 strategy: Two actions to protect the environment. *Mar. Pollut. Bull.* **2021**, *170*, 112590. [CrossRef]
9. Chang, C.-H.; Kontovas, C.; Yu, Q.; Yang, Z. Risk assessment of the operations of maritime autonomous surface ships. *Reliab. Eng. Syst. Saf.* **2021**, *207*, 107324. [CrossRef]
10. Fenton, A.J.; Chapsos, I. Ships without crews: IMO and UK responses to cybersecurity, technology, law and regulation of maritime autonomous surface ships (MASS). *Front. Comput. Sci.* **2023**, *5*, 1151188. [CrossRef]

11. Kurt, I.; Aymelek, M. Operational and economic advantages of autonomous ships and their perceived impacts on port operations. *Marit. Econ. Logist.* **2022**, *24*, 302–326. [CrossRef]
12. Zhang, X.; Wang, C.; Jiang, L.; An, L.; Yang, R. Collision-avoidance navigation systems for Maritime Autonomous Surface Ships: A state of the art survey. *Ocean Eng.* **2021**, *235*, 109380. [CrossRef]
13. Kim, T.-E.; Perera, L.P.; Sollid, M.-P.; Batalden, B.-M.; Sydnes, A.K. Safety challenges related to autonomous ships in mixed navigational environments. *WMU J. Marit. Aff.* **2022**, *21*, 141–159. [CrossRef]
14. Askari, H.R.; Hossain, M.N. Towards utilizing autonomous ships: A viable advance in industry 4.0. *J. Int. Marit. Saf. Environ. Aff. Shipp.* **2022**, *6*, 39–49. [CrossRef]
15. Munim, Z.H. Autonomous ships: A review, innovative applications and future maritime business models. *Supply Chain. Forum Int. J.* **2019**, *20*, 266–279. [CrossRef]
16. Wang, C.; Cai, X.; Li, Y.; Zhai, R.; Wu, R.; Zhu, S.; Guan, L.; Luo, Z.; Zhang, S.; Zhang, J. Research and Application of Panoramic Visual Perception-Assisted Navigation Technology for Ships. *J. Mar. Sci. Eng.* **2024**, *12*, 1042. [CrossRef]
17. Liu, B.; Wang, S.Z.; Xie, Z.; Zhao, J.; Li, M. Ship recognition and tracking system for intelligent ship based on deep learning framework. *TransNav Int. J. Mar. Navig. Saf. Sea Transp.* **2019**, *13*, 699–705. [CrossRef]
18. Leng, X.; Ji, K.; Zhou, S.; Xing, X. Ship detection based on complex signal kurtosis in single-channel SAR imagery. *IEEE Trans. Geosci. Remote Sens.* **2019**, *57*, 6447–6461. [CrossRef]
19. Wang, S.; Zhang, Y.; Huo, R.; Mao, W. A real-time ship collision risk perception model derived from domain-based approach parameters. *Ocean Eng.* **2022**, *265*, 112554. [CrossRef]
20. Zhou, J.; Jiang, P.; Zou, A.; Chen, X.; Hu, W. Ship target detection algorithm based on improved YOLOv5. *J. Mar. Sci. Eng.* **2021**, *9*, 908. [CrossRef]
21. Garcia-Garcia, B.; Bouwmans, T.; Silva, A.J.R. Background subtraction in real applications: Challenges, current models and future directions. *Comput. Sci. Rev.* **2020**, *35*, 100204. [CrossRef]
22. Chen, X.; Wu, H.; Han, B.; Liu, W.; Montewka, J.; Liu, R.W. Orientation-aware ship detection via a rotation feature decoupling supported deep learning approach. *Eng. Appl. Artif. Intell.* **2023**, *125*, 106686. [CrossRef]
23. Guo, Y.; Lu, Y.; Liu, R.W. Lightweight deep network-enabled real-time low-visibility enhancement for promoting vessel detection in maritime video surveillance. *J. Navig.* **2022**, *75*, 230–250. [CrossRef]
24. Liu, R.W.; Lu, Y.; Guo, Y.; Ren, W.; Zhu, F.; Lv, Y. AiOENet: All-in-one low-visibility enhancement to improve visual perception for intelligent marine vehicles under severe weather conditions. *IEEE Trans. Intell. Veh.* **2023**, *9*, 3811–3826. [CrossRef]
25. Wang, C.; Fan, B.; Li, Y.; Xiao, J.; Min, L.; Zhang, J.; Chen, J.; Lin, Z.; Su, S.; Wu, R. Study on the Classification Perception and Visibility Enhancement of Ship Navigation Environments in Foggy Conditions. *J. Mar. Sci. Eng.* **2023**, *11*, 1298. [CrossRef]
26. Li, S.; Cao, X.; Zhou, Z. Research on inshore ship detection under nighttime low-visibility environment for maritime surveillance. *Comput. Electr. Eng.* **2024**, *118*, 109310. [CrossRef]
27. Yan, H.; Hou, Q.; Zhang, J.; Wang, L.; Zhang, G.; Zhu, D. Scheme to implement moving target detection of coastal defense radar in complicated sea conditions. *J. Appl. Remote Sens.* **2022**, *16*, 046510. [CrossRef]
28. Xu, S.; Zhu, J.; Jiang, J.; Shui, P. Sea-surface floating small target detection by multifeature detector based on isolation forest. *IEEE J. Sel. Top. Appl. Earth Obs. Remote Sens.* **2020**, *14*, 704–715. [CrossRef]
29. Rey Charlo, R.E. On-board radio communication and its development in a historical perspective. *Int. J. Marit. Hist.* **2024**, *36*, 140–152. [CrossRef]
30. Xie, C.; Shi, P.; Cao, F.; Qian, Y. Reliability modeling and analysis of ship communication network based on Apriori algorithm. *J. Coast. Res.* **2019**, *93*, 711–716. [CrossRef]
31. Hoole, P.; Ong, S.; Hoole, S. Shore to Ship Steerable Electromagnetic Beam System Based Ship Communication and Navigation. *Appl. Comput. Electromagn. Soc. J.* **2013**, *28*, 747–754.
32. Wang, B.; Ben, K.R.; Hao, Y.X.; Zuo, M.J. SQMCR: Stackelberg Q-learning based Multi-hop Cooperative Routing Algorithm for Underwater Wireless Sensor Networks. *IEEE Access* **2024**, *12*, 56179–56195.
33. Hu, C.; Wu, D.; Liao, Y.; Hu, X. Sliding mode control unified with the uncertainty and disturbance estimator for dynamically positioned vessels subjected to uncertainties and unknown disturbances. *Appl. Ocean Res.* **2021**, *109*, 102564. [CrossRef]
34. Zhao, Z.P.; Zhang, Q. Adaptive self-regulation PID tracking control for the ship course. *Chin. J. Ship Res.* **2019**, *14*, 145–151.
35. Hosseinabadi, P.A.; Abadi, A.S.S.; Mekhilef, S. Fuzzy adaptive finite-time sliding mode controller for trajectory tracking of ship course systems with mismatched uncertainties. *Int. J. Autom. Control* **2022**, *16*, 255–271. [CrossRef]
36. Wang, S.; Er, M.J.; Liu, T.; Gong, H. Path Following Control of Underactuated AUV Based on Improved Model Predictive Control. In Proceedings of the 2023 6th International Conference on Intelligent Autonomous Systems (ICoIAS), Qinhuangdao, China, 22–24 September 2023; pp. 222–227.
37. Abadi, A.S.S.; Hosseinabadi, P.A.; Mekhilef, S. Fuzzy adaptive fixed-time sliding mode control with state observer for a class of high-order mismatched uncertain systems. *Int. J. Control Autom. Syst.* **2020**, *18*, 2492–2508. [CrossRef]
38. Chen, Z.; Qin, B.; Sun, M.; Sun, Q. Q-learning-based parameters adaptive algorithm for active disturbance rejection control and its application to ship course control. *Neurocomputing* **2020**, *408*, 51–63. [CrossRef]
39. Veitch, E.; Alsos, O.A. A systematic review of human-AI interaction in autonomous ship systems. *Saf. Sci.* **2022**, *152*, 105778. [CrossRef]

40. Dlabač, T.; Ćalasan, M.; Krčum, M.; Marvučić, N. PSO-based PID controller design for ship course-keeping autopilot. *Brodogr. Int. J. Nav. Archit. Ocean Eng. Res. Dev.* **2019**, *70*, 1–15. [CrossRef]
41. Volyanskaya, Y.B.; Volyanskiy, S.M.; Onishchenko, O.A.; Shevchenko, V.A.; Trudnev, S.Y. Research of possibilities to increase the exactness of ship stabilizing on a course. Морские интеллектуальные технологии **2019**, *3-3*, 174–181.
42. Volyanskyy, S.; Vorokhobin, I.; Volyanskaya, Y.; Mazur, O.; Onishchenko, O. Marine ship's course stabilization based on an autopilot with a simple fuzzy controller. *Sci. Bull. Mircea Cel Batran Nav. Acad.* **2022**, *25*, 23–35. [CrossRef]
43. Liu, C.; Wang, D.; Zhang, Y.; Meng, X. Model predictive control for path following and roll stabilization of marine vessels based on neurodynamic optimization. *Ocean Eng.* **2020**, *217*, 107524. [CrossRef]
44. Rezaei, A.; Tabatabaei, M. Ship roll stabilization using an adaptive fractional-order sliding mode controller. *Ocean Eng.* **2023**, *287*, 115883. [CrossRef]
45. Sun, M.; Zhang, W.; Zhang, Y.; Luan, T.; Yuan, X.; Li, X. An anti-rolling control method of rudder fin system based on ADRC decoupling and DDPG parameter adjustment. *Ocean Eng.* **2023**, *278*, 114306. [CrossRef]
46. Zhang, Z.; Zhang, X. Course-keeping with roll damping control for ships using rudder and fin. *J. Mar. Sci. Technol.* **2021**, *26*, 872–882. [CrossRef]
47. Qiang, H.; Jin, S.; Feng, X.; Xue, D.; Zhang, L. Model predictive control of a shipborne hydraulic parallel stabilized platform based on ship motion prediction. *IEEE Access* **2020**, *8*, 181880–181892. [CrossRef]
48. You, X.; Li, S.; Liu, J.; Yan, X. Experimental research of the PID tune method for ship path following control. In Proceedings of the ISOPE International Ocean and Polar Engineering Conference, Ottawa, ON, Canada, 19–23 June 2023; p. ISOPE–I-23-364.
49. Nie, J.; Lin, X. FAILOS guidance law based adaptive fuzzy finite-time path following control for underactuated MSV. *Ocean Eng.* **2020**, *195*, 106726. [CrossRef]
50. Shen, C.; Shi, Y. Distributed implementation of nonlinear model predictive control for AUV trajectory tracking. *Automatica* **2020**, *115*, 108863. [CrossRef]
51. Shen, Z.; Bi, Y.; Wang, Y.; Guo, C. MLP neural network-based recursive sliding mode dynamic surface control for trajectory tracking of fully actuated surface vessel subject to unknown dynamics and input saturation. *Neurocomputing* **2020**, *377*, 103–112. [CrossRef]
52. Zhang, H.; Zhang, X.; Cao, T.; Bu, R. Active disturbance rejection control for ship path following with Euler method. *Ocean Eng.* **2022**, *247*, 110516. [CrossRef]
53. Liu, D.; Yao, C.; Yu, J.; Feng, D.; Sun, X. Trajectory Tracking Control of an Intelligent Ship Based on Deep Reinforcement Learning. In Proceedings of the ISOPE International Ocean and Polar Engineering Conference, Rhodes, Greece, 16–21 June 2024; p. ISOPE–I-24-532.
54. Li, S.; Xu, C.; Liu, J.; Xu, Z.; Meng, F. Tracking control of ships based on ADRC-MFAC. *Chin. J. Ship Res* **2023**, *18*, 1–10.
55. Yang, C.-H.; Wu, C.-H.; Shao, J.-C.; Wang, Y.-C.; Hsieh, C.-M. AIS-based intelligent vessel trajectory prediction using bi-LSTM. *IEEE Access* **2022**, *10*, 24302–24315. [CrossRef]
56. Min, B.; Zhang, X. Concise robust fuzzy nonlinear feedback track keeping control for ships using multi-technique improved LOS guidance. *Ocean Eng.* **2021**, *224*, 108734. [CrossRef]
57. Li, H.; Chen, H.; Gao, N.; Aït-Ahmed, N.; Charpentier, J.-F.; Benbouzid, M. Ship dynamic positioning control based on active disturbance rejection control. *J. Mar. Sci. Eng.* **2022**, *10*, 865. [CrossRef]
58. Papadimitrakis, M.; Stogiannos, M.; Sarimveis, H.; Alexandridis, A. Multi-ship control and collision avoidance using MPC and RBF-based trajectory predictions. *Sensors* **2021**, *21*, 6959. [CrossRef]
59. Huang, Y.; Chen, L.; Chen, P.; Negenborn, R.R.; Van Gelder, P. Ship collision avoidance methods: State-of-the-art. *Saf. Sci.* **2020**, *121*, 451–473. [CrossRef]
60. Ding, Z. A ship-motion prediction algorithm based on modified covariance method and neural networks. In Proceedings of the International Conference on Computer Application and Information Security (ICCAIS 2021), Wuhan, China, 18–19 December 2021; pp. 272–278.
61. He, Y.; Li, Z.; Mou, J.; Hu, W.; Li, L.; Wang, B. Collision-avoidance path planning for multi-ship encounters considering ship manoeuvrability and COLREGs. *Transp. Saf. Environ.* **2021**, *3*, 103–113. [CrossRef]
62. Kim, J.H.; Lee, S.; Jin, E.S. Collision avoidance based on predictive probability using Kalman filter. *Int. J. Nav. Archit. Ocean Eng.* **2022**, *14*, 100438. [CrossRef]
63. Abebe, M.; Noh, Y.; Kang, Y.-J.; Seo, C.; Kim, D.; Seo, J. Ship trajectory planning for collision avoidance using hybrid ARIMA-LSTM models. *Ocean Eng.* **2022**, *256*, 111527. [CrossRef]
64. Liu, K.; Yuan, Z.; Xin, X.; Zhang, J.; Wang, W. Conflict detection method based on dynamic ship domain model for visualization of collision risk Hot-Spots. *Ocean Eng.* **2021**, *242*, 110143. [CrossRef]
65. Szlapczynski, R.; Szlapczynska, J. A ship domain-based model of collision risk for near-miss detection and Collision Alert Systems. *Reliab. Eng. Syst. Saf.* **2021**, *214*, 107766. [CrossRef]
66. Liu, R.W.; Huo, X.; Liang, M.; Wang, K. Ship collision risk analysis: Modeling, visualization and prediction. *Ocean Eng.* **2022**, *266*, 112895. [CrossRef]
67. Liu, D.; Shi, G. Ship collision risk assessment based on collision detection algorithm. *IEEE Access* **2020**, *8*, 161969–161980. [CrossRef]

68. Xin, X.; Liu, K.; Yang, Z.; Zhang, J.; Wu, X. A probabilistic risk approach for the collision detection of multi-ships under spatiotemporal movement uncertainty. *Reliab. Eng. Syst. Saf.* **2021**, *215*, 107772. [CrossRef]
69. Shi, J.; Liu, Z. Track pairs collision detection with applications to ship collision risk assessment. *J. Mar. Sci. Eng.* **2022**, *10*, 216. [CrossRef]
70. Perera, L.; Carvalho, J.; Guedes Soares, C. Fuzzy logic based decision making system for collision avoidance of ocean navigation under critical collision conditions. *J. Mar. Sci. Technol.* **2011**, *16*, 84–99. [CrossRef]
71. Park, J.; Jeong, J.-S. An estimation of ship collision risk based on relevance vector machine. *J. Mar. Sci. Eng.* **2021**, *9*, 538. [CrossRef]
72. Chen, P.; Huang, Y.; Papadimitriou, E.; Mou, J.; van Gelder, P. An improved time discretized non-linear velocity obstacle method for multi-ship encounter detection. *Ocean Eng.* **2020**, *196*, 106718. [CrossRef]
73. Tengesdal, T.; Johansen, T.A.; Brekke, E.F. Ship collision avoidance utilizing the cross-entropy method for collision risk assessment. *IEEE Trans. Intell. Transp. Syst.* **2021**, *23*, 11148–11161. [CrossRef]
74. Cheng, Z.; Chen, P.; Mou, J.; Chen, L. Novel collision risk measurement method for multi-ship encounters via velocity obstacles and temporal proximity. *Ocean Eng.* **2024**, *302*, 117585. [CrossRef]
75. Li, S.; Liu, J.; Negenborn, R.R.; Ma, F. Optimizing the joint collision avoidance operations of multiple ships from an overall perspective. *Ocean Eng.* **2019**, *191*, 106511. [CrossRef]
76. Zhu, Z.; Yin, Y.; Lyu, H. Automatic collision avoidance algorithm based on route-plan-guided artificial potential field method. *Ocean Eng.* **2023**, *271*, 113737. [CrossRef]
77. Ren, J.; Zhang, J.; Cui, Y. Autonomous obstacle avoidance algorithm for unmanned surface vehicles based on an improved velocity obstacle method. *ISPRS Int. J. Geo-Inf.* **2021**, *10*, 618. [CrossRef]
78. Zhang, L.; Mou, J.; Chen, P.; Li, M. Path planning for autonomous ships: A hybrid approach based on improved apf and modified vo methods. *J. Mar. Sci. Eng.* **2021**, *9*, 761. [CrossRef]
79. Xu, X.; Lu, Y.; Liu, G.; Cai, P.; Zhang, W. COLREGs-abiding hybrid collision avoidance algorithm based on deep reinforcement learning for USVs. *Ocean Eng.* **2022**, *247*, 110749. [CrossRef]
80. Yang, Z.; Jing, Q.; Li, X. Dynamic Data-Driven Ship Motion Simulation toward Visual-Aided Navigation on Water. *Water* **2023**, *15*, 872. [CrossRef]
81. Chen, X.; Dou, S.; Song, T.; Wu, H.; Sun, Y.; Xian, J. Spatial-Temporal Ship Pollution Distribution Exploitation and Harbor Environmental Impact Analysis via Large-Scale AIS Data. *J. Mar. Sci. Eng.* **2024**, *12*, 960. [CrossRef]
82. Nguyen, D.; Fablet, R. A transformer network with sparse augmented data representation and cross entropy loss for ais-based vessel trajectory prediction. *IEEE Access* **2024**, *12*, 21596–21609. [CrossRef]
83. Zhu, J.; Gao, M.; Zhang, A.; Hu, Y.; Zeng, X. Multi-ship encounter situation identification and analysis based on AIS data and graph complex network theory. *J. Mar. Sci. Eng.* **2022**, *10*, 1536. [CrossRef]
84. Rong, H.; Teixeira, A.; Soares, C.G. Ship collision avoidance behaviour recognition and analysis based on AIS data. *Ocean Eng.* **2022**, *245*, 110479. [CrossRef]
85. Öztürk, Ü.; Boz, H.A.; Balcisoy, S. Visual analytic based ship collision probability modeling for ship navigation safety. *Expert Syst. Appl.* **2021**, *175*, 114755. [CrossRef]
86. He, W.; Lei, J.; Chu, X.; Xie, S.; Zhong, C.; Li, Z. A visual analysis approach to understand and explore quality problems of AIS data. *J. Mar. Sci. Eng.* **2021**, *9*, 198. [CrossRef]
87. Carter, E. Enhancing Maritime Navigation Safety through AIS-Based Visual Augmentation: A Deep Learning Approach to Integrating Real and Virtual Views. *J. Comput. Sci. Softw. Appl.* **2023**, *3*, 21–26.
88. Xia, R.; Chen, J.; Huang, Z.; Wan, H.; Wu, B.; Sun, L.; Yao, B.; Xiang, H.; Xing, M. CRTransSar: A visual transformer based on contextual joint representation learning for SAR ship detection. *Remote Sens.* **2022**, *14*, 1488. [CrossRef]
89. Chen, Z.; Ding, Z.; Zhang, X.; Wang, X.; Zhou, Y. Inshore ship detection based on multi-modality saliency for synthetic aperture radar images. *Remote Sens.* **2023**, *15*, 3868. [CrossRef]
90. Mou, X.; Chen, X.; Guan, J.; Chen, B.; Dong, Y. Marine target detection based on improved faster R-CNN for navigation radar PPI images. In Proceedings of the 2019 International Conference on Control, Automation and Information Sciences (ICCAIS), Chengdu, China, 23–26 October 2019; pp. 1–5.
91. Hsieh, T.-H.; Wang, S.; Gong, H.; Liu, W.; Xu, N. Sea ice warning visualization and path planning for ice navigation based on radar image recognition. *J. Mar. Sci. Technol.* **2021**, *29*, 280–290. [CrossRef]
92. Naus, K.; Wąż, M.; Szymak, P.; Gucma, L.; Gucma, M. Assessment of ship position estimation accuracy based on radar navigation mark echoes identified in an Electronic Navigational Chart. *Measurement* **2021**, *169*, 108630. [CrossRef]
93. Chen, X.; Mu, X.; Guan, J.; Liu, N.; Zhou, W. Marine target detection based on Marine-Faster R-CNN for navigation radar plane position indicator images. *Front. Inf. Technol. Electron. Eng.* **2022**, *23*, 630–643. [CrossRef]
94. Li, Z.; Pan, M.; Hu, J.; Guo, J. Design on ship "video radar" enhanced navigation system based on multi-camera. In Proceedings of the 2022 5th International Conference on Signal Processing and Machine Learning, Dalian, China, 4–6 August 2022; pp. 47–53.
95. Xu, X.; Wu, B.; Xie, L.; Teixeira, Â.P.; Yan, X. A novel ship speed and heading estimation approach using radar sequential images. *IEEE Trans. Intell. Transp. Syst.* **2023**, *24*, 11107–11120. [CrossRef]
96. Zhou, A.; Xie, W.; Pei, J. Background modeling combined with multiple features in the Fourier domain for maritime infrared target detection. *IEEE Trans. Geosci. Remote Sens.* **2021**, *60*, 1–15. [CrossRef]

97. Dong, L.; Ma, D.; Qin, G.; Zhang, T.; Xu, W. Infrared target detection in backlighting maritime environment based on visual attention model. *Infrared Phys. Technol.* **2019**, *99*, 193–200. [CrossRef]
98. Liu, Z.; He, J.; Zhang, T.; Tang, R.; Li, Y.; Waqas, M. Infrared ship video target tracking based on cross-connection and spatial transformer network. In Proceedings of the International Conference on Artificial Intelligence and Security, Qinghai, China, 15–20 July 2022; pp. 100–114.
99. Cao, Y.; Cheng, W.; Wang, X.; Huang, Y. Research on Ship Target Recognition based on Infrared Image Method. In Proceedings of the 2023 4th International Conference on Computing, Networks and Internet of Things, Xiamen, China, 26–28 May 2023; pp. 197–202.
100. Gao, Z.; Zhang, Y.; Wang, S. Lightweight Small Ship Detection Algorithm Combined with Infrared Characteristic Analysis for Autonomous Navigation. *J. Mar. Sci. Eng.* **2023**, *11*, 1114. [CrossRef]
101. Park, J.-H.; Roh, M.-I.; Lee, H.-W.; Jo, Y.-M.; Ha, J.; Son, N.-S. Multi-vessel Target Tracking with Camera Fusion for Unmanned Surface Vehicles. *Int. J. Nav. Archit. Ocean Eng.* **2024**, *16*, 100608. [CrossRef]
102. Li, Y.; Tao, K.; Li, X.; Wang, F. Research on Visual Laser Navigation of Ships. In Proceedings of the 2019 5th International Conference on Transportation Information and Safety (ICTIS), Liverpool, UK, 14–17 July 2019; pp. 191–196.
103. Pan, M.; Liu, Y.; Cao, J.; Li, Y.; Li, C.; Chen, C.-H. Visual recognition based on deep learning for navigation mark classification. *IEEE Access* **2020**, *8*, 32767–32775. [CrossRef]
104. Shao, Z.; Lyu, H.; Yin, Y.; Cheng, T.; Gao, X.; Zhang, W.; Jing, Q.; Zhao, Y.; Zhang, L. Multi-scale object detection model for autonomous ship navigation in maritime environment. *J. Mar. Sci. Eng.* **2022**, *10*, 1783. [CrossRef]
105. Bi, Q.; Wang, M.; Huang, Y.; Lai, M.; Liu, Z.; Bi, X. Ship Collision Avoidance Navigation Signal Recognition via Vision Sensing and Machine Forecasting. *IEEE Trans. Intell. Transp. Syst.* **2023**, *24*, 11743–11755. [CrossRef]
106. Qu, J.; Liu, R.W.; Guo, Y.; Lu, Y.; Su, J.; Li, P. Improving maritime traffic surveillance in inland waterways using the robust fusion of AIS and visual data. *Ocean Eng.* **2023**, *275*, 114198. [CrossRef]
107. Gülsoylu, E.; Koch, P.; Yildiz, M.; Constapel, M.; Kelm, A.P. Image and AIS Data Fusion Technique for Maritime Computer Vision Applications. In Proceedings of the IEEE/CVF Winter Conference on Applications of Computer Vision, Waikoloa, HI, USA, 3–8 January 2024; pp. 859–868.
108. Ding, H.; Weng, J. A robust assessment of inland waterway collision risk based on AIS and visual data fusion. *Ocean Eng.* **2024**, *307*, 118242. [CrossRef]
109. Guo, M.; Guo, C.; Zhang, C.; Zhang, D.; Gao, Z. Fusion of ship perceptual information for electronic navigational chart and radar images based on deep learning. *J. Navig.* **2020**, *73*, 192–211. [CrossRef]
110. Zhang, C.; Fang, M.; Yang, C.; Yu, R.; Li, T. Perceptual fusion of electronic chart and marine radar image. *J. Mar. Sci. Eng.* **2021**, *9*, 1245. [CrossRef]
111. Gao, Z.; Zhu, F.; Chen, H.; Ma, B. Maritime Infrared and Visible Image Fusion Based on Refined Features Fusion and Sobel Loss. *Photonics* **2022**, *9*, 566. [CrossRef]
112. Jeon, R.; Jones, N. Visual and Infrared Detection and Ranging (VAIDAR) for Marine Navigational Hazards. In Proceedings of the OCEANS 2023-MTS/IEEE US Gulf Coast, Biloxi, MI, USA, 25–28 September 2023; pp. 1–6.
113. Sun, S.; Lyu, H.; Dong, C. AIS aided marine radar target tracking in a detection occluded environment. *Ocean Eng.* **2023**, *288*, 116133. [CrossRef]
114. Xu, X.; Wu, B.; Teixeira, Â.P.; Yan, X.; Soares, C.G. Integration of Radar Sequential Images and AIS for Ship Speed and Heading Estimation Under Uncertainty. *IEEE Trans. Intell. Transp. Syst.* **2023**, *25*, 5688–5702. [CrossRef]
115. Wu, Y.; Chu, X.; Deng, L.; Lei, J.; He, W.; Królczyk, G.; Li, Z. A new multi-sensor fusion approach for integrated ship motion perception in inland waterways. *Measurement* **2022**, *200*, 111630. [CrossRef]
116. Xiao, G.; Xu, L. Challenges and Opportunities of Maritime Transport in the Post-Epidemic Era. *J. Mar. Sci. Eng.* **2024**, *12*, 1685. [CrossRef]

Journal of
Marine Science and Engineering

MDPI

Article

Optimizing Ship Draft Observation with Wave Energy Attenuation and PaddlePaddle-OCR in an Anti-Fluctuation Device

Yaoming Wei [1], Huan Du [2,*], Qinyou Hu [1] and Hu Wang [1]

[1] Merchant Marine College, Shanghai Maritime University, Shanghai 201306, China; ymwei@shmtu.edu.cn (Y.W.); qyhu@shmtu.edu.cn (Q.H.); w1632658718@hotmail.com (H.W.)
[2] China Waterborne Transport Research Institute, Beijing 100088, China
* Correspondence: china.duhuan@wti.ac.cn

Abstract: With the development and application of artificial intelligence (AI) in the shipping industry, using AI to replace traditional draft survey methods in bulk carriers can significantly reduce manpower, lower the risks associated with visual observations, improve measurement accuracy, and minimize the impact of human subjective factors. Ultimately, the integration of software and hardware technologies will replace human visual observations with automated draft measurement calculations. A similar anti-fluctuation device described in this article has been used in ship draft observation based on AI-assisted proving, which can ease the fluctuation of the wave inside the pipe. Observers can directly read the water surface inside the pipe and compare it to the ship's draft mark to obtain the final draft, effectively improving draft observation accuracy. However, some surveyors refuse to accept the readings obtained from this device, citing a lack of theoretical basis or the absence of accreditation from relevant technical authorities, leading to the rejection of results. To address these issues, this paper integrates wave energy attenuation theory with PaddlePaddle-OCR recognition to further validate the anti-fluctuation device for accurate ship draft observation. The experimental results are as follows: first, the pipe effectively suppresses the amplitude of external water surface fluctuations by 75%, explaining the fundamental theory that wave heights within the anti-fluctuation device are consistent with external swell heights. When taking a draft measurement, the system dynamically adjusts the position of the main tube in response to the ship's movements, maintaining the stability of the measurement section and significantly reducing the difficulty of observations. Due to the reduction in fluctuation amplitude, there is a noticeable improvement in observation accuracy.

Keywords: ship draft observation; draught survey; wave energy attenuation; wave anti-fluctuation device; PaddlePaddle-OCR

1. Introduction

Traditionally, ship draft is mainly measured by manual observation [1], involving the ship's party and a surveyor, as shown in Figure 1a. It typically involves the use of yachts or standing on the shore side to observe the ship's six draft marks, aiming to achieve an observation angle as parallel to the water surface as possible [2]. The water that immerses the draft mark allows for direct reading of the actual ship's draft [3] in calm water, as shown in Figure 1b. While manual visual observations can yield high precision, fluctuations in the water can result in significant discrepancies between the actual ship draft and the observed draft. The draft mark, where the waterline immerses, indicates the ship's draft, and the corresponding displacement is determined through corrections for apparent draft and trim, draft correction at the perpendicular, displacement adjustments for the center of flotation not being amidships, and the relevant water density. Subsequently, by subtracting the ship's lightship weight and other factors such as fuel and ballast, the constant, and deductibles, the net cargo mass can be calculated, ultimately determining the quantity of cargo loaded [4]. The accuracy of draft readings is crucial in actual work scenarios [5].

Figure 1. The chief officer reading the draft with the surveyor shown in (**a**) and the ship's draft immersed in still water is shown in (**b**).

In practical application scenarios, wave fluctuations can cause water surface readings to be unstable, especially in heavy swell conditions. Significant wave fluctuations complicate the survey process, leading to disputed observation results. For instance, on a cape-size bulk carrier, an observation error of just one centimeter can result in a discrepancy of 120 metric tons of cargo, while for a Panama-size vessel, the error could lead to a discrepancy of 50 metric tons, resulting in disagreements between the ship party and third-party surveyors. In most cases, the shipowner may compromise, leading to potential losses. Although shippers may issue a letter of indemnity to the shipowner in heavy swell conditions to mitigate the risk of inaccurate cargo measurements due to the increased difficulty in taking draft readings, the shipowner may still bear responsibility if there is a significant discrepancy in the actual cargo delivered at the discharge port.

This paper first introduces the core components of the device and explains the function of each component. It then introduces the wave energy attenuation method to explain the fundamental theory of wave heights within the anti-fluctuation device, which is consistent with external swell heights [6]. Finally, it discusses improved observational methods, where draft readings are based on the confirmed original wave height, and utilizes PaddlePaddle-OCR technology [7] and other AI methods to eliminate the uncertainty of subjective human factors [8].

1.1. Literature Review

The paper provides an overview of key research directions in fluid dynamics wave control theory and related fields. Xiao-hui Ren et al. [9] investigated the overlap area and developed a subdivision criterion for interference effects based on experimental velocity ratios. They derived a change curve for interference efficiency, which has practical implications. N. Choudhary et al. [10] introduced a baffle that divides the liquid region into four sections, formulating boundary value problems for each region's potential. By applying matching conditions across virtual interfaces, a system of linear equations was established to determine natural frequencies. Debasish Dutte's et al. [11] research shows that, for a given blockage ratio, the coefficient of inertia decreases with increasing Keulegan–Carpenter (KC) number [12], while the coefficient of pressure increases upstream and decreases downstream of the cylinder. In the review area of auto recognition, Sarkar et al. [13] carry out PaddlePaddle-OCR to automatically recognize unique number plates of vehicles, enabling intelligent traffic and vehicle management, RENATO IVČE et al. [14] proposed a liquid level optical sensor for measuring sea levels in sounding pipes. However, this method suffers from significant errors—up to 120 metric tons for a 3000 dead-weight vessel limiting its practical use. Wei Zhan et al. [15] used the YOLO (You Only Look Once) datasets and unmanned aerial vehicle (UAV) to predict waterline positions, achieving a 0.01 m error margin, though it is effective only in calm wave conditions. Zhong Wang et al. [16] applied edge detection techniques combined with artificial intelligence to observe the ship draft in the near research direction, achieving an error of less than 1cm, but the method is not suitable for rough sea conditions. Yaoming wei et al. [17] review ship draft observation methods, and conclude that an error of one centimeter causes the cargo to be lost on different kinds of bulk carriers. Mevlut Savas Bilican et al. [18] describe draft optimization in voyages. Xinqiang Chen et al. [19] proposed a framework that learns fine-grained personnel features with the help of the object query and transformer encoder–decoder module from the input-port-like image sequences and thus obtains port personnel locations from the input low-visibility images. Hanae Moussaoui et al. [20], after extracting the plate the k-means clustering algorithm, used thresholding techniques and the opening morphological operation to enhance the image and make the characters in the license plate clearer before using addle Paddle OCR. R. C. Hirt et al. [21] used a recurrent neural network (CRNN) to create a DL-based hybrid ID framework that predicts and classifies malicious cyberattacks in the network. Siying Qian et al. [22] demonstrates that MobileNetV3 can obtain a superior balance between efficiency and accuracy for real-life images. Jia Deng et al. [23] illustrate the usefulness of ImageNet through three simple applications in object recognition, image classification and automatic object clustering. Yang He et al. [24] prune filters with smaller norm values in a convolutional neural network. Zeming Li et al. [25] propose a new two-stage detector, Light-Head R-CNN, to address the shortcomings in current two-stage approaches. Eibin Rong et al. [26] proposed two adaptive threshold selection methods based on the mean of image gradient magnitude and standard deviation were put forward for two kinds of typical images. Kristina P. Sinaga [27] proposed a novel unsupervised k-means (U-k-means) clustering algorithm to automatically find an optimal number of clusters without giving any initialization and parameter selection. Boaz Arad [28] found a new, larger-than-ever, natural hyperspectral image data set containing a total of 510 HS images. In terms of date analysis, Çagatay Iris et al. [29] proposed a heuristic algorithm to solve the problem and improve the performance of formulation significantly, and the heuristic efficiently generates high-quality solutions.

1.2. Research Gap

While research on ship draft observation has largely focused on algorithm optimization, there has been a noticeable gap in the exploration of hardware applications combined with AI assistance, particularly under high wave conditions in the near future. This study designs an anti-fluctuation device that improves draft observation accuracy from a hardware perspective, specifically under challenging wave conditions. Previous related research articles are based on observations made in the field with visual observation data used as the base accuracy annotation. In contrast, the data collection conducted in the laboratory for this paper relies on initial wave height judgment, which enhances observation accuracy. The main contributions of this paper can be summarized as follows:

- Despite similar devices being used in various ports by surveyors and seafarers, the anti-fluctuation device has been limited by a lack of theoretical foundation and experimental validation. This paper attempts to establish the theoretical basis as critical. Although some recent studies have explored similar concepts in other fields, such as oil well level measurement, there remains a significant need for more robust research specific to the shipping industry.
- This paper designs an anti-fluctuation device that effectively suppresses the amplitude of water surface fluctuations by 75% inside the pipe. Observers can directly read the surface of the pipe and compare it to the draft mark to obtain the draft observation result. The device dynamically adjusts the position of the main tube in response to the ship's movements, thereby maintaining the stability of the measurement section. This allows the observer to read the water surface relative to the ship's draft mark, significantly reducing the difficulty of observations, improving accuracy, and minimizing

disputes. The cost of the anti-fluctuation device is low, and it can be reused multiple times, making it an ideal candidate for equipping bulk carrier vessels.

- This paper utilizes PaddlePaddle-OCR artificial intelligence recognition as an assistant tool to eliminate the uncertainty of subjective human factors.

2. Research and Development of Anti-Fluctuation Devices

The anti-fluctuation device operates by isolating the water inside the main tube from external wave action. As waves impact the exterior of the tube, the internal baffles and dampers reduce the energy of the waves, resulting in a much calmer water surface inside the tube. Figure 2 illustrates the structure of the anti-fluctuation device and the ship's draft mark system. In this figure, the bottom of the number "2" represents 20 cm, the top of "2" represents 30 cm, and the height between the numbers "2" and "4" is 10 cm. The bottom of "1M" represents 100 cm, while the top represents 110 cm. This stable water surface allows for precise and consistent draft readings.

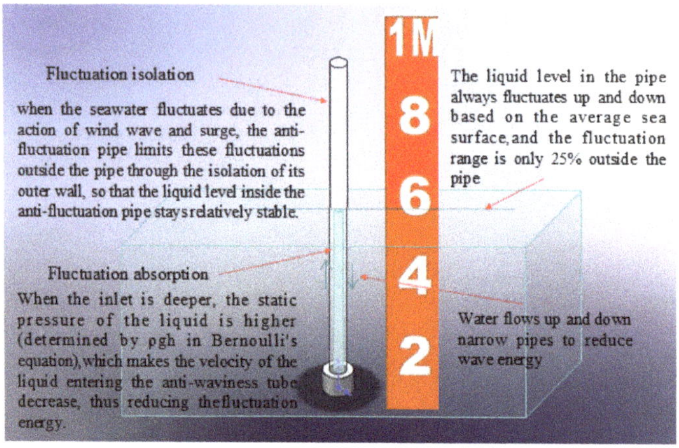

Figure 2. Illustration of structure and material selection of the anti-fluctuation device.

2.1. Basic Principle Introduction

1. Fluctuation isolation component.
   This experiment uses a stainless steel cylinder with a diameter of 10 cm, a height of 10 cm, and a weight of 6.7 kg as the fluctuation isolation component. The lower end of the cylinder has a 1 mm hole to control the speed of water intake, while the upper part is connected to a transparent polyvinyl chloride hose, as shown in Figure 2. The anti-fluctuation device is positioned vertically in the water. When seawater fluctuates due to wind, waves, or surges, the pipe's outer wall isolates these fluctuations, thereby keeping the liquid level inside the pipe relatively stable [30].
2. Fluctuation absorption component.
   The Fluctuation absorption component is made of a transparent polyvinyl chloride hose with a length of 2 m, inner diameter 2 cm, and thickness of 1 mm, which is shown in Figure 2. When the inlet is positioned deeper, the static pressure of the liquid increases (as determined by ρgh in Bernoulli's equation), which slows the velocity of the liquid entering the anti-fluctuation pipe, thereby reducing fluctuation energy. Additionally, a deeper inlet increases resistance and energy loss in the fluid flow (according to fluid dynamics resistance laws) [31], further absorbing and dampening the fluctuation energy [32].

2.2. Basic Equations

The formula demonstrated in this section assumes that the liquid is incompressible and has no viscosity. Based on visual inspection, the water flow in this experiment was found to be unsatisfactory.

Conservation of mass equation:

$$\nabla \cdot (\rho v) = 0 \tag{1}$$

The mass conservation equation, a fundamental continuity equation in fluid mechanics, delineates the principle of mass conservation within a fluid medium. This equation characterizes the temporal and spatial variations in mass density ρ. It is instrumental in analyzing water flow alterations induced by wave action and elucidating the flow field surrounding vessels, thereby facilitating an understanding of volumetric changes in aquatic systems and the governing principles of fluid dynamics [33].

ρ: Fluid density, representing mass per unit volume. t: Time, the temporal dimension for changes in fluid properties. v: Fluid velocity vector, indicating the speed and direction of fluid particles. $\nabla \cdot (\rho v)$: The divergence of the product of density ρ and velocity v, representing the net mass flux out of a control volume.

Navier–Stokes equations.

Conversely, the momentum conservation equation articulates the momentum variation for an inviscid fluid. In practical applications involving complex flow scenarios—such as ship movement influenced by waves—the roles of fluid viscosity and nonlinearity become paramount; thus, it necessitates extension to encompass the generalized Navier–Stokes equations.

$$\rho((v \cdot \nabla) vs.) = -\nabla p + f \tag{2}$$

ρ: Fluid density, representing mass per unit volume. v: Fluid velocity vector, indicating speed and direction. $(v \cdot \nabla)v$: Convective acceleration, describing velocity changes due to fluid movement. $-\nabla p$: Pressure gradient force, the force caused by pressure differences. f: force, external forces like gravity acting on the fluid [34].

The natural oscillation frequency of the liquid depends on the different widths and positions within the pipe, as well as the different structures. When the inlet of the pipe is close to the free liquid surface, it has a greater impact on the frequency. The fluctuation of the liquid level inside the pipe decreases with the increase in the depth of the pipe [35].

Bernoulli's equation.

The interaction of waves with the fluid results in frequent conversions between kinetic energy and pressure. When these waves impinge upon the hull, the kinetic energy of the fluid is transformed into pressure energy, potentially leading to vertical displacement of the hull. The Bernoulli equation serves as a quantitative framework for this energy conversion process [31].

$$p + \frac{1}{2}\rho v^2 + \rho g h = \text{constant} \tag{3}$$

where p is the fluid pressure, v is the fluid velocity, g is the gravitational acceleration, and h his the inlet depth.

Bernoulli's principle states that for an incompressible, non-viscous fluid in a steady flow, the sum of the pressure energy (p), kinetic energy ($\frac{1}{2}\rho v^2$), and potential energy ($\rho g h$) remains constant along a streamline. This implies that an increase in fluid velocity leads to a decrease in pressure and/or height, and vice versa.

Collectively, Equations (1)–(3) comprehensively characterize the motion of the fluid, its energy transfer mechanisms, and its dynamic behavior under wave action. In wave suppression devices, these three equations collaboratively elucidate the impact of waves on both the fluid dynamics and device performance from multiple perspectives.

Wave equation.

The wave equation mainly describes how waves travel through a fluid and the change in energy during wave transport, particularly the energy change in the fluid at the crest and trough of the wave. It plays an important role in the design and analysis of ships and wave suppression devices.

This equation is a linear wave equation that describes the propagation of waves under ideal conditions, assuming that the wave medium is uniform, undamped, and free from external forces. In this case, the propagation speed of the wave c is constant, and the shape of the wave does not change over time.

In the actual physical environment, fluctuations are usually affected by various factors such as damping, nonlinear effects, and more complex boundary conditions. These factors cause the wave behavior to deviate from the ideal linear wave Equation [36]. In order to describe these complex phenomena more accurately, the linear wave equation can be extended to a nonlinear wave equation, and damping and higher-order terms can be introduced [37].

$$\frac{\partial^2 \eta}{\partial t^2} = c^2 \nabla^2 \eta \tag{4}$$

$$\frac{\partial^2 \eta}{\partial t^2} + \gamma \frac{\partial \eta}{\partial t} + \alpha \eta \frac{\partial \eta}{\partial t} + \beta \eta^3 = c^2 \frac{\partial^2 \eta}{\partial x^2} + \epsilon \frac{\partial^4 \eta}{\partial x^4} \tag{5}$$

where η is the offset of the liquid level height and c is the wave velocity.

This equation considers more complex situations, such as viscous resistance ($\gamma \frac{\partial \eta}{\partial t}$), nonlinear effects ($\alpha \eta \frac{\partial \eta}{\partial t}$), and higher-order effects ($\epsilon \frac{\partial^4 \eta}{\partial x^4}$). These factors all lead to a gradual reduction in the wave amplitude.

Darcy–Weisbach equation.

The movement of fluid in the device is influenced by fluctuations, leading to friction. The Darcy–Weisbach equation efficiently calculates pressure loss resulting from interactions between the pipe wall and fluid, based on flow characteristics and geometric parameters within pipes [38].

$$\Delta p = f \frac{L}{D} \frac{\rho v^2}{2} \tag{6}$$

where $\Delta \rho$ is the pressure loss, f is the friction factor, L is the pipe length, and D is the pipe diameter.

The equation shows that the pressure loss is proportional to the square of the velocity (v^2) and depends inversely on the pipe diameter (D). This means that narrower pipes or faster flow rates lead to higher energy losses.

Simultaneous equation of liquid level fluctuation amplitude attenuation.

Simultaneous level fluctuation amplitude attenuation and total energy attenuation, taking into account frequency-dependent and nonlinear effects [39]:

$$\eta(f,t) = \eta_0(f) \exp\left(-\frac{1}{2}\int_0^t \gamma(f)\, dt\right) \cos(\omega(f)t + \phi(f)) \left(1 + \kappa \eta^2(f,t)\right) \tag{7}$$

This equation describes the time evolution of the wave amplitude $\eta(f,t)$ at frequency f and time t. It incorporates several important physical phenomena:

$\eta_0(f)$ is the initial wave amplitude, which is a function of frequency. The exponential term $\exp\left(-\frac{1}{2}\int_0^t \gamma(f)\, dt\right)$ describes the amplitude decay due to the damping coefficient $\gamma(f)$, and this decay is frequency-dependent. $\cos(\omega(f)t + \phi(f))$ represents the oscillatory component of the wave, where $\omega(f)$ is the angular frequency and $\phi(f)$ is the initial phase. $(1 + \kappa \eta^2(f,t))$ accounts for nonlinear effects, implying that the wave amplitude is modified by a term proportional to its square.

$$E(f) \propto S(f) \tag{8}$$

$$E(f,t) = S(f)\exp(-\gamma(f)t) \tag{9}$$

$$E(f,t) = E_0(f)\exp\left(-\int_0^t \gamma(f)\,dt\right) \tag{10}$$

$$E_{\text{total}}(t) = \int_{f_{\min}}^{f_{\max}} S_{\text{eff}}(f)\exp\left(-\int_0^t \gamma(f)\,dt\right)df \tag{11}$$

This Equation [40] describes how the total wave energy $E_{\text{total}}(t)$ varies with time t. The integral is performed over the frequency range $[f_{\min}, f_{\max}]$. $S_{\text{eff}}(f)$ is the effective energy spectrum density at frequency f. The exponential term $\exp\left(-\int_0^t \gamma(f)\,dt\right)$ reflects the energy decay over time, which is also frequency-dependent.

$$\eta_{\text{total}}(t) = \int_{f_{\min}}^{f_{\max}} \eta_0(f)\exp\left(-\frac{1}{2}\int_0^t \gamma(f)\,dt\right)\cos(\omega(f)t + \phi(f))\left(1 + \kappa\eta^2(f,t)\right)df \tag{12}$$

This equation describes the total wave amplitude $\eta_{\text{total}}(t)$ as it evolves over time. The integral is also performed over the frequency range $[f_{\min}, f_{\max}]$. This equation extends the single-frequency component $\eta(f,t)$ from the first equation to account for the total amplitude across the entire frequency range.

2.3. Boundary Conditions

On the inner wall of the fluctuation-resistant tube, the velocity of the fluid perpendicular to the wall is zero in accordance with the no-slip condition. The fluid cannot penetrate the solid wall, and the fluid particles can only slide along the wall [41].

Outlet/inlet boundary above the water surface: the outlet/inlet substance is air, and ρ represents atmospheric pressure at the inlet.

Outlet/inlet boundary below the water surface: outlet/inlet substance is seawater, ρ1.00 g/m^3, and inlet velocity is calculated according to the principle of flow balance in the inlet/outlet anti-fluctuation tube.

2.4. Core Components

2.4.1. Material Selection

The working principle of the equipment relies on its precise physical structure design and floating adjustment mechanism to keep the water level in the measurement area stable under various sea conditions, thereby providing accurate draft data. First, the main tube effectively disperses and absorbs wave energy, significantly reducing water level fluctuations. The ballast system, acting as the device's fluctuation damper, further eliminates minor oscillations, ensuring the water surface in the measurement area remains stable. The floating mechanism senses the movement of the vessel and automatically adjusts the position of the main tube, preventing ship motion or tilting from affecting the water level measurement. This mechanism ensures that even in rough seas, the measurement area remains relatively stationary, preventing reading errors. By integrating these designs and working principles, the device can deliver stable and accurate draft depth data in complex sea conditions, as shown in Figure 3a,b.

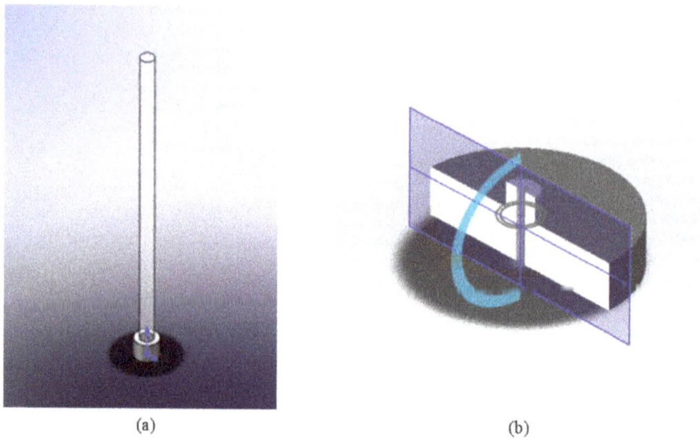

Figure 3. The (**a**) illustrates the direction of water movement into the tube, with the floating mechanism automatically adjusting the position of the main tube. (**b**) explains the inner structure of the fluctuation isolation component.

2.4.2. Core Component Function

The primary function of the device is to effectively resist interference caused by ocean waves, ensuring the accuracy of water level readings. It is widely used for ship draft measurement and other marine observation tasks. The materials and structure are carefully designed to ensure the device's durability, enabling stable performance during long-term operations at sea. The stability and accuracy of the device are crucial for safe navigation, cargo handling, and draft monitoring of vessels.

3. Spatial Color Density Clustering and PaddlePaddle-OCR

This paper introduces an innovative real-time method for ship draft identification that synergizes the Density-Based Spatial Clustering of Applications with Noise (DBSCAN) [42] and PaddlePaddle-OCR within the Paddle platform. PaddlePaddle-OCR, heralded as China's first independently developed deep learning platform, has been open-sourced to the professional community since 2016. It boasts an industrial-grade infrastructure replete with advanced technologies and comprehensive features, encompassing core deep learning frameworks, extensive model libraries, end-to-end development kits, and versatile service platforms. Rooted in real-world industrial applications, PaddlePaddle is dedicated to propelling industrialization and has garnered widespread adoption across diverse sectors, including manufacturing, agriculture, and enterprise services. The PaddlePaddle-OCR system is composed of three key components: a text detector, a direction classifier, and a text recognizer. Figure 4 illustrates the workflow and the optimization techniques employed to enhance model size and performance. By harnessing the sophisticated clustering and noise identification capabilities of DBSCAN, draft marks are adeptly classified into distinct clusters based on their image clustering (RGB) parameters [43], while anomalous points with markedly different RGB values—such as wave crests and hull stains—are effectively flagged as noise. The practical reliability of the anti-surge pipe is evidenced through real-time waterline extraction and a meticulous comparison with water optical character recognition (OCR). This approach extracts the ship's waterline by calculating the spatial geometric distance of different colors, utilizing the positional mapping relationship of color information within the captured images. In the realm of ship draft observation, a cutting-edge ultra-lightweight text detection and recognition system has demonstrated remarkable efficacy in extracting waterlines, recognizing draft marks, and producing precise draft readings. This innovative system harnesses a variety of optimization techniques, including the deployment of lightweight backbone and head networks, the strategic removal of the squeeze and excitation (SE) module [44], and the implementation of cosine learning rate decay along with learning rate warm-up. Additionally, it employs Filter Pruning via Geometric Median (FPGM) pruning [45], all of which culminate in a substantial reduction in the model's size. As a result, these advancements significantly enhance both the efficiency and accuracy of recognition processes, paving the way for more reliable maritime operations.

Based on DBSCAN's clustering and noise identification capabilities, the draft marks are classified into different clusters based on their RGB parameters, while points with significantly different RGB values, such as wave crests and hull stains, are identified as noise. The reliability of the anti-surge pipe in practical applications is demonstrated through real-time extraction of the waterline and a comparison of water gauge reading fluctuations inside and outside the pipe [46].

Ship draft line extraction is one of the key technologies in the automatic interpretation of ship draft lines. The process begins with the prepossessing of the image containing the draft marks, which includes converting the image to Gray-scale and enhancing the contrast to better distinguish the draft marks from the hull. To improve recognition efficiency, a Region of Interest (ROI) is defined [47]. Following this, Density-Based Spatial Clustering of Applications with Noise (DBSCAN) is utilized to detect clusters of varying shapes and sizes from data containing noise and outliers. Finally, the ship draft line is extracted, and the draft is calculated using the recognition capabilities of PaddlePaddle-OCR. Refer to Figure 4 for the ship draft identification flow chart.

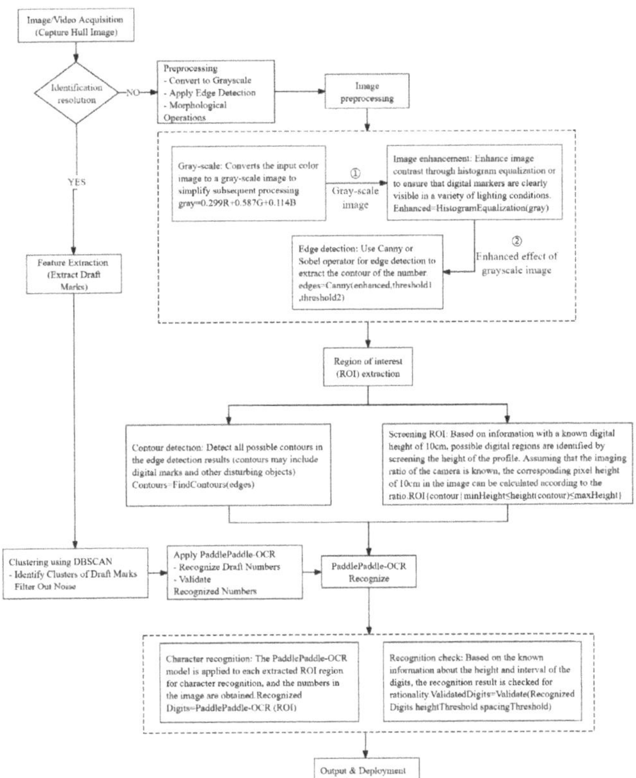

Figure 4. Ship draft identification flow chart.

Figure 5 presents the experimental results, showing the waterline inside and outside the anti-fluctuation tube under calm water conditions. In Figure 5a, the bottom of the number "2" represents 20 cm, the top of the number "2" represents 30 cm, and the distance between "2" and "4" is 10 cm. The draft reading result is 34 cm extracted using the DBSCAN clustering algorithm in combination with noise identification techniques. The result of recognition is shown in the Figure 5b.

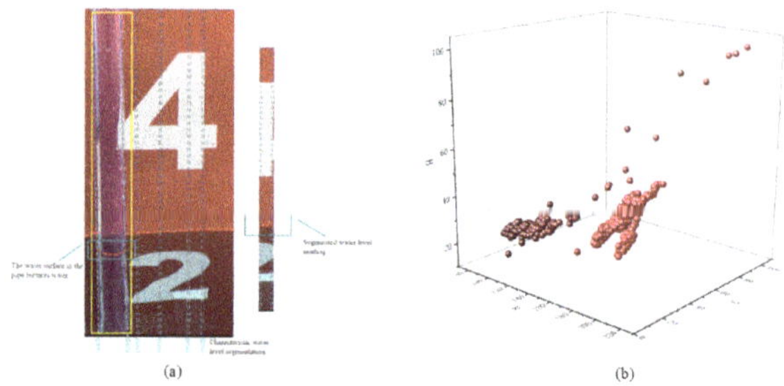

Figure 5. Draught line extraction of the feature column and analysis result.

The DBSCAN algorithm generates clusters through a process known as density divergence, starting from any core point. When a core point is added to the cluster of an already identified core point, the two clusters merge, allowing the cluster to expand further as more core points are included. This merging and expansion process continues until no additional core points can be found within the density-reachability distance. Figure 5b shows the clustering results based on the RGB color distribution. If there are any remaining unvisited core points, the process restarts from one of them, repeating until all core points have been processed. Given m core points, the time complexity of this clustering process is denoted as $T(m)$.

Before the experiment, the height and spacing of the draft marks were precisely defined. This paper utilizes the PaddlePaddle-OCR system, built on Baidu's Paddle deep learning platform, to accurately recognize and interpret the characters on the ship's draft marks. As a tool to help measure the depth of the ship's draft. The height and spacing of the scale and figures are typically 10 cm; this standardized design can ensure the accuracy and reliability of the measurement [48]. The ship's draft mark is usually installed on the hull of the ship. The International Convention on Load Lines (LLC) requires that ships have clearly visible draft marks on both sides of the bow and stern. These marks help to indicate the draft, or depth, of the ship in relation to the waterline, under various loading conditions. These marks should be clearly visible and maintained in good condition. The draft marks are usually in meters or decimeters and should be clearly legible to ensure compliance with draft limitations for safe navigation [49]. Figure 6 illustrates the process and principles of draft character recognition and extraction.

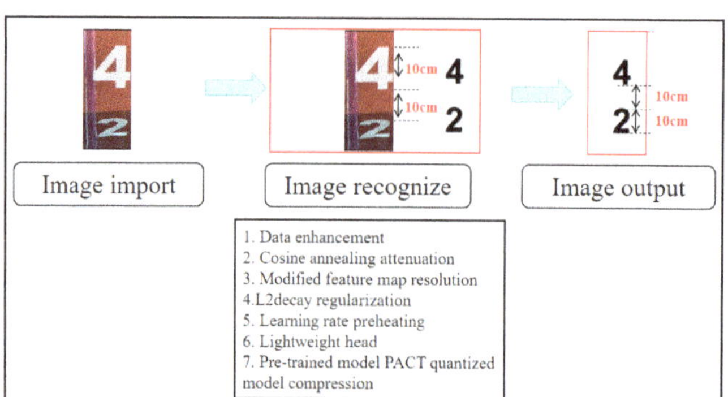

Figure 6. PaddlePaddle-OCR system diagram.

4. Experimental Results and Discussion

4.1. Experimental Presentation

The experiments were carried out in the Wave Simulation Laboratory at Shanghai Maritime University, which is shown in Figure 7. The experiments were conducted in 10 groups at depths of 3 m to 5 m, simulating wave heights of 0 to 0.4 m. The test conditions are shown in Table 1.

Figure 7. Wave Simulation Laboratory

Table 1. Experimental condition.

Size of Site (m)/Model of Simulator	Water Temperature (°C)	Mean Water Density (g/cm^3)	Sea State/Wave Height (m)	Site Depth (m)	Camera
23 × 23 m/KQB-35C	+25.0 °C	1.00 g/cm^3	0~0.4 m	3 m~5 m	I phone 14promax

4.2. The Results of the Experiment

In this paper, the PaddlePaddle-OCR system, based on Baidu's Paddle deep learning platform, is used to identify and read the characters of the ship's water draft. The text recognizer has been enhanced by replacing it with a Convolutional Recurrent Neural Network (CRNN) and a MobileNetV3 lightweight backbone network, improving efficiency.

Meanwhile, a pre-trained model based on the ImageNet dataset is utilized to achieve faster convergence during training for the ship's draft observation recognition scenarios. This pre-trained model also helps improve model accuracy.

In the selected sample pictures, various issues led to certain errors during image collection. These errors include failing to capture the part of the draft line where the ship meets the water surface, recognition failures due to the blurring of the draft mark, and issues arising from the use of handwritten draft marks, among others. These factors caused some captured images to be excluded from the actual draft recognition process.

The accurate recognition rate of the draft can reach 77.2%, as illustrated in Table 2, after filtering out these data. Additionally, waterline extraction, text detection, and text recognition all demonstrated high accuracy.

Table 2. The results of ship draft identification.

Picture Type	Actual Situation	Number
Unavailable image	Fuzzy identification	7
	under illumination	4
	No shot of the waterline	3
	The mark of the water gauge is not standard	9
	total	21
Usable image	Waterline extraction error	3
	Text detection error	2
	Text recognition error	3
	Other unrecognized errors	13
	Identify correctly	58
	total	79
Image availability		79%
Overall recognition rate		61%
Available image recognition rate		77.22%

To avoid subjective interference, the experiment was designed to compare visual observation with AI recognition. Statistical methods were employed to validate that the readings inside and outside the pipe were consistent. It was found that wave fluctuations inside the pipe could be eased by 25%. The results of both visual and AI observations are recorded and discussed in Table 3, which simplifies the draft observation process.

Table 3. Normalized weights.

Date	Outside Visual	Inside Visual	Inside AI (Data Statistical)
1	0.100	0.099	0.099
2	0.096	0.098	0.099
3	0.094	0.100	0.099
	……		
9	0.100	0.101	0.100
10	0.096	0.099	0.099

Based on AI recognition assistance, significant errors in all 79 available experimental diagrams were successfully rectified. A total of 50 sets of data were selected from five groups of experiments. Figure 8 illustrates a subset of the experimental results, with (a) representing the draft parameters under calm water conditions and (b) to (f) depicting the experimental diagrams obtained at the lowest point of liquid level fluctuation under different wave periods. Additionally, (g) to (l) showcase the experimental diagrams acquired at the highest point (wave peak) of liquid level fluctuation under various wave periods.

At the start of the experiment, the water was stable, with an observed still water level of 34.0 centimeters. Next, the anti-fluctuation device was installed in the pool. Once the water level inside the device matched the outside level, waves were generated. Two observers recorded wave heights both outside and inside the instrument, while a camera captured data for AI recognition. Ten sets of experiments were conducted in different areas of the pool to assess the impact of visual observations and to capture higher-quality images.

This paper employs two effective methods to verify the accuracy of the observed results in the wave suppression device. One method involves comparison with human visual observation, while the other uses automated recognition to avoid subjective interference from visual observations. As shown in Figure 9, the data distribution for multiple groups of measurements is presented, including outside, inside visual, and inside AI (average) data.

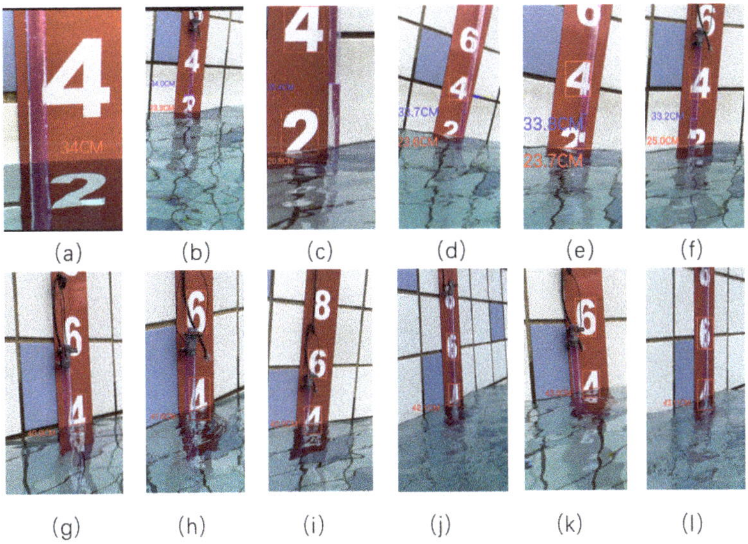

Figure 8. Date sampling of experimental images for recognition and illustrates a subset of the experimental results.

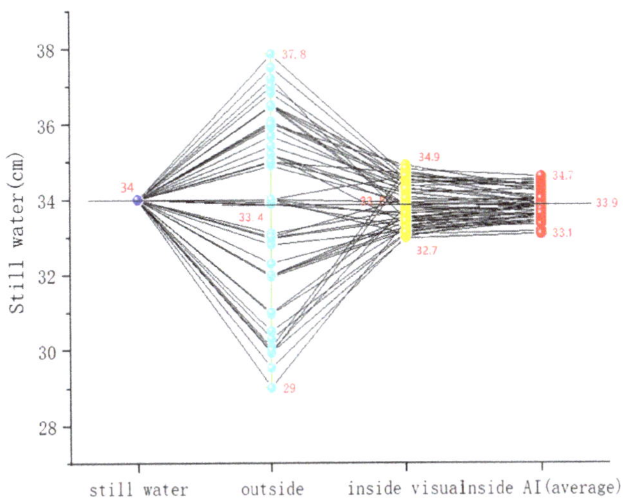

Figure 9. The results of ten groups in the experiment. Blue color represent starting value for "still water". Cyan Color: Indicates data points related to the "outside" Visual results. Yellow Color: Denote the section representing "inside visual" values. Red Color: Represents the "AI (average)" values.

According to Figure 10, the amplitude range outside the anti-fluctuation device can be compared with that inside. The data can be projected into a 3D diagram to more intuitively visualize liquid level fluctuations at different positions (surface locations: 1—outside trough, 2—inside wave, 3—outside wave peak).

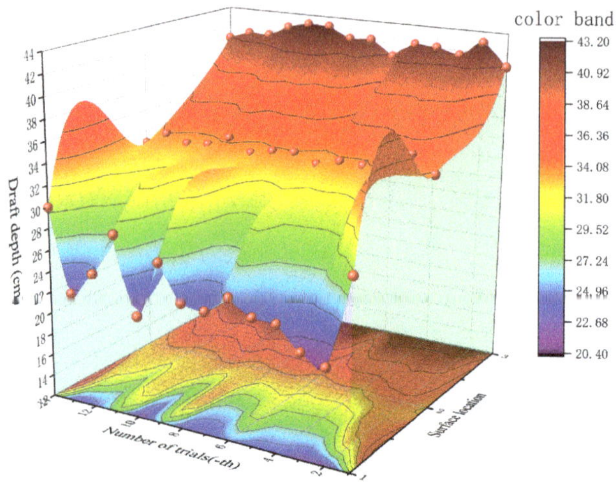

Figure 10. Comparison of the amplitude range outside the anti-fluctuation device with the inside device.

For reference and comparison, the figure illustrates the amplitude range within the tube. It clearly shows that the wave amplitude range inside the anti-fluctuation device is 30 cm to 35 cm, while the range outside the device is 20.8 cm to 43.1 cm. Notably, the anti-fluctuation device can effectively reduce the wave amplitude range by 25.14% compared to the external waves, as expected.

- Pierson–Moskowitz spectrum

$$S(f) = \frac{\alpha g^2}{f^5} e^{-\frac{g}{f^4 U^2}} \tag{13}$$

The Pierson–Moskowitz spectrum describes the energy distribution of ocean waves under constant wind conditions, represented by Equation [50]. $S(f)$ is wave spectral density, α is an empirical constant, g is the acceleration due to gravity, f is frequency, and U is wind speed.

This study conducts a simulation of the variation in liquid levels inside and outside the pipe over time, based on the Pierson–Moskowitz spectrum, which is widely used to model sea waves in fully developed seas. One of the key advantages of the Pierson–Moskowitz spectrum is its ability to accurately describe the energy distribution of waves in conditions where the sea surface is no longer influenced by the wind that originally generated the waves. This makes it particularly suitable for studying long-term wave behavior and wave suppression systems. The spectrum's simplicity and well-established theoretical foundation also make it a reliable tool in various ocean engineering simulations. The results of the simulation are presented in Figure 11. In the figure, the blue line represents the wave variations modeled by the Pierson–Moskowitz spectrum, while the red line illustrates the fluctuations of the liquid level inside the pipe. It is observed that the amplitude of fluctuations inside the pipe is approximately 25% of those outside the pipe. This difference demonstrates the effectiveness of the wave suppression system.

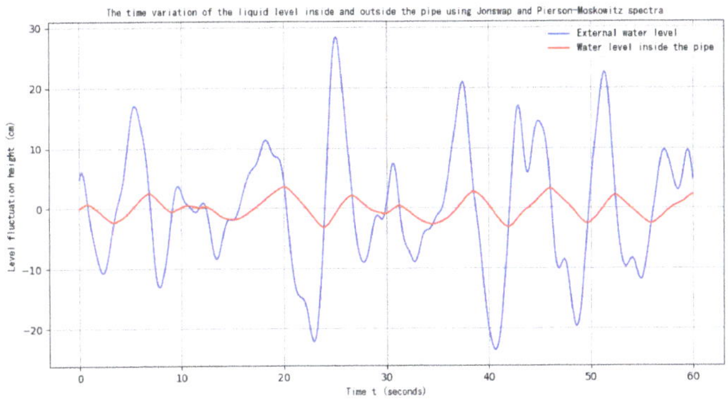

Figure 11. Simulation of the variation in liquid levels inside and outside the pipe over time based on the Pierson–Moskowitz Spectrum.

4.3. Interpretation of Result

Although the wave suppression device has significantly reduced sea surface fluctuations, facilitating the measurement of ship draft readings, obtaining accurate readings from each set of experimental data remains a pressing challenge. Therefore, this paper introduces a weighted averaging method based on energy attenuation to calculate more accurate draft readings.

First, the energy attenuation formula fully accounts for the decay of wave fluctuations at different frequencies over time. This allows for a more accurate reflection of how high-frequency and low-frequency waves differently impact draft readings. Typically, the higher the frequency, the faster the energy attenuates, and thus the less influence it has on the weighted average calculation [51].

Furthermore, traditional weighted averaging methods often overlook the nonlinear effects of waves, such as wave superposition and interaction. However, using the energy attenuation formula allows for a more precise capture of these complex physical phenomena, thereby enhancing the accuracy of the calculation results. The weighted averaging method based on energy attenuation can also effectively reduce the impact of noise and outliers on the final readings. Since the formula assigns smaller weights to readings with larger and faster-decaying fluctuations, the calculation results more accurately reflect the actual draft readings.

More importantly, the weight adjustments in this method are based on the actual physical conditions of each data point, such as frequency and energy, rather than using fixed weight allocations. This dynamic adjustment enables the weighted averaging method to flexibly adapt to the statistical requirements of draft readings in different environments.

- Energy attenuation formula.
  First, we calculate the energy attenuation of the wave over time. Assuming the initial energy is $E_0(f)$, the energy $E(f,t)$ at time t and frequency f is expressed as:

$$E(f,t) = E_0(f)\exp\left(-\int_0^t \gamma(f)\,dt\right) \tag{14}$$

  where $\gamma(f)$ is the damping coefficient dependent on frequency f. $E_0(f)$ is the initial energy at frequency f.
  Simplified as $\gamma(f) = 0.05 \times f$ and $t = 1$, the formula becomes:

$$E(f,t) = \exp(-\gamma(f) \times t) = \exp(-0.05 \times f \times t) \tag{15}$$

- Weight calculation formula.

Based on the energy attenuation formula, the weight w_i for each data point is calculated as:

$$w_i = \frac{1}{1 + \kappa E(f,t)} \quad (16)$$

where κ is a scaling coefficient related to nonlinear effects. $E(f,t)$ is the energy at time t and frequency f for the i-th data point.

- Normalization of weights.
  To ensure the sum of all weights equals 1, the weights w_i are normalized [52]:

$$\text{normalized } w_i = \frac{w_i}{\sum_{i=1}^{n} w_i} \quad (17)$$

- Weighted mean calculation formula.
  Finally, the weighted mean μ_{weighted} is calculated using the normalized weights [53]:

$$\mu_{\text{weighted}} = \sum_{i=1}^{n} \text{normalized } w_i \cdot x_i \quad (18)$$

where μ_{weighted} is the weighted mean. x_i is the i-th data point (such as 'outside_visual', 'inside_visual', 'inside_AI').

Table 3 below shows the normalized w_i of a set of data in the experiment. The value of each data point is multiplied by its corresponding normalized weight, and the weighted results are aggregated to obtain a weighted average.

Table 4 presents the weighted average values from multiple experiments based on energy attenuation. These values demonstrate how energy attenuation has been effectively utilized to calculate more accurate average measurements across different experimental setups.

Table 4. Comparison of wave amplitude measurements.

Group	Still Water (cm)	Outside Visual (cm)	Inside Visual (cm)	Inside AI (Data Statistical) (cm)
1	34.0	35.31	33.7	33.7
2	34.0	34.9	33.72	33.9
3	34.0	33.0	33.7	33.8
4	34.0	32.94	33.9	33.95
5	34.0	33.1	34.06	34.1

The distribution of measurements classified as "outside" shows significantly greater variability compared to "inside visual" and "inside AI (average)", indicating that external measurements may be more susceptible to inconsistencies and potentially less reliable. The "inside AI (average)" readings, which are tightly clustered, suggest that AI processing effectively reduces variability and measurement error. While "inside visual" readings also exhibit reduced variability, they still show a broader range than the AI-processed data.

AI-processed data demonstrate enhanced consistency and accuracy, mitigating the variability seen in raw external measurements and even in internal visual readings. Visual readings taken from within the pipe, although more consistent than external measurements, still exhibit some degree of variability. In contrast, external measurements are more prone to significant errors or fluctuations, likely due to environmental factors.

Figure 12 illustrates the linear distribution of five sets of experimental data. The figure demonstrates that external measurements exhibit a scattered distribution of data points across different groups, with multiple outliers, indicating significant fluctuations and poor reliability. The absence of a clear linear trend confirms the instability of these measurements, aligning with the observation of more outliers. In contrast, internal visual measurements show a more concentrated data distribution, with most points clustering around a clear linear trend, and 95% of the data falling within the confidence intervals, suggesting higher reliability and less variability. Data from different groups also show consistent trends, indicating greater stability in internal measurements. Still water measurements reveal a similarly concentrated distribution, with most data points confined to a narrow range, suggesting consistent but less linear results. The internal visual data points are closely aligned with the fit line, demonstrating higher precision and lower variability, while external measurements are more dispersed, with many points outside the confidence intervals, highlighting their instability. Overall, the figure indicates that internal visual measurements provide more reliable and stable data compared to external measurements, which show more fluctuations and errors. This suggests that the inside visual measurements are more reliable, exhibiting less variability compared to the outside measurements.

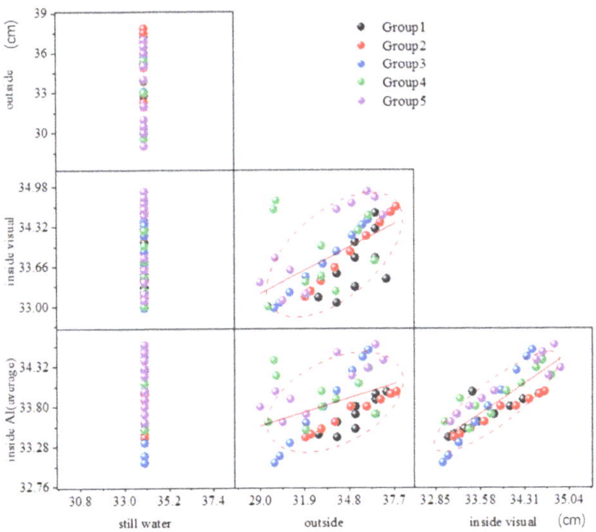

Figure 12. Comparison of the result of five groups in the experiment.

The experimental comparison reveals that the error margin within the anti-fluctuation device ranges from 0 to 0.3 centimeters, while the error margin for outside visual observation is between 0.5 and 1.2 centimeters, representing a 75% improvement in observation accuracy. Additionally, the AI recognition results closely match those obtained inside the anti-fluctuation device.

5. Conclusions

The anti-fluctuation device has demonstrated a remarkable ability to reduce wave amplitude by 25%. It dynamically adjusts the position of the main tube in response to the ship's movements, thereby maintaining the stability of the measurement section. It proves highly valuable in resolving disputes, such as minor overloads of the loading line, where parties may disagree with third-party survey results, significantly enhancing the accuracy of draft observations. This improvement not only contributes to accurate calculations of cargo laden on the ship but also ensures accurate hydrocarbon calculations, thereby protecting the interests of shipowners. Additionally, controlling the ship's emissions is crucial. Moving forward, the next steps involve applying for patent authorization and actively seeking recognition from classification societies to support the device's widespread promotion and industry adoption.

The study has certain limitations. Future research should focus on integrating advanced AI algorithms and expanding the device's functionality to perform effectively in complex scenarios, such as poor visibility and nighttime conditions. Additionally, eliminating minor heaving motions of the ship presents a promising new research direction. This ongoing research prioritizes optimizing the algorithm over relying solely on human visual observation, aiming to enhance accuracy and minimize subjective influences, ultimately advancing observation precision.

Author Contributions: Conceptualization, Y.W. and Q.H.; methodology, Y.W.; software, H.W. and H.D.; validation, Y.W. and H.D.; formal analysis, Y.W.; investigation, Y.W.; resources, Y.W.; data curation, Y.W.; writing—original draft preparation, Y.W.; writing—review and editing, H.D.; visualization, H.W.; supervision, Y.W.; project administration, Y.W. and H.D. All authors have read and agreed to the published version of the manuscript.

Funding: This article received Shanghai University young teachers Fund 2023 .

Institutional Review Board Statement: Not applicable

Informed Consent Statement: Not applicable

Data Availability Statement: The data that support the findings of this study are available from the corresponding author upon reasonable request.

Acknowledgments: The authors want to acknowledge the Wave Simulation Laboratory at Shanghai Maritime University, all the researchers who gave the guideline on the article and all the anonymous reviewers.

Conflicts of Interest: The authors declare no conflicts of interest.

References

1. Jiang, X.; Mao, H.; Zhang, H. Simultaneous optimization of the liner shipping route and ship schedule designs with time windows. *Math. Probl. Eng.* **2020**, *2020*, 3287973. [CrossRef]
2. Dhar, S.; Khawaja, H. Real-Time Ship Draft Measurement and Optimal Estimation Using Kalman Filter. *Int. J. Multiphysics* **2023**, *17*, 407–425.
3. Zhang, B.; Li, J.; Tang, H.; Liu, X. Smart Ship Draft Reading by Dual-Flow Deep Learning Architecture and Multispectral Information. *Sensors* **2024**, *24*, 5580. [CrossRef]
4. International Organization of Legal Metrology (OIML). OIML R 81:2009—Ship Weighing Using Draught Survey. International Recommendation. 2009. Available online: https://www.oiml.org/en/files/pdf_r/r081-e09.pdf (accessed on 28 September 2024).
5. Liu, X.; Wu, J.; Liao, Y.; Tu, X. Column-based waterline detection for lightweight ship draft reading. *IEEE Trans. Instrum. Meas.* **2022**, *71*, 2505010. [CrossRef]
6. Zhang, H.; Li, P.; Jin, H.; Bi, R.; Xu, D. Nonlinear wave energy dissipator with wave attenuation and energy harvesting at low frequencies. *Ocean Eng.* **2022**, *266*, 112935. [CrossRef]
7. Ben-Nun, T.; Hoefler, T. Demystifying parallel and distributed deep learning: An in-depth concurrency analysis. *ACM Comput. Surv. (CSUR)* **2019**, *52*, 1–43. [CrossRef]
8. Chen, X.; Ma, D.; Liu, R.W. Application of Artificial Intelligence in Maritime Transportation. *J. Mar. Sci. Eng.* **2024**, *12*, 439. [CrossRef]
9. Ren, X.h.; Li, P.; Chen, X.; Wang, Y.s.; Liu, Z.; Hao, L.h.; Wang, Y.; Huang, Y.j. Experimental Study on Vortex-Induced Vibration of Marine Riser Model with Coupling Interference Effect Under Combined Internal and External Flow. *China Ocean Eng.* **2023**, *37*, 673–684. [CrossRef]
10. Choudhary, N.; Bora, S.; Strelnikova, E. Study on liquid sloshing in an annular rigid circular cylindrical tank with damping device placed in liquid domain. *J. Vib. Eng. Technol.* **2021**, *9*, 1577–1589. [CrossRef]
11. Dutta, D.; Bihs, H.; Afzal, M.S. Computational Fluid Dynamics modelling of hydrodynamic characteristics of oscillatory flow past a square cylinder using the level set method. *Ocean Eng.* **2022**, *253*, 111211. [CrossRef]
12. Hamelin, J.; Love, J.; Tait, M.; Wilson, J. Tuned liquid dampers with a Keulegan–Carpenter number-dependent screen drag coefficient. *J. Fluids Struct.* **2013**, *43*, 271–286. [CrossRef]
13. Sarkar, O.; Sinha, S.; Jena, A.K.; Parida, A.K.; Parida, N.; Parida, R.K. Automatic Number Plate Character Recognition using Paddle-OCR. In Proceedings of the 2024 International Conference on Innovations and Challenges in Emerging Technologies (ICICET), Nagpur, India, 7–8 June 2024; IEEE: Piscataway, NJ, USA, 2024; pp. 1–7.
14. Ivče, R.; Jurdana, I.; Mohović, R. Determining weight of cargo onboard ship by means of optical fibre technology draft reading. *Promet-Traffic Transp.* **2011**, *23*, 421–429. [CrossRef]

15. Li, W.; Zhan, W.; Han, T.; Wang, P.; Liu, H.; Xiong, M.; Hong, S. Research and Application of U 2-NetP Network Incorporating Coordinate Attention for Ship Draft Reading in Complex Situations. *J. Signal Process. Syst.* **2023**, *95*, 177–195. [CrossRef]
16. Wang, Z.; Shi, P.; Wu, C. A Ship Draft Line Detection Method Based on Image Processing and Deep Learning. In *Proceedings of the Journal of Physics: Conference Series: 5th Annual International Conference on Information System and Artificial Intelligence [ISAI2020], Zhejiang, China, 22–23 May 2020*; IOP Publishing: Bristol, UK, 2020; Volume 1575, p. 012230.
17. Wei, Y. Research Review of Ship Draft Observation Methods. *Am. J. Traffic Transp. Eng.* **2023**, *8*, 33. [CrossRef]
18. Bilican, M.S.; Iris, Ç.; Karatas, M. A collaborative decision support framework for sustainable cargo composition in container shipping services. *Ann. Oper. Res.* **2024**, 1–33. . [CrossRef]
19. Chen, X.; Wu, H.; Han, B.; Liu, W.; Montewka, J.; Liu, R.W. Orientation-aware ship detection via a rotation feature decoupling supported deep learning approach. *Eng. Appl. Artif. Intell.* **2023**, *125*, 106686. [CrossRef]
20. Moussaoui, H.; Akkad, N.E.; Benslimane, M.; El-Shafai, W.; Baihan, A.; Hewage, C.; Rathore, R.S. Enhancing automated vehicle identification by integrating YOLO v8 and OCR techniques for high-precision license plate detection and recognition. *Sci. Rep.* **2024**, *14*, 14389. [CrossRef]
21. Khan, M.A. HCRNNIDS: Hybrid convolutional recurrent neural network-based network intrusion detection system. *Processes* **2021**, *9*, 834. [CrossRef]
22. Qian, S.; Ning, C.; Hu, Y. MobileNetV3 for image classification. In Proceedings of the 2021 IEEE 2nd International Conference on Big Data, Artificial Intelligence and Internet of Things Engineering (ICBAIE), Nanchang, China, 26–28 March 2021; IEEE: Piscataway, NJ, USA, 2021; pp. 490–497.
23. Deng, J.; Dong, W.; Socher, R.; Li, L.J.; Li, K.; Fei-Fei, L. Imagenet: A large-scale hierarchical image database. In Proceedings of the 2009 IEEE Conference on Computer Vision and Pattern Recognition, Miami, FL, USA, 20–25 June 2009; IEEE: Piscataway, NJ, USA, 2009; pp. 248–255.
24. He, Y.; Liu, P.; Wang, Z.; Hu, Z.; Yang, Y. Filter pruning via geometric median for deep convolutional neural networks acceleration. In Proceedings of the IEEE/CVF Conference on Computer Vision and Pattern Recognition, Long Beach, CA, USA, 15–20 June 2019; pp. 4340–4349.
25. Li, Z.; Peng, C.; Yu, G.; Zhang, X.; Deng, Y.; Sun, J. Light-head r-cnn: In defense of two-stage object detector. *arXiv* **2017**, arXiv:1711.07264.
26. Rong, W.; Li, Z.; Zhang, W.; Sun, L. An improved CANNY edge detection algorithm. In Proceedings of the 2014 IEEE International Conference on Mechatronics and Automation, Tianjin, China, 3–6 August 2014; IEEE: Piscataway, NJ, USA, 2014; pp. 577–582.
27. Sinaga, K.P.; Yang, M.S. Unsupervised K-means clustering algorithm. *IEEE Access* **2020**, *8*, 80716–80727. [CrossRef]
28. Arad, B.; Timofte, R.; Ben-Shahar, O.; Lin, Y.T.; Finlayson, G.D. Ntire 2020 challenge on spectral reconstruction from an rgb image. In Proceedings of the IEEE/CVF Conference on Computer Vision and Pattern Recognition Workshops, Seattle, WA, USA, 14–19 June 2020; pp. 446–447.
29. Iris, Ç.; Christensen, J.; Pacino, D.; Ropke, S. Flexible ship loading problem with transfer vehicle assignment and scheduling. *Transp. Res. Part B Methodol.* **2018**, *111*, 113–134. [CrossRef]
30. Liu, Y.; Li, P.; Wang, Y.; Liu, L.h.; Wang, F.; Guo, H.y.; Fu, Q. An experimental study on dynamics features of three side-by-side flexible risers undergoing vortex-induced vibrations in a uniform flow. *China Ocean Eng.* **2020**, *34*, 500–512. [CrossRef]
31. Murad, M.A.S.; Ismael, H.F.; Sulaiman, T.A.; Bulut, H. Analysis of optical solutions of higher-order nonlinear Schrödinger equation by the new Kudryashov and Bernoulli's equation approaches. *Opt. Quantum Electron.* **2024**, *56*, 76. [CrossRef]
32. Hu, T.; Ren, H.; Shen, J.; Niu, Z.; Zhang, M.; Xu, Y.; Song, B.; Sun, T. Experimental investigation on hydrodynamic forces of semi-submerged cylinders in combined steady flow and oscillatory flow. *Ocean Eng.* **2023**, *268*, 113612. [CrossRef]
33. Coppola, G.; Veldman, A.E. Global and local conservation of mass, momentum and kinetic energy in the simulation of compressible flow. *J. Comput. Phys.* **2023**, *475*, 111879. [CrossRef]
34. Wang, W.; Zhang, J.; Chen, S.; Dong, J. Effect of nozzle outlet angle on the fluid flow and level fluctuation in a bloom casting mould. *Chin. J. Eng.* **2007**, *29*, 816–821.
35. Kumar, N.; Choudhary, N. Sloshing reduction in three-dimensional rectangular container using rigid baffles. *Vibroeng. Procedia* **2022**, *42*, 39–44. [CrossRef]
36. Majid, S.Z.; Faridi, W.A.; Asjad, M.I.; Abd El-Rahman, M.; Eldin, S.M. Explicit soliton structure formation for the riemann wave equation and a sensitive demonstration. *Fractal Fract.* **2023**, *7*, 102. [CrossRef]
37. Degtyariov, K.; Kriutchenko, D.; Osypov, I.; Sierikova, O.; Strelnikova, E. Dampers Influence on Sloshing Mitigation in Fuel Tanks of Launch Vehicles and Reservoirs. In Proceedings of the Conference on Integrated Computer Technologies in Mechanical Engineering–Synergetic Engineering, Kharkiv, Ukraine, 12–14 December 2023; Springer: Berlin/Heidelberg, Germany, 2023; pp. 420–430.
38. Dutta, D.; Afzal, M.S.; Alhaddad, S. 3D CFD study of scour in combined wave–current flows around rectangular piles with varying aspect ratios. *Water* **2023**, *15*, 1541. [CrossRef]
39. Ryabov, Y.E.; Puzenko, A. Damped oscillations in view of the fractional oscillator equation. *Phys. Rev. B* **2002**, *66*, 184201. [CrossRef]
40. Kinsman, B. *Wind Waves: Their Generation and Propagation on the Ocean Surface*; Courier Corporation: Mumbai, Maharashtra, 1984.
41. Bauer, H.; Eidel, W. Frictionless liquid sloshing in circular cylindrical container configurations. *Aerosp. Sci. Technol.* **1999**, *3*, 301–311. [CrossRef]

42. Hassanin, M.F.; Hassan, M.; Shoeb, A. DDBSCAN: Different densities-based spatial clustering of applications with noise. In Proceedings of the 2015 International Conference on Control, Instrumentation, Communication and Computational Technologies (ICCICCT), Kumaracoil, India, 18–19 December 2015; IEEE: Piscataway, NJ, USA, 2015; pp. 401–404.
43. Park, S.; Han, S.; Kim, S.; Kim, D.; Park, S.; Hong, S.; Cha, M. Improving unsupervised image clustering with robust learning. In Proceedings of the IEEE/CVF Conference on Computer Vision and Pattern Recognition, Nashville, TN, USA, 19–25 June 2021; pp. 12278–12287.
44. Wei, S.; Qu, Q.; Wu, Y.; Wang, M.; Shi, J. PRI modulation recognition based on squeeze-and-excitation networks. *IEEE Commun. Lett.* **2020**, *24*, 1047–1051. [CrossRef]
45. Gkrispanis, K.; Gkalelis, N.; Mezaris, V. Filter-Pruning of Lightweight Face Detectors Using a Geometric Median Criterion. In Proceedings of the IEEE/CVF Winter Conference on Applications of Computer Vision, Waikoloa, HI, USA, 3–8 January 2024; pp. 280–289.
46. Ram, A.; Jalal, S.; Jalal, A.S.; Kumar, M. A density based algorithm for discovering density varied clusters in large spatial databases. *Int. J. Comput. Appl.* **2010**, *3*, 1–4. [CrossRef]
47. Wang, L.; Fan, S.; Liu, Y.; Li, Y.; Fei, C.; Liu, J.; Liu, B.; Dong, Y.; Liu, Z.; Zhao, X. A review of methods for ship detection with electro-optical images in marine environments. *J. Mar. Sci. Eng.* **2021**, *9*, 1408. [CrossRef]
48. Sutulo, S.; Guedes Soares, C. Review on ship manoeuvrability criteria and standards. *J. Mar. Sci. Eng.* **2021**, *9*, 904. [CrossRef]
49. Shipping, M. (LOAD LINES) REGULATIONS 2019. Ph.D. Thesis, United Nations, New York, NY, USA, 2019.
50. Chen, X.; Wei, C.; Yang, Y.; Luo, L.; Biancardo, S.A.; Mei, X. Personnel trajectory extraction from port-like videos under varied rainy interferences. *IEEE Trans. Intell. Transp. Syst.* **2024**, *25*, 6567–6579. [CrossRef]
51. Gurusamy, S.; Meylan, M.H.; Kumar, D.; Patnaik, B. Frequency dependent decay of water waves due to floating balls with application to simulating wave decay in the marginal ice zone. *J. Fluids Struct.* **2023**, *117*, 103817. [CrossRef]
52. Salimans, T.; Kingma, D.P. Weight normalization: A simple reparameterization to accelerate training of deep neural networks. *Adv. Neural Inf. Process. Syst.* **2016**, *29*.
53. Lo, K.H.; Sun, T.L.; Chen, J.C. A unified weight formula for calculating the sample variance from weighted successive differences. *Discret. Dyn. Nat. Soc.* **2014**, *2014*, 210761. [CrossRef]

Journal of
Marine Science and Engineering

MDPI

Article

A Novel Approach to Enhancing the Accuracy of Prediction in Ship Fuel Consumption

Tianrui Zhou [1,*], Jinggai Wang [2], Qinyou Hu [3] and Zhihui Hu [4]

1 Fujian Chuanzheng Communications College, Fuzhou 350007, China
2 College of Transport and Communications, Shanghai Maritime University, Shanghai 201306, China; wangjinggai01@stu.shmtu.edu.cn
3 Merchant Marine College, Shanghai Maritime University, Shanghai 201306, China; qyhu@shmtu.edu.cn
4 Navigation College, Jimei University, Xiamen 361021, China; huzhihui@jmu.edu.cn
* Correspondence: tianrui_z01@163.com

Abstract: Ship fuel consumption plays a crucial role not only in understanding ships' energy efficiency but also in gaining insights into their emissions. However, enhancing the accuracy of these predictions poses significant challenges due to data limitations and the methods employed. Due to these factors, such as data variability and equipment characteristics, ship fuel consumption exhibits certain fluctuations under specific conditions. Previous fuel consumption prediction methods primarily generate a single specific value, making it difficult to capture the volatility of and variability in fuel consumption. To overcome this limitation, this paper proposes a novel method that integrates Gaussian process prediction with quantile regression theory to perform interval predictions of ship fuel consumption, providing a range of possible outcomes. Through comparative analyses with traditional methods, the possibility of using the method is verified and its results are validated. The results indicate the following: (1) at a 95% confidence level, the proposed method achieves a prediction interval coverage probability of 0.98 and a prediction interval normalized average width of 0.123, which are significantly better than those of the existing backpropagation neural network (BPNN) and gradient boosting decision tree (GBDT) quantile regression models; (2) the prediction accuracy of the proposed method is 92% for point forecasts; and (3) the proposed method is applicable to main datasets, including both noon report and sensor datasets. These findings provide valuable insights into interval predictions of ship fuel consumption and highlight their potential applications in related fields, emphasizing the importance of accurate interval predictions in intelligent energy efficiency optimization.

Keywords: maritime carbon emissions reduction; ship fuel consumption prediction; interval prediction; data-driven

1. Introduction

As a key component of global trade, the shipping industry plays a significant role in the world economy [1]. However, its substantial fuel consumption for transportation results in significant pollutant emissions, posing serious threats to the marine environment and human health [2,3]. Addressing the challenges of saving energy and emission reductions in the shipping sector has become a focal point for governments worldwide. To mitigate carbon emissions from shipping vessels, various countries and relevant organizations have formulated a series of policies aimed at achieving the goal of reducing carbon emissions. The International Maritime Organization (IMO) introduced the Carbon Intensity Indicator (CII) as part of its strategy to reduce greenhouse gas emissions from ships, mandating that ships annually calculate and report their carbon intensity during their operations [4]. Furthermore, if a vessel's carbon emissions do not meet regulatory standards, shipping companies are required to provide an energy efficiency improvement plan. In a similar vein, the European Commission has enacted multiple regulations to compel ship operators to enhance energy efficiency and reduce carbon emissions. For instance, the Monitoring, Reporting, and Verification (MRV) system is applied to enhance the transparency of carbon emissions in the maritime industry and promote the reduction in greenhouse gases [5]. Upon the implementation of the European Union Emissions Trading System (EU ETS) [6], ships will be mandated to hold emission allowances proportional to their carbon emissions. All of these actions are intended to make a positive contribution to the global effort to mitigate climate change within the shipping industry.

To address the stringent environmental pressure for emissions reductions, the maritime industry is accelerating its transition towards greener and low-carbon operations [7]. This compels shipping companies to adopt energy saving and emissions reduction strategies to meet carbon emission regulations, including enhancing ship energy efficiency, optimizing hull cleaning [8], improving propulsion performance [9], operational optimization [10,11], and utilizing alternative fuels [12]. Operational optimization, including route and speed optimization, stands out due to its minimal requirement for additional equipment expenditures [13,14]. The optimization of navigation routes and speed enables vessels to avoid adverse sea conditions, improving their energy efficiency. Consequently, it not only decreases fuel expenses for shipping companies but also ensures compliance with carbon emission regulations [15].

As we all know, the Energy Efficiency Operational Indicator (EEOI) and CII are closely related to a ship's fuel consumption [16]. Additionally, speed optimization and trim optimization are common fuel-saving strategies employed during the operational phase. By accurately predicting a vessel's fuel consumption, it is possible to optimize voyage planning, select the most energy-efficient routes and speeds, and thereby maximize economic benefits. Furthermore, developing an accurate fuel consumption prediction model allows for the continuous monitoring of discrepancies between actual fuel usage and expected consumption, enabling a comprehensive assessment of the vessel's performance and condition. For instance, when the level of fuel consumption exceeds one's expectations, this could indicate potential issues, such as hull fouling or mechanical faults.

The enhancement of ship energy efficiency and reduction in carbon emissions through technological advancements fundamentally depend on the construction of a high-precision ship fuel consumption model [17]. The energy efficiency of a vessel typically comprises two components: the design phase and the operational phase. For a specific vessel, the hull shape, structural configuration, and propulsion system remain fixed over short periods, rendering the energy efficiency determined during the design phase generally stable and unchanged. In contrast, the energy efficiency during the operational phase is influenced by factors [18] such as the navigational route, selected sailing speed, and cargo load, which are subject to significant variability in actual operations. Additionally, during actual navigation, it is essential to first verify the navigability of the chosen route to ensure the safety of the vessel. This involves confirming that the navigational area meets the draft requirements of the ship to avoid grounding and similar incidents. Moreover, it is crucial to ensure that the vessel has sufficient fuel to reach the destination port. Consequently, these factors present significant challenges for accurately predicting a vessel's fuel consumption.

The objective of this study is to develop a highly accurate interval prediction model for ship fuel consumption, considering multiple influencing factors. The proposed model not only accurately predicts the point estimate of ship fuel consumption but also provides a prediction interval for its consumption. Such a model will support the subsequent identification of anomalies in ship energy efficiency. By integrating big data analysis and machine learning techniques [19], this research introduces a novel approach and perspective for predicting ship fuel consumption, thereby promoting energy savings, emission reductions, and green development in the shipping industry.

The structure of this paper is organized as follows: Section 1 introduces the current background of energy conservation and emission reductions in the shipping industry. Section 2 describes the relevant methods for fuel consumption prediction in the previous literature. Section 3 details the data sources and processing methods used in this study. Section 4 presents the construction of the interval prediction model for ship fuel consumption. Section 5 analyzes the model prediction results and compares them with the experimental results of other mainstream existing models. Section 6 summarizes the findings of this study and discusses their practical significance for energy efficiency management.

2. Literature Review

An extensive review of the prior related literature reveals that current ship fuel consumption prediction methods are primarily constructed using either physics-based knowledge or data-driven approaches. Depending on the combination of these two approaches, prediction methods can be classified into three categories: physics-based models, data-driven models, and hybrid models.

2.1. Physics-Based Ship Fuel Consumption Prediction Models

The prediction of ship fuel consumption based on physical models primarily involves calculating the ship's resistance under various wind and wave conditions [20]. By considering the transmission efficiency between the propeller and the main engine, a functional relationship is derived between the ship's fuel consumption and speed under different environmental conditions. The Hotrolp method is commonly used to calculate ship resistance, including wind and wave resistance [21]. Wang et al. employed theoretical formulas to calculate the variation in resistance with speed and subsequently constructed an integrated route–speed optimization model, which was experimentally validated on an ocean-going vessel [22].

A critical aspect of this prediction involves calculating the ship's resistance in various wind and wave conditions. Computational fluid dynamics (CFD) methods have been employed in several studies to numerically simulate the sailing environment and provide precise calculations of ship resistance. La Ferlita et al. utilized CFD software to calculate ship resistance in diverse operational environments and applied automated machine learning algorithms to predict ship performance [23]. Sang et al. analyzed the ship propulsion system and conducted simulation modeling of various operational conditions to develop a fuel consumption prediction model [24], which was experimentally validated on a 38,800 DWT bulk cargo ship, achieving an error margin of less than 5%.

Due to the impact of wind and wave resistance as well as fouling resistance on ship energy consumption, some studies have considered these factors when developing more accurate fuel consumption models. Guo et al. based their ship fuel consumption predictions on their knowledge of physics while accounting for the effects of wind, waves, and fouling on ship resistance [25].

The construction of a physics-driven ship fuel consumption prediction model typically requires complex physical formulas and a series of parameter values to calculate the ship's resistance. Therefore, the accuracy of ship fuel consumption predictions based on this method is easily influenced by these parameters. Due to the integration of physical knowledge and ship mechanism knowledge, the constructed fuel consumption model possesses strong transferability and interpretability.

2.2. Data-Driven Ship Fuel Consumption Prediction Models

The construction of data-driven ship fuel consumption models leverages big data technology and various machine learning techniques to learn from and train on data, thereby outputting ship fuel consumption based on different input variables. The data used for ship fuel consumption prediction include ship maneuvering and navigation data [26,27], marine environmental data [28,29], main engine fuel data [30], and noon report data [31]. These data are either automatically transmitted by ship equipment or manually filled in. Due to the ability of machine learning to learn latent knowledge from large datasets and handle high-dimensional feature data, many studies have utilized machine learning techniques such as multiple linear regressions (MLRs), decision trees (DTs), and artificial neural networks (ANNs) for ship fuel consumption model construction [32].

Many studies assume that ship fuel consumption has a nonlinear functional relationship with voyaging speed, using statistical methods to obtain the relevant coefficients. Du et al. assumed an exponential functional relationship between fuel consumption and speed and derived this relationship from container voyage report data [33]. Additionally, this relationship was used to guide speed and trim optimization.

Due to the ease of understanding and interpreting tree models, many studies have employed them for fuel consumption prediction. Zhou et al. used an RF for constructing a fuel consumption prediction model, utilizing data from two sister ships and publicly available navigation meteorological data [34]. The experimental results showed that the RF model achieved a prediction accuracy of over 92%, surpassing that of an MLR, least absolute shrinkage and selection operator (LASSO), and a Bayesian ridge regression. Yan et al. innovatively proposed a two-stage fuel consumption prediction and optimization method based on the random forest (RF) method [35], which has been validated on a dry bulk carrier.

To address the issue of low model accuracy caused by data mismatches and missing features, some studies utilize publicly available data for data fusion, resulting in more comprehensive and accurate datasets. Li et al. conducted energy efficiency data fusion based on noon report data, automatic identification system (AIS) data, and publicly available data [36]; the experimental results consistently demonstrated that data fusion can improve the accuracy of ship fuel consumption prediction.

To overcome the limitations of single models in fuel consumption prediction, some studies have employed ensemble learning algorithms. Hu et al. proposed a ship fuel consumption prediction model based on ensemble learning [37] and constructed a two-stage trim optimization model, which were validated on an ocean-going vessel, achieving a fuel reduction of 0.69%–1.82%. Lan et al. improved the blending ensemble model (BEM) for fuel consumption predictions of wing-diesel hybrid ships [38], and their experimental results indicated that the improved model reduced prediction errors by 16.99% compared to a support vector machine (SVM) model.

Due to the effectiveness of long short-term memory (LSTM) networks in predicting time series data, models based on LSTM networks are commonly used for real-time fuel consumption prediction. Panapakidis et al. improved the LSTM model to make real-time fuel consumption predictions concerning a passenger ship [39]. Zhang et al. developed a novel bidirectional long short-term memory (Bi-LSTM) ship fuel consumption prediction model based on the LSTM structure and incorporated an attention mechanism [40]. To address the structure and parameters of LSTM models, various bio-inspired evolutionary algorithms are employed for optimization. Han et al. used the improved grey wolf optimization (IGWO) algorithm to optimize the LSTM structure and parameters [41]; their experimental results indicated that these evolutionary algorithms can effectively improve the accuracy of fuel consumption prediction.

2.3. Ship Fuel Consumption Prediction Based on Hybrid Models

Hybrid models, which combine physical models and data-driven approaches, can be classified into parallel and series models based on their connection methods [42]. In parallel models, the physical model and the data-driven model independently predict fuel consumption. The outputs of these two models are then either used together as inputs or combined to form the final prediction result. In series models, the output of the physical model typically serves as the input for the data-driven model, which is trained to learn and predict fuel consumption.

Wang et al. used an RF algorithm combined with a physically derived fuel consumption formula to construct a parallel grey model [43]. Based on this model, they developed a path optimization method using an improved A* algorithm. The effectiveness of the proposed method was validated for an ocean-going vessel, achieving a 5.48% reduction in fuel consumption. Cai et al. developed an LSTM grey fuel consumption prediction model [44] that uses fuel consumption derived from physical formulas as its input. Due to the integration of system knowledge and expert experience, the model's prediction accuracy reached 98.8%.

To address the issue of low fuel consumption prediction accuracies in white-box models for diesel–hybrid ships, Ruan et al. employed six machine learning methods to construct a parallel grey-box model [45]. The effectiveness of this model was validated on a very large ore carrier with a length of 327 m; compared to the white-box model, the proposed model reduced the error by 41.7%.

Due to the integration of prior knowledge from physical models, hybrid models generally exhibit higher fuel consumption prediction accuracies compared to purely physical models. Additionally, in cases of limited data, hybrid models typically outperform data-driven models in terms of prediction accuracy. However, with a large amount of data available, their predictive performance may be inferior to that of data-driven models.

2.4. Research Gaps and Contributions

Despite the extensive research on data-driven ship fuel consumption modeling, there are still several gaps in the existing literature. First, numerous factors influence the energy efficiency of ships. Although some studies have considered various factors, including environmental conditions, propulsion system efficiency, and navigation status, they often neglect the temporal variations of these factors. Therefore, it is essential to account for these temporal changes. Second, the quality and volume of data significantly impact the accuracy of data-driven ship fuel consumption predictions. While some studies have processed data to improve their quality, there remain potential outliers that have not been identified. Thus, the influence of data noise on the uncertainty of ship fuel consumption must be considered. Third, most studies predict ship fuel consumption by outputting a specific value under given input variables, but this approach fails to capture the range of fluctuations in fuel consumption. Therefore, a method is needed to accurately describe fuel consumption variability under various conditions.

To address these gaps, we propose an interval prediction method for ship fuel consumption based on a Gaussian process quantile regression. The main contributions of this research are as follows: (i) Experiments were conducted using noon report data and multi-source sensor data to validate the model's effectiveness and robustness; and (ii) Based on a Gaussian process regression combined with quantile regression theory, a series of quantile intervals were set to construct the interval prediction model for ship fuel consumption. Comparative experiments with quantile regression models based on backpropagation neural networks and gradient boosting decision trees demonstrate that the proposed method achieves a higher accuracy in interval predictions of fuel consumption and more accurately reflects the fluctuations in fuel consumption.

3. Data Description and Preprocessing

3.1. Data Descrption

To verify the effectiveness and robustness of the fuel consumption prediction model, it is necessary to evaluate its performance across various datasets. Currently, mainstream research datasets primarily consist of noon report data and sensor data. Consequently, this study utilizes these two types of datasets for model construction and testing. The noon report data are manually recorded by on duty personnel at noon each day. These datasets include features related to the navigation environment and the main engine's fuel data of the target vessel. Key data features include wind speed, wind direction, current speed, current direction, ship speed, and daily fuel consumption, among others.

Sensor data are collected through various sensor devices installed on the ship. These sensors gather real-time information about the ship's navigation, the main engine's operational status, and the surrounding navigational environment. Compared to noon report data, these sensor data are more comprehensive but also more costly to acquire. For this study, both noon report data and sensor data are used to validate the model's effectiveness and robustness in predicting fuel consumption. The noon report data are sourced from a bulk carrier, while the sensor data are obtained from both a container ship and a bulk carrier. Table 1 provides the basic parameters of the target ships used in this study, Table 2 shows the data collection time range, sampling frequency, and voyage information of the target vessel, and Table 3 shows some of the collected data features.

Table 1. Target ship's basic parameters.

Dataset	Data Type	Ship Type	Ship Length	Ship Width	Design Speed
S1	Sensor data	Container ship	319 m	46 m	24.8 kn
S2	Sensor data	Bulk carrier	361 m	65 m	14.5 kn
S3	Noon report data	Bulk carrier	229 m	32 m	14 kn

Table 2. Sampling information for some features of the target vessel.

Dataset	Time Range	Fuel Consumption Data Frequency	Number of Voyages
S1	14 September 2017–25 September 2018	Every 15 min	34 voyages
S2	2 March 2018–19 April 2018	Every 5 min	1 voyage
S3	5 March 2020–19 September 2021	Daily	10 voyages

Table 3. Some of the collected data features.

Data Number	Ship Speed (kn)	Mean Draft (m)	Wind Speed (m/s)	Wind Direction (Degree)	Current Speed (kn)	Current Direction (Degree)	Fuel Consumption (MT/Day)	…
1#	11.460	22.624	6.023	14.153	0.831	150.693	41.494	…
2#	11.500	22.658	6.016	15.026	0.835	149.892	44.419	…
3#	11.540	22.615	6.008	16.170	0.839	148.762	41.662	…
4#	11.500	22.633	6.000	14.908	0.841	149.967	43.800	…
5#	11.500	22.626	5.993	14.918	0.845	149.966	40.575	…
N#	…	…	…	…	…	…	…	…

3.2. Data Preprocessing

Due to challenges related to data transmission, sensor equipment, and information collection, some of the collected feature data contain errors, omissions, and anomalies. Consequently, these data require preprocessing prior to their utilization in model construction. Leveraging domain knowledge and analyzing the distribution of various features, the data preprocessing for ship energy consumption addresses issues such as duplicate data, missing data, erroneous data, and outliers. The data preprocessing methodology is illustrated in Figure 1.

Based on domain knowledge and a statistical analysis, this study identifies several types of anomalies in energy efficiency data, namely duplicated data points, missing feature data, erroneous data, and outliers. If these anomalous data are used directly for model training, they can interfere with the model's prediction performance. Therefore, it is essential to preprocess the collected data before model construction to obtain cleaner and higher-quality data.

Duplicated data can lead to overfitting; thus, duplicates should be removed, retaining only one instance of each duplicated data point. For missing data, if a specific feature has too many missing values, direct interpolation may bring too many errors. Therefore, a threshold must be set; if the amount of missing data for that feature is below the threshold, linear interpolation is performed using the nearest time points. Otherwise, the data are directly deleted. A similar approach is applied to outlier data, determining whether to interpolate or delete them based on the threshold. For erroneous data, direct interpolation is used for correction.

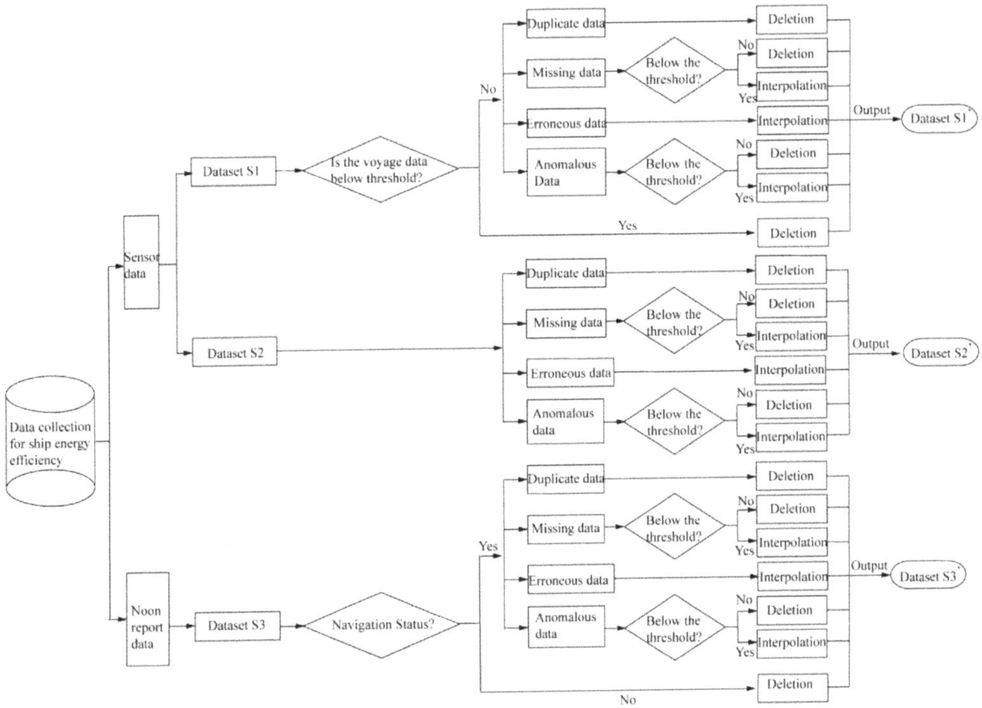

Figure 1. The preprocessing flow for ship energy consumption data.

Through the above preprocessing steps, some obvious outlier data were handled. These preprocessing steps ensure that the data used for model training are of higher quality, thereby improving the model's performance.

3.3. Data Feature Distribution

Due to the differing equipment attributes of each ship, certain features, such as draft and speed, are not collected with the same frequency as fuel consumption data. Consequently, these features must be converted to match the frequency of fuel consumption data collection. The distribution of various features across each dataset is illustrated in Figures 2–4.

As shown in Figures 2–4, the distribution ranges for various features in the datasets for dataset S1 are as follows: the speed distribution range is [5, 25] knots, the average draft (defined as the average of the forward and aft drafts) distribution range is [8, 13] m, the wind speed distribution range is [0, 27] m/s, the current speed distribution range is [0, 1.8] knots, and the ship fuel consumption distribution range is [0, 200] MT. The ranges are as follows for dataset S2: the speed distribution range is [0, 16] knots, and the average draft distribution range is [21, 23.2] m. Since draft data are acquired through measurement instruments, there is an inherent potential for measurement error. Additionally, during ocean voyages, strong winds and waves in the navigational area can induce the rolling and pitching of the vessel, resulting in fluctuations in the fore and aft draft measurements. The wind speed distribution range is [2, 16] m/s, the current speed distribution range is [0, 1.8] knots, and the ship fuel consumption distribution range is [0, 90] MT. The ranges for dataset S3 are as follows: the speed distribution range is [6, 14] knots, the average draft distribution range is [6, 16] m, the wind speed distribution range is [0, 18] m/s, the current speed distribution range is [0, 1] knots, and the ship fuel consumption distribution range is [16, 28] MT.

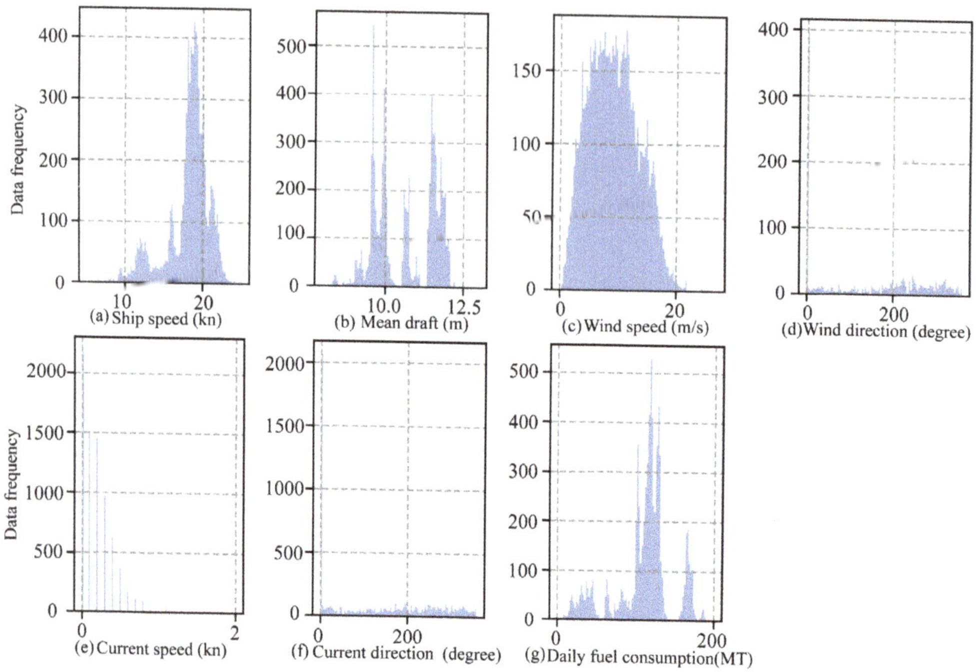

Figure 2. Distribution of various features in dataset S1.

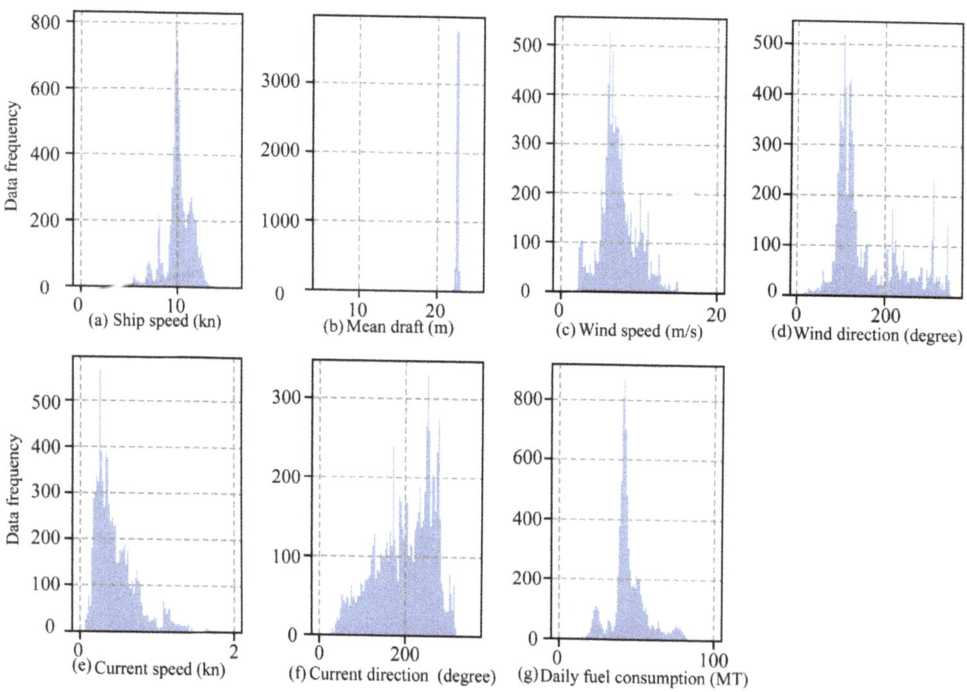

Figure 3. Distribution of various features in dataset S2.

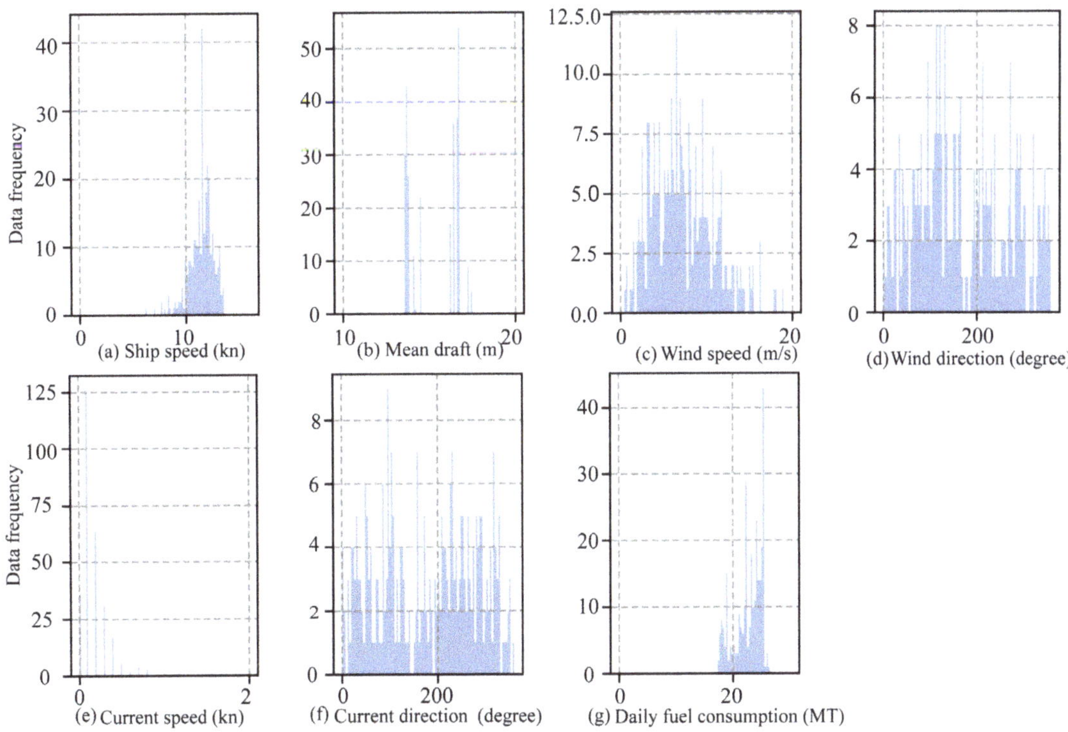

Figure 4. Distribution of various features in dataset S3.

4. Methodology

4.1. Overall Framework

The interval prediction structure for ship fuel consumption in this study consists of three primary components: data collection, data preprocessing, and model construction, as depicted in Figure 5. The data utilized in this research are obtained from multiple sensors and noon reports. Data preprocessing involves applying domain knowledge and statistical methods to address duplicate, abnormal, erroneous, and missing data through interpolation and deletion. For model construction, Gaussian processes and quantile regression models are employed. Ship fuel consumption is influenced by a range of factors, including the ship's navigational status and the maritime environment, both of which are subject to continuous variations throughout the voyage. To effectively capture these dynamic conditions, the model incorporates the ship's fore draft, aft draft, wind speed, wind direction, current speed, and current direction as input variables, while daily fuel consumption is defined as the output variable.

The entire dataset is partitioned into a training dataset and a testing dataset based on a predefined ratio. The training dataset is used for model training and learning, employing Bayesian optimization and cross-validation methods to optimize model parameters, while the testing dataset is reserved for prediction purposes.

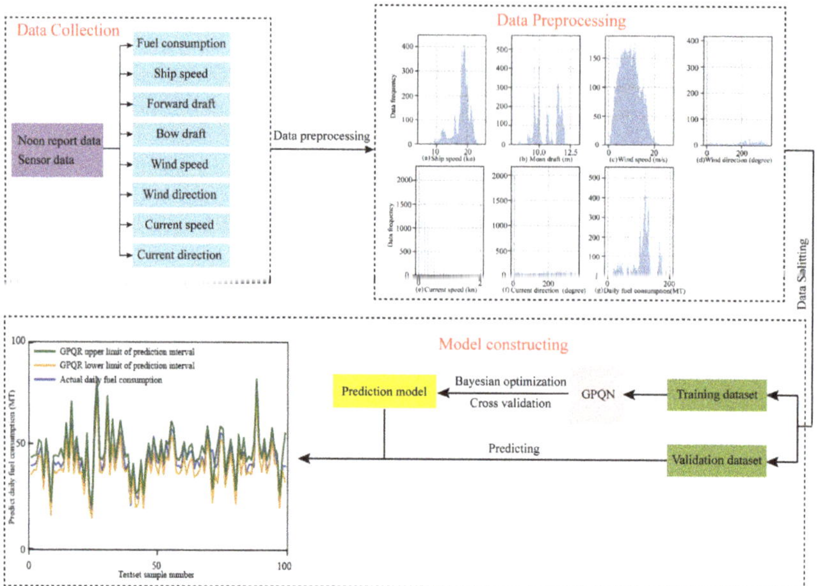

Figure 5. The GPQN framework for the prediction of ship fuel consumption.

4.2. Gaussian Process Regression

Gaussian process regression (GPR) is a machine learning method developed based on statistical learning and Bayesian theory [46]. The method does not require prior knowledge and has strong training and generalization abilities. As a non-parametric method, it is assumed that any subset of the dataset follows a multivariate Gaussian distribution. GPR is represented by mean function $u(x)$ and covariance function $k(x, x_*)$, given test set $T^* = \{(x_i^*, y_i^*), i = n+1, n+2, \dots n+m\}$. $f(x_*)$ is the predicted values for test samples x_*. The Gaussian distribution without noise can be expressed by Equations (1)–(3). The specific information can be found in Appendix A.

$$y = f(x) \sim \mathrm{GP}(u(x), k(x, x_*)) \tag{1}$$

$$u(x) = E[f(x)] \tag{2}$$

$$k(x, x_*) = E\{[f(x) - u(x)]\,|[f(x_*) - u(x)]\} \tag{3}$$

In the prediction of ship fuel consumption, the data typically contain noise. Therefore, it is necessary to model this noise using Gaussian processes for accurate predictions. The Gaussian distribution with noise can be expressed by Equations (4) and (5).

$$y = f(x) + \varepsilon, \qquad \varepsilon \sim N(0, \delta_n^2 I_n) \tag{4}$$

$$cov(y) = k(x, x_*) + \delta_n^2 I_n \tag{5}$$

where ε is noise, following a Gaussian normal distribution, δ_n^2 is noise variance, and I_n is the $n \times n$ unit matrix. After adding noise to the Gaussian process, the joint prior distribution of train set output and test set output can be expressed by Equation (6).

$$\begin{pmatrix} y \\ f_* \end{pmatrix} \sim N\left(u(x), \begin{bmatrix} K + \delta_n^2 I_n & k^T \\ k_* & k(x_*, x_*) \end{bmatrix}\right) \tag{6}$$

The posterior distribution of the test set $f(x_*)$ can be expressed as Equation (7).

$$p(f_*|x,y,x_*) \sim N\left(\overline{f_*}, cov(f_*)\right) \tag{7}$$

wherein the expectations and variances of f_* can be expressed as Equation (8).

$$\overline{f_*} = E(f_*|x,y,x_*) = k(x,x_*)\left(k(x,x)+\delta_n^2 I_n\right)^{-1} y \tag{8}$$

$$cov(\overline{f_*}) = k(x_*,x_*) - k(x,x_*)\left(k(x,x)+\delta_n^2 I_n\right)^{-1} k(x,x_*) \tag{9}$$

$\overline{f_*}$ obeys the standard normal distribution, $E(\cdot)$ is the expected function, $K_* = [k(x_1,x_*), k(x_2,x_*), \ldots, k(x_n,x_*)]$, and K is Gram matrix, represented as Equation (10).

$$K = \begin{bmatrix} k(x_1,x_1) & \cdots & k(x_1,x_*) \\ \vdots & \ddots & \vdots \\ k(x_*,x_1) & \cdots & k(x_*,x_*) \end{bmatrix} \tag{10}$$

The selection of kernel function is crucial for accuracy of Gaussian predictions. The radial basis function is generally chosen as kernel function without much prior knowledge, expressed as Equation (11).

$$k(x_i,x_j) = \delta_f^2\, exp\left(-\frac{(x_i-x_j)^2}{2p^2}\right) \tag{11}$$

δ_f^2 is variance in kernel function, and p^2 is length scale of kernel function.

4.3. Gaussian Process Quantile Regression

Quantile regression is used to study the relationship between the independent variables and the conditional quantiles of the dependent variable [47]. A quantile is defined such that the probability of the random variable α being less than or equal to the quantile τ is $\alpha(\tau)$. The definition is expressed as Equation (12).

$$\tau = P(\alpha \leq \alpha(\tau)) = F((\tau)) \tag{12}$$

The Gaussian process quantile regression (GPQR) model is based on the probabilistic framework of Gaussian processes and integrates quantile regression to output conditional quantiles. GPQR introduces the quantile loss function, as expressed in Equation (13).

$$L_\tau(\alpha_i) = \begin{cases} (\tau-1)\varepsilon_i, & \varepsilon_i < 0 \\ \tau\varepsilon_i, & \varepsilon_i \geq 0 \end{cases} \tag{13}$$

where $\varepsilon_i = y_i - \overline{y_i}$ is the prediction error. By setting different quantiles and minimizing the loss function, the predicted ship fuel consumption at different quantiles can be obtained. Assuming the confidence level is 2β, the upper percentile τ^u is β and the corresponding predicted value is U^β, the lower quantile τ^l is $0.5-\beta$ and the corresponding predicted value is $L^{0.5-\beta}$, and the prediction interval is $[L^{0.5-\beta}, U^\beta]$.

4.4. Bayesian Optimization-Based Gaussian Process Regression

Hyperparameter optimization plays a crucial role in enhancing the predictive accuracy of models. In recent years, methods based on Bayesian algorithms have become prevalent for hyperparameter optimization, owing to their efficacy in navigating various hyperparameter spaces. Unlike traditional grid search and random search techniques, Bayesian optimization is more adept at locating the global optimum, is less prone to local optima, and is more time-efficient.

Bayesian hyperparameter optimization leverages surrogate functions and acquisition functions to iteratively improve the selection of hyperparameters. The acquisition function, which guides the search for the next hyperparameter value, relies on the prior distribution of historical hyperparameters and the objective function. Concurrently, the surrogate function models the relationship between hyperparameter values and the objective function value.

The optimization process is iterative, terminating either when the desired accuracy is achieved or when a predefined number of iterations is completed. The entire Bayesian optimization process can be mathematically expressed in Equation (14).

$$\omega^* = argmin(f(\omega)), \quad \omega \in \overline{\omega} \tag{14}$$

where ω represents the hyperparameters to be optimized. $\overline{\omega}$ is the range of optimized hyperparameters. Bayesian hyperparameter optimization is typically combined with K-fold cross validation for parameter tuning; for the detailed process, please refer to reference [48]. In this study, we employed 10-fold cross validation.

4.5. Error Metric

In the context of single-value prediction for ship fuel consumption, the model's prediction accuracy is evaluated using the determination coefficient (R^2), mean absolute error (MAE), and mean absolute percentage error (MAPE). The specific evaluation metrics can be expressed as Equations (15)–(17).

$$R^2 = 1 - \frac{\sum_{i=1}^{m} (f(x_i) - y_i)^2}{\sum_{i=1}^{m} (y_i - \bar{y})^2} \tag{15}$$

$$MAE = \frac{1}{m}\sum_{i=1}^{m} |f(x_i) - y_i| \tag{16}$$

$$MAPE = \frac{100\%}{m}\sum_{i=1}^{m} \left|\frac{f(x_i) - y_i}{y_i}\right| \tag{17}$$

where $f(x_i)$ is the predicted value of data sample x_i, y_i is actual value of data sample x_i, $\bar{y}$ is the mean of all actual values, and m is the number of data samples. R^2 ranges from 0 to 1, a higher R^2 and lower MAE and MAPE values indicate better model prediction accuracy.

In the context of interval prediction for ship fuel consumption, the prediction interval coverage probability (PICP) and prediction interval normalized average width (PINAW) are employed to assess the performance of the predictive model. The values of these indicators can be determined by Equations (18)–(20).

$$PINAW = \frac{1}{TR^{2\beta}}\sum_{i=1}^{T}(U_i^{\beta} - L_i^{0.5-\beta}) \tag{18}$$

$$PICP = \frac{1}{T}\sum_{i=1}^{T} A(Y_i) \tag{19}$$

$$A_i(Y_i) = \begin{cases} 1, & L_i^{0.5-\beta} \le y_i \le U_i^{\beta} \\ 0, & Y_i > U_i^{\beta},\ y_i < L_i^{0.5-\beta} \end{cases} \tag{20}$$

The variables $L^{0.5-\beta}$ and U^{β} represent the lower and upper limits of interval prediction at a specified confidence level of 2β, respectively. $R^{2\beta}$ represents the interval prediction range at the confidence level of 2β.

According to the definition, a higher PICP indicates that a larger number of samples fall within the predicted confidence interval, and a lower PINAW value signifies smaller prediction intervals. Therefore, effective predictive models should aim for high PICP values and narrow PINAW to ensure both reliability and accuracy in their interval predictions.

5. Results and Discussion

5.1. Analysis of Ship Fuel Consumption Prediction Results

All data prediction experiments in this study were conducted using Python 3.8 version or MATLAB R2012A software. First, the entire dataset is divided into a training set and a validation set. By setting a specific quantile (e.g., the quantile is 0.8), the constructed Gaussian process regression model is used to calculate the ship fuel consumption values. Then, the difference between the actual values and the predicted values is calculated, along with the value of the loss function in Equation (13). Hyperparameter iteration and adjustments are performed based on the loss function value to obtain the model's optimal parameters. Once the model training is complete, it is finally applied to the validation set for predictions. To evaluate the prediction accuracy of the GPR model in comparison with a Backpropagation Neural Network Regression (BPNR) and gradient boosting regression (GBR), each dataset was partitioned into training and testing sets using an 80:20% ratio. Figure 6 presents the ship fuel consumption predictions produced by each model, while Figure 7 illustrates the distribution of the R^2, MAE, and MAPE for each model.

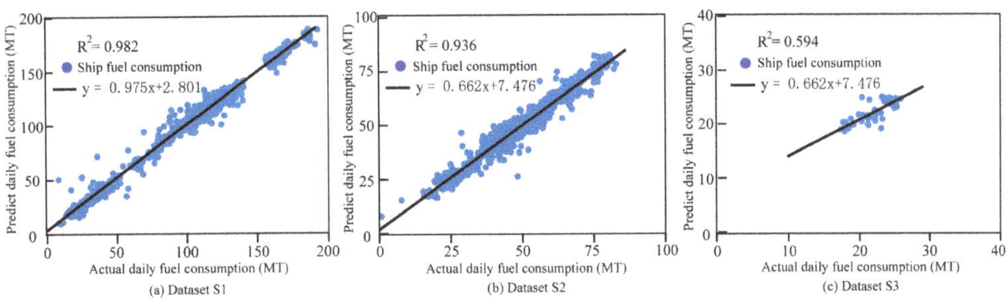

Figure 6. The fuel consumption predictions of each model.

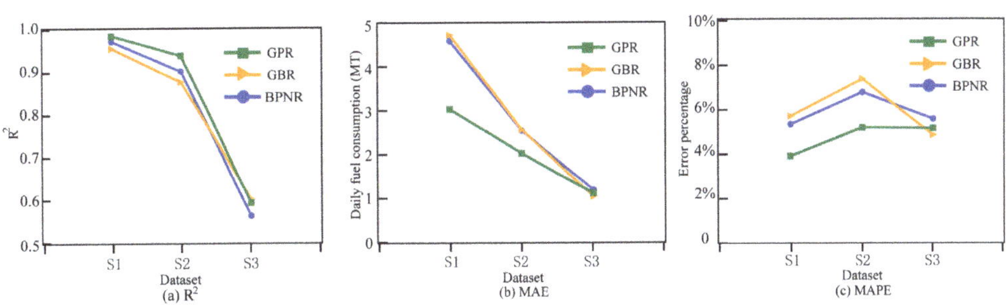

Figure 7. Ship fuel consumption prediction metrics for each model.

As illustrated in Figures 6 and 7, for datasets S1 and S2, the GPR model exhibits the highest R-squared values and the lowest MAE and MAPE values, indicating a superior prediction accuracy for these datasets. In contrast, for dataset S3, the evaluation metrics for each model are relatively similar, suggesting comparable prediction accuracies among the three models for this dataset. Additionally, the prediction accuracy of all models on datasets S1 and S2 was superior to that on dataset S3, indicating that the models performed better with the sensor data than with the noon report data.

5.2. Interval Prediction and Analysis of Ship Fuel Consumption

To investigate the interval predictions of ship fuel consumption at varying confidence levels, the confidence levels were set to 0.8, 0.85, 0.9, and 0.95, respectively. Figure 8 displays the interval prediction distributions of ship fuel consumption based on the GPQR model at these different confidence levels. Due to the large size of the test sets for datasets S1 and S2, only the distributions for 100 testing samples are shown. Conversely, for dataset S3, which has a smaller test set, all interval prediction distributions are presented.

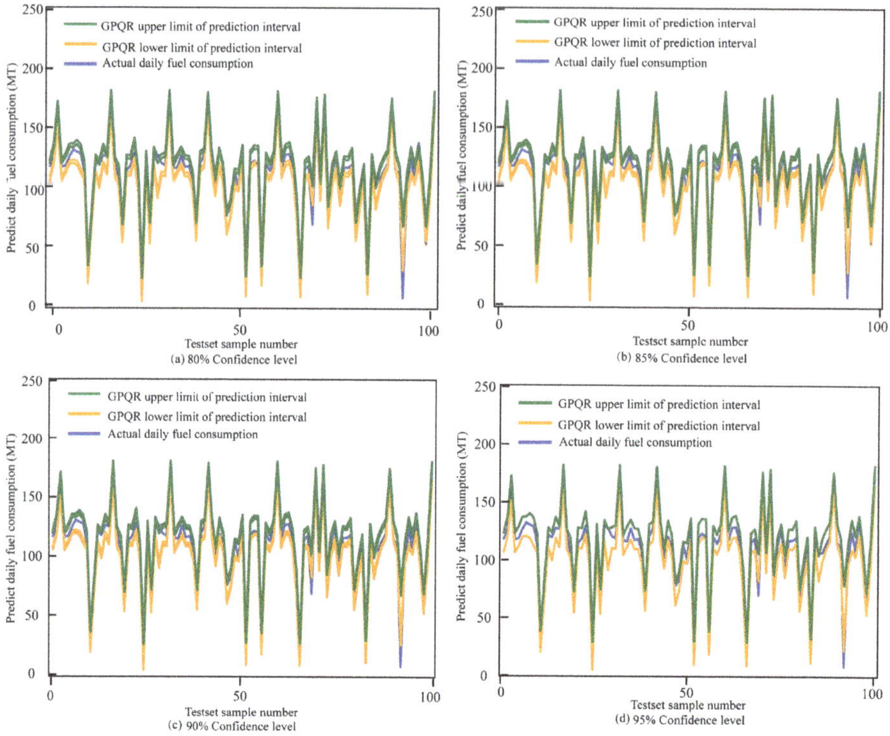

Figure 8. Interval predictions of ship fuel consumption for dataset S1.

As shown in Figures 7–10, the actual values of ship fuel consumption for each dataset mostly fall within the predicted intervals at various confidence levels. Table 4 presents the Gaussian process quantile prediction metrics for each dataset at different confidence levels. As observed in Table 4, for dataset S1, the prediction interval coverage probabilities of the fuel consumption intervals at all four confidence levels exceed the specified confidence level values. For datasets S2 and S3, the value of PICP is nearly equal to the confidence level. Additionally, the value of PINAW for all datasets at the four confidence levels is less than 0.3. Based on the above analysis, it can be concluded that the GPQR model can effectively perform interval predictions of ship fuel consumption.

Table 4. Ship fuel consumption prediction metrics for GPQR model at different confidence levels.

Model	Dataset	Error Metric	80% Confidence Level	85% Confidence Level	90% Confidence Level	95% Confidence Level
GPQR	S1	PICP	0.880	0.905	0.935	0.959
		PINAW	0.142	0.160	0.182	0.217
	S2	PICP	0.831	0.847	0.898	0.949
		PINAW	0.161	0.181	0.206	0.246
	S3	PICP	0.857	0.892	0.927	0.957
		PINAW	0.180	0.202	0.231	0.275

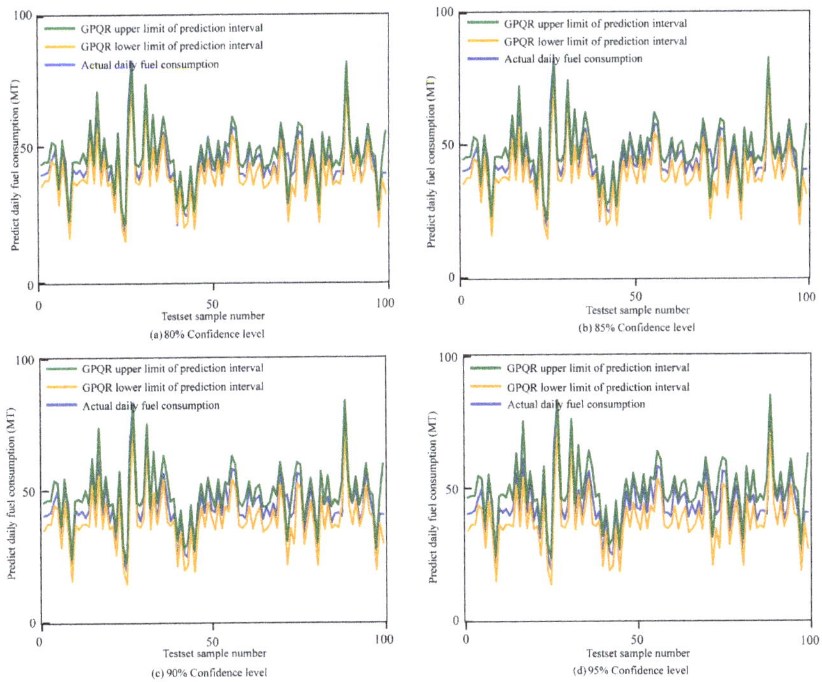

Figure 9. Interval predictions of ship fuel consumption for dataset S2.

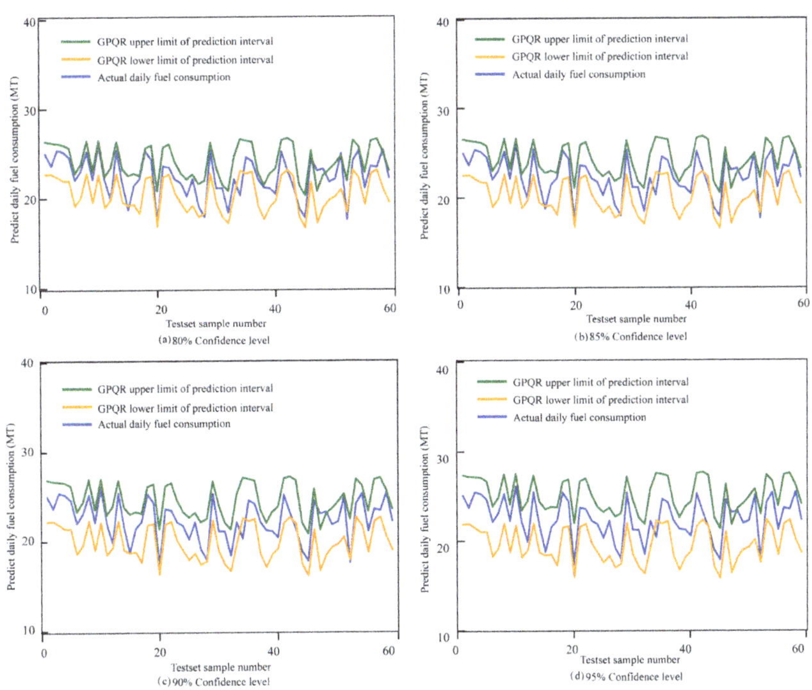

Figure 10. Interval predictions of ship fuel consumption for dataset S3.

5.3. Interval Prediction Results and Analysis for Each Model

To validate the superiority of the GPQR method, a comparison was made with a gradient boosting quantile regression (GBQR) and a backpropagation neural network quantile regression (BPQR). Figures 11–13 present the evaluation metrics for interval predictions of ship fuel consumption using various methods across the three datasets; additionally, the ratio of PICP to PINAW is defined as the prediction interval coverage normalized (PICN) in Figures 11–13.

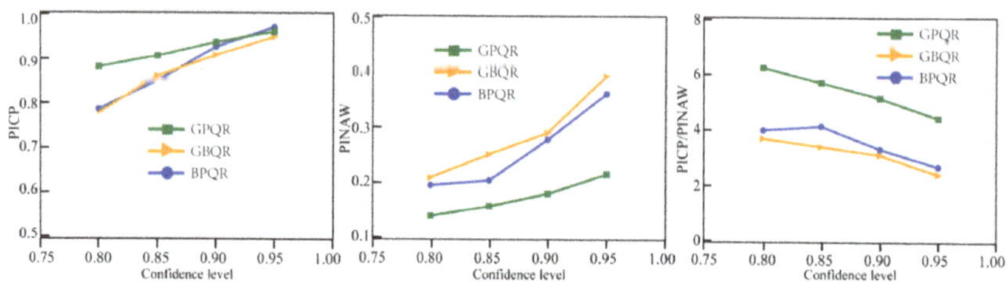

Figure 11. Evaluation metrics for ship fuel consumption interval prediction in dataset S1.

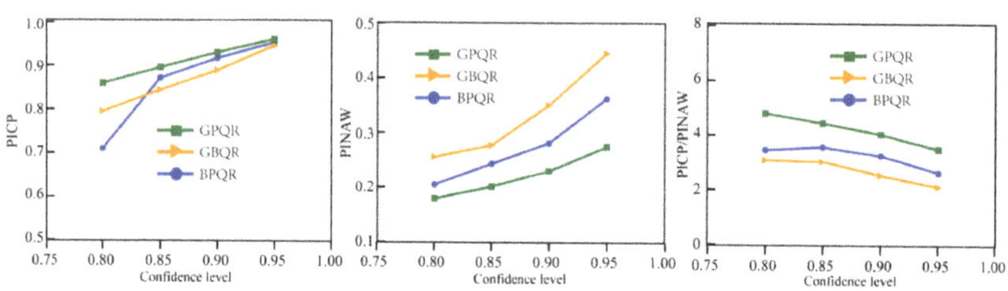

Figure 12. Evaluation metrics for ship fuel consumption interval prediction in dataset S2.

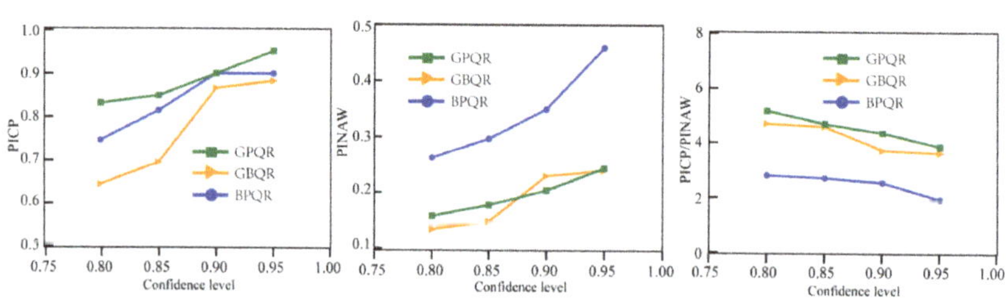

Figure 13. Evaluation metrics for ship fuel consumption interval prediction in dataset S3.

As illustrated in Figures 11–13, the PICP and PINAW values for each model increase with the confidence level. Furthermore, in any dataset, the PICP value of the GPQR is higher than that of the GBQR and BPQR. For datasets S1 and S2, the PINAW value of the GPQR is lower than that of the GBQR and BPQR, indicating that the interval prediction accuracy of the GPQR is superior to that of the GBQR and BPQR on the sensor datasets. Additionally, it is noteworthy that for dataset S3, the GBQR exhibits a lower PINAW compared to the GPQR for confidence levels between 0.80 and 0.85. This phenomenon may be explained by the fact that datasets S1 and S2 consist of sensor data, while dataset S3 contains noon report data. The number of data points used for training and testing in the experiments is 7646, 8263, and 291, respectively. When the dataset size is relatively small, tree-based models may achieve a higher prediction accuracy for ship fuel consumption compared to other models. The PICN value for the GPQR is higher than that for the GBQR and BPQR across all datasets. Based on the above analysis, it can be concluded that the

GPQR outperforms the GBQR and BPQR in terms of accuracy for both sensor data and noon report data.

According to the experimental results, the program's runtime does not exceed five minutes, while the data collection intervals for the experimental datasets S1, S2, and S3 are fifteen minutes, five minutes, and one hour, respectively. As the runtime is shorter than the data collection frequency, the program meets the requirements for real-time fuel consumption prediction for ships. This processing duration is considered acceptable by shipowners and managers.

It should be noted that fuel consumption and emission indices are influenced not only by the amount of fuel consumed but also by the type of fuel used. When calculating gas emissions, it is essential to consider the type of fuel and the associated emission factors.

6. Conclusions

This paper combines a Gaussian process regression and quantile regression to propose a Gaussian process quantile regression method for interval predictions of ship fuel consumption. The method is validated on mainstream ship energy efficiency datasets.

For ship fuel consumption prediction, the mean absolute percentage error of the GPR model for both noon report data and sensor data is below 8%, indicating a prediction accuracy exceeding 92%. The GPR model outperforms the GBR and BPNR across all datasets. This demonstrates that the proposed GPR method is suitable for both sensor and noon report data, and its prediction accuracy surpasses that of the GBR and BPNR, highlighting the superiority of the GPR in ship fuel consumption prediction.

Regarding interval predictions of ship fuel consumption, the prediction accuracy of the GPQR increases with the confidence level. The prediction intervals effectively reflect the volatility of ship fuel consumption under different influencing factors. Whether using sensor data or noon report data, the prediction accuracy of the GPQR is higher than that of the GBQR and BPQR, confirming the superiority of the GPQR in interval predictions of ship fuel consumption.

The proposed method has been primarily applied to three vessels, specifically bulk carriers and container ships. To enhance the generalizability of the method, future research should extend its application to a broader range of vessel types and deadweight tonnages. Such an expansion would be instrumental in validating the method's effectiveness across diverse maritime settings. Additionally, it is crucial to examine the applicability of this method to inland waterway vessels, particularly those operating in the Yangtze River and other inland waterways, to assess its suitability for these unique environments.

Sensor malfunctions leading to data loss or difficulties in data acquisition significantly impact the accuracy and reliability of proposed methods, particularly in complex and variable marine environments. Incomplete data prevent models from fully reflecting real-world conditions, thereby adversely affecting decision making. To enhance the applicability of such methods, future research should focus more on integrating theoretical foundations with practical applications. By incorporating physical formulas and mathematical models, it will be possible to develop more robust and adaptable models tailored to different maritime scenarios, such as varying routes, climate conditions, and vessel types. Additionally, the development of fault-tolerant sensor systems should be considered to minimize the impact of data loss on predictive outcomes, thus improving the overall performance and practical value of these models.

Author Contributions: Conceptualization, T.Z.; methodology, T.Z. and Z.H.; software, T.Z.; validation, T.Z. and J.W.; formal analysis, T.Z. and J.W.; investigation, Q.H.; resources, Q.H.; data curation, Q.H.; writing—original draft preparation, T.Z. and J.W.; writing—review and editing, T.Z. and J.W.; visualization, Z.H.; supervision, Z.H.; project administration, T.Z.; funding acquisition, T.Z. All authors have read and agreed to the published version of the manuscript.

Funding: This research was funded by "Fujian Province Young and Middle-aged Teachers' Education Research Project, JAT231208" and "Doctoral Research Start-up Fund, 20230102".

Institutional Review Board Statement: Not applicable.

Informed Consent Statement: Not applicable.

Data Availability Statement: The data can be accessed by sending an email to the corresponding author.

Conflicts of Interest: The authors declare no conflicts of interest.

List of Abbreviations

Abbreviations	Meaning	Abbreviations	Meaning
BPNN	Backpropagation Neural Network	RF	Random Forest
GBDT	Gradient Boosting Decision Tree	AIS	Automatic Identification System
IMO	International Maritime Organization	BEM	Blending Ensemble Model
CII	Carbon Intensity Indicator	SVM	Support Vector Machine
MRV	Monitoring, Reporting, and Verification	LSTM	Long Short-Term Memory
EUETS	European Union Emissions Trading System	Bi-LSTM	Bidirectional Long Short-Term Memory
EEOI	Energy Efficiency Operational Indicator	GPR	Gaussian Process Regression
CFD	Computational Fluid Dynamics	GPQR	Gaussian Process Quantile Regression
ECA	Emission Control Areas	R^2	Determination Coefficient
EU	European Union	MAE	Mean Absolute Error
MLR	Multiple Linear Regression	MAPE	Mean Absolute Percentage Error
DT	Decision Tree	PICP	Prediction Interval Coverage Probability
ANN	Artificial Neural Network	PINAW	Prediction Interval Normalized Average Width
LASSO	Least Absolute Shrinkage and Selection Operator	BPQR	Backpropagation Neural Network Quantile Regression

Appendix A

Table A1. List of notations used in this study.

Notations	Definitions
x, x_*	Data sample and test data sample.
$u(x)$	Mean function.
$k(x, x_*)$	Covariance function.
$f(x_*)$	The predicted values for test samples x_*.
$E[\cdot]$	Expected function.
$\varepsilon, \delta_n^2, I_n$	Noise, Noise variance, $n \times n$ unit matrix.
K	Gram matrix.
$k(x_i, x_j)$	Kernel function.
δ_f^2, p^2	Variance in kernel function, length scale of kernel function.
τ	Quantile value.
β	Value of confidence level.
$\omega, \omega^*, \overline{\omega}$	Hyperparameters, optimal hyperparameters, the range of optimized hyperparameters.
$R^{2\beta}$	The interval prediction range at the confidence level 2β.
$L^{0.5-\beta}, U^{\beta}$	The lower and upper limits of interval prediction at a specified confidence level 2β.

References

1. Kim, Y.R.; Jung, M.; Park, J.B. Development of a Fuel Consumption Prediction Model Based on Machine Learning Using Ship In-Service Data. *J. Mar. Sci. Eng.* **2021**, *9*, 137. [CrossRef]
2. Chen, X.; Dou, S.; Song, T.; Wu, H.; Sun, Y.; Xian, J. Spatial-Temporal Ship Pollution Distribution Exploitation and Harbor Environmental Impact Analysis via Large-Scale AIS Data. *J. Mar. Sci. Eng.* **2024**, *12*, 960. [CrossRef]
3. Xiao, G.; Wang, Y.; Wu, R.; Li, J.; Cai, Z. Sustainable Maritime Transport: A review of Intelligent Shipping Technology and Green Port Construction Applications. *J. Mar. Sci. Eng.* **2024**, *12*, 1728. [CrossRef]
4. Zhou, T.; Hu, Q.; Hu, Z.; Zhi, J. Theory and Application of Vessel Speed Dynamic Control considering Safety and Environmental Factors. *J. Adv. Transp.* **2022**, *2022*, 5333171. [CrossRef]
5. Ho, D.T.; Bopp, L.; Palter, J.B.; Long, M.C.; Boyd, P.; Neukermans, G.; Bach, L. Monitoring, Reporting, and Verification for Ocean Alkalinity Enhancement. *Sci. Synth. Rep. Assess.* **2023**, *2*, 12.
6. Mao, Z.; Ma, A.; Zhang, Z. Towards Carbon Neutrality in Shipping: Impact of European Union's Emissions Trading System for Shipping and China's Response. *Ocean Coast. Manag.* **2024**, *249*, 107006. [CrossRef]
7. Oloruntobi, O.; Mokhtar, K.; Gohari, A.; Asif, S.; Chuah, L.F. Sustainable Transition Towards Greener and Cleaner Seaborne Shipping Industry: Challenges and Opportunities. *Clean. Eng. Technol.* **2023**, *13*, 100628. [CrossRef]
8. Tadros, M.; Ventura, M.; Guedes Soares, C. Review of the Decision Support Methods Used in Optimizing Ship Hulls Towards Improving Energy Efficiency. *J. Mar. Sci. Eng.* **2023**, *11*, 835. [CrossRef]
9. Karagiannidis, P.; Themelis, N. Data-Driven Modelling of Ship Propulsion and the Effect of Data Pre-Processing on the Prediction of Ship Fuel Consumption and Speed Loss. *Ocean Eng.* **2021**, *222*, 108616. [CrossRef]
10. Bassam, A.M.; Phillips, A.B.; Turnock, S.R.; Wilson, P.A. Artificial Neural Network-Based Prediction of Ship Speed Under Operating Conditions for Operational Optimization. *Ocean Eng.* **2023**, *278*, 114613. [CrossRef]
11. Karatuğ, Ç.; Tadros, M.; Ventura, M.; Guedes Soares, C. Strategy for Ship Energy Efficiency Based on Optimization Model and Data-Driven Approach. *Ocean Eng.* **2023**, *279*, 114397. [CrossRef]
12. Moshiul, A.M.; Mohammad, R.; Hira, F.A. Alternative Fuel Selection Framework Toward Decarbonizing Maritime Deep-Sea Shipping. *Sustainability* **2023**, *15*, 5571. [CrossRef]
13. Jimenez, V.J.; Kim, H.; Munim, Z.H. A Review of Ship Energy Efficiency Research and Directions Towards Emission Reduction in the Maritime Industry. *J. Clean. Prod.* **2022**, *366*, 132888. [CrossRef]
14. Chen, X.; Ma, D.; Liu, R.W. Application of Artificial Intelligence in Maritime Transportation. *J. Mar. Sci. Eng.* **2024**, *12*, 439. [CrossRef]
15. Shih, Y.C.; Tzeng, Y.A.; Cheng, C.W.; Huang, C.H. Speed and Fuel Ratio Optimization for a Dual-Fuel Ship to Minimize Its Carbon Emissions and Cost. *J. Mar. Sci. Eng.* **2023**, *11*, 758. [CrossRef]
16. Fan, A.; Yang, J.; Yang, L.; Wu, D.; Vladimir, N. A Review of Ship Fuel Consumption Models. *Ocean Eng.* **2022**, *264*, 112405. [CrossRef]
17. Smith, A.; Jones, B. A Two-Step Strategy for Fuel Consumption Prediction and Optimization of Maritime Transport. *Transp. Res. Part D Transp. Environ.* **2021**, *93*, 102873.
18. Hu, Z.; Zhou, T.; Osman, M.T.; Li, X.; Jin, Y.; Zhen, R. A Novel Hybrid Fuel Consumption Prediction Model for Ocean-Going Container Ships Based on Sensor Data. *J. Mar.Sci. Eng.* **2021**, *9*, 449. [CrossRef]
19. Chen, X.; Wu, H.; Han, B.; Liu, W.; Montewka, J.; Liu, R.W. Orientation-aware ship detection via a rotation feature decoupling supported deep learning approach. Engineering Applications of Artificial Intelligence. *Eng. Appl. Artif. Intell.* **2023**, *125*, 106686. [CrossRef]
20. Kim, Y.R.; Steen, S.; Kramel, D.; Muri, H.; Strømman, A.H. Modelling of Ship Resistance and Power Consumption for the Global Fleet: The MariTEAM Model. *Ocean Eng.* **2023**, *281*, 114758. [CrossRef]
21. Holtrop, J. A Statistical Re-Analysis of Resistance and Propulsion Data. *Int. Shipbuild. Prog.* **1984**, *31*, 363.
22. Wang, K.; Li, J.; Huang, L.; Ma, R.; Jiang, X.; Yuan, Y.; Mwero, N.A.; Negenborn, R.R.; Sun, P.; Yan, X. A Novel Method for Joint Optimization of the Sailing Route and Speed Considering Multiple Environmental Factors for More Energy Efficient Shipping. *Ocean Eng.* **2020**, *216*, 107591. [CrossRef]
23. La Ferlita, A.; Qi, Y.; Di Nardo, E.; El Moctar, O.; Schellin, T.E.; Ciaramella, A. A Framework of a Data-Driven Model for Ship Performance. *Ocean Eng.* **2024**, *309*, 118486. [CrossRef]
24. Sang, Y.; Ding, Y.; Xu, J.; Sui, C. Ship Voyage Optimization Based on Fuel Consumption under Various Operational Conditions. *Fuel* **2023**, *352*, 129086. [CrossRef]
25. Guo, B.; Liang, Q.; Tvete, H.A.; Brinks, H.; Vanem, E. Combined Machine Learning and Physics-Based Models for Estimating Fuel Consumption of Cargo Ships. *Ocean Eng.* **2022**, *255*, 111435. [CrossRef]
26. Papandreou, C.; Ziakopoulos, A. Predicting VLCC Fuel Consumption with Machine Learning Using Operationally Available Sensor Data. *Ocean Eng.* **2022**, *243*, 110321. [CrossRef]
27. Vasilikis, N.I.; Geertsma, R.D.; Visser, K. Operational Data-Driven Energy Performance Assessment of Ships: The Case Study of a Naval Vessel with Hybrid Propulsion. *J. Mar. Eng. Technol.* **2023**, *22*, 84–100. [CrossRef]
28. Kim, H.S.; Roh, M.I. Interpretable, Data-Driven Models for Predicting Shaft Power, Fuel Consumption, and Speed Considering the Effects of Hull Fouling and Weather Conditions. *Int. J. Nav. Archit. Ocean Eng.* **2024**, *16*, 100592. [CrossRef]

29. Vorkapić, A.; Radonja, R.; Martinčić-Ipšić, S. Predicting Seagoing Ship Energy Efficiency from the Operational Data. *Sensors* **2021**, *21*, 2832. [CrossRef]
30. Yuksel, O.; Bayraktar, M.; Sokukcu, M. Comparative Study of Machine Learning Techniques to Predict Fuel Consumption of a Marine Diesel Engine. *Ocean Eng.* **2023**, *286*, 115505. [CrossRef]
31. Bayraktar, M.; Sokukcu, M. Marine Vessel Energy Efficiency Performance Prediction Based on Daily Reported Noon Reports. *Ships Offshore Struct.* **2024**, *19*, 831–840. [CrossRef]
32. Agand, P.; Kennedy, A.; Harris, T.; Bae, C.; Chen, M.; Park, E.J. Fuel Consumption Prediction for a Passenger Ferry Using Machine Learning and In-Service Data: A Comparative Study. *Ocean Eng.* **2023**, *284*, 115271. [CrossRef]
33. Du, Y.; Meng, Q.; Wang, S.; Kuang, H. Two-Phase Optimal Solutions for Ship Speed and Trim Optimization over a Voyage Using Voyage Report Data. *Transp. Res. Part B Methodol.* **2019**, *122*, 88–114. [CrossRef]
34. Zhou, Y.; Pazouki, K.; Murphy, A.J.; Uriondo, Z.; Granado, I.; Quincoces, I.; Fernandes-Salvador, J.A. Predicting Ship Fuel Consumption Using a Combination of Metocean and On Board Data. *Ocean Eng.* **2023**, *285*, 115509. [CrossRef]
35. Yan, R.; Wang, S.; Du, Y. Development of a Two-Stage Ship Fuel Consumption Prediction and Reduction Model for a Dry Bulk Ship. *Transp. Res. Part E Logist. Transp. Rev.* **2020**, *138*, 101930. [CrossRef]
36. Li, X.; Du, Y.; Chen, Y.; Nguyen, S.; Zhang, W.; Schönborn, A.; Sun, Z. Data Fusion and Machine Learning for Ship Fuel Efficiency Modeling: Part I—Voyage Report Data and Meteorological Data. *Commun. Transp. Res.* **2022**, *2*, 100074. [CrossRef]
37. Hu, Z.; Zhou, T.; Zhen, R.; Jin, Y.; Li, X.; Osman, M.T. A Two-Step Strategy for Fuel Consumption Prediction and Optimization of Ocean-Going Ships. *Ocean Eng.* **2022**, *249*, 110904. [CrossRef]
38. Lan, T.; Huang, L.; Ma, R.; Ruan, Z.; Ma, S.; Li, Z.; Zhao, H.; Wang, C.; Zhang, R.; Wang, K. A Novel Method of Fuel Consumption Prediction for Wing-Diesel Hybrid Ships Based on High-Dimensional Feature Selection and Improved Blending Ensemble Learning Method. *Ocean Eng.* **2024**, *307*, 118156. [CrossRef]
39. Panapakidis, I.; Sourtzi, V.M.; Dagoumas, A. Forecasting the Fuel Consumption of Passenger Ships with a Combination of Shallow and Deep Learning. *Electronics* **2020**, *9*, 776. [CrossRef]
40. Zhang, M.; Tsoulakos, N.; Kujala, P.; Hirdaris, S. A Deep Learning Method for the Prediction of Ship Fuel Consumption in Real Operational Conditions. *Eng. Appl. Artif. Intell.* **2024**, *130*, 107425. [CrossRef]
41. Han, P.; Liu, Z.; Sun, Z.; Yan, C. A Novel Prediction Model for Ship Fuel Consumption Considering Shipping Data Privacy: An XGBoost-IGWO-LSTM-Based Personalized Federated Learning Approach. *Ocean Eng.* **2024**, *302*, 117668. [CrossRef]
42. Leifsson, L.; Sævarsdóttir, H.; Sigurðsson, S.; Vésteinsson, A. Grey-Box Modeling of an Ocean Vessel for Operational Optimization. *Simul. Model. Pract. Theory* **2008**, *16*, 923–932. [CrossRef]
43. Wang, C.; Huang, L.; Ma, R.; Wang, K.; Sheng, J.; Ruan, Z.; Hua, Y.; Zhang, R. A Novel Cooperative Optimization Method of Course and Speed for Wing-Diesel Hybrid Ship Based on Improved A* Algorithm. *Ocean Eng.* **2024**, *302*, 117669. [CrossRef]
44. Cai, Z.; Li, L.; Yu, L.; Li, C.; Sun, M. Diversity, Quality, and Quantity of Real Ship Data on the Black-Box and Gray-Box Prediction Models of Ship Fuel Consumption. *Ocean Eng.* **2024**, *291*, 116434. [CrossRef]
45. Ruan, Z.; Huang, L.; Wang, K.; Ma, R.; Wang, Z.; Zhang, R.; Wang, C. A Novel Prediction Method of Fuel Consumption for Wing-Diesel Hybrid Vessels Based on Feature Construction. *Energy* **2024**, *286*, 129516. [CrossRef]
46. Tartakovsky, A.M.; Ma, T.; Barajas-Solano, D.A.; Tipireddy, R. Physics-Informed Gaussian Process Regression for States Estimation and Forecasting in Power Grids. *Int. J. Forecast.* **2023**, *39*, 967–980. [CrossRef]
47. Abid, N.; Ahmad, F.; Aftab, J.; Razzaq, A. A Blessing or a Burden? Assessing the Impact of Climate Change Mitigation Efforts in Europe Using Quantile Regression Models. *Energy Policy* **2023**, *178*, 113589. [CrossRef]
48. Zhou, T.; Hu, Q.; Hu, Z.; Zhen, R. An Adaptive Hyperparameter Tuning Model for Ship Fuel Consumption Prediction Under Complex Maritime Environments. *J. Ocean Eng. Sci.* **2022**, *7*, 255–263. [CrossRef]

Journal of
Marine Science and Engineering

MDPI

Article

Distributionally Robust Optimal Scheduling of Hybrid Ship Microgrids Considering Uncertain Wind and Wave Conditions

Fang Lu [1,*], Yubin Tian [2], Hongda Liu [3,*] and Chuyuan Ling [1]

[1] College of Intelligent System Science and Engineering, Harbin Engineering University, Harbin 150001, China
[2] Xi'an Aerospace Precision Electromechanical Research Institute, Xi'an 710118, China
[3] Yantai Research Institute, Harbin Engineering University, Yantai 264000, China
* Correspondence: lufang@hrbeu.edu.cn (F.L.); liuhd405@hrbeu.edu.cn (H.L.)

Abstract: A hybrid ship uses integrated generators, an energy storage system (ESS), and photovoltaics (PV) to match its propulsion and service loads, and together with optimal power and voyage scheduling, this can lead to a substantial improvement in ship operation cost, ensuring compliance with the environmental constraints and enhancing ship sustainability. During the operation, significant uncertainties such as waves, wind, and PV result in considerable speed loss, which may lead to voyage delays and operation cost increases. To address this issue, a distributionally robust optimization (DRO) model is proposed to schedule power generation and voyage. The problem is decoupled into a bi-level optimization model, the slave level can be solved directly by commercial solvers, the master level is further formulated as a two-stage DRO model, and linear decision rules and column and constraint generation algorithms are adopted to solve the model. The algorithm aims at minimizing the operation cost, limiting greenhouse gas (GHG) emissions, and satisfying the technical and operational constraints considering the uncertainty. Extensive simulations demonstrate that the expected total cost under the worst-case distribution is minimized, and compared with the conventional robust optimization methods, some distribution information can be incorporated into the ambiguity sets to generate fewer conservative results. This method can fully ensure the on-time arrival of hybrid ships in various uncertain scenarios while achieving expected operation cost minimization and limiting greenhouse gas (GHG) emissions.

Keywords: hybrid ship; distributionally robust optimization; uncertain wind and wave conditions; GHG; energy management

1. Introduction

About 3–5% of global greenhouse gas (GHG) emissions are produced by marine transport [1,2]. If no mitigation measures are taken, GHG emissions from marine transport will rise to 8% of global GHG emissions by 2050 [3,4], so marine transport needs to reduce GHG emissions significantly. To meet the demand for emission reduction, hybrid ships have gained industry attention. Hybrid ships refer to ships driven by different types of engines or electric motors and contain multiple energy sources [5]. In the past, hybrid ships were usually referred to as diesel–electric hybrid ships. With the development of new energy technology, new energy technology represented by photovoltaics (PV) has begun to be applied on ships, and PV has been used when carrying out certain research studies on hybrid ships composed of diesel generators (DGs), energy storage systems (ESSs), and PV [6,7]. In [8], it is verified by the results of actual navigation experiments that the participation of photovoltaic power generation in reasonable scheduling can help the shipping industry meet the "low-carbon navigation" policy.

Energy management is the key to achieving economic objectives and emission reduction on the premise of ensuring power performance. In [9], a deterministic energy management approach is proposed for a cruising ship, to minimize the fuel consumption cost by effectively scheduling the power generation units and voyage. In [10], a coordinated optimal power management method is studied according to the different objectives. The management model is solved by particle swarm optimization. The system is initialized with a population of random solutions and searches for optima by updating generations. At each time step, the velocity of each particle is changed as it heads toward its p-best and l-best locations. Ref. [11] proposes a scheduling model to optimize the economic and environmental objectives for the ship integrated with hybrid BS and energy storage.

However, due to the influence of weather, photovoltaic power generation has fluctuations and uncertainties. Compared to land, ships may encounter more uncertainties when navigating at sea, such as ship pitching and rolling, which greatly increases the prediction error of photovoltaic power generation and affects the accuracy of overall power generation and navigation scheduling. Based on this, in recent years, stochastic optimization and robust optimization have been proposed to deal with uncertain power grid optimization problems. In [12], a stochastic optimization method is adopted to solve the coordinated operation optimization problem of a multiple-energy microgrid under various uncertainties; however, many scenarios need to be generated according to the probability density function in advance, which increases the calculation burden. Compared with stochastic optimization, robust optimization only needs to construct an uncertain set and does not need the probability distribution of uncertain variables which are difficult to obtain, and the solution of robust optimization is robust to all possible scenarios. In [13], a robust optimization method is adopted to address the uncertainty of photovoltaic power generation on a ship to minimize the operating cost under the worst uncertainty scenario while satisfying all the operational constraints.

In addition, most existing studies have not considered the speed loss caused by water waves or wind during ship navigation. Although the speed loss models for different ships may be diverse, various tests have shown that the speed loss induced by water and waves for most ship prototypes is significant and may usually exceed 10% of the nominal speed even in moderate weather conditions. Since the propulsion load has a cubic relationship with the cruising speed, the propulsion power loss led by speed loss is much more significant. In this view, speed loss is always an important issue, and in conventional maritime applications, this problem is referred to as "weather routing" [14]. Therefore, in [14], a robust optimization method is proposed for the joint scheduling of power generation and voyage of a ship considering the uncertain wind and wave, and the results show that the proposed robust optimization model can ensure the punctuality of ships in various uncertain situations. In [15], hydrogen fuel cells are integrated into ship microgrids, and a two-stage robust optimization method is proposed. A two-level column constraint generation algorithm was used to solve the optimization model, aiming to minimize operating costs for the entire voyage in the worst-case scenario.

The above robust optimization methods only consider the worst-case scenario where uncertainty occurs. However, the probability of uncertainty occurring at the extremes is small, so the results of robust optimization are too conservative and often require higher economic costs. Combining the characteristics of stochastic and robust optimization, distributionally robust optimization (DRO) has been proposed in recent years to solve the uncertainty problem [16,17]. DRO is an effective way to handle diverse uncertainties, which incorporates the available probability distribution information into an ambiguity set to characterize the true probability distribution of uncertainties and reduce the solution conservatism. At present, this method has not been used to deal with the uncertainty of ship power systems, which is worthy of further study.

This paper comprehensively considers the uncertainties of photovoltaic output power and wind and wave events and proposes a DRO method to jointly optimize the scheduling of power and voyage speed to minimize the ship operation cost.

This paper contributes to this field of research in the following aspects:

(1) To address diverse uncertainties from photovoltaic systems and wind and wave events, a distributionally robust optimization (DRO) model is proposed to schedule

power generation and voyage. With the proposed method, ships can arrive at each port on time, while ensuring a lower operation cost compared to existing methods.

(2) The original model is decoupled into a bi-level optimization model, the slave level can be solved directly by commercial solvers, the master level is further formulated as a two-stage DRO framework, and linear decision rules are adopted to solve the model, which is suitable for practical applications.

2. Deterministic Optimization Model

As shown in Figure 1, a hybrid ship microgrid consists of DGs, ESSs, and PV for propulsion and service loads.

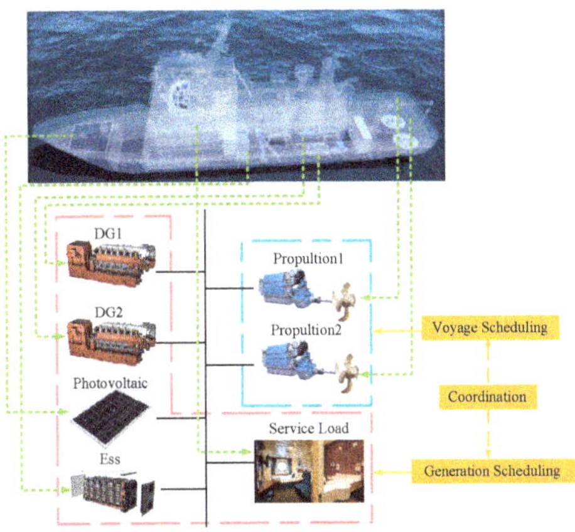

Figure 1. Topology of the ship microgrid.

2.1. Objective

In the optimal scheduling model constructed in this paper, it is assumed that the operation cost of PV and ESSs is zero, and the objective function includes the start–stop cost and the operation cost of the DG units, and the expression is shown as follows:

$$\min(F1 + F2) \tag{1}$$

where $F1$ is the startup and shutdown cost of the DG units; and $F2$ is the operation cost of the DG units.

$F1$ and $F2$ are calculated in (2) and (3), respectively:

$$F1 = \sum_{t=1}^{T}\sum_{i=1}^{N}(C_t^{ion} + C_t^{ioff}) \tag{2}$$

$$F2 = \sum_{t=1}^{T}\sum_{i=1}^{N}C_t^i \tag{3}$$

$$C_t^i = a_{0i}P_t^{i2} + a_{1i}P_t^i + a_{2i} \tag{4}$$

where T is the time period during navigation; t is the time interval; N is the number of onboard DGs; C_t^{ion} denotes the startup cost of the ith generator; C_t^{ioff} denotes the cost of the shutdown of the ith DG; C_t^i denotes the fuel consumption cost of the ith generator; a_{0i}, a_{1i}, and a_{2i} are cost coefficients for the ith generator, respectively; and P_t^i is the output power of the ith DG.

For the convenience of subsequent solving, segmental linearization is used to transform the fuel consumption cost function into a linear function in this paper. By dividing the ith DG to the output P_t^i at the tth time interval into K^i segments, the fuel consumption cost of the DG can be approximated as a linear function in each of the k segments. The expression is shown in Equation (5):

$$\begin{cases} C_t^i = C_0^i + \sum_{k=1}^{K^i} c_{ik} P_t^{ik} \\ P_t^i = P_{\min}^i + \sum_{k=1}^{K^i} P_t^{ik} \\ 0 \leq P_t^{ik} \leq (P_t^i - P_{\min}^i) \end{cases} \tag{5}$$

where c_{ik} is the generation cost coefficient of the ith DG after linearization in section k; P_t^{ik} is the power increment of the ith DG in section k, and C_0^i is the ith DG cost at minimum output power $P_{\min}^i$, which is calculated by the SFC curve and fuel price.

Since the start–stop state of the DG is a discrete variable and the on/off s cost of the unit at each time interval depends on the on/off s state, the start–stop cost is also a discrete function that needs to be linearized as follows:

$$\begin{cases} C_t^{ion} \geq C_{on}^i (o_{t+1}^i - o_t^i) \\ C_t^{ion} \geq 0 \\ C_t^{ioff} \geq C_{off}^i (o_t^i - o_{t+1}^i) \\ C_t^{ioff} \geq 0 \end{cases} \tag{6}$$

where o_t^i is the start–stop state of the ith DG at time t, with 0 indicating that the generator is in the shutdown state and 1 indicating that the generator is in the startup state.

2.2. Constraints

2.2.1. Generation Constraints

1. Power Balance

In a power system, the total power output of the generating unit must be equal to the load power, i.e., the power balance is maintained as follows:

$$\sum_{i=1}^{N} P_t^i + P_t^{Ess} + P_t^{Pv} = P_t^{lp} + P_t^{ls} \tag{7}$$

where P_t^{Ess} is the charging/discharging power of the storage battery t; P_t^{Pv} is the photovoltaic power; P_t^{lp} is the load power of the electric propulsion device; and P_t^{ls} is the load power of the electric service device.

The relationship between ship speed and propulsive load power is as follows:

$$P_t^{lp} = c_1 v^{c_2} \tag{8}$$

where c_1 and c_2 are the proportional and exponential coefficients, respectively; generation scheduling and voyage scheduling are coupled by Equation (8).

2. Generator

The power of DGs is limited by its lower and upper limits, as shown in (9):

$$o_t^i P_{\min}^i \leq P_t^i \leq o_t^i P_{\max}^i \tag{9}$$

where $P_{\min}^i$ is the minimum power; and $P_{\max}^i$ is the rated output power.

The maximal/minimal ramp rates of DGs are limited, as shown in (10):

$$\begin{cases} P_{t+1}^i - P_t^i \leq R_{\max}^i \\ P_t^i - P_{t+1}^i \leq D_{\max}^i \end{cases} \tag{10}$$

3. ESS

State of charge (SOC) refers to the proportion of the remaining capacity to the total capacity, and SOC at any moment is determined by the remaining capacity of the battery at the previous moment and the charging/discharging power, so the dynamic model of the SOC of ESSs at each time step can be described by (12) and (13):

$$SOC_t^{Ess} = \frac{E_t^{Ess}}{E^{Ess}} \tag{11}$$

$$E_t^{Ess} = \begin{cases} E_t^{Ess} - P_t^{Ess}/\eta_{ch}\Delta t & t \neq 1, P_t^{Ess} < 0 \\ E_t^{Ess} - P_t^{Ess}/\eta_{dis}\Delta t & t \neq 1, P_t^{Ess} \geq 0 \end{cases} \tag{12}$$

E_t^{Ess} and E^{Ess} are the remaining capacity and rated capacity of the ESS at the t-th moment, respectively; P_t^{Ess} is the ESS output power at the $t-1$ moment; and η_{ch} and η_{dis} are charging/discharging efficiencies.

The SOC limits during the voyage are expressed as follows:

$$SOC_{\min} \leq SOC_t \leq SOC_{\max} \tag{13}$$

where $SOC_{\min}$ and $SOC_{\max}$ represent the minimal and maximal $SOCs$ of an ESS, respectively.

The power of an ESS is limited by its lower and upper limits, as shown in (14):

$$\begin{cases} P_{ch.\max} < P_{t-1}^{Ess} < 0 & P_{t-1}^{Ess} < 0 \\ 0 \leq P_{t-1}^{Ess} < P_{dis.\max} & P_{t-1}^{Ess} \geq 0 \end{cases} \tag{14}$$

where $P_{ch.\max}$ indicates the maximum charging power of the battery and $P_{dis.\max}$ indicates the maximum discharging power of the battery.

4. Minimum On/Off Time

Once a DG is turned off (or on), it must remain in the same state for a time interval until the next time interval when it is turned on (or off); this is described by (15):

$$t_2 - t_1 \geq T_{on/off} \tag{15}$$

where t_2 and t_1 are any two adjacent on/off state switching (on to off or off to on) times.

5. EEOI

To improve the energy efficiency of ship operation and reduce emissions, EEOI is limited during ship operation as shown in (16) and (17):

$$EEOI = \sum_{i \in N} \sum_{t \in T} \left(\frac{o_i^t \cdot G_{DG}^{i,t} \cdot \Delta t}{m_{AES} Dist_t} \right) \tag{16}$$

$$EEOI \leq EEOI_{\max} \tag{17}$$

where $G_{DG}^{i,t}$ is the carbon dioxide emissions during the operation of the DG, for which the formula is shown in Equation (18); m_{AES} is the weight of the ship cargo, in knots; $Dist_t$ is nautical miles of voyage; and $EEOI_{\max}$ is the maximum allowed.

$$G_{DG}^{i,t}(t) = g_{0,i}(r_{DG}^{i,t}(t))^2 + g_{1,i}(r_{DG}^{i,t}(t)) + g_{2,i} \tag{18}$$

Here, $g_{1,i}$, $g_{2,i}$, and $g_{3,i}$ are the calculation factors for CO_2 emissions from DGs.

6. Action Spinning Reserves Constraints

To ensure the stable operation of the ship's power system, it is necessary to consider the active spinning reserve; the active spinning reserve mathematical model for DGs and $ESSs$ is shown in (19) and (20), and the total active spinning reserve is shown in Equation (21):

$$R_{DG}(t) = \sum\nolimits_{n \in N} (P_{DG,\max} - P_{DG,i}(t)), \forall t \in T \tag{19}$$

$$R_{ESS}(t) = \begin{cases} P_{ESS}^{\max} - P_{ESS}(t), P_{ESS}(t) \geq 0 \\ P_{ESS}^{\max}, P_{ESS}(t) < 0 \end{cases}, \forall t \in T \tag{20}$$

$$\begin{cases} R_R(t) = R_{DG}(t) + R_{ESS}(t) \\ R_R(t) \geq \delta_R \cdot N_{DG} \cdot P_{DG,\max} \end{cases}, \forall t \in T \tag{21}$$

where R_{DG}, R_{ESS}, and R_R are the active spinning reserves of *DGs* and *ESSs* and the total active spinning reserve, respectively; δ_R is the spinning reserve coefficient; and N_{DG} is the number of diesel generators placed.

2.2.2. Voyage Constraints

The ships sail between multiple ports, which contain a variety of operating conditions such as berthing, cruising, and docking, with different speed requirements, and the speed constraints are shown in Figure 2 [14]. The black arrows indicate the speed limits; in each interval, the cruising speeds should be within the upper and lower bounds. The blue arrows indicate the voyage distance limit, meaning that the ship must arrive at the next port within the specified time from the previous port. Equations (22)–(24) are the cruising speed constraints during the cruising intervals and the partial speed intervals.

$$(1 - \delta_v)v_n \leq v_t \leq (1 + \delta_v)v_n \quad \forall t \in T_c \tag{22}$$

$$\eta_v(1 - \delta_v)v_n \leq v_t \leq \eta_v(1 + \delta_v)v_n \quad \forall t \in T_d \tag{23}$$

$$v_t = 0 \quad \forall t \in T_b \tag{24}$$

Here, T_c is the cruising intervals, T_d is the partial speed interval, T_b is the berthing interval; v_n is the rated speed; δ_v is the rate of change of v_n, and when the ship is cruising, the ship speed is shown in (22); and η_v is the ratio between v_t and v_n, and when the ship is sailing at a partial speed, the constraint is shown in (23).

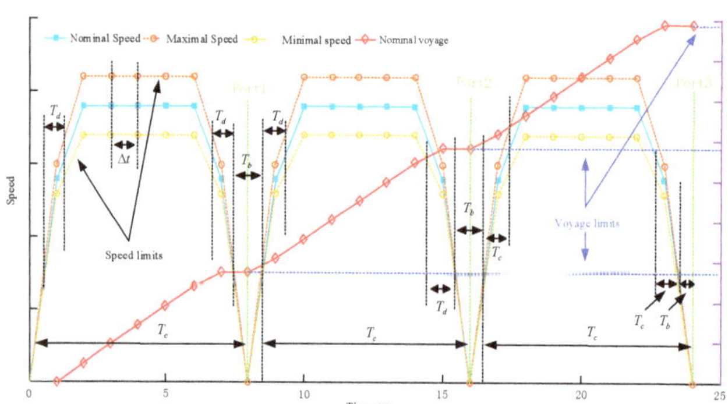

Figure 2. Typical voyage pattern.

Since the ship needs to arrive at the port on time, the voyage distance to the intermediate and terminal ports is required, respectively, as shown in (25) and (26):

$$(1 - \delta_{mid})Dist_n^t \leq Dist^t \leq (1 + \delta_{mid})Dist_n^t \tag{25}$$

$$Dist_n^{nT} \leq Dist^{nT} \leq (1 + \delta_{term})Dist_n^{nT} \tag{26}$$

where *Dist* is the actual sailing distance of the ship; and δ_{mid} and δ_{term} are the maximum permissible distance error for intermediate and terminal ports, respectively.

2.3. Wave and Wind Resistance

Wind resistance and wave resistance, combined with the calm water resistance of the ship, can be used to obtain the speed loss of the ship in different sea conditions.

Calm water resistance is mainly composed of frictional resistance R_f, residual resistance R_r, and attached resistance R_{ap} [18,19], as shown in (27):

$$R_T = R_f + R_r + R_{ap} \tag{27}$$

The International Organization for Standardization (ISO) has constructed a mathematical model for calculating wind resistance and wave resistance based on the data obtained from wind tunnel model tests [20,21], as follows:

$$R_{wind} = \frac{1}{2}\rho_{air}S_t C_{air}\left[(v_t + v_w \cos\theta_t)^2 - v_t^2\right] \tag{28}$$

where R_{wind} is the wind resistance; ρ_{air} is the air density, $\mathrm{kg/m^3}$; S_t is the lateral projected area above the waterline, $\mathrm{m^2}$; C_{air} is the wind resistance coefficient; v_w is the wind speed, m/s; and θ_t is the wind direction.

The wave resistance is calculated using (29):

$$R_{wave} = \frac{1}{L_s}\rho_{water} g h_t^2 B_{\text{int}}^2 C_{D,wat}(\tau_t, \theta_t) \tag{29}$$

where g is the acceleration of gravity, $\mathrm{m/s^2}$; h_t is the wave height, m; B_{int} is the ship's width, m; and $C_{D,wat}$ is the wave resistance coefficient, with the wavelength τ_t and the wind direction θ_t. When the ship is in a certain sea area, the wind direction and wave direction can be approximated to be the same, as both are θ_t.

It should be noted that the variables v_w and h_t also depend on the season and geographic position. For example, in the South China Sea region, v_w may be large and h_t may be tall in winter, thus R_{wind} and R_{wave} are large.

The total resistance (R_c) due to wind and waves can be estimated using Equation (30):

$$R_c = R_T + R_{wind} + R_{wave} \tag{30}$$

The ship speed loss is estimated based on the assumption that the required power P_t at the ship speed in calm water is the same as the required power P_t' in the specific sea condition, as follows:

$$P_t = \frac{R_T \cdot v_t}{\eta_d \eta_s}, P_t' = \frac{R_c \cdot v_t'}{\eta_d \eta_s} P_t = P_t' \tag{31}$$

where η_d and η_s are the propulsion and transmission efficiencies.

Then, the cruising speed with speed loss can be calculated using (32):

$$v_t' = \frac{R_T}{R_c} v_t = \frac{R_T}{R_c} \sqrt[c_2]{\frac{P_t}{c_1}} \tag{32}$$

3. Uncertainty Model

3.1. Uncertainty Variables

In this paper, the uncertain PV power generation P_t^{Pv} and propulsion load power P_t^{lp} associated with speed are expressed as the sum of the expected value and a random term for forecast error $\widetilde{P}_t$, given as

$$P_t^{Pv} = \overline{P}_t^{Pv} + \widetilde{P}_t^{Pv} \tag{33}$$

$$P_t^{lp} = \overline{P}_t^{lp} + \widetilde{P}_t^{lp} \tag{34}$$

where $\overline{P}_t^{Pv}$ and $\overline{P}_t^{lp}$ denote the expected value of PV power generation and propulsion load power at t time interval, respectively, and $\widetilde{P}_t^{Pv}$ and $\widetilde{P}_t^{lp}$ denote the random terms of the forecast error.

The next section will describe the process of a fuzzy set for $\widetilde{P}_t^{Pv}$ and $\widetilde{P}_t^{lp}$. For the convenience of the study, the uncertainty parameter matrix θ is used, where $\theta = [\widetilde{P}_t^{Pv}, \widetilde{P}_t^{lp}]$.

3.2. Fuzzy Set Construction

This section focuses on modeling the random term of PV power and the propulsive load power. According to the historical data of the random variable, as expected, the upper bounds and lower bounds of a random variable are known. The mathematical expression of the fuzzy set Ω constructed in this paper based on the first-order deviation moment function is as follows:

$$\Omega = \left\{ P \in \Im \middle| \begin{array}{c} \mathrm{E}_P\{\theta\} = \mu \\ P\{\theta \in D\} = 1 \\ \mathrm{E}_P\{g_j(\theta)\} \leq \sigma_j, \forall j \in [j] \end{array} \right\} \tag{35}$$

P is the distribution of a random variable, and $\Im$ denotes the set of all probability distributions.

The first line suggests that the expectation of random variable θ is μ, which can be obtained from the historical data; the second line suggests that all the values of random variable θ are within the uncertainty set D, defined the same as the conventional robust optimization. In this paper, the box-type uncertainty set is adopted to limit the range of the fluctuation of the random variable; for $D = \{\underline{\theta} \leq \theta \leq \overline{\theta}\}$, $\overline{\theta}$ and $\underline{\theta}$ are the upper and lower bounds, respectively. $P(g)$ is the probability function; the third line represents the uncertainty of the first-order moment information of θ by adding the first-order deviation moment function $g_j(\theta)$ to the fuzzy set. $g_j(\theta)$ can be approximated to a segmented linear form, $g_j(\theta) = \max\{\theta - \varphi_j, 0\}$, which means the first-order statistical information deviates from a certain given value φ_j, σ_j is a cut-off constant, which can be estimated from historical statistics, j is the index of segments describing the distribution of each random variable, and $[j]$ is a set of segments describing the distribution of each random variable.

Since it is difficult to directly evaluate the expectations for each $g_j(\theta)$ under an uncertain distribution, to facilitate the equivalent transformation of the model, the auxiliary variable u is introduced into the model in this paper [17]. With the extension of the original fuzzy set, this is expressed as follows:

$$\mathrm{G} = \left\{ \mathrm{Q} \in \mathrm{B} \middle| \begin{array}{c} \mathrm{E}_Q\{\theta\} = \mu \\ P\{(\theta, \mathrm{u}) \in H\} = 1 \\ \mathrm{E}_Q\{\mathrm{u}_j\} \leq \sigma_j, \forall j \in [j] \end{array} \right\} \tag{36}$$

$\mathbb{H}$ is the new extended uncertainty set, as shown in (37):

$$\mathbb{H} = \left\{ \begin{array}{c} \overline{\theta} \leq \theta \leq \underline{\theta} \\ u_j \geq \theta - \varphi_j, \forall j \in [j] \\ u_j \geq 0, \forall j \in [j] \end{array} \right\} \tag{37}$$

4. Solution Method

4.1. Matrix Form of the Model

To clearly demonstrate the solution of the proposed distributionally robust nonlinear programming (DRONLP), the compact forms are shown in Equations (38)–(45):

$$Q(x_t, \theta) = \max_{\theta} \min_{x_t} \left[\sum_{t=1}^{T} \sum_{n=1}^{N} (C_t^i + C_t^{ion} + C_t^{ioff}) \right] \tag{38}$$

$$s.t.\ \mathcal{M}_{IEQ} \cdot x_t = m_{IEQ}, \forall t \in \mathrm{T} \tag{39}$$

$$\mathcal{M}_{IEQ} \cdot x_t \leq m_{IEQ}, \forall t \in \mathrm{T} \tag{40}$$

$$C\theta + Du \leq d \tag{41}$$

$$1_a^T \cdot x_t = c_1 \cdot (y_t)^{c_2}, \forall t \in \mathrm{T} \tag{42}$$

$$\mathcal{Y}_{min} \leq \mathcal{Y} = [y_1, \cdots, y_{\mathrm{T}}]^T \leq \mathcal{Y}_{max} \tag{43}$$

$$\mathrm{D_{min}} \leq \mathrm{V}_T \cdot \begin{bmatrix} k_v(u_1) & \mathrm{L} & 0 \\ \mathrm{M} & \mathrm{O} & \mathrm{M} \\ 0 & \mathrm{L} & k_v(u_T) \end{bmatrix} \cdot \mathrm{Y} \leq \mathrm{D_{max}} \tag{44}$$

$$(\theta, u) \in \mathbb{H} \tag{45}$$

where $x_t = [(o_{1,t}^{DG}, \cdots, o_{N,t}^{DG})^T, (P_{1,t}^{DG}, \cdots, P_{N,t}^{DG})^T, P_t^{ESS}, P_t^{Pv}, P_t^{PL}]$ is the generation variable; $y_t = v_t^c$ is the speed variable; and $Q(x_t, \theta)$ is the generator fuel consumption cost and the on/off cost. Equations (39) and (40) represent linear equation and inequality constraints; Equation (41) is the matrix form of the extended fuzzy set; Equation (42) is the relationship between propulsion load and speed; Equation (43) is the sailing speed constraint; Equation (44) is the sailing distance constraint; and Equation (45) is the uncertain parameter set constraint. All of the above coefficient matrices are derived from the original model.

4.2. Bi-Level Formulation of Proposed Model

Equations (38)–(43) are two scheduling problems, i.e., power generation scheduling and voyage scheduling. The two scheduling problems are coupled through Equation (42), i.e., the relationship between propulsion load and cruising speed. If this coupling constraint is removed, the original problem can be decomposed into two levels as shown in Equations (46)–(48) [14]:

$$master : Q(x_t, \theta) \tag{46}$$

$$\begin{gathered} s.t.\ \mathcal{M}_{IEQ} \cdot x_t \leq m_{IEQ}, \forall t \in \mathrm{T} \\ C\theta + Du \leq d \\ 1_a^T \cdot x_t = c_1 \cdot (y_t)^{c_2}, \forall t \in \mathrm{T} \end{gathered} \tag{47}$$

$$slave : y_t^* = \mathrm{argmin}\left(\left| \begin{matrix} D_{\max} - \mathrm{V}_T \cdot K_v \cdot \mathrm{Y} \\ \mathrm{V}_T \cdot K_v \cdot \mathrm{Y} - D_{\min} \end{matrix} \right| : (44), (45) \right) \tag{48}$$

In the above formulation, x_t and y_t are determined in different levels, and $Q(x_t, \theta)$ becomes related to x_t only, while y_t is considered as a constant in the master problem, which is updated by the slave level. For the slave level, y_t is obtained by minimizing the voyage deviation. With this decomposition, Equations (46) and (47) are quadratic programming, and Equation (48) is linear programming. The slave level can be solved directly by commercial solvers; the master level is described in the next section.

4.3. Linearization for the Master Level

4.3.1. Two-Stage Optimization Model

The master level model can be transformed into a two-stage optimization problem, and the variables in the master level are decomposed into first-stage variables and second-stage variables. The first-stage variables are decision variables that do not change with uncertainties, denoted by x_1, and before the power of PV and propulsion loads are determined, the on/off plan of generating units will be formulated first; the second-stage variables will be affected by uncertainties, denoted by x_1, and the power of each generating unit will be optimized according to the actual power of PV and propulsion loads under the constraints.

The first-stage optimization problem can be formulated as shown in (49) and (50):

$$\min c^T x_1 + Q(x_1, \theta) \tag{49}$$

$$s.t.\ A \cdot x_1 \leq b \tag{50}$$

The second-stage objective function $Q(\mathbf{x}_1, \theta)$ can be represented by the second-stage variable x_2, as follows:

$$Q(x_1, \theta) = \min \mathrm{q}^T x_2 \tag{51}$$

$$s.t. \quad Nx_1 + Mx_2 \leq h(\theta) \tag{52}$$

The function $h(\theta)$ contains the random variable θ, which will be affected by uncertainty, as follows:

$$h(\theta) = h^0 + \sum_{s=1}^{S} h_s^\theta \theta_s \tag{53}$$

where h^0 denotes the constant coefficient column vector and h_s^θ denotes the column vector associated with θ_s.

4.3.2. Two-Stage Distributional Robust Optimization Model

Based on the fuzzy sets, the previous two-stage optimization model can be written as a two-stage distributionally robust optimization model, as follows:

$$\begin{gathered} \min c^T x_1 + \max_{\mathbb{Q} \in \mathbb{G}} \mathbb{E}_{\mathbb{Q}} \{Q(x_1, \theta)\} \\ s.t.\ A \cdot x_1 \leq b \end{gathered} \tag{54}$$

The objective function minimizes the expectation of function $Q(x_1, \theta)$ under a distribution P. The objective function minimizes the expectation of the function Q under the probability distribution $\mathbb{Q}$, which is the worst distribution on the fuzzy set $\mathbb{G}$. The function $Q(x_1, \theta)$ represents the economic dispatch cost associated with x_1 under a random variable θ, which can be solved by the second-stage problem.

The second-stage objective function $\max\limits_{\mathbb{Q} \in \mathbb{G}} \mathbb{E}_{\mathbb{Q}} \{Q(x_1, \theta)\}$ can be written in a general matrix form, as follows:

$$\begin{gathered} \max_{\mathbb{Q} \in \mathbb{G}} \mathbb{E}_{\mathbb{Q}} \{Q(x_1, \theta)\} = \min \max_{\mathbb{Q} \in \mathbb{G}} \mathbb{E}_{\mathbb{Q}} \{q^T x_2\} \\ s.t. \quad Nx_1 + Mx_2 \leq h(\theta) \end{gathered} \tag{55}$$

However, solving the second-stage decision variable x_2 based on all realizations θ within fuzzy set G is an NP-hard problem. A linear decision rule is an efficient way to deal with this y intractable problem [22,23], which restricts decision variable x_2 affine dependence on the uncertain parameter θ.

In this paper, a linear decision rule model is adopted which approximates $x_2(\theta, u)$ by a linear affine function of random variables θ and auxiliary variables u.

The linear affine function $x_2(\theta, u)$ can be expressed as (56):

$$x_2(\theta, u) = {x_2}^0 + \sum_{s=1}^{S} x_{2s}^\theta \theta_s + \sum_{j=1}^{J} \sum_{l=1}^{L} {x_2}_{jl}^u u_{jl} \tag{56}$$

By converting the second-stage decision variable x_2 to a linear affine function $x_2(\theta, u)$, the original second-stage optimization problem can also be re-represented as follows:

$$\min \max_{\mathbb{Q} \in \mathbb{G}} \mathbb{E}_{\mathbb{Q}} \left\{ q^T x_2(\theta, u) \right\} \tag{57}$$

$$s.t. \quad Nx_1 + Mx_2(\theta, u) \leq h(\theta) \tag{58}$$

Instead of solving the recourse problem under all outcomes of uncertain parameters, the reformulated problem attempts to search for optimal coefficients of the decision rule $x_2(\theta, u)$ such that the worst-case expectation is minimized under all second-stage constraints. A detailed proof is given in Appendix A in the literature [17].

In the second stage, the inner maximization problem $\max\limits_{\mathbb{Q}\in\mathbb{G}} \mathbb{E}_{\mathbb{Q}}\{q^T x_2(\theta, u)\}$ can be expressed as a semi-infinite optimization problem, the problem is further transformed into a finite-dimensional optimization problem by Lagrange duality, and the original two-layer min-max optimization problem is equivalently reduced to a minimization problem, as shown in Equations (59)–(62):

$$\min \eta + \gamma^T \lambda \tag{59}$$

$$\text{s.t. } \lambda \geq 0 \tag{60}$$

$$\eta + \theta^T \rho + u^T \lambda \geq q^T x_2(\theta, u) \tag{61}$$

$$Nx_1 + Mx_2(\theta, u) \leq h(\theta) \tag{62}$$

After the equivalent transformation of the optimization model, constraints (61) and (62) are still infinite-dimensional linear constraints, they can be transformed based on dual theory, and the model of the second-stage problem is as shown in Equations (63)–(72):

$$\min \eta + \gamma^T \lambda \tag{63}$$

$$s.t. \ \lambda \geq 0 \tag{64}$$

$$\pi_0 \geq 0 \tag{65}$$

$$\pi_m \geq 0 \tag{66}$$

$$\eta - q^T {x_2}^0 + \pi_0^T d \geq 0 \tag{67}$$

$$\pi_0^T C_s = \sum_{n \in N_s^v} q_n x_{2ns}^\theta - \rho_s = q^T x_{2s}^\theta - \rho_s \tag{68}$$

$$\pi_0^T D_i = \sum_{n \in N_i^u} q_n x_{2ni}^u - \lambda_i = q^T x_{2i}^u - \lambda_i \tag{69}$$

$$N_k^T x_1 + M_k^T {x_2}^0 - h_k^0 + d^T \pi_m \leq 0 \tag{70}$$

$$\pi_m^T C_s = \sum_{n \in N_s^v} W_{mn} x_{2ns}^\theta - h_{ms}^\theta = W_m^T x_{2s}^\theta - h_{ms}^\theta \tag{71}$$

$$\pi_m^T D_i = \sum_{n \in N_i^u} W_{mn} x_{2ni}^u = W_m x_{2i}^u \tag{72}$$

After the above series of transformations, the original two-stage distributional robust optimization model is transformed into a mixed-integer linear programming model, objective function is as (73), The constraints include (50) and (64)–(72).

$$\min c^T x_1 + \eta + \gamma^T \lambda \tag{73}$$

As mentioned above, the original problem is decomposed into master and slave levels, where the master level is transformed into a mixed-integer linear programming problem by linear decision rule. The overall process of the proposed solution method is shown in Figure 3.

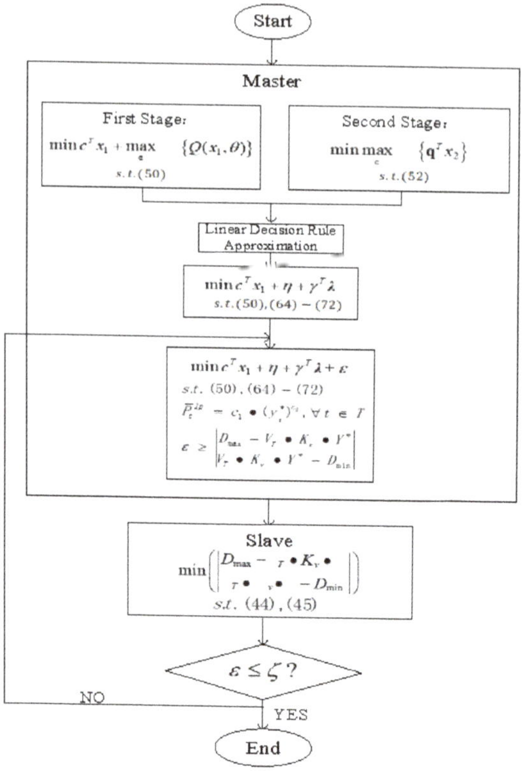

Figure 3. Solution method flow chart.

5. Case Study

5.1. Simulation Parameters

In this study, four DGs, PV, and ESSs were considered for simulations, and the cases had two intermediate ports and one terminal port, as shown in Figure 4. The total voyage is 24 h; ship cruising occurred during t = 1–7, 9–15, and 17–23, and staying in a port occurred during t = 8, 16, and 24, and furthermore when t = 1, 7, 9, 15, 17, and 23,to avoid outage events, a DG remained closed until it needed to be turned on.

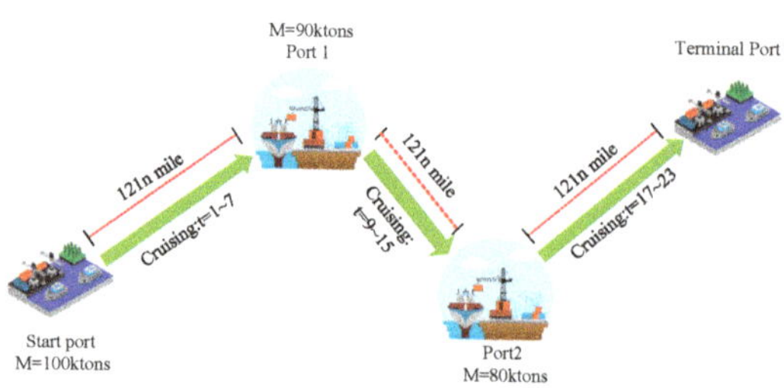

Figure 4. Navigational data for the simulation.

The service load is shown in Table 1 and the basic parameters of DGs and *ESSs* are shown in Table 2 [14,24]. The parameters of the voyage scheduling are shown in Table 3. The predicted mean value and upper and lower bounds of PV power and propulsion power are shown in Figure 5. The capacity of the ESS is 30 MWh, and the rated power is

15 MW. The rated cruising speed is 19 knots. Environmental data along the cruise route were downloaded from the ERA5 global reanalysis data set [25].

Table 1. Electricity service load power.

Time interval	1	2	3	4	5	6	7	8	9	10	11	12
Service load	8.96	8.92	9.43	10.97	8.45	8.52	7.96	8.54	8.01	7.97	8.95	10.82
Time interval	13	14	15	16	17	18	19	20	21	22	23	24
Service load	8.91	7.93	8.99	8.35	8.9	7.46	9.67	10.34	8.51	7.62	7.38	9

Table 2. Diesel generator specifications.

DG	P_{DGn} (MW)	P_{DGmin} (MW)	r_u, r_d (%/Δt)	$a_{0,1,2}$ ($m.u./p.u.$)	$g_{0,1,2}$ ($uCO_2/p.u.$)
DG1	15	4	+/−50%	13.5, 10, 450	386, −2000, 8383
DG2	15	4	+/−50%	13, 12, 430	386, −2000, 8383
DG3	15	4	+/−50%	13.5, 12, 460	363, −650, 950
DG4	15	4	+/−50%	5.6, 58, 390	125, 450, 850
ESS		$\eta_{ch} = 95\%$		$\eta_{dis} = 97\%$	

Table 3. Parameters of voyage scheduling.

$l = 175$ m	$B_{int} = 25.4$ m	$S_t = 274$ m	$\eta_d = 98\%$	$\eta_s = 0.7$	$P_t^{lp} = 0.003{v_t}^3$
$h_t \in [0, 4.5]$ m	$\tau_t \in [0, 1.5]$ m	$v_w \in [0, 15.5]$ m/s	$\theta_t \in [30, 180]$ deg	$C_{air} = 0.025$	$\rho_{air} = 1.29 \times 10^{-3}$ g/cm^3
$\rho_{water} = 1.04$ g/cm^3	$C_{D,wat} = 0.041$ [26]	$g = 9.8$ m/s^2			

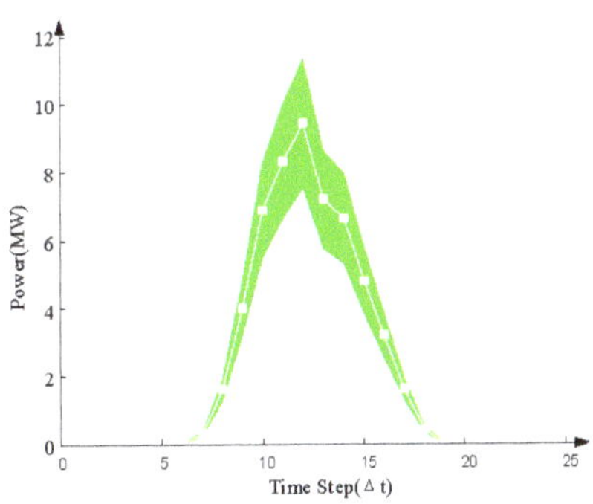

(**a**) predicted PV power

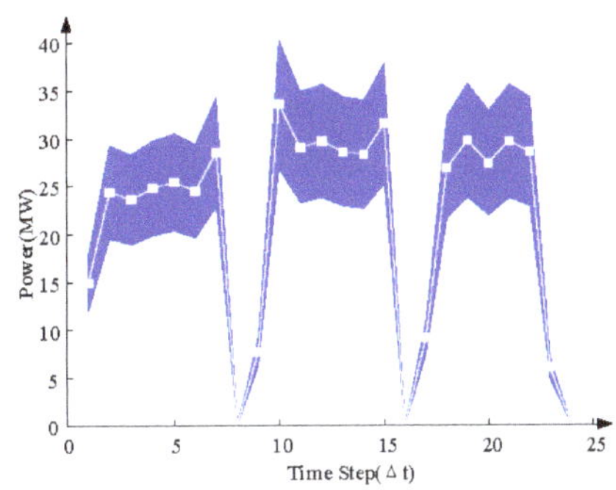

(**b**) predicted propulsion power

Figure 5. Uncertainty variables.

5.2. Analysis of Results

5.2.1. On-Time Rates Under Different Methods

In this study, the speeds of the three methods are analyzed as shown in Figure 6, where RO is robust optimization, DRO is distributionally robust optimization, NR is without considering wind and wave resistance, and Nom denotes the nominal speed. In Figure 6, the speeds of both RO and DRO are significantly higher than those in the case of NR, so both can cover the speed loss caused by wind and waves and ensure that the ship arrives at the port on time. On the other hand, the speed in NR is obviously lower than the rated speed, which makes the ship fail to arrive at the harbor on time.

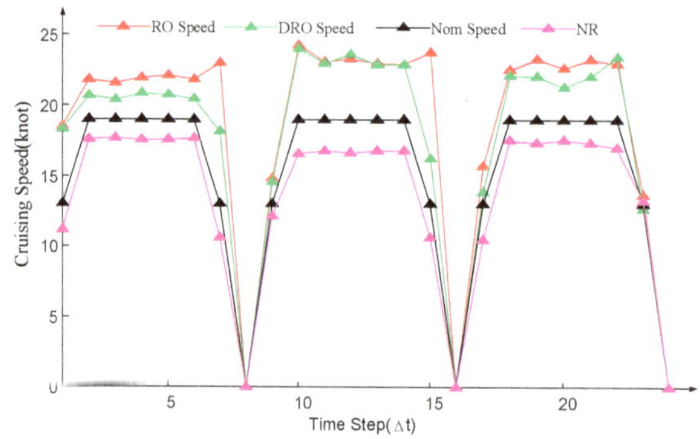

Figure 6. Comparison of speeds for different scenarios.

When t = 1~7 and 17~23, due to DRO not being as conservative as the RO for the uncertainty, the cruise speed is relatively low compared to the RO, and the power of the propulsion load is also lower at these intervals, so the generator output power is reduced accordingly, and so DRO improves the ship's energy-saving and emission reduction capabilities.

5.2.2. Analysis of Economic Costs

Figure 7 shows the fuel consumption for each time step under four different methods, Due to PV being used as supplementary energy during t = 10–15 periods, fuel consumption during t = 10–15 is lower than at other times, so the integration of the PV system can reduce ship operating costs. In NR, the fuel consumption is lowest because the impact of wind and waves is not considered, but it also causes the ship to fail to arrive at the port on time. The RO and DRO consider the uncertainty; therefore, fuel consumption is higher than the Nom Speed scenario. It can also be seen from Figure 8 that the fuel consumption of DRO is always lower than that of RO.

Table 4 shows the total fuel consumption for four different methods, and it can be clearly seen that DRO generally outperforms RO in yielding lower expected costs.

Figure 8 shows that two units are turned on during most of the navigation time for the DRO method since the two units are enough to follow wave, wind, and PV fluctuations at a minimum cost. But for the RO method, three units are turned on for most of the navigation time, which requires more economic costs. Therefore, compared to RO methods, DRO methods greatly reduce the conservatism of decision-making and lower scheduling costs.

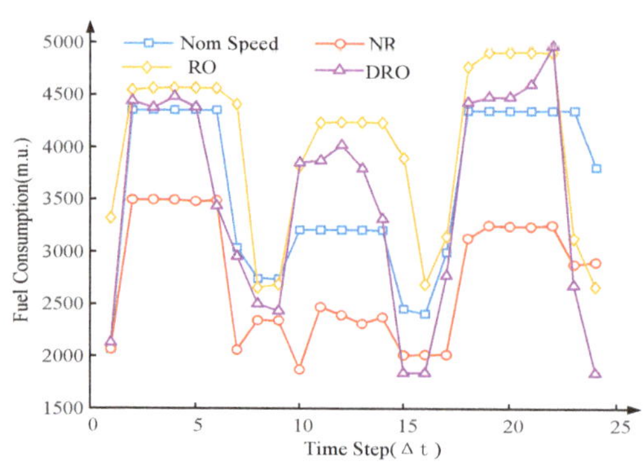

Figure 7. Fuel consumption for 24 time intervals.

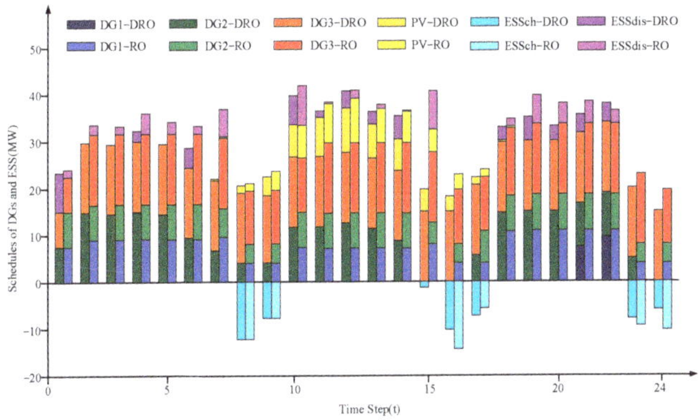

Figure 8. Scheduling schemes of DGs and ESSs.

Table 4. Total fuel consumption for different methods.

	Total Fuel Consumption (m.u.)
DRO	70,222
RO	79,950
NR	54,357
Nom Speed	64,890

5.2.3. Analysis of Sensitivity

When the uncertain variable fluctuates in the range of ±15~±35% of the expected value, it can be seen from Figure 9 that the total cost of DRO is generally lower than that of RO, and with an increase in interval fluctuation, the advantage of DRO is more obvious. This is because the decision made by the traditional RO depends on the worst case, while the DRO can simulate the distribution of uncertain variables according to the probability information contained in the fuzzy set, which reduces the conservative decision.

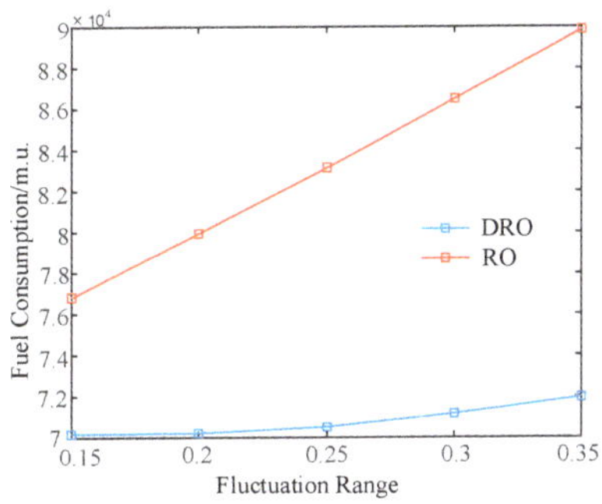

Figure 9. Fuel consumption comparison with an increase in fluctuation.

6. Conclusions

To address the effects of uncertain wave and wind conditions in the navigation route, this paper proposes a DRO formulation for solving the joint scheduling model of power and voyage, aiming at operation cost minimization and GHG emission limitation. The relationship between propulsion load and voyage as influenced by navigation speed and waves is first established, and then a DRO model is established. Due to the complexity of the examined model, the problem is decoupled into a bi-level optimization model, the slave level can be solved directly by commercial solvers, the master level is further formulated as a two-stage DRO model, and linear decision rules and column and constraint generation algorithms are adopted to solve the model. With the exploitation of the proposed method, joint scheduling is achieved with excellent results. The obtained simulation results confirm the effectiveness of the method, and compared with the conventional robust optimization methods, less conservative results are obtained. Expected operation costs are minimized and emissions are reduced while ensuring the on-time arrival of hybrid ships in various uncertain scenarios. Future efforts might be spent on the consideration of incorporating risk into the model to better guide operators.

Author Contributions: Conceptualization, F.L.; methodology, F.L.; software, Y.T.; validation, Y.T.; formal analysis, H.L.; investigation, H.L.; resources, Y.T.; data curation, C.L.; writing—original draft preparation, Y.T.; writing—review and editing, F.L.; visualization, C.L.; project administration, H.L.; funding acquisition, H.L. All authors have read and agreed to the published version of the manuscript.

Funding: This research was funded by the Natural Science Foundation of Heilongjiang Province, China, grant number LH2022E039, the Natural Science Foundation of Shandong Province, China, grant number ZR20210303051, and Key Science and Technology Research Project of Shandong Province (2022CXGC020403).

Institutional Review Board Statement: Not applicable.

Informed Consent Statement: Not applicable.

Data Availability Statement: Data are contained within the article.

Conflicts of Interest: The authors declare no conflicts of interest.

Nomenclature

Acronyms	
DRO	distributionally robust optimization
PV	photovoltaics
GHG	greenhouse gas
SOC	state of charge
ESS	energy storage system
DG	distributed generator
ISO	International Organization for Standardization
DRONLP	distributionally robust nonlinear programming
Sets and indices	
T, t	index and set of time periods
N, i	index and set of generator units
K^i, k	index and set of segments
Parameters	
C_t^{ion}	startup cost of the ith generator at time interval t
C_t^{ioff}	shutdown cost of the ith generator at time interval t
a_{0i}, a_{1i}, a_{2i}	cost coefficients for the ith generator, respectively
c_{ik}	ith DG cost coefficient after linearization in section k
C_0^i	ith DG cost at minimum output power $P_{\min}^i$
o_t^i	start–stop state of the ith DG at time interval t
c_1, c_2	proportional and exponential coefficients
$P_{\min}^i$	minimum power
$P_{\max}^i$	rated output power
E^{Ess}	rated capacity of ESS
η_{chh}, η_{dis}	charging/discharging efficiency
$P_{ch\cdot\max}, P_{dis\cdot\max}$	maximum charging/discharging power of the battery
m_{AES}	weight of the ship cargo
$Dist_t$	miles of voyage
$EEOI_{\max}$	maximum allowed
$g_{1,i}, g_{2,i}, g_{3,i}$	calculation factors for CO_2 emissions from DGs
R_{DG}, R_{ESS}, R_R	active spinning reserve of DGs, ESS, total
δ_R	spinning reserve coefficient

T_c, T_d, T_b	cruising intervals, partial speed intervals, and berthing intervals
v_n	rated speed
δ_v	rate of change of v_n
$Dist$	actual sailing distance of the ship
δ_{mid}, δ_{term}	maximum permissible distance error for intermediate and terminal port
ρ_{air}	air density
S_t	lateral projected area above the waterline, m^2
C_{air}, v_w, θ_t	wind resistance coefficient; wind speed, wind direction
h_t, $C_{D,wat}$, τ_t	wave height, wave resistance coefficient, wavelength
η_d, η_s	propulsion and transmission efficiency
$\widetilde{P}_t^{Pv}$, $\widetilde{P}_t^{lp}$	random term of the forecast error
Variables	
$F1$	startup and shutdown cost of the DG units
$F2$	operation cost of the DG units
P_t^i	output power of the ith DG at time interval t
P_t^{Ess}	charging/discharging power of the storage battery t
P_t^{Pv}	photovoltaic power
P_t^{lp}	load power of the electric propulsion device
P_t^{ls}	load power of the electric service device
E_t^{Ess}	remaining capacity at the t-th time interval
$G_{DG}^{i,t}$	carbon dioxide emissions during the operation of DG
R_f, R_r, R_{ap}, R_{wind}	frictional resistance, residual resistance, attached resistance, and wind resistance
$\overline{P}_t^{Pv}$, $\overline{P}_t^{lp}$	expected value of PV power and propulsion load power

References

1. Wang, Z.; Chen, L.; Wang, B.; Huang, L.; Wang, K.; Ma, R. Integrated optimization of speed schedule and energy management for a hybrid electric cruise ship considering environmental factors. *Energy* **2023**, *282*, 128795. [CrossRef]
2. Chen, X.; Chen, W.; Wu, B.; Wu, H.; Xian, J. Ship visual trajectory exploitation via an ensemble instance segmentation framework. *Ocean Eng.* **2024**, *313*, 119368. [CrossRef]
3. IMO. MEPC.1/Circ.896. In *Guidance on Treatment of Innovative Energy Efficiency Technologies for Calculation and Verification of the Attained EEDI and EEXI*; IMO: London, UK, 2021.
4. Bayraktar, M.; Yuksel, O. A scenario-based assessment of the energy efficiency existing ship index (EEXI) and carbon intensity indicator (CII) regulations. *Ocean Eng.* **2023**, *278*, 114295. [CrossRef]
5. Hou, H.; Gan, M.; Wu, X.X.; Xie, K.; Fan, Z.Y. A review of energy management research on hybrid ships. *China Ship Res.* **2021**, *16*, 14.
6. Wang, X.; Zhu, H.; Luo, X.; Chang, S.; Guan, X. An energy dispatch optimization for hybrid power ship system based on improved genetic algorithm. *Proc. Inst. Mech. Eng. Part A J. Power Energy* **2023**, *238*, 348–361.
7. Wen, S.; Zhao, T.; Tang, Y.; Xu, Y.; Zhu, M.; Huang, Y. A Joint Photovoltaic Dependent Navigation Routing and Energy Storage System Sizing Scheme for More Efficient All-Electric Ships. *IEEE Trans. Transp. Electrif.* **2020**, *6*, 1279–1289. [CrossRef]
8. Yuan, C.; Pan, P.; Sun, Y.; Yan, X.; Tang, X. The evaluating on EEDI and fuel consumption of an inland river 800PCC integrated with solar photovoltaic system. *J. Mar. Eng. Technol.* **2019**, *20*, 77–92. [CrossRef]
9. Kanellos, F.D.; Anvari-Moghaddam, A.; Guerrero, J.M. A cost-effective and emission-aware power management system for ships with integrated full electric propulsion. *Electr. Power Syst. Res.* **2017**, *150*, 63–75. [CrossRef]
10. Kanellos, F.D.; Anvari-Moghaddam, A.; Guerrero, J.M. Smart shipboard power system operation and management. *Inventions* **2016**, *1*, 22. [CrossRef]
11. Fang, S.; Xu, Y.; Li, Z.; Zhao, T.; Wang, H. Two-Step Multi-Objective Management of Hybrid Energy Storage System in All-Electric Ship Microgrids. *IEEE Trans. Veh. Technol.* **2019**, *68*, 3361–3373. [CrossRef]
12. Li, Z.; Xu, Y. Temporally-coordinated optimal operation of a multi-energy microgrid under diverse uncertainties. *Appl. Energy* **2019**, *240*, 719–729. [CrossRef]
13. Li, Z.; Xu, Y.; Fang, S.; Zheng, X.; Feng, X. Robust Coordination of A Hybrid AC/DC Multi-Energy Ship Microgrid with Flexible Voyage and Thermal Loads. *IEEE Trans. Smart Grid* **2020**, *11*, 2782–2793. [CrossRef]
14. Fang, S.; Xu, Y. Multi-objective robust energy management for all electric shipboard microgrid under uncertain wind and wave. *Int. J. Electr. Power Energy Syst.* **2020**, *117*, 105600. [CrossRef]
15. Fan, F.; Aditya, V.; Xu, Y.; Cheong, B.; Gupta, A.K. Robustly coordinated operation of a ship microgird with hybrid propulsion. systems and hydrogen fuel cells. *Appl. Energy* **2022**, *312*, 118738. [CrossRef]
16. Zhai, J.; Zhou, M.; Li, J.; Zhang, X.; Li, G.; Ni, C.; Zhang, W. Hierarchical and robust scheduling approach for vsc-mtdc meshed ac/dc grid with high share of wind power. *IEEE Trans. Power Syst.* **2021**, *36*, 793–805. [CrossRef]

17. Xiong, P.; Jirutitijaroen, P.; Singh, C. A Distributionally Robust Optimization Model for Unit Commitment Considering Uncertain Wind Power Generation. *IEEE Trans. Power Syst.* **2017**, *32*, 39–49. [CrossRef]
18. Krata, P.; Szlapczynska, J. Ship weather routing optimization with dynamic constraints based on reliable synchronous roll prediction. *Ocean Eng.* **2018**, *150*, 124–137. [CrossRef]
19. Shao, S.M.; Zhao, L.E.; Zhu, N.C. *Ship Resistance*; National Defense Industry Press: Beijing, China, 1995.
20. Zhang, H.S.; Wei, Y.B.; Liu, W.Y.; Jiang, S.F. An optimization method of ship's fixed course speed taking into account the wind and wave factors at sea. *Radio Eng.* **2022**, *52*, 724–730.
21. Liu, S.; Shang, B.; Papanikolaou, A. On the resistance and speed loss of full type ships in a seaway. *Ship Technol. Res.* **2019**, *66*, 161–179. [CrossRef]
22. Goh, J.; Sim, M. Distributionally Robust Optimization and Its Tractable Approximations. *Oper. Res.* **2010**, *58*, 902–917. [CrossRef]
23. Lu, X.; Chan, K.W.; Xia, S.; Zhou, B.; Luo, X. Security-constrained multiperiod economic dispatch with renewable energy utilizing distributionally robust optimization. *IEEE Trans. Sustain. Energy* **2019**, *10*, 768–779. [CrossRef]
24. Kanellos, F.D. Optimal Power Management with GHG Emissions Limitation in All-Electric Ship Power Systems Comprising Energy Storage Systems. *IEEE Trans. Power Syst.* **2014**, *29*, 330–339. [CrossRef]
25. Hersbach, H.; Bell, B.; Berrisford, P.; Hirahara, S.; Horányi, A.; Muñoz-Sabater, J.; Nicolas, J.; Peubey, C.; Radu, R.; Schepers, D.; et al. The ERA5 global reanalysis. *Q. J. R. Meteorol. Soc.* **2020**, *146*, 1999–2049. [CrossRef]
26. Kim, Y.R.; Steen, S.; Kramel, D.; Muri, H.; Strømman, A.H. Modelling of ship resistance and power consumption for the global fleet: The MariTEAM model. *Ocean Eng.* **2023**, *281*, 114758. [CrossRef]

Journal of
Marine Science and Engineering

MDPI

Article

Emission Estimation and Spatiotemporal Distribution of Passenger Ships Using Multi-Source Data: A Case from Zhoushan (China)

Xubiao Xu [1], Xingyu Liu [2], Lin Feng [3], Wei Yim Yap [4] and Hongxiang Feng [1,*]

1 Faculty of Maritime and Transportation, Ningbo University, Ningbo 315832, China; 2211120015@nbu.edu.cn
2 Faculty of Governance and Global Affairs, Leiden University, 2311 EZ Leiden, The Netherlands; liuxingyu0423@gmail.com
3 Department of Logistics and Maritime Studies, The Hong Kong Polytechnic University, Hong Kong TU428, China; 23100478d@connect.polyu.hk
4 School of Business, Singapore University of Social Sciences, Singapore 599494, Singapore; wyyap@suss.edu.sg
* Correspondence: fenghongxiang@nbu.edu.cn

Abstract: Quantifying and estimating shipping emissions is a critical component of global emission reduction research and has become a growing area of interest in recent years. However, emissions from short-distance passenger ships operating on inter-island routes and their environmental impacts have received limited attention. This contribution investigated the temporal and spatial distribution characteristics of pollutants emitted by short-distance passenger ships at Zhoushan (China) using Automatic Identification System (AIS) data and the bottom–up emission model integrated with multi-source meteorological data. A year-long emission inventory was investigated. The results indicated that high-speed passenger ships contributed to the largest share of the emissions. The emissions were predominantly concentrated during daytime hours, with the routes between Zhoushan Island and Daishan, Daishan and Shengsi, and Zhoushan Island and Liuheng Island accounting for most of the emissions. Furthermore, intra-port waterways were identified as the primary emission areas for short-distance passenger ships. This study provides essential data support and references for the relevant authorities to understand the emission patterns of short-distance passenger ships, thereby facilitating the formulation of targeted emission reduction strategies for the maritime passenger transport sector.

Keywords: emission estimation; spatiotemporal distribution; automatic identification system (AIS) data; passenger ship; bottom–up emission model

1. Introduction

Maritime activities play a vital role in global exhaust emissions, and the atmospheric pollutants associated with these activities have become significant concerns at both the regional and global levels [1–3]. With increasing attention on climate change, the international community has imposed stricter regulatory requirements on greenhouse gas emissions from the shipping industry. For instance, the International Maritime Organization (IMO) has established carbon emission standards for global shipping and encouraged the adoption of environmentally friendly fuels and technologies, aiming to mitigate global warming by reducing carbon dioxide emissions [4–6]. Some countries have also initiated remote monitoring programs to effectively enforce emission policies and regulations [7,8]. Consequently, the governance of ship emissions has become a critical focus of global regulations.

Effective emission governance requires the support of policy and regulatory measures, and the development of ship emission inventories, which contain data on atmospheric and greenhouse gas emissions as well as their spatial distribution, is crucial to forming the foundation for such actions [9,10]. Most past research on ship emissions has primarily focused on global or regional maritime traffic, particularly emissions from large merchant ships. Although the per-ship emissions of short-distance passenger ships are relatively small, their high frequency of operations, short voyages, and routes located in densely populated coastal areas make their emissions highly impactful on local air quality. These emissions have a direct effect on the health of island and coastal residents [11]. However, studies assessing the environmental and socio-economic impacts of emissions from small passenger ships operating on inter-island routes or ferries navigating urban ports remain insufficient. A key issue is that previous emission inventories have rarely considered short-distance passenger ships within specific regions, resulting in gaps in targeted emission control strategies [12]. Therefore, there is an urgent need to conduct an assessment of the impact of passenger ship emissions in areas such as island cities and island regions where passenger ships sail densely. This led to our research topic: quantifying the pollutant emissions of short-distance passenger ships in the target area, with a particular focus on their spatial and temporal distribution characteristics.

According to IMO regulations [13], all ships with a 300 Gross Tonnage or above engaged in international voyages, cargo ships with a 500 Gross Tonnage or above not engaged in international voyages, and all passenger ships must be equipped with AIS devices. AIS is a navigational system designed to enhance maritime safety and communication between ships and the shore or between ships. It plays a crucial role in maritime services, such as reducing ship collision incidents. AIS data are automatically exchanged at high frequencies between ships and shore stations and recorded, providing geographic information with high temporal precision for ship emission studies. All passenger ships operating within Chinese waters must be equipped with AIS devices, enabling the acquisition of high-precision operational data for short-distance passenger ships with frequent voyages, thereby supporting the development of emission inventories.

The contributions of this research is twofold. First, we propose a framework for estimating emissions from short-distance passenger ships using multi-source data (i.e., AIS data, attribute data, and meteorological and environmental data) and the bottom–up emission model. Secondly, the spatiotemporal characteristics of passenger ship emissions in Zhoushan, China, were investigated and a one-year-long emission inventory was established.

The remainder of this paper is structured as follows: Section 2 reviews the relevant literature; Section 3 outlines the research questions; Section 4 details the methodology, including the model construction and data-processing steps; Section 5 presents an empirical study of short-distance passenger ships in Zhoushan, China; and finally, Section 6 summarises the most important findings.

2. Literature Review

Traditional methods for calculating emission inventories mainly include top–down and bottom–up approaches [14,15]. The top–down method starts from macro-level statistical data, such as fuel consumption and shipping activities, and calculates emissions by multiplying fuel consumption by emission factors [16]. Although simple and efficient in producing emission inventory results, this method may lack the level of detail needed for micro-level analyses due to its broader assumptions and its limited consideration of the actual movement of ships. However, it remains highly effective and accurate for global- or regional-scale studies [17]. In contrast, the bottom–up method can provide more accurate emission estimates for finer-scale analyses [18,19]. It estimates emissions by considering the operational status, engine power, fuel consumption, and corresponding emission factors of ships to calculate emissions for each interval between two consecutive AIS positions [20]. Jalkanen et al. [21] first developed the STEAM (Ship Traffic Emission Assessment) model for ship emissions, using AIS data to calculate pollutant emissions from ships in the Baltic Sea, and constructed an emission inventory. Following this, many studies utilised AIS data to compile emission inventories [22–25]. Chen et al. [26] created a high temporal and spatial resolution emission inventory for ships in China for 2014, filling a gap in the national emission inventory. Toscano et al. [27] developed an emission inventory for the Port of Naples based on 2018 data, assessing its impact on air quality at an urban scale. Ye et al. [28] used emission inventories from 13 case studies, comparing river–sea hybrid models, river–sea models, and mixed models using a bottom–up approach. Li et al. [29] proposed a spatiotemporal STSD iterative correction model based on a random forest algorithm and an ST–DBSCAN-based trajectory segmentation algorithm to estimate CO_2 emissions from ships in China's domestic emission control areas (DECAs). This method is also commonly applied to emission studies of ships in inland river basins. Zhi et al. [30] compiled a 2019 greenhouse gas inventory for China, illustrating the relationship between economic development and emissions. The bottom–up method is particularly suitable for understanding the emission levels of individual ships or specific ship types, such as emissions on particular routes or at ports. Still, it requires a large volume of research data. The influence of meteorological factors on ship emissions has also gained attention, reducing uncertainties in emission models [31]. Moreno-Gutiérrez and Durán-Grados [32] proposed the SENEM model (Ship's Energy Efficiency Model), which, compared to the STEAM model, considers real-time power adjustments influenced by wind, waves, and currents. Studies have consistently shown that bottom–up methods considering operational states are more accurate and have been widely applied to calculate emissions from ships in global waters and port areas.

While numerous studies have explored and calculated ship emission inventories and their impacts, the research gap on short-distance passenger ships still needs to be filled. Significant emissions from large ships have been the primary focus of researchers, with emission calculations typically considering auxiliary engines and boilers. The established emission inventories often include only partial emissions from passenger ships in port areas. Carić et al. [33] assessed the environmental burdens of cruise ships at Croatian ports, highlighting potential human ecological risks, with cruise ships identified as a notable emission source. Dragović et al. [34] investigated cruise ship traffic and found that factors such as berth availability and accessibility, in addition to marine fuel, influence emission levels. Chen et al. [35] studied the activities of passenger ships in the Arctic, revealing adverse impacts on the marine ecosystem and analysing CO_2 emissions under different fuel options, suggesting liquefied natural gas (LNG) as a better choice. Additionally, Chen et al. [36] analysed the impact of COVID-19 on passenger ships in Danish waters, finding significant effects on cruise ships but minimal impact on ferries. For small passenger ships, the research has primarily focused on ferry emissions [37]. Chou et al. [38] estimated the CO_2 emissions of ferries operating on urban port routes and nearby island routes in Taiwan using the Energy Efficiency Index and proposed potential emission reduction policies. Mannarini et al. [39] used a mixed-effect linear model to analyse ferry emissions in the European Economic Area from 2018 to 2020, characterising the impact of COVID-19 on ferry emissions and evaluating the causality between the pandemic and pollutant emissions. Baird and Pedersen [40] calculated the CO_2 emissions from short- and long-distance ferries operating between mainland Scotland and the Orkney Islands, concluding that the optimal solution for medium-sized ferries serving islands is to use or develop the shortest maritime routes to minimise CO_2 emissions from transportation. To optimise emissions, researchers have also explored fuel alternatives for ferries to reduce pollutant emissions [41–43]. However, the results indicate that internal combustion engines remain the dominant hybrid power source, and emissions from short-distance passenger ships require continued attention in the short term. Thus, comprehensive regional evaluations of short-distance passenger ship emissions remain necessary in the absence of alternative power sources.

In summary, bottom–up methods based on dynamic AIS data, which fully consider the operational status of ships and provide precise data for passenger ship emissions, offer higher transparency and traceability compared to model estimates or indirect calculations. This enhances the credibility and reliability of research findings, making them more suitable for calculating emissions from passenger ships. Most studies, however, have concentrated on total emissions from global ocean-going ships or general emissions from port areas [44,45]. Previous research on passenger ship emissions has primarily focused on long-distance recreational cruises and short-distance ferries. In contrast, emissions from short-distance passenger ships operating inter-island routes for transportation purposes require targeted study.

Overall, although emissions from short-distance passenger ships have a relatively smaller global impact compared to ocean-going ships, their high-frequency operations in specific regions significantly affect local environments. This study adopted a bottom–up approach based on dynamic AIS data to estimate and analyse the spatiotemporal characteristics of short-distance passenger ship emissions, uncovering their emission patterns and environmental impacts.

3. Problem Description

Short-distance passenger ships operate according to publicly available schedules, navigating between inter-island passenger terminals and along designated routes specified by regulatory authorities. Although many emission inventories include data on passenger ships, the construction and operation of short-distance passenger ships differ significantly from larger ships such as cruise ships. Previous studies often estimated emissions based on the default main engine power, which likely led to the underestimation of the actual emissions [46].

3.1. Characteristics of Short-Distance Passenger Ships

The development of emission inventories for short-distance passenger ships requires an in-depth understanding of their characteristics. Short-distance passenger ships have the following operational characteristics. First, they are equipped with specialised passenger terminals to facilitate ticket purchases and boarding for residents, which differs from larger ships that dock at general-purpose port berths. Second, their primary purpose is to serve the transportation needs of residents in nearby island regions. These ships are constructed to carry passengers and vehicles, with significant variations in engine power depending on the type and quantity of transported objects as well as the operational purpose of the ship. Since these ships operate exclusively within designated areas, regulatory authorities impose strict schedules for departure, navigation, and berthing.

3.2. Operating Modes of Short-Distance Passenger Ships

According to the definitions in previous emission inventories, the operational modes of large ships can be categorised based on their speed and location [47]. However, the operational modes of short-distance passenger ships differ and can be defined as berthing, unberthing, and cruising. Unlike large ships, idling at berth is not considered for passenger ships since auxiliary engines and boilers are not operational during berthing, and no emissions are generated. The detailed definitions of these operational modes are as follows:

- Berthing: Short-distance passenger ships enter the passenger terminal area from open water, gradually reducing their speed to ensure a safe and stable docking at the designated berth.
- Unberthing: The ship slowly departs from the terminal berth, gradually adapting to navigation conditions in open water.
- Cruising: In this mode, the ship sails along a predetermined route in open water, maintaining a steady speed to ensure a comfortable experience for passengers.

Taking a Roll-on/Roll-off (Ro-Ro) passenger ship as an example, its berthing process, as shown in Figure 1a, involves a gradual deceleration as it approaches the terminal. Around 8:15, the ship's speed decreases from approximately 10 knots to around 3 knots, reducing its kinetic energy to facilitate safe berthing. During this approach, operators adjust the ship's heading based on the wind direction, tidal currents, and other factors, ensuring alignment with the designated berth. Once the alignment is complete, the ship further reduces its speed to below 2 knots, moving slowly into the berth. For unberthing, the ship is carefully unberthed away from the berth at a speed typically below 3 knots, as shown in Figure 1b. Once clear of the berth, operators adjust the ship's heading according to the departure plan, allowing it to transition smoothly to its designated route.

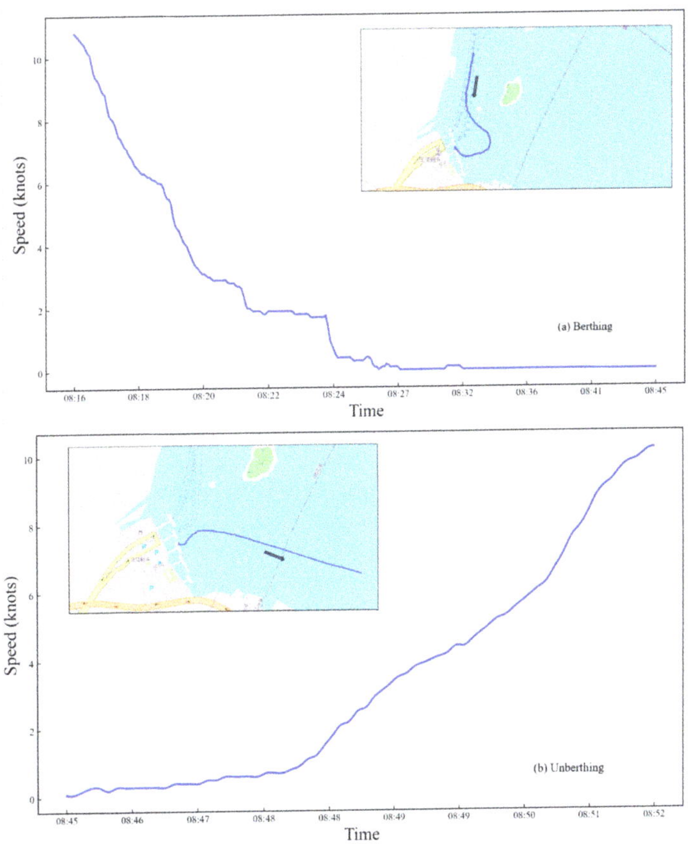

Figure 1. Berthing (**a**) and unberthing (**b**) process of a passenger ship.

4. Methodology

The overall technical workflow of this study is shown in Figure 2. It primarily includes the processing, calculation, and visualisation of the AIS-based ship emission inventory, which consists of five main steps. First, the AIS data underwent pre-processing, including cleaning abnormal information and identifying Maritime Mobile Service Identities (MMSIs) corresponding to passenger ships or high-speed ships, followed by matching the relevant dynamic details. Second, the technical specifications of passenger ships were collected, and corrections to the ship speeds were made based on meteorological factors. Third, the key parameters were determined, including the actual operating speed of the ship, emission factors, main engine load coefficients, as well as the ship's emission location and time stamps. Fourth, the pollutant emissions from passenger ships' main engines were calculated to generate an emission inventory. Finally, visualisation software was used to create spatial distribution maps of the pollutant emissions.

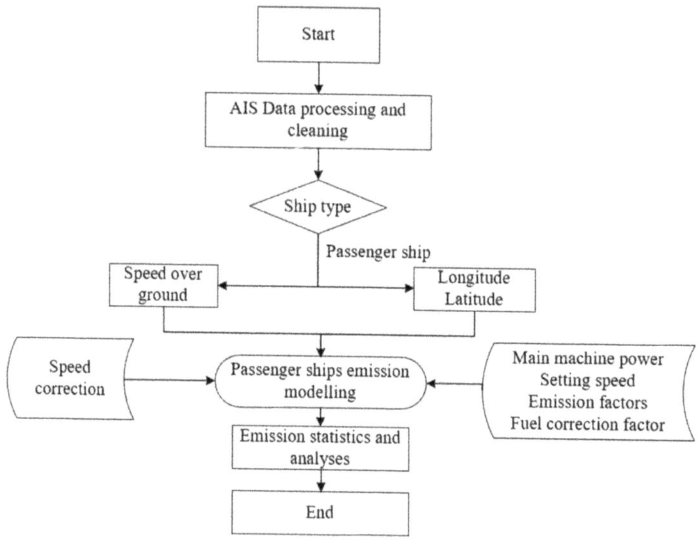

Figure 2. Logic framework.

4.1. Identification Algorithm for Berthing and Unberthing Operations

The identification of operational modes during the berthing and unberthing processes of short-distance passenger ships is a critical foundation for quantifying ship pollutant emissions. This study built upon the algorithm proposed by Chen et al. [48], employing a trajectory-based recognition algorithm. Analysing key parameters such as ship position and speed, the algorithm achieves precise localisation of berthing and unberthing processes. The core of this algorithm lies in constructing a multi-parameter recognition model that integrates ship trajectories, speed variations, and the spatial characteristics of passenger terminals. First, an AIS dataset Si for passenger ships in the terminal area is collected, and complete trajectory information is extracted. Then, based on key parameter definitions, including the distance threshold d and the speed range $(v_{-},\overline{v})$, a refined identification of berthing and unberthing behaviours is conducted. The implementation process is detailed in Algorithm 1.

Algorithm 1 identifies the berthing and unberthing processes of the passenger ships based on trajectory characteristics. The first step involves extracting the trajectory subset Si between the passenger terminal and the berth or vice versa and conducting an initial screening. The second step identifies a collection of continuous low-speed or stationary points approaching the berth, based on predefined distance thresholds and speed ranges, which are marked as the berthing trajectory point set. Subsequently, by detecting trajectory segments where the ship's speed gradually increases and moves away from the berth, the unberthing trajectory point set is further extracted. Using this method, we can identify the berthing and unberthing modes of passenger ships.

Algorithm 1. Framework algorithm for estimating emissions from passenger ships.

Step 1: For the short distance passenger ships trajectory data*set* ***i*** extract the dataset ***Si*** between the passenger terminal boundary to the berth or the berth to the passenger terminal boundary.
Step 2: Determine if the passenger ship trajectory in the ***Si*** data satisfies the following two conditions: i. The speed of the passenger ship trajectory point is between v_- **and** $\overline{v}$. ii. The distance between the passenger ship and the pier is less than ***d***.
Step 3: The trajectory points in the trajectory dataset ***Si*** that continuously approach the terminal are used to construct a set of continuous low-speed\stationary points, which are recorded as the berthing trajectory point set.
Step 4: Detect the changes in the speed and position of the passenger ships in the berthing area. If there are trajectory points that continuously move away from the terminal and the speed gradually increases to $\overline{v}$, this segment of the trajectory is recorded as the unberthing trajectory point set. Step 5: Repeat steps 1–4 to obtain a passenger ship berthing (unberthing) dataset.

4.2. Estimating Emissions of Main Engines

This study adopted a bottom–up approach based on AIS data to calculate the emissions of pollutants such as carbon dioxide (CO_2), nitrogen oxides (NO_x), carbon monoxide (CO), particulate matter ($PM_{2.5}$ and PM_{10}), and sulphur dioxide (SO_2) from passenger ships. The short-distance passenger ships studied here are powered by dual main engines and are not equipped with auxiliary engines or boilers. Thus, the emission calculation in this study focused solely on the main engines. The methodology first estimates the emissions for each trajectory segment between two consecutive AIS messages using the bottom–up calculation approach. These segment-level emissions are then summed to obtain the total emissions of a single ship. Finally, the emissions of all ships in the study region are aggregated. This approach is represented mathematically as Equation (1):

$$E_{m,i,j} = P_m \cdot LF \cdot T_j \cdot EF_{m,i} \cdot LLA \cdot 10^{-6} \quad (1)$$

where

$E_{m,i,j}$ is the emissions of ship i in trajectory segment j (tons);

P_m is the main engine power of the ship (kW);

LF is the load factor of the main engine (dimensionless);

T_j is the running time of trajectory segment j(h);

$EF_{m,i}$ is the emission factor of pollutant i for main engine m (g/kWh);

LLA is the emission adjustment factor, which is equal to 1 if LF is greater than 20% and adjusted otherwise.

$$E_{m,i} = \sum_{j=1}^{N_i} E_{m,i,j} \quad (2)$$

where

$E_{m,i}$ is the total emissions from ship i (tons);

N_i is the total number of trajectory segments of ship i.

$$E = \sum_{i=1}^{M} E_{m,i} \tag{3}$$

where

E is the total emissions of all ships in the study area (tons);

M is the total number of ships in the study area.

4.2.1. Engine Power

The load power of the ship's main engine refers to the actual power output of the main engine during operation, while the rated power of the main engine is the maximum power needed to continuously produce underrated speed and normal maintenance cycles, without external environmental influence. Variations in the power model have a relatively high impact on emission estimations.

In previous studies, researchers have typically extracted data on main engine power characteristics from databases such as the Lloyd's Register, CCS, or IMO database. However, accessing comprehensive local ship databases globally often presents challenges. Many existing studies have attempted to address this data gap by using models to predict missing entries related to ship technical specifications [49]. This approach introduces certain inaccuracies and is one of the largest sources of uncertainty in AIS data-based methods [50,51]. For the short-distance passenger ships under investigation, we adopted a more comprehensive and rigorous data collection strategy to obtain key technical parameters, particularly main engine power. This strategy significantly reduces the amount of missing data and enhances the accuracy of emission calculations.

4.2.2. Emission Factors

Emission factors (EFs) are empirical parameters whose selection is directly related to the engine type, fuel type, and sulphur content, and they are influenced by various external factors [52]. Engines are categorised by speed into slow-speed diesel (SSD) engines and medium-speed diesel (MSD) engines. They are further classified by fuel types, such as heavy fuel oil (HFO), marine diesel oil (MDO), and marine gas oil (MGO). This study used the emission factors recommended under different circumstances by Entec UK Limited [53] and the Starcrest Consulting Group [54], as outlined in Table 1. It is important to note that these studies were based on marine diesel oil with a sulphur content of 0.5% and heavy fuel oil with a sulphur content of 2.7%. For fuels with differing sulphur contents, this study applied specific fuel correction factors to ensure the accuracy and applicability of the emission factors. The specific correction values are detailed in Table 2. The correction factors for the sulphur content were derived from established guidelines and validated through empirical studies. These factors used to adjust the emission estimates to account for variations in fuel sulphur contents. While they improve the accuracy of the estimates, it is important to acknowledge that the sulphur content of fuel may vary across batches and operational conditions, which could introduce uncertainties into the calculations.

Table 1. Emission factors [52,53] (unit: g/kWh).

Machine Type	Engine Type	Fuel Type	Built Year	CO_2	CO	HCs	NO_x	SO_2	$PM_{2.5}$	PM_{10}
ME	MSD	MDO	<1999	649	1.1	0.5	13.2	2.1	0.35	0.38
ME	MSD	MDO	2000–2010	649	1.1	0.5	12.2	2.1	0.35	0.38
ME	MSD	MDO	2011–2015	649	1.1	0.5	10.5	2.1	0.35	0.38
ME	MSD	HFO	<1999	683	1.1	0.5	13	11.5	1.2	1.5
ME	MSD	HFO	2000–2010	683	1.1	0.5	14	11.5	1.2	1.5
ME	MSD	HFO	2011–2015	683	1.1	0.5	11.2	11.5	1.2	1.5

Note: ME refers to the main engine. MSD refers to medium-speed diesel. MDO refers to marine diesel oil. HFO refers to heavy fuel oil.

Table 2. Fuel correction coefficient [55].

Fuel Type	Sulphur Content (%)	CO_2	CO	HCs	NO_x	SO_2	$PM_{2.5}$	PM_{10}
MDO	0.50	1.00	1.00	1.00	0.94	0.18	0.25	0.25
MDO	0.10	1.00	1.00	1.00	0.94	0.04	0.17	0.17

4.2.3. Vessel Speed Correction

Due to the smaller size and scale of short-distance passenger ships, external factors can lead to speed losses [56]. The speed shown in AIS data generally refers to the ship's speed over ground, which already accounts for the influence of external factors such as the wind, current, and waves. To better approximate the actual operating conditions of passenger ships and improve the accuracy of emission estimates, it is necessary to correct the speed over ground to obtain the ship's true output speed. The actual speed of a ship under the influence of wind, waves, and currents is equal to the speed provided by AIS data minus the wind-induced drift speed, current-induced drift speed, and wave-induced drift speed. The speed correction method proposed by Huang et al. [57] was adopted, with the process detailed below.

The wind, wave, and current data used in this study were sourced from the Copernicus Marine Environment Monitoring Service (CMEMS). First, the combined influence of wind and currents on the ship's speed is corrected. Assuming that the ship's pre-affected speed is $\vec{V}_{oss}$, it can be calculated using Equation (4):

$$\vec{V}_{sog} = \vec{V'}_{oss} + \vec{V}_{wnd} + \vec{V}_{cur} \tag{4}$$

Through orthogonal decomposition, Equation (4) can be rewritten as

$$\begin{cases} u_{sog} = \vec{V'}_{oss} + \vec{V}_{cur} * cos\theta_{ct} - \vec{V}_{wnd} * cos\theta_{wt} \\ v_{sog} = \vec{V}_{cur} * sin\theta_{ct} - \vec{V}_{wnd} * sin\theta_{wt} \\ \vec{V}_{sog} = \sqrt{{u_{sog}}^2 + {v_{sog}}^2} \end{cases} \tag{5}$$

where

$\vec{V}_{sog}$ is the speed over ground;

$\vec{V'}_{oss}$ is the engine output speed after the reductions;

$\vec{V}_{wnd}$ is the wind-induced drift speed;

$\vec{V}_{cur}$ is the current speed;

u and v are the rectangular components of the speed;

θ_{ct} is the flow angle between $\vec{V}_{oss}$ and $\vec{V}_{cur}$ at time t;

θ_{wt} is the relative azimuth between $\vec{V}_{org}$ and $\vec{V}_{wnd}$ at time t.

By further processing this formula, a detailed expression for $\vec{V}_{oss}$ can be derived:

$$\vec{V'}_{oss} = \sqrt{\vec{V}_{sog}^{\,2} - \left(\vec{V}_{cur}\sin\theta_{ct} - \vec{V}_{wnd}\sin\theta_{wt}\right)^2} - \vec{V}_{cur}{}^{*}\cos_{\theta_{ct}} + \vec{V}_{wnd}{}^{*}\cos_{\theta_{wt}} \quad (6)$$

Finally, the influence of wave factors is corrected to obtain the original speed output of the ship's engine, $\vec{V}_{org}$, as shown in Equation (7):

$$\vec{V}_{org} = \frac{\vec{V'}_{oss} + (K_1 + K_2h^2 - K_3qh)G}{1 + (K_1h + K_2h^2 - K_3qh)K_4D} \quad (7)$$

where

h is the effective wave height;

q is the relative azimuth;

D is the actual displacement of the ship;

K_1, K_2, K_3, and K_4 are the ship performance parameter values (K_1 = 0.745, K_2 = 0.05015, K_3 = 0.0045, K_4 = 1.35*106);

G is the empirical coefficient that is dependent on ship displacement D:

$$G = \begin{cases} 1.0 & D \leq 10,000t \\ 1.09 & 10,000t < D \leq 20,000t \\ 1.19 & 20,000t < D \leq 40,000t \\ 1.29 & 40,000t < D \leq 60,000t \end{cases} \quad (8)$$

The above model enables the calculation of ships' true engine output speed in the given AIS data and provides information about the wind, waves, and currents in the sailing area.

4.2.4. Load Factors

The load factor (LF) of a ship's main engine is a critical parameter in emission calculations. The relationship between the main engine load and ship speed can be approximately described using the propeller law [58]. Under constant sailing conditions, the ratio of the ship speed to the main engine's rotational speed typically remains unchanged. Thus, the formula for calculating the load factor is as follows:

$$LF = (\vec{V}_{org}/V_{max})^3 \quad (9)$$

where $\vec{V}_{org}$ is the actual output speed of the ship and V_{max} is the design maximum speed of the ship.

In real-world passenger transport operations, the number of passengers and vehicles carried per voyage varies, introducing uncertainty into the load factor of passenger ships. When the engine load of the main engine falls below 20%, there is a tendency for emission factors to increase as the load decreases. This is because diesel engines operate less efficiently at low loads, resulting in incomplete combustion. Therefore, these emission factors should be adjusted as a function of engine load. When the main engine load is less than 20%, a low-load adjustment factor (LLA) must be applied, as detailed in Table 3.

Table 3. Low-load adjustment factor for main engines.

Load	NO_x	SO_2	$PM_{2.5}$	PM_{10}	CO	CO_2	HCs
1%	11.47	5.99	19.17	19.17	19.32	5.82	59.28
2%	4.63	3.36	7.29	7.29	9.7	3.28	21.18
3%	2.92	2.49	4.33	4.33	6.49	2.44	11.68
4%	2.21	2.05	3.09	3.09	4.86	2.01	7.71
5%	1.83	1.79	2.44	2.44	3.9	1.76	5.61
6%	1.6	1.61	2.04	2.04	3.26	1.59	4.35
7%	1.45	1.49	1.79	1.79	2.8	1.47	3.52
8%	1.35	1.39	1.61	1.61	2.45	1.38	2.95
9%	1.27	1.32	1.48	1.48	2.18	1.31	2.52
10%	1.22	1.26	1.38	1.38	1.97	1.25	2.2
11%	1.17	1.21	1.3	1.3	1.79	1.21	1.96
12%	1.14	1.18	1.24	1.24	1.64	1.17	1.76
13%	1.11	1.14	1.19	1.19	1.52	1.14	1.6
14%	1.08	1.11	1.15	1.15	1.41	1.11	1.47
15%	1.06	1.09	1.11	1.11	1.32	1.08	1.36
16%	1.05	1.07	1.08	1.08	1.24	1.06	1.26
17%	1.03	1.05	1.06	1.06	1.17	1.04	1.18
18%	1.02	1.03	1.04	1.04	1.11	1.03	1.11
19%	1.01	1.01	1.02	1.02	1.05	1.01	1.05
20%	1	1	1	1	1	1	1

5. Case Study

5.1. Study Area

Zhoushan is China's fourth largest island (main island) and the largest archipelago, comprising 2085 islands, of which, approximately 105 are inhabited. The region experiences a high level of passenger ship activity. For islands far from the mainland, passenger ships serve as an essential mode of transportation for residents travelling to and from the islands. These ships also fulfil the role of transport vessels, delivering essential supplies to residents.

In this study, we focused on the Zhoushan Port area in Zhejiang Province, China, as the research region. The study area extends northward to Huaniao Town in Shengsi County, Zhoushan, eastward to the Dongji Islands, and southward to the Jiushan Archipelago in Xiangshan County, Ningbo. Using AIS data from the entire year of 2020, we calculated and analysed the pollutant emissions of passenger ships throughout this region. Additionally, we examined the temporal and spatial characteristics of passenger ship emissions within the study area. The research area is shown in Figure 3, which highlights several major passenger terminals.

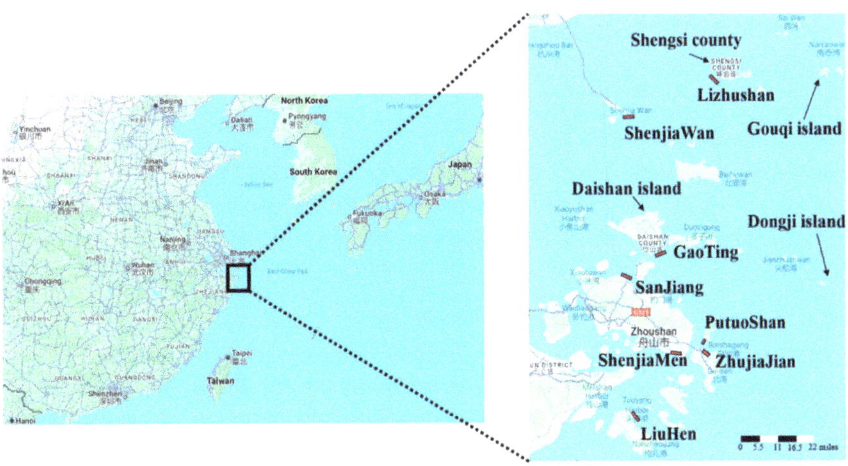

Figure 3. Map of the study area.

5.2. AIS Data

The AIS data used in this study were obtained from the East China Sea Navigation Support Center. Each daily file contained more than 10 million records. Combining the static and dynamic information provided by the AIS data is essential to accurately establish the emission inventory of passenger ships. The static ship information included MMSI (Maritime Mobile Service Identity), ship type code, IMO number, call sign, ship dimensions, estimated time of arrival (ETA), and destination. The dynamic ship information included the timestamp for data transmission and reception (UTC), MMSI, navigational status, rate of turn (ROT), speed over ground (SOG), the ship's latitude and longitude, course over ground (COG), and true heading (HDG), among others [59,60]. It is important to note that ship type codes, which ranged from 60 to 69, represent passenger ships, while codes from 40 to 49 indicate high-speed ships. Some high-speed ships also operate as passenger ships.

5.3. Passenger Ship Characteristics

From the AIS data of one year, we identified 576 ships with type codes classified as passenger ships or high-speed ships. After individually excluding ships with incorrect type codes or those not registered in the Zhoushan area, we identified 153 short-distance passenger ships operating within the study area. Based on the cargo they carry and their operational purposes, these passenger ships were categorised into four types: ordinary passenger ships, Ro-Ro passenger ships, high-speed passenger ships, and car–passenger ferries. The data for these ships were obtained from local passenger transport companies, and the average construction specifications for each type of passenger ship are shown in Table 4. Generally, all types of passenger ships use HFO during their voyages. However, car–passenger ferries differ in that they use MDO while in port and during the 15 min before and after entering or leaving the port, but switch to HFO during open-sea navigation. In the emission calculations, the corresponding emission factors were selected based on the built year, engine type, fuel consumption, and sailing characteristics of each ship.

Table 4. Passenger ship construction specifications.

Ship Type	Average Gross Tonnage	Average Main Engine Power (kW)	Average Passenger Capacity (Person)
Ordinary passenger ship	488.4	994.3	370.4
Ro–Ro passenger ship	1913.9	1902.8	337.2
High-speed passenger ship	215.4	1936	149.6
Car–passenger ferry	308.4	356.4	106.4

5.4. Passenger Ship Trajectory Distribution in the Zhoushan Area

Figure 4 illustrates the trajectory distribution of passenger ships in the Zhoushan area from January to December 2020. The complex trajectory map reflects the frequent movements of ships between the various islands. It can be observed that ordinary passenger ships primarily operated between Zhujiajian Island, Dongji Island, and Shengshan–Gouqi Island on the eastern side, as well as between Shengsi Island and Shenjia Wan. These ships are responsible for transporting passengers over relatively long distances. Ro-Ro passenger ships and high-speed passenger ships mainly navigated between Zhoushan Island, Liuheng, and Daishan. Ferries, on the other hand, were primarily used for short-distance transportation and were seen operating between closely located islands. The most densely trafficked areas were concentrated near Daishan Island, Putuo Island, and Zhoushan Island, where passenger ships frequently entered and left the ports. These locations served as key hubs for docking and transit operations.

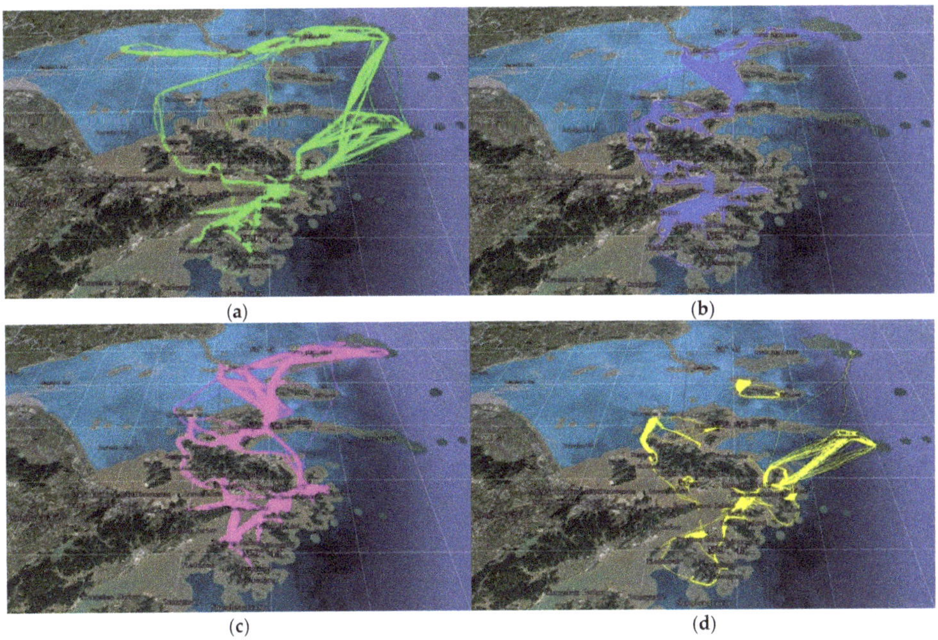

Figure 4. Passenger ship trajectory distribution in Zhoushan area. (Green for ordinary passenger ships, blue for Ro-Ro passenger ships, red for high-speed passenger ships, and yellow for ferries). (**a**) Ordinary passenger ships, (**b**) Ro-Ro passenger ships, (**c**) high-speed passenger ships, and (**d**) ferries.

5.5. Emission Results

Although 2020 was influenced by the COVID-19 pandemic, our analysis suggests that the primary trends and patterns in emissions remained consistent with the objectives of this study. Using the atmospheric pollutant emission model for passenger ships, the emissions of major pollutants in the Zhoushan area were calculated. The results for the total emissions were as follows: 54,895.75 tons for CO_2, 1057.82 tons for NO_x, 101.46 tons for CO, 7.68 tons for $PM_{2.5}$, 8.33 tons for $PM_{2.5}$, 32.02 tons for SO_2, and 50.34 tons for HCs.

5.6. Temporal Features of Emission Inventory

5.6.1. Temporal Distribution

The monthly variation in emissions for the different pollutants is shown in Figures 5 and 6. It could be observed that the emissions in February are significantly low with a narrower and more concentrated distribution. This was attributable to the Chinese Lunar New Year in January and February 2020, as well as the COVID-19 pandemic. The pandemic control measures led to the suspension of most passenger ship operations during this period. The peak monthly emissions of passenger ship pollutants occurred in mid-August and early October. This was likely related to increased passenger traffic during the summer season. As passenger ships are essential for reaching the islands, their operational frequency and load increased during this time, contributing to the highest emissions. In contrast, the emissions in November and December showed consecutive declines, with reductions of 16.3% and 10.5%, respectively, compared to the previous months. We speculate that this decrease was due to seasonal changes and lower temperatures, which reduce the demand for leisure travel, leading to a corresponding decline in emissions.

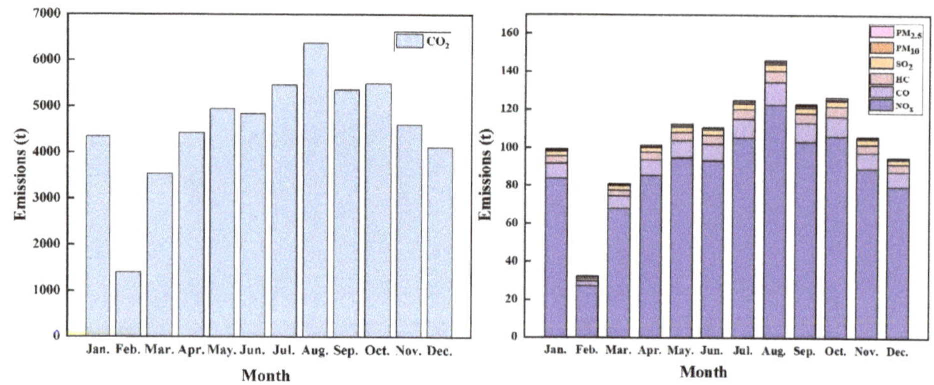

Figure 5. Variation in emissions of individual pollutants over the year (CO_2 on the **left**, other pollutants on the **right**).

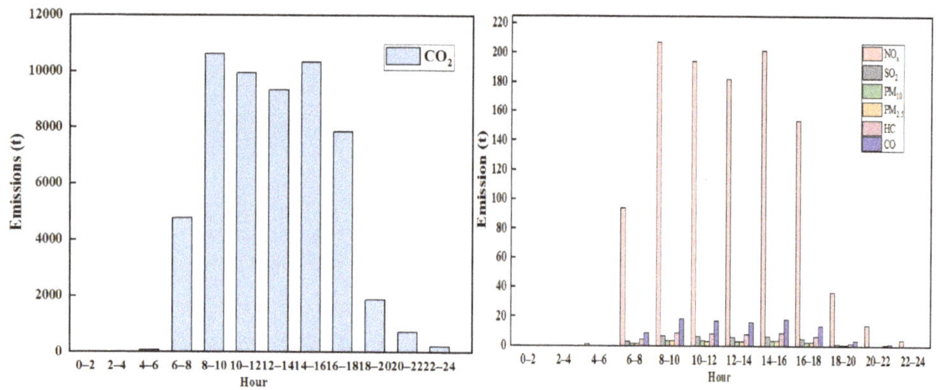

Figure 6. Distribution of emissions throughout the day (CO_2 on the **left**, other pollutants on the **right**).

To further analyse the temporal distribution of the emissions, the data were divided into 12 two-hour intervals, allowing for a detailed examination based on the time of day. Figure 6 illustrates the differences in the emissions across these intervals. The period from 08.00 to 10:00 recorded the highest emissions, with CO_2 emissions reaching 10,402.85 tons. Additionally, 10:00–12:00 and 14:00–16:00 represented two other peak emission periods during the day. These three time slots collectively exhibited higher air pollutant concentrations. For instance, CO_2 emissions during these periods accounted for 17.94% and 18.71% of the total emissions, respectively. Conversely, emissions of air pollutants decrease significantly during the late evening and night-time hours. Between 22:00 and 06:00, the emissions accounted for only 5.32% of the total, representing the lowest emission levels. This reduction was primarily due to the complexity of night-time navigation, poor visibility, and the prioritisation of safety for passenger-focused ships. As a result, only a limited number of short-distance routes operated night services.

5.6.2. The Impact of Severe Weather on Emissions

The annual variation in pollutant emissions from passenger ships over time is shown in Figure 7. The changes in daily CO_2 emissions throughout the year are shown in Figure 8. Certain months exhibit outliers, with notable low-emission days such as August 4, August 25, September 1, and December 30. We hypothesised that these days were affected by severe weather conditions, including typhoons, cold waves, strong winds, and high waves, which can significantly impact navigational visibility. Good visibility is critical for passenger ship navigation, as it enables crew members to accurately identify and follow channels, avoid obstacles, and adjust their course to ensure a safe and timely arrival at the destination. A statistical analysis of severe weather events in the Zhoushan region during 2020 was conducted, with the specific dates summarised in Tables 5 and 6. The timing of these severe weather events closely corresponded to the low-emission days. Typhoons and extreme cold wave conditions were found to have the most pronounced effects on the emissions.

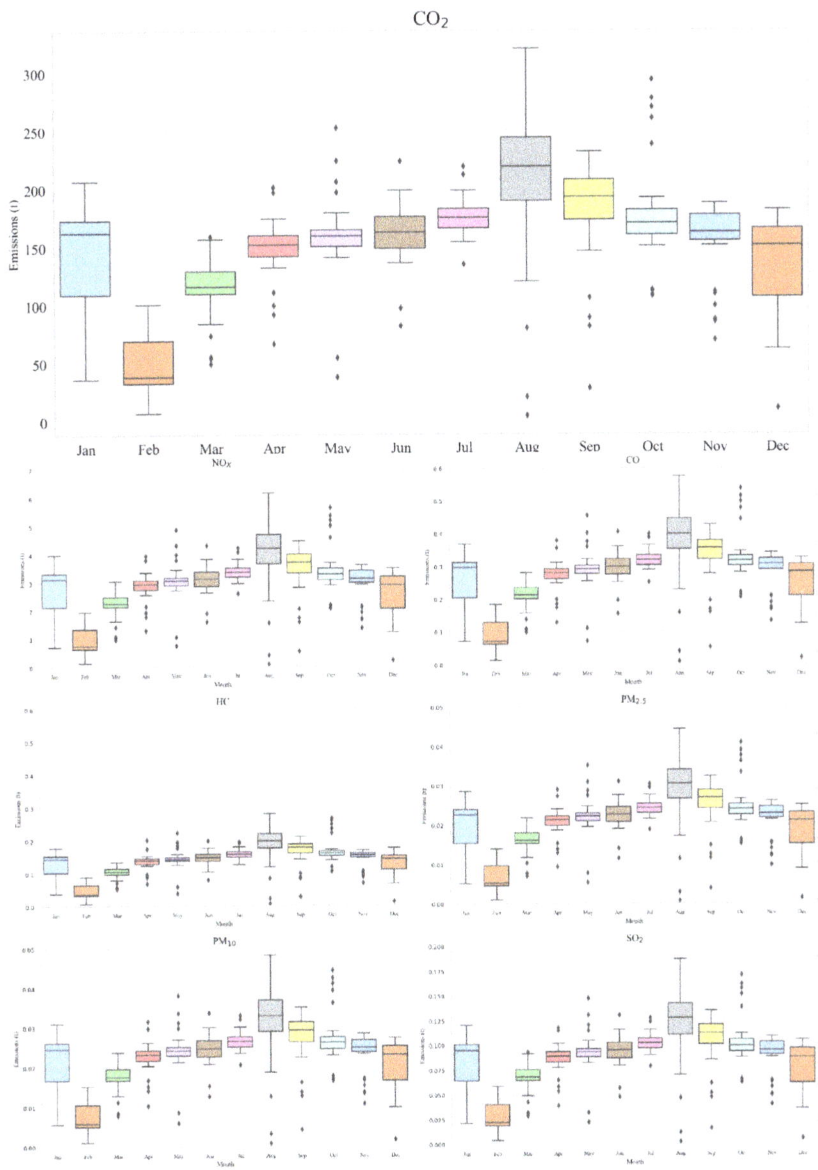

Figure 7. Annual emission distribution of each pollutant.

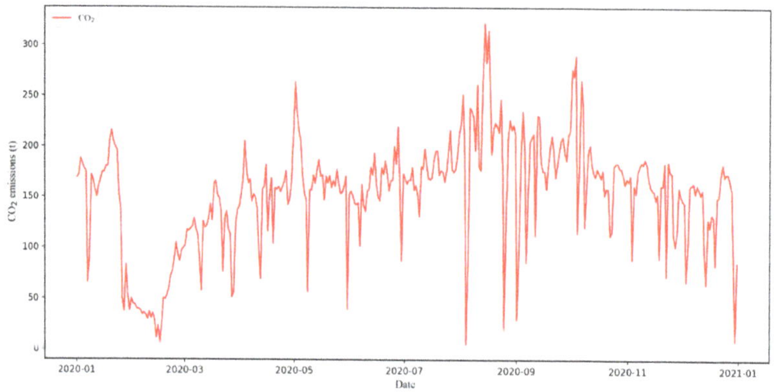

Figure 8. Daily changes in CO_2 emissions.

Table 5. Typhoons affecting Zhoushan area in 2020.

Typhoon No.	Name of the Typhoon	Date of Impact
2004	Hagupit	8.03–8.05
2008	Bavi	8.25–8.26
2009	Maysak	9.01–9.02
2010	Haishen	9.05–9.06

Table 6. Cold waves and gales affecting the Zhoushan area in 2020.

Type of Weather	Elapsed Time of Impact	Level of Warning
Cold wave	4.11–4.12	Severe
Cold wave	12.13–12.15	Moderate
Cold wave	12.28–12.30	Severe
Gale	1.22–1.28	Force 8–9
Gale	3.21	Force 8–11
Gale	4.12	Force 8–10
Gale	9.02–9.04	Force 8–11

During extreme weather events, such as typhoons, passenger ships often suspend operations, particularly in the typhoon-prone months of summer (e.g., August and September). While these suspensions reduce emissions on certain days, compensatory measures around the time of the extreme weather can lead to temporary fluctuations. For instance, after a typhoon, passenger ships may resume operations with increased voyages to recover the lost time, resulting in short-term emission surges. Similarly, some voyages are scheduled during brief weather windows when conditions permit, leading to detours to avoid typhoon paths or increased fuel consumption due to rough sea conditions. These factors collectively contribute to monthly emission outliers, where single-day or single-voyage CO_2 emissions were notably high. The combination of suspended operations during seasonal extreme weather and subsequent compensatory voyages results in significant variability in annual emissions.

5.6.3. Emissions from Different Types of Passenger Ships

The emissions data for different types of passenger ships are shown in Figure 9. The CO_2 emission patterns of the four ship types varied significantly, influenced by the seasonal demand and ship characteristics. Ro-Ro passenger ships exhibited the largest fluctuations in emissions, with their annual emissions ranging from 20 to 100 tons. These ships reach peak emissions during the summer tourism seasons, particularly in August, showing a strong correlation between peak emissions and seasonal demand. Ordinary passenger ships had relatively stable emissions, ranging from 5 to 50 tons. While the emissions increased slightly during summer, the overall fluctuations were minimal, indicating balanced operational characteristics. Car–passenger ferries had the lowest emissions, ranging from approximately 0 to 7 tons. Their seasonal variation was minimal, and the difference between off-season and peak-season emissions was negligible, highlighting their low-emission and steady operational profile. High-speed passenger ships demonstrated significant emission variability, with peaks occurring in January, August, and December, reaching levels of up to 200 tons. This indicates their frequent use during peak passenger periods.

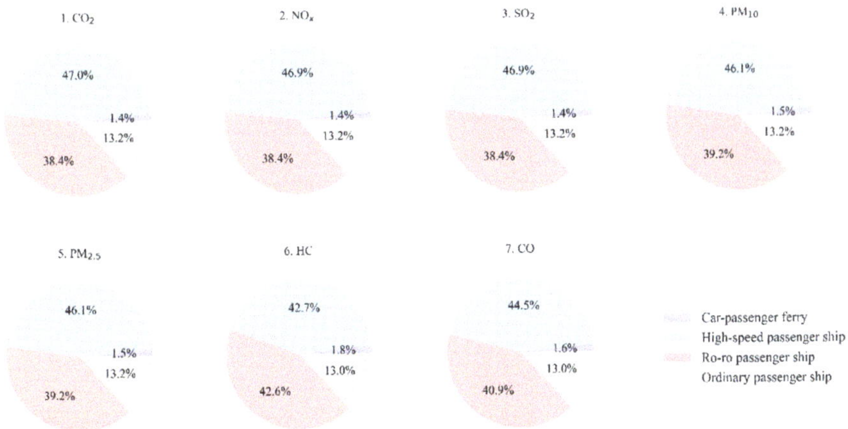

Figure 9. Percentage of emissions of each pollutant from different types of passenger ships.

Figure 9 also illustrates the contribution of each ship type to the pollutant emissions. High-speed passenger ships, which made up 31% of all passenger ships, accounted for 41.7% of the total main engine power and only 8.9% of the total tonnage, yet contributed the highest CO_2 emissions, amounting to 26,054.1 tons. This was largely due to their frequent activities, including multiple port arrivals and departures. Despite regulatory restrictions on night-time operations for high-speed passenger ships, their emissions remained notably high and warrant further attention. Ro-Ro passenger ships ranked second in emissions. Although their average main engine power is comparable to high-speed ships, they carry significantly more passengers and vehicles. With the largest total tonnage among the ship types, they contributed 38.4% of the CO_2 emissions. Ordinary passenger ships, with the highest passenger capacity and being primarily responsible for passenger transport, contributed only about 13.2% of the total CO_2 emissions. Car–passenger ferries had the lowest emissions, with a total CO_2 output of just 814.2 tons, reflecting their low environmental impact.

The similar HC emission contributions of high-speed passenger ships and Ro-Ro passenger ships may have been due to the commonality of the operating characteristics and fuel usage of the two types of ships. First, high-speed passenger ships and Ro-Ro passenger ships are usually equipped with similar medium speed diesel (MSD) engines, which tend to produce a higher proportion of unburned hydrocarbons (HCs) when operating at low loads. Although there were significant differences between the two types of ships in the emission of other pollutants such as CO_2 and NOx, HC emissions are more sensitive to changes in combustion efficiency. Incomplete engine combustion during low-load and idling operation will lead to increased HC emissions, which is a common operating condition during the berthing, unberthing, and low-speed sailing phases of high-speed passenger ships and Ro-Ro passenger ships.

5.7. Emissions from Different Operating Modes

Based on previous studies, the passenger ship emissions in the Zhoushan area were categorised into four operating modes (berthing, anchoring, unberthing, and cruising) to investigate the characteristics of the emissions under the different modes. The proportions of emissions from each mode are shown in Figure 10. The results indicated that, unlike other types of ships such as cargo ships, tankers, and fishing vessels, passenger ships generated 97% to 99% of their emissions during the cruising phase. This aligns with the operational characteristics of passenger ships, which spend most of their operational time in cruising mode and have relatively short anchoring durations. These findings are also consistent with the conclusions of Peng et al. [61]. It is hypothesised that, under normal circumstances, passenger ships powered by dual main engines shut off their engines when approaching passenger terminals. They rely on water currents, inertia, and their residual speed to reach the berth and wait for passengers to board and disembark. This operational behaviour differs from those of other ships entering and leaving ports, leading to especially low emissions during the berthing, anchoring, and unberthing phases.

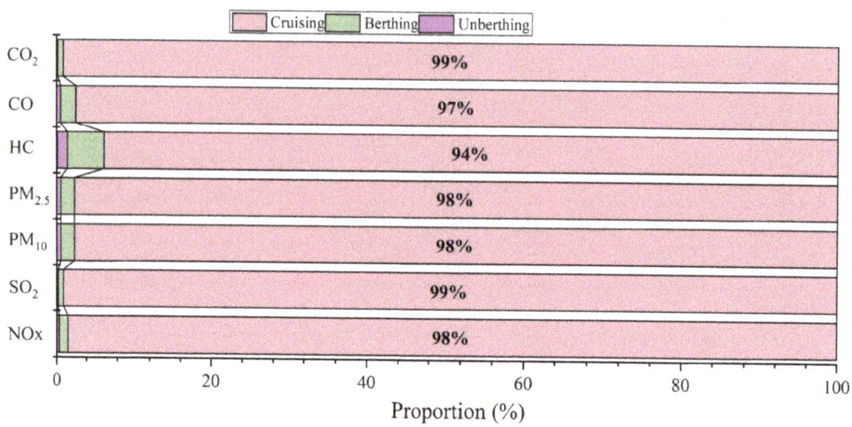

Figure 10. Proportion of emissions from passenger ships during different modes of operation.

5.8. Spatial Distribution

To analyse the spatial distribution of the exhaust emissions, a 0.002° × 0.002° grid was created to rasterise the study area. Based on the geographic coordinates in the AIS data, all the pollutant emissions from passenger ships in the region were aggregated and allocated to the corresponding grid cells. The results are shown in Figure 11.

The spatial distribution of the emissions exhibited clear patterns that correspond to the ships' navigation routes. The emissions along the routes between Daishan Island, Qushan Island, and the Shengsi Archipelago were significantly higher than in other areas. Similarly, there was notable activity on the routes connecting Zhoushan Island with Liuheng Island, Xiazhi Island, and Taohua Island. It is important to note that most passenger ship emissions occurred within designated navigation areas. This can be explained by the nature of passenger ship operations, which primarily serve passenger transport, tourism, and leisure purposes. Before routes are launched, passenger ships will undergo safety risk assessments to ensure compliance with safety requirements. Ships are permitted to only operate in waters that meet passenger safety standards, using routes recommended or designated by administrative authorities, and embark/disembark passengers at designated terminals. In the Zhoushan area, which is crowded with various commercial and fishing vessels, adhering to fixed routes helps reduce the risk of ship collisions and ensures the safety of the passengers onboard.

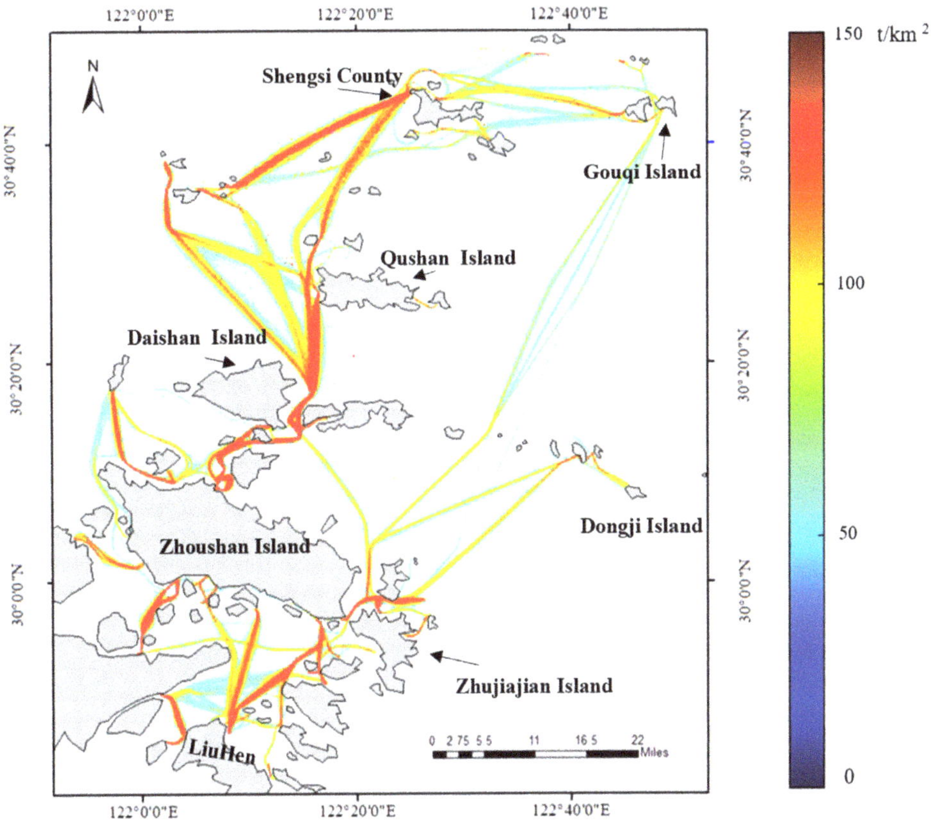

Figure 11. Spatial distribution of annual CO_2 passenger ship emissions.

5.9. Proposed Emission Reduction Measures

Based on the findings of this study, several practical measures can be proposed to reduce emissions from maritime activities. These measures, supported by accurate emissions calculations, could significantly mitigate environmental impacts:

1. Route Optimisation: By analysing vessel traffic patterns and emission hotspots, route optimisation can help minimise unnecessary travel and reduce fuel consumption. Advanced navigation systems and predictive algorithms can identify the most energy-efficient paths, avoiding congested or high-emission zones. This approach not only reduces emissions but also improves operational efficiency.
2. Speed Reduction Programs (Slow Steaming): Lowering vessel speeds, especially in emission-sensitive or congested areas, can significantly reduce fuel consumption and emissions.
3. Training and Awareness Programs: Educating ship operators and crews about energy-efficient practices, such as optimal engine operation and maintenance, can improve compliance and operational efficiency. Awareness programs can highlight the environmental and economic benefits of adopting sustainable practices.

While these measures represent actionable strategies, their successful implementation depends on integrating emissions data into decision-making frameworks. For instance, the granular emission calculations presented in this study can identify high-impact areas for policy intervention or guide investments in cleaner technologies. Future research could explore the cost-effectiveness and practical barriers associated with these measures to provide comprehensive guidance for policymakers and industry stakeholders

6. Conclusions

This study analysed pollutant emissions from short-distance passenger ships in the Zhoushan Archipelago from January to December 2020 using high-resolution AIS data. The analysis incorporated various meteorological factors and detailed passenger ship specifications, employing the bottom–up approach to comprehensively investigate the emission characteristics of passenger ships of different types.

The results indicate that the activity-based bottom–up approach is suitable for analysing frequently operating passenger ships. During the selected timeframe, the total emissions of CO_2, NO_x, CO, $PM_{2.5}$, PM_{10}, SO_2, and HCs were calculated to be 54,895.75 tons, 1057.82 tons, 101.46 tons, 7.68 tons, 8.33 tons, 32.02 tons, and 50.34 tons, respectively. High-speed passenger ships accounted for the highest proportion of the emissions, accounting for approximately 42.1–46.8%, followed by Ro-Ro passenger ships at 37.7–42.2%. The emissions from ordinary passenger ships and car–passenger ferries were relatively lower, contributing 13.8–14.2% and 1.3–1.7%, respectively. When comparing the emissions across different operating modes, cruising accounted for the largest share of pollutant emissions, with the other stages (berthing, anchoring, and manoeuvring) contributing only 1–3%. This was attributed to the unique operational characteristics of passenger ships. The two primary activity areas for passenger ships were between Zhoushan Island, Daishan Island, and the Qushan Archipelago, and between Liuheng Island, Xiazhi Island, and the Taohua Archipelago. These two regions saw the highest emissions due to the frequency of passenger ship activities.

This study utilised high-precision ship attribute data, addressing issues in traditional emission calculations such as inaccuracies related to meteorological factors and main engine power deviations. The findings demonstrate that passenger ship emissions follow a discernible pattern. Clarifying activity-based emission characteristics is crucial for regulatory authorities to formulate effective emission reduction strategies and serves as a reference for passenger transport companies in managing emissions. Additionally, it indirectly supports low-carbon and energy-saving practices in ports, enhancing the resilience of future port and marine environmental management.

However, this study has certain limitations. When applying the model to other regions, the availability of foundational data remains a constraint. Although regulatory authorities mandate that passenger ships use fuel with a sulphur content not exceeding 0.5%, discrepancies between requirements and actual implementation may lead to underestimations of SO_2 emissions. Furthermore, it is challenging to accurately obtain per-voyage fuel consumption data. Future research integrating voyage-specific fuel consumption data could provide a more reliable method for improving precision.

Author Contributions: X.X.: Conception and design of the study, data analysis, interpretation of results, manuscript writing; X.L.: Data collection, data interpretation; L.F.: Literature search, manuscript revision; W.Y.Y.: Statistical analysis, manuscript revision; H.F.: Study design and guidance. All authors have read and agreed to the published version of the manuscript.

Funding: This work was sponsored by K.C. Wong Magna Fund in Ningbo University, China and 111 Project [grant number D21013].

Institutional Review Board Statement: Not applicable.

Informed Consent Statement: Not applicable.

Data Availability Statement: The raw data supporting the conclusions of this article will be made available by the authors on request.

Conflicts of Interest: The authors declare no conflict of interest.

References

1. Wang, L.; Li, Y. Estimation methods and reduction strategies of port carbon emissions-what literatures say? *Mar. Pollut. Bull.* **2023**, *195*, 115451. [CrossRef] [PubMed]
2. Ytreberg, E.; Åström, S.; Fridell, E. Valuating environmental impacts from ship emissions–The marine perspective. *J. Environ. Manag.* **2021**, *282*, 111958. [CrossRef]
3. Kose, S.; Sekban, D. Emphasizing the importance of using cold-ironing technology by determining the share of hotelling emission value within the total emission. *Transp. Saf. Environ.* **2022**, *4*, tdac010. [CrossRef]
4. Gray, N.; McDonagh, S.; O'Shea, R.; Smyth, B.; Murphy, J.D. Decarbonising ships, planes and trucks: An analysis of suitable low-carbon fuels for the maritime, aviation and haulage sectors. *Adv. Appl. Energy* **2021**, *1*, 100008. [CrossRef]
5. Chen, X.; Zheng, J.; Li, C.; Wu, B.; Wu, H.; Montewka, J. Maritime traffic situation awareness analysis via high-fidelity ship imaging trajectory. *Multimed. Tools Appl.* **2024**, *83*, 48907–48923. [CrossRef]
6. Shang, W.-L.; Zhang, J.; Wang, K.; Yang, H.; Ochieng, W. Can financial subsidy increase electric vehicle (EV) penetration---evidence from a quasi-natural experiment. *Renew. Sustain. Energy Rev.* **2024**, *190*, 114021. [CrossRef]
7. Van Roy, W.; Merveille, J.-B.; Van Nieuwenhove, A.; Scheldeman, K.; Maes, F. Policy recommendations for international regulations addressing air pollution from ships. *Mar. Policy* **2024**, *159*, 105913. [CrossRef]
8. Liu, J.; Jiang, X.; Huang, W.; He, Y.; Yang, Z. A novel approach for navigational safety evaluation of inland waterway ships under uncertain environment. *Transp. Saf. Environ.* **2022**, *4*, tdab029. [CrossRef]
9. Kumar, P.; Beig, G.; Sahu, S.; Yadav, R.; Maji, S.; Singh, V.; Bamniya, B. Development of a high-resolution emissions inventory of carbonaceous particulate matters and their growth during 2011–2018 over India. *Atmos. Environ.* **2023**, *303*, 119750. [CrossRef]
10. Zhang, K.; Lin, Q.; Lian, F.; Feng, H. Estimating emissions from fishing vessels: A big Beidou data analytical approach. *Front. Mar. Sci.* **2024**, *11*, 1418366. [CrossRef]
11. Zhai, J.; Yu, G.; Zhang, J.; Shi, S.; Yuan, Y.; Jiang, S.; Xing, C.; Cai, B.; Zeng, Y.; Wang, Y. Impact of ship emissions on air quality in the Greater Bay Area in China under the latest global marine fuel regulation. *Environ. Sci. Technol.* **2023**, *57*, 12341–12350. [CrossRef] [PubMed]
12. Han, Y.; Ma, W.; Ma, D. Green maritime: An improved quantum genetic algorithm-based ship speed optimization method considering various emission reduction regulations and strategies. *J. Clean. Prod.* **2023**, *385*, 135814. [CrossRef]
13. IMO. *Guidelines for the Installation of a Shipborne Automatic Identification System (AIS)*; International Maritime Organization: London, UK, 2002.
14. Nunes, R.; Alvim-Ferraz, M.; Martins, F.; Sousa, S. The activity-based methodology to assess ship emissions-A review. *Environ. Pollut.* **2017**, *231*, 87–103. [CrossRef] [PubMed]
15. You, Y.; Lee, J.C. Activity-based evaluation of ship pollutant emissions considering ship maneuver according to transportation plan. *Int. J. Nav. Archit. Ocean. Eng.* **2022**, *14*, 100427. [CrossRef]
16. Tzannatos, E. Ship emissions and their externalities for Greece. *Atmos. Environ.* **2010**, *44*, 2194–2202. [CrossRef]
17. Corbett, J.J.; Koehler, H.W. Updated emissions from ocean shipping. *J. Geophys. Res. Atmos.* **2003**, *108*, 4650. [CrossRef]
18. Chen, X.; Liu, S.; Liu, R.W.; Wu, H.; Han, B.; Zhao, J.J.O. Quantifying Arctic oil spilling event risk by integrating an analytic network process and a fuzzy comprehensive evaluation model. *Ocean. Coast. Manag.* **2022**, *228*, 106326. [CrossRef]
19. Chen, X.; Xian, J.; Liu, S.; Zhao, J.; Wu, H.; Montewka, J. Autonomous port management based AGV path planning and optimization via an ensemble reinforcement learning framework. *Ocean. Coast. Manag.* **2024**, *251*, 107087. [CrossRef]
20. Li, Y.; Zhang, Y.; Cheng, J.; Zheng, C.; Li, M.; Xu, H.; Wang, R.; Chen, D.; Wang, X.; Fu, X.; et al. Comparative Analysis, Use Recommendations, and Application Cases of Methods for Develop Ship Emission Inventories. *Atmosphere* **2022**, *13*, 1224. [CrossRef]
21. Jalkanen, J.-P.; Brink, A.; Kalli, J.; Pettersson, H.; Kukkonen, J.; Stipa, T. A modelling system for the exhaust emissions of marine traffic and its application in the Baltic Sea area. *Atmos. Chem. Phys.* **2009**, *9*, 9209–9223. [CrossRef]
22. Winther, M.; Christensen, J.H.; Plejdrup, M.S.; Ravn, E.S.; Eriksson, Ó.F.; Kristensen, H.O. Emission inventories for ships in the arctic based on satellite sampled AIS data. *Atmos. Environ.* **2014**, *91*, 1–14. [CrossRef]
23. Goldsworthy, L.; Goldsworthy, B. Modelling of ship engine exhaust emissions in ports and extensive coastal waters based on terrestrial AIS data–An Australian case study. *Environ. Model. Softw.* **2015**, *63*, 45–60. [CrossRef]
24. Pokhrel, R.; Lee, H. Estimation of air pollution from the OGVs and its dispersion in a coastal area. *Ocean. Eng.* **2015**, *101*, 275–284. [CrossRef]

25. Zhou, C.; Huang, H.; Liu, Z.; Ding, Y.; Xiao, J.; Shu, Y. Identification and analysis of ship carbon emission hotspots based on data field theory: A case study in Wuhan Port. *Ocean. Coast. Manag.* **2023**, *235*, 106479. [CrossRef]
26. Chen, D.; Wang, X.; Li, Y.; Lang, J.; Zhou, Y.; Guo, X.; Zhao, Y. High-spatiotemporal-resolution ship emission inventory of China based on AIS data in 2014. *Sci. Total Environ.* **2017**, *609*, 776–787. [CrossRef] [PubMed]
27. Toscano, D.; Murena, F.; Quaranta, F.; Mocerino, L. Assessment of the impact of ship emissions on air quality based on a complete annual emission inventory using AIS data for the port of Naples. *Ocean. Eng.* **2021**, *232*, 109166. [CrossRef]
28. Ye, J.; Chen, J.; Wen, H.; Wan, Z.; Tang, T. Emissions assessment of bulk carriers in China's east Coast-Yangtze River maritime network based on different shipping modes. *Ocean. Eng.* **2022**, *249*, 110903. [CrossRef]
29. Li, H.; Jia, P.; Wang, X.; Yang, Z.; Wang, J.; Kuang, H. Ship carbon dioxide emission estimation in coastal domestic emission control areas using high spatial-temporal resolution data: A China case. *Ocean. Coast. Manag.* **2023**, *232*, 106419. [CrossRef]
30. Zhi, G.; Du, J.; Chen, A.; Jin, W.; Ying, N.; Huang, Z.; Xu, P.; Wang, D.; Ma, J.; Zhang, Y. Progression of an emission inventory of China integrating CO_2 with air pollutants. A chance to learn the influence of development on emissions. *Atmos. Environ.* **2024**, *316*, 120184. [CrossRef]
31. Orlandi, A.; Calastrini, F.; Kalikatzarakis, M.; Guarnieri, F.; Busillo, C.; Coraddu, A. Air quality forecasting of along-route ship emissions in realistic meteo-marine scenarios. *Ocean. Eng.* **2024**, *291*, 116464. [CrossRef]
32. Moreno-Gutiérrez, J.; Durán-Grados, V. Calculating ships' real emissions of pollutants and greenhouse gases: Towards zero uncertainties. *Sci. Total Environ.* **2021**, *750*, 141471. [CrossRef] [PubMed]
33. Carić, H.; Klobučar, G.; Štambuk, A. Ecotoxicological risk assessment of antifouling emissions in a cruise ship port. *J. Clean. Prod.* **2016**, *121*, 159–168. [CrossRef]
34. Dragović, B.; Tzannatos, E.; Tselentis, V.; Meštrović, R.; Škurić, M. Ship emissions and their externalities in cruise ports. *Transp. Res. Part D: Transp. Environ.* **2018**, *61*, 289–300. [CrossRef]
35. Chen, Q.; Lau, Y.-Y.; Ge, Y.-E.; Dulebenets, M.A.; Kawasaki, T.; Ng, A.K. Interactions between Arctic passenger ship activities and emissions. *Transp. Res. Part D Transp. Environ.* **2021**, *97*, 102925. [CrossRef]
36. Chen, Q.; Ge, Y.-E.; Lau, Y.-Y.; Dulebenets, M.A.; Sun, X.; Kawasaki, T.; Mellalou, A.; Tao, X. Effects of COVID-19 on passenger shipping activities and emissions: Empirical analysis of passenger ships in Danish waters. *Marit. Policy Manag.* **2023**, *50*, 776–796. [CrossRef]
37. Huang, H.; Zhou, C.; Xiao, C.; Huang, L.; Wen, Y.; Wang, J.; Peng, X. Effect of seasonal flow field on inland ship emission assessment: A case study of ferry. *Sustainability* **2020**, *12*, 7484. [CrossRef]
38. Chou, C.-C.; Hsu, H.-P.; Wang, C.-N.; Yang, T.-L. Analysis of energy efficiencies of in-port ferries and island passenger-ships and improvement policies to reduce CO_2 emissions. *Mar. Pollut. Bull.* **2021**, *172*, 112826. [CrossRef] [PubMed]
39. Mannarini, G.; Salinas, M.L.; Carelli, L.; Fassò, A. How COVID-19 affected GHG emissions of ferries in Europe. *Sustainability* **2022**, *14*, 5287. [CrossRef]
40. Baird, A.J.; Pedersen, R.N. Analysis of CO_2 emissions for island ferry services. *J. Transp. Geogr.* **2013**, *32*, 77–85. [CrossRef]
41. Cooper, D. Exhaust emissions from high speed passenger ferries. *Atmos. Environ.* **2001**, *35*, 4189–4200. [CrossRef]
42. Mojarrad, M.; Thorne, R.J.; Rødseth, K.L. Technical and cost analysis of zero-emission high-speed ferries: Retrofitting from diesel to green hydrogen. *Heliyon* **2024**, *10*, e27479. [CrossRef] [PubMed]
43. Zhou, C.; Tang, W.; Ding, Y.; Huang, H.; Xu, H. Analysis of Carbon Emission Reduction Paths for Ships in the Yangtze River: The Perspective of Alternative Fuels. *J. Mar. Sci. Eng.* **2024**, *12*, 947. [CrossRef]
44. Chen, X.; Chen, W.; Wu, B.; Wu, H.; Xian, J. Ship visual trajectory exploitation via an ensemble instance segmentation framework. *Ocean. Eng.* **2024**, *313*, 119368. [CrossRef]
45. Shang, W.-L.; Chen, Y.; Yu, Q.; Song, X.; Chen, Y.; Ma, X.; Chen, X.; Tan, Z.; Huang, J.; Ochieng, W. Spatio-temporal analysis of carbon footprints for urban public transport systems based on smart card data. *Appl. Energy* **2023**, *352*, 121859. [CrossRef]
46. Mao, J.; Zhang, Y.; Yu, F.; Chen, J.; Sun, J.; Wang, S.; Zou, Z.; Zhou, J.; Yu, Q.; Ma, W. Simulating the impacts of ship emissions on coastal air quality: Importance of a high-resolution emission inventory relative to cruise-and land-based observations. *Sci. Total Environ.* **2020**, *728*, 138454. [CrossRef] [PubMed]
47. Chen, D.; Zhao, Y.; Nelson, P.; Li, Y.; Wang, X.; Zhou, Y.; Lang, J.; Guo, X. Estimating ship emissions based on AIS data for port of Tianjin, China. *Atmos. Environ.* **2016**, *145*, 10–18. [CrossRef]
48. Chen, S.; Wang, F.; Wei, X.; Tan, Z.; Wang, H.J.T.R.R. Analysis of tugboat activities using AIS data for the Tianjin port. *Transp. Res. Rec. J. Transp. Res. Board* **2020**, *2674*, 498–509. [CrossRef]
49. Tian, Y.; Ren, L.; Wang, H.; Li, T.; Yuan, Y.; Zhang, Y. Impact of AIS data thinning on ship air pollutant emissions inventories. *Atmosphere* **2022**, *13*, 1135. [CrossRef]
50. Chen, X.; Yang, J. Analysis of the uncertainty of the AIS-based bottom-up approach for estimating ship emissions. *Mar. Pollut. Bull.* **2024**, *199*, 115968. [CrossRef]
51. Shu, Y.; Hu, A.; Zheng, Y.; Gan, L.; Xiao, G.; Zhou, C.; Song, L. Evaluation of ship emission intensity and the inaccuracy of exhaust emission estimation model. *Ocean. Eng.* **2023**, *287*, 115723. [CrossRef]

52. Grigoriadis, A.; Mamarikas, S.; Ioannidis, I.; Majamäki, E.; Jalkanen, J.-P.; Ntziachristos, L. Development of exhaust emission factors for vessels: A review and meta-analysis of available data. *Atmos. Environ. X* **2021**, *12*, 100142. [CrossRef]
53. Limited, E.U. *Quantification of Emissions from Ships Associated with Ship Movements Between Ports in the European Community*; European Commission: Brussels, Belgium, 2002.
54. Group, S.C. Port of Los Angeles Inventory of Air Emissions-2012. 2013. Available online: https://kentico.portoflosangeles.org/getmedia/6cbdfeec-49c0-4758-b2b0-8fab74a79578/2012_Air_Emissions_Inventory (accessed on 3 March 2024).
55. Fu, Q.; Shen, Y.; Zhang, J. On the ship pollutant emission inventory in Shanghai port. *J. Saf. Environ.* **2012**, *12*, 57–64.
56. Zhao, J.; Zhang, Y.; Patton, A.P.; Ma, W.; Kan, H.; Wu, L.; Fung, F.; Wang, S.; Ding, D.; Walker, K. Projection of ship emissions and their impact on air quality in 2030 in Yangtze River delta, China. *Environ. Pollut.* **2020**, *263*, 114643. [CrossRef] [PubMed]
57. Huang, L.; Wen, Y.; Geng, X.; Zhou, C.; Xiao, C. Integrating multi-source maritime information to estimate ship exhaust emissions under wind, wave and current conditions. *Transp. Res. Part D Transp. Environ.* **2018**, *59*, 148–159. [CrossRef]
58. Chen, S.; Meng, Q.; Jia, P.; Kuang, H. An operational-mode-based method for estimating ship emissions in port waters. *Transp. Res. Part D Transp. Environ.* **2021**, *101*, 103080. [CrossRef]
59. Lin, Q.; Yin, B.; Zhang, X.; Grifoll, M.; Feng, H. Evaluation of ship collision risk in ships' routeing waters: A Gini coefficient approach using AIS data. *Phys. A Stat. Mech. Its Appl.* **2023**, *624*, 128936. [CrossRef]
60. Feng, H.; Grifoll, M.; Yang, Z.; Zheng, P. Collision risk assessment for ships' routeing waters: An information entropy approach with Automatic Identification System (AIS) data. *Ocean. Coast. Manag.* **2022**, *224*, 106184. [CrossRef]
61. Peng, X.; Ding, Y.; Yi, W.; Laroussi, I.; He, T.; He, K.; Liu, H. The inland waterway ship emission inventory modeling: The Yangtze River case. *Transp. Res. Part D Transp. Environ.* **2024**, *129*, 104138. [CrossRef]

Journal of
Marine Science and Engineering

MDPI

Article

Optimization of Inbound and Outbound Vessel Scheduling in One-Way Channel Based on Reinforcement Learning

Rong Zhen, Meng Sun and Qionglin Fang *

Navigation College, Jimei University, Xiamen 361021, China; zrandsea@163.com (R.Z.); 13462982272@163.com (M.S.)
* Correspondence: fql1437@163.com

Abstract: As the size and number of ships continue to grow, effective management of vessel scheduling has become more and more important for the efficient one-way channel port operation, whose characteristics significantly affect the safety and efficiency of ports. This paper presents a reinforcement-learning-based approach to optimize the scheduling of vessels in a one-way channel, aiming to quickly identify a scheduling solution that enhances port operational efficiency. This method models the vessel scheduling problem in a one-way channel by incorporating navigational constraints, safety requirements, and vessel-specific characteristics. Using the Q-learning algorithm to minimize vessel wait times, it identifies an optimal scheduling solution. Experiments were conducted using real data from the Dayao Bay Pier of Dalian Port to validate the rationality and effectiveness of the proposed model and algorithm. The results show that the reinforcement learning approach achieved approximately a 16% improvement in solution quality compared to the genetic algorithm (GA) while requiring only half the computation time. Additionally, it reduced delay times by over 40% relative to the traditional FCFS strategy, indicating superior overall performance. This research presents an efficient, intelligent approach to vessel scheduling, providing a theoretical foundation for further advancements in this field and enhancing decision support for vessel scheduling in one-way channels with practical implications.

Keywords: inbound and outbound scheduling; one-way channel; reinforcement learning

1. Introduction

As a critical node in global trade, port operational efficiency directly affects the smooth flow of goods and economic benefits [1,2]. However, with the rising number of vessels and channel capacity constraints, port management faces numerous challenges [3]. One key challenge is the scheduling of ship arrivals and departures, with channel arrangements typically supervised by vessel traffic services (VTS) [4,5]. Channel management directly influences the timing of ships' arrivals at berths and departures from the port. In many domestic ports, due to constraints such as natural conditions, the difficulty of channel expansion, and financial limitations, one-way channels remain the primary mode of operation [6,7]. Under this model, ships must strictly follow the port authority's designated time windows for arrivals and departures, which can lead to disorganization in port operations. Inefficient or delayed scheduling may cause port congestion, posing challenges for both ships and the port. By implementing an efficient scheduling system, we can boost the throughput of vessels at ports, enhance operational efficiency, conserve energy resources, and lower carbon emissions, all of which are crucial steps toward meeting China's emission reduction goals [8,9]. Therefore, to alleviate congestion and enhance the safety and efficiency of ship scheduling [10], designing an intelligent and effective scheduling plan is essential for improving port operational capacity and ensuring maritime safety.

The port VTS coverage area typically includes the harbor basin, channels, and anchorages [11,12]. Currently, the most commonly used ship scheduling strategy is the first-come, first-served (FCFS) approach [13], where the VTS management center schedules ships' entry and exit times based on the order of their pre-reported arrival and departure times. However, this approach neglects other valuable information about ships, potentially leading to extended delays during peak arrival periods. This not only increases ship waiting times but also reduces the efficiency of port resource utilization and may even result in traffic congestion and safety risks. In this context, finding an intelligent and flexible scheduling method has become increasingly urgent. To address the above issues, this paper first develops a model that simulates actual port operations and explores the key challenges in one-way channel vessel scheduling. The model takes into account factors such as channel capacity, ship size, and sailing speed, aiming to closely resemble real port environments. Based on this, reinforcement learning techniques are introduced into the field of vessel scheduling, and an intelligent scheduling algorithm suitable for one-way channel environments is designed. This algorithm autonomously learns from real-time information and continuously optimizes scheduling strategies, making the ship entry and exit process more organized and efficient. Finally, through testing with real port data, this paper demonstrates that the reinforcement learning-based scheduling method has significant advantages over traditional approaches. Specifically, the innovations of this paper include:

(1) Method Improvement: By combining reinforcement learning with unidirectional channel ship scheduling problem, we comprehensively consider channel capacity, ship size, sailing speed, and other factors, and we strive to be close to the real port environment and build a more efficient scheduling model.
(2) Algorithm Optimization: A strategy of gradually approaching Q value in reinforcement learning algorithm is adopted to reduce the resource consumption and shorten the running time of the intelligent scheduling algorithm for ships entering and leaving ports in one-way channel.

The remainder of this paper is organized as follows: Section 2 provides a detailed review of the current state of vessel scheduling research, discussing the main achievements and limitations in existing literature and highlighting the focus and trends in current studies. Section 3 presents the model and solution methods for the one-way channel vessel scheduling problem, explaining the reinforcement learning framework and technical details used. Section 4 discusses numerical experiments and results analysis, verifying the effectiveness and superiority of the proposed method through specific case studies and providing a comprehensive evaluation of its performance. Section 5 concludes the paper and suggests future research directions, offering ideas and recommendations for further exploration. The overall research process is illustrated in Figure 1.

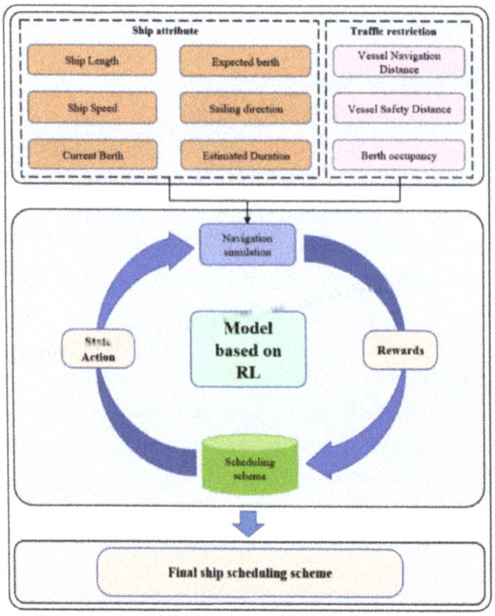

Figure 1. Overall flowchart.

2. Literature Review

2.1. Current Research on Ship Scheduling

Inbound and outbound ship scheduling is a critical process for establishing the queue of ships arriving at and departing from the port, with the primary goal of determining their order. This scheduling process optimizes port operations by ensuring safe and efficient ship movements and maximizing resource utilization. Research on this issue primarily focuses on three areas: scheduling optimization for one-way, two-way, and compound channels, each with distinct requirements and challenges.

Inefficient ship scheduling can lead to severe congestion, disrupting the planned berthing and departure arrangements for inbound and outbound ships, particularly in one-way channels, where it can cause significant delays. To ensure navigational safety, one-way channels prohibit ships from traveling in opposite directions simultaneously. As a result, total waiting time becomes a key metric for measuring the efficiency of port waterways [14]. Existing research mainly focuses on reducing ship waiting times through various optimization methods. Xu et al. [15] proposed an optimal scheduling model for inbound and outbound ships in one-way channels based on onboard AIS data, considering factors such as berth proximity, ship size, type, and draft. Building on this, Zheng et al. [16] considered factors such as the safe navigation distance between ships, alternating entry and exit time windows, and clustered entry and exit rules. They developed a mixed-integer linear programming model aimed at minimizing the total waiting time for inbound ships. Xu et al. [17] considered tidal and water depth requirements and optimized the completion time for all arriving ships. They developed a parallel machine scheduling model for berth allocation with time windows and designed a corresponding heuristic algorithm. Ting et al. [18] proposed a mixed-integer programming model aimed at minimizing the waiting and operating times for arriving ships, addressing the dynamic discrete berth allocation problem. They applied a particle swarm optimization algorithm to solve the model. Mauri et al. [19] developed a mathematical model for berth allocation that accounts for the spatial and temporal constraints required for ship berthing. This model applies to both continuous and discrete berths, and they designed an adaptive large neighborhood search algorithm to solve it. Bai [20] enhanced the artificial fish swarm algorithm to improve search precision and local optimization capability, significantly reducing total delay time due to ship arrival sequencing. Li [21] integrated reinforcement learning with a genetic algorithm to substantially shorten total waiting times for ships in one-way channels while ensuring safe navigation. Zhang et al. [22] developed a scheduling optimization model aimed at minimizing total waiting time and designed a simulated annealing multi-population genetic algorithm suitable for port ship scheduling, validating it through simulation experiments.

In recent years, two-way channels have been widely implemented in major ports to alleviate navigational pressure and enhance efficiency. For two-way channels, Zhang et al. [23] proposed a tidal-influenced ship scheduling model and algorithm to optimize the sequencing of inbound and outbound ships under tidal conditions. This approach reduces waiting times and enhances channel operational efficiency. Building on this, Zhang et al. [24] developed a multi-objective optimization model for two-way port vessel scheduling that incorporates constraints for safety, continuity, and efficiency, taking into account ship attributes and traffic conditions. This model improves scheduling safety and navigational efficiency. Wang [25] proposed a hybrid self-organizing scheduling (HSOS) method for restricted two-way channels that employs a priority mechanism to reduce traffic conflicts, significantly lowering the average delay for large ships under high arrival rates. Meisel et al. [26] addressed the traffic management challenges of the Kiel Canal using multiple optimization models and heuristic algorithms, significantly improving ship transit efficiency and service quality.

Compound channels are relatively advanced, and research on them remains limited both domestically and internationally. Gong [27] proposed an optimized scheduling and tug assignment plan for ports, balancing traffic scheduling rules with tug availability constraints. Zhang et al. [28] developed a ship scheduling model for a typical tidal compound channel, using real tidal data and a genetic algorithm to optimize vessel scheduling in the Tianjin Port channel. Results showed that this approach outperformed other scheduling methods and could be applied to other tide-affected compound channels.

Although extensive research has been conducted on one-way channel ship scheduling, most studies rely on traditional heuristic algorithms, such as mixed-integer linear programming, particle swarm optimization, and genetic algorithms. While these methods can find feasible solutions within relatively short timeframes, they often depend on predefined rules or parameters, making it difficult to guarantee a globally optimal solution [29]. As problem size increases, the computational complexity of traditional algorithms rises sharply, resulting in longer solution times and limiting their suitability for real-time scheduling needs.

2.2. Reinforcement Learning in Combinatorial Optimization

Reinforcement learning is a trial-and-error learning method based on Markov decision processes (MDP) [30,31]. As a key paradigm in machine learning, it enables an agent to optimize its behavior policy through interactions with the environment to achieve specific objectives [32]. Unlike traditional combinatorial optimization methods, reinforcement learning does not rely on precise models, giving it a distinct advantage in addressing complex, hard-to-model problems, particularly in applications with high uncertainty and dynamic environments [33].

In recent years, an increasing number of scholars have sought to apply reinforcement learning to solve combinatorial optimization problems [34]. Among these, the most representative problem is the traveling salesman problem (TSP). Ottoni et al. [35] proposed a version of the TSP with fuel constraints and applied reinforcement learning to solve it. Through multiple case studies, they demonstrated the effectiveness of this approach, achieving optimal results and highlighting the advantages of reinforcement learning in combinatorial optimization problems. Gambardella et al. [36] studied the superior performance of the Ant-Q algorithm in solving problems like the TSP, demonstrating its powerful capabilities in combinatorial optimization. Duan et al. [37] proposed a variable speed control strategy based on Q-learning, significantly improving traffic flow efficiency in bottleneck areas and enhancing overall traffic conditions. The application of reinforcement learning in traffic management has also been expanding. Fotuhi et al. [38] introduced a Q-learning-based model for stacker crane operations, optimizing the service sequence of hauling trucks to minimize waiting times. Huang et al. mentioned that it has also been applied in taxi dispatch in recent years [39]. It is also widely used in traffic trajectory prediction [40]. Xiang et al. [41] developed a reinforcement learning model for outbound flights at multi-airport terminal areas, using it to intelligently sort incoming and outgoing flights, thereby effectively reducing flight delays and the number of delayed aircraft, while lowering average delay times. Zhu et al. [42] applied Q-learning algorithms to optimize air traffic flow, significantly alleviating congestion at flight route intersections with positive results. Wang [43] combined the PPO algorithm with simulation modeling to simulate ship traffic flow in crossing water areas, showcasing the potential of reinforcement learning in managing traffic in complex water environments. Du et al. [44] developed a machine-learning-based model for liner transportation scheduling to address uncertainties in speed adjustments, validating its effectiveness in shipping optimization.

In summary, reinforcement learning, with its ability to learn and adapt in dynamic environments, demonstrates excellent performance in solving complex combinatorial optimization problems, overcoming the limitations of traditional optimization algorithms, such as poor adaptability and high resource consumption. Therefore, this paper applied reinforcement learning to optimize the scheduling of ships entering and exiting a one-way channel, with the goal of reducing ship delay times, improving port scheduling efficiency, and providing innovative technical solutions for port management.

3. Model of Ship Scheduling Problem in One-Way Channels

Port ship operations are a continuous process, and a safe and feasible scheduling plan should be proactive, capable of anticipating potential urgent situations that ships may encounter in the channel, while taking preventive measures in advance [45]. In ports with one-way navigation, the core of ship scheduling lies in the scientific and rational arrangement of the sequence and timing of ships entering and exiting the channel. This not only concerns channel safety but also directly impacts the overall navigation efficiency of the port.

3.1. Construction of One-Way Channel Ship Scheduling Model

The diagram of the one-way channel is shown in Figure 2. The ship arrival and departure process can be divided into three stages: ship arrival, ship departure, and the transition between arrival and departure. For the ship arrival process, when the arriving ship reaches the reporting line, it receives instructions from the VTS [46]. The VTS will develop an arrival plan in advance based on factors such as the ship's scheduled arrival, the distribution of anchorages outside the port, the availability of berths inside the port, and operational information within and outside the port. At the appropriate time, the VTS will issue the arrival instructions to the ship. Upon receiving the instructions, the ship carries out the arrival operation, enters the corresponding harbor via the channel, and reaches the designated berth. For the ship departure process, similar to arrival, after completing operations, the ship needs to submit a departure request to the VTS. After receiving the departure instruction from the VTS, the ship proceeds with steps such as document processing, unmooring, and departure from the berth, then exits the port via the channel [47]. For the arrival-departure transition process, the arrival ship can only begin scheduling after the previous departure ship has left, while the departure ship can only be scheduled after the previous arrival ship reaches the berth, ensuring navigational safety.

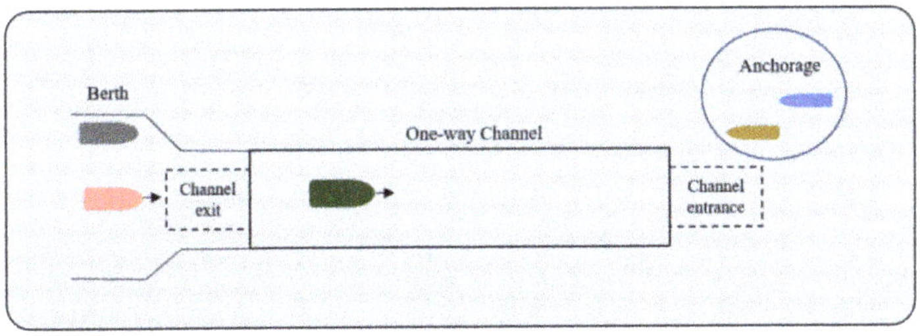

Figure 2. Diagram of a one-way channel.

Several factors influence port ship scheduling, and to simplify the modeling and solution process, the following assumptions are made:

(1) For incoming ships, the application time is considered to be from the moment they apply for entry, and for outgoing ships, the application time is the time when they apply to leave the berth.
(2) The berths for incoming ships have already been pre-assigned.
(3) There are sufficient pilots available.
(4) During the ship scheduling process, factors such as weather, accidents, and other disturbances are not considered.
(5) All ships entering and leaving the port are in the same position near and far from the waterway, that is, the sailing distance of the ship in the waterway is the same.
(6) When applying for entry or exit, pilots and tugboats have already been assigned and are ready.

The parameters are defined as follows: I represents the collection of ships, including incoming and outgoing ships, there are n ships; t_i^s represents the ship's start scheduling time; t_i^a represents the time a ship applies for port entry; i represents the previously scheduled ship; j represents the subsequently scheduled ship; the i and j numbers represent the two ships in a row in the dispatch; S_i represents the distance of ship i from the anchorage; L_i represents the length of ship i; D_i represents the sailing distance of ship i; and v_i represents the average speed of ship i in the channel. The decision variables for subsequent use are defined as follows: B_i represents the berth status of ship i, where 1 indicates that the berth is occupied, and 0 indicates that the berth is available, and IO_i is a binary variable that represents the direction of a ship's movement, where 1 means the ship enters the port and 0 means the ship leaves the port.

Objective function for ship operation delay:

The design of the objective function should be based on the port entry and exit traffic situation of the ships throughout the entire planning period. In the ship scheduling process, aside from the time delays caused by overtaking or crossing in the channel and harbor, the sailing time of the ship in the channel is fixed. Therefore, the optimization focus is on the sum of all ships' waiting times at the berth and the delay times in the channel and harbor, as shown in Equation (1).

$$minT = \sum_{i \in I} (t_i^s - t_i^a) \tag{1}$$

Ship navigation constraints:

(1) Scheduling time constraint

To ensure that ships are effectively scheduled, the start time of the ship's scheduling must be no earlier than its requested scheduling time. This constraint ensures that each ship is scheduled at a reasonable time, avoiding scheduling before the ship's requested time, thus ensuring the feasibility and safety of the scheduling, as shown in Equation (2).

$$t_i^s \geq t_i^a, \forall i \in I \tag{2}$$

(2) Same-direction ship constraint

In traffic flow, the distance and time between vehicles are crucial to improving traffic safety. We need to pay more attention to this in research [48]. As shown in Figure 3, for ships traveling in the same direction in a one-way channel, a safe distance between ships must be maintained. The safe distance s is defined as six times the length of the following ship j [22]. If the starting positions of ship i and ship j are the same, then the scheduled starting time for ship j should be the starting time of ship i plus the safety time interval t_s. Overtaking is prohibited during the navigation of ships to avoid collisions caused by differences in speed or navigational positions. The specific formula is as follows:

$$s = 6 \times L_j \tag{3}$$

$$t_s = s/v_i \tag{4}$$

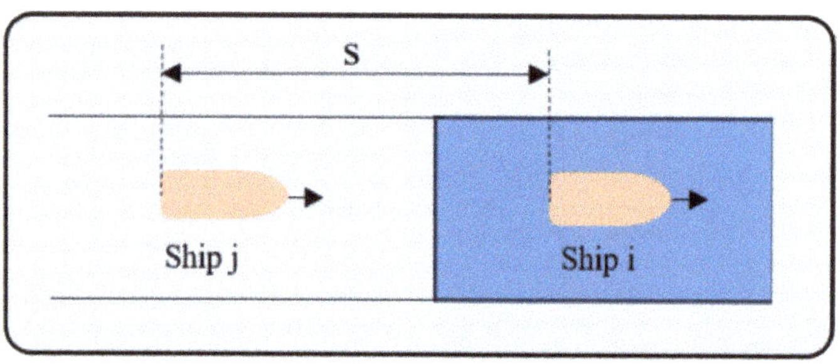

Figure 3. Same-direction sailing of ships in a one-way channel.

To prevent two consecutively scheduled ships from overtaking each other in the channel, specific constraints must be applied when the ships travel in the same direction. For instance, if the leading ship is inbound and the following ship is also inbound, or if both ships are outbound, the scheduling start time of the following ship must be restricted to ensure navigational safety. The specific constraints are detailed in Equation (5).

$$t_j^s = max(t_j^a, t_i^s + t_s), \forall i, j \in I \tag{5}$$

(3) Switching constraints for in-and-out channel process

In a one-way channel, to ensure navigational safety, only ships moving in the same direction are allowed to pass at the same time. Ships moving in opposite directions cannot simultaneously navigate in the channel. Therefore, the inbound and outbound processes must alternate to ensure both the safety and efficiency of the channel. Specifically, if the following ship is moving in the opposite direction to the preceding ship, it must wait until the preceding ship completes its operation before it can begin its own. This constraint ensures smooth traffic flow in the channel and prevents potential collision risks. But this process adds to the overall time and delays subsequent ships. This constraint ensures smooth traffic flow in the channel and prevents potential collision risks. The formula is as follows:

$$t_j^s = t_i^s + S_i/v_i \tag{6}$$

(4) Constraints for Vessels in Opposite Directions

In a one-way channel, when the inbound and outbound processes alternate, it is essential to maintain a safe separation distance between ships. This safe distance, denoted as s', is defined as six times the length of the largest ship. Inbound ships must wait outside the channel entrance at a safe distance until outbound ships have fully passed through the channel. Similarly, outbound ships must remain outside the safe distance from inbound ships. The scheduling start time of the following ship must be delayed by a safety time interval t_s' after the start time of the preceding ship. The formula is as follows:

$$s' = 6 \times max(L_i, L_j) \tag{7}$$

$$t_s' = \frac{s'}{v_j} \tag{8}$$

When the sailing directions of the preceding and following ships are different (e.g., the preceding ship is an inbound ship and the following ship is an outbound ship, or the preceding ship is an outbound ship and the following ship is an inbound ship), the scheduling start time of the following ship must be restricted to ensure safety. The following ship must wait for the preceding ship to complete its entire journey to avoid crossing conflicts. This is shown in Equation (9).

$$t_j^s = max(t_j^a, t_i^s + t_s'), \forall i, j \in I \tag{9}$$

(5) Constraints for the ship following process

To ensure ships maintain a constant speed in the channel and avoid collisions, the speeds of both the leading and following ships must be carefully controlled. Specifically, the speed of the following ship must remain within a safe range to prevent it from catching up to or overtaking the leading ship before it reaches its destination. This speed constraint effectively reduces risks during the navigation process, ensuring a safe distance between ships, preventing rear-end collisions, and improving both the safety and efficiency of the channel. This is shown in Equation (10).

$$\frac{D_i}{v_i} - \frac{D_i}{v_j - v_i} < 0,\ v_j > v_i \tag{10}$$

(6) Constraints to avoid berth conflicts

To ensure that a berth is available when a ship enters the channel, the scheduling of inbound ships must align with berth availability. This ensures that inbound ships can smoothly reach their assigned berths and carry out safe operations, thus avoiding delays and safety risks associated with berth occupancy. The constraint is expressed in Equation (11).

$$(1 - B_i) + (1 - IO_i) > 0 \tag{11}$$

The proposed model seeks to optimize vessel scheduling in ports by minimizing the waiting times of ships within the harbor. To accomplish this objective, a series of constraints have been meticulously designed: ensuring that vessels commence their schedules at or after their requested time; maintaining safe distances between consecutively scheduled vessels traveling in the same direction; regulating the alternating passage of inbound and outbound vessels to ensure orderly traffic flow; preserving adequate separation distances between vessels moving in opposite directions; controlling the speed of following vessels to prevent rear-end collisions; and guaranteeing the availability of berths for incoming vessels to avoid conflicts. Collectively, these measures enhance the safety and efficiency of port operations, mitigate unnecessary delays, and elevate the overall level of shipping management.

3.2. Solving Ship Scheduling Problems Using Reinforcement Learning Algorithms

3.2.1. Algorithm Framework

The port ship scheduling problem, characterized by numerous variables, constraints, and significant time complexity, has been proven to be NP-hard [49]. When the number of ships increases, traditional mathematical methods often fail to find the optimal solution. Consequently, heuristic and intelligent algorithms are typically used to explore the solution space and identify a relatively optimal outcome [50]. Heuristic algorithms are widely applied and studied in related fields, offering feasible, though not always optimal, solutions within a short timeframe. In contrast, reinforcement learning algorithms present unique advantages in complex environments. While they may not consistently guarantee an optimal strategy, particularly for large-scale and intricate problems, reinforcement learning facilitates continuous improvement through autonomous learning and environmental interaction, progressively approaching an optimal solution [51].

The optimization of ship scheduling for inbound and outbound operations in a one-way channel is a complex problem involving multivariable planning and multi-stage decision making. The objective is to minimize ship waiting times by determining the order and timing of ship operations. This multi-stage decision problem can be effectively modeled and analyzed as a Markov decision process (MDP). In the decision-making process, the MDP is represented by a tuple of five elements: $\langle S, A, R, P, \gamma \rangle$. In this context, the state space variable S represents the set of possible ship operation sequences; the action space A represents the set of actions available to agents in the MDP; the reward space R represents the interactions between the agent and the environment, providing an evaluation mechanism to guide the adjustment of action strategies; the state transition probability P represents the environmental changes resulting from the agent's actions, determining the likelihood of selecting a specific ship; the discount factor γ reflects the agent's emphasis on future rewards and adjusts the action selection process accordingly. In an MDP, the construction of the reward space is crucial. The reward space is often a cumulative process, representing the long-term accumulated reward. It can be expressed as

$$G_t = R_{t+1} + \gamma R_{t+2} + \gamma^2 R_{t+3} + \cdots = \sum_{k=0}^{K} \gamma^k R_{t+k+1} \tag{12}$$

where G_t is the accumulated reward at time t, R_{t+1} is the reward at time $t+1$, and K is the accumulation duration.

The goal of reinforcement learning is to guide the optimal action strategy by learning the interaction between the intelligent system and the environment, which is achieved by maximizing the long-term accumulated reward for the optimization and iteration of the agent. In MDP, the optimization strategy of the agent is evaluated by constructing state value functions and action value functions, which can be expressed as [52]

$$V^{\pi}(s) = \max_{a \in A}(R_s^a + \gamma \sum_{s_{t+1} \in S} P_{ss_{t+1}}^a V(s_{t+1})) \tag{13}$$

$$Q(s,a) = \max_{a \in A}(R_s^a + \gamma \sum_{s_{t+1} \in S} P_{ss_{t+1}}^a \max_{a \in A} Q(s_{t+1}, a_{t+1})) \quad (14)$$

where s_{t+1} is the state at time $t+1$, $P_{ss_{t+1}}^a$ is the probability of transitioning to state s_{t+1} from state s when action a is taken, and $Q(s,a)$ is the cumulative reward for selecting action a in state s. The current optimal state value function and optimal action value function can be expressed by the optimal state value function and optimal action value function of the next state. Through iteration, the convergent value function is obtained, which in turn determines the optimal strategy. Q-learning is one of the commonly used reinforcement learning models [53]. It determines which action should be taken in each state to maximize cumulative rewards by constructing and updating a Q-table, ultimately finding the optimal or near-optimal solution. It is a simple and effective reinforcement learning method. The basic model of reinforcement learning is shown in Figure 4.

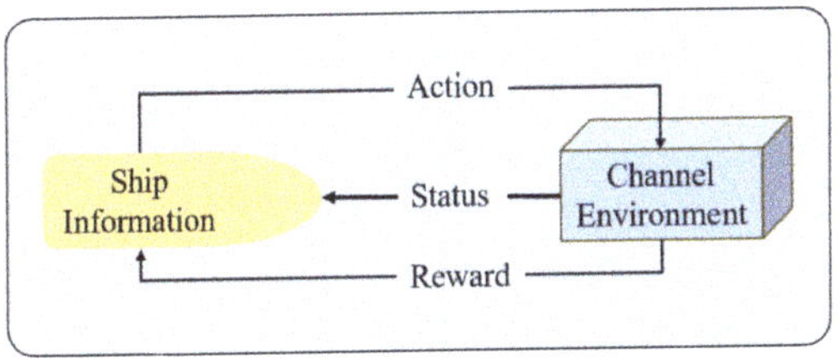

Figure 4. Reinforcement learning model.

3.2.2. Q-Learning-Based Ship Scheduling Optimization Model Design

The reinforcement learning model in this paper consists of a set of states, actions, and a reward function [53–56]. The proposed structure is as follows:

(1) State Space Design: The scheduling of ship arrivals and departures can be viewed as a traversal decision of nodes in a graph, which can be considered a variation of the traveling salesman problem (TSP). The ship traversal process is shown in Figure 5. The state in the ship scheduling model is represented by the sequence of all visited ships. The number of states varies depending on the number of nodes in the instance. All ship sequences and their actual start times together form the state set, denoted as S.

$$S = [ship_1, ship_2, \cdots, ship_n] \quad (15)$$

where n represents the number of ships:

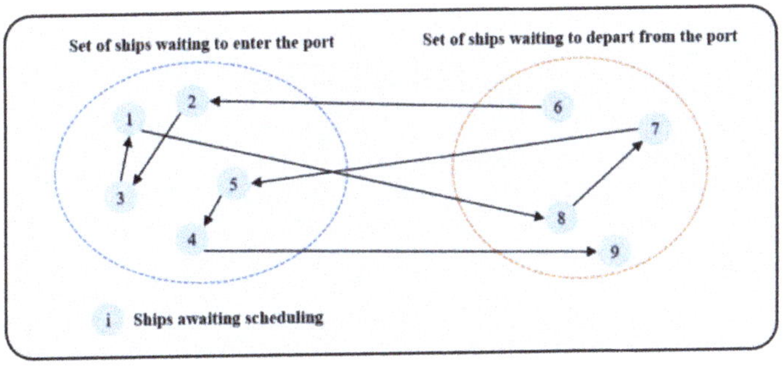

Figure 5. Illustration of ship traversal.

(2) Action Space Design: In the ship scheduling model for port entry and exit, adjusting each ship's position and actual start time is considered an action. This adjustment

ensures that the safety distance between ships is maintained and that no ship operates before its designated application time. The set of all selectable ship IDs constitutes the action set in the algorithm, denoted by A.

$$A = [1, 2, \cdots, n] \tag{16}$$

(3) Action Selection Strategy: In this paper, the Rollout algorithm was used, and the most basic and commonly used ε-greedy strategy in reinforcement learning was selected as the action selection strategy. The ε-greedy strategy balances exploration and exploitation. Specifically, the action with the highest value function is exploited, while other suboptimal actions are still selected with a certain probability.

$$a = \begin{cases} max(Q(s,a)) \\ random(A) \end{cases} \tag{17}$$

(4) Mask Design: When an action fails to satisfy the constraint conditions, it is assigned a value of 0, effectively masking the original action set and thereby blocking or selecting specific actions. The mask construction incorporates constraints such as ship speed limitations and berth conflict restrictions [38]. The mask is denoted by M, and its length is equal to the number of actions.

$$M = [1, 1, \cdots, 1, 1] \tag{18}$$

when an action $a \in A$ does not satisfy the constraint conditions, the mask $M(a) = 0$; otherwise, $M(a) = 1$. The final valid action set is

$$A' = A \cdot M \tag{19}$$

(5) Design of the Wait Time Minimization Reward Function: Based on the established ship scheduling model for port entry and exit, the Q-learning algorithm in reinforcement learning is employed to minimize the delay in the start time of each ship's operation relative to its requested time at the designated position. In order to conform to the Q-learning process, this algorithm improves the reward function accordingly and considers the delay of each ship and the overall delay of all ships. Moreover, the result is de-logarithmic to ensure that the calculation result has no negative value and enhances the stability of the algorithm. The corresponding objective function is formulated and can be computed using Formula (8).

$$R = log\left(\left(\frac{\overline{t^{int}}}{\sum_{i \in u}\left(t_i^s - t_i^a\right)}\right)^{\beta} * \left(\frac{\overline{t^{int}}}{t_i^{delay}}\right)^{\alpha}\right) \tag{20}$$

where α and β are used to adjust the reward function. $\overline{t^{int}}$ represents the average time interval for all ships. Increasing α amplifies the impact of each ship's delay on the reward, while increasing β amplifies the total delay's impact on the reward.

(6) Penalty Function Design: When the resulting sequence fails to meet all ship scheduling requirements, a penalty mechanism is introduced to adjust the outcome. This penalty mechanism can be expressed mathematically as follows:

$$F(s') = f(s') + P(s') \tag{21}$$

Here, the scheduling sequence is denoted as S', the objective function as $f(S')$, and the penalty term as $P(S')$. In this study, the penalty P is set to a significantly large value of 5000 s.

(7) Q-Table Design: In reinforcement learning, the Q-table stores the value of each state-action pair, guiding the agent in selecting the optimal action for a given state. In this paper, the Q-table represents the likelihood of each ship being selected at each decision-making step.

$$Q = \begin{pmatrix} Q(s_1,a_1) & Q(s_2,a_2) & \cdots & Q(s_1,a_n) \\ Q(s_2,a_1) & Q(s_2,a_2) & \cdots & Q(s_2,a_n) \\ \cdots & \cdots & \cdots & \cdots \\ Q(s_n,a_1) & Q(s_n,a_2) & \cdots & Q(s_n,a_n) \end{pmatrix} \tag{22}$$

The update method can be constructed according to Equation (14), as shown below:

$$Q(s_i,a_j) = Q(s_i,a_j) - lQ(s_i,a_j) + lQ(s_i',a_j') \tag{23}$$

where $Q(s_i',a_j')$ represents the Q-table for the next state, and l denotes the learning rate.

3.2.3. Algorithm Process

First, the separation distances between ships are calculated based on factors such as inbound and outbound statuses, speeds, ship lengths, and channel width for different scenarios. This process generates a delay time matrix between ships. Subsequently, this paper employed the rollout method as the core mechanism for implementing the scheduling strategy. In each scheduling iteration, the algorithm first selects an initial ship based on the current policy. It then determines the next ship to be scheduled by following specific policy rules, ensuring that the pre-calculated safety delay times are not violated. To enhance exploration, the ϵ-greedy strategy is implemented, allowing ships to be selected randomly with a certain probability while following the current optimal policy in all other cases. Furthermore, the algorithm assigns a probability to each decision and records both the decisions and their probabilities for subsequent learning updates. After completing a round of ship scheduling, the algorithm evaluates the quality of the scheduling result based on the predefined reward function (total delay time). If the new scheduling solution is better than the known optimal solution, it is updated as the new optimal solution. Through repeated iterations of the rollout process, the algorithm gradually optimizes and learns more effective ship scheduling strategies for specific scenarios. The specific process is illustrated in Figure 6, In the figure, Y represents the condition that is met, and N represents the condition that is not met..

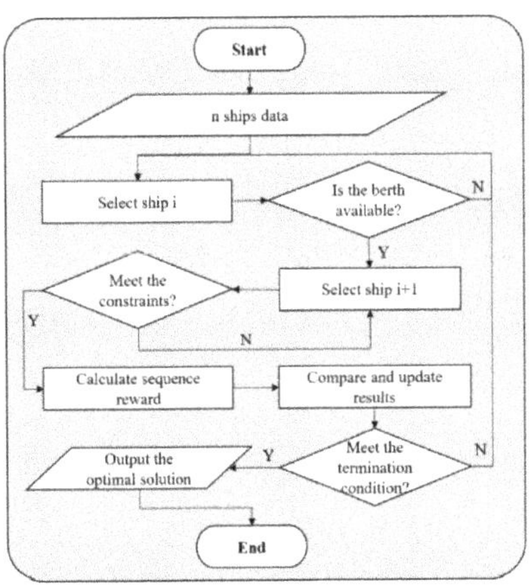

Figure 6. Algorithm process.

4. Experimental Case Analysis

4.1. Parameter Input

For model validation and analysis, this paper took the Dayao Bay Pier of Dalian Port in Liaoning Province as an example (see Figure 7) [57]. The Dayao Bay Pier is located in the northern part of the Port of Dalian, featuring a single-lane waterway that is 1481.6 m long, as shown in Figure 8a. Under normal circumstances, the anchorage areas include the Dayao Bay Anchorage and the Outer Sea Cargo Anchorage, as illustrated in Figure 8b. The data for the ships anchored in the anchorages on a particular day is shown in Table 1 [57], which includes information such as vessel ID, direction of entry and exit, berth number, vessel length, average speed, requested entry/exit time, and berth distance. This is assuming that the first vessel, whether entering or leaving the port, starts its scheduling at 7:00 a.m., and that berth allocation has already been completed by the terminal operator, meeting the mooring requirements of the vessels.

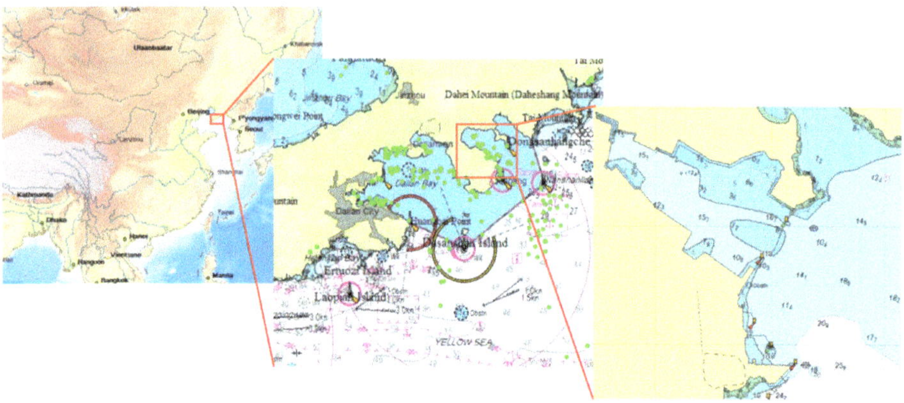

Figure 7. Dayao Bay Pier of Dalian Port.

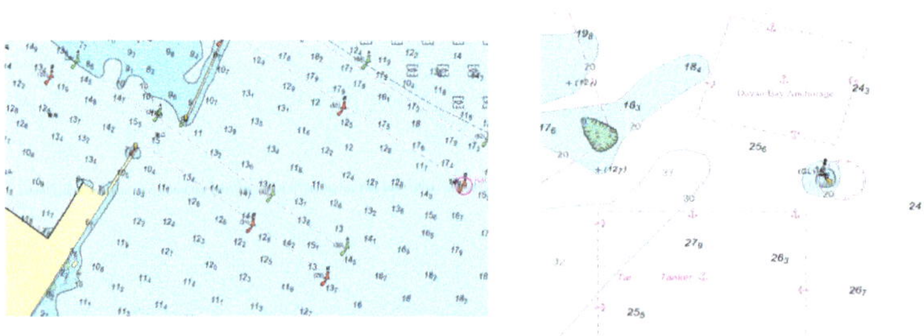

(**a**) (**b**)

Figure 8. Dayao Bay illustration, (**a**) is the channel, (**b**) is the anchorage.

The experimental simulation in this paper was conducted on a Windows platform with an Intel i5-8250U CPU and 8 GB of memory. The reinforcement learning development environment included Python 3.10, PyCharm 2022, and Anaconda 3.4.1. Relevant parameters for the reinforcement learning algorithm were set as follows: learning rate $l = 0.85$ [55], delay impact index $\alpha = 1.5$ and $\beta = 2$, and exploration rate $\varepsilon = 0.01$, and the initial Q-table was set to 0. The maximum number of iterations was set to 500.

Table 1. Basic data of ships.

Ship Number	Direction	Length (m)	Berth Number	Speed (kn)	Sailing Distance (n mile)	Setup Time (hh:mm)
1	In	148	3	8.0	1.5	7:08
2	In	142.7	6	14.0	1.8	7:01
3	In	84.4	1	8.8	1.3	7:07
4	Out	144.8	5	9.0	1.7	7:18
5	Out	278.7	14	8.3	2.6	7:04
6	In	241.3	12	7.1	2.4	7:40
7	In	115.8	9	11.2	2.1	7:56
8	Out	97.2	13	9.5	2.5	7:49
9	Out	150.4	6	15.5	1.8	7:36
10	In	88.8	13	7.7	2.5	7:23
11	In	130	17	10.7	2.9	7:24
12	Out	108.4	10	11.7	2.2	8:29
13	In	142.7	5	11.5	1.7	7:12
14	In	72	2	9.6	1.4	7:47
15	In	166.2	7	12.3	1.9	7:38
16	In	147.5	11	10.9	2.3	7:09
17	In	334.1	14	12.9	2.6	7:03
18	In	85.5	10	8.6	2.2	7:01
19	Out	145.1	3	16.5	1.5	7:11
20	In	95.2	4	7.8	1.6	7:00

4.2. Results Analysis

The optimal scheduling scheme for the 20 ships, as determined by the proposed reinforcement learning algorithm, is presented in Table 2. The iteration process is illustrated in Figure 9, where it is evident that the algorithm reached stabilization within 500 iterations. This indicates the algorithm's ability to identify near-optimal solutions in a relatively short time, thus improving scheduling efficiency. Table 2 presents the optimal scheduling scheme for the ships' port entry and exit, along with the exact start time for scheduling and the delay time for each ship.

Table 2. Ship delay table.

Ship Number	Direction	Setup Time (hh/mm)	Start Time (hh/mm)	Delay Time (min)
5	Out	7:04	7:04	0
3	In	7:07	7:28	21
16	In	7:09	7:32	21
19	Out	7:11	7:44	28
4	Out	7:18	7:46	0
9	Out	7:36	7:49	25
7	In	7:56	7:59	3
14	In	7:47	8:00	42
6	In	7:40	8:05	13
11	In	7:24	8:09	87
20	In	7:00	8:10	45
17	In	7:03	8:19	33
8	Out	7:49	8:31	101
2	In	7:01	8:49	13
10	In	7:23	8:50	105
13	In	7:12	8:53	23
12	Out	8:29	9:02	76
18	In	7:01	9:16	135
1	In	7:08	9:19	33
15	In	7:38	9:23	105

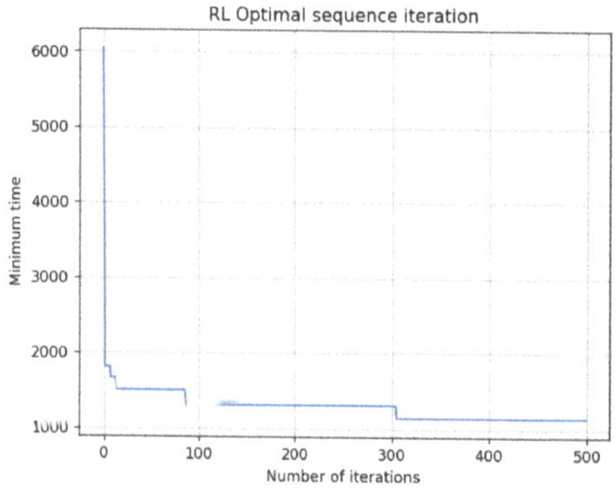

Figure 9. Reinforcement learning optimization iteration process.

(1) Reasonableness verification

Based on the analysis of the anchored ship data, both Ship 1 and Ship 19 are associated with Berth 3. Ship 19 has already docked and is preparing to depart, while Ship 1 is applying to enter the port, resulting in a berth conflict. According to the berth conflict constraint, Ship 19 must complete its departure plan before Ship 1 enters the port. As shown in Table 2, the optimal scheduling plan prioritizes the departure of Ship 19, followed by the scheduling of Ship 1's entry, ensuring the rationality and feasibility of the scheduling plan. To further illustrate the scheduling results of the 20 ships in Table 2, a Gantt chart, as shown in Figure 10, was created. The diagram clearly illustrates the optimal entry and exit schedule for the 20 ships, providing a visual representation of each ship's waiting and sailing times. It also details the sequence of entry and exit for each vessel. The timeline begins at 7:00 a.m. on the day of scheduling and is measured in minutes, offering a comprehensive view of the entire scheduling process, from the ship's entry application to its departure completion.

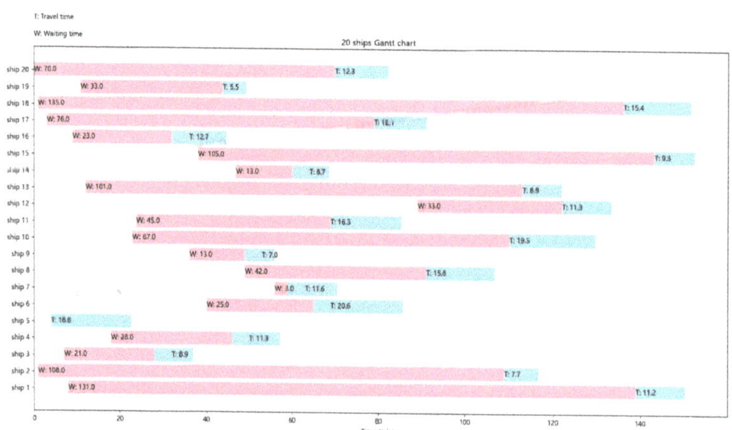

Figure 10. Optimal scheduling diagram.

(2) Algorithm Superiority

From the perspective of optimal scheduling efficiency, for a one-way channel, since ships cannot sail in opposite directions simultaneously, it is necessary to ensure that there are no ships in the channel before changing direction. Excessive direction changes lead to unnecessary time waste. In theory, for longer one-way channels, reducing the number of direction changes helps shorten the ship scheduling time. Table 3 compares 20 ship scheduling schemes based on first-come-first-served (FCFS) rule, reinforcement learning (RL) algorithm, and basic genetic algorithm (GA). According to the table, the optimal scheduling plan using the reinforcement learning algorithm requires 7 changes of direction, while the FCFS rule requires 11 changes, which significantly reduces the ship's delay time. In addition, the table also shows the comparison between reinforcement learning algorithm and basic genetic algorithm in scheduling results. The results show that the reinforcement learning algorithm can reduce the delay time by 16% compared with the genetic algorithm. Figure 11 presents the iteration curves for both algorithms, clearly showing that the reinforcement learning approach converges faster and results in shorter waiting times. Figure 12 displays the average runtime of both algorithms over 10 iterations, with the genetic-algorithm-based scheduling optimization taking nearly 700 s, while the reinforcement learning-based scheduling optimization requires only half that time. Figure 13 illustrates the range of optimal values achieved by both algorithms under the same ship scheduling model across 10 iterations. The graph reveals that the Q-Learning algorithm has a smaller difference between the best and worst solutions compared to the genetic algorithm, indicating greater stability. These results demonstrate that the reinforcement learning algorithm is able to identify more efficient and superior scheduling solutions more quickly and consistently.

Table 3. Comparison of FCFS and RL scheduling schemes for 20 ships.

FCFS				RL				GA			
Ship Number	Direction	**Setup Time (hh:mm)**	**Start Time (hh:mm)**	Ship Number	Direction	**Setup Time (hh:mm)**	**Start Time (hh:mm)**	Ship Number	Direction	**Setup Time (hh:mm)**	**Start Time (hh:mm)**
20	In	7:00	7:00	5	Out	7:04	7:04	19	Out	7:11	7:11
18	In	7:01	7:12	3	In	7:07	7:28	16	In	7:09	7:19
5	Out	7:04	7:27	16	In	7:09	7:32	20	In	7:00	7:21
17	In	7:03	7:46	19	Out	7:11	7:44	1	In	7:08	7:24
3	In	7:07	7:58	4	Out	7:18	7:46	9	Out	7:36	7:36
16	In	7:09	8:07	9	Out	7:36	7:49	4	Out	7:18	7:37
19	Out	7:11	8:20	7	In	7:56	7:59	2	In	7:01	7:51
1	In	7:08	8:25	14	In	7:47	8:00	3	In	7:07	7:52
4	Out	7:18	8:36	6	In	7:40	8:05	14	In	7:47	7:53
13	In	7:12	8:48	11	In	7:24	8:09	12	Out	7:23	8:29
11	In	7:24	8:56	20	In	7:00	8:10	15	In	7:38	8:42
9	Out	7:36	9:13	17	In	7:03	8:19	7	In	7:56	8:44
2	In	7:01	9:20	8	Out	7:49	8:31	6	In	7:40	8:48
15	In	7:38	9:27	2	In	7:01	8:49	8	Out	7:49	9:09
6	In	7:40	9:37	10	In	7:23	8:50	5	Out	7:04	9:14
14	In	7:47	9:57	13	In	7:12	8:53	17	In	7:03	9:38
8	Out	7:49	10:06	12	Out	8:29	9:02	13	In	7:01	9:40
10	In	7:23	10:22	18	In	7:01	9:16	11	In	7:24	9:43
7	In	7:56	10:41	1	In	7:08	9:19	18	In	7:01	9:44
12	Out	8:29	10:52	15	In	7:38	9:23	10	In	7:23	9:46
Total delay (h)		30.37				18.20				21.90	

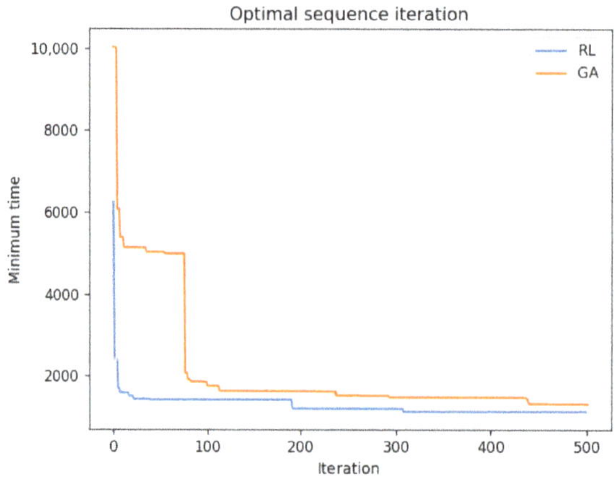

Figure 11. Convergence curves of RL and GA algorithms.

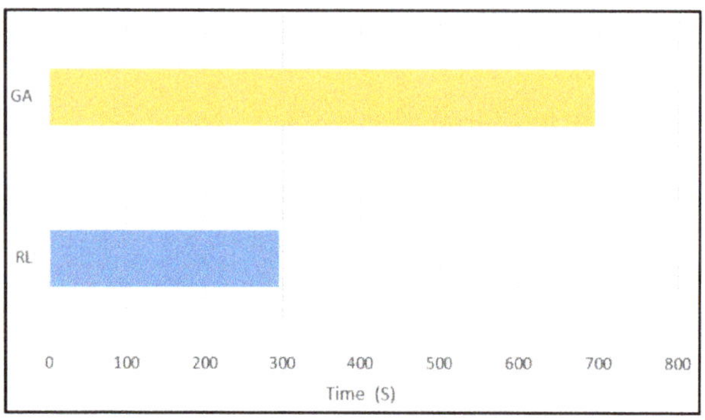

Figure 12. The average time for both algorithms across 10 identical scheduling problems.

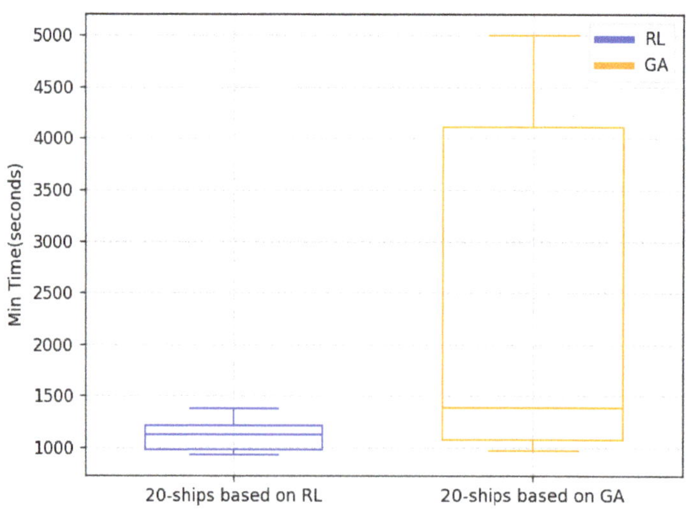

Figure 13. The best result range for both algorithms over 10 identical scheduling problems.

5. Conclusions

To address the ship scheduling optimization problem for a one-way channel, this paper proposes a ship scheduling optimization model based on reinforcement learning algorithms, with the objective of minimizing waiting times for ships in the port. The model incorporates multiple constraints, including initialization constraints, traffic conversion constraints, time-slot allocation constraints, and berth conflict resolution constraints, all based on the traffic flow characteristics of different port areas. A reinforcement learning algorithm was designed and implemented using Python to solve the problem. Targeted simulation scenarios were created to validate the proposed model. The computational results demonstrate that, compared to the FCFS scheduling method, the proposed model reduced the number of ship entry and exit transitions and decreases the total scheduling time by over 40%. Furthermore, when compared to the genetic algorithm (GA) method, the proposed model achieved more efficient and reasonable scheduling within the same time frame. Consequently, the approach presented in this paper ensures more efficient ship entry and exit at the port. However, the model is simplified, and future research should aim to develop more accurate and adaptable models. Ship scheduling is a complex problem, and this study does not consider dynamic factors such as fluctuating arrival and departure times of ships. With the upsizing trend of ships, the tidal factor will become a major influence on ship scheduling, and the weather will also increase the delay time of ships. At the same time, the allocation of port resources (such as the allocation of pilots, tugs and berths) will also be a major influence. From the perspective of algorithms, due to the limitation of Q-learning algorithm, as the number of ships increases, the computing power requirements will also increase, and the computing speed will be affected, which can be further improved to adapt to more scenarios and obtain better performance.

Author Contributions: Conceptualization, R.Z. and M.S.; literature search, R.Z. and M.S.; data curation, M.S. and Q.F.; investigation, R.Z., M.S., and Q.F.; writing—original draft, R.Z. and M.S.; methodology, R.Z. and M.S., writing—review and editing, R.Z.; visualization, R.Z., M.S., and Q.F.; funding acquisition, Q.F. All authors have read and agreed to the published version of the manuscript.

Funding: This research was funded by the National Natural Science Foundation of China (grant number 52001134) and the Fujian Provincial Natural Science Foundation (grant numbers 2024J01102 and 2023J01326).

Institutional Review Board Statement: Not applicable.

Informed Consent Statement: Not applicable.

Data Availability Statement: The original contributions presented in this study are included in the article. Further inquiries can be directed to the corresponding author.

Conflicts of Interest: The authors declare no conflicts of interest.

References

1. Lalla-Ruiz, E.; Shi, X.; Voss, S. The Waterway Ship Scheduling Problem. *Transport. Res. Part D-Transport. Environ.* **2018**, *60*, 191–209. [CrossRef]
2. Zhen, R.; Shi, Z.; Gu, Q.; Yang, S. A Novel Deterministic Search-Based Algorithm for Multi-Ship Collaborative Collision Avoidance Decision-Making. *Ocean Eng.* **2024**, *292*, 116524. [CrossRef]
3. Tong, Y.; Zhen, R.; Dong, H.; Liu, J. Identifying Influential Ships in Multi-Ship Encounter Situation Complex Network Based on Improved WVoteRank Approach. *Ocean Eng.* **2023**, *284*, 115192. [CrossRef]
4. Jiang, X.; Zhong, M.; Shi, G.; Li, W.; Sui, Y. Vessel Scheduling Model with Resource Restriction Considerations for Restricted Channel in Ports. *Comput. Ind. Eng.* **2023**, *177*, 109034. [CrossRef]
5. Liu, B.; Li, Z.-C.; Wang, Y. A Two-Stage Stochastic Programming Model for Seaport Berth and Channel Planning with Uncertainties in Ship Arrival and Handling Times. *Transp. Res. Part E Logist. Transp. Rev.* **2022**, *167*, 102919. [CrossRef]

6. Zhang, P. Optimization Study of Ship Scheduling in Unidirectional Channel Considering Ship Extension. Master's Thesis, Dalian Maritime University, Dalian, China, 2023.
7. Lei, K. Integrated Scheduling Optimization of One-Way Channel Vessel Entry and Exit and Tugboat Distribution. Master's Thesis, Dalian Maritime University, Dalian, China, 2022.
8. Shang, W.-L.; Zhang, J.; Wang, K.; Yang, H.; Ochieng, W. Can Financial Subsidy Increase Electric Vehicle (EV) Penetration---Evidence from a Quasi-Natural Experiment. *Renewable and Sustainable Energy Reviews* **2024**, *190*, 114021. [CrossRef]
9. Shang, W.-L.; Chen, Y.; Yu, Q.; Song, X.; Chen, Y.; Ma, X.; Chen, X.; Tan, Z.; Huang, J.; Ochieng, W. Spatio-Temporal Analysis of Carbon Footprints for Urban Public Transport Systems Based on Smart Card Data. *Appl. Energy* **2023**, *352*, 121859. [CrossRef]
10. Zhen, R.; Lv, P.; Shi, Z.; Chen, G. A Novel Fuzzy Multi-Factor Navigational Risk Assessment Method for Ship Route Optimization in Costal Offshore Wind Farm Waters. *Ocean Coast. Manag.* **2023**, *232*, 106428. [CrossRef]
11. Dong, H.; Zhen, R.; Gu, Q.; Lin, Z.; Chen, J.; Yan, K.; Chen, B. A Novel Collaborative Collision Avoidance Decision Method for Multi Ship Encounters in Complex Waterways. *Ocean Eng.* **2024**, *313*, 119512. [CrossRef]
12. Huang, L.; Wan, C.; Wen, Y.; Song, R.; van Gelder, P. Generation and Application of Maritime Route Networks: Overview and Future Research Directions. *IEEE Trans. Intell. Transp. Syst.* **2025**, *26*, 620–637. [CrossRef]
13. Wang, D.; Liao, F. Analysis of First-Come-First-Served Mechanisms in One-Way Car-Sharing Services. *Transp. Res. Part B Methodol.* **2021**, *147*, 22–41. [CrossRef]
14. Xia, Z.; Guo, Z.; Wang, W.; Jiang, Y. Joint Optimization of Ship Scheduling and Speed Reduction: A New Strategy Considering High Transport Efficiency and Low Carbon of Ships in Port. *Ocean Eng.* **2021**, *233*, 109224. [CrossRef]
15. Xu, G.; Guo, T.; Wu, Z. Optimum Scheduling Model for Ship in/Outbound Harbor in One-Way Traffic Fairway. *J. Shanghai Marit. Univ.* **2008**, *34*, 50–153, 157.
16. Zheng, H.; Liu, B.; Deng, C.; Feng, P. Ship Scheduling Optimization in One-Way Channel Bulk Harbor. *Oper. Res. Manag. Sci.* **2018**, *27*, 28–37.
17. Xu, D.; Li, C.-L.; Leung, J.Y.-T. Berth Allocation with Time-Dependent Physical Limitations on Vessels. *Eur. J. Oper. Res.* **2012**, *216*, 47–56. [CrossRef]
18. Ting, C.-J.; Wu, K.-C.; Chou, H. Particle Swarm Optimization Algorithm for the Berth Allocation Problem. *Expert Syst. Appl.* **2014**, *41*, 1543–1550. [CrossRef]
19. Mauri, G.R.; Ribeiro, G.M.; Lorena, L.A.N.; Laporte, G. An Adaptive Large Neighborhood Search for the Discrete and Continuous Berth Allocation Problem. *Comput. Oper. Res.* **2016**, *70*, 140–154. [CrossRef]
20. Bai, X.; Li, B.; Xu, X. Sequencing of Ships Entering a Port Based on an Improved Artificial Fish Swarm Algorithm. *J. Shanghai Marit. Univ.* **2021**, *42*, 85–90.
21. Li, R.; Zhang, X.; Li, J.; Jiang, L. Application of Self-Learning Genetic Algorithm Based on Reinforcement Learning in Ship Scheduling. *J. Dalian Marit. Univ.* **2022**, *48*, 20–30. [CrossRef]
22. Zhang, X.; Lin, J.; Guo, Z.; Chen, X. Vessel Scheduling Optimization Based on Simulated Annealing and Multiple Population Genetic Algorithm. *Navig. China* **2016**, *39*, 26–30.
23. Zhang, B.; Zheng, Z.; Wang, D. A Model and Algorithm for Vessel Scheduling through a Two-Way Tidal Channel. *Marit. Policy Manag.* **2020**, *47*, 188–202. [CrossRef]
24. Zhang, X.; Li, R.; Lin, J. Vessel Scheduling Optimization in Two-Way Traffic Ports. *Navig. China* **2018**, *41*, 36–40.
25. Wang, H.; Tian, W.; Zhang, J.; Li, Y. A Hybrid Self-Organizing Scheduling Method for Ships in Restricted Two-Way Waterways. *Brodogradnja* **2020**, *71*, 15–30. [CrossRef]
26. Meisel, F.; Fagerholt, K. Scheduling Two-Way Ship Traffic for the Kiel Canal: Model, Extensions and a Matheuristic. *Comput. Oper. Res.* **2019**, *106*, 119–132. [CrossRef]
27. Gong, H. Research on Integrated Dispatching Optimization of Compound Channel and Tugboat. Master's Thesis, Dalian Maritime University, Dalian, China, 2022.
28. Zhang, B.; Zheng, Z. Model and Algorithm for Vessel Scheduling Optimisation through the Compound Channel with the Consideration of Tide Height. *Int. J. Shipp. Transp. Logist.* **2021**, *13*, 445–461. [CrossRef]
29. Wang, W.; Ding, A.; Cao, Z.; Peng, Y.; Liu, H.; Xu, X. Deep Reinforcement Learning for Channel Traffic Scheduling in Dry Bulk Export Terminals. *IEEE Trans. Intell. Transp. Syst.* **2024**, *25*, 17547–17561. [CrossRef]
30. Chen, X.; Wu, H.; Han, B.; Liu, W.; Montewka, J.; Liu, R.W. Orientation-Aware Ship Detection via a Rotation Feature Decoupling Supported Deep Learning Approach. *Engineering Applications of Artificial Intelligence* **2023**, *125*, 106686. [CrossRef]
31. Lin, C.; Zhen, R.; Tong, Y.; Yang, S.; Chen, S. Regional ship collision risk prediction: An approach based on encoder-decoder LSTM neural network model. *Ocean Eng.* **2024**, *296*, 117019. [CrossRef]
32. Jonathan, H. Connell and Sridhar Mahadevan. In *Robotica*; Cambridge University Press: Cambridge, UK, 1999; Volume 17, pp. 229–235. [CrossRef]
33. Lopes Silva, M.A.; de Souza, S.R.; Freitas Souza, M.J.; Bazzan, A.L.C. A Reinforcement Learning-Based Multi-Agent Framework Applied for Solving Routing and Scheduling Problems. *Expert Syst. Appl.* **2019**, *131*, 148–171. [CrossRef]

34. Hottung, A.; Kwon, Y.-D.; Tierney, K. Efficient Active Search for Combinatorial Optimization Problems. Available online: https://arxiv.org/abs/2106.05126v3 (accessed on 16 September 2024).
35. Ottoni, A.L.C.; Nepomuceno, E.G.; de Oliveira, M.S.; de Oliveira, D.C.R. Reinforcement Learning for the Traveling Salesman Problem with Refueling. *Complex Intell. Syst.* **2022**, *8*, 2001–2015. [CrossRef]
36. Gambardella, L.M.; Dorigo, M. Ant-Q: A Reinforcement Learning Approach to the Traveling Salesman Problem. In *Machine Learning Proceedings 1995*; Prieditis, A., Russell, S., Eds.; Morgan Kaufmann: San Francisco, CA, USA, 1995; pp. 252–260. ISBN 978-1-55860-377-6.
37. Duan, H.; Liu, P.; Li, Z.; Tang, D. Variable Speed Limit Control at Freeway Merge Bottlenecks Based on Reinforcement Learning. *J. Transp. Syst. Eng. Inf. Technol.* **2015**, *15*, 55–61.
38. Fotuhi, F.; Huynh, N.; Vidal, J.M.; Xie, Y. Modeling Yard Crane Operators as Reinforcement Learning Agents. *Res. Transp. Econ.* **2013**, *42*, 3–12. [CrossRef]
39. Huang, H.; Fang, Z.; Wang, Y.; Tang, J.; Fu, X. Analysing Taxi Customer-Search Behaviour Using Copula-Based Joint Model. *Transp. Saf. Environ.* **2022**, *4*, tdab033. [CrossRef]
40. Meng, X.; Tang, J.; Yang, F.; Wang, Z. Lane-Changing Trajectory Prediction Based on Multi-Task Learning. *Transp. Saf. Environ.* **2023**, *5*, tdac073. [CrossRef]
41. Xiang, Z.; Zhou, D.; Sun, H.; Chu, T. Research on Departure Sequencing in Multi- airport Terminal Area Based on Reinforcement Learning. *Aeronaut. Comput. Tech.* **2023**, *53*, 30–34.
42. Zhu, C.; Zhang, Z. Application of Q-Learning Agent in Aircraft Sequencing at En-Route Intersection. *Aeronaut. Comput. Tech.* **2015**, *45*, 68–70.
43. Wang, Y. Ship Traffic Organization in Intersection Waters Based on Reinforcement Learning. Master's Thesis, Dalian University of Technology, Dalian, China, 2022.
44. Du, J.; Zhao, X.; Guo, L.; Wang, J. Machine Learning-Based Approach to Liner Shipping Schedule Design. *J. Shanghai Jiaotong Univ. (Sci.)* **2022**, *27*, 411–423. [CrossRef]
45. Xin, X.; Liu, K.; Zhang, J.; Chen, S.; Wang, H.; Cheng, Z. A Self-Organizing Grouping Approach for Ship Traffic Scheduling in Restricted One-Way Waterway. *Mar. Technol. Soc. J.* **2019**, *53*, 83–96. [CrossRef]
46. Jiang, J. Analysis of Domestic Intelligent Ship Traffic Management System. *Mar. Equip./Mater. Mark.* **2023**, *31*, 26–28. [CrossRef]
47. Liu, C.; Liu, J.; Zhou, X.; Zhao, Z.; Wan, C.; Liu, Z. AIS Data-Driven Approach to Estimate Navigable Capacity of Busy Waterways Focusing on Ships Entering and Leaving Port. *Ocean Eng.* **2020**, *218*, 108215. [CrossRef]
48. Liu, X.; Tang, J.; Yuan, C.; Gao, F.; Ding, X. Examining the Characteristics between Time and Distance Gaps of Secondary Crashes. *Transp. Saf. Environ.* **2024**, *6*, tdad014. [CrossRef]
49. Zheng, H.; Xu, H.; Liu, B.; Cao, H. One-Way Channel Ship Inbound Order and Berth Allocation Collaborative Optimization. *Oper. Res. Manag. Sci.* **2017**, *26*, 37–45.
50. Naderi, E.; Pourakbari-Kasmaei, M.; Cerna, F.V.; Lehtonen, M. A Novel Hybrid Self-Adaptive Heuristic Algorithm to Handle Single- and Multi-Objective Optimal Power Flow Problems. *Int. J. Electr. Power Energy Syst.* **2021**, *125*, 106492. [CrossRef]
51. Naeem, M.; Rizvi, S.T.H.; Coronato, A. A Gentle Introduction to Reinforcement Learning and Its Application in Different Fields. *IEEE Access* **2020**, *8*, 209320–209344. [CrossRef]
52. Clifton, J.; Laber, E. Q-Learning: Theory and Applications. *Annu. Rev. Stat. Its Appl.* **2020**, *7*, 279–301. [CrossRef]
53. Wang, T.; Zhang, Y.; Hu, X. A Q-Learning Based Hyper-Heuristic Scheduling Algorithm with Multi-Rule Selection for Sub-Assembly in Shipbuilding. *Comput. Ind. Eng.* **2024**, *197*, 110567. [CrossRef]
54. Bianchi, R.A.; Ribeiro, C.H.; Costa, A.H. On the Relation between Ant Colony Optimization and Heuristically Accelerated Reinforcement Learning. In Proceedings of the 1st International Workshop on Hybrid Control of Autonomous System, Pasadena, CA, USA, 13 July 2009; AAAI: Palo Alto, CA, USA, 2009; pp. 49–55.
55. Ottoni, A.L.C.; Nepomuceno, E.G.; de Oliveira, M.S. A Response Surface Model Approach to Parameter Estimation of Reinforcement Learning for the Travelling Salesman Problem. *J. Control Autom. Electr. Syst.* **2018**, *29*, 350–359. [CrossRef]
56. Júnior, F.C.D.L.; Neto, A.D.D.; De Melo, J.D. Hybrid Metaheuristics Using Reinforcement Learning Applied to Salesman Traveling Problem. In *Traveling Salesman Problem, Theory and Applications*; IntechOpen: Rijeka, Croatia, 2010; ISBN 953-307-426-4.
57. Zhang, X.; Chen, X.; Ji, M.; Yao, S. Vessel Scheduling Model of a One-Way Port Channel. *J. Waterw. Port Coast. Ocean Eng.* **2017**, *143*, 04017009. [CrossRef]

Journal of
Marine Science and Engineering

MDPI

Article

Effects of High-Frequency Vibration on Residual Stress and Microstructure of Carbon Steel for Marine Structures: Comparative Analysis with Tempering

Guanhua Xu [1,2,*] and Feilong Liu [3]

1 State Key Laboratory of Fluid Power & Mechatronic Systems, College of Mechanical Engineering, Zhejiang University, Hangzhou 310027, China
2 Key Laboratory of 3D Printing Process and Equipment of Zhejiang Province, College of Mechanical Engineering, Zhejiang University, Hangzhou 310027, China
3 School of Mechanical Engineering, Tongji University, Shanghai 200092, China; feilongliu@tongji.edu.cn
* Correspondence: xuguanhua@zju.edu.cn; Tel.: +86-18550185607

Abstract: To improve the safety and service life of carbon steel used in marine structures, appropriate regulation of residual stress in carbon steel is required. This paper investigates the effects of high-frequency vibratory stress relief (VSR) and tempering on the residual stress, microstructure, and surface hardness of 45 steel. After the high-frequency VSR and tempering at 200 °C for 30 min treatment, the microstructure is still tempered martensite. When the 45 steel experimental specimens were tempered at 600 °C for 30 min, the microstructure changed from tempered martensite to tempered sorbite, and the residual stress regulation effect of 45 steel experimental specimens was significantly improved. However, its surface hardness decreased significantly, which reduces the mechanical properties of marine structural components. Comparatively, high-frequency VSR is an effective method to regulate residual stress while ensuring that the microstructure of marine structural components does not undergo drastic changes. This study provides technical and theoretical support for the residual stress regulation treatment of 45 steel in marine engineering.

Keywords: vibratory stress relief; tempering; residual stress; surface hardness; microstructure

1. Introduction

The marine environment hosts numerous large-scale equipment and structural components [1]. During the manufacturing process of these equipment and components, residual stress is inevitably introduced due to the production techniques [2,3]. Residual stress can present compressive, neutral, or tensile stress fields. Existing studies have shown that both residual tensile stress and residual compressive stress significantly affect the stability and service life of metal structures [4]. Residual tensile stress, in particular, has a significant influence on the mechanical strength and fatigue strength of metals [5]. In order to ensure the safety of personnel in the marine environment, on the one hand, it is possible to avoid potential hazards during application through technologies such as machine vision and mathematical modeling [6–8]. On the other hand, it is necessary to improve the reliability, mechanical strength, and service life of equipment and structural components during the manufacturing process [9,10]. Consequently, reducing residual stress in marine equipment and structural components to enhance their safety and service life has garnered significant attention. Currently, widely used stress regulation methods include thermal stress relief (TSR) technology and vibratory stress relief (VSR) technology [11,12]. The TSR technology promotes a homogenization of residual stress in the whole volume of the part, as seen for additive manufactured 316L after the annealing process [13]. But TSR requires substantial energy consumption during implementation, incurs high costs, and poses challenges in controlling certain key parameters during the process, depending on the material, geometry, and size, among others. In contrast, VSR technology offers simpler equipment, shorter treatment durations, and an easier on-site application compared to the TSR. It has now become a highly promising stress regulation technology. As a result, the VSR technology has been extensively studied and has yielded significant achievements.

Wang et al. [14] carried out VSR treatment on AA6061 aluminum alloy, in which the frequency and amplitude were 46 Hz and 3.5 g. It is found that higher residual stress relieving can be obtained when the AA6061 aluminum alloy has a larger initial residual stress. VSR treatment reduces the stored strain energy and dislocation density so that the residual stresses are released. When the VSR treatment was performed for 20 min, the residual stress relaxation rate reached 40%. Chuvas et al. [15] carried out the VSR treatment on welded DP600 hot-rolled steel. The results show that the longitudinal residual stress relieving rate of the weld is about 40%, and the transverse residual stress relieving rate of the weld is about 20%. Gao et al. [16,17] found that fatigue life decreases with increasing amplitude. Although the residual stress relieving effect at a high amplitude is good, and the surface residual stress relieving rate is 50%, the fatigue life will be reduced. Low amplitude VSR can effectively reduce the residual stress, but the fatigue life is almost unchanged. Gong et al. [18] investigated the effect of VSR on the deformation and residual stress in 7075 aluminum alloy thin-walled components. The results show that the VSR significantly improves the shape and dimensional stability of the material by relieving residual stress in the thin-walled components. Moraes et al. [19] carried out the VSR treatment and the heat treatment experiments on HSLA Domex700 steel. The results show that the effect of the VSR on relieving residual stress is obviously worse than that of heat treatment, but the fracture toughness, yield strength, and dimensional stability of HSLA Domex700 steel treated by the VSR are better than those of HSLA Domex700 steel treated by the heat treatment. Song et al. [20] proposed a segmented thermal–vibration stress relief (STVSR) process for large rings and carried out related experiments on 2219 aluminum–copper alloy rings. The results show that the circumferential and axial residual stress relieving rates are 44.43% and 45.14%, respectively. The size and shape of the grains and the coarse Al_2Cu particles are not obviously changed by the STVSR, which indicates that the STVSR can only cause micro-plastic deformation within the grain interior, but not macro-plastic deformation. Tu et al. [21] studied the effects of VSR on the microstructure, residual stress, hardness, and magnetic properties of TiC-reinforced steel matrix composite materials. The results show that the VSR treatment significantly released the residual stress and improved the Vickers hardness of the composite while also promoting the precipitation of fine carbides in the matrix and the microstructure transformation of metastable retained austenite to relatively stable martensite. Song et al. [22] carried out VSR with different parameters, aiming to explain the microscopic mechanism of VSR from the perspective of molecular dynamics. Gao et al. [23] carried out VSR on the welded joints of dissimilar steels and carried out corresponding numerical simulations. Their results show that VSR reduces the residual stress introduced by welding and improves the impact toughness.

Low-frequency VSR technology is suitable for regulating the residual stress of large metal structures. In order to expand the application range of VSR technology, high-frequency VSR technology has been proposed and studied [24–26]. Jiang et al. [24] proposed a multi-degree-of-freedom system model of metal particles based on microscopic dynamics theory, and the mechanism of high-frequency VSR was discussed. He et al. [25] carried out the high-frequency VSR treatment on the small quenched specimens of the $Cr_{12}MoV$ steel. It is found that the maximum residual stress relieving rates are, 66.0% and 62.9% along the x-axis and y-axis directions of the specimens, respectively. Gu et al. [26] designed a high-frequency vibration energy amplification device and carried out the high-frequency VSR on the quenched $Cr_{12}MoV$ steel. It is found that the effect of the high-frequency VSR can be improved since the high-frequency vibration energy imposed into the metal materials can be amplified by the high-frequency vibration energy amplification device. Moreover, the high-frequency VSR can decrease the dislocation density.

VSR is a technique for the reduction and homogenization of residual stress by mechanical vibration. In terms of effectiveness in the reducing residual stress, TSR is excellent compared to VSR for the regulation of residual stress in metal structures. Song et al. [27] studied 2219 aluminum–copper alloy monolithic thin-walled parts with the TSR treatment. The results show that the axial residual stress relieving rate and circumferential residual stress relieving rate could reach 86.37% and 85.77%, respectively, after the TSR treatment. Xiao et al. [28] investigated the effect of annealing Al_2O_3 diffusion-reinforced alloy at 600 °C for 6 h on the residual stress. The results show that after creep annealing, the residual stress relieving was 80.3%. Jiang et al. [29] investigated the effect of heat treatment on the residual stress in selective laser melting $AlSi_{10}Mg$, and the results show that the residual stress relieving can be effectively relieved. By comparing the effects of VSR and TSR, it can be found that TSR can obtain more significant effects than VSR. In addition, scholars are increasingly paying attention to energy consumption, and whether the environmental friendliness of treatment technologies is one of the factors that needs to be considered [30–32]. In order to be safely applied in marine structural components, it is necessary to study in-depth the treatment methods for regulating residual stresses and to analyze their mechanisms. Therefore, the mechanism of TSR for efficiently regulating residual stress should be analyzed, which will also help to explore the mechanism of high-frequency VSR. The existing research on the mechanism of the VSR is mainly based on the plastic deformation caused by micro-scale dislocation motion, and there are few studies on the mechanism of the VSR based on other new perspectives. As a result, there is no systematic and complete explanation of the mechanism of VSR, which limits the large-scale application of VSR technology in the field of marine engineering.

Based on the extensive application of 45 steel in marine equipment and structural components, along with the lack of systematic comparative experimental analysis of high-frequency VSR and tempering heat treatment, this study designed corresponding experiments. This research investigates and compares the effects of high-frequency VSR and tempering heat treatment on residual stress, microhardness, and microstructure. Furthermore, the mechanism of high-frequency VSR was explored through the lens of microstructural inheritance and evolution.

2. Experiments

2.1. Mechanism of High-Frequency VSR

In order to study the influences of the high-frequency VSR for the 45 steel specimens, the experimental system of high-frequency VSR is established as shown in Figure 1 [25]. The host computer system controls the signal generator to output a sinusoidal exciting signal with the adjustable frequency and amplitude. The exciting signal amplified by the power driver is used to control the electromagnetic exciter to generate the high-frequency vibration. The high-frequency vibration energy amplification device is used to amplify the vibration energy output by the electromagnetic exciter. The specimen is clamped on the upper surface of the high-frequency vibration energy amplification device in order to act the higher vibration energy on the specimen. Meanwhile, the acceleration sensor is clamped on the upper surface of the high-frequency vibration energy amplification device in order to obtain the vibration amplitude acting on the specimen. The vibration waveform is displayed on the oscilloscope and transmitted back to the host computer system as well. The determination of excitation frequency for high-frequency VSR using the sweep frequency method was conducted as follows: the acceleration sensor was mounted on the upper surface of the high-frequency vibration energy amplification device. A frequency sweep range of 1000–5000 Hz was implemented with a step increment of 10 Hz, while the host computer monitoring system recorded the excitation amplitude at each discrete frequency. Experimental results demonstrated that the maximum excitation amplitude was captured at 2160 Hz by the acceleration sensor. The high-frequency VSR conducted on the specimen at this frequency allows the specimen to acquire greater vibration energy, thereby achieving improved residual stress elimination effects.

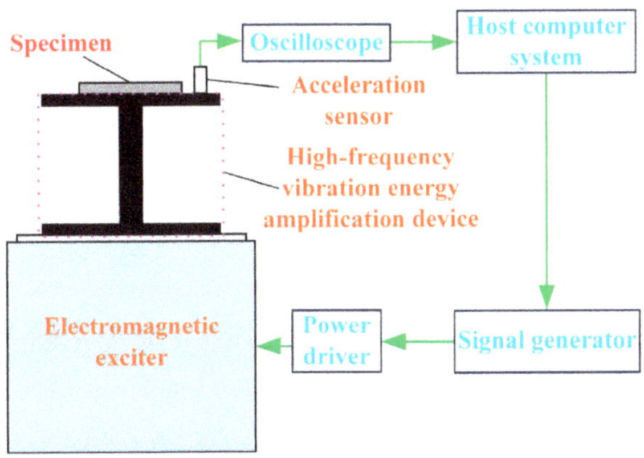

Figure 1. The experimental system of the high-frequency VSR [25].

Based on the dislocation activation model, it can be observed that regardless of whether it is mechanical energy or thermal energy, as long as the energy injected into the material is sufficient to activate dislocation motion, the purpose of reducing residual stress can be achieved. Dislocation activation inevitably leads to processes such as dislocation accumulation and annihilation, resulting in changes in the internal dislocation density of metallic components. At the same time, the changes in dislocation density can lead to variations in flow stress, which in turn affects dislocation activation motion. Therefore, further research is needed on dislocation evolution to clarify the evolution after dislocation activation and elucidate the microscopic mechanism by which external energy injection eliminates residual stress in metallic components.

Figure 2 illustrates the evolution after dislocation activation, including two main processes: dislocation accumulation and annihilation. As shown in Figure 2a, assuming a Frank–Read dislocation source on the slip plane releases a series of dislocations outward under the shear stress τ_m, they accumulate in front of strong obstacles (such as grain boundaries and forest dislocations), increasing the local dislocation density in metal components. Dislocation activation induces both accumulation and annihilation processes. For simplicity, the annihilation mechanism is demonstrated using edge dislocations. Edge dislocations are divided into positive and negative categories. In Figure 2b, positive and negative edge dislocations activated on the slip plane interact, resulting in annihilation and a subsequent decrease in local dislocation density within the metal structure.

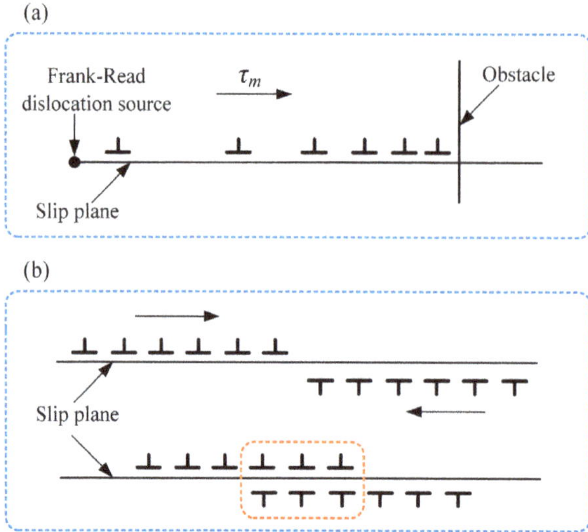

Figure 2. Dislocation evolution process: (**a**) dislocation accumulation; (**b**) dislocation annihilation.

From the above analysis, it can be observed that changes in dislocation density in metal components include increases due to dislocation accumulation and decreases due to dislocation annihilation, which can generally be expressed as follows [33]:

$$\frac{d\rho}{d\gamma^p} = \frac{d\rho^+}{d\gamma^p} - \frac{d\rho^-}{d\gamma^p} \tag{1}$$

in which $\frac{d\rho}{d\gamma^p}$ represents the change in dislocation density during the plastic deformation process after dislocation activation; $\frac{d\rho^+}{d\gamma^p}$ denotes the portion of dislocation density increase caused by dislocation accumulation at obstacles under external energy input; $\frac{d\rho^-}{d\gamma^p}$ signifies the portion of dislocation density decrease due to dislocation annihilation under external energy input.

Consequently, the dislocation density evolution during plastic deformation of metal components can be modeled by the following equation [33]:

$$\begin{cases} \frac{d\rho^+}{d\gamma^p} = \alpha_1\sqrt{\rho} \\ \frac{d\rho^-}{d\gamma^p} = \alpha_2\rho \\ \frac{d\rho}{d\gamma^p} = \alpha_1\sqrt{\rho} - \alpha_2\rho \end{cases} \tag{2}$$

In the above equation, α_1 is the dislocation storage coefficient, while α_2 is the coefficient for dynamic recovery coefficient. Previous studies and related research findings indicate that high-frequency VSR above 1 kHz reduces the internal dislocation density of metal components [26,34,35]. Similarly, tempering also leads to a decrease in dislocation density [36]. To analyze the mechanism of external energy's influence on dislocation evolution, the effects of external energy on dislocation accumulation and annihilation were incorporated into the dislocation density evolution control Equation (2), resulting in the following equation:

$$\frac{d\rho}{d\gamma^p} = \alpha_1(1+\lambda_{\alpha_1})\rho^{1/2} - \alpha_2(1+\lambda_{\alpha_2})\rho^{1/2} \tag{3}$$

in which $\lambda_{\alpha 1}$ and $\lambda_{\alpha 2}$ represent the ratios of change in α_1 and α_2, respectively, under the effect of external energy. Based on the dislocation density evolution control Equation (3), it can be concluded that during plastic deformation, metal components undergo two fundamental processes: dislocation accumulation and annihilation. Previous studies and findings suggest that high-frequency VSR reduces dislocation density in metal components. This suggests that when high-frequency mechanical energy is introduced, dislocation accumulation and annihilation occur, with annihilation prevailing, leading to reduced dislocation density and lower residual stress.

In the field of marine engineering, carbon steel is widely used in various structures due to its excellent mechanical properties and low cost [37–39], such as in offshore drilling platforms, oil and gas pipelines [40], as well as smaller components like bolts and rods [41]. Therefore, 45 steel, a high-quality carbon steel, was selected for experimentation in this study. To visually represent the distribution of high-frequency vibration energy in metal components, ANSYS 15.0 software was employed to perform numerical modal analysis on 45 steel, and the energy distributions of different modes were obtained. The specific steps are as follows:

(1) Construct the solid model and set material parameters: The 3D solid model of 45 steel was constructed with dimensions consistent with subsequent experimental studies, namely, 60 mm in length, 20 mm in width, and 3 mm in thickness. For numerical modal analysis, the material properties were defined as follows: density $\rho = 7850\,\mathrm{kg/m^3}$, elastic modulus $E = 200\,\mathrm{GPa}$ and Poisson's ratio $\nu = 0.28$;

(2) Build the finite element model: The element type was set to SOLID95. Since the 3D solid model of 45 steel is a regular rectangular block, the mapped method was used for mesh generation, resulting in regular hexahedral meshes;

(3) Solve: In the Solution module, the solution type was set to Modal, and the solution algorithm was chosen as Block Lanczos. The number of modes to solve was set to the first 50. To obtain strain modes, the "Calculate elem results" option was selected;

(4) Results analysis: Using the General Postprocessing module, the natural frequencies, displacement modes, and strain modes of the model were extracted.

Figures 3 and 4, respectively, show the displacement modes and strain modes corresponding to some natural frequencies, which are 4299 Hz, 11,722 Hz, 22,526 Hz, and 35,817 Hz.

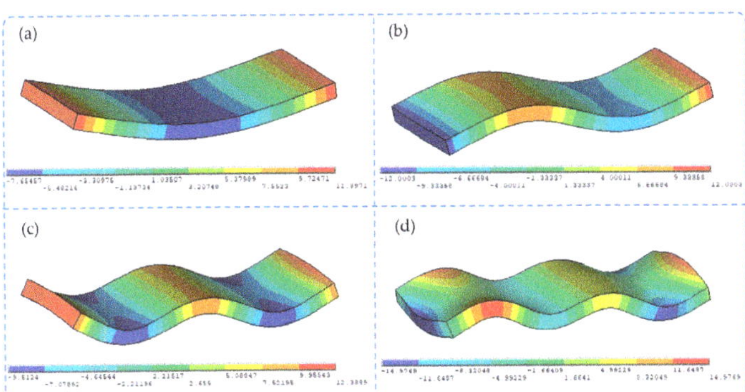

Figure 3. Displacement modes: (**a**) 4299 Hz; (**b**) 11,722 Hz; (**c**) 22,526 Hz; (**d**) 35,817 Hz.

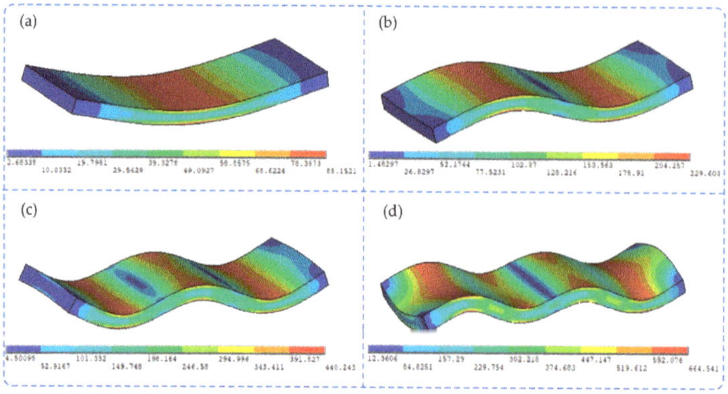

Figure 4. Strain modes: (**a**) 4299 Hz; (**b**) 11,722 Hz; (**c**) 22,526 Hz; (**d**) 35,817 Hz.

From Figures 3 and 4, it can be observed that at higher natural frequencies, metal components exhibit more peak regions, indicating a greater amount of high-frequency mechanical energy injected into the interior of the components. Compared to low-frequency VSR, high-frequency VSR can stimulate more intense vibrations among metal particles. Based on the dislocation density evolution control Equation (3), high-frequency VSR facilitates the acquisition of greater mechanical energy in more localized regions of metallic components, aiding dislocation motion activation.

2.2. Experimental Methods

The marine environment is relatively complex, so ensuring the safety of marine structures is the most important goal. The introduction of residual stress is unavoidable during the manufacturing and construction of structural components. Without proper treatment, residual stress can severely affect the performance of structural components and shorten their service life [42]. Thus, treatments are required to release residual stress, improve the performance of structural components, extend their service life, and ensure safety [37,43]. To study the effectiveness of the two methods for relieving residual stress, quenched 45 steel specimens were subjected to high-frequency VSR and tempering, analyzing the evolution of residual stress introduced by quenching under external field effects.

The 45 steel specimens are prepared by annealing at 850 °C for 3 min, followed by water quenching for a high cooling rate. The Rockwell hardness for the quenched 45 steel can be as high as about 60 HRC, which makes it difficult to measure the residual stress by the hole-drilling strain-gage method [44] using the electric hand drill. In this case, the quenched 45 steel specimens are tempered at 200 °C for 30 min to decrease the surface hardness for the 45 steel specimens, which still have the large initial residual stress. In this case, drilling on the 45 steel specimens during the process of the residual stress measuring is facilitated by the hole-drilling strain-gage method [44]. The tempering temperature and tempering time for the initial specimens, which are named 1# specimens, are 200 °C and 30 min, respectively. The 1# specimens were prepared as the initial specimens. Tempering and high-frequency VSR were applied based on these specimens.

The effects of key parameters of tempering heat treatment, such as different times and temperatures, were comprehensively analyzed. In order to study the influences of the tempering time on the 45 steel specimens, three batches of the 1# specimens are randomly selected for the tempering experiments. The tempering temperature for the 2# specimens, 3# specimens, and 4# specimens are 200 °C, while the tempering time are selected as 60, 90, and 120 min, respectively. In addition, in order to study the influences of the tempering temperature on the 45 steel specimens, a batch of the 1# specimens are randomly selected for the tempering experiments. The tempering temperature and tempering time for the 5# specimens are 600 °C and 30 min, respectively.

In the context of marine structures, high-frequency VSR equipment is easier to operate on-site and provides more flexibility and convenience in regulating residual stress compared to tempering. In order to study the influences of the high-frequency VSR on the 45 steel specimens, five batches of the 1# specimens are randomly selected for the high-frequency VSR experiments. The exciting frequency and the amplitude of the high-frequency VSR are 2160 Hz and 10 μm, respectively. The vibration time is selected as 5, 10, 15, 20, and 25 min, respectively. All specimen numbers and treatment details are shown in Table 1.

Table 1. Specimen number and processing conditions.

Specimen	Treatment	Condition
1#	Tempering	200 °C, 30 min
2#	Tempering	200 °C, 60 min
3#	Tempering	200 °C, 90 min
4#	Tempering	200 °C, 120 min
5#	Tempering	600 °C, 30 min
6#	High-frequency VSR	5 min
7#	High-frequency VSR	10 min
8#	High-frequency VSR	15 min
9#	High-frequency VSR	20 min
10#	High-frequency VSR	25 min

The hole-drilling strain-gage method was used to measure the residual stress before and after treatment, which is a widely used and recognized method for residual stress measurement [45,46]. This method allows for an accurate assessment of the effectiveness of both treatments, thereby ensuring the safety of marine structures. The hole-drilling strain-gage method, follows the ASTM E837-08 standard [44]. Figure 5 is the schematic of the attached B-type triaxial strain rosette on the residual stress measuring point. The B-type triaxial strain rosette with the type of the BX120-1CA shown in Figure 5 is used in this study. The strain grid 1 and the strain grid 3 coincide with the x-axis and the y-axis, respectively. The angle between the strain grid 2 and the x-axis is 45°. The drilling depth and diameter are 2 mm and 1.5 mm, respectively. The residual stress measuring point shown in Figure 5 is located on the center of the upper surface of the 45 steel specimens, in which d_0 is the diameter of the drilled hole and d is the diameter of the gage circle. The drilling equipment was an ASMD5-8 Strainmeter (Jinan Sigmar Sci-tech Co., Ltd., Jinan, China).

The schematic diagram of the wire connection for the residual stress measuring by the hole-drilling strain-gage method [44] is shown in Figure 5. The triaxial strain rosette for measuring the released strains around the drilled hole is attached on the 45 steel specimens as shown in Figure 5. Meanwhile, the compensating strain rosette is attached on the compensating specimen with the material of the 45 steel to eliminate the effects of the ambient temperature on the released strains as shown in Figure 5. According to the standard of the hole-drilling strain-gage method [44], the x axial and y axial residual stress around the drilled hole of the 45 steel specimens can be calculated by the following the equation

$$\begin{cases} \sigma_x = \frac{E}{2}\left(\left(\frac{1}{(1+v)\overline{a}} + \frac{1}{\overline{b}}\right)\varepsilon_1 + \left(\frac{1}{(1+v)\overline{a}} - \frac{1}{\overline{b}}\right)\varepsilon_3\right) \\ \sigma_y = \frac{E}{2}\left(\left(\frac{1}{(1+v)\overline{a}} - \frac{1}{\overline{b}}\right)\varepsilon_1 + \left(\frac{1}{(1+v)\overline{a}} + \frac{1}{\overline{b}}\right)\varepsilon_3\right) \end{cases} \quad (4)$$

in which σ_x is the x axial residual stress, σ_y is the y axial residual stress, E is the elastic modules, v is Poisson's ratio, ε_1 is the x axial released strain, ε_3 is the y axial released strain,

$\bar{a}$ is the calibration constant for the isotropic stress, and $\bar{b}$ is the calibration constant for the shear stress.

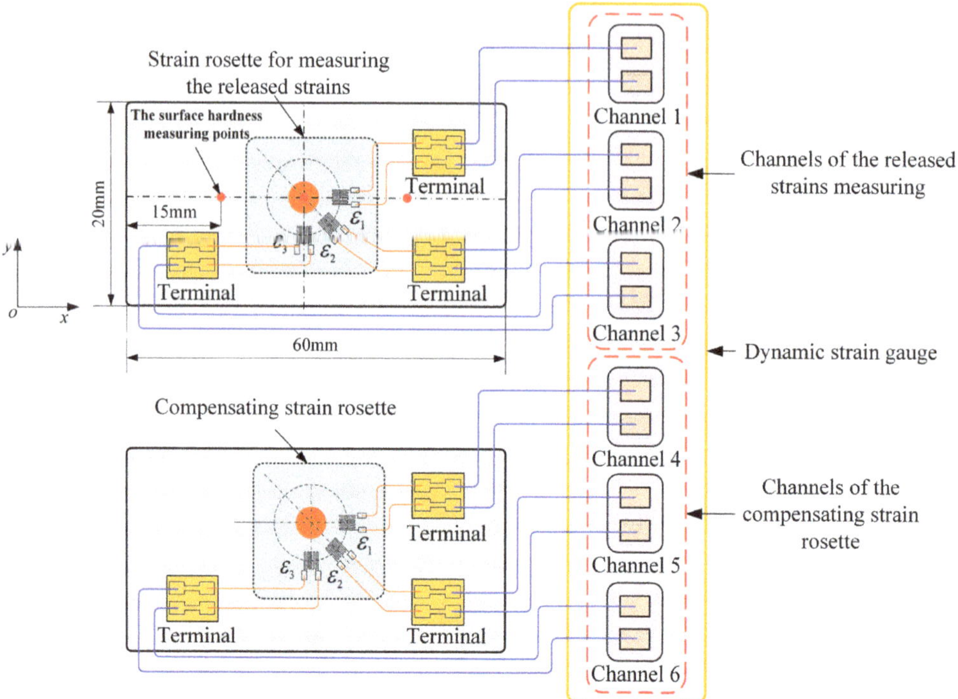

Figure 5. Diagram of residual stress testing.

The HRS-150AT (Laizhou Laihua Testing Instrument Factory, Laizhou, China) Rockwell hardness (HRC) tester is used to measure the Rockwell hardness and the LCW-1K (SAIKASI, Beijing, China) Vickers hardness tester is used to measure the Vickers hardness for the 45 steel specimens. The load for the hardness test is 9.8 N. As shown in Figure 5, three equally spaced measuring points on the symmetrical center line in the horizontal direction of the 45 steel specimens are selected as the surface hardness measuring points. Each measuring point was measured once for hardness. The distance between the surface hardness measuring points is 15 mm. Furthermore, the average value of the surface hardness at the measuring point is regarded as the surface hardness value of the 45 steel specimens under the different treatment conditions.

The OLYMPUS GX53 (Olympus Co., Ltd., Tokyo, Japan) Optical Microscope (OM) is used to analyze the microstructure of the 45 steel specimens. The central 10 × 10 mm^2 area of the 45 steel specimens is cut to observe the microstructure. In addition, the area needs to be polished and etched. The etch solution is 4% nitric acid alcohol solution, and the etch time is 3~5 s.

3. Results

3.1. Residual Stress of the 1# Specimens

In order to compare the effect of two residual stress modulation treatments on marine structural components, the residual stress results of 45 steel tempered for 30 min after quenching need to be obtained as a basis. Also to ensure the reliability of the data, three random specimens were used for residual stress measurements. To reasonably assess the residual stress, equivalent stress is used for comparison. The calculation method for equivalent stress is as follows:

$$\sigma_{eq} = \sqrt{\sigma_x^2 + \sigma_y^2} \quad (5)$$

in which σ_{eq} is the equivalent stress, σ_x is the x axial residual stress, and σ_y is the y axial residual stress. The average value of the residual stress is 346.4 MPa. The standard deviation is 7.4 MPa. The average values of the residual stress are regarded as the initial residual stress for the 1# specimens. This measurement is more similar to the experimental results of Schubnell et al. [1]. It can be a more realistic representation of the magnitude of residual stresses introduced in marine structural components because of the manufacturing process, which ensures the reliability and reasonableness of the subsequent research in this paper.

3.2. Effects of Tempering Time on Residual Stress

Time and temperature are the two most critical parameters for tempering structural components in the marine environment. Within these two parameters, it is relatively easy to increase the tempering time because higher temperatures place higher demands on the power of the equipment used to perform the tempering process. Therefore, the effect of increasing the tempering time at lower temperatures was investigated first. Residual stresses were measured on different tempered specimens. Figure 6 shows in detail the effects of the tempering time on the residual stress for the 45 steel specimens, in which the tempering temperature is maintained at 200 °C. It can be roughly found from Figure 6 that the tempering can reduce the residual stress for the 45 steel specimens.

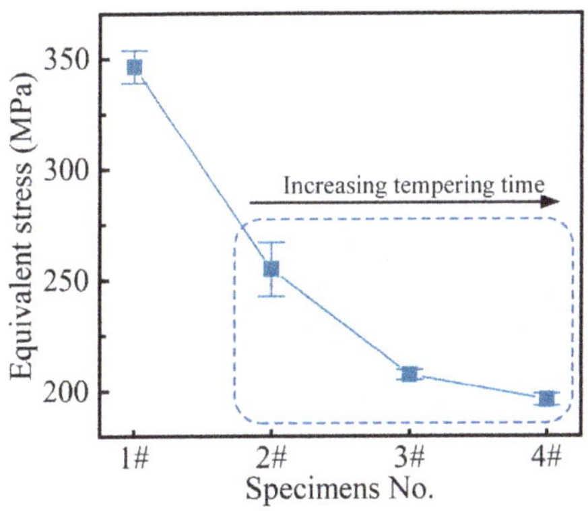

Figure 6. The effects of the tempering time on the equivalent stress.

To reasonably evaluate the treatment effect, the residual stress relieving rate was adopted as the assessment metric. The residual stress relieving rate is calculated according to the following equation:

$$\eta = \frac{|\sigma_{1\#} - \sigma_{i\#}|}{|\sigma_{1\#}|} \times 100\% \quad (6)$$

in which η is the residual stress relieving rates, $\sigma_{1\#}$ is the residual stress for the 1# specimens, and $\sigma_{i\#}$ is the residual stress for the i# specimens (i = 2, 3, . . ., 10).

When the tempering time is 60 min, the residual stress relieving rates for the 2# specimens are 26.4%. The residual stress relieving rates can reach 40.1% when the tempering time is 90 min. Meanwhile, the residual stress relieving rates for the 4# specimens with the tempering time of 120 min are 43.3%. It can be concluded that the residual stress relieving effect can be improved by increasing the tempering time. However, the residual stress relieving effect is better in the initial stage of the tempering, and the residual stress relieving effect has gradually become stable when the tempering time reaches 90 min as shown in Figure 6. Even if the subsequent tempering time is increased, the residual stress relieving effect has not been significantly improved. In this section, the shortest tempering time is 60 min.

For large structural components commonly found in marine engineering, this treatment time would occupy significant time and space. Extending the treatment time may yield better results, but it would consume more energy and provide minimal improvement. It indicates that the residual stress relieving effect is relatively insignificant when tempering is carried out at a lower temperature. In this case, it is necessary to increase the tempering temperature to improve the regulation of residual stresses and enhance the safety and service life of marine structures.

3.3. Effects of Tempering Temperature on Residual Stress

Tempering at lower temperatures did not provide sufficiently good regulation of residual stress. The tempering temperature was raised to 600 °C to significantly regulate the residual stresses in marine structural components, thereby improving their safety and service life.

Figure 7 shows the effects of the tempering temperature on the residual stress for the 45 steel specimens, in which the tempering time is maintained at 30 min. In this condition, the residual stress relieving rates for the 5# specimens are 92.8%. It can be seen that when the tempering time is 30 min and the tempering temperature reaches 600 °C, the residual stress of the 5# specimens is almost in a stress-free state. It implies that the residual stress relieving rates can be significantly improved by increasing the tempering temperature.

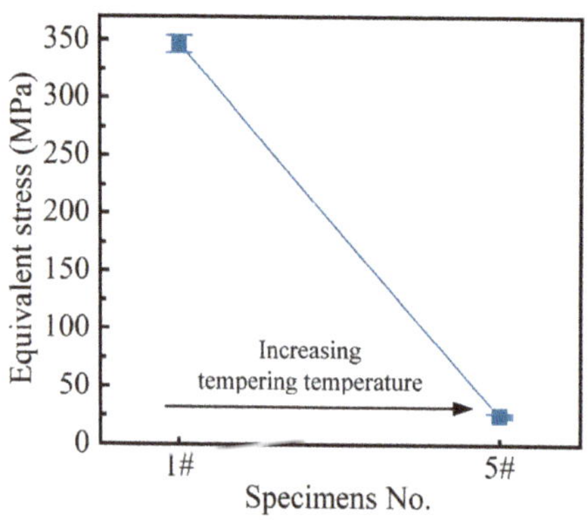

Figure 7. The effects of the tempering temperature on the equivalent stress.

Combined with the experimental results in the previous section, elevating the tempering temperature is more effective than elevating the time. It can be enough to obtain significant residual stress relieving, which may be a great enhancement for structural components applied in marine engineering. Further research and analysis of the mechanism of high temperature tempering to regulate the residual stresses in marine structural components is needed to ensure practical application.

3.4. Effects of High-Frequency Vibration on Residual Stress

High-frequency VSR can be conveniently applied to marine structures and consumes less energy than tempering, and the effect of different treatment times was investigated first. Figure 8 shows the effects of the vibration time on the residual stress for the 45 steel specimens, in which the frequency and the amplitude of the high-frequency VSR are maintained at 2160 Hz and 10 μm, respectively. It can be found from Figure 8 that the high-frequency VSR can reduce residual stress for the 45 steel specimens. When the vibration time is 5 min, the residual stress relieving rates for the 6# specimens are 19.8%. Meanwhile, residual stress relieving rates for the 8# specimens are 56.7%, in which the vibration time reaches 15 min. It can be concluded that the residual stress relieving rates increase approximately linearly with the increasing of the vibration time when the range of the vibration time is 5 to 15 min as shown in Figure 8. Furthermore, the residual stress relieving rates for the 10# specimens are 63.0% when the vibration time increases to 25 min. It indicates that the residual stress relieving rates are gradually stable when the vibration time reaches 15 min.

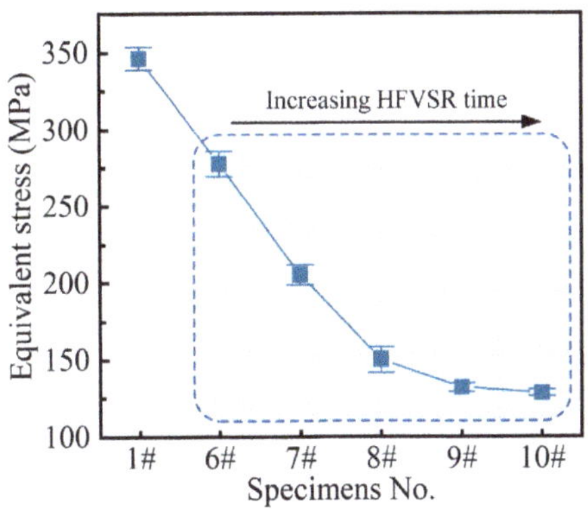

Figure 8. The effects of the vibration time on the equivalent stress.

High-frequency VSR has a clear advantage in this complex marine application environment. Even large structural components do not require the massive energy consumption that heat treatment does. The most obvious advantage is that high-frequency VSR requires much less time. In this section, the longest treatment time designed was 25 min, still shorter than the shortest heat treatment time of 60 min, yet the residual stress elimination effect is similar. In less time, high-frequency VSR can significantly regulate residual stress in marine structures, improving their safety and service life.

3.5. Effects of High-Frequency Vibration and Tempering on Hardness

To further analyze the effect of high-frequency VSR and tempering on the mechanical properties of the marine structural components, the hardness of the specimens was measured before and after treatment. The results of the Rockwell hardness and the Vickers hardness for the 45 steel specimens under each condition are shown in Figure 9. It can be found from Figure 9 that the Rockwell hardness and the Vickers hardness for the 2# specimens, 3# specimens, and 4# specimens are basically unchanged under the lower temperature tempering at 200 °C. In this case, it can be concluded that the effects of the tempering time on the surface hardness for the 45 steel specimens can be neglected. The Rockwell hardness and the Vickers hardness for the 5# specimens with the tempering temperature of 600 °C and time of 30 min will decrease to 22.8 HRC and 265.3 HV. Figure 10 is the Vickers indentations of the center point on the 45 steel specimens under each condition. It can also be found from Figure 10 that the area of the Vickers indentation for the 5# specimens is larger, which indicates that the Vickers hardness is smaller in this treatment condition. It indicates that the Rockwell hardness and the Vickers hardness for the 45 steel specimens decrease significantly with the increasing temperature. Moreover, the Rockwell hardness and the Vickers hardness for the 10# specimens with the vibration time of 25 min are 50.8 HRC and 531.5 HV. In this case, it can be concluded that the effects of the vibration time on the surface hardness are negligible.

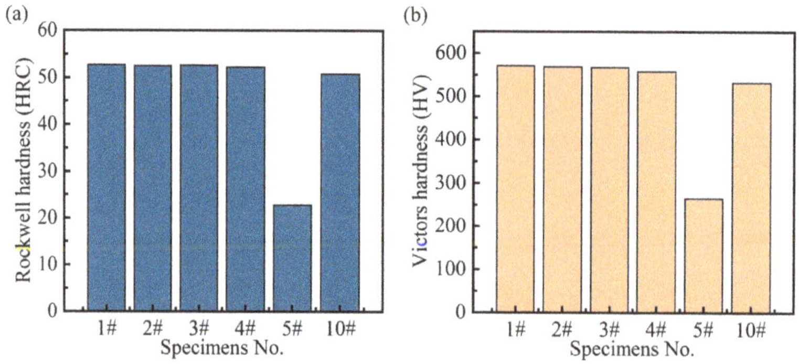

Figure 9. The Rockwell hardness and the Vickers hardness testing results for the 45 steel specimens: (**a**) the Rockwell hardness; (**b**) the Vickers hardness.

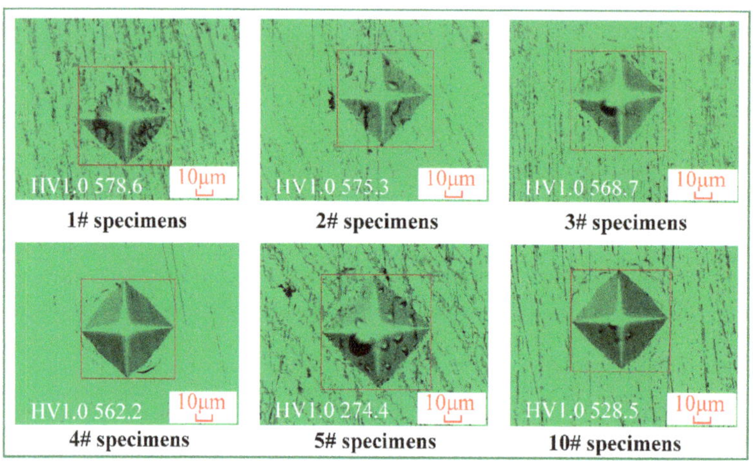

Figure 10. Vickers indentations under each condition.

Studies have shown that there is a certain relationship between hardness and tensile properties, where higher hardness generally correlates with stronger tensile properties [47,48]. The experimental results above indicate that after tempering at 600 °C, hardness significantly decreases. Based on the relationship between hardness and tensile properties, it can be inferred that the tensile properties will be severely weakened after this type of tempering. The decrease in tensile properties directly lowers the safety of the marine structure. Even though the residual stress elimination rate is the highest under this condition, from a safety perspective, this is unacceptable.

3.6. Effects of High-Frequency Vibration and Tempering on Microstructure

The microstructure can directly indicate the composition of the organization of the marine structural components and can also contribute to the analysis of the mechanism of the high-frequency VSR and tempering treatments to regulate the residual stresses and assist in evaluating the effect of the regulation, so the microstructure before and after the treatments were observed. Figure 11a is the OM photo for the 45 steel specimens treated by the quenching at 850 °C for 3 min. It can be clearly observed that the microstructure of the 45 steel specimens, which are quenched at 850 °C for 30 min, consists of martensite. Figure 11b is the OM photo for the 1# specimens, showing that the microstructure of the 1# specimens tempered at 200 °C for 30 min is tempered martensite. Figure 11c is the OM photo for the 4# specimens treated by the tempering at 200 °C for 120 min, showing that the microstructure of the 4# specimens subjected to tempering at 200 °C for 120 min consists of tempered martensite. This microstructure is in good agreement with that observed in Figure 11b, suggesting excellent inheritance of the microstructure. The microstructure for the 5# specimens treated by the tempering at 600 °C for 30 min are the tempered sorbite as shown in Figure 11d. In comparison with the microstructure of the 1# specimens presented in Figure 11b, the current microstructure has undergone substantial alterations, which implies that genetic mutations in the microstructure have taken place. As shown in Figure 11e, the microstructure for the 10# specimens treated by the high-frequency VSR for 25 min are the tempered martensite, which is consistent with the microstructure observed in Figure 11b,c, indicating that the microstructure morphology is well inherited.

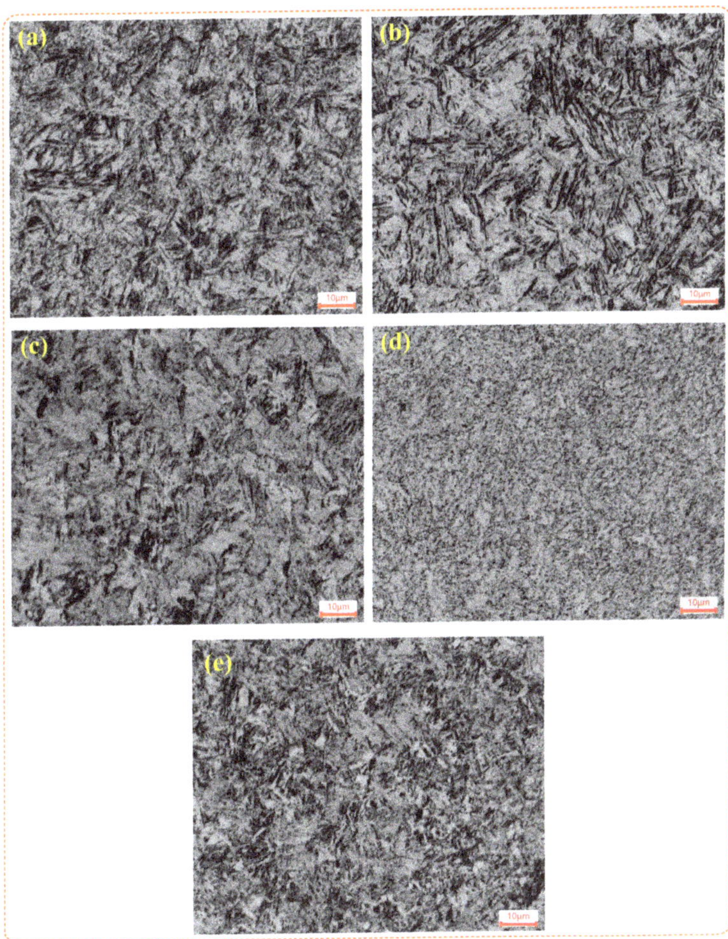

Figure 11. (**a**) The OM photos for the 45 steel specimens treated by quenching at 850 °C for 3 min; (**b**) the OM photos for the 1# specimens; (**c**) the OM photos for the 4# specimens treated by tempering at 200 °C for 120 min; (**d**) the OM photos for the 5# specimens treated by tempering at 600 °C for 30 min; (**e**) the OM photos for the 10# specimens treated by the high-frequency VSR for 25 min.

The microstructure results further support the hardness measurement results from the previous section. The tempering at 600 °C altered the microstructure of marine structures, reducing the tensile properties of the marine structures, and thus lowering their safety.

4. Discussion

The complex marine environment imposes strict safety requirements on structural components. The regulation of residual stress must be considered. However, during the residual stress relieving process, the inherent mechanical properties of the marine structural components should not be weakened, and at the very least, they should be maintained at the same level as before treatment. In this chapter, the effects of tempering and high-frequency VSR on the microstructure will be discussed in detail, and the feasibility of both treatments in marine structural components will be analyzed.

As shown in Figure 11, the microstructure for the quenched 45 steel is martensite, which is the supersaturated solid solution of carbon atoms in the bcc iron also known as $\alpha-$Fe. In this case, the martensite is in the high-energy metastable state, and the large residual stress for the quenched 45 steel is produced owing to the serious lattice distortion of $\alpha-$Fe. Under the action of external energy, the martensite will change from the high-energy metastable state to the low-energy stable state. Meanwhile, the residual stress will decrease during the state transition process. Generally, the martensite in the high-energy metastable state will decompose, and the supersaturated carbon atoms will precipitate in the form of carbides when the tempering temperature is 100~200 °C. In this case, the supersaturation degree of the martensite is decreased. The precipitated carbides are distributed in a network on the martensite matrix, which is known as tempered martensite. As shown in Figure 11b, the microstructure for the quenched 45 steel treated by tempering at 200 °C for 30 min is the tempered martensite. The residual stress and the surface hardness for the 45 steel will decrease since the precipitation of supersaturated carbon atoms decreases the degree of the lattice distortion. In this case, the surface hardness for the 1# specimens decreases from 58.2 HRC to 53.1 HRC as shown in Figure 9. Meanwhile, the average residual stress for the 1# specimens is 346.4 MPa. In other words, the tempered 45 steel 1# specimens still have large residual stress, and the surface hardness decreases to a certain extent relative to the quenched 45 steel specimens, which is conducive to the development of the hole-drilling strain-gage method experiment.

In the case of 4# specimens tempered at 200 °C for 120 min, more carbides were precipitated due to the longer tempering time, but due to the lower tempering temperature, the precipitated carbides were still distributed in a reticular pattern on the martensite matrix and would not be transformed into carburite, resulting in good inheritance of the microstructure of tempered martensite without changes in morphology with the extension of tempering time. In this case, the residual stress relieving effect was limited, and the Vickers hardness was basically unchanged. A comparison with 2#, 3#, and 4#, the residual stress relieving effect gradually stabilized as the tempering time increased. This is due to the good genetic inheritance of microstructure compared to the 1# specimens. As shown in Figure 6, the residual stress relieving rate was 43.3%, where the tempering time was 120 min. At the same time, the Vickers hardness decreased by only 2.26%. It can be found that the microstructure based on tempering at 200 °C for 120 min was essentially unchanged compared to 1# specimens. In this case, the residual stress relieving effect was limited, and the Vickers hardness remained essentially unchanged.

During the high-frequency VSR process, the temperature did not rise significantly, indicating that the high-frequency vibration energy injected into the 10# specimens was not sufficient to have a significant effect on the precipitation process of supersaturated carbon atoms. In this case, when the 10# specimens were subjected to high-frequency

VSR treatment, no transformation of tempered martensite occurred and the microstructure morphology was well inherited with a high-frequency VSR time, vibration amplitude, and excitation frequency of 25 min, 10 μm, and 2160 Hz, respectively, as shown in Figure 11e. Moreover, the residual stress relieving effect shown in Figure 8 was gradually stabilized with the increasing vibration time. The residual stress elimination rate of the 10# specimen reached 63.0%. At the same time, the Vickers hardness of 10# specimens were reduced by 6.90% compared to 1# specimens. Compared with the 1# specimens, the microstructure of the 4# and 10# specimens did not change significantly and were well inherited, resulting in a limited residual stress relieving effect. In contrast, the microstructure of 5# specimens was genetically mutated, and the microstructure was changed, resulting in a significant increase in the residual stress relieving effect.

When the high-frequency vibration energy is injected into the metal materials, the dislocation density of the metal materials will decrease, which indicates that the dislocation annihilation process dominates in the high-frequency VSR process [35]. In this case, high-frequency VSR can reduce the dislocation density of metal materials, thereby reducing the residual stress inside metal materials. In addition, as shown in Figure 11e, during the high-frequency VSR process, the tempered martensite morphology of the 10# specimens remained almost unchanged compared to the 1# specimens, and good inheritance was obtained. In this case, the residual stress elimination mechanism of high-frequency VSR is the decrease in dislocation density caused by the dominant dislocation annihilation process.

The heat energy produced by tempering can activate the dislocation movement, similar to the high-frequency vibration energy, which can also affect the activation of dislocation movement. The dislocation pile-up process and dislocation annihilation process also occur inside the metal materials. In this case, the heat energy produced by tempering has a great influence on the dislocation annihilation process [36,49], resulting in a decrease in residual stress, thereby reducing the dislocation density. In addition, during the tempering process at 200 °C, the microstructure of the tempered martensite of the 4# specimens remained almost unchanged, and compared with the microstructure of the 1# specimens, it was well inherited. In this case, the residual stress elimination mechanism of tempering at 200 °C is also the decrease in dislocation density caused by the dominant dislocation annihilation process.

When the tempering temperature reaches 600 °C, the microstructure heredity of tempered martensite changes abruptly and granular cementite aggregates and spheroidizes, forming a mechanical mixture of polygonal ferrite and spherical cementite, as shown in Figure 11d, which is tempered sorbite. As a result, the residual stress shown in Figure 7 and the surface hardness shown in Figure 9 are significantly reduced. When the tempering temperature is 600 °C and the tempering time is 30 min, the residual stress elimination rate of the 5# specimen is 92.8%. At the same time, the Rockwell hardness of the 5# specimens decreased to 22.8 HRC, and the Vickers hardness decreased to 265.3 HV. Ferrite is the interstitial solid solution of carbon atoms in bcc iron, with good plasticity, low strength, and microhardness. However, tempered martensite is a supersaturated solid solution of carbon atoms in bcc iron, which induces severe lattice distortion and a solid solution strengthening effect. At this time, the tempered martensite has a large hardness as shown in Figure 9. In addition, with the increasing tempering temperature, the dislocation density decreases significantly, because the higher the tempering temperature, the greater the influence on the dislocation annihilation process. In this case, the residual stress and surface hardness will decrease significantly with the increasing tempering temperature. Marchese et al. [50] found that when the nickel-based superalloy Inconel 718 was heat treated at 1065 °C, the microstructure morphology produced an obvious dendritic structure, and the residual stress was obviously eliminated. Park et al. [51] performed 800 °C heat treatment on the cold-rolled Invar alloy, and the microstructure was recrystallized. The residual stress elimination rate measured along the rolling direction was 73%, and the residual stress measured along the transverse direction was almost zero. This indicates that the considerable release of residual stress can be attributed to partial recrystallization and recovery. Chao et al. [52] annealed 316L austenitic stainless steel based on additive manufacturing at 1100 °C. After prolonging the solution annealing time at this temperature, the recrystallization of columnar grains and the coarsening of nano-inclusions occurred, which led to the decrease in yield strength and ductility, making the specimens exist in an almost a stress-free state.

In conclusion, for marine structural components, tempering at 200 °C and high-frequency VSR can both regulate residual stress while maintaining the original microstructure. High-frequency VSR achieves a significant residual stress relieving rate with only a slight decrease in microhardness, outperforming tempering at 200 °C. Additionally, the residual stress relieving rate of high-frequency VSR increases steadily with treatment time, which is an advantage that tempering at 200 °C cannot provide. If the temperature is raised to 600 °C, although an excellent residual stress relieving rate is achieved, it changes the microstructure of the marine structures, reduces their tensile properties and significantly weakens their safety. In terms of treatment time, the maximum treatment time for high-frequency VSR is 25 min, which is still shorter than the shortest tempering time of 60 min. For marine structural components, high-frequency VSR not only efficiently relieves residual stress in a shorter time but also preserves the original microstructure, thus, improving the safety and service life of the marine structural components.

5. Conclusions

Based on carbon steel commonly used in marine structural components, comparative experiments of high-frequency VSR and tempering are designed, and the effects of both processes on residual stress regulation in marine structural components as well as the evolution mechanisms of the two processes are analyzed.

(1) The high-frequency VSR has a good residual stress regulation effect, and the effect gradually stabilizes with increasing vibration time. When the vibration time is 25 min, the residual stress relieving rate of 45 steel specimens is 63.0%. With increasing tempering time at 200 °C, the elimination effect of residual stress is improved. The residual stress relieving rate of 4# specimens tempered at 200 °C for 120 min is 43.3%. When the tempering temperature is 600 °C, the residual stress is significantly eliminated, and the relieving rate reaches 92.8%.

(2) High-frequency VSR and the tempering treatment at 200 °C have almost the same effect on the microstructure and microhardness. The microstructure after these two processes inherits well from the original state, remaining tempered martensite. The microhardness of both processes shows a slight decrease. After tempering at 600 °C, the microstructure mutates, eventually evolving into tempered sorbite, accompanied by a significant reduction in hardness.

(3) The mechanical energy provided by high-frequency VSR and the thermal activation energy provided by 200 °C tempering make the dislocation density of the initial specimens decrease, which is the main reason for the elimination of residual stress. However, since the microstructure remains essentially the same as before the processes, there is no substantial effect on the microhardness. When the temperature reaches 600 °C, the microstructure undergoes abrupt changes as carbon atoms precipitate from the center. This process reduces supersaturation and diminishes the solid solution

strengthening effect, ultimately leading to a significant decline in both residual stress and microhardness.

(4) Comparing the microstructure and microhardness of the initial specimens, those after high-frequency VSR, and those after tempering, it can be concluded that causing a large change in microstructure genetics is an effective way to obtain better residual stress.

(5) In the process of regulating residual stress in marine structural components, high-frequency VSR was recommended. High-frequency VSR can achieve an excellent residual stress relieving rate while maintaining the microstructure of the component with only a slight reduction in hardness. Comparative analysis reveals that tempering at 200 °C similarly maintains microstructure and hardness characteristics yet exhibits inferior stress relieving performance compared with high-frequency VSR. Raising the temperature to 600 °C provides an excellent residual stress relieving rate, but it significantly alters the microstructure, affecting the hardness and tensile performance, thereby weakening structural safety. Furthermore, considering the time, space, and energy required for both processes, high-frequency VSR shows more advantages for regulating residual stress in marine structures.

Author Contributions: G.X.: Conceptualization, Methodology, Validation, Resources, Data Curation, Writing—Review and Editing, Supervision, Project Administration, and Funding Acquisition. F.L.: Writing—Original Draft Preparation, Software, Formal Analysis, Investigation, Visualization. All authors have read and agreed to the published version of the manuscript.

Funding: This research was sponsored by the National Natural Science Foundation of China for Young Scholars with the Grant No. 51805477, which was chaired by the corresponding author.

Institutional Review Board Statement: Not applicable.

Informed Consent Statement: Not applicable.

Data Availability Statement: Data will be made available on request.

Conflicts of Interest: The authors declare no conflict of interest.

References

1. Schubnell, J.; Carl, E.; Widerspan, V.; Collmann, M. Determination of Loading and Residual Stresses on Offshore Jacket Structures by X-Ray Diffraction. *J. Mar. Sci. Eng.* **2023**, *11*, 1304. [CrossRef]
2. Wang, D.; Gao, F.; Shen, Z.; Li, Y.; Zheng, Z.; Zhou, M.; Yang, F.; Zhang, S.; Zhang, Z. Study on Type IV Cracking Behavior of T92/Super304H Dissimilar Steel Welded Joints in Long-Term Service. *Materials* **2024**, *17*, 5888. [CrossRef]
3. Gu, B.; Wang, Y.; Gao, L.; Xu, G.; Liu, L.; Sun, S. Effects of ultrasonic impact treatment on the tensile properties of DH36 steel welded joints. *Eng. Fract. Mech.* **2025**, *315*, 110798. [CrossRef]
4. Chen, B.-Q.; Liu, K.; Xu, S. Recent Advances in Aluminum Welding for Marine Structures. *J. Mar. Sci. Eng.* **2024**, *12*, 1539. [CrossRef]
5. Gu, B.; Zhuo, J.; Xu, G.; Gao, L. Study on the effects of ultrasonic impact treatment on the local mechanical properties of welded joints based on the digital image correlation. *Measurement* **2025**, *239*, 115441. [CrossRef]
6. Chen, X.; Chen, W.; Wu, B.; Wu, H.; Xian, J. Ship visual trajectory exploitation via an ensemble instance segmentation framework. *Ocean Eng.* **2024**, *313*, 119368. [CrossRef]
7. Chen, X.; Zheng, J.; Li, C.; Wu, B.; Wu, H.; Montewka, J. Maritime traffic situation awareness analysis via high-fidelity ship imaging trajectory. *Multimed. Tools Appl.* **2024**, *83*, 48907–48923. [CrossRef]
8. Liu, J.; Jiang, X.; Huang, W.; He, Y.; Yang, Z. A novel approach for navigational safety evaluation of inland waterway ships under uncertain environment. *Transp. Saf. Environ.* **2022**, *4*, tdab029. [CrossRef]
9. Xing, X.; Duan, X.; Jiang, T.; Wang, J.; Jiang, F. Ultrasonic Peening Treatment Used to Improve Stress Corrosion Resistance of AlSi10Mg Components Fabricated Using Selective Laser Melting. *Metals* **2019**, *9*, 103. [CrossRef]
10. Gu, B.; Wang, Y.; Xu, G.; Gao, L. Effect of ultrasonic impact on the microstructure and properties of 30%WC-Ni ceramic coatings by laser cladding. *Ceram. Int.* **2025**, *51*, 4677–4692. [CrossRef]

11. Xiu, L.; Wu, J.; Liu, Z.; Ma, J.; Ji, H.; Zhang, Z. Application of Vibration Stress Relief in CFETR Vacuum Vessel Welding. *Fusion Eng. Des.* **2021**, *171*, 112561. [CrossRef]
12. Gao, H.; Wu, S.; Wu, Q.; Li, B.; Gao, Z.; Zhang, Y.; Mo, S. Experimental and Simulation Investigation on Thermal-Vibratory Stress Relief Process for 7075 Aluminium Alloy. *Mater. Des.* **2020**, *195*, 108954. [CrossRef]
13. Rodopho, F.V.; Vladimir, L.; Filomena, S.; Giovani, G.R.; Julian, A.; Vicente, A.; Javier, S.; Irene, G.C. The Effect of the Deposition Strategy and Heat Treatment on Cold Spray Additive Manufactured 316L Stainless Steel. *Adv. Eng. Mater.* **2024**, *26*, 2302156.
14. Wang, J.S.; Hsieh, C.C.; Lin, C.M.; Chen, E.C.; Kuo, C.W.; Wu, W. The Effect of Residual Stress Relaxation by the Vibratory Stress Relief Technique on the Textures of Grains in AA 6061 Aluminum Alloy. *Mater. Sci. Eng. A* **2014**, *605*, 98–107. [CrossRef]
15. Chuvas, T.C.; Castello, D.A.; Cindra Fonseca, M.P. Residual Stress Relief of Welded Joints by Mechanical Vibrations. *J. Braz. Soc. Mech. Sci. Eng.* **2016**, *38*, 2449–2457. [CrossRef]
16. Gao, H.-J.; Zhang, Y.-D.; Wu, Q.; Song, J. Experimental Investigation on the Fatigue Life of Ti-6Al-4V Treated by Vibratory Stress Relief. *Metals* **2017**, *7*, 158. [CrossRef]
17. Gao, H.; Zhang, Y.; Wu, Q.; Song, J.; Wen, K. Fatigue Life of 7075-T651 Aluminium Alloy Treated with Vibratory Stress Relief. *Int. J. Fatigue* **2018**, *108*, 62–67. [CrossRef]
18. Gong, H.; Sun, Y.; Liu, Y.; Wu, Y.; He, Y.; Sun, X.; Zhang, M. Effect of Vibration Stress Relief on the Shape Stability of Aluminum Alloy 7075 Thin-Walled Parts. *Metals* **2019**, *9*, 27. [CrossRef]
19. De Moraes, A.G.; Clarke, T.G.R.; Diehl, I.L. Quantification of CTOD Fracture Toughness in Welded Joints to Evaluate the Efficacy of Vibration Stress Relief Compared to Thermal Stress Relief. *Mater. Res.* **2020**, *23*, e20200157. [CrossRef]
20. Song, H.; Gao, H.; Wu, Q.; Zhang, Y. Effects of Segmented Thermal-Vibration Stress Relief Process on Residual Stresses, Mechanical Properties and Microstructures of Large 2219 Al Alloy Rings. *J. Alloys Compd.* **2021**, *886*, 161269. [CrossRef]
21. Tu, X.X.; Xiao, L.R.; Cai, Z.Y.; Peng, Z.W.; Ren, P.H.; Zeng, D.L.; Zhao, X.J. Effects of Vibration Aging on Residual Stress and Performance of Instrument-Grade TiC Reinforced Steel Matrix Composite. *Mater. Lett.* **2022**, *325*, 132829. [CrossRef]
22. Song, H.; Gao, Z.; Zhou, X.; Zhang, Q.; Zhang, B. Research on the evolution law and control mechanism of residual stress under vibration stress relief. *Mater. Today Commun.* **2024**, *40*, 110072. [CrossRef]
23. Gao, L.; Yang, Y.; Gu, B.; Li, S.; Xu, G.; Yu, Y. Experimental and simulation analysis on the effect of vibratory stress relief on impact toughness of Q345/316 L dissimilar steel welded joints. *Mater. Today Commun.* **2024**, *38*, 108084. [CrossRef]
24. Jiang, G.; He, W.; Zheng, J.Y. Mechanism and experimental research on high-frequency VSR. *J. Zhejiang Univ. (Eng. Sci.)* **2009**, *7*, 1269–1272. (In Chinese)
25. He, W.; Gu, B.; Zheng, J.; Shen, R. Research on High-Frequency Vibratory Stress Relief of Small Cr12MoV Quenched Specimens. *Appl. Mech. Mater.* **2012**, *157–158*, 1157–1161. [CrossRef]
26. Gu, B.P.; Hu, X.; Lai, J.T.; Jin, Z.D.; Zhou, H.; Yang, Z.S.; Pan, L. Effects of High-Frequency Vibration on Quenched Residual Stress in Cr12MoV Steel. *J. Mater. Res.* **2016**, *31*, 3588–3596. [CrossRef]
27. Song, H.; Gao, H.; Wu, Q.; Zhang, Y. Residual Stress Relief Mechanisms of 2219 Al-Cu Alloy by Thermal Stress Relief Method. *Rev. Adv. Mater. Sci.* **2022**, *61*, 102–116. [CrossRef]
28. Xiao, Z.; Huang, Y.; Chen, C.; Li, Z.; Gong, S.; Huang, Y.; Zhang, C.; Zhang, X. Effects of Thermal Treatments on the Residual Stress and Micro-Yield Strength of Al2O3 Dispersion Strengthened Copper Alloy. *J. Alloys Compd.* **2019**, *781*, 490–495. [CrossRef]
29. Jiang, X.; Xiong, W.; Wang, L.; Guo, M.; Ding, Z. Heat treatment effects on microstructure-residual stress for selective laser melting AlSi10Mg. *Mater. Sci. Technol.* **2020**, *36*, 168–180. [CrossRef]
30. Xiao, G.; Tong, H.; Shu, Y.; Ni, A. Spatial-temporal load prediction of electric bus charging station based on S^2TAT. *Int. J. Electr. Power Energy Syst.* **2025**, *164*, 110446. [CrossRef]
31. Xiao, G.; Pan, L.; Lai, F. Application, opportunities, and challenges of digital technologies in the decarbonizing shipping industry: A bibliometric analysis. *Front. Mar. Sci.* **2025**, *31*, 1523267. [CrossRef]
32. Kose, S.; Sekban, D. Emphasizing the importance of using cold-ironing technology by determining the share of hotelling emission value within the total emission. *Transp. Saf. Environ.* **2022**, *4*, tdab010. [CrossRef]
33. Fang, X.F.; Dahl, W. Strain hardening of steels at large strain deformation. Part I: Relationship between strain hardening and microstructures of b.c.c steels. *Mater. Sci. Eng. A* **1995**, *203*, 14–25. [CrossRef]
34. Gu, B.P.; Hu, X.; Lai, J.T.; Zhang, C.; Jin, Z.D.; Xu, G.H.; Yang, Z.S.; Pan, L. Effects of High Frequency Vibration on the Residual Stress in AISI 1045 Steel Subjected to Pulsed Laser Beam Surface Irradiation. *Lasers Eng.* **2018**, *39*, 265.
35. Gu, B.P.; Hu, X.; Xu, G.H.; Lai, J.T.; Pan, L. Microcosmic Mechanism of High-Frequency Vibratory Stress Relief Based on Dislocation Density Evolution. *Rare Met. Mater. Eng.* **2018**, *47*, 2477–2482. (In Chinese)
36. Haiko, O.; Kaijalainen, A.; Pallaspuro, S.; Hannula, J.; Porter, D.; Liimatainen, T.; Kömi, J. The Effect of Tempering on the Microstructure and Mechanical Properties of a Novel 0.4C Press-Hardening Steel. *Appl. Sci.* **2019**, *9*, 4231. [CrossRef]
37. Jin, L.; Zhang, X.; Liang, X.; Wang, Z.; Wu, Q. Mechanical Properties of Corroded Carbon Steel Based on Random Pit Corrosion in Marine Environment. *Ocean Eng.* **2022**, *260*, 111759. [CrossRef]

38. Grinon-Echaniz, R.; Paul, S.; Thornton, R.; Refait, P.; Jeannin, M.; Rodriguez, A. Prediction of Thermal Spray Coatings Performance in Marine Environments by Combination of Laboratory and Field Tests. *Coatings* **2021**, *11*, 320. [CrossRef]
39. Gu, B.; Chu, J.; Wang, Y.; Xu, G.; Gao, L.; Yang, Y.; Hu, Y.; Zhang, H. Effects of ultrasonic impact treatment on the corrosion resistance of laser-cladded CrMnFeCoNi high-entropy alloy coatings. *Surf. Coat. Technol.* **2024**, *489*, 131102. [CrossRef]
40. Tian, X.; Pei, J.; Rong, J. Dynamic Mechanical Properties and Damage Parameters of Marine Pipelines Based on Johnson–Cook Model. *J. Mar. Sci. Eng.* **2023**, *11*, 1666. [CrossRef]
41. Zhou, Z.; Ding, Y. A Study on the Fatigue Performance and Corrosion Resistance of 304/45 Bimetallic Composite Bolts. *Materials* **2023**, *16*, 4454. [CrossRef]
42. Liu, C.; Xu, Q.; Liu, X.; Fan, X.; Li, Y. Comparative Study of Marine Steel and Welding Joint in Artificial Seawater Based on Stress Corrosion Cracking and Crack Growth. *J. Mater. Res. Technol.-JMRT* **2024**, *33*, 2683–2691. [CrossRef]
43. Wang, R.; Chu, G.; Zhang, J.; Liu, R.; Wang, Y.; Sun, L.; Cao, Z.; Song, L.; Ma, F. Influence of Stress on Corrosion Behavior and Evolution Model of Q235 Steel in Marine Environments. *Int. J. Press. Vessel. Pip.* **2025**, *214*, 105388. [CrossRef]
44. *ASTM E837–08;* Standard Test Method for Determining Residual Stresses by the Hole-Drilling Strain-Gage Method. ASTM International: West Conshohocken, PA, USA, 2008.
45. Pei, J.; Wang, S.; Qin, S.; Yu, Z.; Gao, J.; Xiao, J.; Li, Z. Using Iterative Correction to Improve the Accuracy of the Blind-Hole Welding Residual Stress Test. *Sensors* **2024**, *24*, 6243. [CrossRef] [PubMed]
46. Gu, B.; Wang, C.; Wang, Y.; Gao, L.; Xu, G.; Yang, Y.; Zhu, X. Effect of ultrasonic impact treatment on surface residual stress, microstructure and electrochemical properties of the hot-rolled S355 steel. *Surf. Coat. Technol.* **2024**, *494*, 131468. [CrossRef]
47. Ceschini, L.; Morri, A.; Morri, A.; Pivetti, G. Predictive Equations of the Tensile Properties Based on Alloy Hardness and Microstructure for an A356 Gravity Die Cast Cylinder Head. *Mater. Des.* **2011**, *32*, 1367–1375. [CrossRef]
48. Tiryakioğlu, M.; Campbell, J.; Staley, J.T. On Macrohardness Testing of Al-7 Wt.% Si-Mg Alloys II. An Evaluation of Models for Hardness-Yield Strength Relationships. *Mater. Sci. Eng. A* **2003**, *361*, 240–248. [CrossRef]
49. Takuro, M.; Shohei, U.; Kyosuke, M.; Setsuo, T.; Toshihiro, T. Analysis of Dislocation Density by Direct-Fitting Modified Williamson-Hall (DF MWH) Method in Tempered Low-Carbon Martensitic Steel. *Tetsu-Hagane/J. Iron Steel Inst. J.* **2020**, *106*, 826–834. (In Japanese)
50. Marchese, G.; Atzeni, E.; Salmi, A.; Biamino, S. Microstructure and Residual Stress Evolution of Laser Powder Bed Fused Inconel 718 under Heat Treatments. *J. Mater. Eng. Perform.* **2021**, *30*, 565–574. [CrossRef]
51. Park, S.J.; Jo, S.-H.; Kim, J.G.; Kim, J.; Lee, R.; Oh, Y.-S.; Kim, S.-J.; Lee, H.W.; Kang, S.-H.; Jung, J. Evolution of Microstructure, Mechanical Properties and Residual Stress of a Cold Rolled Invar Sheet Due to Heat Treatment. *Metals* **2022**, *12*, 110. [CrossRef]
52. Chao, Q.; Thomas, S.; Birbilis, N.; Cizek, P.; Hodgson, P.D.; Fabijanic, D. The Effect of Post-Processing Heat Treatment on the Microstructure, Residual Stress and Mechanical Properties of Selective Laser Melted 316L Stainless Steel. *Mater. Sci. Eng. A* **2021**, *821*, 141611. [CrossRef]

Journal of
Marine Science and Engineering

MDPI

Article

ANFIS-Based Course Controller Using MMG Maneuvering Model

Yu Guo [1], Rui Yang [1], Zhiheng Zhang [1,*] and Bing Han [2]

1 Merchant Marine College, Shanghai Maritime University, Shanghai 201306, China; yuguo@shmtu.edu.cn (Y.G.); 202330110258@shmtu.edu.cn (R.Y.)
2 Shanghai Ship and Shipping Research Institute Co., Ltd., Shanghai 200135, China; han.bing@coscoshipping.com
* Correspondence: zhzhang@shmtu.edu.cn

Abstract: In the domain of course control, traditional methods such as proportional–integral–derivative (PID) control often exhibit limitations when addressing complex nonlinear systems and uncertain disturbances. To mitigate these challenges, the adaptive neuro-fuzzy inference system (ANFIS) has been integrated into course control strategies. The primary objective of this study is to investigate the course control characteristics of vessels governed by the ANFIS controller under both normal and severe sea conditions. A three-degree-of-freedom (3-DOF) maneuvering model set (MMG) was employed and validated through sea turning tests. The design of the ANFIS controller involved a combination of the backpropagation algorithm with the least square method. Training data for the ANFIS control system were derived from a linear control framework, followed by simulation tests conducted under normal and severe sea conditions to assess control performance. The simulation results indicate that in normal sea conditions, ANFIS has more stable heading control (smaller A_ψ), but at the cost of more energy consumption (larger I_δ). Notably, response time is reduced by approximately 36.7% compared to that of the linear controller. Conversely, during severe sea conditions, ANFIS exhibits an increase in response time by about 33.3% relative to the linear controller while maintaining a smaller I_δ. In the whole course control stage, the stability is better than the linear controller, and it has better energy-saving characteristics. Under scenarios involving small and large course alterations, A_ψ values for ANFIS are approximately 11.28% and 13.97% higher than those observed with the best-performing linear controller (λ_ψ = 60), respectively. As the propeller speed increases, the A_ψ value of the ANFIS controller decreases significantly, to about 62.71%, indicating that the energy efficiency is improved and the course stability is also enhanced. In conclusion, it can be asserted that the implementation of an ANFIS controller yields commendable performance in terms of controlling vessel courses effectively.

Keywords: heading control; ANFIS; MMG model; simulation

1. Introduction

Maritime transport safety has become an important factor restricting the rapid development of the shipping industry [1]. In addition to the burning process of the ship's main engine and auxiliary engine, a large number of pollutants will be produced, affecting the safety of maritime transportation [2]. Ship course control is also the main content of maritime transportation security, and the intelligent development of control systems is the main trend of navigation technology development. For ocean-going ships, course control is essential to enhance maritime traffic safety, especially in the era of autonomous ships [3].

The traditional controller is considered an effective way to realize ship course keeping. The common method is proportional–integral–differential (PID) control. Mochammad S et al. [4] proposed a ship control system using a PID controller, realized the parameter identification algorithm of the control system based on gradient approximation, and achieved good ship maneuverability. Sutulo S et al. [5] proposed a conventional PID controller for course keeping. In the range of water depth, lateral distance, and speed, the hydrodynamic interaction and maneuvering speed response of two ships during overtaking are simulated. When the ratio of water depth to the ship's draft depth is less than 2.0, the PID controller is still effective for overtaking boats. Bayezit L et al. [6] proposed a strategically coupled system based on a sensor model and PID controller for course angle control to understand the motion dynamics of ocean vessels and demonstrated the practical feasibility of the system through the "vehicle in the loop" test, demonstrating its potential to improve the efficiency of recreational ocean navigation.

Conventional PID controllers are not robust to external interference, in contrast to fraction-order controllers, sigmoid PID controllers, BELBIC PID controllers, and multi-node hormone neuroendocrine PID controller, which have received attention. Liu L et al. [7] proposed an FOPID controller based on the three-dimensional stable region analysis method, which improved the robustness of the title control of unmanned ships. Although the robustness of the controller has been improved to some extent, the complexity of determining the parameters of the FOPID controller results in a long control response time, which is obviously insufficient in the face of transient title control situations. Suid MH et al. [8] proposed the parameters of the SPID controller obtained using an enhanced self-tuning heuristic optimization method called the nonlinear sine-cosine algorithm (NSCA) to achieve a better dynamic response. TAO S et al. [9] used (BELBIC PID) to control the heading of autonomous ground vehicles. Then, in order to achieve better path tracking performance and drive stability, the PSO algorithm was used to optimize the parameters, and the results were good. To sum up, the PID controller has the characteristics of a simple structure in the control field, but it needs manual adjustment of parameters and has poor adaptability.

In order to improve the shortcomings of PID control, robust controllers are introduced into this topic. Muzammal M et al. [10] proposed nonlinear control algorithms based on variable structure, namely integral sliding mode, double integral sliding mode, terminal sliding mode, and super spiral sliding mode controller. The controller switches between different control laws depending on the system state, ensuring robust performance in different sea conditions. Hosseinabadi P et al. [11] designed a new finite-time robust controller for ship heading (heading) control systems with unknown mismatched external disturbances and uncertainties. A fuzzy adaptive finite time sliding mode control (FAFSMC) scheme is proposed. By combining the concept of fuzzy controller, adaptive time stability, and sliding mode control (SMC) scheme, the advantages of these two schemes are utilized and the shortcomings of single applications are compensated. Rezaei A et al. [12] designed an adaptive fractional-order sliding mode controller (AFOSMC) to reduce the adverse effects of ship roll motion. Through this adaptive control method, the designed controller is robust to the ship's lateral motion mechanics and the uncertain parameters in the fin actuator. Xu Y et al. [13] designed a global fast terminal synovial controller to track and compensate ship heave motion data. The results show that the tracking compensation effect of the improved sliding mode controller is better than that of the traditional PID control, and the tracking error is smaller than that of the traditional PID control when the improved sliding mode controller is introduced. Global fast terminal sliding mode control has good robustness and can be used in a ship heave compensation system. To sum up, the course control accuracy and stability of the robust controller are stronger, but it lacks intelligence.

In the current field of course control, in order to assist ship driving more safely, effectively reduce the danger to the crew, and improve the efficiency of maritime traffic [14], intelligent controllers are widely concerned. Both artificial neural network control and fuzzy control belong to intelligent control. NIM N et al. [15] developed a nonlinear autoregressive model based on a neural network by combining an artificial neural network with fuzzy logic control. The results show that the two-wheeled wheelchair system combining fuzzy logic control and a neural network can improve the stability and performance. Aghaseyedabdollah M et al. [16] proposed a fuzzy adaptive synovial control scheme for cable robots. In the proposed scheme, the intelligent method is combined with the traditional synovial control to achieve the optimal adjustment of control parameters. Arifi A et al. [17] proposed an ASP modeling and control method based on Takagi–Sugeno (TS) and successfully implemented the method for carbon removal. The results show the effectiveness and superiority of the proposed control method based on TS fuzzy for dealing with complex and nonlinear biochemical processes. Han B et al. [18] proposed an improved fuzzy control method based on the integrated line-of-sight (ILOS) guidance principle in order to meet the needs of autonomous navigation and high-precision ship trajectory control. Simulation results show that the algorithm has good following performance and can maintain smooth rudder angle output. The research results provide a reference for ship path tracking control. Li Y et al. [19] proposed a berthing decision method for very large ships based on fuzzy logic. The research results provide theoretical and practical insights for the development of human-like decision-making methods for autonomous navigation in port waters and maritime safety management in the shipping industry. The model can also be further applied to develop more widely applicable autonomous navigation decision systems in narrow waters. Du L et al. [20] proposed a model combining a computational fluid dynamics solver (ANN) to calibrate the efficiency and precision weights in high-dimensional problems of ship resistance optimization. The adaptive mechanism of the model reduced the calculation cost from 1638 min to 456 min, and the optimization efficiency increased by 72.16%. The results show that the proposed adaptive mechanism can dynamically rebalance the efficiency and accuracy of the framework and provide theoretical and technical support for ship design optimization. Unar S et al. [21] proposed an artificial neural network controller for course and position control systems and selected a mathematical model with four effective thrusters to test the performance of the proposed controller. Rohit D et al. [22] proposed a deep Q-learning method based on an artificial neural network to solve the ship heading control problem. This method can make optimal decisions based on sufficient learning experience and realize the interaction between the numerical model and waypoint tracking. Wakita K et al. [23] proposed a system identification method using recurrent neural networks and free-running model testing to generate a low-speed maneuvering model, paying special attention to low-speed maneuvering in the last stage of berthing, so as to achieve automatic berthing course control. The results show that this method can accurately represent the low-speed motion of the ship. Bouaiss O et al. [24] proposed a novel nested control strategy based on adaptive radial basis function neural networks (RBFNNs) and embedded integrators (IBS) for neural network supervised control to reduce modeling uncertainty, perceived noise, and external bounded interference, with robustness and effectiveness superior to PID controllers. Zhao et al. [25] used radial basis neural networks to compress unknown system terms and external disturbances into unknown parameters, aiming at the problem of course control and maintenance with unknown environmental interference and model uncertainty. Finally, numerical simulation verified the effectiveness of the algorithm. Le et al. [26] studied the application of artificial neural networks in ship heading control systems. A two-layer multi-layer feedforward neural network heading control system is proposed. The simulation results show that the course control system can maintain the predefined direction under various sea conditions, and the proposed method can be used to develop and apply the design of a real ship automatic driving system. To sum up, the combination of neural networks and fuzzy logic control can effectively resist the interference of the external environment. It has stability and generalization. This kind of controller meets the requirements of current navigation practice.

ANFIS combines the advantages of artificial neural networks (ANNs) and fuzzy inference systems (FISs). This hybrid control has the advantages of both neural network and fuzzy inference algorithms to achieve better control performance [27]. Compared with the sigmoid PID controller, BELBIC PID controller, and other PID controllers mentioned above, the ANFIS controller can automatically adjust fuzzy rules and parameters through training data, and has strong learning ability without manual intervention. The sigmoid PID controller lacks adaptive learning ability and requires manual parameter adjustment. Although the BELBIC PID controller has certain adaptability, its learning mechanism is relatively complex and depends on the accuracy of the motion model. Therefore, the ANFIS controller can automatically adjust the control strategy by combining fuzzy rules and a neural network to adapt to different sea conditions and ship states. This integration enables the handling of intricate nonlinear relationships, enhancing control accuracy and robustness while maintaining ship course stability in complex maritime environments.

PID controllers require manual adjustment of parameters and cannot control complex nonlinear models. Robust controllers (SMC) are very sensitive to parameter selection, and small changes in parameters can lead to significant changes in controller performance. Therefore, using an ANFIS controller can overcome the shortcomings of the previous two.

In this study, the ANFIS controller based on the MMG model realizes stable ship course control. Rudder angle and propeller speed are used as input parameters to analyze the response performance of the controller. The main contributions can be summarized as follows:

(1) By establishing an MMG model considering the influence of waves, the accuracy of the MMG model is verified by the rotating circle test of the self-propelled model. A large amount of training data comes from a linear control system based on the MMG model, which takes propeller speed and rudder angle as two control inputs and yaw angle speed as the control output. After training, ANFIS takes the difference between the actual course and the expected course, yaw angular speed as input, and rudder angle as output. Course control is executed under both normal and adverse sea conditions. The effectiveness of the proposed ANFIS controller is verified. The MMG model and ANFIS controller can accurately describe the ship's motion characteristics and provide a reliable theoretical basis for the design and verification of the ship's course controller.

(2) By combining the learning ability of a neural network and the reasoning ability of fuzzy logic, a heading controller without manual parameter adjustment is designed. This kind of controller can realize the effective and high-precision control of the desired course and improve the automation level of ship course control.

(3) Simulation tests are carried out under normal sea conditions and bad sea conditions, and the results show that the ANFIS controller is superior to the traditional linear controller in course control performance. In normal sea conditions, the ANFIS controller is able to maintain course with greater accuracy, despite consuming slightly more energy. In harsh sea conditions, the ANFIS controller automatically obtains the best control response through the learning ability of the neural network, showing better robustness. Finally, the influence of propeller speed variation on ship course and rudder angle response is analyzed, which provides theoretical support for ship control in actual sailing.

2. Ship Motion Simulation Model

2.1. Coordinate System

To study the mathematical model of ship motion, it is essential to establish a coordinate system. Two right-handed rectangular coordinate systems are considered, as illustrated in

Figure 1. In the presence of waves in a complex ocean environment, denoted by χ as the wave angle, The *O-XYZ* is the global coordinate system. Point *O* is usually chosen as the position of the ship's center of gravity at the initial moment. In this system, the *OX* axis points due north, *OY* points due east, and *OZ* points toward the center of the earth. The *o-xyz* is the body-fixed coordinate system, with *o* as the midpoint of the line connecting the ship's bow and stern. In this system, *ox* points to the bow along the ship's centerline, and *oy* points to the port side. The heading angle ψ is referenced from the north direction, the rudder angle δ is positive for the right rudder, and *r* is positive in the clockwise direction. In many ship motions and control scenarios, heaving, pitching, and rolling motions are often negligible, simplifying the problem to a plane motion with only 3 degrees of freedom.

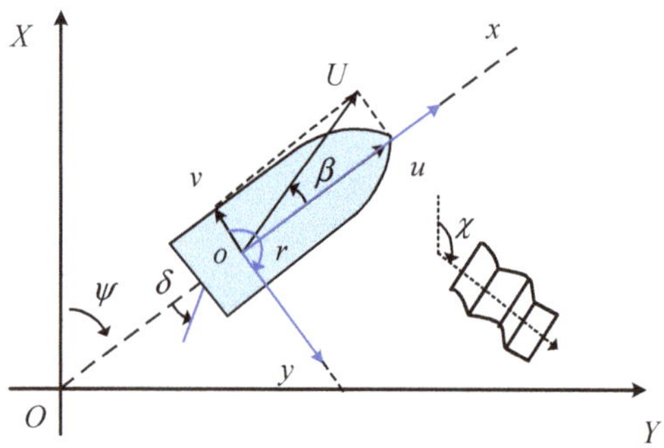

Figure 1. Describes the motion of the ship in the inertial coordinate system and the attached coordinate system.

2.2. 3-DOF MMG Model

The MMG model primarily focuses on decomposing the fluid forces and moments acting on a ship into components related to the bare hull, open water propeller, and open water rudder, as well as their interactions [28]. This decomposition allows for a clearer understanding of the physical forces at play and can be easily supported by experimental data. By establishing a mathematical model of ship motion that considers the interactions between these components, the MMG model addresses issues such as path tracking and heading control. The model simplifies the analysis by only considering the ship's movement with three degrees of freedom on the horizontal plane, omitting rolling, pitching, and heaving motions. Following the separated modeling approach, the study divides the hydrodynamic forces and moments into contributions from the hull, propeller, and rudder. The model expresses the external forces and torques affecting the ship's motion, *X*, *Y*, and *N*, as fluid inertia forces, fluid viscosity forces acting on the hull, propeller thrust, and rudder forces. The basic structure of the MMG model is captured in Equation (1).

$$\begin{cases} m(\dot{u} - vr) = X_I + X_H + X_P + X_R \\ m(\dot{v} + ur) = Y_I + Y_H + Y_P + Y_R \\ I_{zz}\dot{r} = N_I + N_H + N_P + N_R \end{cases} \tag{1}$$

The subscripts *H*, *P*, and *R* denote the low-frequency hydrodynamic forces on the hull, propeller, and rudder, respectively. The subscript *I* denotes the hull fluid inertial force. When a ship changes speed or turns in a fluid, it induces acceleration or deceleration in the surrounding medium, resulting in a reaction force called fluid inertia force. This force, associated with acceleration, hinders the non-inertial motion of the ship in the respective direction.

The fluid inertia force acting on the hull in Equation (1) can be expressed as the following Equation (2).

$$\begin{cases} X_I = -(m_x\dot{u} - m_y vr) \\ Y_I = -(m_y\dot{v} + m_x ur) \\ N_I - -J_{zz}\dot{r} \end{cases} \quad (2)$$

Substituting Equation (2) into Equation (1), the MMG model can be written as follows

$$\begin{cases} (m + m_x)\dot{u} - (m + m_y)vr = X_H + X_P + X_R \\ (m + m_y)\dot{v} + (m + m_x)ur = Y_H + Y_P + Y_R \\ (I_{ZZ} + J_{ZZ})\dot{r} = N_H + N_P + N_R \end{cases} \quad (3)$$

From Equation (3), it is evident that the fluid inertia force results in an increase in the mass and inertia moment of the hull by a specific value. where m is the mass of a ship. m_x and m_y are the added masses along the x and y axes, respectively. I_{zz} and J_{zz} denote inertia moments and the added inertia moments about the Z axis, respectively.

2.2.1. Hydrodynamic Forces Acting on the Hull

Hydrodynamic forces acting on hull can be expressed as follows

$$\begin{cases} X_H = 0.5\rho L^2 V^2 X'_H(v', r') - R(u_0) \\ Y_H = 0.5\rho L^2 V^2 Y'_H(v', r') \\ N_H = 0.5\rho L^3 V^2 N'_H(v', r') \end{cases} \quad (4)$$

where $R(u_0)$ means the ship resistance. X_H, Y_H, and N_H, represent the surge force, sway force, and yaw moment, respectively. The total speed V is expressed as $V = \sqrt{v^2 + u^2}$. L is the length between perpendiculars. The non-dimensional forms of r and v are $r\prime = r(L/u)$ and $v\prime = v/V$.

The hydrodynamic coefficients for dimensionless maneuvering are represented in polynomial form with respect to v' and r'

$$\begin{cases} X'_H(v, r') = X'_{vv}v'^2 + X'_{vr}vr' + X'_{rr}r'^2 \\ Y'_H(v, r') = Y'_v v' + Y'_r r' + Y'_{vvv}v'^3 + Y'_{vvr}v'^2 r' + Y'_{vvv}v'^2 + Y_{rrr}r'^3 \\ N'_H(v, r') = N'_v v' + N'_r r' + N'_{vvv}v'^3 + N'_{vvr}v'^2 r' + N'_{\mathrm{vrr}}v' r'^2 + N_{rrr}r'^3 \end{cases} \quad (5)$$

The non-dimension coefficients $X\prime_{vv}$, $X\prime_{vvv}$, $N\prime_{vvv}$, etc., denote hydrodynamic derivatives with respect to physical variables, and they can be obtained by either the planar motion mechanism test or the circular motion test. The hull force is represented as a third-order polynomial, as depicted in Equation (5). The maneuvering coefficients $X\prime_{vv}$, $Y\prime_{vvv}$, and $N\prime_{vvv}$ utilized in this study were obtained through planar motion mechanism (PMM) experiments [29]. This formula illustrates that the hull force can be approximated by a third-order polynomial [30].

2.2.2. Hydrodynamic Forces on the Propeller

Due to the minimal impact of the lateral force and moment of the propeller, accurately measuring their values poses a challenge. In the MMG model, these forces are typically considered in conjunction with the hydrodynamic force of the steering hull, thus being incorporated into Equation (3) while torque is omitted. As a result, the calculation focuses solely on the longitudinal hydrodynamic force of the propeller, specifically the thrust. The propeller thrust is commonly determined using the following formula.

$$X_p = (1 - t_p)\rho n_p^2 D_p^4 K_T \quad (6)$$

In Equation (6), t_p, n_p, and D_p represent the thrust derating coefficient, propeller speed, and propeller diameter, respectively. The thrust coefficient K_T is a function of the advance speed coefficient and is expressed as a second-order polynomial.

$$K_T(J_p) = j_0 + j_1 J_p + j_2 J_p^2 \tag{7}$$

The coefficients j_0, j_1, and j_2 in Equation (7) can be determined through propeller open water testing, and the advance speed coefficient J_p is expressed as

$$J_p = \frac{u(1-w_p)}{nD} \tag{8}$$

Equation (8) defines w_p as the wake coefficient at the propeller, with its value typically varying during maneuvering motion. This article employs a specific formula to estimate w_p.

$$w_p = w_{p0} \cdot \exp(-8.0\beta_p^2) \tag{9}$$

In Equation (9), w_{p0} represents the wake fraction at the propeller when the ship is sailing straight, and β_p represents the inflow angle at the propeller during maneuvering motion.

$$\beta_p = \beta - x'_p r' = \tan^{-1}(-v/u) - x'_p r' \tag{10}$$

In Equation (10), $\beta = \tan^{-1}(-v/u)$ represents the drift angle, and $x\prime_p$ represents the dimensionless x-coordinate of the propeller.

2.2.3. The Rudder Forces

For ship maneuverability, the calculation of rudder force is of paramount importance. The rudder forces are evaluated as

$$\begin{cases} X_R = -(1-t_R)F_N \sin\delta \\ Y_R = -(1+a_H)F_N \cos\delta \\ N_R = -(x_R + a_H x_H)F_N \cos\delta \end{cases} \tag{11}$$

where F_N is the rudder normal force. t_R, a_H and x_H represent the interaction coefficients between ship hull and rudder. x_R denotes the x coordinate of the center of the rudder.

The rudder normal force F_N is evaluated by

$$F_N = \frac{1}{2}\rho \frac{6.13\Lambda}{\Lambda + 2.25} A_d U_R^2 \sin\alpha_R \tag{12}$$

where ρ is the water density. Λ and A_d are the rudder aspect ratio and the rudder area, respectively. U_R and α_R represent the resultant rudder inflow velocity and the inflow angle, respectively, and they can be estimated as follows

$$\begin{cases} \alpha_R = \delta - \tan^{-1}(v_R/u_R) \\ u_R = u_p \varepsilon \sqrt{1 + 8\kappa K_T/(\pi J_P^2)} \\ u_p = u\left[(1-w_p) + \tau\beta_p^2\right] \\ v_p = \left(\gamma v' + C_{Rr} r' + C_{Rrrr} r'^3 + C_{Rrrv} r'^2 v'\right)U \\ U_R^2 = u_R^2 + v_R^2 \end{cases} \tag{13}$$

where the coefficients ε, κ, τ and γ represent the interaction among the hull, propeller, and rudder. v_R is the inflow velocities of the rudder in the x direction. C_{Rr}, C_{Rrrr} and C_{Rrrv} are experimental constants for expressing v_R accurately [31]. u_R and u_P are the inflow velocities to the rudder and the propeller, respectively. K_T, w_p, and J_p are the propeller thrust, the wake coefficient, and the propeller advanced ratio, respectively.

2.3. Wave Disturbance

Ships navigating the sea are inevitably influenced by waves, and a precise description of the wave disturbances experienced by ships is crucial for accurate analysis in ship maneuvering control research. To predict the wave disturbance affecting a ship, this study employs the modified Pearson–Moscowitz (MPM) spectrum, which is detailed as follows

$$S_{\zeta\zeta}(\omega) = \frac{4\pi^3 H_s^2}{T_z^4} w^{-5} \exp(-\frac{16\pi^3}{T_z^4}\omega^{-4}) \tag{14}$$

where H_s is the significant wave height. T_z is the average zero-crossings period. One 2nd-order linear shape function W_r is utilized to approximate the wave disturbance acting on course. It is expressed as follows

$$W_r = h_r(s) \cdot W(s) \cdot \sin(\psi - \chi) \tag{15}$$

where $W(s)$ is the zero-mean Gaussian white noise and its power spectrum1.0. One 2nd-order wave transfer function $h_r(s)$ is approximately described as

$$h_r(s) = \frac{2\xi_0\omega_0\sigma_r s}{s^2 + 2\xi_0\omega_0 s + \omega_0{}^2} \tag{16}$$

where σ_r denotes the dominated wave strength coefficients of Yaw. ξ_0 and ω_0 are the damping coefficient and the dominating wave frequency, respectively. The ξ_0 is chosen as 0.26 in the present study [32]. Substituting $s = \bar{j}\omega(\bar{j}^2 = -1)$ into Equation (11) yields

$$h_r(\bar{j}\omega) = \frac{2\bar{j}\xi_0\omega_0\sigma_r\omega}{(\omega_0^2 - \omega^2) + 2\bar{j}\xi_0\omega_0\omega} \tag{17}$$

The $P_{yy}(\omega)$ is showed as follows

$$P_{yy}(\omega) = \left|h_r(\bar{j}\omega)\right|^2 P_{\omega\omega}(\omega) = \frac{4(\xi_0\omega_0\sigma_r)^2\omega^2}{(\omega_0^2 - \omega^2)^2 + 4(\xi_0\omega_0\omega)^2} \tag{18}$$

where $P_{\omega\omega}(\omega)$ is selected as 1.0. P_{yy} reflects the energy distribution of $S_{\xi\xi}(\omega)$ in the actual frequency range. When $\omega = \omega_0$, the maximum value of $P_{yy}(\omega)$ is obtained, which is

$$\max P_{yy}(\omega) = P_{yy}(\omega_0) = \sigma_r^2 \tag{19}$$

when $P_{yy}(\omega)$ obtains the maximum value, use $P_{yy(\omega)}$ to approximate the MPM spectrum. At this time, the expressions of ω_0 and σ_r can be obtained

$$\begin{gathered} (\frac{dS_{\zeta\zeta}(\omega)}{d\omega})_{\omega=\omega_0} = 0 \\ P_{yy}(\omega_0) = S_{\zeta\zeta}(\omega_0) = \sigma_r^2 \end{gathered} \tag{20}$$

According to Formula (20), we can get

$$\begin{gathered} \omega_0 = \sqrt[4]{\frac{4}{5} \times \frac{16\pi^3}{T_z^4}} \\ \sigma_r = H_s \times \sqrt{0.0201 T_z} \end{gathered} \tag{21}$$

when the wave spectrum changes, the expressions of ω_0 and σ_r also need to be rederived according to Equation (20). When the encounter frequency ω replaces the wave frequency ω_0 and is substituted into Equation (17), considering infinite water depth and the influence of the ship's forward speed, the resulting output can be obtained.

$$h_r(s) = \frac{2\xi_0\omega\sigma_r s}{s^2 + 2\xi_0\omega s + \omega^2} \tag{22}$$

In summary, the three-degree-of-freedom MMG model has a high degree of nonlinearity and is suitable for ship course control research. The relationship between this model and the wave force model is shown in Figure 2 below.

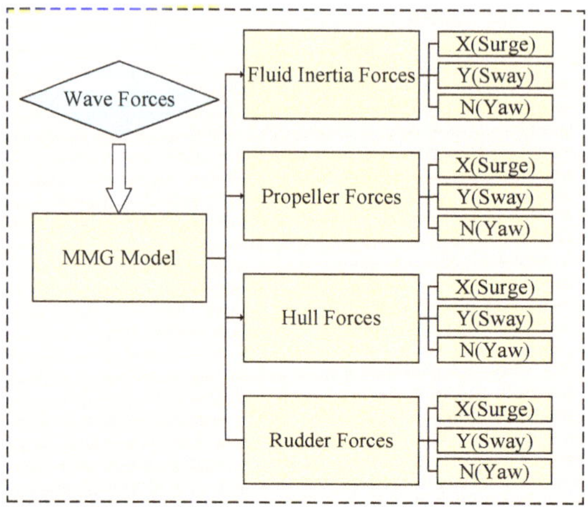

Figure 2. The relationship between this model and the wave force model.

3. Adaptive Neuro-Fuzzy Inference System

The adaptive neuro-fuzzy inference system aims to integrate the fuzzy inference system, which is built on the Takagi–Sugeno model's characteristics and structure, with the learning abilities of neural networks. The Takagi–Sugeno model is a valuable tool for modeling uncertain systems, especially for approximating nonlinear functions and complex dynamic processes.

ANFIS has garnered interest for its innovative approach to system modeling, effectively mapping intricate relationships between input and output data through the integration of learning data and adjustments to membership functions and fuzzy rule parameters. This process involves a neural network-like structure, where the parameters of the fuzzy controller are iteratively optimized based on feedback until the control effect meets a predetermined standard. ANFIS shows promising potential in ship heading control applications.

Figure 3 shows the structure of the ANFIS control system based on the MMG model. The ANFIS controller calculates the rudder angle correction instruction according to the deviation between the set course and the actual course, and the instruction acts on the ship through the actuator to adjust the course. The MMG model simulates the dynamic response of a ship under the action of rudder angle and considers the influence of wave interference on ship motion. System outputs, including swing distance and heading, are displayed graphically to facilitate analysis of the ship's navigational performance. The purpose of the entire system is to keep the ship's course stable by automatically adjusting the rudder angle, even in the case of external disturbances such as waves.

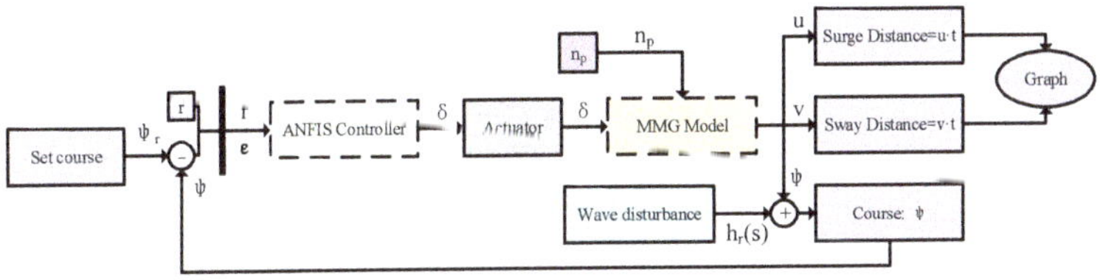

Figure 3. The structure of ANFIS control system based on MMG model.

The training data [$Y1$, $Y2$, $Y3$] were added to the MATLAB working area. When the training period was set to 50 times, the convergence state was finally presented, and the training error was about 0.04°, as shown in Figure 4, indicating that the accuracy of the training process was satisfactory.

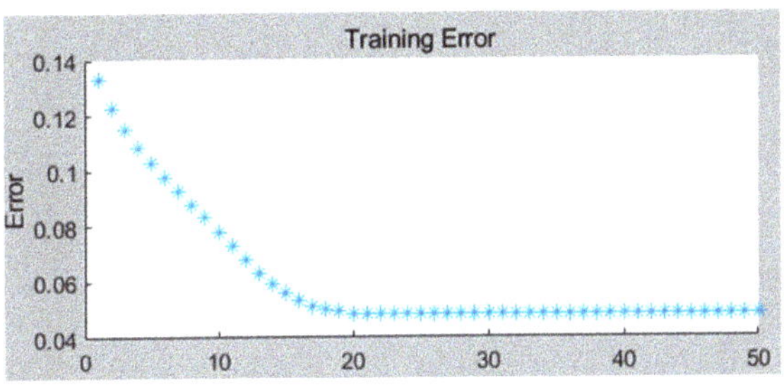

Figure 4. ANFIS training error ("*" represents the training error for each time step).

3.1. ANFIS Ship Heading Controller Structure

The ship's control system operates through a range of sensors, such as gyroscopes, heading sensors, speed sensors, rudder angle sensors, and wave sensors. The sensor feeds some data into the controller and then outputs it.

The ANFIS described in this study is a feedforward fuzzy neural network consisting of 5 layers, as illustrated in Figure 5. It incorporates artificial neural network elements into a fuzzy neural system to create an artificial neural network and applies the learning algorithm of artificial neural networks to derive the inference rules of the fuzzy system.

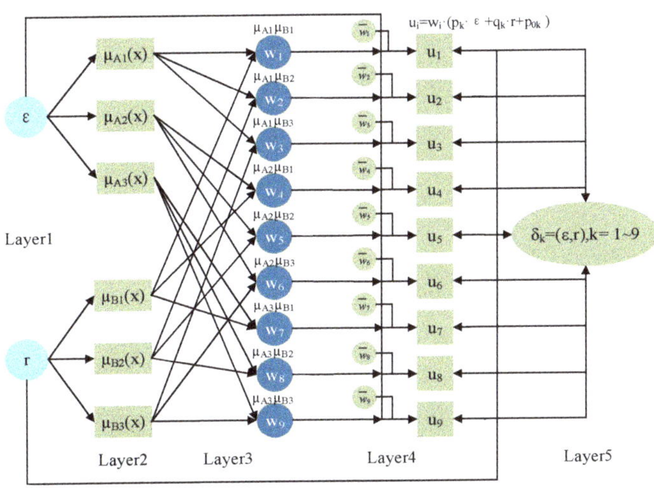

Figure 5. ANFIS with nine rules for two inputs and one output.

Layer1: Input variable layer, the two input variables are the heading error $\varepsilon = \psi_r - \psi$ and the head angular velocity $r = \dot{\psi}$. The first layer simply transmits the input values to the next layer.

Layer2: The input language layer is responsible for fuzzifying the input signal using two input languages ε and r, corresponding to membership functions $\mu_{A1(x)}$, $\mu_{A2(x)}$, $\mu_{A3(x)}$ and $\mu_{B1(x)}$, $\mu_{B2(x)}$, $\mu_{B3(x)}$, respectively. Each square neuron node represents a membership function, and the output of the fuzzification layer is the corresponding membership function. Data clustering technology is utilized to analyze sample data and determine the clustering center of input variables, leading to fuzzy partitioning based on the clustering results.

Layer3: The excitation intensity of each rule is calculated by controlling the rule layer and applying the formula $w_i = \mu_{Ai1}\mu_{Bi2}$, where i ranges from 1 to 3. In this system, 9 Sugeno-type reasoning rules are designed. The rule structure is defined as follows: if ε corresponds to m_{f1} and r also corresponds to m_{f1}, then $\delta = w_k$, where k spans from 1 to 9.

Layer4: The output language layer represents the language value of the controller's output language variable. The front end of the output language layer is the normalized result of all rule strengths and then calculates the output of each rule u_i. Each neural network node corresponds to a specific membership function, which is responsible for delineating the fuzzy boundaries of the language value. It is worth noting that the membership function used adopts the Sugeno model, which is a first-order linear function. This is because of the efficient performance of the Sugeno model. In model construction, if the preconditions are fuzzy, the subsequent results will be clear quantities to ensure the output results' accuracy and reliability.

Layer5: Output variable layer, representing the output variable value of the controller.

3.2. ANFIS Ship Heading Control Optimization Algorithm

In this study, the backpropagation learning algorithm was integrated with the least square method to optimize control parameters using the time backpropagation algorithm, resulting in a hybrid learning algorithm for ANFIS.

The network adjusts its internal parameters by comparing the output with the expected trajectory in each period and fine-tunes itself through error feedback to learn and approximate the desired model behavior. Jang introduced an adaptive neuro-fuzzy inference system by integrating time-backward propagation (TBP), enhancing the traditional backward propagation (BP) learning algorithm. By training the neural network with the TBP mechanism, the controlled system's output can be more effectively guided to accurately track the desired path, leading to a significant enhancement in control system performance. The learning process of the ship heading ANFIS control system using the TBP learning algorithm is illustrated as a closed-loop dynamic simulation behavior in Figure 6. The optimization criterion is to minimize the control performance metric, which is usually expressed as the error between the actual heading and the desired heading.

In Figure 6, Δ represents the time shift factor, $\psi_r(k)$ denotes the heading command, the desired trajectory is formed based on the set heading $\psi_r(j)$, and all $\psi(j)$ is stored to create the actual motion trajectory. The expression of the closed-loop control performance index is an optimization function, as shown in Equation (23). The consequent parameters $\{p_{Ak}, q_{Bk}, p_{0k}\}$ of the Sugeno-type fuzzy rule in ANFIS are determined through the learning law provided by the BP algorithm, as shown in Equation (24).

$$E = \sum_{k=0}^{n} \delta^k (r - \overline{\psi})^2 \tag{23}$$

$$\Delta a_i = -\eta \frac{\partial E}{\partial a_i} = -\eta \sum_{k=1}^{n} \frac{\partial E(j)}{\partial a_i} \tag{24}$$

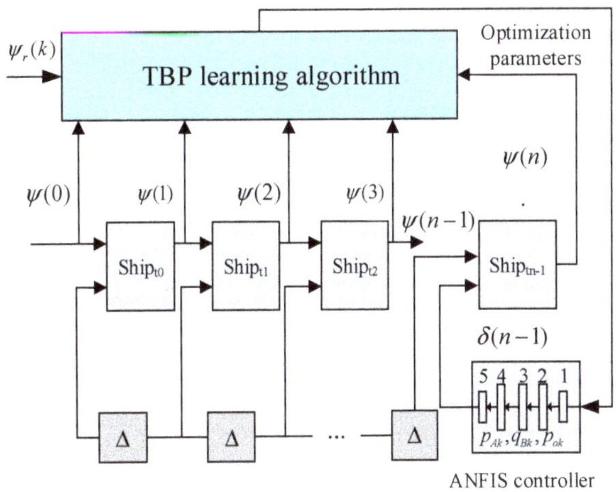

Figure 6. TBP learning process of ship heading ANFIS controller.

Utilizing Equation (24) of the TBP algorithm to optimize the parameter ai, the ANFIS controller can be updated with these parameters and subsequently applied in the next control cycle.

ANFIS's TBP parameter optimization algorithm consists of two main processes: forward calculation and reverse calculation. The forward calculation process addresses the storage and accumulation of data needed for learning through direct calculations, while the reverse calculation process utilizes the TBP algorithm to adjust parameters.

(1) Forward calculation of ANFIS

Layer1: The first layer determines that the two inputs are, respectively, the deviation ε between the ship's current course and the expected course and the heading angular speed r.

$$\varepsilon = \psi_r - \psi; r = \overline{\psi} \tag{25}$$

Layer2: The generalized bell-shaped membership function is determined by the parameters a, b, and c in Formula (26), with a and b typically being positive values. The parameter c is utilized to specify the center of the curve.

$$\mu_{Ai/Bi}(x,a,b,c) = \frac{1}{1 + \left|\frac{x-c}{a}\right|^{2b}}; i = 1,2,3 \tag{26}$$

Layer3: Calculate the excitation intensity of each rule, as depicted in Equation (27). Subsequently, normalize each excitation intensity to obtain Equation (28).

$$w_k = \mu_{Ai_1}\mu_{Bi_2}; i_1, i_2 \in \{1,2,3\}; k = 1,2,\cdots,9 \tag{27}$$

$$\overline{w}_k = \frac{w_k}{\sum\limits_{k=1}^{9} w_k} \tag{28}$$

Layer4: The fourth layer is the output of fuzzy rules. The output of each rule is a linear combination of input variables.

$$\overline{w_k}u_k = \overline{w_k}(p_{Ak}\varepsilon_k + p_{Bk}r_k + p_{0k}) \tag{29}$$

Layer5: The output is the command rudder angle given by the ANFIS autopilot, which is affected by the consequent parameters $\{p_i, q_i, p_{0k}\}$.

$$\delta_k = \sum_{k=1}^{9} \overline{w_k}u_k \tag{30}$$

(2) Reverse calculation of ANFIS

During a sampling period, the parameters in Equation (24) are represented with simplified symbols: $E(j) = E_j$, $\varepsilon(j) = \varepsilon_j$, $\psi(j) = r_j$, $\delta r(j) = \delta rj$, $w_k = w^k$, $\overline{w_k} = \overline{w^k}$, $\overline{w_A}(k) = \overline{w^k_A}$, $\overline{w_B}(k) = \overline{w^k_B}$,$u_k = u^k$, $u_k(j) = u^jk$.

$$\frac{\partial E}{\partial a_k} = 2\sum_{j=1}^{n} \varepsilon_j t_j \frac{\partial \varepsilon_j}{\partial a_k} \tag{31}$$

$$\frac{\partial \varepsilon_j}{\partial a_k} = \frac{\partial \varepsilon_j}{\partial \psi_j} \cdot \frac{\partial \psi_j}{\partial \delta_j} \cdot \frac{\partial \delta_j}{\partial \delta_{rj}} \cdot \frac{\partial \delta_{rj}}{\partial a_k} = \frac{\partial \psi_j}{\partial \delta_j} \cdot \frac{\partial \delta_j}{\partial \delta_{rj}} \cdot \frac{\partial \delta_{rj}}{\partial a_k} \tag{32}$$

From Equation (33), Equation (34) can be derived.

$$\delta_{rj} = \sum \overline{w^k_j} u^k_j \tag{33}$$

$$\frac{\partial \delta_{rj}}{\partial a_k} = \begin{cases} \overline{w^k_j}\varepsilon_j, a_k = p_{Ak} \\ \overline{w^k_j} r_j, a_k = p_{Bk} \\ \overline{w^k_j}, a_k = p_{0k} \end{cases} \tag{34}$$

By comprehensively applying Equations (23) to (34), we can solve the problem of updating the controller consequent parameters $\{p_{Ak}, q_{Bk}, p_{0k}\}$.

To sum up, the ANFIS structure shown in Figure 5 needs to be learned based on the TBP learning algorithm. The control object is the MMG model, which constitutes the control system shown in Figure 2.

4. Model Validation

4.1. Free Self-Propelled Model Test

In general, the more exact a control plant is established, the more accurate control performances can be evaluated. In the present study, the control plant is obtained from the free-running model tests, which were conducted at the ocean engineering wave basin in Shanghai Jiao Tong University. The dimension of the basin is 50 m × 30 m. A high-fidelity MMG model mentioned in Section 2.2 is adopted for the S175 ship, whose main particulars are described in Table 1. The model test and numerical simulations are carried out based on the ship model with a scale ratio of 1:57.686.

According to the turn test defined in Figure 7, the tactical diameter and forward distance need to be determined. The International Maritime Organization (IMO) requires a tactical diameter of less than 5.0 L and a forward distance of less than 4.5 L.

Table 1. Principal particulars of ship models.

	Items	Units	Full Scale	Model
Hull	Length between perpendiculars, L	m	175.0	3.0337
	Beam, B	m	25.4	0.4403
	Draft, D	m	9.5	0.1647
	Displacement, ∇	m3	24,742	0.127
	Block coefficient, C_B	/	0.57	0.57
	Radius of gyration in roll, k_{xx}	/	0.33 B	0.33 B
	Radius of gyration in pitch, k_{yy}	/	0.25 L	0.25 L
	Radius of gyration in yaw, k_{zz}	/	0.269 L	0.269 L
Propeller	Diameter, D_p	m	6.5064	0.1128
	Pitch ratio	/	0.915	0.915
	Number of blades	/	5	5
Rudder	Area, A_d	m^2	32.46	0.0098
	Height, H	m	7.7	0.1335
	Aspect ratio, Λ	/	1.8268	1.8268

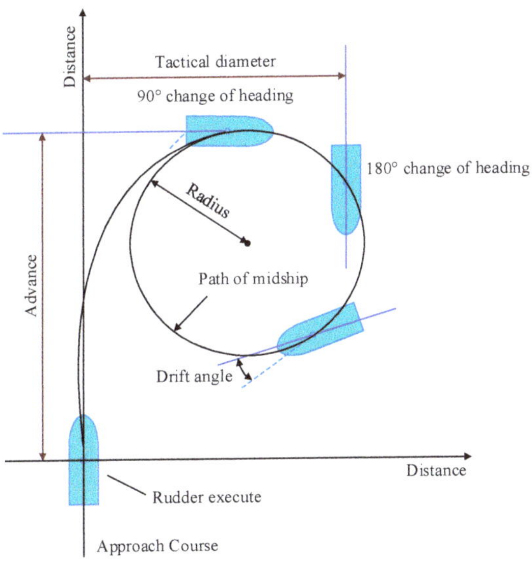

Figure 7. Notations and criteria in turning test.

4.2. Numerical Simulation of Rotary Motion in Still Water

Zou Z.J.'s team at Shanghai Jiao Tong University conducted a self-propelled model rotation test in still water to obtain reliable data. The interference coefficients and hydrodynamic derivatives for the ship-propeller-rudder system were computed and are presented in detail in Table 2. The shape factor k_f and residual resistance coefficient C_r for the hull were estimated using empirical formulas provided. Additional coefficients can be found in the PMM test data of Yasukawa [33].

The rotation trajectory of the S175 container ship model in still water is illustrated in Figure 8. The dimensions of the ship model, initial velocity, propeller speed, and rudder speed were all kept constant throughout the test. Upon comparing the numerical simulation findings with the experimental rotation trajectory in still water, a high level of agreement was observed.

Table 2. Hydrodynamic derivative and interference coefficient in S175 maneuvering motion model.

	Symbol	Value	Symbol	Value	Symbol	Value
Hull	m'_x	0.000238	Y'_v	−0.0116	N'_v	−0.00385
	m'_y	0.007049	Y'_r	0.00242	N'_r	−0.00222
	J'_{xx}	0.0000034	Y'_{vvv}	−0.109	N'_{vvv}	0.001492
	J'_{zz}	0.000419	Y'_{vvr}	0.0214	N'_{vvr}	−0.0424
	X'_{vv}	−0.00386	Y'_{vrr}	−0.0405	N'_{vrr}	0.00156
	X'_{vr}	−0.00311	Y'_{rrr}	0.00177	N'_{rrr}	−0.00229
	X'_{rr}	0.0002	z_H/D	0.5	N_{10}	0.0082
	C_r	0.001	kf	1.04		
Propeller	t_P	0.175	1-wp0	0.816	x_P	−0.47
	j_0	0.5179	j_1	−0.1179	j_2	−0.3618
Rudder	t_R	0.29	x'_H	−0.48	x'_R	−0.5
	a_H	0.237	ε	0.921	γ	0.088 ($v' < 0$)
	K	0.631	z_H/D	0.7		0.193 ($v' \geq 0$)
	C_{Rr}	−0.156	C_{Rrr}	−0.275	C_{Rrvv}	1.96

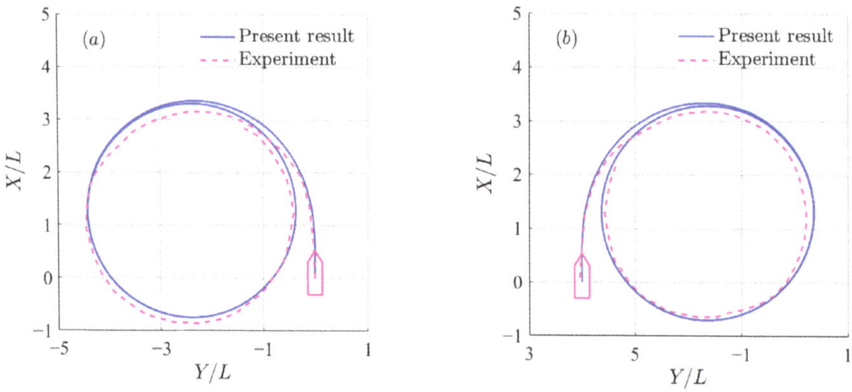

Figure 8. Turning tests in clam water at F_n = 0.15: (**a**) port turning (δ = −35°); (**b**) starboard turning (δ = 35°).

Table 3 presents a comparison of the results for rotation diameter, tactical diameter, advance distance, and horizontal distance between numerical simulation and experiment. The relative error E_r is calculated using Equation (35):

$$E_r = \frac{|V_n - V_t|}{V_t} \times 100\% \tag{35}$$

where V_n and V_t denote the values of numerical results and the test results, respectively.

Table 3. Comparison of numerical results and test trajectories.

Parameter Name	Numerical	Test	Relative Error
Tactical diameter (δ = −35°)	4.5 L	4.4 L	2.27%
Tactical diameter (δ = 35°)	4.5 L	4.2 L	7.14%
Advance (δ = −35°)	3.4 L	3.2 L	6.25%
Advance (δ = 35°)	3.4 L	3.1 L	9.68%

The errors of tactical diameter and the advance are about 2.27% and 6.25%, respectively. In addition, the numerical results of the tactical diameter and the advance are 4.5 L and

3.4 L, and they are less than the requirements of 5 L and 4.5 L, respectively. It shows that the MMG model is effective to describe the maneuverability of S175. The results indicate that the MMG model utilized in this study is effective, with satisfactory calculation accuracy. The maneuverability of the S175 container ship is specifically applied in practical navigation scenarios.

5. Results

5.1. Performance Analysis Metrics

The ship uses a controller to control the heading needed to continue sailing through the waves. For further quantitative analysis, the two criteria A_ψ and I_δ evaluate the control performance as follows:

$$\begin{aligned} A_\psi &= \frac{1}{t_\infty - t_0}\int_{t_0}^{t_\infty} |S_\psi - T_\psi| dt \\ I_\delta &= \frac{1}{t_\infty - t_0}\int_{t_0}^{t_\infty} |U_\delta| dt \end{aligned} \tag{36}$$

where S_ψ denotes the desired heading; T_ψ is the actual heading; U_δ is the input rudder; A_ψ represents the mean absolute error of course keeping to measure the response performances of course keeping; I_δ is the mean absolute input of rudder to evaluate energy consumption of rudder engine.

5.2. Simulation of Navigation Trajectory in the Still Sea

The ANFIS controller is utilized to model the ship's sailing trajectory during a simulation lasting 860 s. The ship is traveling north with a speed of F_n = 0.15, while the propeller speed is limited to n_p = 20rps. The wave disturbance caused by Beaufort No. 6 wind is incorporated to introduce white noise with a power of 0.001. In the simulation, λ_ψ = 100 controller is taken as an example to illustrate the control performances. Various desired heading values of 40°, 60°, 90°, 270°, 300°, and 320° are input into the controller. The navigation trajectory is depicted in Figure 9. The outcomes reveal that under normal sea conditions, the actual heading of the navigation trajectory aligns closely with the expected heading, indicating a successful control performance.

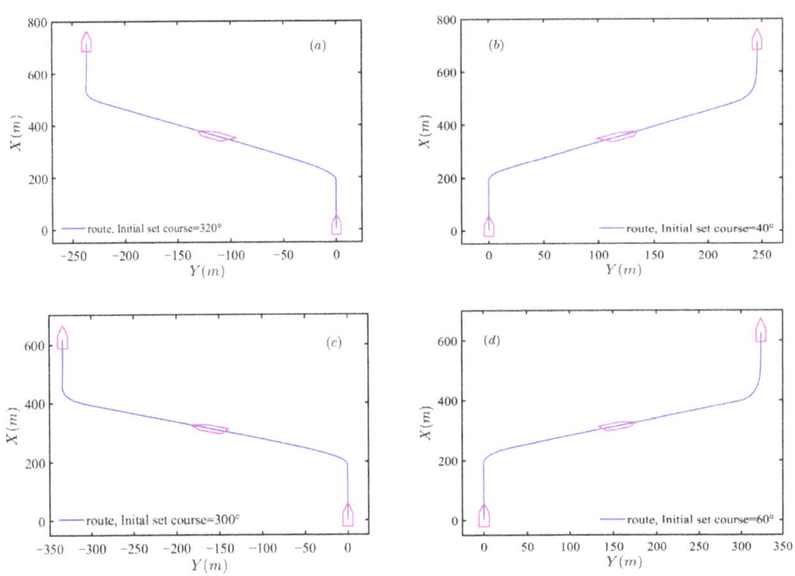

Figure 9. *Cont.*

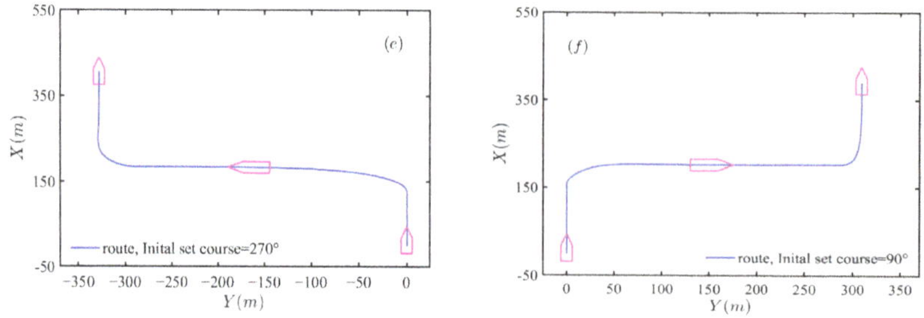

Figure 9. Navigation trajectories of different setting course in clam water conditions: (**a**) 320°; (**b**) 40°; (**c**) 300°; (**d**) 60°; (**e**) 270°; (**f**) 90°.

5.3. Course and Roll Stabilization Performances in the Still Sea

The parameters of the linear controller for generating training data are selected from the literature [34]. The performance parameters of the linear controller can be adjusted to λ_ψ to quantitatively meet different control requirements. The performance parameters can reflect the response characteristics and stability of the controller.

As depicted in Figure 10, the ship initially departs from the due north direction (0°) under normal sea conditions. At 200 s into the simulation, the ship begins to turn and eventually reaches the desired heading, maintaining it thereafter. By 600 s, the ship gradually returns to its initial heading. The simulation can be divided into different stages: 0–200 s for the step duration, 200–400 s for significant heading changes, 400–600 s for heading maintenance, 600–900 s for heading recovery, and 900–1000 s for heading maintenance. Setting the desired headings at 40°, 60°, 90°, 270°, 300°, and 320° initially, both controllers effectively manage the heading throughout the simulation.

As depicted in Figure 11, the simulation experiment enforces a limitation on the ship's rudder speed to a range of [−5°/s, 5°/s], with the rudder angle saturation limit set at $\delta \in [-35°, 35°]$. When there is a deviation in the desired heading, both the controller and linear controller in ANFIS exhibit similar changes in the rudder angle. The simulation results confirm the precision of the training data and demonstrate the effective control performance of both controllers.

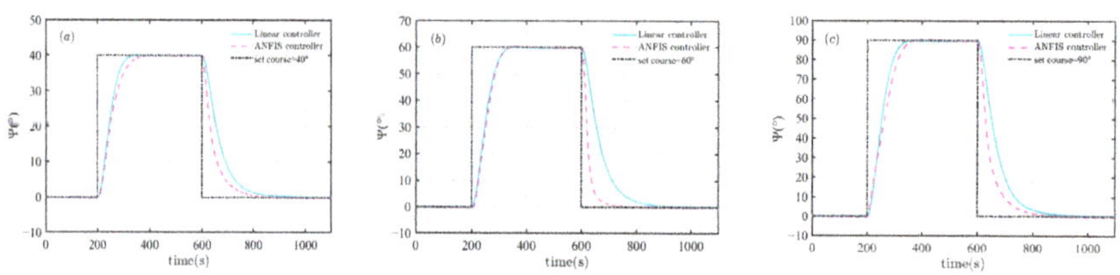

Figure 10. The effect of linear controller and ANFIS controller on ship heading control: (**a**) set course = 40°; (**b**) set course = 60°; (**c**) set course = 90°.

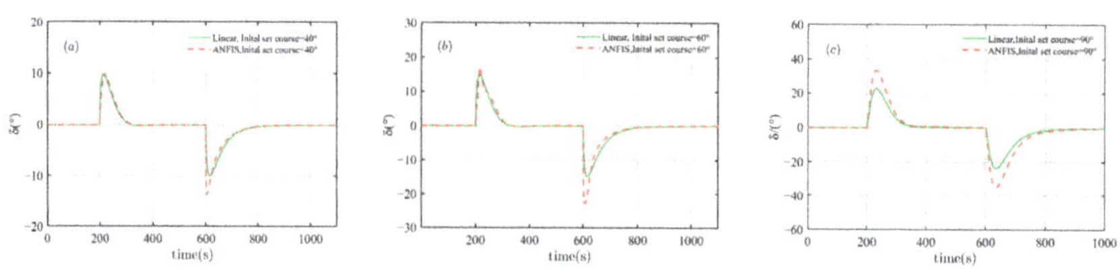

Figure 11. Linear controller and ANFIS controller control ship's rudder angle: (**a**) set course = 40°; (**b**) set course = 60°; (**c**) set course = 90°.

Based on the control effects illustrated in Figures 10 and 11, and in conjunction with Equation (36), the values of A_ψ and I_δ were calculated with λ_ψ set to 100. Tables 4 and 5 present a comparative analysis of the performance between linear controllers and ANFIS controllers during starboard turn operations at initial moments, with expected courses of 40°, 60°, and 90°. For the linear controller, the A_ψ values are 3.02°, 4.62°, and 6.96°, while the corresponding I_δ values are 0.51°, 0.73°, and 1.09°. In contrast, for the ANFIS controller, the A_ψ values are 3.00°, 4.32°, and 5.94°, and the I_δ values are 1.01°, 2.15°, and 2.99°, respectively. When the ship needs to recover its initial course and perform a left turn, the A_ψ values for the linear controller are 4.15°, 6.31°, and 9.16°, and the I_δ values are 0.67°, 0.98°, and 1.47°, respectively. For the ANFIS controller, the A_ψ values are 4.02°, 5.98°, and 7.54°, and the I_δ values are 1.59°, 2.64°, and 3.61°, respectively. The results indicate that both controllers exhibit increasing A_ψ and I_δ values as the expected course increases. While the ANFIS controller demonstrates more stable heading control (smaller A_ψ), this comes at the expense of higher energy consumption (larger I_δ).

Table 4. The analysis of control performance (right rudder).

Symbol	$A_\psi(\psi = 40°)$	$A_\psi(\psi = 60°)$	$A_\psi(\psi = 90°)$	$I_\delta(\psi = 40°)$	$I_\delta(\psi = 60°)$	$I_\delta(\psi = 90°)$
Linear	3.02°	4.62°	6.96°	0.51°	0.73°	1.09°
ANFIS	3.00°	4.32°	5.94°	1.01°	2.15°	2.99°

Table 5. The analysis of control performance (left rudder).

Symbol	$A_\psi(\psi = 40°)$	$A_\psi(\psi = 60°)$	$A_\psi(\psi = 90°)$	$I_\delta(\psi = 40°)$	$I_\delta(\psi = 60°)$	$I_\delta(\psi = 90°)$
Linear	4.15°	6.31°	9.16°	0.67°	0.98°	1.47°
ANFIS	4.02°	5.98°	7.54°	1.59°	2.64°	2.99°/3.61°

As illustrated in Figure 12 and Table 6, when the expected heading is set to 40°, 60°, and 90°, the maximum response rudder angle differences between the ANFIS controller and the linear controller are 0°, 3°, and 11°, respectively. For expected headings of 270°, 300°, and 320°, these differences increase to 3°, 8°, and 29°, respectively. These findings indicate that the maximum response rudder angle of the ANFIS controller exceeds that of the linear controller, with the disparity becoming more pronounced as the expected heading increases. Consequently, the ANFIS controller imposes a higher load on the steering gear. Additionally, while the response times for both controllers are comparable at expected headings of 40°, 60°, and 90°, the ANFIS controller demonstrates significantly faster response times—approximately 28.57%, 43.75%, and 37.78% quicker—at expected headings of 270°, 300°, and 320°, respectively. This evidence underscores the superior response speed of the ANFIS controller compared to the linear controller. In summary, the larger rudder angle responses suggest that the ANFIS controller can more rapidly adjust the ship's control status under typical sea conditions, thereby enhancing the overall dynamic performance of the system.

Table 6. Comparison of the maximum rudder angle and response time of the ship turning.

Symbol	$\psi = 40°$	$\psi = 60°$	$\psi = 90°$	$\psi = 270°$	$\psi = 300°$	$\psi = 320°$
Linear (δ/t)	10°/130 s	14°/145 s	22°/170 s	−10°/350 s	−14°/400 s	−22°/450 s
ANFIS (δ/t)	10°/135 s	17°/135 s	33°/170 s	−13°/250 s	−22°/225 s	−51°/280 s

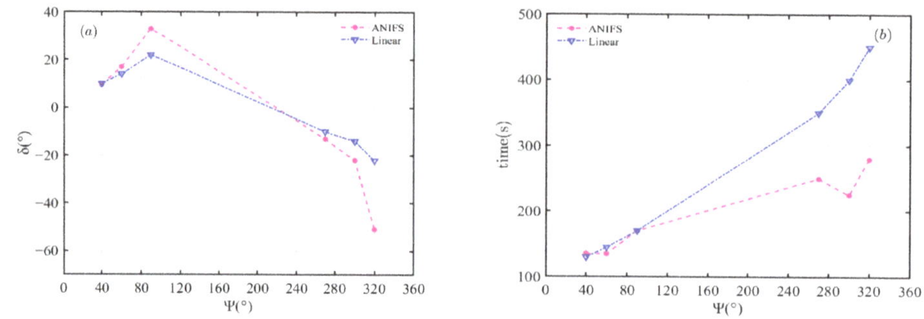

Figure 12. Comparison of the maximum rudder angle and response time of the ship when turning: (**a**) the maximum rudder angle; (**b**) response time.

5.4. Course and Roll Stabilization Performances in the Heavy Sea

According to Figure 13 and Table 7, it can be seen that under bad sea conditions, when the expected heading is 40° and λ_{ψ} = 100, the response time of the linear controller and ANFIS controller is 100 s and 180 s, respectively, and the response time is reduced by about 33.3%. The A_{ψ} is 5.24° and 5.85°, and the I_{δ} is 1.84° and 1.54°, respectively. When the left rudder operation is performed, A_{ψ} is 7.05° and 6.96°, and I_{δ} is 1.96° and 1.71°, respectively. The results show that the response time of ANFIS is longer than that of the linear controller, but its I_{δ} index is smaller. The course stability of the two controllers is basically the same regardless of whether the rudder is turned right or left. It can be concluded that the response time of ANFIS is affected in bad sea conditions, but the stability of ANFIS is better than that of the linear controller during the whole course-keeping phase, and it has better energy-saving characteristics.

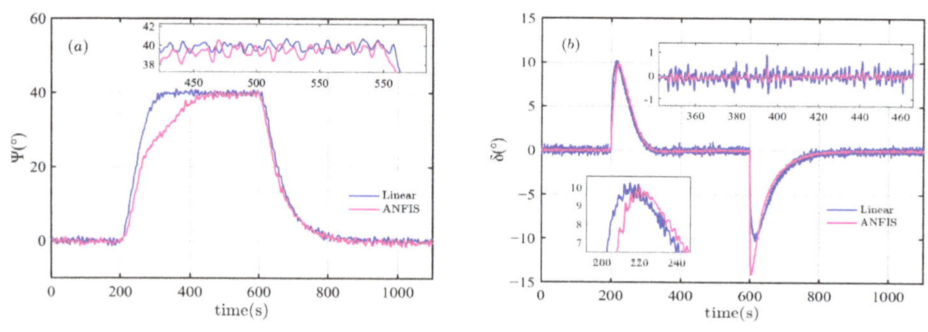

Figure 13. Comparison of steering stability between ANFIS controller and linear controller in the heavy sea: (**a**) course control; (**b**) rudder angle.

Table 7. The analysis of control performance (ψ = 40°).

Symbol	Response Time	$A_{\psi}(\psi = 40°$, Right Rudder)	$A_{\psi}(\psi = 40°$, Left Rudder)	$I_{\delta}(\psi = 40°$, Right Rudder)	$I_{\delta}(\psi = 40°$, Left Rudder)
Linear	100s	5.24°	7.05°	1.84°	1.96°
ANFIS	180s	5.85°	6.96°	1.54°	1.71°

Figure 14 and Table 8 show small amplitude course control and rudder angle performance (ψ = 60°). It is worth mentioning that course performance is determined by 100–400 s of course change and 400–1000 s of course hold, respectively. As shown in Figure 14a and Table 8, the A_{ψ} using the λ_{ψ} = 60, λ_{ψ} = 90, and λ_{ψ} = 120 controllers are 2.57°,

3.19°, and 4.14°, respectively, increasing by 24.12% and 29.78%. The controller (λ_ψ = 60) has the fastest response time but has a slight overshoot. The controller (λ_ψ = 120) has the longest response time. The A_ψ of the ANFIS controller is 2.86°, which is only 11.28% higher than that of the linear controller with the best directional stability (λ_ψ = 60), has satisfactory stability and safety, and does not produce an overshoot phenomenon. Figure 14b and Table 8 show that the controller (λ_ψ = 60) has a larger control input (maximum rudder angle) than the other controllers, but its I_δ of 0.79° is the smallest among the three linear controllers with different parameters, due to its fast response speed. The I_δ of the ANFIS controller is 1.12°, which not only has a good energy-saving effect but also has no overshoot.

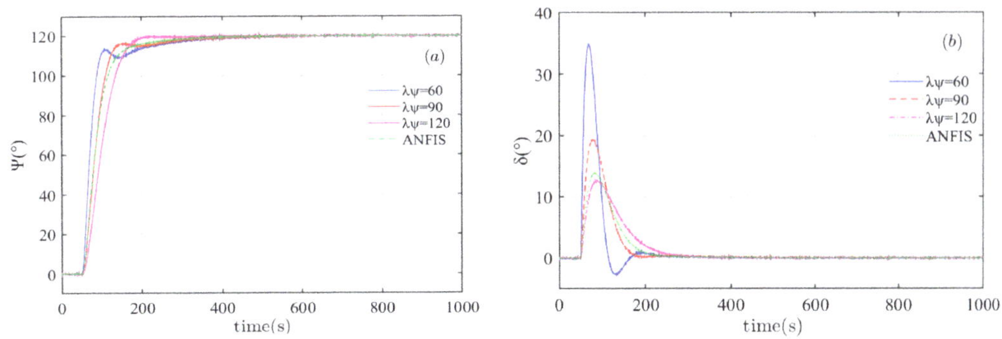

Figure 14. Comparison of course control and rudder angle in the heavy sea between ANFIS controller and linear controller (ψ = 60°): (**a**) course control; (**b**) rudder angle.

Table 8. The analysis of control and rudder performance in the heavy sea between ANFIS controller and linear controller (ψ = 60°/120°).

Symbol	A_ψ $\psi = 60°/\psi = 120°$	$A_\psi(\lambda_\psi = 60)$ $\psi = 60°/\psi = 120°$	$A_\psi(\lambda_\psi = 90)$ $\psi = 60°/\psi = 120°$	$A_\psi(\lambda_\psi = 120)$ $\psi = 60°/\psi = 120°$	I_δ $\psi = 60°/\psi = 120°$	$I_\delta(\lambda_\psi = 60)$ $\psi = 60°/\psi = 120°$	$I_\delta(\lambda_\psi = 90)$ $\psi = 60°/\psi = 120°$	$I_\delta(\lambda_\psi = 120)$ $\psi = 60°/\psi = 120°$
Linear	/	2.57°/5.51°	3.19°/7.01°	4.14°/8.52°	/	0.79°/1.76°	1.48°/1.52°	1.46°/1.50°
ANFIS	2.86°/6.28°	/	/	/	1.12°/1.46°	/	/	/

Figure 15 and Table 8 show large course control and rudder angle performance (ψ = 120°). Figure 15a and Table 8 show that the A_ψ using the λ_ψ = 60, λ_ψ = 90, and λ_ψ= 120 controllers are 5.51°, 7.01°, and 8.52°, respectively. They increased by 27.22 percent and 21.54 percent, respectively. The A_ψ of the ANFIS controller is 6.28°, which is second only to the λ_ψ = 60 controller and 13.97% higher. It has satisfactory stability and security. As shown in Figure 15b and Table 8, the I_δ of the linear controller decreases as λ_ψ increases. Notably, the I_δ of the ANIFS controller is measured at 1.46°, which is lower than the minimum value of λ_ψ = 120 for the linear controller. This indicates a superior energy-saving effect achieved by the ANIFS controller.

A linear controller and an adaptive neural fuzzy inference system (ANFIS) controller are designed based on the validated MMG model. A propeller speed limit ($0 \leq n_P \leq 40$ rps) was added to the simulation. The linear controller can quantitatively adjust the performance parameters λ_ψ to meet different control requirements. Any controller must be compared to a typical controller to account for its performance. In this paper, a linear controller is established with λ_ψ = 60, λ_ψ = 90, and λ_ψ = 120 as examples. The ANFIS controller operates on the basis of the training data and performs a small heading control (ψ = 60°) and a large heading control (ψ = 120°), respectively. For fair comparison, simulation parameters, wave interference (bad sea conditions), and propeller speed (n_P = 20 rps) were the same for all three controllers.

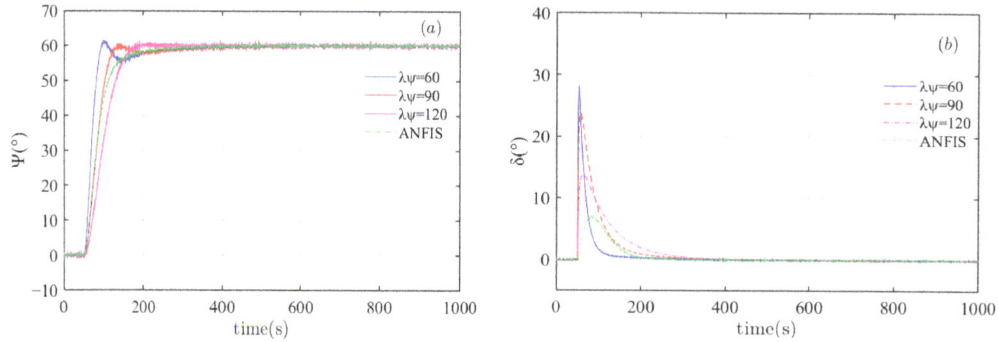

Figure 15. Comparison of course control and rudder angle in the heavy sea between ANFIS controller and linear controller (ψ = 120°): (**a**) course control; (**b**) rudder angle.

The conclusion is that the increase in the parameter λ_ψ improves the response speed of the linear controller, enhances the energy-saving effect of the controller, and increases the volatility of the controller. This makes it impossible to get the best of both worlds by manually adjusting parameters. Although the smallest λ_ψ controller has strong stability and a good energy-saving effect in the course-keeping stage, it will cause overshoot in the response stage, which will cause damage to the steering gear and is not conducive to the safe navigation of the ship. Adaptive neural fuzzy inference system (ANFIS) can be based on the coupling of a neural network and an adaptive fuzzy controller, without manual adjustment, and can automatically adjust the performance parameters to achieve the control effect that the linear controller cannot achieve, so as to adapt to the interference of the external environment. The overshoot phenomenon is overcome, and the stability and energy saving of heading control are effectively enhanced, which shows that the neural network has strong generalization ability and the fuzzy controller is robust.

Figure 16a shows that when the controller performance parameter λ_ψ (λ_ψ = 100) and environmental disturbance remain unchanged, the increase in propeller speed will accelerate the system response, but there will be slight overshooting. Figure 16b shows that the maximum rudder angle remains basically constant at different speeds, while the system oscillation increases slightly with increasing speed. Table 9 lists the A_ψ and I_δ values of the linear controller and ANFIS controller at different propeller speeds. For the linear controller, when the propeller speed is n_P = 10 rps, n_P = 20 rps, and n_P = 30 rps, the A_ψ value is 5.24°, 2.79° and 1.99°, respectively, with a decrease of about 62.02%. For ANFIS controllers, the A_ψ values at these speeds are 5.31°, 2.92°, and 1.98°, respectively, representing a reduction of about 62.71%. The results show that the performance of both controllers is improved with the increase in propeller speed. Linear controller in n_P = 10 rps, n_P = 20 rps, and n_P = 30 rps when I in the delta value is 3.04°, 1.71°, and 1.31°, respectively. Similarly, at these speeds, the I_δ values of the ANFIS controller are 3.03°, 1.71°, and 1.30°, respectively. The results show that the two controllers can achieve a comparable energy-saving effect, and the best energy efficiency is observed at n_P = 30 rps.

Table 9. The analysis of control and rudder performance in the heavy sea between ANFIS controller and linear controller (ψ = 40°).

Symbol	$A_\psi(n_P = 10$ rps)	$A_\psi(n_P = 20$ rps)	$A_\psi(n_P = 30$ rps)	$I_\delta(n_P = 10$ rps)	$I_\delta(n_P = 20$ rps)	$I_\delta(n_P = 30$ rps)
Linear	5.24°	2.79°	1.99°	3.04°	1.71°	1.31°
ANFIS	5.31°	2.92°	1.98°	3.03°	1.71°	1.30°

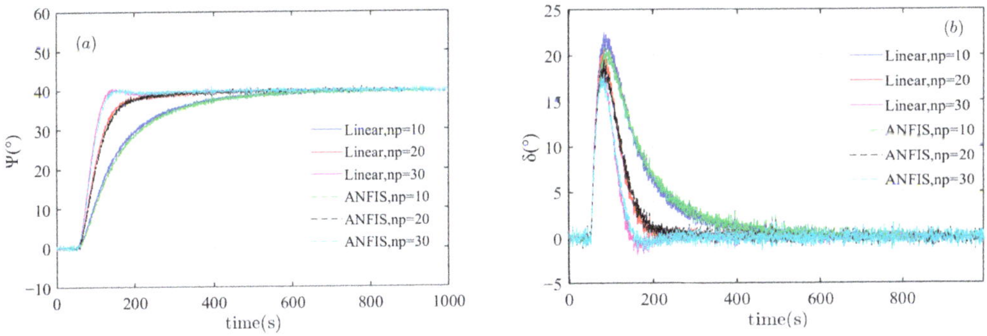

Figure 16. Comparison of course control and rudder angle in the heavy sea between ANFIS controller and linear controller (λ_ψ = 100): (**a**) course control; (**b**) rudder angle.

The conclusion is that appropriate np is necessary when environmental disturbances and other performance parameters are constant because excessive overshoot and errors should be avoided when ships are performing rescue missions or sailing in limited waters. Under the influence of different speeds, the response characteristics of the ANFIS controller are almost the same as those of the linear controller, which shows that the ANFIS controller using the combination of a neural network and fuzzy control has good practicability and generalization.

6. Conclusions

The three-degree-of-freedom MMG model was adopted and verified by the data of the S175 free model turning circle. In order to achieve efficient and stable course control performance, the ANFIS controller is designed by combining the backpropagation algorithm and the least square method. The MMG model is controlled by the linear control method and the ANFIS method, respectively, in normal sea states and bad sea states. Heading error, output rudder angle, and yaw angular speed in the linear controller are the training data sources of ANFIS. ANFIS trains and controls the MMG model through the powerful learning ability of the neural network, simulating and analyzing the performance of the two controllers.

In normal sea conditions, for course control performance, ANFIS can output the correct course trajectory according to the input rudder angle of the MMG model, and A_ψ is lower than the linear controller, which has stronger course stability performance and faster response speed. The maximum output rudder angle of the ANFIS controller is greater than that of the linear controller, and the energy-saving effect is slightly weaker than that of the linear controller.

In harsh sea conditions, the ANFIS controller achieved an I delta improvement of nearly 16% and 13%, respectively, over the linear controller, both on the left and right rudder, while maintaining excellent course stability. On the premise of keeping the propeller speed unchanged, different control performance can be obtained by manually adjusting the performance parameter λ_ψ in the linear controller to meet different engineering requirements. It is worth mentioning that the ANFIS controller can rely on the strong learning ability of the neural network to control whether the desired course is a large course (ψ = 120°) or a small course (ψ = 60°), it can maintain course stability and improve energy efficiency, even in harsh sea conditions. This is achieved without the need to manually adjust parameters and has significant advantages over traditional linear control methods. Keeping the performance parameters and the environment unchanged, changing only the propeller speed can improve the control performance of the controller, and the maximum steering angle does not change with the speed. The ANFIS-based course controller offers a promising ship course control scheme that strikes a balance between energy efficiency and control performance. Its adaptability and robustness make it suitable for engineering applications, with the potential to lead to safer, more economical, and environmentally friendly navigation practices.

Currently, our research focuses on the motion response performance of ships with three degrees of freedom. The number of degrees of freedom will be increased in the future to more accurately describe the motion of ships in complex sea conditions. Considering the coupling effect of wind, wave, current, and other environmental factors, the structure number of the controller and the accuracy of the model are further improved.

Author Contributions: Conceptualization, Y.G., R.Y., Z.Z. and B.H.; methodology, Y.G.; software, R.Y.; validation, Y.G. and R.Y.; formal analysis, R.Y.; investigation, B.H.; resources, Z.Z.; data curation, Y.G.; writing—original draft preparation, Y.G. and R.Y.; writing—review and editing, R.Y.; visualization, R.Y.; supervision, Z.Z.; project administration, Y.G.; funding acquisition, Y.G.; Z.Z. and B.H. All authors have read and agreed to the published version of the manuscript.

Funding: The present work was supported by the National Natural Science Foundation of China (Grant No. 52071200). We gratefully acknowledge the financial support of the Shanghai Commission of Science and Technology Project (Grant No. 23010501900).

Institutional Review Board Statement: Not applicable.

Informed Consent Statement: Not applicable.

Data Availability Statement: Data are contained within the article.

Conflicts of Interest: Author Bing Han is employed by Shanghai Ship and Shipping Research Institute, others declare no conflict of interest.

References

1. Liu, J.; Jiang, X.; Huang, W.; He, Y.; Yang, Z. A novel approach for navigational safety evaluation of inland waterway ships under uncertain environment. *Transp. Saf. Environ.* **2022**, *4*, tdab029. [CrossRef]
2. Kose, S.; Sekban, D.M. Emphasizing the importance of using cold-ironing technology by determining the share of hotelling emission value within the total emission. *Transp. Saf. Environ.* **2022**, *4*, tdac010. [CrossRef]
3. Chen, X.; Zheng, J.; Li, C.; Wu, B.; Wu, H.; Montewka, J. Maritime traffic situation awareness analysis via high-fidelity ship imaging trajectory. *Multimed. Tools Appl.* **2024**, *83*, 48907–48923. [CrossRef]
4. Sahal, M.; Gamayanti, N.; Cendikiarani, M.T. Surge Speed and Yaw Angle Control System Design on Unmanned Surface Vehicle (USV) Using PID-Fuzzy-Genetic Algorithm. In Proceedings of the 2024 International Seminar on Intelligent Technology and Its Applications (ISITIA), Mataram, Indonesia, 10–12 July 2024; IEEE: New York, NY, USA, 2024.
5. Xu, C.; Zhou, X.; Ren, H.; Sutulo, S.; Soares, C.G. Real-time simulation of ship overtake maneuvering at different water depths with a ship motion controller. *Ocean Eng.* **2024**, *311*, 119021. [CrossRef]
6. Bayezit, I.; Jambak, A.I.; Kinaci, O.K.; Fridan, B. Sensor data-based system identification and modular autopilot control instrumentation of marine craft. *Ocean Eng.* **2025**, *317*, 119793. [CrossRef]
7. Liu, L.; Zhang, L.; Pan, G.; Zhang, S. Robust yaw control of autonomous underwater vehicle based on fractional-order PID controller. *Ocean Eng.* **2022**, *257*, 111493. [CrossRef]
8. Suid, M.H.; Ahmad, M.A. Optimal tuning of sigmoid PID controller using Nonlinear Sine Cosine Algorithm for the Automatic Voltage Regulator system. *ISA Trans.* **2022**, *128*, 265–286. [CrossRef]
9. Tao, S.; Ju, Z.; Zhang, H.; Dong, X.; Chen, J. Path following of autonomous vehicles with an optimized brain emotional learning–based intelligent controller. *SAE Int. J. Connect. Autom. Veh.* **2023**, *6*, 241–250. [CrossRef]
10. Islam, M.M.; Siffat, S.A.; Ahmad, I.; Liaquat, M. Supertwisting and terminal sliding mode control of course keeping for ships by using particle swarm optimization. *Ocean Eng.* **2022**, *266*, 112942. [CrossRef]
11. Hosseinabadi, P.A.; Abadi, A.S.S.; Mekhilef, S. Fuzzy adaptive finite-time sliding mode controller for trajectory tracking of ship course systems with mismatched uncertainties. *Int. J. Autom. Control* **2022**, *16*, 255–271. [CrossRef]
12. Rezaei, A.; Tabatabaei, M. Ship roll stabilization using an adaptive fractional-order sliding mode controller. *Ocean Eng.* **2023**, *287*, 115883. [CrossRef]

13. Xu, Y.; Zhao, X.; Li, G. Design and analysis of 3-UPU wave compensation platform based on improved sliding mode control. *Ships Offshore Struct.* **2024**, *19*, 2040–2049. [CrossRef]
14. Chen, X.; Wu, H.; Han, B.; Liu, W.; Montewka, J.; Liu, R.W. Orientation-aware ship detection via a rotation feature decoupling supported deep learning approach. *Eng. Appl. Artif. Intell.* **2023**, *125*, 106686. [CrossRef]
15. Nasir, N.M.; Ghani, N.M.A.; Nasir, A.N.K.; Ahmad, M.A.; Tokhi, M.O. Neuro-modelling and fuzzy logic control of a two-wheeled wheelchair system. *J. Low Freq. Noise Vib. Act. Control* **2024**, 14613484241287608. [CrossRef]
16. Aghaseyedabdollah, M.; Abedi, M.; Pourgholi, M. Interval type-2 fuzzy sliding mode control for a cable-driven parallel robot with elastic cables using metaheuristic optimization methods. *Math. Comput. Simul.* **2024**, *218*, 435–461. [CrossRef]
17. Arifi, A.; Bouallègue, S. Takagi–Sugeno fuzzy-based approach for modeling and control of an activated sludge process. *Int. J. Dyn. Control* **2024**, *12*, 3123–3138. [CrossRef]
18. Han, B.; Duan, Z.; Peng, Z.; Chen, Y. A Ship Path Tracking Control Method Using a Fuzzy Control Integrated Line-of-Sight Guidance Law. *J. Mar. Sci. Eng.* **2024**, *12*, 586. [CrossRef]
19. Li, Y.; Song, G.; Yip, T.-L.; Yeo, G.-T. Fuzzy Logic-Based Decision-Making Method for Ultra-Large Ship Berthing Using Pilotage Data. *J. Mar. Sci. Eng.* **2024**, *12*, 717. [CrossRef]
20. Du, L.; Wu, Q.; Shu, Y.; Li, G. The effects of online-training artificial neural network mechanism and multi-stage parametric modeling method on simulation-based design system for ship optimization. *Ocean Eng.* **2024**, *309*, 118284. [CrossRef]
21. Unar, S.; Unar, A.M.; Memon, A.Z.; Narejo, S. A neural network-based heading and position control system of a ship. *Mehran Univ. Res. J. Eng. Technol.* **2022**, *41*, 172–177. [CrossRef]
22. Deraj, R.; Kumar, R.S.; Alam, M.S.; Somayajula, A. Deep reinforcement learning based controller for ship navigation. *Ocean Eng.* **2023**, *273*, 113937. [CrossRef]
23. Wakita, K.; Maki, A.; Umeda, N.; Miyauchi, Y.; Shimoji, T.; Rachman, D.M.; Akimoto, Y. On neural network identification for low-speed ship maneuvering model. *J. Mar. Sci. Technol.* **2022**, *27*, 772–785. [CrossRef]
24. Bouaiss, O.; Mechgoug, R.; TalebAhmed, A.; Brikel, A.E. Adaptive neural network-based compensation control of quadrotor for robust trajectory tracking. *Int. J. Adapt. Control Signal Process.* **2023**, *37*, 2772–2793. [CrossRef]
25. Zhao, Z.; Deng, L.; Yu, W.; Wang, C.; Wang, X.; Zhang, X. Adaptive Neural Event-Based Course-Keeping Control of Marine Vessel. In Proceedings of the 2024 Second International Conference on Cyber-Energy Systems and Intelligent Energy (ICCSIE), Shenyang, China, 17–19 May 2024; IEEE: New York, NY, USA, 2024; pp. 1–5.
26. Le, T.T. Ship heading control system using neural network. *J. Mar. Sci. Technol.* **2021**, *26*, 963–972. [CrossRef]
27. Karaboğa, D.; Kaya, E. Training ANFIS by using the artificial bee colony algorithm. *Turk. J. Electr. Eng. Comput. Sci.* **2017**, *25*, 1669–1679. [CrossRef]
28. Yasukawa, H.; Sakuno, R. Application of the MMG method for the prediction of steady sailing condition and course stability of a ship under external disturbances. *J. Mar. Sci. Technol.* **2020**, *25*, 196–220. [CrossRef]
29. Zhang, G. Calculation of Hydrodynamic Derivative of Ship Maneuverability Based on Numerical PMM Test. Ph.D. Thesis, Dalian Maritime University, Dalian, China, 2023.
30. Yasukawa, H.; Hirata, N.; Yamazaki, Y. Effect of bilge keels on maneuverability of a fine ship. *J. Mar. Sci. Technol.* **2018**, *23*, 302–318. [CrossRef]
31. Zhang, W.; Zou, Z.; Deng, D. A study on prediction of ship maneuvering in regular waves. *Ocean Eng.* **2017**, *137*, 367–381. [CrossRef]
32. Fossen, T.I. *Handbook of Marine Craft Hydrodynamics and Motion Control*; John Wiley & Sons: London, UK, 2011.
33. Yasukawa, H. Simulations of a ship maneuvering in waves. *J. Jpn. Soc. Nav. Archit. Ocean. Eng.* **2006**, *4*, 127–136. [CrossRef]
34. Zhang, Z.; Zou, Z.; Yang, J.; Zhang, X. Nonlinear hydrodynamic model-based course control and roll stabilization by taking rudder and propeller actions. *Ocean Eng.* **2022**, *263*, 112377. [CrossRef]

Journal of
Marine Science and Engineering

MDPI

Article

Risk Assessment of Hydrogen Fuel System Leakage in Ships Based on Noisy-OR Gate Model Bayesian Network

Gen Li [1,*], Haidong Zhang [1], Shibo Li [1] and Chunchang Zhang [1,2,3]

[1] Merchant Marine College, Shanghai Maritime University, Shanghai 201306, China; 202330110101@stu.shmtu.edu.cn (H.Z.); sbli@shmtu.edu.cn (S.L.); cczhang@shmtu.edu.cn (C.Z.)
[2] National Engineering Research Center of Special Equipment and Power System for Ship and Marine Engineering, Shanghai 201306, China
[3] Shanghai Engineering Research Center of Ship Intelligent Maintenance and Energy Efficiency Control, Shanghai 201306, China
* Correspondence: gli@shmtu.edu.cn

Abstract: To mitigate the risk of hydrogen leakage in ship fuel systems powered by internal combustion engines, a Bayesian network model was developed to evaluate the risk of hydrogen fuel leakage. In conjunction with the Bow-tie model, fuzzy set theory, and the Noisy-OR Gate model, an in-depth analysis was also conducted to examine both the causal factors and potential consequences of such incidents. The Bayesian network model estimates the likelihood of hydrogen leakage at approximately 4.73×10^{-4} and identifies key risk factors contributing to such events, including improper maintenance procedures, inadequate operational protocols, and insufficient operator training. The Bow-tie model is employed to visualize the causal relationships between risk factors and their potential consequences, providing a clear structure for understanding the events leading to hydrogen leakage. Fuzzy set theory is used to address the uncertainties in expert judgments regarding system parameters, enhancing the robustness of the risk analysis. To mitigate the subjectivity inherent in root node probabilities and conditional probability tables, the Noisy-OR Gate model is introduced, simplifying the determination of conditional probabilities and improving the accuracy of the evaluation. The probabilities of flash or pool fires, jet fires, and vapor cloud explosions following a leakage are calculated as 4.84×10^{-5}, 5.15×10^{-5}, and 4.89×10^{-7}, respectively. These findings highlight the importance of strengthening operator training and enforcing stringent maintenance protocols to mitigate the risks of hydrogen leakage. The model provides a valuable framework for safety evaluation and leakage risk management in hydrogen-powered ship fuel systems.

Keywords: hydrogen-powered ship fuel system; Bayesian network; Bow-tie model

1. Introduction

International shipping carries over 80% of global trade by volume, making it a cornerstone of the global economy. Safety [1,2], intelligence, and environmental friendliness are the development direction of the shipping industry. Faced with the challenges of climate change and the global energy crisis, reducing greenhouse gas emissions has become a key priority for nations worldwide. In 2023, the International Maritime Organization set a target for international shipping, aiming for at least 5% of energy to come from near-zero emissions by 2030. Hydrogen has emerged as a promising solution for maritime propulsion. Research indicates that replacing fossil fuels with hydrogen could reduce CO_2 emissions from international shipping by 34% [3]. Therefore, developing hydrogen-powered vessels is crucial to achieving carbon neutrality. While hydrogen fuel cells remain the dominant technology, hydrogen internal combustion engines (ICEs) offer a viable alternative. They require less pure hydrogen and can leverage existing manufacturing infrastructure developed for petroleum-based engines, significantly lowering costs and investment [4,5].

However, hydrogen's low density, flammability, and low ignition energy present significant safety concerns compared with conventional fuels. The release of hydrogen in confined or semi-confined spaces, such as within the fuel supply system or vessel enclosures, poses critical risks. In the event of a hydrogen leak, especially in confined areas, the gas may accumulate rapidly, creating a flammable or explosive atmosphere. Major hydrogen leaks can result in catastrophic outcomes, such as large-scale explosions or fires, depending on the amount of hydrogen released and the confinement conditions [6]. Conversely, small-scale hydrogen releases, often referred to as fugitive releases, are harder to detect due to their lower volumes and the difficulty in pinpointing their origin. These smaller leaks, while less immediately dangerous, can still lead to hazardous accumulations of hydrogen that, if ignited, may cause delayed explosions or fires [7]. Therefore, ensuring hydrogen safety is the foremost concern in the use of hydrogen-fueled ships. Current academic research has largely focused on the safety risk assessment of hydrogen fuel cell vessels, while the risk assessment of hydrogen internal combustion engine vessels' fuel supply systems remains limited. Mao et al. focused on a hydrogen fuel cell ship, conducting simulations to analyze hydrogen leakage risks and the diffusion process, as well as the explosion impacts under different ignition scenarios [8]. Duong et al. studied the ammonia fuel gas supply system and analyzed ammonia leakage risks in the engine room using FLACS-CFD [9]. Byongug et al. used event tree and fault tree analysis methods to investigate the explosion risks of the LNG fuel supply system in the fuel preparation room of a 300,000 DWT ore carrier [10]. While there is extensive research on the safety risks associated with alternative fuel systems, such as methanol and natural gas-powered internal combustion engines, studies specifically addressing the hydrogen ICE fuel supply system remain limited. Given this context, the objective of this paper is to develop a reliable approach conducting a systematic assessment of the hydrogen leakage risk in the fuel supply system of hydrogen-powered internal combustion engine vessels.

In earlier research, Bayesian networks (BNs) have gained significant popularity in risk assessment because of their clear graphical structure and the ability to quantitatively model the relationships between various influencing factors using probabilistic methods. They are increasingly recognized as a standard approach to determine failure probabilities and to evaluate pipeline failure risks [11] and oil degradation of marine engine lubrication [12]. Singh et al. [13] combined fuzzy fault tree analysis and BNs to propose a new approach. This approach is used to obtain the latest prior probabilities of basic and top events in the system fault tree when new information is available. Weng et al. [14] proposed a three-layer BN model of "pipeline failure–cause of failure–influencing factors". This innovative approach aims to address the challenges of excessive subjectivity and diminished accuracy often associated with assessing the likelihood of pipeline failures. Hassan et al. [15] used BNs to predict the main failure modes of pipelines, considering failure causes as variables.

BNs excel in handling uncertain situations. This means they are capable of reasoning and learning the associations between information items using conditional probabilities under settings of limited, imperfect, and unreliable information. However, obtaining conditional probability tables (CPTs) for BNs is not easy. Currently, the application of hydrogen-powered ships remains limited, resulting in a lack of complete failure case datasets for hydrogen internal combustion engine gas supply systems. An alternative approach to obtain CPTs is through a combination of expert judgment and specific algorithms, where CPTs are obtained based on evaluations from experts with relevant experience and knowledge. By incorporating expert opinions into the Bayesian network, Mahmood [16] et al. effectively addressed its inherent limitation—the inability to accurately capture statistical dependencies and causal relationships. However, the probabilities for root nodes and the CPTs derived from expert opinions may involve subjective factors. Additionally, the analysis of logical gates is definitive, and failure events do not result in a straightforward binary outcome (e.g., "1" for occurrence and "0" for non-occurrence). To overcome these limitations, this paper introduces a combined BN and Noisy-OR Gate model to address the absolute nature of CPT descriptions [17]. Wang et al. [18] adopts a Leaky Noisy-OR Gate model to dynamically enhance and compute conditional probabilities, effectively addressing the rigidity of traditional conditional probability representations. Zhang et al. [19] used a fuzzy fault tree approach based on natural gas stations to establish a risk assessment model for micro-leakage in bolt–washer–flange connections. Due to the high number of micro-leakages, which could have serious consequences and are difficult to obtain complete information for in practice, this study ultimately enhanced the effectiveness of expert input and reduced the workload by utilizing this method. Zhao et al. [20] employed Noisy-OR Gate BNs to assess the likelihood of navigation accidents, demonstrating that their proposed model effectively addressed the rigid relationships typically found between basic and intermediate events in BNs.

While some studies have successfully applied BNs to assess the safety of fuel systems, there remains ongoing academic discussion regarding whether internal combustion engines or fuel cells should dominate future technological pathways. Some researchers emphasize the superior environmental performance of fuel cells, whereas others highlight the technical maturity and cost-effectiveness of internal combustion engines. Thus, developing a comprehensive safety assessment framework for internal combustion engine systems is essential to fill a gap in the existing body of research.

Against this backdrop, the current research seeks to establish a comprehensive risk assessment framework for hydrogen fuel systems in internal combustion engines. The Bow-tie model is utilized to pinpoint the elements that may lead to hydrogen leaks and to evaluate the possible repercussions. The Bow-tie model is applied to identify the factors contributing to hydrogen leakage and to analyze the potential consequences. A hybrid approach combining the BN model and the Noisy-OR Gate model is employed to overcome challenges associated with complex conditional probability computations. By utilizing BN-based diagnostic inference, the study identifies key causes of leakage incidents and estimates the probabilities of hydrogen leakage and corresponding accident outcomes. In cases where failure data for basic events (risk factors) are unavailable or difficult to obtain, a fuzzy logic approach based on expert linguistic judgments is introduced.

2. Research Methodology

Figure 1 illustrates the risk assessment framework for hydrogen leakage in the fuel supply system of marine hydrogen internal combustion engines, which primarily consists of five steps: (1) data collection, review of the literature, and study of fundamental theories to understand the composition and operating principles of the fuel supply system, as well as analysis of the influencing factors of hydrogen leakage incidents; (2) risk identification; (3) development of the Bow-tie model; (4) construction of the BN model. The fault tree structure in the Bow-tie model is transformed according to transformation rules to establish a BN model. The fuzzy comprehensive evaluation method is used to convert expert qualitative judgments into probabilities of basic events, and the Noisy-OR Gate model is employed to generate the CPTs for intermediate events; and (5) risk assessment and mitigation.

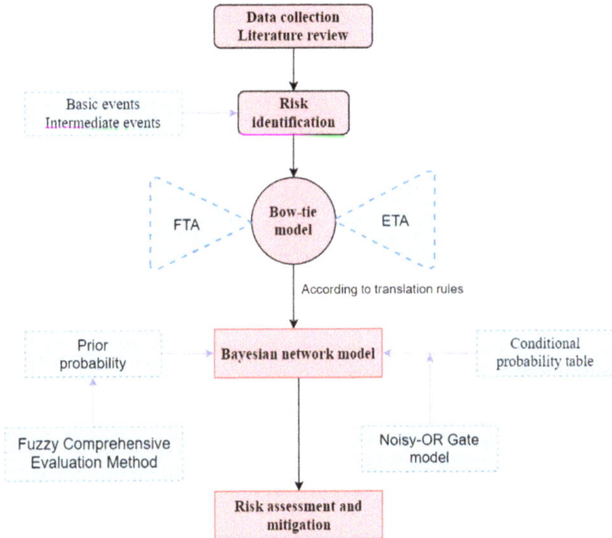

Figure 1. The risk assessment framework for hydrogen leakage.

2.1. Bow-Tie Model

The Bow-tie analysis method is a general model used to conceptualize and analyze risks. It posits that any hazardous event (central node) involves risks and uses a combination of fault tree analysis (FTA) and event tree analysis (ETA) to help analyze and prevent or mitigate these risks. FTA is used to identify the root causes of a hazardous event (defined by specific risks), where the causes can be categorized as either basic or intermediate, and relationships between events are represented using logical "AND" and "OR" gates. An "AND" gate implies that all related causes must occur for the event to happen, while an "OR" gate means that the occurrence of any related cause will trigger the event. ETA, on the other hand, is used to describe the scenario of consequence evolution following the hazardous event. The Bow-tie model is chosen for its ability to integrate both FTA and ETA, providing a unified approach to assess the causes and consequences of hazardous events. This combined methodology is particularly suited for complex systems, such as hydrogen internal combustion engine fuel supply systems, where it allows for a thorough risk assessment by addressing both the underlying failure mechanisms and the potential sequence of events.

2.2. Bayesian Network

The BN is chosen for its robust capability in modeling the complex probabilistic relationships among variables, which is particularly advantageous in systems with interdependent factors. A BN is a type of probabilistic model that employs a directed acyclic graph to show the relationships among variables. Each node in this structure corresponds to a random variable, and the directed edges indicate direct dependencies between the variables. Every node is associated with a probability distribution; non-root nodes are linked to conditional probability distributions, while root nodes correspond to marginal distributions. BNs allow for both qualitative and quantitative assessments. From a qualitative perspective, they illustrate the dependencies and independencies between variables through the graph's structure. Quantitatively, the conditional probability distributions capture the statistical correlations among the variables [21]. This dual capability—qualitative insight and quantitative rigor—makes the BN model an invaluable tool for conducting comprehensive risk assessments, where understanding both the relationships and the uncertainties between the variables is crucial.

2.3. Fuzzy Comprehensive Evaluation Method

Using historical data is an effective approach for estimating the failure probability of fundamental events. However, conducting statistical analysis becomes challenging when data are limited. To address this issue, this study employs the fuzzy comprehensive evaluation method, combining semantic judgments of incomplete intuitionistic fuzzy preference relations with fuzzy set theory to capture the uncertainty of fundamental events and quantify their failure probabilities. The fuzzy comprehensive evaluation method requires transforming experts' fuzzy linguistic levels into fuzzy numerical sets, which are then converted into failure probabilities [22]. In a fuzzy set, variables are obtained from expert ratings, transforming subjective perceptions into fuzzy numbers to express ambiguous or unclear information. Semantic judgments can be represented in a number of ways using fuzzy numbers, including triangular and trapezoidal variations. For the purposes of this research, trapezoidal fuzzy numbers are employed to develop the membership function. A trapezoidal fuzzy number is denoted as $\theta(a, b, c, d)$, where its membership function, $\mu_\theta(x)$, is defined in Equation (1).

$$\mu_\theta(x) = \begin{cases} \frac{x-a}{b-a}, & a \leq x \leq b \\ 1, & b \leq x \leq c \\ \frac{d-x}{d-c}, & c \leq x \leq d \\ 0, & others \end{cases} \tag{1}$$

As shown in Table 1, the transformation scale includes seven linguistic terms used to estimate the likelihood of basic events. The failure probability is expressed using seven fuzzy linguistic terms through an evaluation scoring process, with experts' judgments used as weighted criteria, transforming them into fuzzy numbers suitable for fuzzy calculations. Figure 2 depicts the membership function curves of the seven fuzzy numbers, including "very low, low, relatively low, medium, relatively high, high, very high" fuzzy scales, which map expert judgments to a quantitative level.

Table 1. Fuzzy linguistic terms and their corresponding fuzzy numbers.

Fuzzy Linguistic Term	Fuzzy Number
Very Low (VL)	(0, 0, 0.1, 0.2)
Low (L)	(0.1, 0.2, 0.2, 0.3)
Relatively Low (RL)	(0.2, 0.3, 0.4, 0.5)
Medium (M)	(0.4, 0.5, 0.5, 0.6)
Relatively High (RH)	(0.5, 0.6, 0.7, 0.8)
High (H)	(0.7, 0.8, 0.8, 0.9)
Very High (VH)	(0.8, 0.9, 1.0, 1.0)

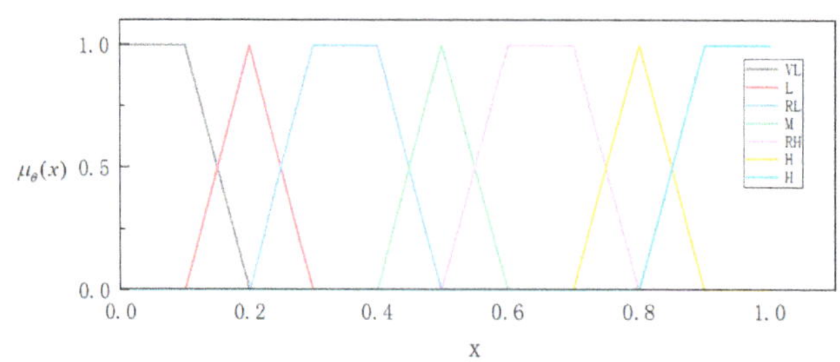

Figure 2. Diagram of a simple BN.

Since experts may have differing views on the probability of basic events, it is necessary to merge or consolidate those views into a single perspective. This research utilizes the linear opinion pool method [23], with the combined fuzzy number P_j depicted in Equation (2).

$$P_j = \sum_{i=1}^{n} (W_i \times P_{ij}), j = 1, 2, 3, \cdots, m \quad (2)$$

In this context, W_i denotes the weight attributed to the i-th expert, and $\sum_{i=1}^{n} W_i = 1$, while P_{ij} signifies the linguistic assessment offered by expert i regarding event j. Here, n represents the aggregate number of experts involved, and m indicates the overall count of fundamental events.

Considering each expert's confidence level and knowledge, expert weights are used to handle uncertainties related to expert judgments. The criteria for assigning weights to experts are shown in Table 2. Expert weights are calculated using Equation (3).

$$W_i = WS_i / \sum_{i=1}^{n} WS_i \quad (3)$$

where WS_i denotes the weight score assigned to the i-th expert, calculated as $WS_i = PPS_i + STS_i + ELS_i$, PPS_i, STS_i, and ELS_i correspond to the scores for professional position, service duration, and education level.

Table 2. Expert weighting criteria.

Professional Position		Service Time		Education Level	
Category	Score	Category	Score	Category	Score
Professor/Captain	5	≥30 years	5	PhD	5
Associate Professor /Chief Engineer	4	20–29 years	4	Master	4
First Officer/First Engineer	3	10–19 years	3	Bachelor	3
Engineer/Surveyor	2	6–9 years	2	Vocational education	2
Lecturer/Worker	1	≤5 years	1	High school	1

Expert weights are assigned based on factors such as professional position, service duration, and educational background, reflecting the expert's level of expertise and experience in the relevant domain. Although the weighting system involves some subjectivity, the criteria are carefully designed to minimize bias and ensure that contributions are proportional to the expert's qualifications. However, expert judgment bias may arise if the experts selected do not represent a diverse range of perspectives, potentially skewing the results. To mitigate this, experts are chosen from varied backgrounds with extensive, relevant experience. Additionally, the validity of subjective probability inputs is ensured by using cross-validation techniques, such as comparing expert judgments with aggregating opinions from multiple experts. These measures help reduce bias and enhance the overall reliability of the aggregated probabilities.

The subsequent stage involves transforming the collected fuzzy numbers into a fuzzy possibility score (FPS), a process referred to as defuzzification. By applying the centroid method for defuzzification, one can derive the defuzzied value of the trapezoidal fuzzy number $\theta(a, b, c, d)$ as outlined in Equation (4) [24].

$$X^* = \frac{\int_a^b \frac{x-a}{b-a} x dx + \int_b^c x dx + \int_c^d \frac{d-x}{d-c} x dx}{\int_a^b \frac{x-a}{b-a} dx + \int_b^c dx + \int_c^d \frac{d-x}{d-c} dx} = \frac{1}{3} \times \frac{(d+c)^2 - dc - (a+b)^2 + ab}{(d+c-a-b)} \tag{4}$$

Therefore, *FPS* = X^*, where *FPS* is the fuzzy possibility score.

Using the function provided by Onisawa, the *FPS* is converted into a fuzzy rate (*FR*). Specifically, the *FPS* serves as a measure of the degree of possibility within a fuzzy system, while the *FR* provides a normalized representation of this possibility, making it more interpretable for analysis. The relationship between the *FPS* and the *FR* lies in the transformation process facilitated by Onisawa's function, which adjusts the raw possibility score into a more practical rate for decision making and analysis. The fuzzy rate (*FR*) can be represented using Equations (5)–(7).

$$FR = \begin{cases} \frac{1}{10^C}, & FPS \neq 0 \\ 0, & FPS = 0 \end{cases} \tag{5}$$

$$C = \frac{1}{K} \times \left(\frac{1 - FPS}{FPS}\right)^{\frac{1}{3}} \tag{6}$$

$$K = \frac{1}{\log_{10}\left(\frac{1}{P_{SC}}\right)} \tag{7}$$

Here, K is a constant, and P_{SC} represents a safety rate, set at 5×10^{-3}, which give $1/K = 2.301$.

2.4. Noisy-OR Gate Model

The Noisy-OR Gate model is employed to depict the relationships among multiple variables and their corresponding child node Y. This model is particularly useful in risk assessment when multiple factors may influence the occurrence of an event. However, the event is unlikely to occur unless one or more of these factors are present, which makes this model ideal for situations where the occurrence of an event depends on the presence of certain conditions. In this study, the Noisy-OR Gate model is selected to address certain limitations of traditional methods. Specifically, root node probabilities and conditional probability tables derived from expert opinions can be influenced by subjectivity. Additionally, logical gate analysis is often binary and absolute, making it inadequate for scenarios where the occurrence of failure events is not simply a binary outcome. To overcome these challenges, this study combines the Bayesian network model with the Noisy-OR Gate model to simplify the acquisition of conditional probabilities and improve the accuracy of risk assessments. For this combined model to be valid, it must meet two criteria [20]:

(1) The variables are independent of each other;
(2) If a single variable X_i occurs while others do not, the probability of Y occurring is given by $P = P(Y = 1 \mid \overline{X}_1, \ldots, X_i, \overline{X}_{i+1}, \ldots, \overline{X}_m)$,with the CPT represented as $P_1, \ldots, P_i, \ldots, P_m$, as shown in Equation (8).

$$P(Y|X_p) = 1 - \prod_{i:X_i \in X_p} (1 - P_i) \tag{8}$$

In cases where X_p is an empty set, $P(Y \mid X_p) = 0$, indicating no dependency between Y and its parent node X_p. This limitation mirrors the traditional fault tree analysis. However, in real-world scenarios, additional unknown factors, apart from X, may influence child node Y, and these cannot always be easily identified. To address this, the combined Leaky Noisy-OR Gate BN is introduced. This extended model attributes the unidentified risk factors to an auxiliary set X_0, with a leakage probability $P_0 = P(Y \mid \overline{X}_1, \overline{X}_2, \ldots, \overline{X}_m, X_0)$. This probability represents the chance that Y enters a failure state when X_L fails, while the other parent nodes remain unaffected.

In this scenario, let us assume that node Y has precisely two parent nodes: C_0 and C_{all}, with C_{all} signifying the collective influence of factors aside from C_0. The associated probabilities are denoted as P_0 and P_{all}, respectively, and their connection is articulated through Equations (9) and (10).

$$P(Y|C_0) = 1 - (1 - P_0)(1 - P_{all}) = P_0 + P_{all} - P_0 P_{all} \tag{9}$$

$$P(Y|\overline{C}) = P_{all} \tag{10}$$

Subsequently, the relationship can be derived as shown in Equation (11).

$$P_0 = \frac{P(Y|C_0) - P(Y|\overline{C})}{1 - P(Y|\overline{C})} \tag{11}$$

In the presence of a leakage node X_0, the conditional probability for node Y is expressed in Equation (12).

$$P(Y|X_p, X_0) = 1 - (1 - P_0)\prod\nolimits_{i:X_i \in X_p}(1 - P_i) \tag{12}$$

3. Hydrogen Leakage Risk Assessment Model

3.1. Fuel System Overview

The schematic of the diesel–hydrogen dual-fuel engine setup, shown in Figure 3, illustrates the critical components and their integration within the fuel system, highlighting how hydrogen and diesel are combined to optimize engine performance. This fuel system includes hydrogen storage tanks, diesel storage tanks, hydrogen flow meters, an engine control unit (ECU), combustion analysis devices, and various types of auxiliary systems and safety equipment. By proportionally mixing hydrogen and diesel, the system injects the mixture into the engine cylinders for combustion, providing a more efficient and environmentally friendly power source. The system also features real-time monitoring of hydrogen through high-pressure storage tanks and flow meters, with the ECU controlling the mixture's injection to ensure stability and efficiency. Auxiliary safety measures, such as leak detection and emergency shutdown systems, are integrated to guarantee safe and reliable operation.

Hydrogen is monitored in real time through high-pressure storage tanks and flow meters and is mixed with air in the intake system with injection controlled by the ECU. Diesel is supplied through an independent fuel system to ensure combustion stability. The system uses a low flow safety system to regulate fuel temperature, pressure, and flow, and it dynamically adjusts the fuel ratio according to operating conditions to optimize combustion efficiency. Multiple safety devices, such as leak detection and emergency shutdown systems, are in place to ensure that protective measures are automatically triggered in the event of abnormalities.

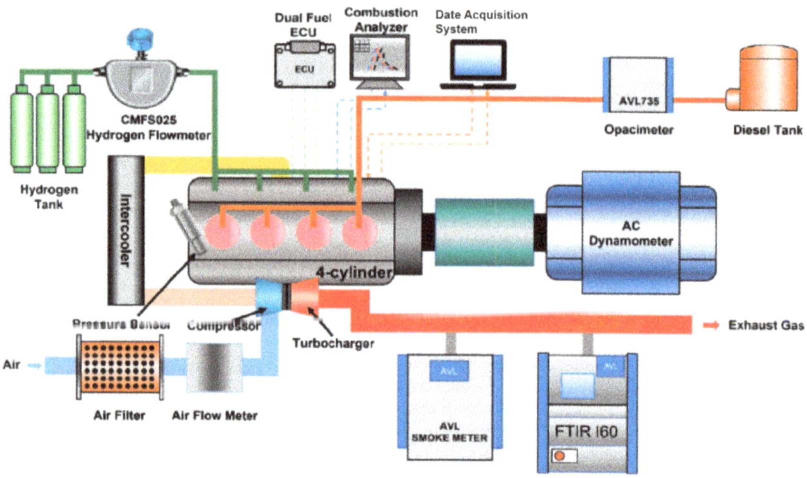

Figure 3. Diagram of a simple BN schematic of diesel–hydrogen dual-fuel engine setup.

3.2. Bow-Tie Model Construction

Hydrogen leakage is considered the top event (i.e., hazardous event), and the causes of hydrogen leakage are analyzed from three aspects: equipment failure, human factors, and external factors. This comprehensive analysis aims to identify all possible risk paths leading to hydrogen leakage. Although hazardous events can occur in countless scenarios, only a limited number of representative cases have been selected for study. Once a hazardous event has been determined, its consequences can be classified into different levels. The Bow-tie model of the hydrogen leakage incident is shown in Figure 4, and descriptions of the various events involved are provided in Table 3. Different subscripts of M represent intermediate events, while different subscripts of X represent basic events.

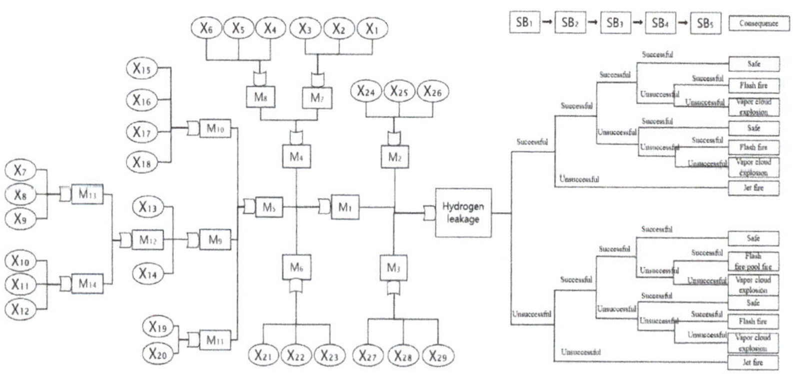

Figure 4. Bow-tie model of hydrogen leak accident.

The safety barriers (SBs) included in the Bow-tie model for hydrogen leakage incidents are as follows:

Leak Prevention Barrier (SB_1): This barrier is implemented by establishing an emergency shutdown system and equipping the system with high-sensitivity hydrogen detection devices. The strategic placement of these sensors is critical for early detection of potential leaks. The placement is designed to maximize coverage in high-risk areas where hydrogen leaks are most likely to occur, such as near joints, valves, and hydrogen storage locations.

Upon detecting an excessive hydrogen concentration or leakage, the sensors automatically trigger the emergency shutdown system, immediately cutting off the hydrogen supply and initiating predefined protective measures.

Table 3. Description of each event.

Event	Event Description	Event	Event Description
X_1	Mechanical impact	X_{23}	Heat exchanger leakage
X_2	Fire	X_{24}	Insufficient operator training
X_3	Internal tank coating detachment	X_{25}	Inadequate operating procedures
X_4	Safety valve failure	X_{26}	Improper maintenance procedures
X_5	Relief valve failure	X_{27}	Ship collision
X_6	Leak at sensor opening	X_{28}	Ship fire
X_7	Material defects (cracks, crystal defects, etc.)	X_{29}	Heavy object impact
X_8	Welding defects	M_1	Equipment failure
X_9	Internal residual stress	M_2	Human factors
X_{10}	High hydrogen concentration	M_3	External factors
X_{11}	High-pressure hydrogen environment	M_4	Leakage from cryogenic hydrogen storage tank
X_{12}	External mechanical stress (uneven flange bolt preload; external stress due to pipe elbows; internal stress due to temperature differences inside and outside the pipeline, etc.)	M_5	Piping failure
X_{13}	Internal pipe coating detachment	M_6	Other equipment failure
X_{14}	Mechanical vibration, periodic pressure fluctuations	M_7	Tank leakage
X_{15}	Aging of flange gasket	M_8	Tank accessory leakage
X_{16}	Improper flange installation or fastening	M_9	Pipe rupture
X_{17}	External damage to flange	M_{10}	Flange failure
X_{18}	Gasket damage	M_{11}	Valve failure
X_{19}	Valve stem seal failure	M_{12}	Hydrogen embrittlement
X_{20}	External valve damage	M_{13}	Internal factors
X_{21}	Emergency shutdown device failure	M_{14}	External factors
X_{22}	Filter failure (poor sealing or damage)		

Immediate Ignition Barrier (SB_2): This barrier involves immediately igniting hydrogen if a leak occurs or if ignition is triggered by an incident. The primary aim is to control heat sources and prevent electrical sparks, reducing the risk of ignition. However, hydrogen has a very low Minimum Ignition Energy (MIE) of 0.019 M_J [25], which makes it highly susceptible to ignition. Therefore, relying solely on the immediate ignition barrier is insufficient. It should be combined with other preventive measures such as electrical isolation to prevent sparks, heat source isolation to avoid contact with hot surfaces, and automated fire suppression systems (e.g., gas extinguishing or water mist systems) to quickly mitigate any fire risk. Real-time hydrogen leakage monitoring and automatic shutdown systems are also essential for early intervention, minimizing the risk of ignition. By integrating these measures, a more comprehensive safety strategy can be achieved.

Diffusion Prevention Barrier (SB_3): This barrier is used to prevent the diffusion of hydrogen leaks, limiting the impact scope and duration of potential consequences. The barrier can be categorized into three types: passive, active, and procedural. Passive barriers, such as walls or structural enclosures, work continuously without requiring activation. They physically isolate hydrogen, preventing its diffusion and containing it within a specific area. Active barriers, such as water mist spray systems and high-expansion foam, are designed to be activated when hydrogen leakage is detected. These systems reduce the spread and duration of hydrogen leaks by quickly responding to the situation. Lastly, procedural barriers involve safety protocols and operational practices that help manage hydrogen leakage risks. These include regular maintenance, monitoring systems, emergency response plans, and personnel training to ensure the proper functioning of both active and passive barriers, thus contributing to overall safety. To ensure effectiveness, the design and implementation of such barriers should align with Atmosphères Explosibles (ATEX) directives, which establish safety requirements for explosive atmospheres [26]. ATEX standards provide guidance on risk assessment, material compatibility, and system integrity, ensuring that preventive barriers not only minimize hydrogen dispersion but also mitigate ignition risks in compliance with established safety regulations.

Delayed Ignition Barrier (SB_4): Through rational layout design, this barrier ensures that there are no potential ignition sources near the leakage source and uses fire prevention systems, such as automatic extinguishing devices, to reduce the risk of hydrogen vapor cloud ignition in areas away from the leakage source.

Environmental Condition Barrier (SB_5): This barrier can prevent vapor cloud explosions. The environmental condition barrier is primarily achieved by maintaining good ventilation and an open environment. To ensure effective ventilation, a real-time monitoring system should monitor environmental conditions. When an increase in hydrogen concentration is detected, the ventilation system should immediately increase airflow or adjust the ventilation strategy to control hydrogen accumulation and dispersion, ensuring that hydrogen does not accumulate in hazardous concentrations and minimizing the risk of explosion.

The consequences of hydrogen refueling leakage are classified into five categories, denoted as C_1, C_2, C_3, C_4, and C_5, as shown in Table 4.

Table 4. Consequences of hydrogen leakage.

Consequence	Consequence Factors				
	Physical Consequence	**Casualties**	**Economic Loss**	**Material Loss**	**Expected Probability**
C_1	Safe	None	None	None	Low
C_2	Leak without fire and safe dispersion	Small	Moderate	Small	Very High
C_3	Flash fire/pool fire	Significant	Significant	Significant	Low
C_4	Jet fire	Significant	Major	Major	Low
C_5	Vapor cloud explosion	Major	Major	Major	Very Low

3.3. Construction of the BN Model

The Bow-tie model illustrating the hydrogen leakage incident, depicted in Figure 5, has been transformed into a BN model. Within this framework, the top event, intermediate events, and basic events from the fault tree are represented as child nodes, intermediate nodes, and parent nodes in the BN, respectively. Logic gates define the conditional relationships between nodes in the BN. The CPT for node *A* is shown in Table 5.

Similarly, the CPTs for the "AND" and "OR" gates are derived from their logical relationships. Figure 5 illustrates this transformation, showing the mapping of these gates from the fault tree to the BN model, along with their respective conditional probabilities. Additionally, the failure probabilities of safety barriers in the event tree are mapped to the BN model, reflecting the consequences of these failures.

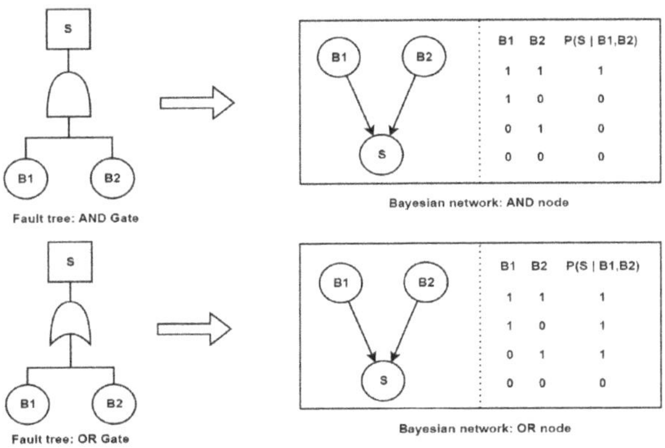

Figure 5. The "AND" and "OR" gate in fault tree and Bayesian network representation.

Table 5. Conditional probability table of the "OR" gate.

X_1	X_2	A (Not Occurred)	A (Occurred)
Not Occurred	Not Occurred	1	0
Not Occurred	Occurred	0	1
Occurred	Not Occurred	0	1
Occurred	Occurred	0	1

3.4. Acquisition of Prior Failure Probability for Basic Events

Currently, the application of hydrogen internal combustion engines in marine settings is minimal, resulting in a lack of comprehensive failure case datasets. Therefore, a combination of expert scoring and fuzzy set theory is used to calculate the prior failure probability of basic events. A group of nine experts was selected to evaluate the fundamental events of the fault tree. Their assessment was based on a combination of factors, including their professional expertise, knowledge, and available technical reports, as well as both the on-site and environmental conditions. These experts included three chief engineers, one first engineer, one associate professor, two engineers, one lecturer, and one official from the Shanghai Maritime Bureau responsible for ship inspection. The expert scoring results are shown in Table 6. Expert weights, calculated using Equation (3), are presented in Table 7.

Table 6. Prior and posterior probability calculation results for basic events.

Basic Event	Expert Evaluation Level Result									Prior Probability	Posterior Probability
X_1	VL	VL	L	RL	RL	RH	VL	VL	RL	4.86×10^{-4}	1.16×10^{-3}
X_2	M	RL	L	M	RH	H	VL	M	RL	3.30×10^{-3}	1.02×10^{-2}
X_3	VL	RL	VL	VL	M	VL	L	L	VL	1.38×10^{-4}	3.69×10^{-4}
X_4	L	M	L	M	RL	H	H	M	VL	3.14×10^{-3}	5.66×10^{-3}
X_5	RL	L	L	M	M	RL	L	M	L	1.23×10^{-3}	2.24×10^{-3}
X_6	VL	L	RL	M	L	H	VL	M	RL	1.38×10^{-3}	1.96×10^{-3}
X_7	VL	VL	RL	RH	RH	VH	VL	L	L	1.47×10^{-3}	1.59×10^{-3}
X_8	VL	VL	RL	RL	RH	H	VL	L	RL	1.15×10^{-3}	1.22×10^{-3}
X_9	VL	L	L	RH	RH	RH	VL	RL	RL	1.73×10^{-3}	1.81×10^{-3}
X_{10}	VL	L	L	RH	M	H	M	M	L	2.48×10^{-3}	2.65×10^{-3}
X_{11}	L	VL	RL	RH	RH	H	M	M	L	3.24×10^{-3}	3.41×10^{-3}
X_{12}	M	RL	RL	RL	RH	VH	H	RH	M	7.36×10^{-3}	7.63×10^{-3}
X_{13}	L	RL	RL	RL	L	RL	RH	RL	RL	1.44×10^{-3}	4.84×10^{-3}
X_{14}	RH	M	RL	M	RL	H	RH	RH	RH	8.05×10^{-3}	1.56×10^{-2}
X_{15}	H	M	M	M	RL	H	RH	VL	M	6.16×10^{-3}	2.69×10^{-2}
X_{16}	RH	M	M	RL	RH	VH	RH	VL	RL	5.87×10^{-3}	2.20×10^{-2}
X_{17}	RL	L	M	RL	M	RH	RH	VL	M	2.93×10^{-3}	8.28×10^{-3}
X_{18}	H	L	M	RL	M	H	RH	L	M	4.94×10^{-3}	9.79×10^{-3}
X_{19}	RH	RL	M	RL	RL	H	RH	RL	RH	5.59×10^{-3}	1.86×10^{-2}
X_{20}	L	VL	M	RL	RL	L	RH	VL	RL	8.99×10^{-4}	2.50×10^{-3}
X_{21}	L	VL	RL	M	RL	L	L	L	L	4.95×10^{-4}	7.80×10^{-3}
X_{22}	L	L	RL	L	RL	L	M	RL	L	6.62×10^{-4}	2.89×10^{-3}
X_{23}	RH	VL	M	RH	RL	L	M	RH	L	2.56×10^{-3}	3.63×10^{-3}
X_{24}	RH	M	M	H	M	VL	RH	RL	RH	6.05×10^{-3}	2.53×10^{-1}
X_{25}	H	RH	M	H	M	L	H	RL	RH	9.06×10^{-3}	2.16×10^{-1}
X_{26}	RH	M	M	H	RH	L	H	RH	RH	9.89×10^{-3}	1.69×10^{-1}
X_{27}	L	VL	RL	RL	RH	M	RL	RL	RL	1.54×10^{-3}	1.12×10^{-2}
X_{28}	L	L	M	M	RH	RH	L	M	RL	2.76×10^{-3}	1.44×10^{-2}
X_{29}	L	VL	RL	M	RL	M	L	VL	RL	8.77×10^{-4}	4.20×10^{-3}

Table 7. Expert weights.

Expert	Expert 1	Expert 2	Expert 3	Expert 4	Expert 5	Expert 6	Expert 7	Expert 8	Expert 9
Weight	0.084211	0.126316	0.094737	0.136842	0.105263	0.136842	0.094737	0.084211	0.136842

3.5. Establishment of Conditional Probability Tables Between Nodes

In the BN model, the CPTs may incorporate subjective factors, since logical gate analysis tends to be excessively inflexible. Failure events do not occur solely in a binary fashion, which deviates from real-world situations and does not adequately represent the relationship between intermediate and basic events. To address this, the Leaky Noisy-

OR Gate model is introduced. Taking the intermediate node "Valve Failure" (M_2) as an example, it has two parent nodes, X_{19} and X_{20}, from which the conditional probability of M_2 can be calculated, as shown in Table 8. For instance, if $X_{19} = 1$, it indicates that event X_{19} definitely occurs; if $X_{19} = 0$, it indicates that event X_{19} definitely does not occur. The closer the value of X_{19} is to 1, the higher the probability of event X_{19} occurring.

Table 8. Conditional probability for valve failure (M_2).

Conditional Probability	Probability Value	Conditional Probability	Probability Value
$P(M_2 = 1 \mid X_{19} = 1)$	0.67	$P(M_2 = 1 \mid X_{19} = 0)$	0.33
$P(M_2 = 1 \mid X_{20} = 1)$	0.65	$P(M_2 = 1 \mid X_{20} = 0)$	0.35

According to Equation (4), the joint probabilities are $P(X_{19}) = 0.67$ and $P(X_{20}) = 0.20$. Assuming the unknown factor is X_0 and setting $P(X_0 = 1) = 0.1$, substituting these probability values into Equation (5) yields the conditional probabilities of M_2 under different circumstances, as shown in Table 9. By following a similar approach, the conditional probabilities for all intermediate events can be obtained.

Table 9. Corrected conditional probability using the Leaky Noisy-OR Gate model for valve failure (M_2).

Event	X_{19}	X_{20}	$M_2 = 1$	$M_2 = 0$
Probability	1	0	0.56	0.44
	0	1	0.51	0.49
	1	1	0.76	0.24

4. Results and Analysis Using Bayesian Network Model in GeNle

4.1. Forward Reasoning Analysis

Using data from Sections 3.3–3.5, a BN model was constructed with GeNle Academic 4.1 software (see Figure 6). In this model, Y denotes the occurrence of an event, while N represents the non-occurrence. After inputting the relevant prior and conditional probabilities, the calculated likelihood of a leakage accident in the hydrogen dual-power ship fuel system is 4.73×10^{-4}. This suggests that, under the current conditions and safety measures, the probability of system leakage is relatively low and meets the safety standards defined by the ALARP principle [27].

Based on the probabilities of safety barrier failures and hydrogen leakage, the occurrence probabilities of consequences C_1 to C_5 were calculated as 0.9995, 4.27×10^{-4}, 2.23×10^{-5}, 2.37×10^{-5}, and 2.25×10^{-7}, respectively. These results indicate that consequence C_1, which represents relatively minor damage, is the most likely outcome following a hydrogen leakage incident. The key nodes influencing this outcome include the failure rates of the safety barriers, particularly the immediate ignition barrier and the diffusion prevention barrier. These nodes have a significant impact on the severity of the consequences, as they directly control the diffusion of hydrogen and the likelihood of ignition. The model's calculation suggests that with effective safety barriers in place, the risk of catastrophic events like explosions is minimized, as indicated by its extremely low probability of 2.25×10^{-7}.

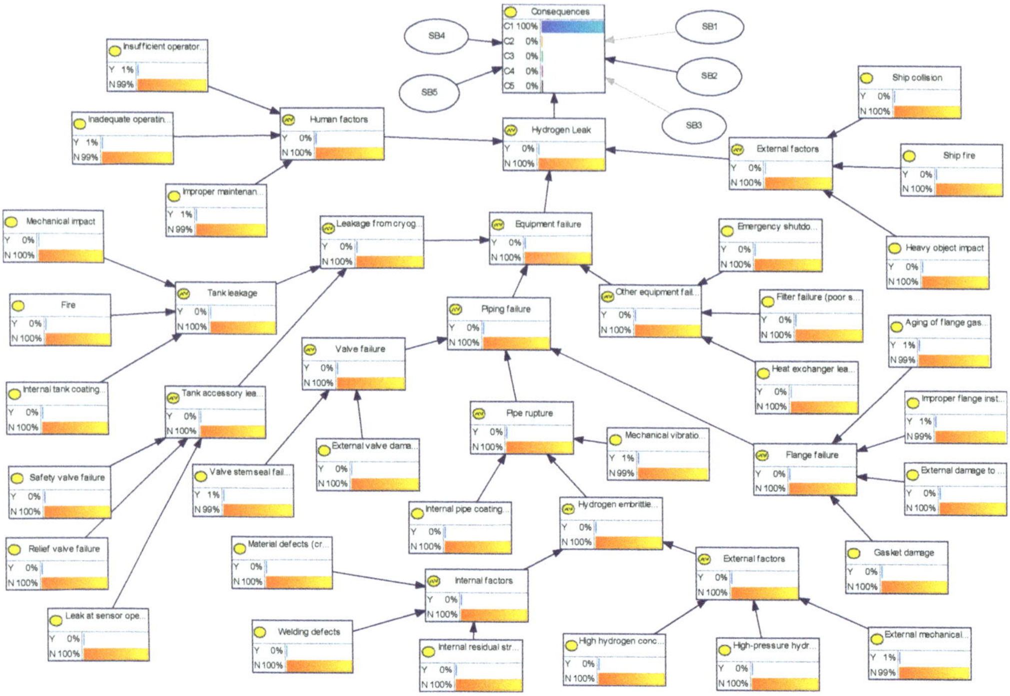

Figure 6. BN model for evaluating leakage risk in hydrogen-powered ship fuel systems using GeNIe.

Further analysis of the safety barriers reveals that the immediate ignition barrier and the diffusion prevention barrier are critical in mitigating the severity of hydrogen leakage incidents. These barriers effectively limit hydrogen diffusion, thereby preventing the escalation of physical damage and reducing the risk of casualties. Therefore, maintaining the reliability of these key barriers, along with regular maintenance and testing, is essential for effective accident consequence control.

4.2. Backward Inference

BNs are capable of performing backward inference, meaning that once the occurrence of a top event is known, the posterior probabilities of related basic events can be updated. This process, known as Bayesian updating, involves using prior probabilities and conditional dependencies to adjust the probabilities of events based on new evidence. As illustrated in Figure 7, this allows for the determination of the posterior probabilities of key contributing factors, such as X_{26} (improper maintenance procedures), X_{25} (insufficient operational protocols), X_{24} (inadequate operator training), X_{19} (valve stem seal failure), X_{15} (aging of flange gaskets), X_{14} (mechanical vibration and periodic pressure fluctuations), and X_{16} (improper flange installation or fastening), which all have relatively high posterior probabilities, making them significant factors contributing to hydrogen leakage. Notably, the human-related factors, including X_{24}, X_{25}, and X_{26}, have much higher posterior probabilities compared with other factors. This suggests that when hydrogen leakage (the top event) occurs, shortcomings in operational management are often the primary drivers. Specifically, the higher posterior probabilities indicate that these factors directly influence the occurrence of hydrogen leakage. Notably, human-related factors such as X_{24}, X_{25}, and X_{26} have much higher posterior probabilities compared with other factors, suggesting that shortcomings in operational management, especially in maintenance procedures, operational protocols, and operator training, are often the primary drivers behind hydrogen leakage.

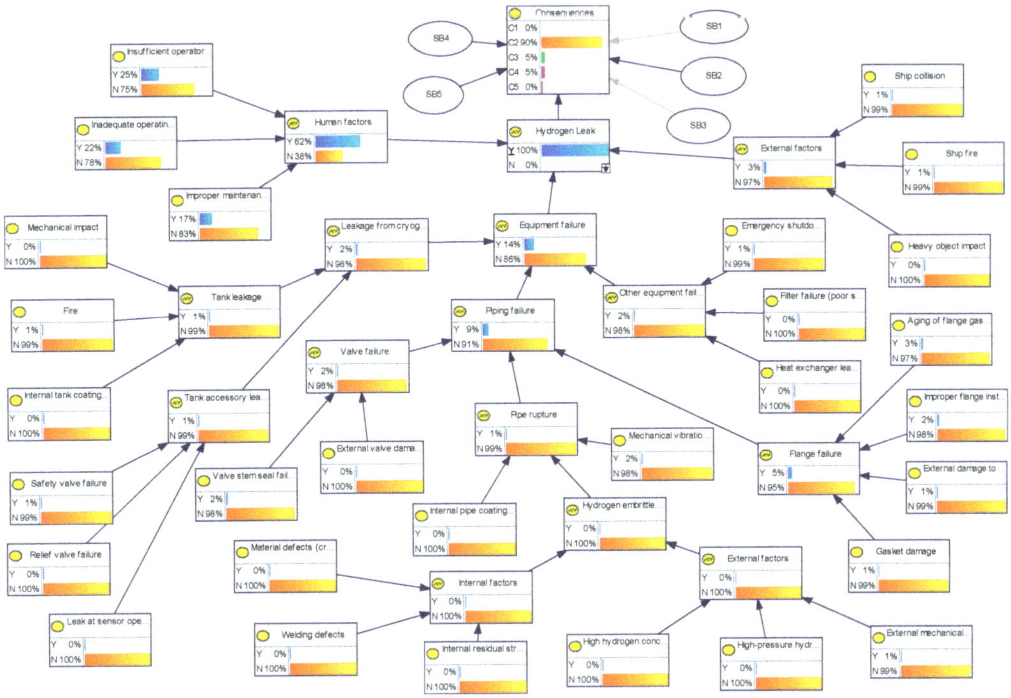

Figure 7. Posterior probability update of the model.

4.3. Sensitivity Analysis

Bayesian sensitivity analysis is used to quantify the impact of changes in uncertainty variables on the occurrence of hydrogen leakage incidents. In this analysis, the sensitivity of nodes is used to measure the magnitude of their influence on accident occurrence. A comparative analysis of prior and posterior probabilities (as shown in Figure 8) further reveals the sensitivity of each basic event to hydrogen leakage incidents. Specifically, events X_{24}, X_{25}, and X_{26} exhibit notably high sensitivity when such incidents occur, indicating that operational and management failures are key drivers of hydrogen leakage. While these factors may be infrequent during normal operations, their occurrence significantly raises the risk of severe accidents. Therefore, these high-sensitivity factors should be prioritized in risk management and control measures. This aligns with previous findings, reinforcing that enhancing maintenance protocols, refining operational standards, and improving personnel training are essential to effectively mitigate accident risks.

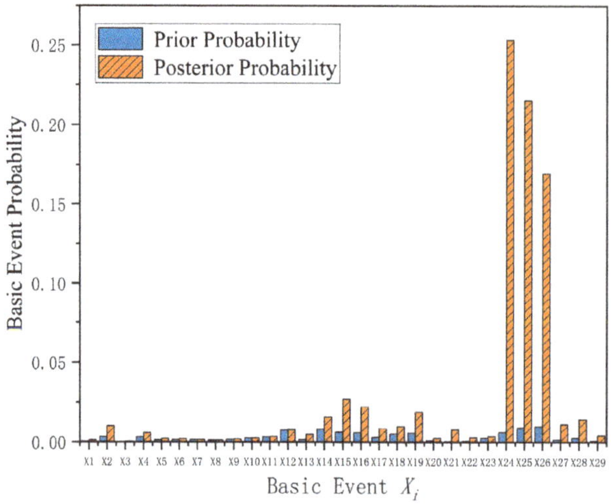

Figure 8. Comparison of prior and posterior probabilities of basic events.

5. Conclusions

This study employed a risk assessment approach that integrates the Bow-tie model with a BN model to systematically analyze the primary hazards and potential consequences of leakage incidents in the hydrogen dual-power ship fuel system. The main findings are summarized below:

(1) The probability of a leakage occurring in the hydrogen dual-power ship fuel system is calculated to be 4.73×10^{-4}. This value is within the typical range observed in industrial risk assessments for LNG systems. For instance, the average leakage probability for hoses in LNG systems is 5.48×10^{-4}. This figure reflects typical conditions for hoses used in industrial settings, considering the presence of safety systems such as pull-away prevention and mitigation systems. In contrast, loading arms exhibit a slightly higher leakage probability of 3.50×10^{-3}, reflecting differences in equipment design and operational environments [28]. It is important to account for the differences in measurement units, as some probabilities are reported per operation, while others are annualized per year. Thus, the hydrogen leakage probability in hydrogen-powered internal combustion engines is comparable to those observed in other sectors, though differences in reporting conventions and operational contexts must be carefully considered. Following such an incident, the probabilities of a flash fire or pool fire, a jet fire, and a vapor cloud explosion are estimated at 2.23×10^{-5}, 2.37×10^{-5}, and 2.25×10^{-7}, respectively.

(2) Human factors such as improper maintenance procedures, inadequate operational standards, and insufficient operator training are the primary contributors to leakage accidents. Therefore, it is essential to focus on strengthening both operator training and the enforcement of stringent maintenance protocols. However, it is also important to recognize that human error cannot be entirely eliminated, which points to the need for designing more automated systems and fail-safe mechanisms to further reduce the impact of such errors. One key opportunity for improving safety lies in the integration of real-time monitoring systems and advanced predictive maintenance technologies. These technologies can provide early warnings of potential failures, allowing for more proactive risk management. Additionally, improvements in the design and layout of

critical system components, such as valves, pipelines, and hydrogen storage tanks, can help mitigate leakage risks by reducing the likelihood of failures.

While the model presented here provides a robust foundation for risk assessment, it has certain limitations. The accuracy of the risk predictions is dependent on the quality and completeness of the input data, which in this study were based on the available literature and estimated probabilities. Real-world variations, such as equipment degradation, extreme environmental conditions, and the effect of operational practices, were not fully incorporated into the model. Future work will involve gathering more real-world data from hydrogen dual-fuel ships, which will help refine the model and ensure that it better reflects operational realities. This will include case studies, data from publicly available risk databases, and collaborations with research institutions. Moreover, extending the analysis to include additional risk scenarios, such as extreme weather events or equipment failures, will help improve the robustness of the safety measures proposed.

In conclusion, while the study provides valuable insights into the risks associated with hydrogen dual-power ship fuel systems, further research is needed to address the identified limitations and explore innovative solutions to improve safety in this emerging technology.

Author Contributions: Conceptualization, C.Z. and S.L.; methodology, H.Z.; software, H.Z.; validation, G.L.; formal analysis, G.L.; investigation, S.L.; data curation, H.Z.; writing—original draft preparation, H.Z.; writing—review and editing, G.L.; visualization, H.Z.; supervision, C.Z.; project administration, C.Z.; funding acquisition, C.Z. All authors have read and agreed to the published version of the manuscript.

Funding: This research was funded by National Key R&D Program of China, grant number 2022YFB4301400.

Institutional Review Board Statement: Not applicable.

Informed Consent Statement: Not applicable.

Data Availability Statement: The data presented in this study are available on request from the corresponding authors.

Conflicts of Interest: The authors declare no conflict interest.

References

1. Chen, X.; Wu, H.; Han, B.; Liu, W.; Montewka, J.; Liu, R.W. Orientation-aware ship detection via a rotation feature decoupling supported deep learning approach. *Eng. Appl. Artif. Intell.* **2023**, *125*, 106686. [CrossRef]
2. Chen, X.; Zheng, J.; Li, C.; Wu, B.; Wu, H.; Montewka, J. Maritime traffic situation awareness analysis via high-fidelity ship imaging trajectory. *Multimed. Tools Appl.* **2024**, *83*, 48907–48923. [CrossRef]
3. Palmén, M.; Lotrič, A.; Laakso, A.; Bolbot, V.; Elg, M.; Banda, O.A.V. Selecting Appropriate Energy Source Options for an Arctic Research Ship. *J. Mar. Sci. Eng.* **2023**, *11*, 2337. [CrossRef]
4. Fernández-Ríos, A.; Santos, G.; Pinedo, J.; Santos, E.; Ruiz-Salmón, I.; Laso, J.; Lyne, A.; Ortiz, A.; Ortiz, I.; Irabien, Á.; et al. Environmental sustainability of alternative marine propulsion technologies powered by hydrogen—A life cycle assessment approach. *Sci. Total Environ.* **2022**, *820*, 153189. [CrossRef]
5. Soni, G.; Neto, R.C.; Moreira, L. Hydrodynamic Simulation of Green Hydrogen Catamaran Operating in Lisbon, Portugal. *J. Mar. Sci. Eng.* **2023**, *11*, 2273. [CrossRef]
6. Yang, N.; Deng, J.; Wang, C.; Bai, Z.; Qu, J. High pressure hydrogen leakage diffusion: Research progress. *Int. J. Hydrog. Energy* **2023**, *50*, 1029–1046. [CrossRef]
7. Palazzi, E.; Currò, F.; Fabiano, B. Low rate releases of hazardous light gases under semi-confined geometry: A consequence based approach and case-study application. *J. Loss Prev. Process Ind.* **2020**, *63*, 104038. [CrossRef]
8. Mao, X.; Ying, R.; Yuan, Y.; Li, F.; Shen, B. Simulation and analysis of hydrogen leakage and explosion behaviors in various compartments on a hydrogen fuel cell ship. *Int. J. Hydrogen Energy* **2021**, *46*, 6857–6872. [CrossRef]
9. Duong, P.A.; Lee, J.; Ryu, B.R.; Kang, H. A preliminary safety assessment of fuel gas supply system in the engine room of the ammonia fuelled ship. *J. Mar. Eng. Technol.* **2025**, 1–20. [CrossRef]

10. Jeong, B.; Lee, B.S.; Zhou, P. Quantitative risk assessment of fuel preparation room having high-pressure fuel gas supply system for LNG fuelled ship. *Ocean. Eng.* **2017**, *137*, 450–468. [CrossRef]
11. Ventikos, N.P.; Louzis, K. Risk dynamics for marine systems: Towards a bio-inspired framework for dynamic risk assessment. *Transp. Saf. Environ.* **2022**, *4*, 34–50. [CrossRef]
12. Ventikos, N.P.; Sotiralis, P.; Annetis, E. A combined risk-based and condition monitoring approach: Developing a dynamic model for the case of marine engine lubrication. *Transp. Saf. Environ.* **2022**, *4*, tdac020. [CrossRef]
13. Singh, K.; Kaushik, M.; Kumar, M. Integrating α-cut interval based fuzzy fault tree analysis with Bayesian network for criticality analysis of submarine pipeline leakage: A novel approach. *Process Saf. Environ. Prot.* **2022**, *166*, 189–201. [CrossRef]
14. Weng, Y.; Sun, X.; Yang, Y.; Tao, M.; Liu, X.; Zhang, H.; Zhang, Q. An Improved Prediction Method for Failure Probability of Natural Gas Pipeline Based on Multi-Layer Bayesian Network. *Processes* **2024**, *12*, 2930. [CrossRef]
15. Hassan, S.; Wang, J.; Kontovas, C.; Bashir, M. An assessment of causes and failure likelihood of cross-country pipelines under uncertainty using bayesian networks. *Reliab. Eng. Syst. Saf.* **2022**, *218*, 108171. [CrossRef]
16. Mahmood, Y.; Chen, J.; Yodo, N.; Huang, Y. Optimizing natural gas pipeline risk assessment using hybrid fuzzy bayesian networks and expert elicitation for effective decision-making strategies. *Gas Sci. Eng.* **2024**, *125*, 205283. [CrossRef]
17. Xie, S.; Huang, Z.; Wu, G.; Luo, J.; Li, L.; Ma, W.; Wang, B. Combining precursor and Cloud Leaky noisy-OR logic gate Bayesian network for dynamic probability analysis of major accidents in the oil depots. *Reliab. Eng. Syst. Saf.* **2024**, *241*, 109625. [CrossRef]
18. Wang, L.; Zhang, J.; Wang, H.; You, Q.; Zhuo, J.; Zhang, S.; Qiao, J.; Wei, J. Hydrogen leakage risk assessment of HECS based on dynamic bayesian network. *Int. J. Hydrogen Energy* **2024**, *78*, 256–267. [CrossRef]
19. Zhang, Z.; Guo, Y.; Wang, D.; Li, G.; Peng, D. A Noisy-OR gate based fuzzy fault tree approach for micro-leakage evaluation of bolt-gasket-flange connection (BGFC). *J. Loss Prev. Process Ind.* **2021**, *71*, 104521. [CrossRef]
20. Zhao, C.; Yip, T.L.; Wu, B.; Lyu, J. Use of fuzzy fault tree analysis and Bayesian network for occurrence likelihood estimation of navigational accidents in the Qinzhou Port. *Ocean. Eng.* **2022**, *263*, 112381. [CrossRef]
21. Wang, Q.-A.; Chen, J.; Ni, Y.; Xiao, Y.; Liu, N.; Liu, S.; Feng, W. Application of Bayesian networks in reliability assessment: A systematic literature review. *Structures* **2025**, *71*, 108098. [CrossRef]
22. Clemen, R.T.; Winkler, R.L. Combining Probability Distributions From Experts in Risk Analysis. *Risk Anal.* **2022**, *50*, 187–203. [CrossRef]
23. Soule, D.; Grushka-Cockayne, Y.; Merrick, J. A Heuristic for Combining Correlated Experts When There Are Few Data. *Manag. Sci.* **2024**, *70*, 6637–6668. [CrossRef]
24. Yu, J.; Ding, H.; Yu, Y.; Wu, S.; Zeng, Q.; Xu, Y. Risk assessment of liquefied natural gas storage tank leakage using failure mode and effects analysis with Fermatean fuzzy sets and CoCoSo method. *Appl. Soft Comput.* **2024**, *154*, 111334. [CrossRef]
25. Vairo, T.; Cademartori, D.; Clematis, D.; Carpanese, M.P.; Fabiano, B. Solid oxide fuel cells for shipping: A machine learning model for early detection of hazardous system deviations. *Process Saf. Environ. Prot.* **2023**, *172*, 184–194. [CrossRef]
26. Cheng, Y.-F.; Wu, H.-B.; Liu, R.; Yao, Y.-L.; Su, J.; Wang, W.-T.; Shu, C.-M. Combustion behaviors and explosibility of suspended metal hydride TiH2 dust. *Int. J. Hydrogen Energy* **2020**, *45*, 12216–12224. [CrossRef]
27. Aziz, M.A.A.; Said, M.S.M. ALARP demonstration in management of change using quantitative Bowtie analysis risk assessment tool for an offshore gas platform. *Process. Saf. Prog.* **2022**, *42*, 310–327. [CrossRef]
28. Gerbec, M.; Aneziris, O. Uncertainties in failure rates in the LNG bunkering risk assessment. *Saf. Sci.* **2022**, *152*, 105774. [CrossRef]

Journal of
Marine Science and Engineering

MDPI

Article

Tramp Ship Routing and Scheduling with Integrated Carbon Intensity Indicator (CII) Optimization

Haiying Yang [1], Feiyang Ren [2], Jingbo Yin [1,*], Siqi Wang [1] and Rafi Ullah Khan [1]

[1] State Key Laboratory of Ocean Engineering, Department of Transportation Engineering, School of Ocean and Civil Engineering, Shanghai Jiao Tong University, Shanghai 200240, China; yanghaiying@sjtu.edu.cn (H.Y.); iqisgnaw@sjtu.edu.cn (S.W.); asaduetian1@gmail.com (R.U.K.)
[2] COSCO Shipping Technology Co., Ltd., Shanghai 200135, China; ren.feiyang@coscoshipping.com
* Correspondence: jingboyin@sjtu.edu.cn; Tel.: +86-135-8588-9263

Abstract: In response to growing environmental concerns and regulatory pressures, reducing carbon emissions in maritime transport has become a priority. Shipping companies face the challenge of balancing profitability objectives with the imperative to minimize their environmental footprint. This study addresses the tramp ship routing and scheduling problem by incorporating the carbon intensity indicator (CII) into the optimization framework. A bi-objective optimization model is developed, with two objective functions aimed at maximizing fleet profit and improving CII ratings. The Gale–Shapley algorithm is employed to achieve stable vessel–cargo matching, and the genetic algorithm is adopted for iterative optimization. This computational study, based on real historical data, verifies the effectiveness of the proposed model and algorithm. The results demonstrate notable improvements in fleet efficiency and environmental performance, increasing profitability by 4.38% while maintaining favorable CII ratings. The findings provide valuable theoretical guidance for shipping companies navigating increasingly stringent CII regulations.

Keywords: tramp ship routing and scheduling; carbon intensity indicator; bi-objective optimization model; Gale–Shapley algorithm; genetic algorithm

1. Introduction

As an important mode of freight transportation, maritime shipping accounts for more than 80% of the global freight trade [1]. In 2022, the international seaborne trade volume reached 12,027 million tons, with a world fleet capacity of 2273 million deadweight tons [2]. According to Clarkson's statistics, in 2022, the international trade in dry bulk cargoes was 5252 million tons, making up about 50% of seaborne trade and occupying an important position in the shipping market. Additionally, because of its uncertainties, including uncertain cargoes, calling ports, routes, timetables, and freight rates, the schedule of tramp shipping is more complex compared with liner shipping and industrial shipping, and research on tramp ship scheduling remains relatively rare [3].

Although maritime shipping plays a crucial role in boosting world trade and connecting the global economy, it is also a major energy consumer, one of the main sources of greenhouse gas (GHG) emissions, accounting for 3% of total global GHG emissions [4]. With the rapid growth in seaborne trade volume, shipping CO_2 emissions have increased from 977 million tons in 2012 to 1056 million tons in 2018, and international shipping emissions are likely to increase by 40% based on the 2008 level by 2050 [4].

In response to climate change, the IMO has established GHG reduction targets. In 2018, the IMO adopted the "Initial Strategy on Reduction of GHG Emissions from Ships", which sets out quantified targets for emission reductions in the short, medium, and long term [5]. Each ship receives a CII rating annually based on its carbon emissions intensity for that year. CII is a metric that evaluates the carbon emissions intensity of a ship engaged in transport work. Ships with poor ratings may face challenges in obtaining a CII compliance statement. This could adversely affect their normal operations. This implies that ships need to strike a balance between increasing profits and reducing carbon emissions. There are several methods shipping companies can use to improve their ships' CII ratings, thereby avoiding incurring penalties. These methods can be divided into technical and operational aspects.

One effective technical measure is to switch from fossil fuel to alternative low-carbon or zero-carbon fuel, although high investment costs may make this a long-term goal [6]. Chemical absorption has been identified as the commercially viable mainstream approach despite challenges in high energy consumption, low regeneration efficiency, and spatial constraints. The development and application of carbon capture and storage (CCS) technology will provide more possibilities to achieve carbon reduction targets for the shipping industry [7]. The IMO lists several applications that can improve vessel energy efficiency, such as energy-saving devices (ESDs), engine power limitation (EPL), waste heat recovery systems (WHRS), solid-oxide fuel cells (SOFCs), and improvements to marine equipment, such as propellers and rudder blades [8–11]. Regular maintenance and timely inspections also help optimize mechanical performance and reduce energy consumption. Adland et al. [12] pointed out that periodic hull cleaning and dry-docking can significantly reduce fuel consumption. Different emission allowance allocation policies significantly influence shipowners' adoption of emission reduction technologies, with grandfathering effectively reducing emissions but potentially hindering advanced technology adoption, while non-free allocation, though cost-neutral, impedes low-carbon technology uptake, emphasizing the need for adaptive policymaking to optimize emission reductions in evolving market and technological landscapes [13,14]. Operational solutions are generally more cost-effective and easier to implement than most technical measures [15]. Controlling sailing speed and optimizing routes are key strategies for reducing carbon emissions. Slow steaming is widely used in commercial shipping to lower fuel consumption, but it could result in cargo delivery delays and reduced transport commitments, thereby impacting costs and reputation [16]. Wang et al. [17] developed a mixed-integer linear programming model to optimize ship deployment, sailing speed, and fuel selection, balancing cost minimization with compliance through the strategic use of traditional and renewable fuels in designated areas. These practices cut operational costs and align with sustainability goals.

In these circumstances, maximizing economic benefits while meeting the rating requirements of CII is a crucial concern in the current marine transportation industry. The main contribution of this paper is the bi-objective mixed-integer programming model, aiming to increase the profits of the entire fleet while simultaneously achieving high CII ratings. The model also includes self-owned ships, chartered ships, and cargo time windows. As vessel–cargo matching involves both shipowners and cargo owners, matching must be performed after considering and selecting options for both sides. Gale and Shapley [18] introduced a stable matching model that enables agents from both sides to achieve optimal pairing. Furthermore, the genetic algorithm (GA) has been widely recognized as an effective solution for vehicle routing problems [19]. This paper uses the Gale–Shapley algorithm for vessel–cargo matching, followed by iterative speed optimization using the GA to output the final fleet scheduling plan. The effectiveness of the method is demonstrated through a case study.

The remainder of this paper is organized as follows. Section 2 presents the literature review, detailing previous research and methodologies related to ship routing and scheduling, along with carbon intensity metrics in maritime operations. Section 3 describes the specific problem of tramp ship routing and scheduling with consideration of CII. We propose a mathematical optimization model and outline methods for solving this problem based on the Gale–Shapley algorithm and the GA. Section 5 provides a detailed case study to assess the applicability and effectiveness of the proposed model. Section 6 concludes this paper by summarizing the findings and suggesting potential directions for future research.

2. Literature Review

The optimization of ship routing and scheduling is a typical operations research problem and a complex dynamic issue. Carriers must allocate their own ships rationally based on shippers' transport demands and shipping capabilities, creating transport plans for cargo assignments. This study focuses on two aspects of the research literature: ship routing and scheduling and the integration of carbon emissions, including CII considerations.

Traditional ship routes and schedules are manually planned based on past experience. Recently, there has been in-depth research on this issue. Researchers have explored more sophisticated approaches dealing with carrier selection, many-to-many matching, and two-sided matching. These studies aim to optimize the allocation of vessels to cargoes, considering the preferences and constraints of shipowners and cargo owners. Since 1995, the dry bulk transportation market has faced an overcapacity of vessels, leading to a continuous decline in vessel freight rates in the shipping market. This situation prompted Kim and Lee [20] to develop a ship scheduling decision-making system based on a generalized elastic ship capacity segmentation model. The intention was to optimize the static one-stage decision-making optimization problem for a single voyage or a limited number of voyages, a single ship's scheduling, and a cargo-determined ship-model-based DSS in Ship Scheduling (MoDiss). On this basis, Fagerholt [21] established a corresponding model for tramp ship scheduling, using a hybrid heuristic algorithm, adding a search engine interface to help shipowners and cargo owners search in the system for the ships or cargoes of direct interest to them. The aims were to increase the chances of reaching a cooperative deal between the two parties, to reduce the voyage time of empty ships and the waiting time of ships in line at ports, to shorten the voyage mileage of empty ships, and to reduce the additional costs incurred by empty ships. Fagerholt and Christiansen [22] presented a bulk ship scheduling problem that is a combined multi-ship pickup and delivery problem with time windows (m-PDPTW) and a multi-allocation problem. Zeng et al. [23] developed an integer linear programming model to improve coal transportation efficiency and designed a two-stage heuristic tabu search algorithm to optimize fleet planning and route selection. Numerical experiments showed that the proposed method could reduce unit transportation costs and average ship delays and improve the reliability of the coal transport system.

Compared with liner shipping, tramp shipping needs to find cargo sources in a highly uncertain market. Bronmo et al. [24] considered variable cargo sizes and quantities, introducing flexible cargo sizes into tramp ship scheduling and routing problems. They proposed a realistic mixed-integer programming model and used a set partitioning method to solve multi-vessel loading and unloading problems with time windows and flexible cargo sizes, aiming to maximize profits. Lin and Liu [25] simultaneously addressed the problems of ship allocation, freight assignment, and ship routing in tramp shipping. They constructed sub-networks for transportation and navigation based on the port locations and transportation time of cargoes, transforming the ship routing problem into a network arc flow problem in operations research. Vilhelmsen et al. [26] combined the voyage separation requirements to minimize the time intervals between voyages, designed a dynamic programming algorithm, and created a time window branching scheme. Computational results showed that this algorithm converged better and could find optimal solutions more effectively compared with previous ship routing optimizations. In addition to the uncertainty of cargo demand, seasonal fluctuations and weather-related unseaworthy conditions are difficulties in tramp ship scheduling. Yu et al. [27] considered all the factors, constructed a mixed-integer programming model to maximize profits and proposed a heuristic method. Zhao et al. [28] proposed a robust multi-objective optimization model, integrating chance-constrained programming and NSGA-II to address voyage duration uncertainties for ore transportation, demonstrating enhanced robustness.

The sailing speed of ships is a crucial parameter in ship design and operation, determining the time window, affecting fuel consumption, and thereby determining emissions. Consequently, it considerably impacts shipping companies' ship scheduling and routing, operational efficiency, and profitability. In recent years, increasing concerns about carbon emissions and continuous fluctuations in fuel prices, mean that speed optimization has become a vital cost-control decision for shipping companies. Ma et al. [29] considered emission control areas (ECA) regulations and weather conditions, investigating the optimization of ship routes and speed solved by the Non-dominated Sorting Genetic Algorithm (NSGA-II), minimizing sailing cost and time. Carbon emissions are directly affected by fuel consumption. Wang et al. [30] considered factors including ship carbon emissions, as well as demurrage, dispatch in voyage chartering, and proposed speed decision models under three forms of carbon emission taxation. Addressing a multiple-ship scenario, Wen et al. [31] derived ship scheduling plans and optimal speeds under different objective functions, such as the shortest sailing time, minimum total cost, and lowest carbon emissions. As the IMO enforces increasingly stringent regulations on ship emissions, more scholars believe that a comprehensive analysis of ship emissions is a critical step in devising effective strategies to reduce air pollution. Weng et al. [32] estimated the emissions from vessels in the Yangtze River Estuary based on AIS data, examining the impact of vessel type, operational mode, emission equipment, time, and location on emissions. Based on noon report data, Yan et al. [33] proposed a two-stage model for predicting and reducing ship fuel consumption. In the first stage, a random forest regression model was developed to predict fuel consumption for dry bulk carriers under different conditions, such as sailing speed, cargo load, sea state, and weather. In the second stage, a mixed-integer programming model was built based on the prediction model to optimize sailing speed and minimize total fuel consumption over a voyage segment. Numerical experiments demonstrated that this model could effectively reduce fuel consumption by 2–7%, thereby reducing CO_2 emissions. Increasingly stringent regulations on ship energy-saving and emission-reduction technologies promoted Sun et al. [34] to consider wind-assisted devices such as rotors and sails to harness renewable wind energy. They proposed an improved ship weather routing optimization framework based on the A* algorithm, using ship design, weather forecasts, and historical navigation information to determine the optimal route and operation of ships equipped with wind-assisted rotors. Several improvements to the classic A* algorithm, including directional search and three-dimensional expansion, have been introduced to enhance optimization efficiency and effectiveness. Considering ship speed and payload, Li et al. [35] proposed a mixed-integer nonlinear programming model, introducing a discretized speed variable and employing a branch-and-price algorithm, the research successfully solves large-scale problem instances, demonstrating the model's effectiveness in reducing ship fuel consumption. Yuan et al. [36] proposed an optimization method that integrates speed and route planning to enhance the operational efficiency of ship fleets while ensuring compliance with CII regulations. Focusing on the Northeast Passage, Tsai et al. [37] found that the CII system could significantly influence future shipping routes as it aligns with global efforts to reduce carbon emissions in maritime shipping.

Although an increasing number of scholars have proposed various models and algorithms for optimizing ship routing and scheduling, much of the existing research focuses on maximizing economic benefits. While prior studies have incorporated sailing speed and carbon emissions into optimization models, few studies have taken the CII into account. This study addresses this gap by integrating the CII as both one of the objective functions and a constraint in the research framework. In contrast to conventional manual matching, this study introduces a two-sided matching mechanism that explicitly balances the interests of shipowners and cargo owners. Furthermore, rather than focusing on individual vessels, the research adopts a fleet-centric perspective, ensuring that each vessel remains profitable and meets carbon intensity standards, thereby achieving overall fleet optimization. In short, the main objective of this study is to develop a bi-objective optimization framework for tramp ship scheduling that simultaneously maximizes the fleet's overall profit and optimizes the CII rating, which can provide guidance for operational decision-making.

3. Mathematical Model

The tramp shipping market, characterized by the extensive presence of numerous carriers and shippers, exhibits a perfectly competitive market structure. Within this environment, the scheduling problem, taking carbon emission intensity into account, is a complex task that requires strategic consideration. The problem has two objectives: to maximize the total operational profit and to reduce the unit carbon emission. These objectives are often in conflict. In this section, we examine the specifics of the bi-objective optimization model, exploring the mathematical formulations and potential solutions for addressing tramp ship routing and scheduling, taking profit and CII objectives into consideration.

The first objective is to optimize the fleet schedule to increase the total operational profits. The shipping company must strategically allocate ships in a manner commensurate with cargo demands to optimize operational efficiency and enhance service quality. The second objective is to improve the CII rating. Sailing speed determines fuel consumption, fuel cost, and carbon emissions. However, lower energy consumption may reduce economic benefits. Therefore, when executing freight tasks, it is necessary to adjust the speed to maximize fleet profits while achieving optimal CII ratings for the vessels. Striking a balance between economic objectives and environmental responsibilities is a challenging prospect. Therefore, this paper focuses on improving the profit of a fleet and achieving a higher CII rating.

3.1. Problem Description

A voyage charter is commonly used in tramp shipping, particularly for meeting one-time or short-term cargo transportation needs. The flexibility of a voyage charter makes it a crucial option for shippers when dealing with volatile market conditions and fluctuating cargo demands. Shipping companies and cargo owners frequently enter into contracts of affreightment (COAs), which are long-term agreements between the carrier and shipper. These contracts provide both parties with enhanced certainty and stability, facilitating more predictable and secure operational planning over extended periods. A COA stipulates the carrier's commitment to transport specified quantities of cargo between designated ports for the shipper. The specific details regarding the routes, time window, and freight rate are usually agreed upon in the contract. For carriers, COAs provide a steady source of cargo volume, allowing for better planning and utilization of the fleet. Shippers can secure transportation capacity at a known cost, thereby reducing exposure to freight rate volatility.

Let us take the example of a shipping company operating a heterogeneous fleet of ships consisting of dry bulk carriers with specific ship characteristics, including ship capacity, cost structure, and original location. As a carrier, the shipping company is obligated to execute designated transportation tasks within the planning period. The company must carefully assign vessels to each shipment, taking into account cargo volume, destination, and the time windows for loading and unloading. Several types of dry bulk cargo are scheduled for transport, including COA cargo and spot cargo. Each vessel within the fleet is limited to carrying a single shipment per voyage, precluding the possibility of loading or unloading additional cargo en route. If the capacity of the fleet is insufficient to handle all cargoes during the planning horizon, additional vessels may be chartered on the spot market to meet transportation demand. The revenue generated from these shipments is determined by the quantity of cargo and the negotiated freight rate.

3.2. Model Assumptions

To facilitate the process of calculation, the following assumptions are made:

- The tramp ships in the fleet are heterogeneous in terms of size, loading capacity, cruising speed, draft, fuel efficiency, and some other indicators;
- Each ship can only transport one cargo on one voyage, and each cargo can only be transported by one ship on one voyage; there are no split loads or trans-shipments;
- The shipping company can choose its own ships or charter ships on the spot market if necessary;
- There are no unforeseen delays due to weather conditions or port congestion. Details of each cargo demand, including its specific weight, loading port, discharging port, and time window, are fixed and known in advance;
- Ships sail at a specific constant speed on each leg;
- Ships have enough fuel to reach their destinations, and fuel prices do not change over time;
- Carriers and shippers are rational.

3.3. Model Formulation

3.3.1. Cost Calculation

When making informed decisions regarding freight assignments, cost is one of the key factors that needs to be considered. Because tramp ships do not have fixed sailing routes and schedules, each voyage has different costs and revenues based on known information, such as cargo quantities, ports of call, sailing distances, and ship parameters. In addition, if the current fleet capacity is insufficient to meet all transportation needs, there will be extra expenses for chartering ships on the spot market. The freight assignments must be carefully analyzed and calculated to ensure higher profits.

Voyage costs are mainly made up of two components: fuel cost and port cost. The specific calculation methods are as follows:

Fuel Cost:

The daily fuel consumption of the ship is given by a function $f(v,w)(tons/day)$. We use the realistic closed-form solution of Psaraftis and Kontovas [38].

$$f(v,w) = \mu v_{ijk}^3 (w_k + A_k)^{\frac{2}{3}} \tag{1}$$

where v is the ship speed, w is the ship payload, A is the lightship weight, and μ is a constant.

Let p_f be heavy fuel price. The shipping cost for ship k between nodes i and j can be defined as

$$C_{ijk}^s = p_f \mu v_{ijk}^3 (w_k + A_k)^{\frac{2}{3}} \frac{d_{ijk}}{24 v_{ijk}} \tag{2}$$

Port Cost:

During operation and berthing at port, ships will load and discharge cargoes, inevitably queuing at anchorages or ports, incurring fuel costs at port C_{ik}^{pf}:

$$C_{ik}^{pf} = p_f \gamma \{t_{ik} + \max[(T_{i\min} - T_{ik}), 0]\} \tag{3}$$

where γ is the fuel consumption of the ship at port(ton/h), t_{ik} is the total berth time at node i for ship k, $T_{i\min}$ and T_{ik} denote the real times for arriving at node i for ship k and the earliest arrival time for the ship at node i expected by the shipper, respectively.

Ships also need to use channels, berths, docks, and anchorages and apply for various services provided by the port authority, such as pilotage and tug assistance. Port disbursement is a crucial component of the voyage cost for tramp ships. It consists of port charges, stevedoring charges, cleaning holds, and other items. In this paper, let P_{ik} be the port disbursement, and let C_{ik}^{p} denote the port cost of ship k at node i.

$$C_{ik}^{p} = p_f \gamma \{t_{ik} + \max[(T_{i\min} - T_{ik}), 0]\} + P_{ik} \tag{4}$$

3.3.2. CII Calculation Method

According to the IMO, the CII determines the annual reduction factor needed to ensure continuous improvement of a ship's operational carbon intensity within a specific rating level. The actual annual operational CII achieved must be documented and verified against the required annual operational CII. This enables the operational carbon intensity rating to be determined.

A ship rated D for three consecutive years, or E for one year, has to submit a corrective action plan to show how the required index of C or above will be achieved [39].

According to the IMO, the steps for CII rating of ships are as follows [40–44]:

1. Calculation of CII (G1 CII calculation methods)

The attained annual operational CII is a measure used to evaluate the efficiency of a ship's CO_2 emissions over a certain period, typically on an annual basis. This is calculated by dividing the total amount of CO_2 emitted by the ship (M) by the transport work completed (W):

$$Attained\ CII = \frac{M}{W} = \frac{\sum_j FC_j \times C_{Fj}}{Capacity \times D_t} \tag{5}$$

where M is the sum of CO_2 emissions, j is the fuel type, FC_j is the total mass(in grams) fuel consumption of type j, C_{Fj} represents the fuel oil mass of type j to CO_2 mass conversion factor, D_t is the total distance traveled (in nautical miles), and W is the transport work completed by the ship, defined as the product of the ship's capacity and D_t. For bulk carriers, deadweight tonnage should be used as $Capacity$.

2. Reference Lines for CII (G2 reference lines guideline)

The 2023 IMO Strategy on Reduction of GHG Emissions from Ships set a goal of reducing the CO_2 emission intensity of shipping by 40% relative to 2008. Because of the unavailability of data in 2008, the IMO decided to base the development on fuel consumption data of operating ships in 2019. The reference guideline of CII, CII_{ref}, is a function of capacity. Equation (6) indicates the computation of CII_{ref}.

$$CII_{ref} = aCapacity^{-c} \tag{6}$$

G2 gives the values of parameters a and c for each ship type. In this paper, for bulk carriers, a equals 4745 and c equals 0.622.

3. Operational Carbon Intensity Reduction Factors Relative to Reference Lines (G3 reduction factors)

The required CII can be calculated using CII_{ref} and reduction factor Z. The annual operational carbon intensity reduction rate is defined as the percentage by which the required annual operational CII value for a given year is lower than the baseline value.

$$Required\ CII = \left(1 - \frac{Z}{100}\right) \times CII_{ref} \tag{7}$$

Z is determined as 5, 7, 9, and 11 from 2023 to 2026, respectively. The values for the years beyond 2026 will be further studied and strengthened.

4. Operational Carbon Intensity Rating of Ships (G4 rating guidelines)

The CII rating is determined by comparing the Attained CII with the Required CII. It can be classified into one of the five performance levels (A, B, C, D, E), representing excellent, good, average, fair, or unsatisfactory, respectively. There are four boundaries to separate the levels: superior, lower, upper, and inferior. The formulas for the four boundary values are as follows:

$$\begin{aligned} \text{Superior boundary} &= \exp(d_1) \times \text{Required CII} \\ \text{Lower boundary} &= \exp(d_2) \times \text{Required CII} \\ \text{Upper boundary} &= \exp(d_3) \times \text{Required CII} \\ \text{Inferior boundary} &= \exp(d_4) \times \text{Required CII} \end{aligned} \tag{8}$$

The value of $\exp(d_i)$ is different for each ship type. For bulk carriers,

$$\begin{aligned} \text{Superior boundary} &= 0.86 \times \text{Required CII} \\ \text{Lower boundary} &= 0.94 \times \text{Required CII} \\ \text{Upper boundary} &= 1.06 \times \text{Required CII} \\ \text{Inferior boundary} &= 1.18 \times \text{Required CII} \end{aligned} \tag{9}$$

The ship's rating is determined by the range of its Attained CII value for the corresponding year, as Figure 1 shows.

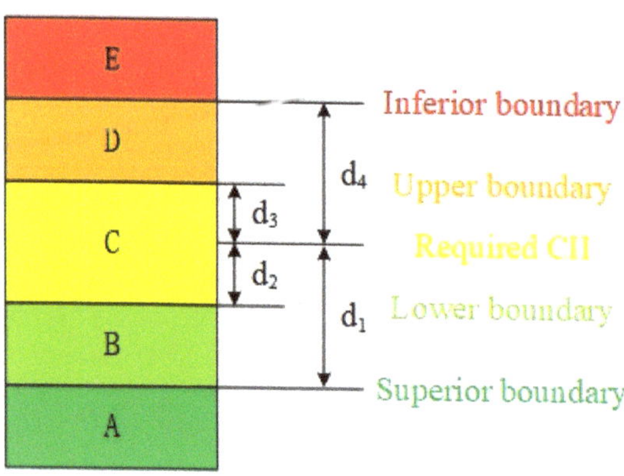

Figure 1. Schematic diagram of the CII ratings and boundaries.

3.3.3. Optimization Model

This section presents the mathematical formulation of the optimization model for the tramp ship scheduling problem considering CII rating issues.

Tables 1–3 show the notations used in the model.

Table 1. Indices and sets.

Sets	Description
N	The set of loading ports, discharge ports, and the initial position of each ship in the fleet.
K	The set of ships, each element is denoted as k. $K = \{1, 2, \ldots, k, \ldots, m\}$.
I	The set of cargoes, each element is denoted as i. $I = \{1, 2, \ldots, i, \ldots, n\}$.
N_L	The set of loading ports associated with cargo i is a loading port i. $N_L = \{1, 2, 3 \ldots, n\}$.
N_D	The set of discharge ports associated with cargo i is a discharge port $n + i$. $N_D = \{n+1, n+2, n+3 \ldots, 2n\}$.
N_k	The set of nodes that can be visited by ship k.
N_k^L	The set of loading ports that ship k may visit.
N_k^D	The set of discharge ports that ship k may visit.
Ω_k	The set of arcs that are traversable by ship k. $\Omega_k = \{(i, j) : i, j \in N_L \cup N_D, i \neq j\}$.

Table 2. Parameters.

Parameters	Description
$o(k)$	The origin node of the ship k.
$d(k)$	The artificial destination node of ship k.
Q_k	The load capacity of ship k (tons).
D_i	The total quantity of cargo i (tons).
q_i	The quantity of cargo i that has been actually transported (tons).
t_{ik}	The total berth time at node i for ship k, including loading and discharging time, waiting for service time, etc (days).
T_{ik}	The time for arrival node i for ship k (days).
$T_{i\min}$	The earliest arrival time for the ship at node i expected by the shipper.
$T_{i\max}$	The latest time for the start of service at node i.
$[T_{i\min}, T_{i\max}]$	The expected time window associated with node i.
d_{ij}	The sailing distance from node i to node j (n miles).
γ	The amount of fuel consumption when the ship is at port (tons/hour)

Table 3. Decision variables.

Decision Variables	Description
w_{ik}	The exact load on board at node i.
x_{ijk}	If ship k sails from node i to node j, $x_{ijk} = 1$; otherwise, $x_{ijk} = 0$.
y_i	If cargo i is transported by a charter ship in the spot market, $y_i = 1$; otherwise, $y_i = 0$.
v_{ijk}	The sailing speed of ship k from node i to node j.

Based on these, the model can be stated as follows:

The objective functions are as follows:

$$\max \sum_{i \in I} R_i q_i - \sum_{k \in K} \sum_{(i.j) \in \Omega_k} x_{ijk} C_{ijk}^s - \sum_{k \in K} \sum_{(i.j) \in \Omega_k} x_{ijk} C_{ik}^p - \sum_{i \in I} y_i C_i \tag{10}$$

$$\min AttainedCII = \frac{\sum_j FC_j \times C_{Fj}}{C \times D_t} \tag{11}$$

The constraint conditions are as follows:

$$\sum_{k \in K} \sum_{j \in N_k} x_{ijk} + y_i = 1, \forall i \in I \tag{12}$$

$$\sum_{k \in K} x_{i(n+i)k} \leq 1, \forall i \in I \tag{13}$$

$$\sum_{i \in I} x_{ijk} \leq 1, \ \forall k \in K, \ j \in N_k \tag{14}$$

$$x_{ijk} = 0, \forall k \in K, i \in N_k^L, j \in N_k^L \tag{15}$$

$$x_{ijk} = 0, \forall k \in K, i \in N_k^D, j \in N_k^D \tag{16}$$

$$\sum_{j \in N_k^L \cup \{d(k)\}} x_{o(k)jk} = 1, \ \forall k \in K \tag{17}$$

$$\sum_{j \in N_k^D \cup \{o(k)\}} x_{id(k)k} = 1, \ \forall k \in K \tag{18}$$

$$\sum_{j \in N_k} x_{ijk} - \sum_{j \in N_k} x_{jik} = 0, \forall k \in K, i \in N_k \backslash \{o(k), d(k)\} \tag{19}$$

$$\sum_{j \in N_k} x_{ijk} - \sum_{j \in N_k} x_{j(n+i)k} = 0, \forall k \in K, i \in N_k^L \tag{20}$$

$$x_{i(n+i)k} w_{(n+i)k} = 0, \forall k \in K, i \in N_k^L, (i, n+i) \in \Omega_k \tag{21}$$

$$x_{(n+i)jk}\left(q_j - w_{jk}\right) = 0, \forall k \in K, i, j \in N_k^L, (n+i, j) \in \Omega_k \tag{22}$$

$$\sum_{j \in N_k} x_{ijk} D_i (1-\mu) \le q_i \le \sum_{j \in N_k} x_{ijk} D_i (1+\mu), \forall i \in N_k^L \tag{23}$$

$$0 \le w_{ik} \le Q_k, \forall k \in K, i \in N_k \tag{24}$$

$$x_{ijk}\left[\max(T_{ik}, T_{i\min}) + t_{ik} + \frac{d_{ij}}{24 v_{ijk}} - T_{jk}\right] = 0, \forall k \in K, (i,j) \in \Omega_k \tag{25}$$

$$\max(T_{ik}, T_{i\min}) + t_{ik} + \frac{d_{i(n+i)}}{24 v_{i(n+i)k}} = T_{(n+i)k}, \forall k \in K, (i,j) \in \Omega_k \tag{26}$$

$$T_{i\min} \le T_{ik} \le T_{i\max}, \forall k \in K, i \in N_k \tag{27}$$

$$\frac{AttainedCII}{RequiredCII} \le 1.06 \tag{28}$$

$$\underline{v_{bk}} \le v_{ijk} \le \overline{v_{bk}}, \forall k \in K, i \in N_k \backslash N_k^L, j \in N_k^L \tag{29}$$

$$\underline{v_{lk}} \le v_{i(n+i)k} \le \overline{v_{lk}}, \forall k \in K, i \in N_k^L \tag{30}$$

$$x_{ijk} \in \{0,1\}, \forall k \in K, (i,j) \in \Omega_k \tag{31}$$

$$y_i \in \{0,1\}, \forall i \in I \tag{32}$$

Objective (10) is to maximize the total profit of the fleet over the planning period. The total cost includes fuel cost, port cost, and charter cost.

Objective (11) is to minimize Attained CII, which means optimizing the CII rating.

Constraint (12) guarantees that each consignment should be transported by a ship in the fleet or by a charter ship.

Constraints (13) and (14) indicate that each consignment should be shipped in the whole ship transport and cannot be mixed.

Constraints (15) and (16) indicate that the ship cannot consecutively dock at two loading ports or two discharging ports, respectively.

Constraint (17) ensures the ship leaves its initial position only once.

Constraint (18) ensures that the ship arrives at the virtual destination only once.

Constraints (19) and (20) describe the flow of the sailing route for ship k. Constraint (19) indicates that the number of ships arriving and leaving any port is the same. Constraint (20) ensures the consistency of the scheduling plan for ship k when transporting cargo i, ship k must visit loading node i before discharging at node $n+i$.

Constraints (21) and (22) describe the relationship between the quantity of cargo i on board and loading and discharging ports, respectively.

Constraint (23) indicates that the actual quantity of cargo i on board should be within the specified range.

Constraint (24) ensures that the load on board cannot exceed the capacity of ship k.

Constraint (25) states the relationship between the sailing route and the sailing time.

Constraint (26) indicates that the discharging node for cargo i must be visited after its loading node.

Constraint (27) ensures that the ship must meet the time window of cargo i.

Constraint (28) ensures that the CII rating can be C and above.

Constraints (29) and (30) are the lower and upper bounds for ballast and loading speed variables, respectively.

Constraints (31) and (32) are binary restrictions.

4. Algorithm Design

In this paper, the Gale–Shapley algorithm is used to solve the stable vessel–cargo matching problem. Shipowners evaluate available cargoes, and cargo owners compile preference lists for vessels, leading to multiple vessel–cargo matches. The one-to-one matching result is obtained by using the dynamic bilateral matching of ship and cargo. Then, through the iteration of the GA, ship scheduling and sailing speed are optimized, resulting in the fleet's optimal transportation scheduling scheme.

4.1. Gale–Shapley Algorithm

The Gale–Shapley algorithm, also known as the deferred-acceptance algorithm and originally developed for the stable marriage problem, is a method used to solve the problem of finding stable matching between sets of elements [18]. In vessel–cargo matching, considering attributes such as cargo type, volume, port specification, and time window, each vessel has a ranking of cargoes based on their preferences, and each cargo ranks vessels similarly. Vessels propose to their most preferred cargo or to one that prefers them over their current match. Cargoes can reject proposals except for those from their most preferred vessel. This process continues until no vessel or cargo would benefit from switching partners, achieving stable matching where all preferences are optimally aligned. This algorithm can be applied to efficiently match vessels with available cargo in a way that maximizes overall efficiency, satisfaction, and profitability.

When applying the Gale–Shapley algorithm to the problem of vessel–cargo matching, the steps can be summarized as follows:

1. Initialize vessel data and cargo data.
2. For each vessel, call the vessel-to-cargo matching algorithm to obtain the result set of matches sorted by voyages; this could be the preference list ranking the potential matches.
3. Extract the optimal solution for each vessel to form an initial array of solutions.
4. Extract the part of the initial solution that does not have conflicting cargoes and save it to the final result set.
5. Extract the cargo involved to eliminate unlikely candidates before the next round of the cycle.
6. Remove from backward to forward the initial solution that has entered the final set of matching results.
7. Rank the corresponding solutions according to their profits for the optimal set of outcomes where multiple ships are present simultaneously for one cargo.

8. Add the conflicting scenario with the best net benefit to the final matching result and list the other conflicting scenarios as deletions.
9. Loop steps 5, 6, 7, and 8 until the initial solution is empty.
10. For the set of vessels that have not yet been matched, a new initial result set is formed by selecting the suboptimal solution with no cargo conflicts based on the gains from the initial matching results.
11. Output the final stable matching scheme.

4.2. Genetic Algorithm

GAs are adaptive search heuristics that mimic the process of natural selection and genetic recombination. They are used to generate optimal solutions to optimization and search problems by iteratively evolving a potential population solution continuously optimizing the fitness of individuals. In GA, each potential solution is represented as an individual, called a chromosome, within a population of possible solutions. The algorithm starts with a randomly generated initial population. Each individual is then evaluated based on a fitness function, which assigns a fitness value based on how well the individual satisfies the objective of the optimization problem. Operations such as selection, crossover, and mutation are applied to evolve and improve these solutions iteratively. In short, they solve problems by simulating the survival of the fittest among individuals of consecutive generations.

When applied to ship routing and scheduling problems, GAs are effective because of their ability to handle large search spaces and multiple constraints efficiently. To solve the bi-objective optimization model proposed in this paper, considering the total profit of the fleet and the CII, the GA iteratively improves the allocation of cargo to vessels while simultaneously optimizing the speed of each sailing leg.

The specific steps of the GA are as follows:

1. Initialization:

Generate a set of chromosomes to form an initial population of solutions randomly or by using heuristics. Each individual in the population represents a potential scheduling solution, including a sequence of ports to visit, the allocation of cargo to ships, and sailing speed.

2. Evaluation:

Develop a fitness function to calculate and evaluate the fitness value of each individual in the population, potentially reflecting the performance of each scheme. The fitness function should be based on multiple criteria. In this paper, the values of the fleet profit and Attained CII are deformed and weighted as the fitness function. We have assigned them equal weights (1:1) in the fitness function to reflect their operational parity.

3. Selection:

Select chromosomes with higher fitness values from the current generation to be parents for the next generation. There are several selection methods, such as roulette wheel selection, tournament selection, or rank selection. In this paper, roulette wheel selection is employed.

4. Crossover:

Crossover is a genetic operator used to combine the genes of parents to generate new offspring. This could involve swapping segments of two different routes to create a new one.

5. Mutation:

Introduce random mutations to some chromosomes to maintain the diversity of the population, thereby avoiding becoming stuck in local optima.

6. Replacement:

Repeat steps 2 to 5 to generate the new offspring, which will replace some individuals with lower fitness scores in the previous population, ensuring the population size remains constant. This can help preserve diversity and improve scheduling schemes.

7. Termination:

Set an appropriate termination condition, such as reaching a preset maximum number of iterations, reaching a satisfactory solution, and running out of computational resources. Repeat the selection, crossover, mutation, and replacement steps until the termination condition is met. The algorithm halts and outputs the optimal scheme for the current population.

5. Computational Study

A computational study was conducted in this section to validate the effectiveness of the proposed mathematical model and evaluate the performance of the algorithms for considering carbon intensity in the tramp ship scheduling problem. We applied the proposed model and algorithm to a historical instance, comparing the optimal solution with actual operational data from past tramp ship voyages for a comprehensive evaluation. This study used Python 3.10 to simulate a dry bulk shipping fleet, thereby providing a robust framework for analyzing performance and efficiency under real-world conditions.

5.1. Data Source and Description

To conduct a comparative analysis with the optimized results, this study constructs an empirical analysis case based on accessible historical operational data, selecting actual freight data from a specific period in China's coastal region as the research sample. Data for the computational study were sourced from AIS and COSCO Shipping Technology Co., Ltd. The dataset mainly covers details on 30 cargo bookings along the coastal regions of China and information on a fleet of 15 bulk carriers transporting dry bulk commodities, such as iron ore, coal, and grain.

The fleet is composed of 15 ships with varying specifications, as Table 4 shows. We numbered these 15 ships from 1 to 15 as Ship ID. The second column, MMSI, is the acronym for Maritime Mobile Service Identity; it is a unique nine-digit numerical assigned to a vessel for digital identification and communication in maritime tracking systems to ensure safety, navigation, and emergency coordination. Deadweight tonnage is the maximum weight a ship can safely carry, including cargo, fuel, ballast water, crew, and supplies, while maintaining its designated draft and structural integrity. Lightship weight is the total weight of a ship in its empty operational state, including its hull, machinery, and permanent equipment but excluding all cargo, fuel, crew, and consumable supplies. The Min speed and Max speed columns define the lower and upper bounds of the vessel's navigational speed. The initial location column refers to the vessel's location at the start of the planning period, expressed in longitude and latitude coordinates. Positive values denote east longitude or north latitude, while negative values indicate west longitude or south latitude. The last two columns represent the daily fuel consumption of the vessel during port anchoring and berthing.

Table 5 provides information on transportation volumes, loading and discharging ports, time windows, and other cargo information. We numbered these 30 cargos from 1 to 30 as Cargo ID. The quantity column refers to the specific shipment volume assigned to each task. Columns 3 and 4 represent the loading port and discharging port for each freight task, respectively. The last column indicates the lay days of each shipment; the vessel must arrive at the designated port and be ready to load within the period. Table 6 provides full names corresponding to the port code used in Table 5.

Table 4. Information on 15 ships.

Ship ID	MMSI	Deadweight Tonnage (ton)	Lightship Weight (ton)	Min Speed (kn)	Max Speed (kn)	Initial Location		Fuel Consumption (ton/day)	
						Longitude (°)	Latitude (°)	Anchoring	Berthing
1	354643000	76,546	40,896	10.239	15.359	4.049	51.937	6.114	2.867
2	354879000	55,676	30,953	9.888	14.900	113.486	23.082	5.080	3.099
3	356760000	56,907	33,044	10.765	16.084	73.907	7.153	7.030	4.120
4	412343000	55,566	30,963	9.888	14.900	127.735	34.907	4.408	2.995
5	412523920	32,034	19,995	11.510	17.201	140.670	35.922	4.553	2.998
6	412775000	57,617	33,511	11.063	16.594	123.337	34.771	5.121	1.987
7	413219590	52,818	29,995	10.481	15.660	102.003	2.170	6.396	3.271
8	414139000	64,542	38,384	12.529	18.793	118.012	38.386	6.719	2.892
9	414166000	49,998	31,893	11.312	16.968	118.455	38.987	5.625	2.582
10	414775000	49,256	28,714	10.981	16.407	122.471	31.273	5.514	2.510
11	414783000	49,281	28,714	10.981	16.407	118.188	38.950	5.232	2.454
12	477043400	63,261	36,378	10.732	16.189	121.374	31.501	6.610	3.423
13	477595900	76,364	40,896	10.239	15.359	−43.930	−23.128	5.185	2.698
14	477961500	56,687	32,899	10.765	16.084	29.458	40.763	5.976	2.785
15	563023600	52,395	30,053	8.117	13.529	171.427	41.853	7.411	3.836

Source: Compiled by the authors from data provided by COSCO Shipping Technology Co., Ltd.

Table 5. Information on cargoes.

Cargo ID	Quantity (ton)	Loading Port	Discharging Port	LAN/CAN
1	49,000	CNQHD	CNZHE	1–3 July 2022
2	47,500	CNQHD	CNZJG	4–6 July 2022
3	57,000	CNQHD	CNNBO	4–6 July 2022
4	58,500	CNTJN	CNZHE	5–7 July 2022
5	76,550	CNHUA	CNNJG	5–7 July 2022
6	44,500	CNJIT	CNZJG	6–8 July 2022
7	59,000	CNHUA	CNDAL	6–8 July 2022
8	47,000	CNQHD	CNYZH	6–8 July 2022
9	57,200	CNQHD	CNGUA	6–8 July 2022
10	46,600	CNJIT	CNZJG	6–8 July 2022
11	46,600	CNJIT	CNZJG	6–8 July 2022
12	47,500	CNZHA	CNJIA	7–9 July 2022
13	47,000	CNCFD	CNNJG	7–9 July 2022
14	43,500	CNJIT	CNZHE	8–10 July 2022
15	53,000	CNJIT	CNTZH	8–10 July 2022
16	47,000	CNHUA	CNNJG	8–10 July 2022
17	70,500	CNQHD	CNGUA	9–11 July 2022
18	70,500	CNQHD	CNGUA	9–11 July 2022
19	47,000	CNCFD	CNYZH	10–12 July 2022
20	50,000	CNJIT	CNZJG	10–12 July 2022
21	47,800	CNCFD	CNYZH	10–12 July 2022
22	47,000	CNHUA	CNYZH	10–12 July 2022
23	45,000	CNZHA	CNJIJ	10–12 July 2022
24	65,000	CNHUA	CNZHA	11–13 July 2022
25	57,500	CNCFD	CNGUA	12–14 July 2022
26	49,300	CNJIT	CNZJG	12–14 July 2022
27	58,300	CNJIT	CNZJG	12–14 July 2022
28	70,000	CNHUA	CNMEZ	12–14 July 2022
29	69,000	CNCFD	CNGUA	12–14 July 2022
30	46,000	CNJIT	CNNJG	15–17 July 2022

Source: Compiled by the authors from data provided by COSCO Shipping Technology Co., Ltd.

Table 6. Port code cross-reference table.

Port Code	Port Name
CNCFD	Caofeidian Port
CNDAL	Dalian Port
CNGUA	Guangzhou Port
CNHUA	Huangpu Port
CNJIA	Jiangyin Port
CNJIJ	Jingjiang Port
CNJIT	Jingtang Port
CNMEZ	Meizhou Port
CNNBO	Ningbo Port
CNNJG	Nanjing Port
CNQHD	Qinhuangdao Port
CNTJN	Tianjin Port
CNTZH	Taizhou Port
CNYZH	Yangzhou Port
CNZHA	Zhanjiang Port
CNZHE	Zhenjiang Port
CNZJG	Zhangjiagang Port

The freight rate in this paper is based on the freight rate of several important coal shipping routes along China's coast over the past three months. The fuel price is $589.6/ton.

5.2. Computational Results and Analysis

Cargo owners compile preference lists for ships, and shipowners evaluate available cargoes, leading to multiple cargo–vessel matches. Transitioning from one-to-many matching, where one party has multiple options, to many-to-many matching, where both parties have several options, requires reconciling potential conflicts. A comprehensive assessment of preference ranking lists for shipowners and cargo owners is crucial for a stable, mutually beneficial outcome. First, the original dataset is preprocessed to generate a preference list based on the characteristics of the ships and cargoes. Second, this list is inputted into the model for computational analysis. We leverage the Gale–Shapley algorithm to output the final stable matching scheme, as Figure 2 shows.

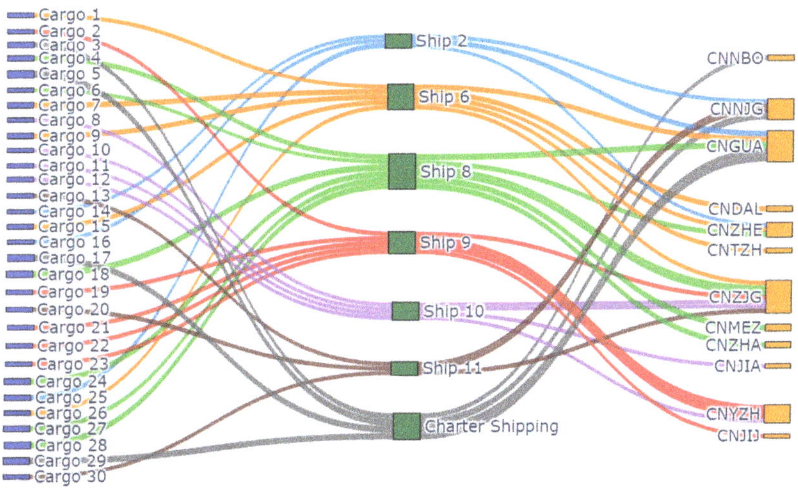

Figure 2. The Result of Vessel—Cargo Matching.

In Figure 2, the left column represents the cargo, the width of the blocks, and the thickness of the connecting lines indicate the volume of the cargo; thicker lines denote a larger volume of the respective cargo. Each shipment is linked to the matching vessel; the right column represents the destination ports of the cargo. Upon reviewing the calculation results, it was consistently observed that not all self-owned ships in the fleet were assigned to cargo transport tasks. Additionally, a few of the 30 shipments required the chartering of external ships. Specifically, Ships 1, 3, 4, 5, 7, 12, 13, 14, and 15 have no cargo to transport; Ship 2 transports Cargoes 14, 16, and 25; Ship 6 transports Cargoes 1, 7, 9, 15, and 26; Ship 8 transports Cargoes 4, 6, 18, 24, 27, and 28; Ship 9 transports Cargoes 2, 19, 21, 22, and 23; Ship 10 transports Cargoes 8, 10, 11, and 12; Ship 11 transports Cargoes 13, 20, and 30. Unfortunately, there is no suitable ship for Cargoes 3, 5, 17, and 29.

Factors contributing to these outcomes include the following:

- A mismatch between ship capabilities and cargo requirements: The type, capacity, or other characteristics of some ships may not align with cargo needs.
- Distance to departure ports: Certain ships were too far from the departure ports to meet the time windows, making chartering ships necessary to avoid time constraint violations.
- Impact on CII ratings: Some voyages may increase unit carbon emissions, thereby lowering the CII rating below acceptable levels.
- Flexibility of chartering on the spot market: Charter shipping possesses considerable flexibility and enables rapid adjustment of shipping capacity and routes. In some cases, charter shipping can be more economical and efficient.

The results indicate that ships in the fleet have been partially laid up, while several cargoes are proposed to be transported by chartered ships. The ships assigned to transportation tasks have been renumbered for clarity, with the original Ship 2 now designated as 001, Ship 6 as 002, Ship 8 as 003, Ship 9 as 004, Ship 10 as 005, and Ship 11 as 006.

Based on the vessel–cargo matching results, we take Ship 2 as an example to evaluate the performance of the Genetic Algorithm (GA), Simulated Annealing (SA), and Particle Swarm Optimization (PSO) for optimizing profit and CII rating. The comparative outcomes are presented in Table 7.

Table 7. Performance comparison of GA, SA, and PSO.

Metric	**Genetic Algorithm (GA)**	**Simulated Annealing (SA)**	**Particle Swarm Optimization (PSO)**
Profit (USD)	205,001.19	183,200.45 (−10.6%)	192,450.70 (−6.1%)
CII Rating	A (0.69)	A (0.82)	C (0.95)
Convergence Iterations	430	680	590
Speed Optimization Strategy	Dynamic speed adjustment	Single-speed optimization	Locally Optimal Speed Combinations
Time Window Satisfaction Rate	100%	92%	88%

As summarized in Table 7, the GA outperforms SA and PSO in economic efficiency, environmental performance, and computational effectiveness.

- Economic efficiency: The profit achieved by GA is $205,001.19, exhibiting 10.6% and 6.1% improvements over SA ($183,200.45) and PSO ($192,450.70), respectively. By dynamically optimizing sailing speeds, GA balances fuel costs and transportation efficiency. In contrast, SA and PSO fail to identify superior speed combinations because of insufficient local search capabilities.

- CII rating: GA achieves an A-level CII rating with the Attained CII to Required CII ratio of 0.69, significantly better than the A rating (0.82) of SA and the C rating (0.95) of PSO. The speed strategies of SA and PSO result in higher carbon emissions per transport work, failing to meet the IMO's stringent standards.
- Convergence Efficiency and Global Search Capability: GA converges after 430 iterations, which is faster than PSO (590 iterations), while also delivering superior solution quality. SA requires 680 iterations because of its conservative cooling strategy and is prone to local optima. GA's crossover and mutation mechanisms effectively maintain population diversity, preventing premature convergence.
- Practical Constraint Satisfaction: The transportation schedules optimized by GA achieve full compliance with time window constraints (100%), whereas SA and PSO exhibit lower compliance rates of 92% and 88%, respectively, which significantly escalates operational risks.

This validation confirms the superior performance of GA in addressing complex ship scheduling challenges, attributed to its global search capabilities and effective multi-objective optimization strategies.

Based on the aforementioned results, the final fleet routing and scheduling plan is determined by the GA. We set the maximum number of iterations as 1000, the crossover probability as 0.8, and the mutation probability as 0.003. Table 8 shows the scheme and calculation results.

Table 8. Optimized tramp ship routing and scheduling scheme.

Ship	MMSI	Route	Attained CII/ Required CII	CII Rating	Profit ($)	Speed (Knot)
001	354879000	14-16-25	0.69	A	205,001.19	[11.86, 14.32]-[14.21, 10.10]-[12.28, 10.72]
002	412775000	1-15-7-26-9	1.05	C	429,955.47	[14.36, 16.49]-[14.71, 16.55]-[13.88, 13.75]-[11.94, 15.04]-[15.61, 14.69]
003	414139000	6-27-4-24-18-28	0.90	B	259,211.12	[16.94, 13.64]-[14.90, 13.97]-[17.36, 17.85]-[18.50, 18.17]-[12.82, 15.39]-[14.18, 15.58]
004	414166000	2-21-19-22-23	0.87	B	290,509.53	[14.30, 12.34]-[11.88, 13.18]-[13.09, 13.14]-[15.46, 16.44]-[14.03, 14.62]
005	414775000	8-10-11-12	0.79	A	294,967.22	[16.17, 11.91]-[14.92, 11.42]-[11.57, 12.15]-[13.06, 13.25]
006	414783000	13-30-20	0.90	B	220,333.49	[15.57, 11.21]-[13.26, 13.74]-[14.84, 13.68]
C	-	3-5-17-29	-	-	-	-

In Table 8, the first column is the updated ship number. The Route column specifies the cargo assignment and transport sequence for each ship. For example, Ship 001 is scheduled to first load Cargo 14 and transport it to the destination port, followed by Cargo 16, and finally, to complete its tasks by delivering Cargo 25. The Attained CII/Required CII column represents the ratio of the Attained CII to the Required CII. The CII Rating column provides the ship's CII rating result after optimizing. The Profit column reflects the net profit for each ship within the planning period, calculated as the difference between the revenue earned from completing transportation tasks and the associated operational costs. The Speed column shows the sailing speed of each ship in each leg. In this context, we refer to both the voyage for picking up cargo and delivering it as a leg. Similarly, using Ship 001 as an example, the optimized results show that the ship sails at a speed of 11.86 knots from its initial position to the loading port of Cargo 14 and at 14.32 knots to the discharge port. The sailing speed of the leg to pick up Cargo 16 is 14.21 knots, while the loaded transport speed is 10.10 knots. From the discharge port of Cargo 16 to the next Cargo 25, the speeds are 12.28 knots and 10.72 knots for the respective legs.

Lowering sailing speed is a prevalent strategy for cutting fuel consumption and enhancing energy efficiency, thereby directly lowering the numerator in the CII calculation formula. Some carriers also curtail fuel expenditure by idling their ships or selecting shorter routes; this is indeed the case with Ship 004 and Ship 005. Figure 3 illustrates the optimized shipping routes of these two vessels, where green and orange lines denote CII ratings of A and B, respectively. Ship 004 has completed five shipments and achieved a B rating; Ship 005 has transported four shipments with an A rating.

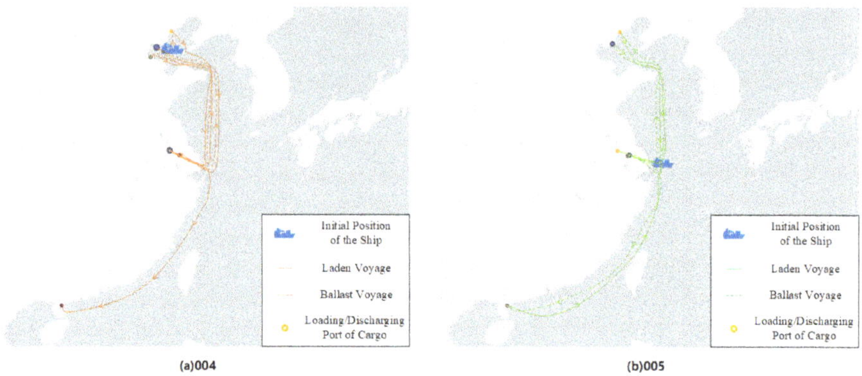

Figure 3. Optimized Shipping Routes of Ship 004 and Ship 005.

However, simply slowing down or reducing voyage frequency does not necessarily optimize the CII rating. Although higher speed increases fuel consumption, it also raises the transport workload of the ship. This corresponds to the denominator in the formula by allowing the ship to complete more cargo tasks efficiently, thereby potentially improving overall CII performance.

Taking Ship 002 and Ship 003 as examples, Ship 002 has undertaken five cargo shipments, with a ratio of Attained CII to Required CII being 1.05, resulting in the CII rating of C, indicated by red lines. Meanwhile, Ship 003 has completed six tasks with a ratio of 0.90, achieving a B rating. This indicates that, despite increased transportation tasks, optimizing scheduling and improving operational efficiency can effectively reduce carbon intensity and improve the CII rating. Figure 4 illustrates the optimized shipping routes of these two vessels.

Similarly, as Figure 5 shows, Ship 001 and Ship 006 both transport three batches of cargo. Ship 001 achieves an A rating, while Ship 006 is rated B. The cargo volume transported by these two ships is similar, but the sailing routes of 006 are much shorter. This requires attention to the relationship between route length, transportation efficiency, and ship scheduling. The results validate the model's effectiveness, demonstrating that reasonable scheduling and route optimization can achieve reduced carbon intensity and improved CII ratings.

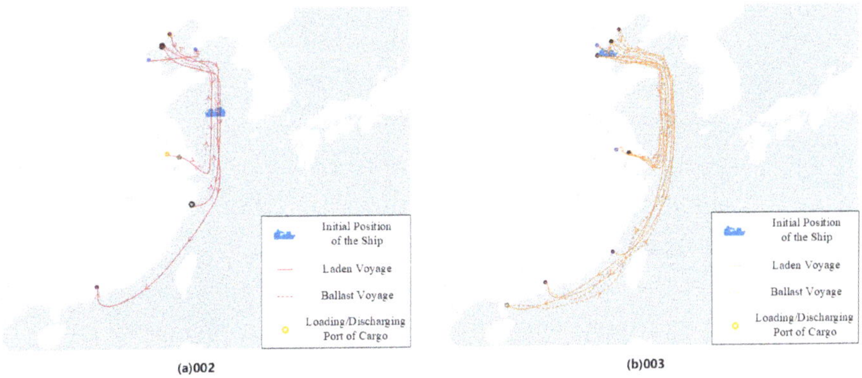

Figure 4. Optimized Shipping Routes of Ship 002 and Ship 003.

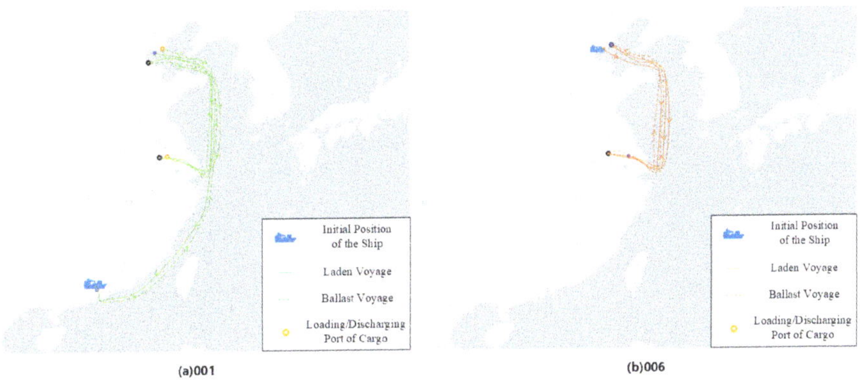

Figure 5. Optimized Shipping Routes of Ship 001 and Ship 006.

After optimization, all six ships achieve a CII rating of C or higher. A comparative analysis between actual data and optimized outcomes is presented in Table 9. Five of the six ships show improvements in their CII ratings. Specifically, Ship 001 and Ship 003 see considerable upgrades, transitioning from noncompliant D and E ratings to A and B, respectively. However, these environmental enhancements are accompanied by trade-offs in profitability, with profit margins decreasing by 16.00% and 10.90% for Ship 001 and Ship 003, respectively. Additionally, Ship 004 and Ship 006 both upgrade from C to B, achieving profit increases of 12.82% and 11.19%, respectively. While Ship 005 improves from C to A, with a 7.09% reduction in profit. These results suggest that through reasonable allocation and optimization of routing and speed, these ships achieve improvements in transportation efficiency and CII ratings, validating the effectiveness of the proposed methodology. In contrast, the rating of Ship 002 decreases from B to C while exhibiting a 34.12% profit surge. After optimization using the proposed model and algorithm, the total fleet profit has increased by 4.38% compared with actual operational data. This result indicates that achieving greater economic efficiency may involve trade-offs in which some environmental benefits are sacrificed while compliance with carbon intensity standards is still ensured.

Furthermore, by monitoring and updating the real-time ship position, the proposed method effectively matches ships with nearby cargoes, allowing for the timely allocation of new transportation assignments. This approach improves ship utilization and accelerates cargo flow and delivery. These enhancements result in a more reliable service, fostering greater satisfaction among cargo owners and strengthening the shipping company's competitiveness.

Table 9. Changes in CII ratings and profit before and after optimization.

Ship	Actual CII Rating	Optimized CII Rating	Actual Profit ($)	Optimized Profit ($)	Optimized Profit Change
001	D	A	244,058.00	205,001.19	−16.00%
002	B	C	320,563.99	429,955.47	+34.12%
003	E	B	290,911.29	259,211.12	−10.90%
004	C	B	257,488.50	290,509.53	+12.82%
005	C	A	317,483.07	294,967.22	−7.09%
006	C	B	198,159.95	220,333.49	+11.19%
Total	-	-	1,628,664.80	1,699,978.02	+4.38%

6. Conclusions

The increasing global emphasis on sustainability and stringent regulations on carbon emissions make it imperative to integrate CII into the optimization problem. While considerable research efforts have been dedicated to optimizing ship routing and scheduling, and some have considered carbon emissions, there remains a notable gap in integrating CII into these problems. This paper has introduced CII into the tramp ship routing and scheduling problem. We aimed to address the dual objectives of maximizing economic benefits and improving the CII rating while comprehensively considering the complex relationships between sailing speed, CII ratings, fleet operating costs, and shipment revenues. As a result, we established a bi-objective mixed-integer programming model. The model effectively balances economic returns and environmental impacts by optimizing resource allocation and operational strategies. The Gale–Shapley algorithm was employed to solve the cargo–vessel matching problem, ensuring optimal matching between cargoes and vessels. This approach enhances scheduling efficiency and resource utilization, avoiding unnecessary empty voyages and resource wastage and achieving a win–win situation for shippers and carriers. A GA was also used to optimize ship routing and scheduling and sailing speed, further improving operational efficiency and CII ratings.

The computational results validate the effectiveness of our methodology, demonstrating a 4.38% increase in the fleet's overall profit through model and algorithmic optimizations, with all vessels achieving CII compliance and significant improvements in environmental performance. This assists stakeholders in evaluating the trade-offs between economic returns and CII regulations. In an increasingly competitive market with stringent low-carbon requirements, the proposed model provides critical insights for tramp shipping operations, demonstrating considerable practical implications.

However, this study has limitations that could be considered in future studies. The current research results are mainly focused on coastal routes in China. Expanding the scope of this study to include a broader range of routes could yield more comprehensive conclusions. Additionally, shipowners need to ensure that their vessels meet CII standards to avoid penalties. In future research, it would be valuable to incorporate the CII requirements of chartered vessels into the optimization model.

Author Contributions: Conceptualization, H.Y. and J.Y.; methodology, H.Y., F.R. and J.Y.; software, H.Y. and F.R.; validation, H.Y., F.R. and R.U.K.; formal analysis, H.Y.; investigation, H.Y., F.R. and S.W.; resources, F.R.; data curation, H.Y., F.R. and J.Y.; writing—original draft preparation, H.Y.; writing—review and editing, H.Y., J.Y., S.W. and R.U.K.; visualization, H.Y. and F.R.; supervision, J.Y.; funding acquisition, J.Y. All authors have read and agreed to the published version of the manuscript.

Funding: This research was funded by the National Natural Science Foundation of China (Grant No. 52472362) and the Open Research Fund of the State Key Laboratory of Maritime Technology and Safety (Grant No. W24CG000037).

Data Availability Statement: The original contributions presented in this study are included in the article. Further inquiries can be directed to the corresponding author.

Acknowledgments: The authors want to acknowledge Xinbo Wang for project administration, Yi Han for conceptualization, and Shaohan Wang for providing essential resources.

Conflicts of Interest: Feiyang Ren was employed by COSCO Shipping Technology Co., Ltd. The remaining authors declare that the research was conducted in the absence of any commercial or financial relationships that could be construed as potential conflicts of interest.

References

1. UNCTAD. *Review of Maritime Transport*; UNCTAD: Geneva, Switzerland, 2021.
2. UNCTAD. *Review of Maritime Transport*; UNCTAD: Geneva, Switzerland, 2023.
3. Christiansen, M.; Fagerholt, K.; Nygreen, B.; Ronen, D. Ship routing and scheduling in the new millennium. *Eur. J. Oper. Res.* **2013**, *228*, 467–483. [CrossRef]
4. IMO. *Fourth IMO GHG Study*; IMO: London, UK, 2020.
5. IMO. *2023 IMO Strategy on Reduction of GHG Emissions from Ships*; IMO: London, UK, 2023.
6. Ammar, N.R.; Seddiek, I.S. Eco-environmental analysis of ship emission control methods: Case study RO-RO cargo vessel. *Ocean Eng.* **2017**, *137*, 166–173. [CrossRef]
7. Hua, W.S.; Sha, Y.S.; Zhang, X.L.; Cao, H.F. Research progress of carbon capture and storage (CCS) technology based on the shipping industry. *Ocean Eng.* **2023**, *281*, 114929. [CrossRef]
8. Bayraktar, M.; Yuksel, O. A scenario-based assessment of the energy efficiency existing ship index (EEXI) and carbon intensity indicator (CII) regulations. *Ocean Eng.* **2023**, *278*, 114925. [CrossRef]
9. IMO. *Resolution MEPC.308(73)—2018 Guidelines on the Method of Calculation of the Attained Energy Efficiency Design Index (EEDI) for New Ships*; IMO: London, UK, 2018.
10. Gianni, M.; Pietra, A.; Coraddu, A.; Taccani, R. Impact of SOFC Power Generation Plant on Carbon Intensity Index (CII) Calculation for Cruise Ships. *J. Mar. Sci. Eng.* **2022**, *10*, 1478. [CrossRef]
11. Chen, X.Q.; Zheng, J.B.; Li, C.F.; Wu, B.; Wu, H.F.; Montewka, J. Maritime traffic situation awareness analysis via high-fidelity ship imaging trajectory. *Multimed. Tools Appl.* **2023**, *83*, 48907–48923. [CrossRef]
12. Adland, R.; Cariou, P.; Jia, H.Y.; Wolff, F.C. The energy efficiency effects of periodic ship hull cleaning. *J. Clean. Prod.* **2018**, *178*, 1–13. [CrossRef]
13. Zhou, R.; Luo, M.; Jiang, C.; Yang, D.; Wang, K. How does emission allowance allocation affect shipowners' selection of emission reduction technologies? *Transp. Res. Part D Transp. Environ.* **2025**, *139*, 104573. [CrossRef]
14. Chen, X.Q.; Chen, W.P.; Wu, B.; Wu, H.F.; Xian, J.F. Ship visual trajectory exploitation via an ensemble instance segmentation framework. *Ocean Eng.* **2024**, *313*, 119368. [CrossRef]
15. Hua, R.; Yin, J.B.; Wang, S.H.; Han, Y.; Wang, X.B. Speed optimization for maximizing the ship's economic benefits considering the Carbon Intensity Indicator (CII). *Ocean Eng.* **2024**, *293*, 116712. [CrossRef]
16. Aydin, N.; Lee, H.; Mansouri, S.A. Speed optimization and bunkering in liner shipping in the presence of uncertain service times and time windows at ports. *Eur. J. Oper. Res.* **2017**, *259*, 143–154. [CrossRef]
17. Wang, H.Q.; Liu, Y.; Wang, S.A.; Zhen, L. Optimal Ship Deployment and Sailing Speed under Alternative Fuels. *J. Mar. Sci. Eng.* **2023**, *11*, 1809. [CrossRef]
18. Gale, D.; Shapley, L.S. College Admissions and the Stability of Marriage. *Am. Math. Mon.* **1962**, *69*, 9–15. [CrossRef]
19. Al-Hamad, K.; Al-Ibrahim, M.; Al-Enezy, E. A genetic algorithm for ship routing and scheduling problem with time window. *Am. J. Oper. Res.* **2012**, *2*, 22440. [CrossRef]
20. Kim, S.H.; Lee, K.K. An optimization-based decision support system for ship scheduling. *Comput. Ind. Eng.* **1997**, *33*, 689–692. [CrossRef]
21. Fagerholt, K. A computer-based decision support system for vessel fleet scheduling—Experience and future research. *Decis. Support Syst.* **2004**, *37*, 35–47. [CrossRef]
22. Fagerholt, K.; Christiansen, M. A combined ship scheduling and allocation problem. *J. Oper. Res. Soc.* **2000**, *51*, 834–842. [CrossRef]
23. Zeng, Q.C.; Yang, Z.Z.; VanAlbada, G.D. Model integrating fleet design and ship routing problems for coal shipping. In Proceedings of the Computational Science—ICCS 2007, Beijing, China, 27–30 May 2007; pp. 1000–1003.

24. Bronmo, G.; Christiansen, M.; Nygreen, B. Ship routing and scheduling with flexible cargo sizes. *J. Oper. Res. Soc.* **2007**, *58*, 1167–1177. [CrossRef]
25. Lin, D.Y.; Liu, H.Y. Combined ship allocation, routing and freight assignment in tramp shipping. *Transp. Res. Part E-Logist. Transp. Rev.* **2011**, *47*, 414–431. [CrossRef]
26. Vilhelmsen, C.; Lusby, R.M.; Larsen, J. Tramp ship routing and scheduling with voyage separation requirements. *OR Spectr.* **2017**, *39*, 913–943. [CrossRef]
27. Yu, B.; Wang, K.M.; Wang, C.; Yao, B.Z. Ship scheduling problems in tramp shipping considering static and spot cargoes. *Int. J. Shipp. Transp. Logist.* **2017**, *9*, 391–416. [CrossRef]
28. Zhao, K.; Zhang, D.; Jin, J.G.; Dong, G.X.; Lee, D.H. Vessel voyage schedule planning for maritime ore transportation. *Ocean Eng.* **2024**, *291*, 116503. [CrossRef]
29. Ma, W.H.; Lu, T.F.; Ma, D.F.; Wang, D.H.; Qu, F.Z. Ship route and speed multi-objective optimization considering weather conditions and emission control area regulations. *Marit. Policy Manag.* **2021**, *48*, 1053–1068. [CrossRef]
30. Wang, C.X.; Xu, C.Y. Sailing speed optimization in voyage chartering ship considering different carbon emissions taxation. *Comput. Ind. Eng.* **2015**, *89*, 108–115. [CrossRef]
31. Wen, M.; Pacino, D.; Kontovas, C.A.; Psaraftis, H.N. A multiple ship routing and speed optimization problem under time, cost and environmental objectives. *Transp. Res. Part D-Transp. Environ.* **2017**, *52*, 303–321. [CrossRef]
32. Weng, J.X.; Shi, K.; Gan, X.F.; Li, G.R.; Huang, Z. Ship emission estimation with high spatial-temporal resolution in the Yangtze River estuary using AIS data. *J. Clean. Prod.* **2020**, *248*, 119297. [CrossRef]
33. Yan, R.; Wang, S.A.; Du, Y.Q. Development of a two-stage ship fuel consumption prediction and reduction model for a dry bulk ship. *Transp. Res. Part E-Logist. Transp. Rev.* **2020**, *138*, 101930. [CrossRef]
34. Sun, W.Y.; Tang, S.Y.; Liu, X.Y.; Zhou, S.A.; Wei, J.F. An Improved Ship Weather Routing Framework for CII Reduction Accounting for Wind-Assisted Rotors. *J. Mar. Sci. Eng.* **2022**, *10*, 1979. [CrossRef]
35. Li, L.Z.; Ji, B.; Yu, S.S.; Zhou, S.Q.; Fang, X.P. Branch-And-Price Algorithm for the Tramp Ship Routing and Scheduling Problem Considering Ship Speed and Payload. *J. Mar. Sci. Eng.* **2022**, *10*, 1811. [CrossRef]
36. Yuan, Q.; Wang, S.; Peng, J. Operational efficiency optimization method for ship fleet to comply with the carbon intensity indicator (CII) regulation. *Ocean Eng.* **2023**, *286*, 115487. [CrossRef]
37. Tsai, Y.M.; Lin, C.Y. Effects of the Carbon Intensity Index Rating System on the Development of the Northeast Passage. *J. Mar. Sci. Eng.* **2023**, *11*, 1341. [CrossRef]
38. Psaraftis, H.N.; Kontovas, C.A. Ship speed optimization: Concepts, models and combined speed-routing scenarios. *Transp. Res. Part C-Emerg. Technol.* **2014**, *44*, 52–69. [CrossRef]
39. IMO. *Resolution MEPC.328(76)—2021 Amendments to the Annex of the Protocol of 1997 to Amend the International Convention for the Prevention of Pollution from Ships, 1973, as Modified by the Protocol of 1978 Relating Thereto 2021 Revised MARPOL Annex VI*; IMO: London, UK, 2021.
40. IMO. *Resolution MEPC.352(78)—2022 Guidelines on Operational Carbon Intensity Indicators and the Calculation Methods (CII Guidelines, G1)*; IMO: London, UK, 2022.
41. IMO. *Resolution MEPC.352(78)—2022 Guidelines on the Reference Lines for Use with Operational Carbon Intensity Indicators (CII Reference Lines Guidelines, G2)*; IMO: London, UK, 2022.
42. IMO. *Resolution MEPC.338(76)—2021 Guidelines on the Operational Carbon Intensity Reduction Factors Relative to Reference Lines (CII Reduction Factors Guidelines, G3)*; IMO: London, UK, 2021.
43. IMO. *MEPC.354(78)—2022 Guidelines on the Operational Carbon Intensity Rating of Ships (CII Rating Guidelines, G4)*; IMO: London, UK, 2022.
44. IMO. *MEPC.355(78)—2022 Interim Guidelines on Correction Factors and Voyage Adjustments for CII Calculations (CII Guidelines, G5)*; IMO: London, UK, 2022.

Journal of
Marine Science and Engineering

MDPI

Article

An AIS-Based Study to Estimate Ship Exhaust Emissions Using Spatio-Temporal Approach

Akhahenda Whitney Khayenzeli [1], Woo-Ju Son [2], Dong-June Jo [1] and Ik-Soon Cho [3,*]

1 Department of Maritime AI & Cyber Security, Graduate School of National Korea Maritime & Ocean University, Busan 49112, Republic of Korea; akhahenda@g.kmou.ac.kr (A.W.K.); dongjune.jo@g.kmou.ac.kr (D.-J.J.)
2 Research Institute of Medium & Small Shipbuilding, Changwon 51965, Republic of Korea; wjson@rims.re.kr
3 Division of Maritime AI & Cyber Security, National Korea Maritime & Ocean University, Busan 49112, Republic of Korea
* Correspondence: ischo@kmou.ac.kr; Tel.: +82-51-410-5072

Abstract: The global shipping industry facilitates the movement of approximately 80% of goods across the world but accounts for nearly 3% of total greenhouse gas (GHG) emissions every year, and other pollutants. One challenge in reducing shipping emissions is understanding and quantifying emission characteristics. A detailed method for calculating shipping emissions should be applied when preparing exhaust gas inventory. This research focused on quantifying CO_2, NO_x, and SO_x emissions from tankers, containers, bulk carriers, and general cargo in the Republic of Korea using spatio-temporal analysis and maritime big data. Using the bottom-up approach, this study calculates vessel emissions from the ship engines while considering the fuel type and operation mode. It leveraged the Geographic Information System (GIS) to generate spatial distribution maps of vessel exhausts. The research revealed variability in emissions according to ship types, sizes, and operational modes. CO_2 emissions were dominant, totaling 10.5 million tons, NO_x 179,355.2 tons, and SO_x 32,505.1 tons. Tankers accounted for about 43.3%, containers 33.1%, bulk carriers 17.3%, and general cargo 6.3%. Further, emissions in hoteling and cruising were more significant than during maneuvering and reduced speed zones (RSZs). This study contributes to emission databases, providing a basis for the establishment of targeted emission control policies.

Keywords: exhaust gas; spatio-temporal; big data; GIS; emission database; international maritime organization (IMO)

1. Introduction

Maritime transportation plays a vital role in the development of global trade and economic growth. However, it is also a remarkable contributor to atmospheric pollution [1]. Approximately 90% of the global trade volume is carried by ship, leading to high volumes of exhaust emissions, which have negative impacts [2]. The Second International Maritime Organization (IMO) Greenhouse Gas study [3] suggests that shipping greenhouse emissions accounted for approximately 3.3% of global emissions, with projections indicating a continued rise if mitigation strategies are not intensified. In the IMO GHG study, carbon dioxide was the highest emitted greenhouse gas in terms of quantity and global warming potential. A ship's engine combustion emits 450 different air pollutants [4], but the key ship sources of air pollutants are greenhouse gases, carbon monoxide, NO_x, SO_x, and particulate matter [5,6]. Around 70% of these vessel emissions are discharged within a radius of

https://doi.org/10.3390/jmse13050922

400 km from the coastline, and the emissions travel further through the atmosphere as ships maneuver near densely inhabited coastal areas [7]. As a result, around 230 million people living in the top 100 world ports are directly impacted by marine pollution [8].

To address this, international bodies such as the IMO have set increasingly ambitious targets. The IMO, through its legislation, the International Convention on the Prevention of Pollution from Ships (MARPOL) Annex VI, has regulations to mitigate maritime pollution. Since 2018, the IMO has actively analyzed shipping emissions to establish long-term strategies to control marine pollution, including the IMO 2020 sulfur limit [9]. In the NO_x ECAs, tiered emission limits are imposed based on the ship's engine construction date and type. Carbon dioxide is one of the largest contributors to global warming; hence the revised 2023 IMO GHG strategy to minimize shipping carbon intensity by up to 40% by 2030 and 70% by 2050 compared to 2008 [9]. IMO's ambitious strategy to achieve net-zero GHG emissions in international shipping by around 2050 has encouraged the formulation of policies such as the use of alternative green fuels and the establishment of green shipping corridors as pilots for systemic change. These goals align with broader global frameworks like the Paris Agreement, which emphasizes rapid decarbonization across all sectors. The Paris Agreement aims to maintain global average temperatures to below 2 °C and limit its increase to 1.5 °C above pre-industrial levels. In the agreement, all parties were required to report the national GHG inventory and information on mitigation policies and support framework to the United Nations [10]. According to the IPCC [11], global greenhouse gas emissions need to be cut by at least 40% below 1990 levels by 2030, to net-zero emissions by 2050 and move towards negative emissions thereafter to achieve the Paris Agreement goals.

Aside from the IMO pursuing stricter regulations to cut GHG emissions, individual countries and regions are also striving to contribute to the decarbonization of the shipping industry. The European Union (EU) expanded its emissions trading system (ETS) to include the maritime sector in 2024, requiring vessels over 5000 GT calling ports within the European Economic Area (EEA) to purchase EU allowances for GHG monitoring in line with the EU Monitoring, Reporting and Verification (EU MRV) regulation [12]. Ships that do not comply with this regulation are subject to pecuniary penalties and may be denied entry into ports within the EEA territory. Further, the United States enacted the Inflation Reduction Act (IRA) in 2022, which has provisions for direct support of electrification to reduce port emissions and the hydrogen tax incentive that aims to significantly contribute to making the ammonia supply chain greener [12]. China continues to be a global leader in designing and building ammonia-ready vessels, in addition to implementing its short-term measures to decrease vessel GHG emissions [12]. All these milestones indicate the motivation of nations to achieve low-carbon transformation in their shipping operations. Similarly, South Korea has also made great strides in developing maritime carbon reduction goals to achieve carbon neutrality by 2050.

Shipping forms the backbone of South Korea's economy, as nearly all of its imports and exports, totaling 99.7%, rely on maritime transport [12]. Internationally, South Korea is a prominent player in the maritime industry, ranking seventh globally in vessel ownership, fourth in container port traffic, and first in shipbuilding [2]. Being a maritime nation and a global leader in shipbuilding, South Korea is directly impacted by the maritime pollution issue, thus actively responding to these environmental challenges. Studies estimate South Korea's GHG equivalent emissions to be around 701.3 million tons of CO_2-equivalent, with the transportation sector, including the shipping industry, accounting for approximately 14% of the total emissions [10]. The Ministry of Oceans and Fisheries in Korea has reported that Korea is responsible for less than 1% of national emissions; however, the estimations do not cover the bulk of emissions arising from international emissions. As a result, it fails to give an accurate depiction of the entire shipping industry's pollution potential since GHG emissions from international shipping are intensive [10]. The Korean government launched multiple initiatives, such as the "Air Quality Management Basic Plan", to cut maritime emissions and protect people from harmful air pollutants. In 2019, the Ministry of Oceans and Fisheries [13] introduced VSR programs at key ports of Korea (Busan/Incheon/Ulsan/Yeosu–Gwangyang), requiring vessels to reduce speeds within designated zones to lower exhaust gases. In 2020, Korea announced its goal to achieve carbon neutrality in domestic shipping by 2050. Following the carbon neutrality declaration, the Ministry of Oceans and Fisheries (MOF), which oversees the shipping sector, introduced a decarbonization plan for domestic shipping and fisheries [13]. The government passed the Eco-Friendly Ship Act which establishes a basis for pushing the manufacture and distribution of eco-friendly ships that use green fuels. Based on this act, in 2021, the Ministry of Ocean and Fisheries announced the Greenship-K Promotion Strategy with the aim to produce low-CO_2 vessels and incentivize the use of alternative fuels to create eco–friendly ships [13]. On an international scale, South Korea is taking part in green corridor development to accelerate shipping decarbonization. During COP27, Korea announced an intent to explore the feasibility of creating the Busan–Seattle/Tacoma green corridor between Busan port and Seattle/Tacoma port [14]. Other policy recommendations include setting the zero-emissions "At Berth" policy by 2030 to mandate the use of shore power in Korean ports to reduce harmful emissions.

The Natural Resources Defense Council [15] reports that numerous research articles have studied and proven the adverse impacts of shipping emissions on global climate and the environment. Additionally, along with causing poor air quality [16] and climate change, plenty of studies have related these pollutants to severe health impacts [17,18]. Researchers have focused on preparing ship emission inventories to evaluate the impacts of pollutants on air quality [19,20]. According to [21], studying exhaust gas characteristics is fundamental to guiding the implementation of sustainable policies. To support maritime emission reduction strategies, numerous studies have developed ship emission estimation models, referring to the top-down and bottom-up approaches in the guidelines provided by the Inter-Governmental Panel on Climate Change [11]. The top-down approach utilizes statistically analyzed marine fuel consumption data and fuel-related emission factors to estimate emissions [7,22], but the results are inaccurate since it does not take into account the vessel's operating conditions. The method would be ideal since fuel consumption and emissions discharged from the engine combustion have a positive correlation. However, [23] found a huge discrepancy between fuel sale statistics and the actual fuel used by the global fleet. In Korea, the Ministry of Oceans and Fisheries, the Ministry of Environment, and the Korea Maritime Institute (KMI) applied a top-down approach to estimate emissions and determine the amount of air pollutant emissions in ports, but the exhaust gas data were inaccurate [24]. In contrast, the bottom-up method estimates emissions based on vessel activity and technical parameters such as engine specifications, speed profiles, and activity time derived from extensive maritime data such as AIS [25]. In an attempt to differentiate the two methods, [26] calculated the quantity of pollutants from ships. The bottom-up method was reliable for understanding emission patterns and intensity, leading to the establishment of effective green policies for air quality management [27–29]. However, the approach could introduce uncertainties for global applications due to the diverse input parameters required, model assumptions, and data anomalies that introduce uncertainties when used in large-scale applications without validation [30–32].

Since IMO mandated the installation of AIS receivers on passenger and merchant ships of over 300 gross tonnages (GT) on international voyages, many scholars have used AIS data to create emission assessment models for different world regions [33]. A few authors developed the Ship Emission Inventory Model (SEIM) to evaluate the impacts of ship pollution in East Asia [34]. Ref. [35] estimated carbon emissions in global high seas using a proposed Geographic Emission Estimation Model (GEEM) based on geographic location. Ref. [24] estimated emissions from ships at Busan Port, highlighting the hoteling mode and CO_2 as major contributors, but their model lacked vessel-specific temporal granularity. Ref. [36] also conducted a study to estimate emissions from vessels operating at Yeosu-Gwangyang port in Korea using the bottom-up approach and established that CO_2 and the hoteling mode accounted for the highest emission levels. Further, the authors reported that containers and tankers together emitted the highest volume of exhaust gases. However, their estimations did not factor in the updated policies, such as the IMO 2020 sulfur cap [9]. Refs. [19,37] calculated the exhaust emissions of oceangoing ships in Hong Kong and Busan Port, respectively, obtaining an emission contribution rate of different ship types. On the other hand, [38] conducted a comparative analysis of the estimation methods of ship pollutants and proposed that data statistics and management are essential foundations for research. Such inventories provide a basis for further refining the emission estimation models and enhancing emission databases for implementing macro-emission policies.

Despite these advancements, few studies have mapped emissions at grid-level resolution using spatial occupancy metrics and aggregated grid spatial emissions. This study responds to the global efforts to reduce maritime emissions and utilizes a comprehensive, data-driven emission estimation framework for vessels operating within the Republic of Korea's coastal waters. Over a period of 12 months (September 2021–August 2022), exhaust gases of CO_2, NO_x, and SO_x were calculated for four major vessel types: tankers, container ships, bulk carriers, and general cargo vessels across four distinct operational modes, such as cruising, RSZ, maneuvering, and hoteling. This study integrates Automatic Identification System (AIS) data, IHS ship registry data, and spatial grid modeling (0.5 km $\times$ 0.5 km) to generate detailed, high-resolution maps of exhaust emissions, all visualized using ArcGIS Pro. What sets this study apart is the application of the spatial-temporal density method, derived from the EMODnet framework, enabling the estimation of vessel occupancy time within grid cells, which is then used to calculate emissions based on the US EPA's bottom-up modeling formula. Additionally, this research contributes to the creation of a robust maritime emissions database, a resource that is increasingly critical for data-driven decision-making to promote sustainable development. The increasing global efforts to decarbonize shipping, with the IMO targeting net zero by around 2050, presents a pressing need for transparent, traceable, and high-resolution emissions data. Maritime stakeholders, including governments, port authorities, and environmental regulators, require insights into when, where, and why emissions occur, categorized per vessel type, operational mode, engine profile, and pollutant class. By combining operational data, this study provides detailed information, thereby availing opportunities for sustainable management of shipping pollution. Figure 1 is an illustration of the study background, the research gaps in current research, and how this study aims to enhance the current emission databases.

The remaining sections are divided as follows: In Section 2, details of the data preprocessing steps, emission estimation, and spatial methods are provided. Section 3 presents the emission characterization results for different categories. Section 4 highlights the discussions of the study results. In Section 5, we summarize the study and describe its limitations and future research tasks.

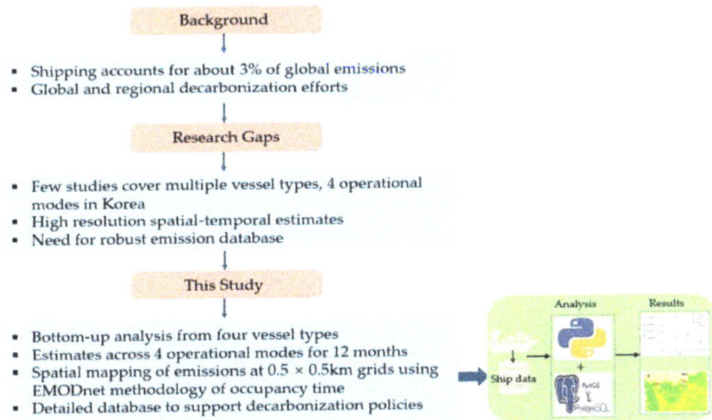

Figure 1. A visual summary of the study's framework.

2. Materials and Methods

This study was conducted in the Republic of Korea to analyze the exhaust gas emissions generated by vessels operating in the region. The Republic of Korea's entire coastal area was designated as the area of interest (AOI) for the study, as shown in Figure 2a. Additionally, grids of 0.5 km by 0.5 km were created over the Republic of Korea's area, as illustrated in Figure 2b. The grid cell size was chosen to allow for higher spatial resolution when visualizing the emissions in each grid. As shown in Table 1, the AIS data collection period is 12 months, from September 2021 to August 2022. The data volume sums up the values for the tankers, containers, bulk carriers, and general cargo ships at all speeds of navigation.

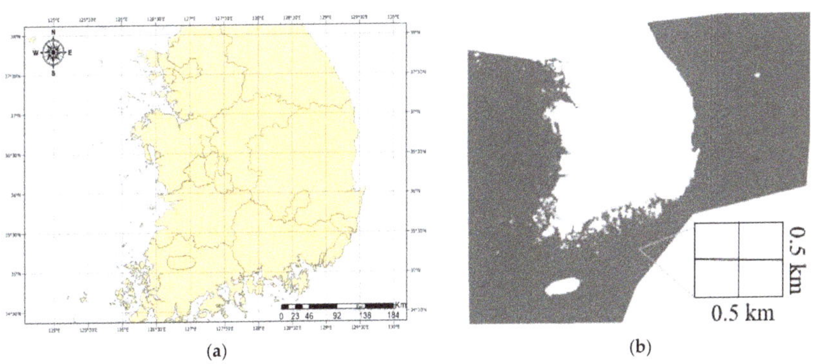

Figure 2. Setting spatial extent (area of interest): (**a**) Area of interest and (**b**) grid cells.

Table 1. Description of study data.

Categorization	Automatic Identification System
Period	1 September 2021–31 August 2022 (12 months)
Data Collection Area	Republic of Korea
Volume of Data	Approximately 112 Gigabytes (GB)
Ship Speed	All speeds

2.1. Area of Study

Figure 3 is a flowchart of the analysis steps conducted in this study. First, we input the data from various sources, such as AIS and IHS, for the study period and area. The data preprocessing steps entailed data cleaning, filtering, merging, and enrichment to ensure all columns needed for emission calculation are available, and finally, imputation to infill all missing values. Then, the emission modeling process involved the creation of an emission matrix area analysis to define an area of interest and define the grids. The emission matrix was generated based on vessel length and speed to group the emissions by case and then the occupancy time calculation for each grid. Finally, emission analysis entailed calculating the hourly emissions for each equipment and multiplying the hourly emissions by occupancy to obtain total emissions in the grids. Additionally, statistical maps of exhaust gases were done using ArcGIS Pro version 3.0 software, manufactured by ESRI with headquarters in Redlands, CA, USA.

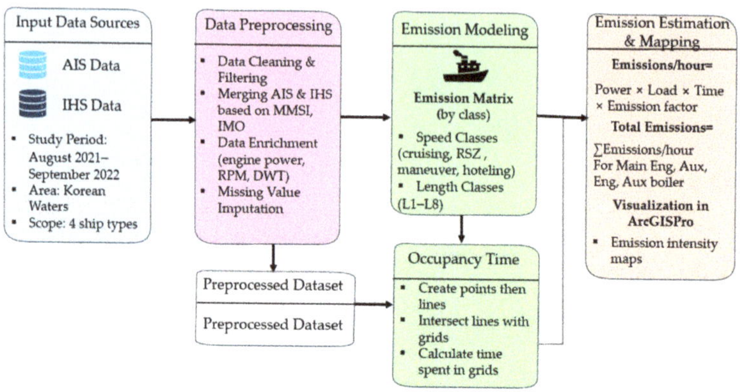

Figure 3. Flow diagram of research methodology.

2.2. Data and Characteristics

To prevent collisions at sea, the IMO established a regulation that required vessels to install AIS instruments on all international ships with gross tonnages (GT) of more than 300 GT, on domestic cargo ships with 500 GT and above, and on all passenger ships regardless of size [33]. AIS transmits static data in real time, allowing adjacent ships to recognize other vessels' statuses and movement. AIS transponders have satellite receivers capable of collecting AIS data over the open sea that are also available [39]. The data contains vessels' static information, including the ship name, Maritime Mobile Service Identity (MMSI), ship type, length, width, and draught, while the dynamic information provides the navigation status, speed, Course over Ground (COG), Heading over Ground (HOG), and position [40]. Nonetheless, AIS data are limited to ship technical data such as engine power that is required in estimating ship emissions. Therefore, the IHS database from Lloyd's Register is used to complement the AIS data. The IHS database covers all seagoing merchant vessels of over 100 GT and has valid IMO numbers. The data are continuously updated with newly built ships and provide a range of ship information, including the IMO number, deadweight (DWT), engine power, and revolutions per minute (RPM).

2.3. Data Preprocessing

Data directly taken from the source will likely have inconsistencies and errors and, hence are not ready to be used for a data analysis process [41]. The preprocessing steps of the study data are shown in Figure 4. First, the static and dynamic AIS data are combined and then filtered according to IMO's defined ship-type numbers to obtain the ship types. However, AIS data only provide generalized cargo data, so to obtain the specific vessel types under cargo data, the AIS database was matched to the IHS database based on IMO or MMSI numbers to distinguish container, bulk carriers, and general cargo ships. For tankers, the AIS data were sufficient to identify the fleet as defined by IMO. The data were filtered according to the ship-type profile, and then the necessary columns required for analysis were obtained. These columns are the date, latitude, longitude, MMSI, LOA, deadweight, service speed, SOG, ship type, and main engine power.

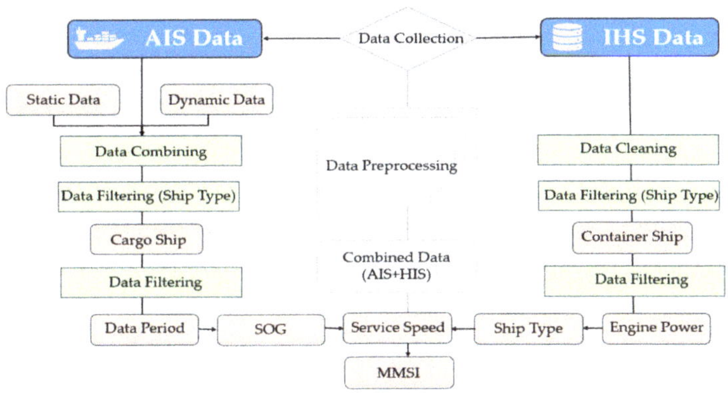

Figure 4. Data preprocessing steps.

Missing value imputation is a basic solution method for incomplete dataset problems, particularly those where some data samples contain missing attribute values [42]. The study data had a lot of missing parameters, including deadweight, length overall, main engine power, speed over ground (SOG), service speed, and maximum speed. Rows with incomplete values for deadweight were deleted since they were a smaller percentage. Then, applying correlation analysis between parameters (length vs. deadweight and power vs. deadweight), polynomial regression analysis was applied to infill the missing length and power values. Fitted curves of the regression analysis are shown in Figure 5 for all four vessel types.

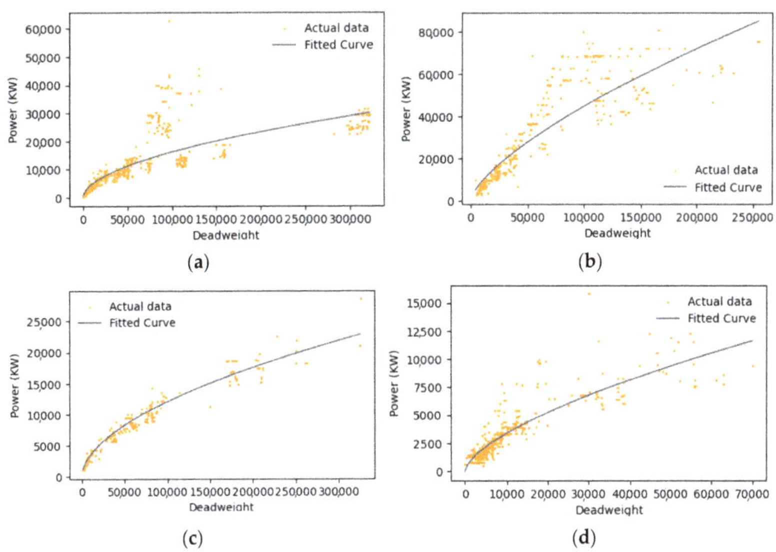

Figure 5. Correlation analysis results for DWT vs. main engine output: (**a**) Tanker, (**b**) container, (**c**) bulk carrier, and (**d**) general cargo.

The analysis generated r-squared values for tankers at 0.71, containers at 0.81, bulk carriers at 0.96, and general cargo at 0.76. While polynomial regression models demonstrated strong fits, especially for bulk carriers (R^2 = 0.96) and containers (R^2 = 0.81), there remains inherent uncertainty in the imputed values. Factors such as the vessel age, engine retrofits, or hull design may influence the power output in ways not captured by deadweight alone. Therefore, these estimates, though statistically supported, may not perfectly reflect the actual values for all vessel classes, thereby introducing errors in the results. For maximum speed and service speed, the missing parameters were backfilled by considering the characteristics of ships with similar lengths and speeds. Each of the vessel types was divided into emission matrices based on their speed and length, and then the mean values calculated within each class were assigned to the missing data rows.

2.4. Occupancy Time Analysis

Occupancy time refers to the duration a ship spends at a specific location. Occupancy time is critical in estimating maritime pollution, as it describes vessel activity within space. Vessel activity can be determined through the spatial-temporal density technique developed by the European Marine Observation and Data Network (EMODnet) [43]. Before calculating the occupancy time of the vessels under study, ArcGIS Pro software was used to designate the area of analysis and create grids of 0.5 km by 0.5 km over the relevant sea area. Grid cells of 0.5 km by 0.5 km were chosen to enable higher spatial resolution since smaller grid cells allow a more detailed representation of emissions in space. Ships' hourly emissions vary throughout their voyage depending on the operational mode (speed), type, and size. As a result, the combined preprocessed data are classified using the Python program version 3 as an emission matrix. Python is an open-source language developed by Guido van Rossum and maintained by the Python Software Foundation, whose initial registered office is in Delaware, USA. The grid cells defined in ArcGIS and the emission analysis matrix are input into the Structured Query Language (SQL) database using the PostgreSQL and PostGIS programs version 16. PostgreSQL is derived from a POSTGRES package written at the University of California at Berkeley, whereas PostGIS is developed by Refractions Research Inc. based in Victoria, British Colombia and Canada. The occupancy time of the ships in the grid cells is calculated using the spatial-density analysis method, which applies an SQL syntax. The method considers vessel movement distance with the concept of space and time simultaneously. The spatial-temporal density method creates lines by connecting two consecutive positional points and then calculates the time and distance of the connected line [43]. It also calculates the time within each cell, referred to as the occupancy time for the generated line. The obtained occupancy time is multiplied by the hourly emissions to generate total exhaust gas emissions in each grid cell. Finally, the results are exported as a polygon layer to the ArcGIS program, where the calculated values are constructed as space–time statistical maps. The detailed procedure is demonstrated in Figure 6.

There is a wide range of methods for analyzing the density of vessel passage, including point, line, and vessel density. However, in this study, the occupancy time of the ships in the grid cells is calculated using the spatial-density analysis method. The method considers vessel movement distance with the concept of space and time simultaneously. The spatial-temporal density method creates lines by connecting two consecutive positional points and then calculates the time and distance of the connected line. The procedure also calculates the time within each cell, referred to as the occupancy time for the generating line, as shown in Figure 7 below. Figure 7 is an example of a voyage, whereby the green line's occupancy time in the cell is determined as 92.5 s by dividing the length covering the grid (370 m) by the entire length (600 m) and then multiplying by the line time given as 150 s. In this study, the occupancy time refers to the duration in which emissions occur within a space range at the time the ship is in the cell. Therefore, the emissions in the grid can be determined by multiplying the occupancy time and the ship's activity in the cell, which depends on its operational mode and type.

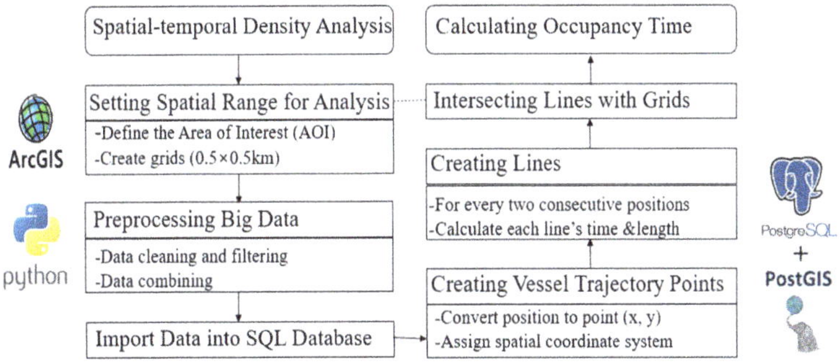

Figure 6. Process of calculation of occupancy time.

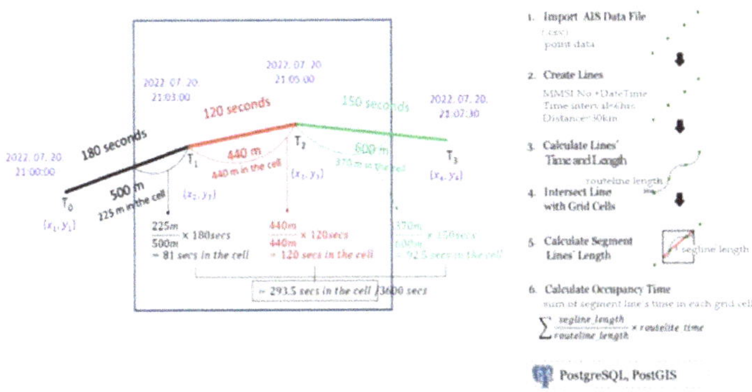

Figure 7. An overview of the calculation of occupancy time.

2.5. Emission Output Modeling

Maritim studies commonly apply two approaches for emissions modeling: top-down and bottom-up. In this study, we adopted the bottom-up approach as recommended by the US EPA, which estimates vessel exhaust emissions based on actual ship activity data such as the ship's speed, engine power, load factor, and activity time. This model for emissions output entails the estimation of quantitative pollutant data from the oceangoing vessels, enabling localized, high-resolution assessments of exhaust gases at the vessel level.

The bottom-up approach for ship emission estimation is widely regarded for its high spatial and temporal resolution, enabling detailed analysis of emissions based on vessel-specific operational characteristics such as engine power, load factors, and activity time [36], unlike the top-down approaches, which rely on aggregate fuel consumption data and are less suited for local-scale analysis. However, despite its advantages, the bottom-up approach is sensitive to the quality and completeness of data input. Missing technical information, such as engine power, or imprecise vessel specifications, such as operational mode, introduce uncertainty through emission calculations. Moreover, assumptions regarding constant load factors or vessel behavior during certain operational modes may not reflect real-world variability [34]. Model outputs are especially affected by inaccuracies in AIS-derived speed data and imputed technical values, which are often estimated through regression models. Furthermore, discrepancies can arise when matching AIS data to fleet registries, particularly in large datasets with older vessels or incomplete identifiers. Hence, although the bottom-up method is robust for localized emission mapping, authors must account for uncertainties tied to data imputation, operational assumptions, and technical variability. The modeling process applied in this study incorporates emission formulas based on engine specifications and operational phases recommended by the US EPA. Further, an emission matrix based on the ship's speed and length is utilized to obtain emission results based on the operational characteristics of the ships.

2.5.1. Emission Matrix

Ship size is a critical factor in maritime operations and emissions calculations. The most important indicators of vessel size are deadweight, gross tonnage, length overall (LOA), beam, and draft. The ship size index used in this study was selected by considering the Spearman correlation between the size index, navigation speed, and maximum power output. The results revealed a positive correlation between the ship's overall length, navigation speed, and maximum power output. Therefore, based on these results, vessel length was divided into eight length categories with an incremental length of 50 m. Figure 8 shows graphs of the correlation results.

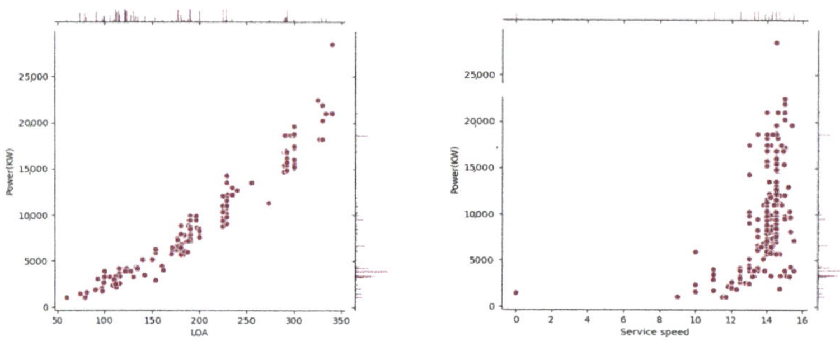

(a) (b)

Figure 8. Correlation analysis of ship variables: (**a**) power in Kw vs. LOA and (**b**) power in kW vs. service speed.

For navigation speed, the operational mode of the ship was considered, as highlighted in Table 2. While in service, vessels operate in these phases: hoteling, maneuvering, and cruising. According to [7], ships move inside the port boundary with all engines running and at service speed when cruising, while during maneuvering, they transit between the breakwaters with a slower speed maintained as the vessel approaches the pier/wharf/dock. At the reduced speed zone, vessels move at speeds less than cruise but greater than maneuvering. South Korea introduced the VSR program, where ships are mandated to reduce their speeds within certain levels to reduce emissions from OGVs as they enter the ports. In the VSR program in Korea, vessels operate at speeds between 8–10 knots, depending on the vessel type [13]. During hoteling, ships may be at anchorage waiting for a berth call, or they may be at berth with the propulsion engine off, and only the auxiliary engine and boiler operate at peak loads to provide power onboard and for loading and off-loading equipment. In ports where cold ironing is applied, ships use shore power instead of auxiliary engines. In this study, vessels are assumed to operate auxiliary engines during the hoteling mode since the ship data used (2021–2022) were at a period when cold ironing had not been implemented in most Korean ports. The operating phase can determine a ship's power demand, which is essential when estimating exhaust gases.

Table 2. Classification of ship's operation mode based on speed.

Operating Mode	Cruising	RSZ	Maneuvering	Hoteling
Ship Speed	v > 12 kts	8 kts < v ≤ 12 kts	3 kts < v ≤ 8 kts	v ≤ 3 kts

2.5.2. Emission Estimation Method and Parameters

The research applies the bottom-up method, based on AIS and IHS data, to estimate exhaust gas from ships. The methodology reflects the ship's real-time movements and demands the use of detailed specifications data; hence, the results are more accurate [37,44,45]. The three main sources of exhaust gases are the main engine, auxiliary engine, and auxiliary boiler [46]. Ref. [47] mentions that the working conditions of the power equipment vary frequently for different ship movements, leading to significant differences in emissions. When sailing, the ship's main and auxiliary engines operate at peak loads and generate many waste gases. However, while berthing or mooring, the auxiliary engine and auxiliary boiler are operational; thus, they are the primary sources of pollutants. While some ports worldwide offer shore-side electrical power to allow vessels to switch off their auxiliary engines at berth, this research, which is based in South Korea, assumes auxiliary engines are always on during berthing at South Korean ports. South Korea has initiated the implementation of shore power, also known as Alternative Maritime Power (AMP), at several major ports, but the rollout is ongoing and not yet comprehensive or mandatory for all vessels [48]. Therefore, the use rate of shore power in South Korean ports remains low, and most ships at berth, especially those not equipped or required to use shore power, continue to rely on their auxiliary engines for electricity generation. The Korean government plans to build AMP infrastructure in 230 berths by 2030 in the major ports of Busan, Incheon, Gwangyang, Pyeongtaek-Dangjin, Ulsan, and Pohang [10]. Several pilot projects are ongoing; for example, in January 2025, HMM's LNG-ready container ship utilized AMP in Gwangyang port in the Republic of Korea, resulting in approximately 16 tons of GHG [49]. With regard to this, the assumption that the vessels at Korean ports utilized auxiliary engines rather than shore is substantiated by the currently existing data and infrastructure. The generalized bottom-up formula for emission determination recommended by US EPA [46] is as in Equation (1):

$$E = P \times LF \times A \times EF \quad (1)$$

where E represents the emission of the exhaust gas in grams (g), P is the installed power of the main engine, also called Maximum Continuous Rating (MCR) in kilowatts (kW), LF is the load factor, A is the ship activity time, and EF is the emission factor in grams per kilowatt-hour (g/kWh). Equations (2) and (3) provide the generalized formula, while Equations (4)–(6) are detailed emission formulas for each equipment that emits pollutants as follows:

$$E_{Total\ Emission} = E_{Emission\ per\ Hour} \times A_{Occupancy\ Time} \quad (2)$$

$$E_{Emission\ per\ Hour} = E_{ME} + E_{AE} + E_{AB} \quad (3)$$

$$E_{Main\ Engine(ME)} = MCR_{ME} \times LF_{ME} \times EF_{ME} \quad (4)$$

$$E_{Auxiliary\ Engine(AE)} = MCR_{AE} \times LF_{AE} \times EF_{AE} \quad (5)$$

$$E_{Auxiliary\ Boiler(AB)} = Energy\ Demand_{AB} \times EF_{AB} \quad (6)$$

Based on Equation (1), the components for emission estimation are engine power, load factor, emission factor, and activity time. Figure 9 is an illustration of the bottom-up approach framework applied in the study for emission estimation.

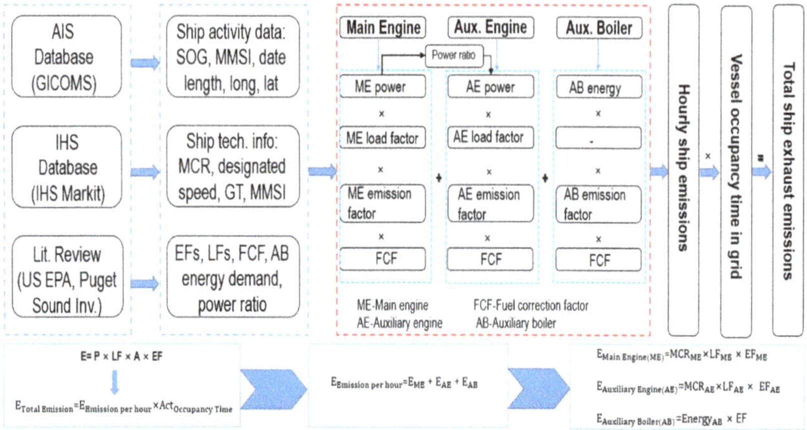

Figure 9. A bottom-up approach framework for emission estimation.

1. Engine power (MCR)

Ships typically have three engine types: a main engine for propulsion, auxiliary engine for electricity generation, and auxiliary boiler for steam generation. The vessel-operating phase determines the engine power. Data on ship specifications, including main engine power, maximum design speed, vessel type, engine type, gross tonnage, and deadweight, were obtained from the Lloyd Register database (IHS Markit), as described in Section 2.2 in this study. The installed engine power of a ship is essential for emission estimation; however, some values were missing from the IHS database. To estimate the engine power, a correlation between the ship's length and deadweight was established, and polynomial regression was applied to infill missing data. The main propulsion values were determined by averaging the MCR values of distinct lengths in all the operating modes. Table 3 highlights a positive correlation between length and power in that longer ships demand more engine power to maintain speed. Container ships have higher power requirements compared to other ship types because they operate at higher speeds to meet their fixed schedules. Tankers and bulk carriers have relatively slower speeds and lower main power values because they have flexible schedules that entail longer waiting times at the ports or berth, while general cargo has moderate power requirements.

Table 3. Average main engine power values obtained from the IHS database.

Ship Type	Tanker	Container	Bulk Carrier	General Cargo
L1	629.59	-	-	692.58
L2	1864.55	3007.02	1935.5	1668.92
L3	3854.4	7178.22	3594.64	3393.37
L4	8303.32	13,872.77	7368.57	7883.44
L5	12,687.92	22,879.86	10,335.06	19,400.45
L6	23,712.13	43,859.86	16,774.80	-
L7	26,752.60	58,764.16	20,312.90	-
L8	-	59,235.68	27,011.10	-

On the other hand, the auxiliary engine power installed information is not reported on Lloyd's database. Therefore, an estimation of the auxiliary engine power was done based on the main engine power adopted from the US EPA report [36]. The values are obtained from a survey conducted by the California Air Resources Board (ARB) on oceangoing ships totaling 327 and can be applied to mid-tier inventory development if no auxiliary power data are available [46]. The ratio of the propulsion engine to auxiliary engine provided by the US EPA [46], as shown in Table 4, was used to determine auxiliary engine power.

Table 4. The power ratio of main propulsion to auxiliary engine (ARB Survey).

Ship Type	Auxiliary to Propulsion Engine Power Ratio
Auto Carrier	0.266
Bulk Carrier	0.222
Container Ship	0.220
Cruise Ship	0.278
General Cargo	0.191
Tanker	0.211

2. Boiler energy demand

In addition to auxiliary engines, ships also have auxiliary boilers. Boiler energy demand is the amount of energy required to operate boilers on a ship. Boilers are essential for purposes of heating heavy fuel oil, lubricating oil, or for maintaining specific engine parts at required temperatures. Boilers also provide steam for propulsion in steam-powered vessels. In modern vessels, steam is supplied to auxiliary equipment for electricity or freshwater generation and hot water generation for crew needs. During cruising and the reduced speed zone, boilers are typically off because most ships are equipped with exhaust heat recovery systems that utilize heat from the main engine exhaust for hot water needs [46]. When the main engine's temperature falls below an adequate amount for waste heat recovery, such as during maneuvering and hoteling, the boilers can be operated. For boiler energy demand, the values recommended by the US EPA [46] shown in Table 5 were applied for each vessel type and activity.

Table 5. Energy boiler demand according to operation mode.

Ship Type	Cruising	RSZ	Maneuvering	Hoteling
Auto Carrier	0	0	371	371
Bulk Carrier	0	0	109	109
Container Ship	0	0	506	506
Cruise Ship	0	0	1000	1000
General Cargo	0	0	106	106
Tanker	0	0	371	3000

3. Load factor

The shipload factor is the percentage of the actual output power to the installed capacity of the ship's power equipment. It determines how efficiently fuel is burned, affecting the composition of pollutants [50]. Using the Propeller Law, the main engine's load factor of the ship main engine can be obtained. The calculation formula is in Equation (7):

$$LF = \frac{AS^3}{MS} \tag{7}$$

where AS represents the actual speed of the ship obtained from the AIS database, and MS represents the maximum speed of the ship in knots extracted from the IHS database. The load factor of the main engines calculated using the Propeller Law is determined for the four vessel types at different operational modes. On the other hand, the auxiliary load factor is extracted from the research report of the US EPA [46], as illustrated in Table 6. The US EPA provides guidance on assumptions of the load factors as part of their methodology for estimating emissions from marine vessels. The load factor assumptions from the US EPA report vary based on ship type and operation mode.

Table 6. Auxiliary engine load factor.

Ship Type	Cruising	RSZ	Maneuvering	Hoteling
Auto Carrier	0.15	0.30	0.45	0.26
Bulk Carrier	0.17	0.27	0.45	0.10
Container Ship	0.13	0.25	0.48	0.19
Cruise Ship	0.80	0.80	0.80	0.64
General Cargo	0.17	0.27	0.45	0.22
Tanker	0.24	0.28	0.33	0.26

4. Emission factor

Emission factors vary for engines with different speeds. When the load factor of the main engine is greater than 20%, the emission factor is considered constant, but it increases as the load factor decreases to less than 20%. There is limited data from which emission factors can be obtained, as emission testing of OGVs is relatively difficult and expensive. The emission factors provided in the US EPA report are from an analysis published by Entec highlighting individual factors for slow-speed engines, medium-speed engines, high-speed engines, steam turbines, gas turbines, and the three fuel types (HFO, MDO, and MGO) [46]. The US EPA [46] report has provided emission factors for different pollutants and fuel types for each of the combustion equipment; the values are as indicated in Table 7. In this study, it is assumed that all the vessels used MDO with a 0.5% sulfur content outside ECAs, and a MGO with a 0.1% sulfur content within the ECAs as per IMO regulations.

Table 7. Emission factors for each pollutant.

	CO_2			NO_X			SO_X		
	HFO 2.7%	MDO 0.5%	MGO 0.1%	HFO 2.7%	MDO 0.5%	MGO 0.1%	HFO 2.7%	MDO 0.5%	MGO 0.1%
Main Eng.	677.9	646.0	646.0	14.0	13.2	13.2	11.2	1.9	0.4
Aux. Eng.	722.5	690.7	690.7	14.7	13.9	13.9	11.9	2.1	0.4
Aux. Boiler	970.7	922.9	922.9	2.1	2.0	2.0	16.1	2.8	0.5

3. Results

This section highlights the results of emission calculation for the vessel types operating in the Republic of Korea for a period of one year calculated using the methodology mentioned above, with the values rounded off to one decimal point.

3.1. Results of Occupancy Time Analysis

Table 8 is a summary of the total occupancy time in hours for each vessel in each mode. From the results, bulk carriers have the highest total occupancy time. These ships transport bulky raw materials such as coal and grain that may require complex and lengthy loading and unloading times compared to containerized cargo, which are handled using standardized equipment. On the other hand, tankers have the highest occupancy time in the Reduced Speed Mode (RSZ) compared to other ship types because of operational and regulatory factors. Tankers transport large quantities of oil and gas, making them subject to stricter environmental regulations. Tankers may need to travel at much slower speeds for longer periods when entering certain regions to minimize the risk of accidents such as oil spills. In the cruising phase, containers had the highest occupancy time of all other ship types because of their operational patterns that resulted in frequent movements. Overall, container, bulk carriers, and general cargo had the highest occupancy time in the hoteling phase compared to other operation modes due to potentially longer waiting times at the ports. However, tankers had the lowest occupancy during hoteling because they dock at specialized terminals designed to handle specific liquid bulk cargo, making it less likely for them to wait for longer hours for an available berth.

Table 8. Calculated occupancy time in hours.

Ship Type	Cruising	RSZ	Maneuvering	Hoteling	Total
Tanker	412,009.9	668,095.1	294,201.5	157,924.9	1,532,231
Container	476,316.2	191,979.0	109,114.9	518,229.6	1,295,639
Bulk Carrier	270,434.0	365,208.4	102,559.8	1,072,769	1,810,971
General Cargo	107,002.0	403,859.7	150,189.3	708,553.2	1,369,604

3.2. Ship Emissions per Each Equipment

The main engine emits the largest pollutants for all vessels and pollutant types because the main engine operates at higher loads. However, it is worth noting that for tankers, emissions from the main engine and the auxiliary boilers are almost at the same level, while the emission levels from these two pieces of equipment were significantly different for the other ship types. Tables 9–11 present the results of emissions for the three equipment, rounded off to one decimal point. For the CO_2 emissions in Table 9, tankers had the highest total emissions (4,543,196 tons), with significant contributions from auxiliary boilers; for the NO_x emissions in Table 10, tankers and container vessels emitted similar levels of approximately 64,000 tons each, dominated by main engine exhaust. Table 11 of the SO_x emissions results highlights tankers again as the highest contributors, with about 13,937.7 tons, followed by container vessels and bulk carriers.

Table 9. Results of emission calculation (equipment/CO_2 in tons).

Equipment	Tanker	Container	Bulk Carrier	General Cargo
Main Engine	2,186,217	2,157,368	1,426,695	451,487.8
Aux. Engine	800,277.8	1,000,683	370,935.9	127,447.9
Aux. Boiler	1,556,701	316,650.6	118,118.2	83,926.62
Total	4,543,196	3,474,702	1,797,631	662,862.3

Table 10. Results of emission calculation (equipment/NO_x in tons).

Equipment	Tanker	Container	Bulk Carrier	General Cargo
Main Engine	44,671.9	44,082.4	29,193.0	9225.40
Aux. Engine	16,121.5	20,144.7	7471.14	1833.77
Aux. Boiler	3376.79	686.87	256.22	182.05
Total	64,170.2	64,913.8	36,920.3	11,974.8

Table 11. Results of emission calculation (equipment/SO_x in tons).

Equipment	Tanker	Container	Bulk Carrier	General Cargo
Main Engine	6700.78	6612.36	4375.8	1383.8
Aux. Engine	2458.82	3072.43	1139.1	391.57
Aux. Boiler	4778.16	971.73	362.55	257.60
Total	13,937.7	10,656.7	5877.6	2032.9

Figure 10 presents the main engine emission inventory. Tankers and containers had the highest emissions, with tankers slightly leading because they have larger engines with higher fuel requirements. Container ships also consume large amounts of fuel due to their need to maintain high speeds and operation frequency. Bulk carriers have moderate emissions, while general cargo ships have the lowest due to their small size and lower fuel consumption.

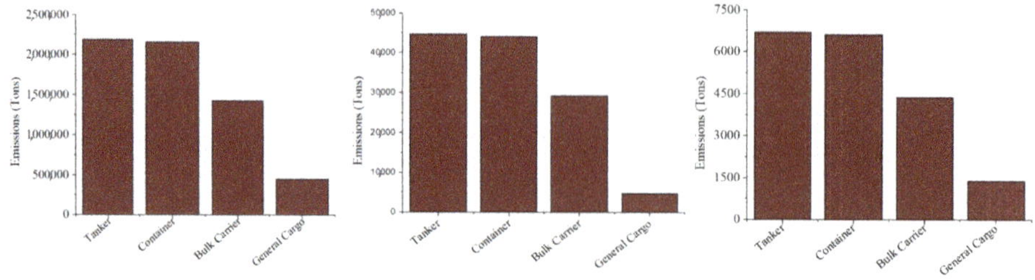

(**a**) (**b**) (**c**)

Figure 10. Ship inventory from the main engine: (**a**) CO_2, (**b**) NO_X, and (**c**) SO_X.

For the auxiliary engine shown in Figure 11, the results are different from the main engine inventory. Here, container ships emit slightly higher CO_2, NO_X, and SO_X exhaust gases compared to tankers. Auxiliary engines on container ships have higher operational demands during port stays, which explains the high emissions. Bulk carriers and general cargo had lower emissions, which is consistent with their lower auxiliary engine usage compared to tankers and containers.

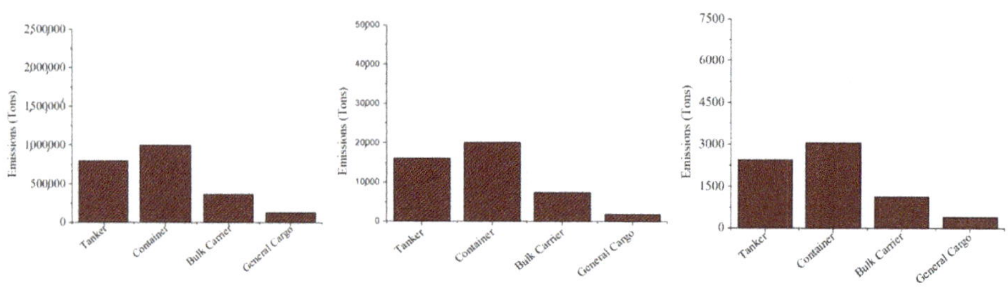

(**a**) (**b**) (**c**)

Figure 11. Ship emission inventory from the auxiliary engine: (**a**) CO_2, (**b**) NO_X, and (**c**) SO_X.

The auxiliary boiler emission inventory in Figure 12 revealed the highest CO_2, NO_X, and SO_X emissions from tanker vessels with significant differences compared to other ships. Auxiliary boilers on tankers are frequently used for cargo operations, such as heating oil for viscosity control and other processes that demand significant fuel consumption. Containers, bulk carriers, and general cargo have minimal emissions, reflecting that they are less reliant on boilers. Only CO_2 and SO_X emissions are significantly high from tankers, but NO_X pollutants are very low. The disparity occurred because CO_2 emissions directly correlate with the amount of fuel consumed during operation. Similarly, SO_X exhaust gases directly relate to sulfur fuel, and tankers use high-sulfur fuels. However, NO_X formation is more dependent on combustion temperatures rather than fuel amount or sulfur content.

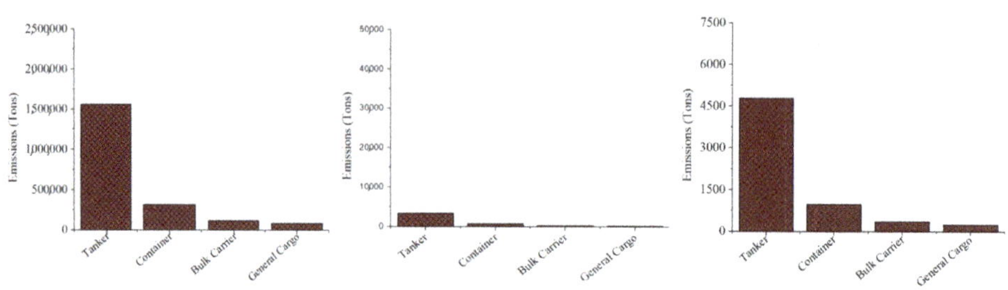

(**a**) (**b**) (**c**)

Figure 12. Ship emission inventory from the auxiliary boiler: (**a**) CO_2, (**b**) NO_X, and (**c**) SO_X.

3.3. Ship Emissions by Vessel Type

The summarized emissions per ship type are presented in Table 12, with the values rounded off to one decimal point. Tankers recorded the largest emissions, accounting for approximately 4,601,007.1 tons, followed by containers at 3,546,082.8 tons. General cargo had the least amount of emissions, totaling around 676,690.3 tons, because of their small number in operation and the use of small engines that consume less fuel compared to the other ship types.

Table 12. Results of emissions per ship type in tons.

Ship Type	CO_2	NO_x	SO_x	Total
Tanker	4,543,196	64,170.1	13,937.7	4,601,007
Container	3,474,702	64,913.8	10,656.7	3,546,082
Bulk Carrier	1,797,631	36,920.3	5877.6	1,971,715
General Cargo	662,862.7	11,974.8	2032.9	676,690.3

3.4. Ship Emissions by Operation Mode

Tables 13–15 present detailed results for emissions categorized by the operational mode. Container ships generated peak emissions during cruising and maneuvering. However, overall, the tankers had the highest total emissions combined in all phases due to significantly large pollutants released during the hoteling mode. Although tanker ships had shorter activity times during hoteling, they were the largest emitters of CO_2, illustrated in Table 13, because of high energy demands when stationery for oil heating, operation of cargo pumps, and other safety equipment for handling its cargo during loading or unloading operations. Crude oil carriers demand a continuous supply of heat to maintain an optimal temperature for the oil. Additionally, tankers still use less efficient fuels while in ports; therefore, although they spend less time in ports or at berth, their emissions are disproportionately higher than other vessels that use systems that are more efficient and cleaner fuels.

Table 13. Results of CO_2 emissions in tons.

Ship Type	Cruising	RSZ	Maneuvering	Hoteling	Total
Tanker	1,448,052	966,954.2	233,074.8	1,874,819	4,543,196
Container	2,088,042	449,536.4	275,091.1	737,842.7	3,474,702
Bulk Carrier	918,186.9	648,491.1	91,535.80	257,535.5	1,797,631
General Cargo	219,072.6	256,795.3	58,327.00	128,487.8	662,862.3

Table 14. Results of NO_x emissions in tons.

Ship Type	Cruising	RSZ	Maneuvering	Hoteling	Total
Tanker	29,558.1	19,707.3	3002.1	11,902.6	64,170.1
Container	40,973.0	9134.80	4659.1	10,146.9	64,913.8
Bulk Carrier	18,773.0	26,401.4	1666.5	3250.00	36,920.3
General Cargo	4473.5	5236.70	918.60	1346.00	11,974.8

Table 15. Results of SO_x emissions in tons.

Ship Type	Cruising	RSZ	Maneuvering	Hoteling	Total
Tanker	4439.0	2965.0	730.7	5802.8	13,937.7
Container	6156.3	1379.1	848.5	2275.3	10,656.7
Bulk Carrier	2817.7	1985.7	281.0	790.9	5877.6
General Cargo	671.50	787.30	179.1	394.9	2032.9

Bulk carriers and general cargo vessels showed moderate but non-negligible emissions across all modes. General cargo ships produced the lowest emissions in all modes due to their smaller size and shorter cruising time. Tankers and containers have the highest NO_x emissions, particularly when cruising when their engines operate at higher loads and higher temperatures. Bulk carriers showed notable NO_x pollutants when operating in the RSZ, as shown in Table 14. All the vessels had the lowest NO_x emission during maneuvering since engines typically operate at lower loads during port entry or exit. Containers emitted the highest SO_x emissions in the cruising mode, while tankers were the largest contributors of SO_x during hoteling, as indicated in Table 15. This can likely be explained by the large fuel consumption by containers and tankers when sailing. During hoteling, tankers use heavy fuel oil, which has a higher sulfur content than marine diesel oil or marine gas oil utilized by other ship types, leading to high SO_x emissions.

3.5. Ship Emissions by Pollutant

Table 16 consolidates the exhaust gases by pollutant (CO_2, NO_x, and SO_x) across all vessel types. CO_2 dominated the total emissions, accounting for a higher majority of total tonnage. The total CO_2 emissions from tankers, containers, bulk carriers, and general cargo vessels were approximately 10,478,392.7 tons. Tankers accounted for about 43.3%, containers 33.1%, bulk carriers 17.3%, and general cargo 6.3%. For NO_x, tankers emitted 33.6%, containers 34%, bulk carriers 26.2%, and general cargo 6.3%. On the other hand, SO_x emissions were 42.9% for tankers, 32.8% for containers, 18.1% for bulk carriers, and 6.3%, as highlighted in Table 16. Similarly, Figure 13 is a visualization of the variations in emissions across operational modes and vessel types for CO_2, NO_x, and SO_x. Distinct peaks are observed for container ships during cruising, while tanker emissions peak sharply during hoteling operations.

Table 16. Results of emissions per pollutant in tons.

Ship Type	CO_2	NO_x	SO_x	Total
Tanker	4,522,900	64,170.1	13,936.8	4,601,007
Container	3,470,512	64,913.8	10,656.6	3,546,082
Bulk Carrier	1,915,749	50,090.9	5875.30	1,971,715
General Cargo	662,682.7	11,974.8	2032.80	676,690.3
TOTAL	10,478,392	191,149.6	32,501.50	10,702,043

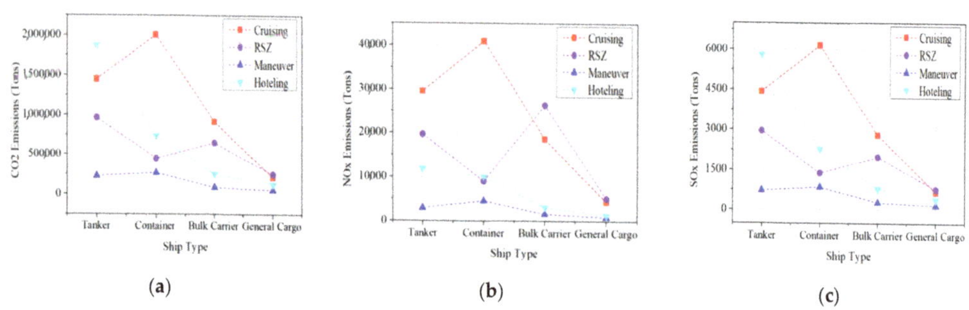

Figure 13. A line graph illustrating emission variations: (**a**) CO_2, (**b**) NO_x, and (**c**) SO_x.

3.6. Emission Distribution

This section provides the spatial visualization of the total emissions over a period of 1 month (1–31 January 2022) in a major port in South Korea from the four ship types using a Geographic Information System. After estimating the vessel density for each grid cell through spatio-temporal analysis in PostgreSQL/PostGIS, the data were combined with vessel-specific hourly emission values that were calculated for the four ship types and four operational modes to estimate the total emissions. The emissions output was then imported into ArcGIS Pro using PostGIS to generate detailed emission intensity maps using a 0.5 km by 0.5 km grid system. The spatial maps illustrate the emissions distribution with significant spatial differences based on the ship-type operation in the region. These maps provide a visualization of where the emissions are most concentrated, providing a basis for environmental monitoring and the establishment of pollution control policies.

The emission distribution maps in Figures 14–16 highlight distinct spatial differences across vessel types. Overall, container vessels exhibit high emission densities in the approach channels and anchorage zones near Busan Port, corresponding to their frequent maneuvering and prolonged hoteling. Tankers show more dispersed emissions patterns, reflecting their activity across outer harbor zones and anchoring areas. In contrast, general cargo ships show moderate and more spatially uniform emission footprints, particularly along secondary routes, while bulk carrier emissions are minimal. The visualized emission patterns directly correlate with the operational behavior and port usage characteristics of each vessel type. Detailed discussions of the spatial distribution maps for all the pollutant types are done below.

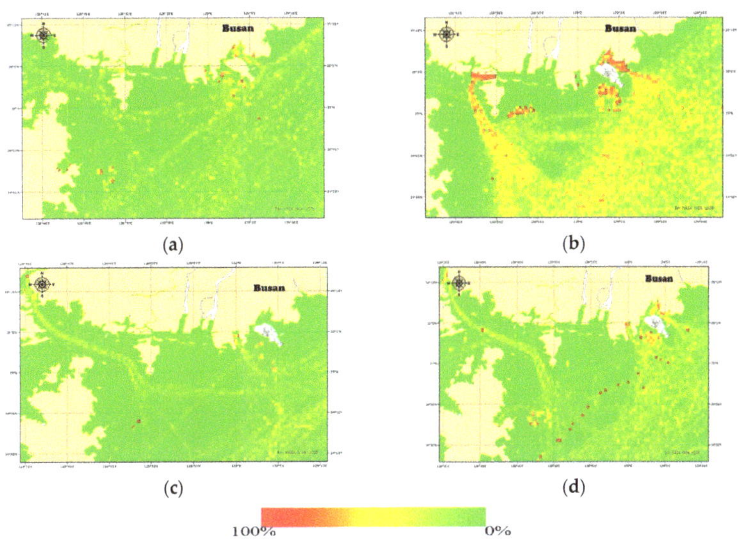

Figure 14. CO_2 emission distribution in Busan: (**a**) Tanker, (**b**) container, (**c**) bulk carrier, and (**d**) general cargo.

Based on the emission maps in Figure 14a, emission clusters are observed in Busan North Port and its anchorage area near Yeongdo–gu. Busan North Port is a major hub for various maritime operations, including handling oil and chemical products. The emissions result from the use of boilers during cargo operations as tankers use this zone. The west of the North Port also has an emission cluster, as it could be an outer anchorage area used by tankers awaiting berth assignments or clearance to enter regulated port waters. In contrast, container vessel CO_2 emissions show the highest concentrations symbolized by red clusters in the North Port and New Port and in their anchorage areas, as shown in Figure 14b. Busan is the largest container handling port in South Korea, processing millions of twenty-foot Equivalent Units (TEUs) annually. There are concentrated emissions near the container terminals in the North Port, New Port, and anchorage areas. The dense red clusters highlight intense port and anchorage activities with emissions being generated during maneuvering and hoteling modes. Bulk carrier CO_2 emissions are significantly less in Busan and primarily located in anchorage zones due to their longer hoteling times during loading and unloading of cargo. For general cargo in Figure 14d, noticeable emissions are seen on the vessel tracks approaching/leaving the North Port anchorage area. General cargo vessels transport a wide variety of cargo, typically in smaller quantities, and their emissions are dispersed, aligning with operational routes.

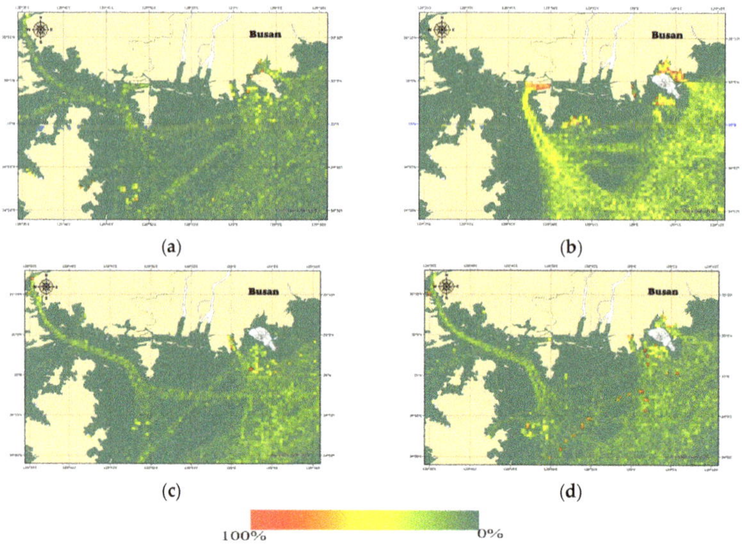

Figure 15. NO_x emission distribution in Busan: (**a**) Tanker, (**b**) container, (**c**) bulk carrier, and (**d**) general cargo.

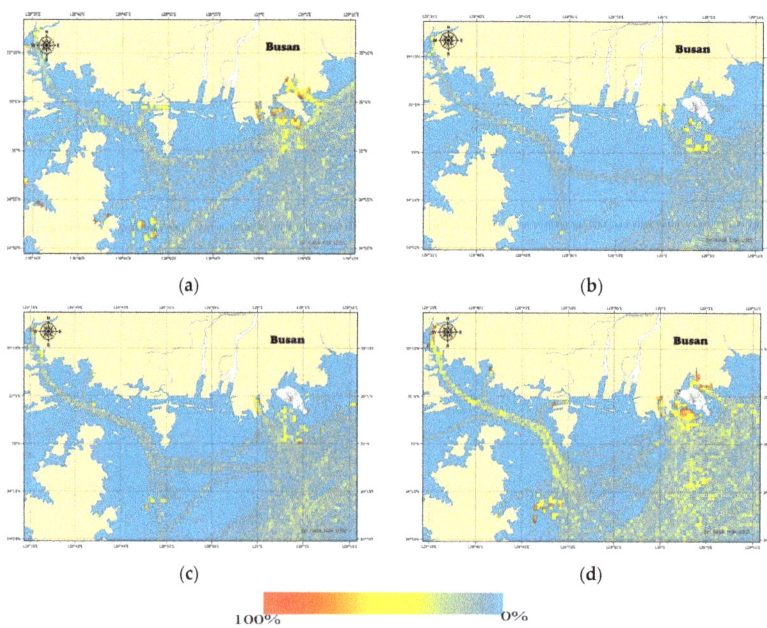

Figure 16. SO_x emission distribution in Busan: (**a**) Tanker, (**b**) container, (**c**) bulk carrier, and (**d**) general cargo.

For tankers, low quantities of NO_x exhaust, likely resulting from maneuvering and low-speed operations near anchorage zones, can be observed around the North Port, as shown in Figure 15a. NO_x emissions from container ships, in Figure 15b, show the most significant concentrations in Busan, especially in the North Port and New Port. Medium–intensity emission tracks can be seen in the navigation channels approaching Busan ports. The constant flow of large container ship traffic in Busan leads to high NO_x exhaust during port entry and exit. The clustered patterns of medium intensity near the terminals indicate high activity during maneuvering. Container vessels operate at higher engine loads during cruising and maneuvering modes, resulting in high temperatures that cause substantial NO_x formation. Figure 15c above is the spatial result of the emissions from bulk carriers in the Republic of Korea. The emissions are modest and more spatially dispersed, primarily occurring in anchorage zones rather than on transit paths. On the other hand, distinct NO_x clusters along the navigation routes approaching the Busan Old Port anchorage can be observed for general cargo ships, as illustrated in Figure 15d. These emissions result from continuous low-speed movements during the approach or departure of ships.

Tanker ships emitted the greatest SO_x emissions compared to other vessel types. SO_x pollutants are emitted when sulfur-based fuel is burnt. Therefore, SO_x emissions directly correlate to the sulfur levels in the fuel and partly to the quantity of fuel. Figure 16a shows high-intensity emission clusters in the port areas and anchorage areas despite the fuel sulfur limits imposed by the IMO on ships when entering or exiting the ports. Low-intensity emission tracks can also be observed on the fairways, indicating cruise-related exhausts. Despite the high container traffic, SO_x emissions are the lowest compared to other vessel types, as seen in Figure 16b. This could be due to the strict enforcement of the 0.1% sulfur fuel limit or the use of scrubbers applied to ECAs. Additionally, large container vessels likely have efficient systems installed that lead to reduced pollution. Clusters of medium-intensity emissions are only visible in the Busan Old Port anchorage, and this could be due to ships waiting in the anchorage zones. Bulk carriers also had lower SO_x emissions, as shown in Figure 16c, consistent with their slower operations and less frequent port calls at Busan Port. General cargo also emitted notable SO_x exhaust with several high-intensity clusters illustrated in Figure 16d. This could be due to general cargo vessels being older fleets that use less efficient engines or longer waiting times during hoteling and maneuvering. The yellow color signifies medium-intensity pollutants emitted from ships cruising slowly when approaching or departing the port.

4. Discussion

This study aligns with the global and national efforts to reduce maritime pollution by providing a data-driven emission inventory and a detailed analysis of current emissions levels using a bottom-up and spatio-temporal approach. Through the integration of AIS, IHS, and GIS technologies, vessel emissions across different ship types, operational modes, and equipment categories were quantified and mapped for Korean coastal waters, identifying key areas where compliance with international regulations can further minimize the environmental footprint of maritime activities.

This research revealed significant differences in emissions levels from the four vessel types. Tankers were the largest emitters of CO_2, NO_x, and SO_x emissions, followed by container vessels, bulk carriers, and general cargo ships. As shown in Table 12, tankers accounted for the largest share of CO_2, nearly 43.3%, contributing around 4.5 million tons annually, corresponding to [51]'s research that found tanker ships as the highest emitters at Incheon port. Container ships contributed slightly lower CO_2 amounts, totaling 3.5 million tons, but emitted comparable levels of NO_x, while general cargo had the least value at 6.3%. The dominance of tanker emissions can be attributed to their larger engine sizes, leading to higher fuel consumption, extended hoteling periods, and intensive auxiliary boiler operations during cargo loading and unloading operations. Additionally, there are many tanker ships, accounting for a significant percent of the global fleet; hence, generally, CO_2, NO_x, and SO_x emissions were highest for tanker ships and lowest for general cargo. Tables 9–11 showing equipment-specific results emphasize the primary sources of emissions for CO_2 and NO_x as the main engines, while auxiliary boilers significantly contributed to SO_x emissions, particularly for tankers and general cargo ships. The operational mode results presented in Tables 13–15 show that the cruising mode generated the highest emissions for container vessels, particularly for CO_2 and NO_x, compared to other ship types. In contrast, hoteling was a major emission contributor for tankers and bulk carriers due to long idle times at anchorage using auxiliary engines and boilers. This emphasizes the need to expedite the adoption of shore power at major ports like Busan. In the line graphs in Figure 13, emissions during maneuvering and RSZ operations were non-negligible and varied significantly between vessel types, indicating the operational profiles' critical impact on total emissions.

Overall, this study highlights tankers and containers as major contributors to maritime emissions. Tankers were the largest contributors to the total emissions in Korean coastal waters. Container ships, while contributing slightly lower overall CO_2 emissions, exhibited high NO_x emissions, particularly during the maneuvering and cruising phases, reflecting their frequent port entries and high-load engine operations. Bulk carriers and general cargo vessels produced lower emission volumes but showed characteristic patterns linked to extended hoteling periods and slow-speed operations near port approaches. These vessel categories, often older and operating less efficiently, contributed a disproportionately higher amount of SO_x relative to their CO_2 emissions, suggesting the potential use of higher-sulfur fuels. Additionally, vessels generated varying exhaust gases depending on their sizes and operational modes, suggesting that distinct operational characteristics of ships played a critical role in emission levels. The high amounts of CO_2 emissions across all ship types suggest that the use of high-quality fossil fuels has been insufficient in minimizing GHG emissions. Therefore, all ship types should be strictly required to adhere to IMO's energy efficiency standards. Additionally, emphasis should be directed towards the adoption of clean energy sources such as green ammonia or hybrid propulsion options.

Moreover, this study revealed significant spatial variations in emissions in Busan, highlighting the influence of vessel types and operational modes on emission volumes. The emission hotspots identified through GIS mapping correspond to anchorage zones and the major shipping lanes. Anchorage zones and ship terminals and berths had high-intensity emissions, especially for CO_2. In the study, Busan had the highest CO_2 emissions clustering from container vessels near the North Port and the New Port, aligning with its position as among the busiest container terminals in the world. However, SO_x emissions were notably lower in Busan, primarily concentrated around anchorage zones due to stringent fuel regulations in ECAs. These results highlight the concentration of container shipping activities in Busan Port and the need for localized maritime pollution control strategies. Additionally, the research revealed the need to apply different emission control measures for different vessel types, regions, and shipping activities. For example, tanker and general cargo vessels' emissions can be significantly reduced by adopting shore power in port areas that experience the highest vessel activities. Promoting the use of shore power in these zones will reduce the environmental impact of stationary vessels at the port. Port authorities should expand the shore power networks and charging infrastructure to allow ships to switch off the auxiliary engines while at ports. Furthermore, improving communication and cooperation between ports will contribute to environmentally friendly port management.

On the other hand, container vessels' cruising emissions can be controlled by finding a balance between speed reduction and maintaining their schedules.

Generally, CO_2 emissions were the highest, totaling approximately 10.5 million tons for a period of 1 year. In 2019, Korea's annual shipping GHG emissions accounted for around 27.4 million tons of CO_2-equivalent, representing approximately 3.9% of the total GHG emissions in Korea (about 701.3 million tons of CO_2-equivalent) [10]. Additionally, according to the Greenhouse Gas Inventory and Research Center of Korea (GIR), the Republic of Korea's total GHG emissions in 2022 was reported as 642.8 million tons of CO_2-equivalent [52]. Focusing specifically on CO_2 from fuel combustion, the total emissions was around 549.31 million tons in 2022 [53]. To contextualize the results of this study, the 10.5 million tons of CO_2 emissions estimated from the four vessel types represent approximately 1.9% of the Republic of Korea's total CO_2 emissions from the fuel combustion generated in 2022. The value only represents a percentage of emissions from the shipping sector since this study only covers the four dominant contributors of ship exhaust gases (tankers, containers, bulk carriers, and general cargo). Although this proportion seems modest, it is significant in Korea's national emissions reduction target since the maritime industry is integral to the region's development. The prolonged reliance on fossil fuels is a huge hindrance to decarbonization in the Republic of Korea, with research indicating that by 2021, the transport sector will be mostly dominated by fossil fuels [54]. In this study, all the ships used fossil fuels, specifically MDO outside ECAs and MGO within the ECAs. The share of low-carbon fuels in the transport mix needs to increase to between 40–60% by 2040 if the Republic of Korea wants to largely reduce national GHG emissions [55]. Consequently, MOF released the 2050 Net-Zero Roadmap, which sets target emission reductions for Nationally Determined Contributions (NDCs), which include domestic shipping emissions at 40% by 2030 and 70% by 2050 compared to 2018 levels, whereas, for international shipping, the goal aligns with IMO's 50% GHG reduction in emissions. Furthermore, the Republic of Korea implemented several policies to facilitate emission reductions in its shipping industry. These include a green fleet transition that entails conversion to eco-friendly ships, the use of alternative fuels, the establishment of zero-emission shipping routes, and international corporations [10]. With increasing global pressure to decarbonize and the Republic of Korea's vision to be among the leaders in shipping and shipbuilding to realize net-zero emissions by 2050, a comprehensive database of current emission levels from Korean ships is paramount. This study complements the existing emission inventories and avails emission data that are vital in promoting the development of targeted emission reduction strategies to achieve the Republic of Korea's broader climate goal and maintain its competitiveness in global shipping.

The bottom-up methodology combined with the spatio-temporal technique applied in this study generated detailed emission values by vessel type, operational mode, and equipment type, thus contributing to existing emission databases to help mitigate maritime pollution [55]. However, the bottom-up approach still generates uncertainties in the results due to the imputation of missing technical data, load factors, vessel operational behaviors, and lack of integration of dynamic environmental conditions that greatly affect fuel consumption and emissions [32]. Future work should address these uncertainties by incorporating probabilistic modeling, real-time metocean data, and machine learning techniques to refine predictive capabilities. Further, the differentiated emissions by vessel type, operational mode, pollutant type, and spatial location offer actionable insights for maritime stakeholders, including port authorities, regulators, and policymakers. From the findings, port authorities are mandated to expand shore power systems at terminals where hoteling emissions are dominant. Korea has already implemented vessel speed programs at its major ports; however more should be done to further mitigate pollution in its port areas. The government should also increase incentivization for fuel switching or provide retrofit investments based on ship emission profiles as part of its maritime pollution policies. In addition, emission monitoring systems integrating AIS, remote sensing, and onboard reporting should be mandated for all ships to enable South Korea to pursue its carbon neutrality targets along with IMO's net-zero targets.

To conclude, this study underlines the use of a bottom-up and spatio-temporal analysis approach as a basis for conducting ship-type and mode-specific emission inventories. This has implications for tracking emission changes and understanding its patterns at various levels, which is fundamental for targeted emission reduction strategies that will contribute towards addressing the adverse impacts of ship pollutants in the Republic of Korea's coastal regions. These findings provide actionable insights for developing differentiated policies aimed at specific vessel categories and operational behaviors, supporting the Republic of Korea's broader maritime decarbonization goals aligned with the IMO's net-zero strategy.

5. Conclusions

Shipping is a pivotal industry for global trade and is particularly significant for countries like South Korea that heavily rely on exports. Data-driven decisions play a vital role in the formulation of effective policies with the growing demand for sustainability and IMO's tightened regulatory frameworks to achieve net-zero emissions in the shipping sector. This study on emissions from ships operating in Korean waters offers a comprehensive analysis of vessel exhaust gases and highlights findings regarding the distribution and intensity of CO_2, NO_x, and SO_x emissions by vessel types, operational modes, and pollutants. The main conclusions are summarized below:

1. Tankers were identified as the largest contributors to CO_2 (about 4.5 million tons), NO_x (around 64,000 tons), and SO_x (approximately 13,900 tons) emissions, followed by container vessels, bulk carriers, and general cargo ships. Further, analysis of operational mode pollution showed that container ships generated the most emissions during cruising, while tankers' emissions were highest during hoteling.
2. Main engines were the dominant source of CO_2 and NO_x emissions, whereas auxiliary boilers were significant emitters of SO_x emissions, particularly during hoteling for tanker ships.
3. The spatial distribution maps highlighted emission hotspots from container vessels at Busan's North Port and New Port, emphasizing the influence of vessel density and operational behaviors on pollution patterns.
4. CO_2 emissions were the highest, accounting for approximately 1.9% of the total CO_2 emissions from fuel combustion in Korea. Therefore, maritime emissions in Korea represent a substantial share of the national CO_2 emissions, highlighting an urgent need for more sector-specific mitigation measures.
5. Furthermore, this study validates the significance of spatio-temporal analysis combined with a bottom-up approach for vessel-specific inventories to implement effective decarbonization policies.

This study leveraged maritime big data combined with emission models to enable the accurate spatial mapping of shipping emissions, with the aim of supporting global efforts for minimizing maritime pollution. Thus, it contributed to the existing emission databases that highlight emission characteristics to help formulate effective pollution control policies. This research had implications for tracking emission changes and understanding its patterns according to the mentioned criteria; this is fundamental for targeted emission reduction strategies that will contribute towards addressing the adverse impacts of ship pollutants in the Republic of Korea's coastal regions. Policy recommendations based on this study include:

1. Expediting the expansion of shore power (AMP) infrastructure at key port terminals in Korea to curb hoteling emissions.
2. Implementing other vessel speed reduction programs at key cruising shipping lanes.
3. Intensifying incentivization programs to promote the operation of low-emission vessels.

In conclusion, this study contributes to the existing research on developing detailed emission databases that provide a foundation for emission reduction strategies contributing to the broader goal of achieving net-zero emissions in shipping. Nevertheless, this study was not sufficient due to the following limitations:

1. First, ships are greatly affected by weather and environmental factors such as wind and waves when sailing that affect fuel consumption and the dispersion of pollutants [50]. In the future, dynamic weather and environmental data should be integrated to make the research more comprehensive.
2. Second, future research should integrate GIS tools with machine learning techniques for dynamic visualizations to help uncover complex trends in emission distributions.
3. Third, although regression-based imputation allowed recovery of missing technical parameters, these models introduced error uncertainties in the overall results. Future work could incorporate probabilistic approaches to better capture this uncertainty or conduct a scenario analysis to assess the influence of imputation on emission results.
4. Lastly, future work should consider the use of satellite emission tracking alongside AIS data to correlate emissions with specific vessels. These improvements will increase the viability of these emission results by enhancing transparency and traceability.

Author Contributions: Conceptualization, I.-S.C., W.-J.S., D.-J.J. and A.W.K.; methodology, D.-J.J.; software, A.W.K. and W.-J.S.; validation, I.-S.C., W.-J.S. and A.W.K.; formal analysis, A.W.K.; investigation, A.W.K.; resources, A.W.K.; data curation, A.W.K.; writing—original draft preparation, A.W.K.; writing—review and editing, I.-S.C., W.-J.S. and A.W.K.; visualization, A.W.K.; supervision, I.-S.C.; project administration, I.-S.C.; funding acquisition, W.-J.S. All authors have read and agreed to the published version of the manuscript.

Funding: This research was funded by the National Research Foundation of Korea (NRF) grant funded by the Korean government (Ministry of Science and ICT (MSIT)), grant number RS-2025-00554193.

Data Availability Statement: This study was conducted using data provided by the Ministry of Oceans and Fisheries (MOF, Republic of Korea) for ship traffic route research.

Conflicts of Interest: The authors declare no conflicts of interest.

Abbreviations

The following abbreviations are used in this manuscript:

GHG	Greenhouse gas gases
CO_2	Carbon dioxides
NOx	Nitrogen oxides
SOx	Sulfur oxides
GIS	Geographic Information System
RSZ	Reduced Speed Zone
IMO	International Maritime Organization
MARPOL	International Convention on Maritime Pollution
IPCC	International Panel on Climate Change
MOF	Ministry of Ocean Fishery

AIS	Automatic Information System
GT	Gross tonnage
SEIM	Ship Emission Inventory Model
ECA	Emission Control Area
IHS	Information Handling System
EMODnet	European Marine Observation and Data Network
AOI	Area of interest
USEPA	United States Environmental Protection Authority
GB	Gigabyte
MMSI	Maritime Mobile Service Identity
HOG	Heading Over Ground
COG	Course Over Ground
DWT	Deadweight
RPM	Revolutions Per Minute
SOG	Speed Over Ground
SQL	Structured Query Language
LOA	Length overall
VSR	Vessel speed reduction
OGV	Ocean Going Vessels
AMP	Alternative Maritime Power
HMM	Hyundai Merchant Marine
LNG	Liquefied natural gas
MCR	Maximum Continuous Rate
LF	Load factor
EF	Emission factor
ME	Main engine
AE	Auxiliary engine
AB	Auxiliary boiler
FCF	Fuel Correction Factor
ARB	Air Resources Board
HFO	Heavy fuel il
MDO	Marine diesel oil
MGO	Marine gas oil
TEU	Twenty-foot Equivalent Units
GEEM	Geographic Emission Estimation Model
KMI	Korea Maritime Institute
GIR	Greenhouse Gas Inventory and Research
NDCs	Nationally Determined Contributions

References

1. Kwon, Y.; Lim, H.; Lim, Y.; Lee, H. Implication of activity-based vessel emission to improve regional air inventory in a port area. *Atmos. Environ.* **2019**, *203*, 262–270. [CrossRef]
2. United Nations Conference on Trade and Development (UNCTAD). *The UNCTAD Handbook of Statistics 2021*; UNCTAD: New York, NY, USA, 2021.
3. IMO. *Second International Maritime Organization GHG Study 2009: Update of the 2000 IMO GHG Study*; IMO: London, UK, 2009.
4. Bilgili, L.; Celebi, U.B. Emission routing in maritime transportation. In *Energy, Transportation and Global Warming*; Springer: Berlin/Heidelberg, Germany, 2016; pp. 837–849.
5. Carlton, J.S.; Danton, S.D.; Gawen, R.W. *Marine Exhaust Emissions Research Programme*; Lloyd's Register Engineering Services: London, UK, 1995.
6. Eyring, V.; Köhler, H.W.; Lauer, A.; Lemper, B. Emissions from international shipping: 2. Impact of future technologies on scenarios until 2050. *J. Geophys. Res. Atmos.* **2005**, *110*, 1–23. [CrossRef]
7. Song, S. Ship emissions inventory, social cost and eco-efficiency in Shanghai Yangshan port. *Atmos. Environ.* **2014**, *82*, 288–297. [CrossRef]
8. Merk, O. *Shipping Emissions in Ports*; International Transport Forum: Paris, France, 2014.

9. IMO. *Fourth IMO Greenhouse Gas Study 2020: Safe, Secure and Efficient Shipping on Clean Oceans*; International Maritime Organization (IMO): London, UK, 2021.
10. Yum, J. Korea's Green Shipping Pathways: The Korean Shipping Landscape and Policy Recommendations for Ocean-Climate Leadership in Shipping. In *Solutions for Our Climate*; PACIFIC ENVIRONMENT: Seoul, Republic of Korea, 2022.
11. IPCC. *Intergovernmental Panel on Climate Change Guidelines for National Greenhouse Gas Inventories*; Institute for Global Environmental Strategies (IGES): Kanagawa, Japan, 2006.
12. Yoon, Y. *Policy for Maritime Industry Development*; National Assembly Secretariat: Seoul, Republic of Korea, 2009.
13. Ministry of Oceans and Fisheries. *Vessel Speed Reduction (VSR) Program to Start December This Year*; Ministry of Oceans and Fisheries: Sejong-si, Republic of Korea, 2019.
14. Office of the Spokesperson. U.S Announcements Under the Green Shipping Challenge at COP27. November 2022. Available online: https://2021-2025.state.gov/u-s-announcements-under-the-green-shipping-challenge-at-cop27/ (accessed on 25 April 2025).
15. Natural Resources Defense Council. *Harboring Pollution Strategies to Clean Up U.S. Ports*; Natural Resources Defense Council: New York, NY, USA, 2004.
16. Vutukuru, S.; Dabdub, D. Modeling the effects of ship emissions on coastal air quality: A case study of southern California. *Atmos. Environ.* **2008**, *42*, 3751–3764. [CrossRef]
17. Andersson, C.; Bergström, R.; Johansson, C. Population exposure and mortality due to regional background PM in Europe–Long-term simulations of source region and shipping contributions. *Atmos. Environ.* **2009**, *43*, 3614–3620. [CrossRef]
18. Corbett, J.J.; Winebreak, J.J.; Green, E.; Kasibhatla, P.; Eyring, V.; Lauer, A. Mortality from Ship Emissions: A Global Assessment. *Environ. Sci. Technol.* **2007**, *41*, 8512–8518. [CrossRef]
19. Yau, P.S.; Lee, S.C.; Cheng, Y.; Huang, Y.; Lai, S.; Xu, X. Contribution of ship emissions to the fine particulate in the community near an international port in Hong Kong. *Atmos. Res.* **2013**, *124*, 61–72. [CrossRef]
20. Ye, S.; Zheng, J.; Pan, Y.; Wang, S.; Lu, Q.; Zhong, L. Marine emission inventory and its temporal and spatial characteristics in Guangdong Province. *Acta Sci. Circumstantiae* **2014**, *34*, 537–547.
21. Zheng, C.M.; Hoffmann, M.R.; Mauzerall, D.L.; Gan, J.; Song, L. Expanding our horizons on the Earth's sustainable future. *Sustain. Horiz.* **2021**, *1*, 100001. [CrossRef]
22. Nunes, R.A.O.; Alvim-Ferraz, M.C.M.; Martins, F.G.; Sousa, S.I.V. The activity-based methodology to assess ship emissions—A review. *Environ. Pollut.* **2017**, *231*, 87–103. [CrossRef]
23. Corbett, J.J.; Koehler, H.W. Updated emissions from ocean shipping. *J. Geophys. Res. Atmos.* **2003**, *8*, 87–107. [CrossRef]
24. Kwon, J.; Yeo, S.; Lee, W. Assessment of Shipping Emissions on Busan Port of South Korea. *J. Mar. Sci. Eng.* **2023**, *11*, 716. [CrossRef]
25. Yang, J.; Yin, P.; Ye, S.; Wang, S.; Zheng, J.; Ou, J. Marine emission inventory and its temporaland spatial characteristics in the city of Shenzhen. *Environ. Sci. J. Integr. Environ. Res.* **2015**, *36*, 1217–1226.
26. Xing, H.; Duan, S.; Liang, B.; Liu, Q. Methods to estimate exhaust emissions from sea-going ships according to operational data. *Navig. China* **2016**, *39*, 93–98.
27. Villalba, G.; Gemechu, E.D. Estimating GHG emissions of marine ports-the case of Barcelona. *Energy Policy* **2011**, *39*, 1363–1368. [CrossRef]
28. McCollum, D.; Yang, C. Achieving deep reductions in US transport greenhouse gas emissions: Scenario analysis and policy implications. *Energy Policy* **2009**, *37*, 5580–5596. [CrossRef]
29. Tichavska, M.; Tovar, B. External costs from vessel emissions at port: A review of the methodological and empirical state of the art. *Transp. Rev.* **2017**, *37*, 383–402. [CrossRef]
30. Maragkogianni, A.; Papaefthimiou, S.; Zopounidis, C. Current Methodologies for the Estimation of Maritime Emissions. In *Mitigating Shipping Emissions in European Ports*; Briefs in Applied Sciences and Technology; Springer: Berlin/Heidelberg, Germany, 2016; pp. 25–35.
31. Psaraftis, H.N.; Kontovas, C.A. CO_2 emission statistics for the world commercial fleet. *WMU J. Marit. Aff.* **2009**, *8*, 1–25. [CrossRef]
32. Miola, A.; Ciu, B. Estimating air emissions from ships: Meta-analysis of modelling approaches and available data sources. *Atmos. Environ.* **2011**, *45*, 2242–2251. [CrossRef]
33. Mou, N.; Zhang, X.; Yang, T.; Xu, H.; Zheng, Y.; Wang, J.; Niu, J. Carbon footprints: Uncovering multilevel spatiotemporal changes of ship emissions during 2019–2021 in the US. *Sci. Total Environ.* **2024**, *912*, 169395. [CrossRef]
34. Liu, H.; Fu, M.; Jin, X.; Shang, Y.; Shindell, D.; Faluvegi, G.; Shindell, C.; He, K. Health and climate impacts of ocean-going vessels in East Asia. *Nat. Clim. Change* **2016**, *6*, 1037–1041. [CrossRef]
35. Li, Y.; Jia, P.; Jiang, S.; Kuang, H.; Hong, Y.; Wang, S.; Zhao, X.; Guan, D. The climate impact of high seas shipping. *Natl. Sci. Rev.* **2023**, *10*, nwac279. [CrossRef]
36. Kim, H.; Bui, H.D.; Hong, S.S. Estimating of Air Pollution from Ships in Port Area: A Case Study of Yeosu and Gwangyang Ports in Korea. *Atmosphere* **2022**, *13*, 1890. [CrossRef]

37. Song, S.K.; Shon, Z.H. Current and future emission estimates of exhaust gases and particles from shipping at the largest port in Korea. *Environ. Sci. Pollut. Res.* **2014**, *21*, 6612–6622. [CrossRef]
38. Maragkogianni, A.; Papaefthimiou, S.; Zopounidis, C. *Mitigating Shipping Emissions in European Ports: Social and Environmental Benefits*; Springer International Publishing: Berlin, Germany, 2016.
39. IMO. *Third International Maritime Organization Greenhouse Gas Study*; IMO: London, UK, 2014.
40. Shu, Y.; Daamen, W.; Ligteringen, H.; Wang, M.; Hoogendoorn, S. Calibration and validation for the vessel maneuvering prediction (VMP) model using AIS data of vessel encounters. *Ocean Eng.* **2018**, *169*, 529–538. [CrossRef]
41. Gracia, S.; Luengo, J.; Herrera, F. *Data Preprocessing in Data Mining*; Springer International Publishing: Cham, Switzerland, 2015; p. 72.
42. Lin, W.; Tsai, C. Missing Value Imputation: A review and analysis of the literature (2006–2017). *Artif. Intell. Rev.* **2020**, *53*, 1487–1509. [CrossRef]
43. EMODnet. Human Activities; Making Use of Our Oceans. EU Vessel Density Map Detailed Method. 2019. Available online: https://www.emodnet-humanactivities.eu/documents/Vessel%20density%20maps_method_v1.5.pdf (accessed on 30 April 2025).
44. Lee, H.; Park, D.; Choo, S.; Pham, H. Estimation of the Non-Greenhouse Gas Emissions Inventory from Ships in the Port of Incheon. *Sustainability* **2020**, *12*, 8231. [CrossRef]
45. Liu, T.; Sheu, H.; Tsai, J. Sulfur dioxide emission estimates from merchant vessels in a port area and related control strategies. *Aerosol Air Qual. Res.* **2014**, *14*, 413–421. [CrossRef]
46. U.S. Environmental Protection Agency. *Current Methodologies in Preparing Mobile Source Port-Related Emission Inventories*; ICF International: Ottawa, ON, Canada, 2009.
47. Weng, J.; Shi, K.; Gan, X.; Li, G.; Huang, Z. Ship emission estimation with high spatial-temporal resolution in the Yangtze River estuary using AIS data. *J. Clean. Prod.* **2020**, *248*, 119297. [CrossRef]
48. Yum, J.; Han, Y. Port Decarbonization: Focusing on South Korean Five Major Ports; Solutions for Our Climate. 2023. Available online: https://content.forourclimate.org/files/research/XuvmFUe.pdf (accessed on 30 April 2025).
49. Ajsa, H. HMM's LNG-Ready Containership Trials Shore Power at South Korean Port. 2025. Available online: https://www.offshore-energy.biz/hmms-lng-ready-containership-trials-shore-power-at-south-korean-port/ (accessed on 30 April 2025).
50. Jalkanen, J.; Brink, A.; Kalli, J.; Pettersson, H.; Kukkomen, J.; Stipa, T. A modelling system for the exhaust emissions of marine traffic and its application in the Baltic Sea area. *Atmos. Chem. Phys.* **2009**, *9*, 9209–9223. [CrossRef]
51. Lee, H.; Pham, H.; Chen, M.; Choo, S. Bottom-Up Approach Ship Emission Inventory in Port of Incheon Based on VTS Data. *J. Adv. Transp.* **2021**, *2021*, 5568777. [CrossRef]
52. Greenhouse Gas Inventory & Research Center of Korea. 2022 GHG Inventory. 2022. Available online: https://www.gir.go.kr/eng/ (accessed on 30 April 2025).
53. International Energy Agency. Total CO_2 Emissions from Energy. In Summit on the Future of Energy Security. Available online: https://www.iea.org/countries/korea/emissions (accessed on 30 April 2025).
54. Climate Transparency. Climate Transparency Report: Comparing G20 Climate Action. 2022. Available online: https://www.climate-transparency.org/wp-content/uploads/2022/10/CT2022-South-Korea-Web.pdf (accessed on 30 April 2025).
55. Bojić, F.; Gudelj, A.; Bošnjak, R. An Analytical Model for Estimating Ship-Related Emissions in Port Areas. *J. Mar. Sci. Eng.* **2023**, *11*, 2377. [CrossRef]

Journal of
Marine Science and Engineering

MDPI

Article

Smart Maritime Transportation-Oriented Ship-Speed Prediction Modeling Using Generative Adversarial Networks and Long Short-Term Memory

Xinqiang Chen [1], Peishi Wu [1], Yajie Zhang [1], Xiaomeng Wang [2,*], Jiangfeng Xian [1,*] and Han Zhang [3]

1 Institute of Logistics Science and Engineering, Shanghai Maritime University, Shanghai 201306, China; chenxinqiang@stu.shmtu.edu.cn (X.C.); 202330510045@shmtu.edu.cn (P.W.); zhangyajie@stu.shmtu.edu.cn (Y.Z.)
2 College of Transport and Communications, Shanghai Maritime University, Shanghai 201306, China
3 College of Ocean Science and Engineering, Shanghai Maritime University, Shanghai 201306, China; 202440410003@stu.shmtu.edu.cn
* Correspondence: wangxm@shmtu.edu.cn (X.W.); jfxian@shmtu.edu.cn (J.X.)

Abstract: Ship-speed prediction is an emerging research area in marine traffic safety and other related fields, occupying an important position with respect to these areas. At present, the effectiveness of techniques used in in time-series forecasting methods in ship-speed prediction is poor, and there are accumulated errors in long-term forecasting, which is limited in its processing of ship-speed information combined with multi-feature data input. To overcome this difficulty and further optimize the accuracy of ship-speed prediction, this research proposes a new deep learning framework to predict ship speed by combining GANs (Generative Adversarial Networks) and LSTM (Long Short-Term Memory). First, the algorithm takes an LSTM network as the generating network and uses the LSTM to mine the spatiotemporal correlation between nodes. Secondly, the complementary characteristics linked between the generative network and the discriminant network are used to eliminate the cumulative error of a single neural network in the long-term prediction process and improve the prediction accuracy of the network in ship-speed determination. To conclude, the Generator–LSTM model advanced here is used for ship-speed prediction and compared with other models, utilizing identical AIS (automatic identification system) ship-speed information in the same scene. The findings indicate that the model demonstrates high accuracy in the typical error measurement index, which means that the model can reliably better predict the ship speed. The results of the study will assist maritime traffic participants in better taking precautions to prevent collisions and improve maritime traffic safety.

Keywords: smart ship; AIS data; ship-speed prediction; ensemble GAN network; maritime traffic situation awareness; smart ship

1. Introduction

1.1. Introductory Context

Over the past few years, with recent vigorous advancements in maritime transportation and trade, the investigation into intelligent ships has emerged as a key area of research in the shipping field. One crucial aspect of this research is the accurate prediction of ship speed, which plays a vital role in ensuring the continuous growth and stability of the shipping sector [1]. Therefore, it is necessary to make good predictions of ship speed.

Researchers worldwide have conducted extensive studies on the prediction of ship navigation dynamics, including factors such as speed and trajectory [2]. As the relevant applied technologies have advanced, there have been significant changes in forecasting methods at the technical level, which have transitioned from empirical formulas and physical approaches to data-driven models [3]. Early researchers constructed empirical formulas for speed within a specific tonnage range by fitting a collection of operational constants established from actual data, and some of these methods are still in use today. Li et al. utilized AIS (automatic identification system) data to model and predict ship trajectories. For this task, they employed five machine learning techniques: Kalman Filter (KF), Support Vector Regression (SVR), Backpropagation Network (BP), Gaussian Process Regression (GPR), and Random Forest (RF) [4]. In addition, seven deep learning approaches were used: Recurrent Neural Networks (RNNs), Long Short-Term Memory (LSTM), Gated Recurrent Unit (GRU), Bidirectional LSTM (Bi-LSTM), Sequence-to-Sequence (Seq2seq), Bidirectional GRU (Bi-GRU), and Transformer. The effectiveness of each method was demonstrated under various maritime conditions.

AI has been crucial in advancing the autonomous navigation of unmanned vessels, and collision risk has become a major concern as the shipping industry expands, and the number of ships rises, especially autonomous vessels functioning in intricate environments [5]. Aristov et al. focused on maritime ports and their associated terminals, using several ports along the east coast of the United States as case studies. They traced the historical routes of container ships and accurately mapped the areas in which ships docked within terminals. This research provided valuable insights into ship scheduling and offered significant value to stakeholders in the supply chain industry, contributing to both theoretical and practical applications in maritime logistics [6]. In addition, some scholars have used the existing knowledge of ship dynamics and fluid mechanics to build a physical model of a ship's sailing motion in wind and waves, based on the descriptions of ship propulsion force and resistance, aiming to predict ship speed.

This paper presents a new deep learning framework for ship-speed prediction by combining GANs (Generative Adversarial Networks) and LSTMs. Unlike other deep learning models such as the GRU, LSTM, and WAGCN, this model takes geospatial dependence into consideration. Compared with the GAN–GRU and GAN–WAGCN combined with GAN, the proposed model can improve the prediction effectiveness by enhancing the diversity of models, capturing more complex data patterns, and improving the accuracy and stability of the prediction. First, the algorithm takes an LSTM network as the generating network and uses an LSTM to mine the spatiotemporal correlation between nodes. Secondly, the complementary characteristics linked between the generative network and the discriminant network are used to eliminate the cumulative error of a single neural network in the long-term prediction, optimizing and boosting the prediction accuracy of the network in ship-speed determination. Ultimately, the proposed Generator–LSTM model is used for vessel speed prediction, and compared with other models, using the same AIS ship-speed information in the same scene. The key contributions of this paper are following three aspects: (1) a ship-speed-prediction model which enhances model diversity is proposed; (2) more spatial features and complex data patterns can be captured by introducing GAN models; (3) the Generator–LSTM is verified using the data of two different types of ship scenarios. The organization of the paper is as follows: Section 2 covers the related research in speed prediction. Section 3 covers the specifics of the approach in this study. Section 4 presents the experimental settings and analyzes the results. Finally, Section 5 summarizes the research.

As sensor reception capabilities and computational power continue to improve, data-driven, technology-based speed-prediction methods have become increasingly feasible [7]. Previously, various relevant ship trajectory prediction tactics and techniques for ship-speed prediction have been proposed, among which deep learning methods have attracted significant attention due to the ability of this approach to take into account key factors and adjust to complex situations; nevertheless, the use of this technique in practice has been limited due to the insufficient extraction of environmental factors and feature data. Pensado et al. presented a real-time trajectory optimizer for shore-to-ship operations utilizing Unmanned Aerial Vehicles (UAVs), aiming to enhance the efficiency of the transportation system by leveraging UAVs for parcel deliveries to offshore ships, streamlining logistics and reducing delivery times [8].

1.2. Literature Review

With recent increases in maritime traffic flow and frequent navigation activities, ship collisions, as well as groundings and other accidents, have been occurring. In helping to mitigate such safety concerns, high-precision and stable ship trajectory prediction plays a vital role in the development of accident prevention technologies for the maritime IoT industry. Magalhães et al. considered ship trajectory prediction a challenge due to the difficulties that traditional deep learning techniques have in capturing complex changes in trajectories. To overcome this obstacle, they used the deep learning network encoded by Geohash to predict ship trajectory using clustering AIS data in an approach developed for the IoT industry, and applied a density-based, noise-aware spatial clustering algorithm to perform cluster analysis on the trajectories. Trajectory data across different categories were entered into the LSTM optimization network for the purpose of Geohash coding training [9]. A majority of the available research has indicated that deep learning-based methods for ship-speed prediction have been incorporated into the industry, significantly enhancing prediction accuracy and aiding maritime professionals in identifying potential safety risks. Padmavathi et al. proposed the use of the Kolmogorov–Arnold Networks (KAN) algorithm for prediction, which has effectively improved accuracy. The performance of the proposed method is evaluated using various metrics, including MAE, MSE, RMSE, and MAPE, based on the AIS dataset [10].

Accurate prediction using deep learning methods based on AIS data has become a key focus in modern maritime transportation research. Donandt et al. focused on inland Vessel Traffic Planning (VTP) and applied Gaussian Mixture Models to a fused dataset of AIS and discharge measurements. This approach generated multi-modal distribution curves that capture typical lateral vessel positioning in the fairway, as well as dislocation speeds along the waterway [11]. Data-driven models can integrate unmodeled physical phenomena into physical models, resulting in improved predictions, provided that the data and algorithms are used rationally. With the developments in the capabilities of computers, the utilization of AIS data allows for the estimation of ship speed. AIS is a ship-based proportional communication device utilized for both navigation and collision prevention. Some scholars have proposed a Support Vector Regression model based on adaptive chaotic differential evolution; this approach selected speed, course, time, latitude, and longitude as sample features from the AIS data, aiming to predict ship trajectory [12]. Owing to its extensive coverage and abundant information, including static information, dynamic information, and navigation-related information, AIS is widely used in ship-speed prediction. AIS is considered to be a key data source, in that it represents maritime traffic spatially and temporally in the context of marine spatial planning; it is widely used in ship-speed prediction research, and can offer informational support for navigation risk assessment.

Sigillo et al. tackled the challenging problem of vessel route forecasting using a vast number of AIS records. They proposed a novel deep learning architecture, SeaFormer, which leveraged transformer modules to capture long-term dependencies in the data. This approach enabled the forecasting of vessel routes even several hours ahead, improving the accuracy and timeliness of maritime navigation predictions [13]. Gülsoylu et al. addressed the problem by developing a technique that fused AIS data with data describing the vessels detected in the images to create enriched datasets. This fusion combined visual data with vessel-related information, such as type, size, speed, and direction, providing a more comprehensive understanding of vessel behavior and improving analysis capabilities [14]. In addition, AIS data are also used to predict ship tracks under uncertain conditions, extract ship-movement patterns, and identify abnormal ship behavior, in order to reduce threats to other nearby ships. Before using AIS data, it should be noted that because the communication channels of AIS are concentrated in traffic-intensive waterways, the transmission rates associated with the AIS data will change. Based on the navigation status of a given ship, there will be anomalies in the original AIS data, so data cleaning and interpolation are needed [15].

In recent years, the study of ship trajectory prediction has been developed rapidly due to the rich ship-movement information (including ship position, speed, course, and heading) provided by AIS [16]. Various data-driven approaches have been developed, including SVM (Support Vector Machines), BP, and LSTM. Ship track prediction contributes significantly to the maintenance of traffic safety in busy waterways. Conventional forecasting techniques often ignored the complex spatiotemporal interactions and the built-in collision prevention actions that take place during encounters between multiple vessels, resulting in lack of precision in the trajectory forecasting associated with the interaction. Pappagallo et al. focused on implementing a 'Port Classifier,' evaluating several models, including Conv1D, MLP, and LSTM, on AIS trajectories labeled through heuristic algorithms. Their study showed promising results, with Conv1D demonstrating superiority in port classification tasks. They also conducted an Exploratory Data Analysis (EDA) to gain a deeper understanding of the data, further improving the classification performance [17]. As described in case studies, neural networks combined with GANs have demonstrated the capability to develop navigation rules beyond the current regulations, and provide valuable insights for improving safety in offshore operations.

Ship speed is a critical factor in the decision-making processes of maritime transportation, such as determining ETA (estimated time of arrival), estimating fuel usage, planning collision-free routes, and so on. Nevertheless, the speed of the ship is occasionally anomalously reduced because of increased wind/wave friction and decreased propulsion efficiency, so accurate ship-speed forecasting demands a thorough understanding of drag, propulsion, machinery, and more. Various models have been created for the prediction of ship speed; these could be classified as data-driven or semi-data-driven formulas, including computational liquid dynamic, as well as experimental, simulations [18]. In addition to the above methods, dynamic prediction utilizing time-series data has also found important uses in the shipping industry. Certain researchers have used AIS data and artificial intelligence approaches to forecast ship speed, and the prediction accuracy has been improved [19].

There are, occasionally, obvious present or potential trends in ship sailing states and marine meteorological conditions, and constructing time series for relevant predictions may achieve good results [20]. Comprehending the dynamics between a ship and its surroundings allows for optimal path prediction, which is essential for enhancing the safe navigation of autonomous vessels. Predicting future trajectories is a highly complex task because of the inherent uncertainties and the intricate temporal and spatial relationships between various ships. Nonetheless, current approaches disregard the continuity and cross-border aspects of ship-to-ship impacts. Chen et al. proposed a novel ship route planning method that took into account various factors, such as weather conditions and carbon emissions. Their research findings aimed to assist the maritime community in making ship routing decisions that were more reasonable and environmentally conscious under different navigation conditions, enhancing efficiency and sustainability in maritime operations [21]. Ourc et al. introduced a practical framework for predicting risky close encounters and gray-zone interactions, using historical AIS data and LSTM networks. Their methodology developed an extensive encounter model that identified ship-to-ship interactions and classified them as risky, gray-zone, or non-risky encounters, based on ship domain violations. This approach enhanced safety by providing early detection and classification of potential risks in maritime navigation [22]. Shin et al. studied port ship trajectory forecasting using a deep learning technique method for maritime data preprocessing and berthing-side integration, and this maritime special data preprocessing technology improved the performance of the model [23]. The model was trained using AIS data and the mean square error was used as an evaluation index, aiming to forecast future outcomes for sailing ship location.

Yang et al. combined the Transformer model, which was popular within the domain of artificial intelligence, with the traditional KF model, and proposed a new flight-path prediction model [24]. The experimental results showed that, compared to the use of a conventional method, this approach offered improved prediction accuracy and superior parallel processing of sequence data. Mehri et al. introduced a ship trajectory prediction approach utilizing RNN. For the data preprocessing stage, a method based on symmetric segmented path distance was developed to mitigate the impacts of redundant data and noise. For the prediction process, an RNN model based on the gated cycle unit was designed, attempting to achieve effective and timely prediction of ship position details [25].

In addition, the sailing sea area, loading state, and ship-type of the ship are obviously different for each voyage, which causes inconsistent data distribution and makes it difficult to extend the trained model to the new data [26]. Xian et al. studied a received signal strength (RSS)-based three-dimensional (3D) drowning target localization that simultaneously considered the absorption effect, uncertain transmission power (UTP), and the time-varying path loss exponent (PLE) for underwater search and rescue missions, induced by the frequent occurrence of maritime accidents [27]. Researchers in similar areas have been working on developing a series of ship-speed prediction models that are more accurate and robust, aiming to effectively monitor the future navigation statuses of ships. These models aim to enhance the safety and efficiency of maritime navigation by providing timely and reliable predictions, allowing for better decision-making in real-time navigation scenarios [28,29].

2. Materials and Methods

2.1. Research Area

It is known that relative to the world-wide carrying capacity for tankers, the present number of vessels constitutes less than 30%, while the counterpart for container ships is 13.5%. More specifically, both tankers and container ships are commonly used for the transfer of cargo goods within the maritime industry. Thus, tankers transport liquid cargoes, especially crude oil and chemicals, and their maritime safety is particularly important. Ship speed is a key parameter of maritime navigation safety. The ship-speed predictions for oil tankers and container ships play an important role in ensuring maritime safety [30]. Accurate ship-speed prediction can help the crew to perform better navigation planning, reduce risks, and improve efficiency [31]. Weather conditions, ship load, marine environment, and ship performance will directly or indirectly affect ship speed.

Actions relating to modern tankers and container ships can also be predicted using artificial intelligence and machine learning algorithms. These technologies build complex predictive models by analyzing large amounts of historical data, such as ship speed, weather, sea state, etc. AI can automatically recommend the best ship speed based on real-time input of weather data and information relating to the marine environment.

2.2. Methodology

2.2.1. AIS Data Preprocessing

The data we used is derived from AIS information, which contains some ship-related information, such as MMSI (Maritime Mobile Service Identification), location, heading and speed. Depending on the speed and route, the intervals between messages sent varies from 2 s to 3 min. The AIS data used in this study were located at https://marinecadastre.gov/ais/ (accessed on 24 March 2025), and were download from the website [32].

In terms of speed prediction, the original AIS information was gathered first. Then, five steps, comprising data outlier detection, speed statistics extraction, speed information for data interpolation determination, uniform interval processing, and data standardization, were carried out. Speed information was used for data interpolation, uniform interval processing, and data standardization. The data preprocessing workflow is illustrated in Figure 1.

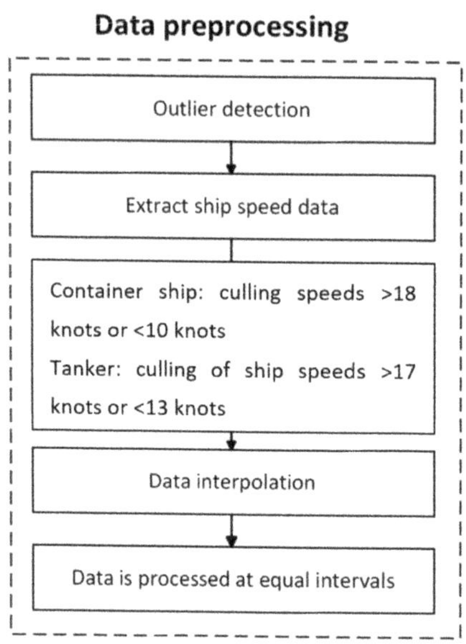

Figure 1. Data preprocessing flow chart.

After assigning the data as the input label for model training, anomaly detection is initially applied to identify deviant samples prior to executing de-duplication procedures. Following the initial data acquisition, velocity vectors are computationally derived, with subsequent application of threshold-based filtering to eliminate non-compliant kinematic measurements then performed. We used a low-pass filter to smooth out noise, especially in cases of drastic changes in speed or position, in order to reduce the interference of short-term abnormal fluctuations in model training. Furthermore, we have also enhanced the robustness of the model through data augmentation techniques, enabling it to better handle errors caused by noise.

AIS data play an important role in ship monitoring and navigation analysis, especially in applications such as ship-speed prediction. AIS data have some inherent limitations, such as missing values, transmission errors, and other problems, all of which would affect the accuracy and reliability of ship-speed prediction. Missing values in AIS data may lead to the discontinuity of the time series, affecting the training and prediction accuracy of the model. In this paper, the missing values are filled through linear interpolation, and when using time-series prediction models (such as the LSTM, GRU, etc.), the processing of missing data is taken as a specific input feature in optimizing the model. Transmission errors may lead to outliers or inconsistencies in AIS data, especially data that are abnormal in terms of speed and heading. This kind of error will directly affect the training of the ship-speed prediction model and may lead to the deviations in the prediction results. This paper identified and corrected the transmission errors by performing outlier detection on the data.

In the container-ship experiment scenario, due to the high loading and unloading speeds and short port-stop times of container ships, most of them utilize high speeds. In recent years, in order to save energy, a more economical speed was generally adopted, which was about 18 nautical miles per hour, while the speed of a container ship in short-distance coastal navigation is only about 10 nautical miles per hour. Therefore, data with velocity values outside the interval $10 \leq v \leq 18$ knots were removed in the data preprocessing for the container-ship experiment scenarios. In the tanker experiment scenario, since the speed of the tanker is generally 13 to 17 knots, the data associated with speed values outside the interval $13 \leq v \leq 17$ knots were removed in the data preprocessing for the tanker experiment scenarios. Next, the cubic spline interpolation technique was applied to set the interval between each data point to 10s, which met the need for a uniform time interval associated with the prediction task. Finally, in order to smooth the training and converge the network, normalization was required; the normalization value is expressed in Equation (1).

$$X_{normalized} = \frac{X_i - X_{min}}{X_{max} - X_{min}} \tag{1}$$

where $X_{normalized}$ is the eigenvector after normalization, X_i is the value prior to normalization, X_{min} represents the minimum value in the sample data, and X_{max} is the maximum data point in the sample.

2.2.2. Basic Research Ideas

The framework uses the Generator–LSTM model to achieve the ship-speed prediction task, as built upon AIS dynamic information. The Generator–LSTM model is composed of two components: the LSTM generator and the discriminator. The ship-speed prediction algorithm framework introduced in this research is shown in Figure 2. The research idea mainly includes four processes: the initial step is data preparation and preprocessing, the second process is network model design, the third process is model training and optimization adjustment, and the fourth process is model performance evaluation.

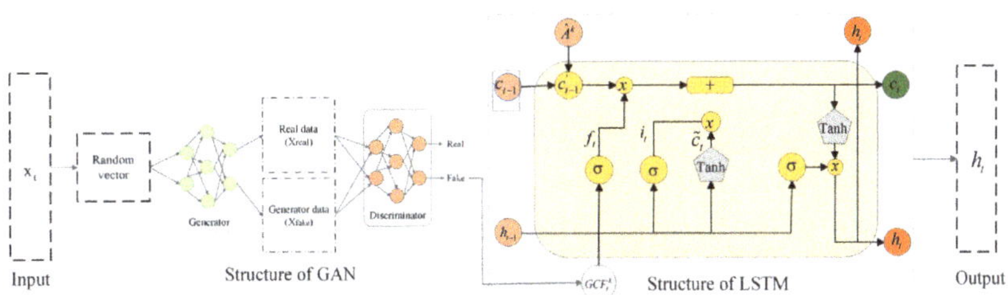

Figure 2. Basic block diagram.

In the first process, data preparation is first carried out to obtain ship AIS data, including the ship's latitude and longitude, course, and speed, as well as other information. These data are usually time-series data, and are subsequently organized and processed in chronological order. Then, AIS data is preprocessed, including data cleaning (removal of outliers and error data) and normalization processing, to verify that the data format is suitable for the input of the LSTM model.

In the second process, since an LSTM is appropriate for handling time-series data and capable of capturing the long-term dependencies in the data, an LSTM neural network model is structured to learn the time-series patterns in the AIS data. The designed LSTM model is embedded in the generator of a GAN network to produce sequential data for the prediction of ship speed.

In the third process, the processed AIS data are input into the LSTM generator, which learns the distribution of the provided data in order to generate sequential data to predict the ship speed. The role of the discriminator is to differentiate the generated ship-speed data from the real AIS data, so that the generator can learn to generate realistic ship-speed data. Then a GAN network is trained by using the GAN loss function, learning rate and Adam optimizer. In this study, the generator optimizer is Adam, with a relatively low initial learning rate of 10^{-4}, and momentum parameters set at $\beta_1 = 0.5$ and $\beta_2 = 0.999$ to ensure stable gradient updates. The discriminator optimizer also uses Adam, but with a slightly higher learning rate of 5×10^{-4}; this is aimed at accelerating the improvement of the discriminator's ability. Additionally, a smaller learning rate of 10^{-5} is used for Adam to prevent overfitting and improve the model's generalization ability. To prevent gradient explosion during GAN training, a concern which is particularly important when handling long-term dependencies in an LSTM, the gradient norm is limited to a threshold of 0.01. This approach ensures a more stable and effective training of the model.

Based on the training process, the model's hyperparameters (e.g., the number of hidden units and layers of the LSTM) are adjusted to enhance both prediction accuracy and the quality of the generated data. In the fourth process, the MSE (Select Mean Square Error), RMSE (Root Mean Square Error), MAPE (Mean Absolut Percentage Error) and MAE (Mean Absolute Error) evaluation indices were used to quantify the accuracy of the model's predictions and the fidelity of generated data, and to further evaluate the differences between the ship-speed data generated by the GAN network and the real data. Through the above steps, a GAN network combined with an LSTM as a generator can effectively learn and generate ship-speed forecasting using AIS data, and improve the efficiency and accuracy of ship transportation and navigation management.

2.2.3. Generator–LSTM Network

Figure 3 shows the GAN network architecture diagram. A GAN is a deep learning model founded on the concept of adversarial training and comprising two opposing neural networks, a generator and a discriminator, the competition between which enables them to replicate any data distribution. The generator's role is to generate simulated data so that the discriminator cannot distinguish between the generated data and the actual data. The discriminator's role is to differentiate between real data and generated data. The generator network should continuously refine the data it generates to make it difficult for the discriminant network to evaluate the data, and the discriminator network should also refine itself in order to enhance the accuracy of its judgment [33]. The connection between them is antagonistic, so it is called an antagonistic network. Generators have no tags and are unsupervised networks; the discriminator has a label and a supervisory network, and its label states are "false" and "true" (0 and 1).

When processing sequence data (like text and time-series data), an LSTM can be used as a generator; this research method can be utilized in the application scenario of ship-speed prediction research. The following is a description of the GAN network architecture utilized in this study, using an LSTM as generator.

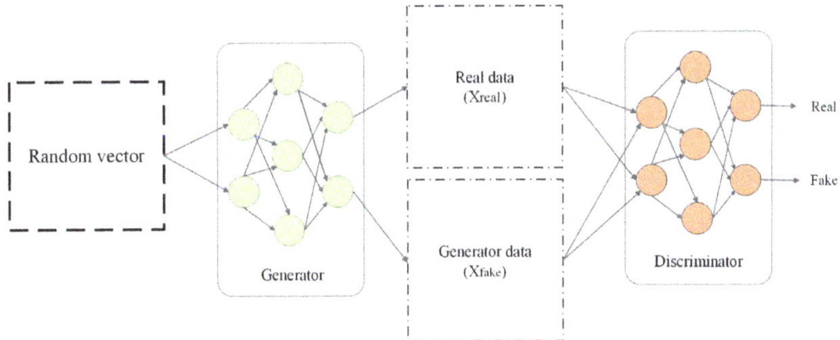

Figure 3. GAN network architecture diagram.

The generator's task is to generate a sequence similar to the training data from random noise. In this study, the LSTM network is used to build the generator. An LSTM is a unique variant form of recurrent neural network that can capture long-term dependencies and is ideal for analyzing and forecasting significant events with extended intervals and delays in the associated time-series data. Figure 4 illustrates the fundamental architectural diagram of an LSTM, which consists of the input gate, forget gate, and output gate, which work together to update the unit state and produce the hidden-state output. In this study, the model consists of 3 LSTM layers. The bottom LSTM layer is designed to capture short-term time dependencies, while the higher-level LSTM layers are responsible for learning the characteristics of long-term time series. Each LSTM layer contains 256 hidden units, allowing the model to effectively capture both short-term and long-term patterns in the data, improving the overall performance of the prediction task. This architecture strikes a balance between learning fine-grained short-term dynamics and understanding broader temporal trends in the data.

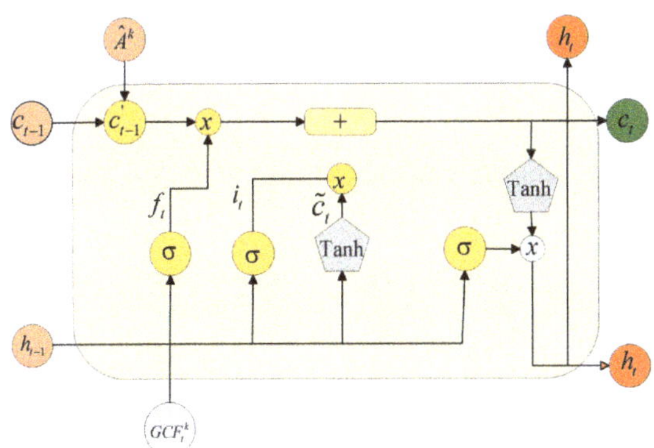

Figure 4. LSTM: basic structure diagram.

The input gate controls whether the input should be added to the memory. The formula is expressed in Equation (2).

$$i_t = \sigma(W_i[h_{t-1}, x_t] + b_i) \quad (2)$$

where σ represents the sigmoid function, W_i denotes the weight matrix of the input gate, h_{t-1} represents the hidden layer state at time $t-1$, x_t represents the input quantity at time t, and b_i represents the bias information of the input gate. The $[h_{t-1}, x_t]$ represents the splicing of the tensors h_{t-1} and x_t into a single tensor, using the cat function.

The forget gate controls whether the previous memory is retained in the present memory. This is expressed in Equation (3).

$$f_t = \sigma\left(W_f[h_{t-1}, x_t] + b_f\right) \tag{3}$$

where W_f represents the weight matrix of the forget gate, and b_f represents the bias term of the forget gate.

The output gate controls the output at the current time. This is expressed in Equation (4).

$$o_t = \sigma(W_o[h_{t-1}, x_t] + b_o) \tag{4}$$

where W_o represents the weight matrix of the forget gate, and b_o represents the bias term of the forget gate.

The cell status update can be carried out based on i_t and f_t, and the formula is expressed in Equations (5) and (6).

$$\widetilde{C}_t = \tanh(W_C[h_{t-1}, x_t] + b_C) \tag{5}$$

$$C_t = f_t * C_{t-1} + i_t * \widetilde{C}_t \tag{6}$$

where $\widetilde{C}_t$ is the candidate memory unit state, W_C is the weight matrix associated with the updated memory unit state, b_C is the offset entry of the revised memory unit state, and C_t is the memory unit state of the present time-step.

Finally, the hidden state at the present time-step can be generated based on the output gate and the memory cell state. The hidden-state output formula of an LSTM's basic module is expressed in Equation (7).

$$h_t = o_t * \tanh\ (C_t) \tag{7}$$

where h_t is the hidden state of the current time-step.

The loss function of the generated network is used to estimate the error, expressed as the difference between the predicted value of the model and the true value. The formula is expressed in Equation (8).

$$L_G = H(1, D(G(z))) \tag{8}$$

where G represents the generation network, D represents the discriminant network, H denotes the cross entropy, and z is the input random data. $D(G(z))$ is the discriminator output of the posterior probability, for which 1 indicates that the data is absolutely true and 0 means that the data is absolutely false. $H(1, D(G(z)))$ represents the distance between the judgment result and 1. Obviously, for the generation network to obtain good results, it must be ensured that the discriminator can distinguish the generated data from the true data.

The discriminator is used to distinguish between real data and generated data. When using an LSTM as a generator, the discriminator is also a network capable of processing sequence data. The loss function is expressed in Equation (9).

$$L_D = H(1, D(x)) + H(0, D(G(z))) \tag{9}$$

where x is the real data, $H(1, D(x))$ is the distance between the real data and 1, and $H(0, D(G(z)))$ represents the distance between the generated data and 0. Obviously, if the discriminant network obtained good results, this will make the distance between the real data and 1 small, and the distance between the generated data and 0 small.

2.2.4. Evaluation Index

In order to analyze the prediction accuracy of each model, a comparison experiment was conducted on the test datasets associated with two ships: a container and a tanker. MSE, RMSE, MAPE, and MAE, were the four indicators used to analyze the degree of sample-fitting of the prediction model. Lower values for the four indicators indicate that the predicted values of the model are closer to the true observed value. The specific formula is expressed in Equations (10)–(13).

$$MSE = \frac{1}{m}\sum_{j=1}^{m}\left(\tilde{x}_j - x_j\right)^2 \tag{10}$$

$$RMSE = \sqrt{\frac{1}{m}\sum_{i=1}^{m}\left(\tilde{x}_j - x_j\right)^2} \tag{11}$$

$$MAE = \frac{1}{m}\sum_{i=1}^{m}\left|\tilde{x}_j - x_j\right| \tag{12}$$

$$MAPE = \frac{100\%}{m}\sum_{i=1}^{m}\left|\frac{\tilde{x}_j - x_j}{x_j}\right| \tag{13}$$

where $\tilde{x}_j$ and x_j represent the forecasted and actual values of latitude and longitude, respectively, which can show the prediction strength of the model.

3. Experiments and Results

3.1. Data Description

In this study, the first type of ship is a container ship, and the second ship-type is an oil tanker; the rest are, severally, one bulk carrier, one passenger ship, one fishing boat, and one dredger. The motion information relating to the six ships is used to predict the ship speeds. In this study, four-dimensional historical data comprising the longitude, dimensions, headings, and speeds of the six ships under six sailing scenarios were selected for ship-speed prediction research. The data were uniformly sampled for 10 s in the experiment. For these six ships, the AIS data are filtered by using their ship position data, including longitude and dimension, and then only the ship-speed information is included; that is, the ship speed is predicted by the AIS speed information. In addition, in the task of speed prediction, the robustness of the proposed model is further verified due to the different variation trends associated with ship speed in the different scenarios.

Table 1 shows the relevant information for the six different ship-types. (1) The first vessel, MMSI 255805778, is a container ship named PANDA 009 (Portugal) and is sailing in the southeastern waters near Hortaherkclep. Figure 5 illustrates the port location of this ship. The original AIS data selected for the experiment began at 1:12 a.m. on 8 February 2016 and ended at 5:08 a.m. on 13 February 2016. (2) The second MMSI is 235085368. It is of the oil tanker type and named KOHZAN MARU (UK). Figure 6 shows its navigation in the southeast sea near the city of Shima. The original AIS data selected for the experiment began from 6:03 a.m. on the 1 February 2016 and ended from 3:52 a.m. on the 7 February 2016. (3) The third vessel, MMSI of 214182516, is a bulk carrier named SARA REGINA (Moldova). Figure 7 shows its navigation area near Miami Beach. The start and end times of the original AIS data selected for the experiment were 0:15 a.m. on 7 February 2016 and 22:47 p.m. on 12 February 2016, respectively. (4) The MMSI of the fourth vessel is 246506000. It is a passenger ship named VEENDAM (Netherlands). Figure 8 shows its navigation area near Katakron. The start and end times of the original AIS data selected for the experiment were 0:03 a.m. on 13 March 2016 and 23:21 p.m. on March 31, 2016, respectively. (5) The MMSI of the fifth vessel is 257294000, and its type is a fishing boat, called INTER SCOTIA (Norway). Figure 9 shows its navigation area. The start and end times of the original AIS data selected for the experiment were 23:46 p.m. on 12 March 2016 and 5:20 a.m. on 31 March 2016, respectively. (6) The MMSI of the sixth vessel is 210921000, and its type is a dredger, called SHOREWAY (Cyprus). Figure 10 shows its navigation area near Taichung City. The start and end times of the original AIS data selected for the experiment were 22:02 p.m. on 8 March 2016 and 21:29 p.m. on 31 March 2016, respectively.

Table 1. Details associated with each ship's specific information.

Ship Experimental Data Information					
MMSI	**Ship Type**	**Navigation Area**	**Data Length**	**Ship Length**	**Ship Breadth**
255805778	Container ship	Coastal navigation	3869	277 m	40 m
235085368	Oil tanker	Coastal navigation	4961	180 m	32 m
214182516	Bulk cargo ship	Coastal navigation	2217	91 m	14 m
246506000	Passenger ship	Coastal navigation	3658	220 m	32 m
257294000	Fishing vessel	Internal navigation	6059	76 m	17 m
210921000	Sand dredger	Coastal navigation	1554	98 m	22 m

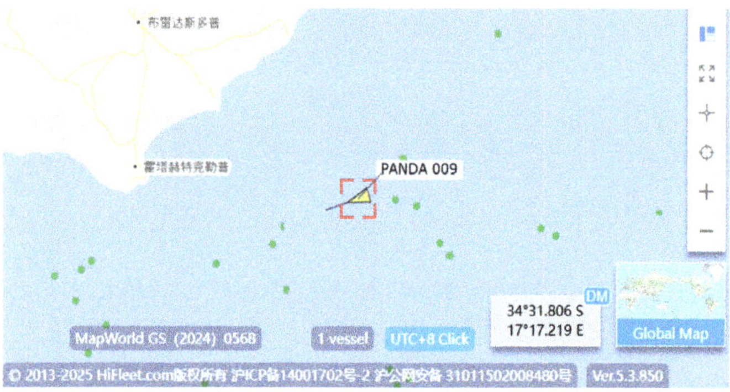

Figure 5. The container ship.

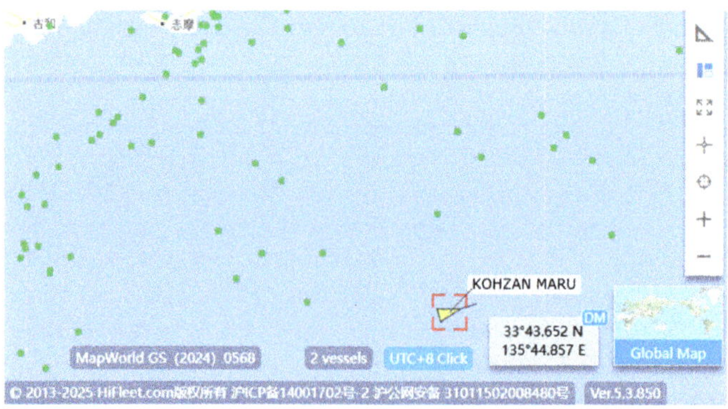

Figure 6. The oil tanker.

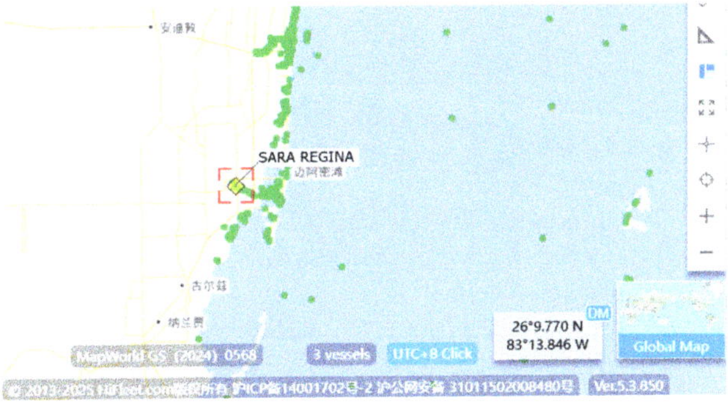

Figure 7. The bulk cargo ship.

Figure 8. The passenger ship.

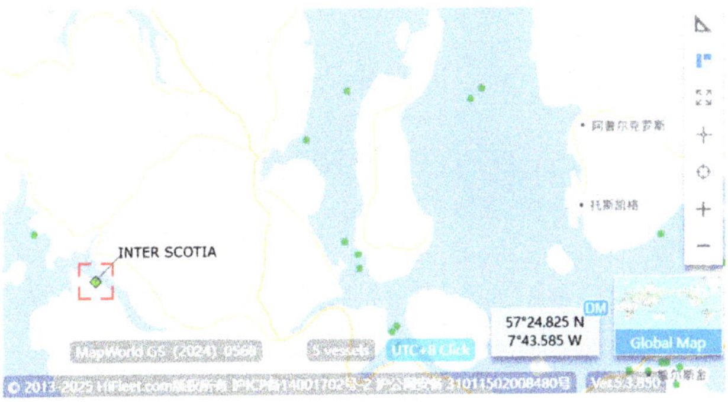

Figure 9. The fishing vessel.

Figure 10. The sand dredger.

Figures 5 and 6 show the geographical locations and travel areas of the container ship and the oil tanker, respectively.

3.2. Experimental Setup

In order to verify the capacity of the proposed model, the error of the ship-speed prediction model is evaluated against the errors of standard time-series prediction models, including the LSTM and GRU, and against those of the WAGCN. On the basis of this comparison test, for further assessment of the effectiveness of a GAN network in ship-speed prediction and to verify that Generator–LSTM has the most accurate ship-speed prediction results, the generator of the GAN network is replaced with the GRU and WAGCN, respectively, and two comparison experiments are conducted.

All models were implemented on the PyTorch 3.9.18 server with NVIDIA driver version 545.92 installed. The dataset is divided into three parts: training set, verification set, and test set, and the ratio is 7:1:2. We trained the model by minimizing MSE, because this value is widely used in ship-speed prediction research, facilitating the use of optimization algorithms such as gradient descent to train the model, determine the robustness of MSE to outliers, and the find direct correlations between MSE and prediction error. By trying different batch-size values and evaluating their performance on the validation set, it was found that batch sizes greater than 16 would cause performance degradation because each update would make it difficult for the model to escape the local minimum; therefore, it was decided to set the batch size to 16. The training period used in the experiment was 200 epochs. To prevent overtraining and overfitting, early stopping criteria were established as follows: ① monitoring indicators, verify losses; ② patience value, ten cycles; and ③ improvement threshold (min Delta), 0.001. These strategies ensured that training halts when no further significant improvement is observed, thus avoiding overfitting and ensuring a more generalizable model. The learning rate is the small step-size of each parameter update, which directly affects how fast the model's parameters change during training. The learning rate for the initial setting was 10^{-3}; increasing the learning rate during training will lead to the instability of the training process, although the convergence speed is accelerated. As a result, the learning rate was set at 10^{-3}.

3.3. Experimental Results

3.3.1. Experimental Results for the Container Ship Scenario

Table 2 gathers the evaluation index results for the Generator–LSTM model and the other five models in the container ship scenario. As can be seen in Table 2, the Generator–LSTM model has the smallest value for each index among all models, and preceded other models, which indicated that it achieved the best speed-prediction effect, compared to other models. It can be found that, considering the time dependence, compared to the GRU model, the LSTM model belonging to the recurrent neural network is more effective in capturing and processing the long-term dependence relationships in the sequential data, and with higher prediction accuracy. Compared with the WAGCN model, an LSTM has better prediction effectiveness and reliability due to its unique time-series modeling ability and structural advantages. For example, the RMSE value for the LSTM model is 0.1036, which is about 6.3% lower than the GRU model and about 3.8% lower than the WAGCN model.

In contrast to the single neural network models such as the LSTM, GRU, and WAGCN, the MSE value of the Generator–LSTM model is 0.33%, 2.02%, and 1.26% lower than those of the LSTM, GRU, and WAGCN. RMSE values are lower by 1.75%, 8.01%, and 5.53%, respectively. The MAPE values were lower by 0.15%, 0.37%, and 0.21%, respectively. The MAE values are lower by 2.05%, 4.6%, and 2.66%, respectively, indicating that on the one hand, it is difficult for a single neural network model to capture complex spatiotemporal correlations in the data; on the other hand, compared with a single neural network model, the GAN–LSTM model demonstrates higher performance in ship-speed prediction and is more effective in capturing spatiotemporal correlations.

Table 2. Comparison of results for speed-prediction accuracy of the models in the container-ship scenario.

Model	MSE	RMSE	MAPE (%)	MAE
LSTM	1.07×10^{-2}	1.03×10^{-1}	6.34×10^{-3}	8.15×10^{-2}
GRU	2.76×10^{-2}	1.66×10^{-1}	8.52×10^{-3}	1.07×10^{-1}
WAGCN	1.20×10^{-2}	1.41×10^{-1}	6.92×10^{-3}	8.76×10^{-2}
Generator–GRU	1.37×10^{-2}	1.17×10^{-1}	6.28×10^{-3}	7.98×10^{-2}
Generator–WAGCN	2.70×10^{-2}	1.64×10^{-1}	9.46×10^{-3}	1.20×10^{-1}
Generator–LSTM	7.42×10^{-3}	8.61×10^{-2}	4.81×10^{-3}	6.10×10^{-2}

Compared with the GAN network model, the generators of which are the GRU and WAGCN, the MSE values of Generator–LSTM are 0.63% and 1.95% lower, the RMSE values are 3.08% and 7.82% lower, and the MAPE values are 0.15% and 0.47% lower, respectively, in the four error indicators. The MAE values are lower by 1.88% and 5.93%, respectively, which means that the raised GAN–LSTM model maintains the optimal speed-prediction effectiveness among the three GAN network models. This suggests that using Generative Adversarial Network (GAN) structures, in which the generator part is an LSTM, can improve prediction performance by increasing model diversity, capturing more complex data patterns, and improving prediction accuracy and stability.

For the container-ship scenario, the comparison of the results for the ship-speed prediction accuracy of each experimental model are shown in Table 2.

In order to understand the effectiveness of the raised model in predicting ship speed more intuitively, the subset segment of the current container-ship condition was used to analyze the results. Figure 11 exhibits the ship-speed predictions of the WAGCN, GRU, LSTM, and LSTM models, as well as the Generator–WAGCN, Generator–GRU, and Generator–LSTM models, with reference to the same actual ship speeds in the same time periods. As can be seen, the results in Figure 11 are in line with those in Table 2. The offset between the forecasted values and the observed values for the GRU and Generator–WAGCN models is maximal in all models, as illustrated in Figure 11b,d. The LSTM model has an optimal ship-speed prediction effectiveness relative to all models except the Generator–LSTM model, but the speed value predicted by Generator–LSTM is nearer to the true value than that predicted by the LSTM, and the true value and predicted value fit better, as shown in Figure 11c,f. This shows that the proposed model can utilize the available data, while the features of the dataset are enhanced by capturing higher-order statistical features of the data and the prediction performance is improved. Thus, it is rational to think that the Generator–LSTM model is superior to the other models.

In order to more clearly represent the ship-speed prediction results of all models within a subset segment of the current general situation of container ships, they are plotted on a coordinate axis, as illustrated in Figure 12. The solid black line represents the original input data, and the solid red line represents the forecast data of the Generator–LSTM model. It can be seen that with respect to the corresponding predicted ship-speed line segments of all models, the red solid line (i.e., the Generator–LSTM model) fits better with the black solid line (i.e., the original input data); that is, the model has better ship-speed prediction performance.

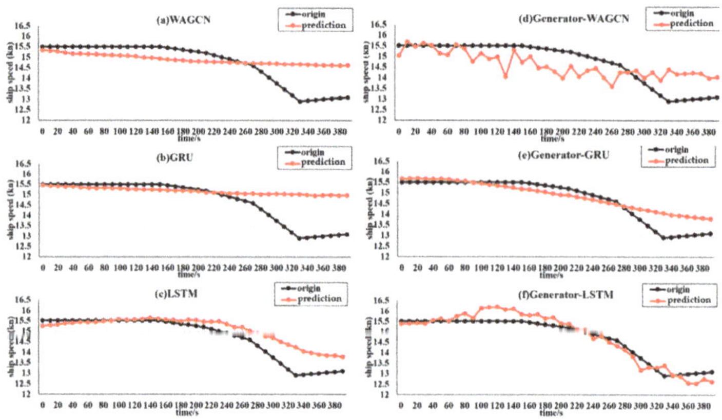

Figure 11. Comparison of model ship-speed prediction results in the tanker scenario.

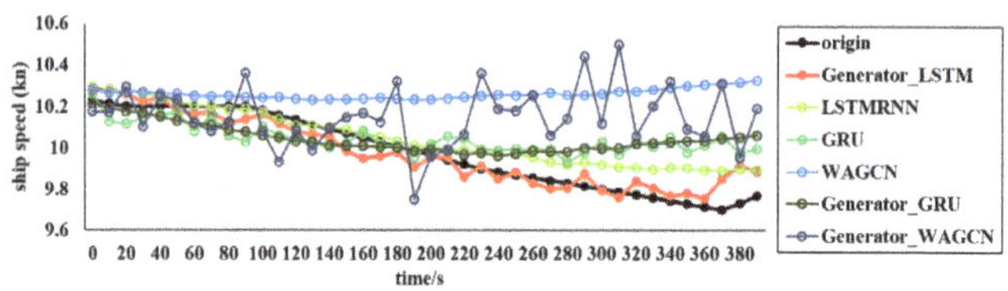

Figure 12. Comparison of overall prediction results of 6 models under container ship scenario.

This study proposes that this model demonstrates optimal performance in ship-speed prediction, which is reflected not only in the excellent degree of fit between the predicted data and the original input data, but also in each error index. As shown in Figure 13, the abscissa represents the error index value, and the ordinate is used to represent the name of each model. Each model corresponds to four types of error indicators. Values for the same class of error indicators are represented by the same color. In observing the figure, it can be seen that the Generator–LSTM model has the smallest area of the histogram areas for the four kinds of error indicators; that is, the Generator–LSTM model has the lowest values for the four kinds of error indicators and the best performance in ship-speed prediction.

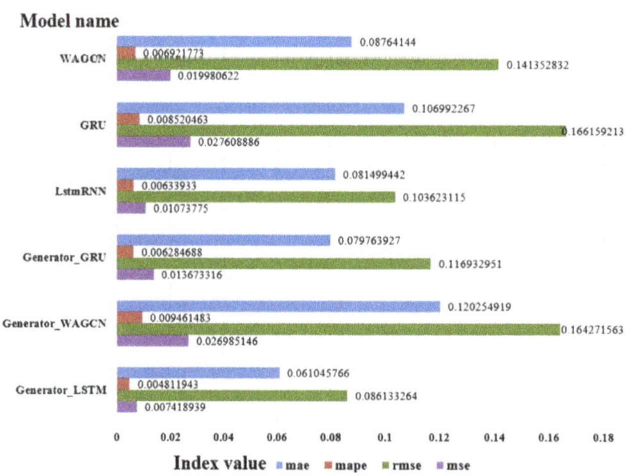

Figure 13. Comparison of the model indicators in the container-ship scenario.

3.3.2. Experimental Results for the Tanker Scenario

In addition, the ship-speed prediction model introduced in this paper can also be applied to other contexts. Table 3 shows the distribution of indicators in the tanker scenario. Similar to the experimental results for the container-ship scenario, from the perspectives of MSE, RMSE, MAPE, and MAE, the precision of LSTM prediction is vastly lower than that of the GRU and WAGCN model, possibly because LSTM introduces a more complex memory unit structure, including cell state and three gates (forgetting gate, input gate, and output gate). These structures enable an LSTM to arrest and maintain long-term dependencies more efficiently for sequential data such as ship speed.

Table 3. Speed-prediction error for the different models in the tanker scenario.

Model	MSE	RMSE	MAPE (%)	MAE
LSTM	7.43×10^{-2}	2.73×10^{-1}	1.15×10^{-2}	1.67×10^{-1}
GRU	2.13×10^{-1}	4.62×10^{-1}	2.23×10^{-2}	3.25×10^{-1}
WAGCN	1.65×10^{-1}	4.06×10^{-1}	1.88×10^{-2}	2.74×10^{-1}
Generator–GRU	1.29×10^{-1}	3.59×10^{-1}	1.70×10^{-2}	2.43×10^{-1}
Generator–WAGCN	1.70×10^{-1}	4.12×10^{-1}	2.15×10^{-2}	3.10×10^{-1}
Generator–LSTM	5.83×10^{-2}	2.41×10^{-1}	1.14×10^{-2}	1.65×10^{-1}

By comparing the four indices for the Generator–LSTM and the LSTM model, it is evident that the prediction accuracy of the Generator–LSTM model is considerably lower compared to the LSTM model, indicating that due to the introduction of a GAN structure, the competitive learning process between generator and discriminator can prompt the generator to generate prediction results closer to the real data. Specifically, in the analysis of ship-speed forecasting, the accuracy and stability of the prediction are improved.

Compared with the GAN network model, the generators of which are the GRU and WAGCN, the MSE values of Generator–LSTM are 7.04% and 11.13% lower, the RMSE values are 11.73% and 17.03% lower, and the MAPE values are 0.56% and 1.01% lower, respectively, for the four error indicators. MAE values are lower by 7.82% and 14.49%, respectively, which means that the introduced Generator–LSTM maintains an optimal speed-prediction effectiveness among the three GAN network models. This suggests that using Generative Adversarial Network GAN structures, where the generator part is an LSTM, can improve prediction performance by increasing model diversity, capturing more complex data patterns, and improving prediction accuracy and stability.

Similarly, in order to more clearly understand the efficiency of the suggested model for ship-speed forecasting, the consequent of a sub-trajectory segment describing the current tanker situation is included as an illustration. Figure 14 illustrates the ship-speed predictions of the WAGCN, GRU, LSTM, and LSTM models, as well as those of the Generator–WAGCN, Generator–GRU, and Generator–LSTM models, corresponding to the same actual ship speeds in the same time periods. As can be seen, the results in Figure 14 are consistent with those in Table 3. Similar to the experimental results for the container-ship scenario, the gap between the predicted and actual values of the GRU model is maximal in all models, as illustrated in Figure 14b. The LSTM still has an optimal effectiveness in ship-speed prediction with respect to all models other than the Generator–LSTM model, as shown in Figure 14c. The Generator–LSTM model is the model with highest precision, which is to say, optimal prediction, compared to all models in the experiment. As shown in Figure 14f, its predicted values are nearer to the true values, as compared to the results from the LSTM, and the true values and predicted values fit better.

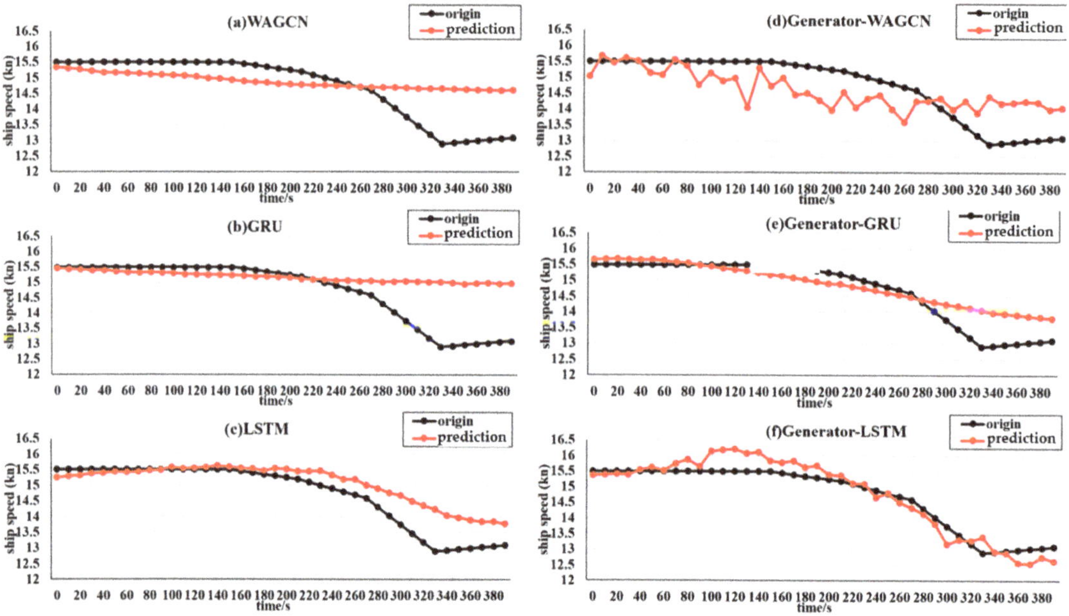

Figure 14. Comparison of model ship speed prediction results in tanker scenario.

By contrasting the ship-speed forecasting accuracy of each model in the two different scenarios, it can be verified that the proposed model exceeds the other models in ship-speed forecasting. In the ship-speed-prediction task, an LSTM as a generator can help the model to better capture and simulate the complex time-series characteristics and change laws.

In order to more clearly represent the ship-speed prediction results of all the models of this subset segment within the current tanker scheme, they are drawn in a coordinate axis, as depicted in Figure 15. The solid black line represents the original input data, and the solid red line represents the forecast data of the Generator–LSTM. It is clear that within the predicted data of each model, the red solid line (i.e., the Generator–LSTM model) fits better with the black solid line (i.e., the original input data); that is, the model has better ship-speed prediction performance.

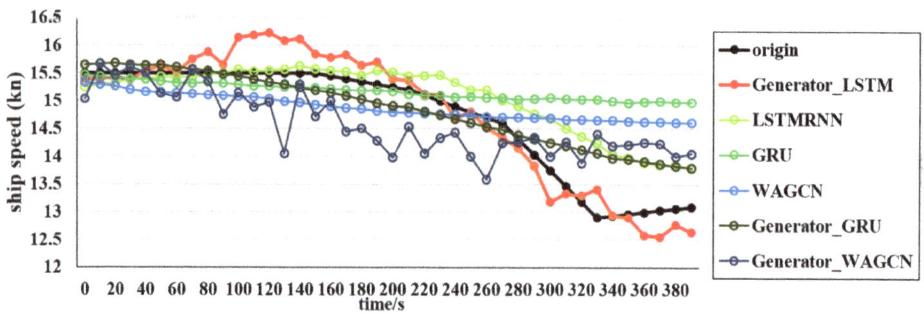

Figure 15. The overview of the prediction results from the six models in the tanker scenario.

Similarly, observe the model-error histogram in this scenario. As shown in Figure 16, the Generator–LSTM model has the smallest area in the histogram area for the four types of error indicators; that is, the Generator–LSTM model has the lowest value for the four types of error indicators and the best ship-speed prediction performance.

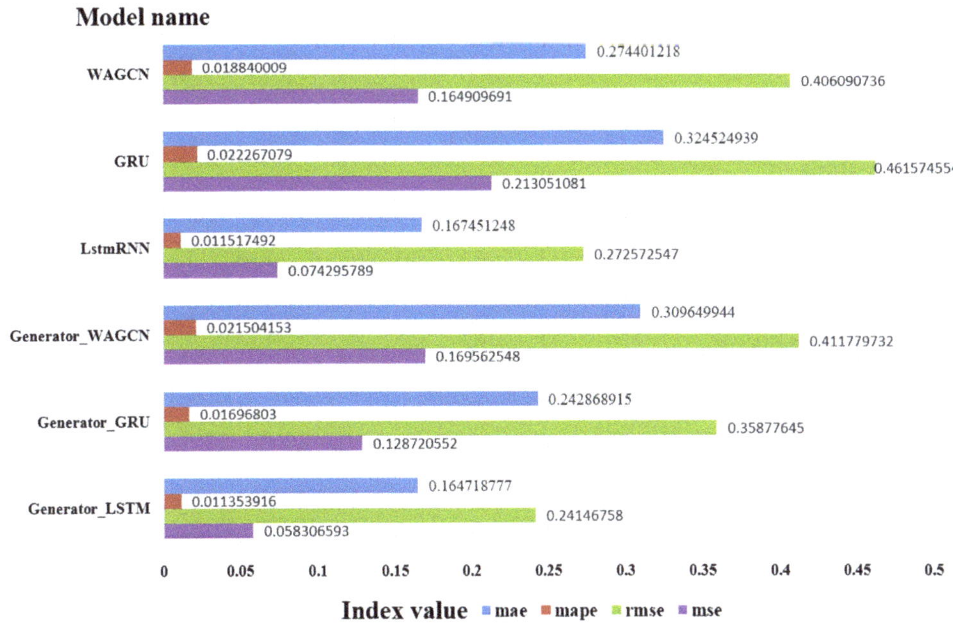

Figure 16. Comparison of model indices in the tanker scenario.

3.3.3. Experimental Results in the Bulk Cargo Ship, Passenger Ship, Fishing Ship and Sand Dredger Scenarios

The experimental results of the first two scenarios, namely, the container experimental scenario and the oil tanker experimental scenario, indicate that the GAN–LSTM has a better effectiveness in ship-speed prediction. To expand the substantive nature of the experimental verification, four different types of experimental scenarios, namely, bulk carriers, passenger ships, fishing boats, and dredgers, have been added. And benchmark tests were conducted for two advanced models, namely, Transformer and GNN, as well as the target model GAN–LSTM. Tables 4–7, respectively, show the error index results of the seven comparison models in the experimental scenarios of bulk carriers, passenger ships, fishing boats, and dredgers.

Table 4. Comparison of results as to experimental model speed-prediction accuracy in the bulk cargo ship scenario.

Model	MSE	RMSE	MAPE (%)	MAE
LSTM	2.64×10^{-2}	2.14×10^{-1}	3.63×10^{-2}	3.76×10^{-1}
GRU	4.76×10^{-2}	1.90×10^{-1}	6.17×10^{-2}	4.85×10^{-1}
WAGCN	3.07×10^{-2}	1.54×10^{-1}	4.60×10^{-2}	4.06×10^{-1}
Generator–GRU	2.36×10^{-2}	1.85×10^{-1}	3.48×10^{-2}	3.54×10^{-1}
Generator–WAGCN	4.44×10^{-2}	1.66×10^{-1}	5.73×10^{-2}	5.21×10^{-1}
Transformer	9.59×10^{-2}	3.10×10^{-2}	1.96×10^{-2}	1.53×10^{-1}
GNN	1.77×10^{-1}	1.33×10^{-1}	7.26×10^{-2}	1.94×10^{0}
Generator–LSTM	1.81×10^{-2}	2.50×10^{-2}	1.47×10^{-2}	1.06×10^{-1}

Table 5. Comparison of results for experimental model speed-prediction accuracy in the passenger ship scenario.

Model	MSE	RMSE	MAPE (%)	MAE
LSTM	3.59×10^{-3}	5.99×10^{-2}	4.84×10^{-2}	3.18×10^{-1}
GRU	1.22×10^{-2}	1.10×10^{-1}	1.41×10^{-1}	6.70×10^{-1}
WAGCN	5.72×10^{-3}	7.57×10^{-2}	1.00×10^{-1}	5.26×10^{-1}
Generator–GRU	4.52×10^{-3}	6.72×10^{-2}	6.75×10^{-2}	4.32×10^{-1}
Generator–WAGCN	5.22×10^{-3}	7.22×10^{-2}	8.89×10^{-2}	5.28×10^{-1}
Transformer	2.73×10^{-3}	1.65×10^{-2}	1.37×10^{-2}	8.06×10^{-2}
GNN	8.74×10^{-3}	2.96×10^{-2}	2.27×10^{-2}	1.34×10^{-1}
Generator–LSTM	2.50×10^{-3}	3.82×10^{-3}	1.10×10^{-2}	1.70×10^{-2}

Table 6. Comparison of results for experimental model speed-prediction accuracy in the fishing ship scenario.

Model	MSE	RMSE	MAPE (%)	MAE
LSTM	1.45×10^{-3}	1.21×10^{-1}	6.98×10^{-2}	8.58×10^{-2}
GRU	1.07×10^{-2}	3.28×10^{-1}	1.71×10^{-1}	2.09×10^{-1}
WAGCN	3.52×10^{-3}	1.88×10^{-1}	1.03×10^{-1}	1.25×10^{-1}
Generator–GRU	2.89×10^{-3}	1.70×10^{-1}	9.19×10^{-2}	1.12×10^{-1}
Generator–WAGCN	6.48×10^{-3}	2.55×10^{-1}	1.34×10^{-1}	1.63×10^{-1}
Transformer	5.04×10^{-3}	2.24×10^{-2}	1.86×10^{-2}	1.95×10^{-1}
GNN	6.14×10^{-3}	2.48×10^{-2}	1.70×10^{-2}	1.77×10^{-1}
Generator–LSTM	1.25×10^{-3}	1.20×10^{-2}	1.50×10^{-2}	7.97×10^{-2}

Table 7. Comparison of results for experimental model speed-prediction accuracy in the sand dredger scenario.

Model	MSE	RMSE	MAPE (%)	MAE
LSTM	9.95×10^{-3}	9.98×10^{-2}	5.40×10^{-2}	7.69×10^{-2}
GRU	5.69×10^{-2}	2.39×10^{-1}	1.29×10^{-1}	1.85×10^{-1}
WAGCN	3.92×10^{-2}	1.98×10^{-1}	9.78×10^{-2}	1.42×10^{-1}
Generator–GRU	1.73×10^{-2}	1.31×10^{-1}	6.43×10^{-2}	9.28×10^{-2}
Generator–WAGCN	6.02×10^{-2}	2.45×10^{-1}	1.39×10^{-1}	1.98×10^{-1}
Transformer	2.11×10^{-2}	1.45×10^{-1}	8.86×10^{-2}	2.27×10^{0}
GNN	1.66×10^{-2}	1.29×10^{-1}	7.02×10^{-2}	2.02×10^{0}
Generator–LSTM	6.77×10^{-3}	8.23×10^{-2}	3.71×10^{-2}	5.29×10^{-2}

The results of the model comparison error index for the bulk-carrier experimental scenario show that the minimum MSE value is that of GAN–LSTM, 1.81×10^{-2}; this model also returns the minimum RMSE value, 2.50×10^{-2}; the minimum MAPE value, 1.47×10^{-2}; and the minimum MAE value, 1.06×10^{-1}. The four error indicators of GAN-LSTM are the lowest indicators for the benchmark tests, including Transformer and GNN, relative to the remaining five models, indicating that it is the most effective in ship-speed prediction.

The results of the comparison of the model error index for the passenger ship experimental scenario show that the minimum value of MSE is that of GAN–LSTM, which is 2.50×10^{-3}; this model also returns the minimum RMSE value, 3.82×10^{-3}; the minimum MAPE value, 1.10×10^{-2}; and the minimum MAE value, 1.70×10^{-2}. The four error indicators of GAN–LSTM have the lowest indicators among the benchmark tests, including Transformer and GNN, relative to the remaining four models, indicating that it is the most effective in ship-speed prediction.

The results of the model comparison error index for the fishing ship experimental scene show that the minimum MSE value is that of GAN–LSTM, 1.25×10^{-3}; this model also returns the minimum RMSE value, 1.20×10^{-2}; the minimum MAPE value, 1.50×10^{-2}; and the minimum MAE value, 7.97 × 10^{-2}. The four error indicators of GAN–LSTM have the lowest indicators among the benchmark tests, including Transformer and GNN, relative to the remaining five models, indicating that it is the most effective in ship-speed prediction.

The results of the model comparison error index for the dredger experimental scene show that the minimum MSE value is that of GAN–LSTM, at 6.77 × 10^{-3}; this model also reports the lowest values for RMSE, at 8.23 × 10^{-2}; MAPE, at 3.71 × 10^{-2}, and MAE, at 5.29 × 10^{-2}. The four error indicators of GAN–LSTM have the lowest indicators among the benchmark tests, including Transformer and GNN, relative to the remaining five models, indicating that it is the most effective in ship-speed prediction.

4. Discussion

4.1. An In-Depth Critical Discussion of the Experimental Results

The experimental scenarios selected six types of ships, namely, container ships, oil tankers, bulk carriers, passenger ships, fishing boats, and dredgers; these can represent the ship scenarios within a certain range. The experimental comparison models include the LSTM, GRU, WAGCN, GAN–GRU, and GAN–WAGCN. Meanwhile, the benchmark tests of the target model GAN–LSTM on Transformer and GNN are included, and have certain substantive significance for the experimental verification. The comparison of the experimental index results for the six types of ships shows that GAN–LSTM has the best effectiveness in predicting ship speed. However, for the purpose of a more in-depth and critical discussion of the experimental results, it encompasses three aspects: limitations, error analysis, and practical significance. The following is a detailed elaboration of these dimensions.

4.1.1. Limitations

(1) Discussion relating to the limitations of GAN–LSTM in predicting ship speed

In terms of data quality and quantity, the performance of the GAN–LSTM model usually depends on the quality and diversity of the training data. If the data contain noise, or are incomplete or biased, the generated prediction results may not be accurate. In the application of ship-speed prediction, collecting complete and accurate historical data is a challenge, especially due to the influence of factors such as having different types of ships, different routes, and weather conditions. A GAN–LSTM usually requires a large amount of historical data to train the model. For certain specific areas or special types of ships, it may be difficult to collect sufficient training data, which limits the performance of the model.

In terms of the complexity of model training, a GAN–LSTM requires the training of both the generator and the discriminator simultaneously, and the LSTM part also needs to handle long-term series data. This may lead to the model training process becoming extremely complex, requiring a large amount of computing resources and time for convergence. The generator in a GAN may experience pattern collapse; that is, the generated prediction results might lack diversity, resulting in the model's inability to accurately capture the complex changes in ship speed, especially when ships are affected by various factors such as weather and sea conditions. Both the GAN and the LSTM themselves have many hyperparameters. The selection of these parameters will affect the performance of the model, and the parameter tuning process requires a large number of experiments. If the model is overly dependent on the training data, it may lead to overfitting, that is, performing poorly on new data. Especially in the prediction of ship speed, the different navigation conditions of ships may lead to gaps between the training data and the actual scenarios, which will affect the generalization ability of the model.

In addition, the speed of a ship is not only affected by historical speed data, but also by many external environmental factors, such as wind speed, ocean currents, weather, and tides [34]. Although a GAN–LSTM can capture certain patterns of ship speed through historical data, it is difficult for it to directly simulate these complex external factors, resulting in limitations to its prediction accuracy.

(2) Discussion relating to the respective limitations of the remaining experimental models compared as to ship-speed prediction

As to the limitations of an LSTM, when dealing with long-term series data, although the LSTM represents an improvement, compared to traditional RNN, with respect to extremely long series of data, it may encounter problems such as vanishing or exploding gradients, resulting in a decline in its ability to capture long-term dependencies. The computational overhead of an LSTM is relatively high, especially when dealing with complex time-series data, which may reduce the real-time performance of the model. This is a limitation for application scenarios with high real-time requirements, such as ship-speed prediction [35].

As to the limitations of a GRU, the GRU performs slightly worse, relative to the LSTM, when modeling long-term and short-term dependencies. Although it has fewer parameters and higher computational efficiency, when capturing very long-term dependencies, the GRU performs worse, relative to the LSTM. The structure of the GRU is simpler than that of the LSTM, which also means that it is not as flexible as an LSTM when dealing with complex time-series data, especially under complex navigation conditions.

As to the limitations of a WAGCN, the WAGCN is a graph convolutional network, and is usually used to handle graph-structured data. In the context of ship-speed prediction, although it can effectively handle information with graph structures, if the task data do not have a clear graph structure, the WAGCN cannot function effectively. A WAGCN requires that the input data have a good graph-based informational structure, or that the relationships between ships needed to be modeled, which would increase the complexity of data processing and the cost of preprocessing.

As to the limitations of a GAN–GRU, similar to a GAN–LSTM, the GAN–GRU also faces the problem of generator pattern collapse, which affects the accuracy of the model. The GRU has fewer parameters compared to an LSTM, but its ability to handle long-term series is poor and it cannot effectively capture long-term dependencies in the changes of ship speed.

As to the limitations of a GAN–WAGCN, the GAN–WAGCN combines the advantages of graph convolution and Generative Adversarial Networks, but is also limited to graph-structured data. If there is no clear graph structure of the relationships between ships, it would be difficult to model, and the effectiveness of the model would be affected. A GAN–WAGCN faces a training instability problem similar to those of other GAN models and requires a good training strategy to avoid the problems of generator pattern collapse and the inability of the discriminator to be effectively trained.

(3) Discussion of the limitations of the benchmark tests for the GAN–LSTM, the Transformer and GNN

The discussion and analysis of the limitations of a GAN–LSTM have been elaborated specifically in "(1) Discussion on the limitations of a GAN–LSTM in predicting ship speed" in this section and will not be repeated here.

The limitations of Transformer are mainly manifested in the following four aspects:

High memory consumption occurs when processing long sequences. Although Transformer is more efficient than are the RNN and LSTM when dealing with long sequences, its self-attention mechanism requires calculating the interaction information for all input sequences, which leads to high computational and memory overhead. For particularly long time series, the problem of low computational efficiency may be faced.

Transformer has a strong dependence on data quality. The Transformer model performs very well on high-quality and large-scale data. However, if the data quality is poor or the data volume is insufficient, Transformer may not achieve the expected effect, and this may even lead to overfitting due to the instability of the training data [36].

The lack of time-dependent modeling makes it difficult to handle the sequential nature of time series. Although the self-attention mechanism of Transformer can capture the dependencies at long distances in the sequence, it itself lacks an inherent mechanism for handling temporal order, which makes it less direct and effective than an LSTM when dealing with tasks that strictly rely on temporal order, such as ship-speed prediction.

Real-time prediction of delay proves difficult. Since Transformer needs to perform global calculations on the entire sequence when calculating self-attention, it will encounter computational bottlenecks in real-time or low-latency prediction tasks. Especially in application scenarios that require rapid response, Transformer cannot complete the prediction in a timely manner.

The limitations of GNN are mainly manifested in the following four aspects:

The construction of graph data is complex and the construction of graph structure is difficult. GNN mainly relies on graph structures to represent the relationships between data points. For ship-speed prediction, constructing an appropriate graph structure can be very complex, especially when the relationships between data are relatively implicit or difficult to model clearly.

It is difficult to capture global dependencies in the balance between local information and global information. Although GNN can model the relationships between local nodes well, it is not as direct and effective as an RNN or LSTM when dealing with long-term global dependencies. Ship-speed prediction involves rather complex global relationships in terms of time and space, and GNN performs poorly in this aspect.

The computational complexity of graph networks is high. When graph neural networks handle large-scale graphs, the computational and memory overheads will also increase rapidly, especially when the graph structure needs to be updated frequently and complex graph convolution needs to be performed, which will lead to a decrease in the efficiency of the model in the training and inference stages [37].

It is prone to overfitting small-scale data. In the practical application of ship-speed prediction, the dataset is not particularly large. Especially in the case of missing data or a substantial amount of noise, GNN is prone to overfitting, resulting in a poor generalization ability of the model.

4.1.2. Analysis of Errors

(1) Sources of error

The first source of error is in the AIS data quality issues. The AIS system itself may be affected by environmental factors such as weather, sea conditions, or terrain, resulting in errors or inaccuracies during data collection. For example, ships may lose signals, or signals may be interfered with, resulting in positional errors. Inaccurate position and velocity information will affect the training and prediction results of the GAN and LSTM models. In addition, AIS data usually have a time delay, especially when the vessel is far away or the signal coverage is poor. The AIS system sometimes loses data due to hardware or communication issues. The delay or absence of historical data can lead to the inability to obtain complete information during model training, which may result in the inability to accurately capture the changing trends in ship speed during the training process, thereby affecting the prediction accuracy.

The second source of error is in the dynamic environmental factors. The speed of a ship not only depends on the operation of the vessel itself, but is also affected by the marine environment. These external factors may not have been fully considered or might be difficult to quantify accurately. If these external factors are not fully modeled or reflected in the AIS data, the prediction results of the model may not accurately reflect the actual changes in ship speed.

The third source of error is in the complexity of time-series characteristics. The variations in ship speed may present complex nonlinear and periodic characteristics. Although an LSTM is suitable for processing time-series data, it may be difficult to fully capture all time-varying patterns associated with ship speed. If the LSTM model fails to fully capture these complex time-series patterns, this may lead to an increase in prediction errors, especially during long-term predictions, in which the errors will accumulate over time.

(2) Biases of datasets

The first dataset bias is the time synchronization bias. AIS data typically contain the timestamps and positions of ships, but due to network delays, communication issues or equipment failures, the reporting times of ships may not be completely synchronized. Therefore, there may be a certain deviation between the timestamp and the actual ship speed. Time synchronization deviation can lead to the disruption of the temporal order of data, thereby affecting the learning effectiveness of the LSTM model for the time series.

The second dataset bias consists of data noise and outliers. AIS data may sometimes contain noise or outliers, which may be caused by factors such as sensor failure and signal interference. Data noise and outliers can make it difficult for the LSTM model to learn effective patterns from normal data, affecting the training effectiveness of the model.

The third dataset bias is in the nonlinear relationship between ship speed and position. The relationship between the speed of a ship and its position may be a complex nonlinear one, and the speed of a ship may be affected by factors such as waterways, traffic density, and wind speed. In AIS data, the relationship between ship speed and position is not a simple linear one, but has strong dynamic changes. An LSTM is capable of handling nonlinear relationships in a time series. However, if the nonlinear relationships in AIS data are not modeled correctly, the LSTM model may have difficulty accurately capturing the true laws of ship-speed changes.

4.1.3. Practical Significance

The first element of practical significance is that a GAN–LSTM can improve the prediction accuracy of ship-speed prediction in practical applications, enhance the robustness of the model, and achieve ship-speed optimization.

Traditional time-series prediction methods may have limitations when dealing with complex ship-speed data, especially when the data are affected by multiple factors such as the marine environment, climate change, and ship-types. By combining LSTM with a GAN, the time-series characteristics and potentially nonlinear relationships in the ship-speed data can be effectively captured. Meanwhile, training samples that are more diverse are generated through the GAN to enhance the generalization ability of the model, thereby improving the prediction accuracy. Combining the generation ability of a GAN and the time-series modeling ability of an LSTM can enhance the model's adaptability to various forms of complex and nonlinear data. By generating synthetic samples, a GAN can expand the training dataset, especially in the case of scarce data, enhancing the robustness of the model, enabling the model to better cope with incomplete or variable data in practical applications. In addition, accurate prediction of ship speed not only helps to enhance the safety of shipping, but also plays a positive role in optimizing routes, saving fuel, and reducing shipping costs. Through more accurate ship-speed prediction, ship managers can better plan the voyage and avoid unnecessary speed fluctuations, and thereby improve operational efficiency.

The second element of practical significance is that the GAN–LSTM can achieve automated and real-time prediction in predicting ship speed in practical applications, improving shipping logistics and achieving the goal of energy conservation and emission reduction.

The GAN–LSTM model can achieve automated ship-speed prediction in shipping systems. With the advancement of self-driving ships, real-time prediction of ship speed is of great significance in aspects such as navigation control, collision avoidance, and channel selection. The GAN–LSTM model can be utilized to provide real-time speed adjustment suggestions for ships, helping crew members better cope with varying sea conditions. In the shipping and logistics industry, accurate prediction of ship speed not only helps ships better control their voyage time, but also enables logistics companies to schedule ships more efficiently, optimizing the cargo transportation process, reducing delays, and thereby enhancing overall logistics efficiency and customer satisfaction. Accurate ship-speed prediction can help shipping companies adjust the speed of a ship based on its real-time conditions, thereby optimizing fuel consumption and reducing carbon emissions. This is of positive significance for achieving the green transformation and energy conservation and emission reduction goals of the shipping industry.

Finally, deploying the ship-speed prediction model based on a GAN–LSTM in the actual maritime traffic management system has significant practical significance.

Firstly, it can enhance navigation safety, optimize the voyage range, and reduce navigation delays, thereby improving traffic flow management and strengthening the real-time response capability of the system. Accurate prediction of ship speed helps ensure that ships can navigate as planned and reduces potential safety hazards caused by excessive fluctuations in ship speed. By using the GAN–LSTM model, the real-time speed of ships can be accurately predicted, helping the maritime traffic management system to monitor and adjust the navigation routes in real time, avoiding accidents such as collisions and stranding, and ensuring navigation safety. Traditional voyage planning usually relies on historical data and preset speeds, while the GAN–LSTM model can conduct dynamic prediction in combination with real-time data. This means that shipping companies and maritime traffic management institutions can adjust routes and speeds in real time based on accurate ship-speed predictions and optimize voyages, thereby reducing delays, improving the operational efficiency of vessels, and enhancing the overall logistics operation speed. A GAN–LSTM combines the advantages of Generative Adversarial Networks and Long Short-Term Memory networks and can quickly adapt to the changes in ship-speed prediction in the unpredictable marine environment.

Secondly, it can enhance fuel efficiency and reduce emissions, improve the efficiency of resource allocation, and provide significant support for automated navigation and intelligent shipping. In maritime traffic management, the reasonable prediction of ship speed can help ships better control their speed and avoid sailing too fast or too slow. A reasonable speed can not only save fuel but also reduce carbon emissions, which is in line with the green and environmental protection goals increasingly emphasized by the modern shipping industry. With the gradual application of automation technology, the role of accurate ship-speed prediction in automated navigation is particularly crucial [38]. The ship-speed prediction based on the GAN–LSTM model can provide accurate real-time data for the automated navigation system, helping ships automatically adjust the speed and route without manual intervention to ensure the safety and efficiency of navigation.

4.2. Construction of Marine Ship Intelligence and a Self-Navigation System

With the advancement of smart vessels, the automated navigation systems on ships must handle real-time dynamic data processing, such as weather conditions, ocean currents, hull status, and more [39]. A GAN can generate ship-speed data that match these conditions, help automated systems predict ship speed in real time, and adjust the speed according to the predicted results. This is vital for the automated operation of ships, especially in the absence of human operators, where the reliability of ship-speed forecasts directly affects the safety and efficiency of navigation [40]. In the self-navigation system of intelligent ships, a GAN can combine real-time data to predict ship-speed changes, especially the ship-speed adjustments made in bad weather or complex sea conditions.

When actually deploying the GAN–LSTM model to predict the speeds of oil tankers and container ships, the computational efficiency is associated with high requirements for real-time performance, accuracy, and resource consumption [41].

(1) In terms of training time: The GAN–LSTM model combines the advantages of a GAN and an LSTM, but at the same time requires a longer training time. Especially when it comes to complex time-series data (such as ship speed, position, etc.), the training process can be very time-consuming. Adversarial training in a GAN requires repeated iterations of the generator and discriminator, and an LSTM needs to effectively learn the long-term dependencies in the time series [42]. The datasets of oil tankers and container ships usually include multi-dimensional time-series information (such as ship speed, heading, longitude and latitude, etc.), and a large amount of computing time and resources are consumed during the training process.
(2) In terms of the required resources: The GAN–LSTM model requires a large amount of computing resources during the training process. Since the model contains a generator, a discriminator, and an LSTM module, all of these parts require high computational power in terms of calculation [43]. In actual deployment, this high demand may limit this application under the condition of limited resources. Speed prediction for oil tankers and container ships involves a large amount of data and usually requires real-time calculation. Therefore, the computing resource requirements during training may place higher demands on hardware platforms (such as GPU, TPU) [44]. High demands for memory and computing power may lead to performance degradation when deployed on low-configuration hardware.

4.3. *Collaborative Optimization of Smart Ports and Ship-Speed Prediction*

With the development of the maritime industry, the research on ship-speed prediction cannot ignore the necessity of the collaborative optimization of smart ports and routes. Smart ports can leverage IoT and Big Data technologies to monitor a wide range of dynamic information both inside and outside the port in real time, including ship locations, cargo loading and unloading progress, port weather, and more [45,46]. A GAN can optimize coordination between ships and ports through real-time ship-speed prediction and navigation data generation. For example, by predicting the accurate arrival times of ships, ports can better arrange berths, scheduling, and cargo handling plans, thereby reducing port congestion and waiting times and improving port operational efficiency [47]. In addition, the ship-speed prediction not only affects the ship sailing process but also has an important impact on the port logistics chain. Therefore, precise ship-speed prediction is a key prerequisite for optimizing the port logistics chain. The logistics management system of a smart port can use the ship-speed data generated by a GAN to optimize the cargo handling efficiency of all links from the ship to the port [48,49]. By accurately predicting ship speed, the port can reasonably arrange the flow of goods and the times at which ships enter and exit the port, and adjust the equipment and personnel configuration in advance, reducing the overall logistics cost.

With the improvements in environmental protection requirements, ship energy optimization has become the key [50]. By utilizing a GAN for ship-speed prediction, ships can more precisely manage their speed to achieve minimum fuel consumption and optimal energy use. Ship speed and fuel consumption are related in a nonlinear fashion. A GAN can generate predictive models for ship speed and fuel consumption under various sea conditions and loads, assisting ships in adjusting their speed to achieve green shipping objectives and reduce carbon emissions [51].

The GAN model is essentially a complex deep learning model that lacks sufficient interpretability. This makes the model's predictions difficult to understand and verify, especially when key decisions require transparency and interpretability (e.g., port scheduling decisions, channel optimization) [52]. In this study, GAN architectures with better stability, such as the LSTM–GAN or WAGCN–GAN, are used, which can reduce the phenomenon of training instability by improving the loss function and regularization method. Additionally, the study began with simple tasks used to train the GAN models and progressively increased the complexity of the models. The model's capabilities are enhanced through incremental training, facilitating easier convergence [53].

Ship-speed prediction not only aims to accurately predict ship speed, but also involves the collaborative optimization of multiple objectives, such as port scheduling, channel planning, etc., some of which may conflict with each other [54,55]. In this research, according to the importance of different target data, different types of data are prioritized and weighted, and some important targets are given priority in the training process in order to balance the conflicts between different targets [56,57]. By introducing the multi-objective optimization function into the generator and discriminator, the model can not only generate ship-speed prediction but also consider many factors such as port resource utilization efficiency and channel safety.

5. Conclusions

A GAN can be used to generate a richer training dataset. In cases where traditional models have less data or unbalanced data, a GAN can be combined in certain scenarios to enhance the training process and improve the generalization ability of the model by using the data generated by a GAN. However, the combination of traditional models and GANs can lead to unstable training, high demand for computing resources, unreliable quality of generated data, complex parameter adjustment, and difficulties in model fusion, etc. An LSTM is a kind of neural network specifically used for processing time-series data (such as the change in ship speed over time), and it can capture the long-term dependencies in the time series. Compared with traditional machine learning models, an LSTM can better capture the temporal characteristics and dynamic patterns of ship speed when processing time-series data.

In this study, a GAN–LSTM model used for ship-speed prediction is proposed, one which combines a GAN network and an LSTM model. First, the algorithm takes an LSTM network as the generating network and utilizes the LSTM to capture spatiotemporal dependencies between nodes. Secondly, the complementary characteristics linked between the generative network and the discriminant network are used to eliminate the cumulative error of a single neural network over the long-term prediction process, improving the accuracy of the network in ship-speed determination. Ultimately, the Generator–LSTM model is utilized for predicting ship speed and compared with other models, using the same AIS ship-speed information, and in the same scene. The results show that the model achieves high accuracy in typical error metrics, which means that the model can more accurately predict ship speed. In addition, the performance of the Generator–LSTM model shows the superiority of this method in the task of speed prediction. This can be attributed to the Generator–LSTM model's ability to realize the task of speed prediction by using the spatiotemporal correlations in velocity-related data. In addition, the Generator–LSTM model has a high prediction accuracy, indicating that good temporal and spatial correlation is a crucial component of a model for accurate ship-speed prediction.

This study still has some limitations, such as the limited nature of the dataset. Although we used datasets from multiple sources, these data may not fully cover all possible actual situations. Therefore, under certain specific conditions, the prediction accuracy of the model may be reduced. Secondly, there is also a limitation related to the universality of the model. The neural network (or LSTM) model used in the research performs well in certain specific scenarios, but as to situations involving extreme or uncommon situations, the robustness of the model still needs to be further verified. Furthermore, due to the limitations in computing resources, this study failed to conduct more hyperparameter optimizations or train on larger-scale datasets. Therefore, the performance of the model may not have reached its optimum level.

In future research, it is important to further investigate the performance of this model in long-term speed-prediction tasks, and research on the expression of water-traffic situational knowledge based on AIS data is an area also worthy of further study.

Author Contributions: Conceptualization, X.C., P.W. and J.X.; methodology, X.C., Y.Z. and X.W.; writing—original draft preparation, P.W., Y.Z. and H.Z.; writing—review and editing, P.W. and X.W.; funding acquisition, X.C. All authors have read and agreed to the published version of the manuscript.

Funding: This work was jointly supported by National Natural Science Foundation of China (52331012, 52472347), and Shanghai Science and Program of Shanghai Academic/Technology Research Leader (23XD1431000).

Data Availability Statement: Data can be accessed by sending an email to the corresponding author.

Conflicts of Interest: The authors declare no conflicts of interest.

References

1. Goncharov, V.K.; Klementieva, N.Y. Parameterization of the propeller thrust for modelling ship braking in ice channel behind icebreaker. *Transp. Saf. Environ.* **2023**, *5*, tdac042. [CrossRef]
2. Goncharov, V.K.; Chugaeva, A.E.; Kviatkovskaia, E.S.; Zueva, E.S. Analysis of the impact of Arctic environment conditions on the psychophysical status of navigators and the safety of ice navigation. *Transp. Saf. Environ.* **2024**, *6*, tdae019. [CrossRef]
3. Korupoju, A.K.; Kapadia, V.; Vilwathilakam, A.S.; Samanta, A. Ship Collision Risk Evaluation using AIS and weather data through fuzzy logic and deep learning. *Ocean. Eng.* **2025**, *318*, 120116. [CrossRef]
4. Li, H.; Jiao, H.; Yang, Z. AIS data-driven ship trajectory prediction modelling and analysis based on machine learning and deep learning methods. *Transp. Res. Part E Logist. Transp. Rev.* **2023**, *175*, 103152. [CrossRef]
5. Klemm, J.; Niemi, A.; Torres, F.S. Predicting Ship Tracks in Waterways Using Long Short-Term Memory. In Proceedings of the 7th International Conference on Transportation Information and Safety, Halifax, NS, Canada, 23–26 September 2024. [CrossRef]
6. Aristov, N.; Li, Z.; Koch, T.; Dugundji, E.R. Elucidating US Import Supply Chain Dynamics: A Spatial-Temporal Graph Neural Network Approach. In Proceedings of the IEEE High Performance Extreme Computing Conference, Wakefield, MA, USA, 23–27 September 2024. [CrossRef]
7. Lee, G.; Kim, D.; Park, S.; Sim, S.; Liu, L.; Bae, H. Neural Bezier Interpolation with Manifold Learning for Reliable Vessel Trajectory Prediction. In Proceedings of the IEEE 6th International Conference on Cognitive Machine Intelligence, Washington, DC, USA, 28–31 October 2024. [CrossRef]
8. Pensado, E.A.; López, F.V.; Jorge, H.G.; Pinto, A.M. UAV Shore-to-Ship Parcel Delivery: Gust-Aware Trajectory Planning. *IEEE Trans. Aerosp. Electron. Syst.* **2024**, *60*, 6213–6223. [CrossRef]
9. Magalhães, A.; Estima, J.; Cardoso, A. Vessel Voyage Trajectory Extrapolation: Comparing the Performance of Kalman Filters. In Proceedings of the 2023 6th Experiment@ International Conference, Évora, Portugal, 5–7 June 2023. [CrossRef]
10. Padmavathi, A.; Parameshachari, B.D.; Fallah, M.H.; Pushpalakshmi, G.; Saranya, N.N. A Marine Traffic Pattern Prediction System Based on Graph Convolutional and Kolmogorov-Arnold Networks. In Proceedings of the International Conference on Integrated Intelligence and Communication Systems, Kalaburagi, India, 22–23 November 2024. [CrossRef]

11. Donandt, K.; Söffker, D. Incorporating Navigation Context into Inland Vessel Trajectory Prediction: A Gaussian Mixture Model and Transformer Approach. In Proceedings of the 2024 27th International Conference on Information Fusion (FUSION), Venice, Italy, 8–11 July 2024. [CrossRef]
12. Ivanovic, P.; Windmann, A.; Neumann, P. SmartShip: Data Decoding Optimization for Onboard AI Anomaly Detection. In Proceedings of the 2024 23rd International Symposium on Parallel and Distributed Computing (ISPDC), Chur, Switzerland, 8–10 July 2024. [CrossRef]
13. Sigillo, L.; Marzilli, A.; Moretti, D.; Grassucci, E.; Greco, C.; Comminiello, D. Sailing the Seaformer: A Transformer-Based Model for Vessel Route Forecasting. In Proceedings of the 2023 IEEE 33rd International Workshop on Machine Learning for Signal Processing (MLSP), Rome, Italy, 17–20 September 2023. [CrossRef]
14. Gülsoylu, E.; Koch, P.; Yıldız, M.; Constapel, M.; Kelm, A.P. Image and AIS Data Fusion Technique for Maritime Computer Vision Applications. In Proceedings of the IEEE/CVF Winter Conference on Applications of Computer Vision Workshops, Waikoloa, HI, USA, 1–6 January 2024. [CrossRef]
15. Nitish, R.; Prabhat, K.; Divyanshi, G. Predicting Vessel Speed Using Ensemble MLR and RF Models on AIS Data. *Def. Sci. Inf. Doc. Cent.* **2025**, *75*, 250–256. [CrossRef]
16. Oleg, T.; Danila, E.; Alex, T. A benchmark study on ship speed prediction models in arctic conditions: Machine learning, process-based and hybrid approaches. *Ocean. Eng.* **2024**, *311*, 118884. [CrossRef]
17. Pappagallo, A.; Ortame, F.; Massacci, G.; Sisti, F.; Pugliese, F. *Deep Learning for the Classification of Ports in Maritime Transport Statistics via AIS Data*; Lecture Notes in Computer Science; Springer: Cham, Switherland, 2025; pp. 318–332. [CrossRef]
18. Kumar, R.H.; Ramanarayanan, C.P.; Bharath, R. *Machine Learning Based Automated Process for Predicting the Anomaly in AIS Data*; Lecture Notes on Data Engineering and Communications Technologies; Springer: Singapore, 2023; pp. 303–314. [CrossRef]
19. Marta, S.; Ireneusz, C. VAR vs XGBoost: Performance of both models in AIS data reconstruction. *Procedia Comput. Sci.* **2024**, *246*, 5084–5093. [CrossRef]
20. Xian, J.; Li, Z.; Wu, H. Novel Polarization Construction Method and Synchronization Algorithm for Underwater Acoustic Channel Under T-Distribution Noise Environment. *J. Mar. Sci. Eng.* **2025**, *13*, 362. [CrossRef]
21. Chen, X.; Hu, R.; Luo, K.; Wu, H.; Biancardo, S.A.; Zheng, Y.; Xian, J. Intelligent Ship route planning via an A* search model enhanced double-deep Q-Network. *Ocean. Eng.* **2025**, *327*, 120956. [CrossRef]
22. Ourc, M.F.; Altan, Y.C. Deep learning for risky encounter prediction: Leveraging LSTM networks and AIS data. In *Advances in the Collision and Grounding of Ships and Offshore Structures*; Taylor & Francis: Abingdon, UK, 2023; pp. 83–89. [CrossRef]
23. Shin, G.-H.; Yang, H. Vessel trajectory prediction in harbors: A deep learning approach with maritime-based data preprocessing and Berthing Side Integration. *Ocean. Eng.* **2025**, *316*, 119908. [CrossRef]
24. Yang, Y.; Liu, Y.; Li, G.; Zhang, Z.; Liu, Y. Harnessing the power of Machine learning for AIS Data-Driven maritime Research: A comprehensive review. *Transp. Res. Part E Logist. Transp. Rev.* **2024**, *183*, 103426. [CrossRef]
25. Mehri, S.; Alesheikh, A.A.; Basiri, A. A context-aware approach for vessels' trajectory prediction. *Ocean. Eng.* **2023**, *282*, 114916. [CrossRef]
26. Liao, W.; Wu, Y.; Zhou, P.; Wang, H.; Yu, W.; Zhang, C.; Luo, C. Dynamic navigation: Integrating GL-STGCNN and MPC for collision avoidance with future Awareness. *Ocean Eng.* **2024**, *309 Pt 1*, 118416. [CrossRef]
27. Xian, J.; Ma, J.; Chen, X. Robust and Precise Drowning Target Localization under Dual-parameters Uncertainty. *Reg. Stud. Mar. Sci.* **2025**, *85*, 104131. [CrossRef]
28. Waltz, M.; Paulig, N.; Okhrin, O. 2-level reinforcement learning for ships on inland waterways: Path planning and following. *Expert Syst. Appl.* **2025**, *274*, 126933. [CrossRef]
29. Alam, M.M.; Spadon, G.; Etemad, M.; Torgo, L.; Milios, E. Enhancing short-term vessel trajectory prediction with clustering for heterogeneous and multi-modal movement patterns. *Ocean. Eng.* **2024**, *308*, 118303. [CrossRef]
30. Li, Y.; Liang, M.; Li, H.; Yang, Z.; Du, L.; Chen, Z. Deep learning-powered vessel traffic flow prediction with spatial-temporal attributes and similarity grouping. *Eng. Appl. Artif. Intell.* **2023**, *126 Pt B*, 107012. [CrossRef]
31. Chen, X.; Wu, S.; Shi, C.; Huang, Y.; Yang, Y.; Ke, R. Sensing Data Supported Traffic Flow Prediction via Denoising Schemes and ANN: A Comparison. *IEEE Sens. J.* **2020**, *20*, 14317–14328. [CrossRef]
32. Chen, X.; Ling, J.; Yang, Y.; Zheng, H.; Xiong, P.; Postolache, O.; Xiong, Y. Ship Trajectory Reconstruction from AIS Sensory Data via Data Quality Control and Prediction. *Math. Probl. Eng.* **2020**, *2020*, 1–9. [CrossRef]
33. Peel, G.J.; Gedikli, E.D.; Hendrikse, H. Navigating Arctic Waters: A Summary of Ship Activities and Ice-Ship Interactions in Alaskan Waters. In Proceedings of the International Conference on Port and Ocean Engineering Under Arctic Conditions, Glasgow, UK, 12–16 June 2023.
34. Chen, X.; Wei, C.; Xin, Z.; Zhao, J.; Xian, J. Ship Detection under Low-Visibility Weather Interference via an Ensemble Generative Adversarial Network. *J. Mar. Sci. Eng.* **2023**, *11*, 2065. [CrossRef]
35. Wang, D.; Jing, Y. Ship collision risk analysis in port waters integrating GRA algorithm and BPNN. *Transp. Saf. Environ.* **2025**, *7*, tdaf012. [CrossRef]

36. Liu, X.; Qiu, L.; Fang, Y.; Wang, K.; Li, Y.; Rodríguez, J. Event-Driven Based Reinforcement Learning Predictive Controller Design for Three-Phase NPC Converters Using Online Approximators. *IEEE Trans. Power Electron.* **2025**, *40*, 4914–4926. [CrossRef]
37. Gao, X.; Dai, W.; Yu, L.; Yu, Q. Study on factors contributing to severity of ship collision accidents in the Yangtze River estuary. *Transp. Saf. Environ.* **2024**, *6*, tdae014. [CrossRef]
38. Liu, X.; Qiu, L.; Fang, Y.; Wang, K.; Li, Y.; Rodríguez, J. Combining Data-Driven and Event-Driven for Online Learning Predictive Control in Power Converters. *IEEE Trans. Power Electron.* **2025**, *40*, 563–573. [CrossRef]
39. Huu, K.D. A Speed Prediction Model Using A HIL Simulator and Neural Network for Cargo Ships. In Proceedings of the International Russian Smart Industry Conference, Sochi, Russia, 25–29 March 2024. [CrossRef]
40. Zhu, Q.; Xi, Y.; Weng, J.; Han, B.; Hu, S.; Ge, Y.-E. Intelligent ship collision avoidance in maritime field: A bibliometric and systematic review. *Expert Syst. Appl.* **2024**, *252 Pt A*, 124148. [CrossRef]
41. Chen, J.; Zhuang, C.; Shi, J.; Jiang, H.; Xu, J.; Liu, J. Risk factors extraction and analysis of Chinese ship collision accidents based on knowledge graph. *Ocean Eng.* **2025**, *322*, 120506. [CrossRef]
42. Karachalios, T.; Moschos, P.N.; Fanariotis, A.; Orphanoudakis, T.; Leligou, H.-C. Integration of AIS Data for Efficient Drone Navigation in Maritime Real-Time Ship Inspection Missions. In Proceedings of the IEEE/SICE International Symposium on System Integration, Munich, Germany, 21–24 January 2025. [CrossRef]
43. Dai, Y.; He, Y.; Zhao, X.; Xu, K. Testing method of autonomous navigation systems for ships based on virtual-reality integration scenarios. *Ocean Eng.* **2024**, *309 Pt 2*, 118597. [CrossRef]
44. Han, Z.; Wu, D.; Zhang, J.; Huan, T.; Han, Q.-L.; Zhang, M. COLREGs-Adaptive trajectory planning and decision-making in maritime autonomous surface ships. *Ocean Eng.* **2024**, *312 Pt 3*, 119308. [CrossRef]
45. Purcell, E.; Nejad, A.R.; Böhm, A.; Sapp, L.; Lund, J.; Von Bock und Polach, F.; Nickerson, B.M.; Bekker, A.; Gilges, M.; Saleh, A.; et al. On Methodology for a digital twin of ship propulsion under harsh environmental conditions. In Proceedings of the International Conference on Offshore Mechanics and Arctic Engineering, Singapore, 9–14 June 2024. [CrossRef]
46. Zhen, R.; Lv, P.; Shi, Z.; Chen, G. A novel fuzzy multi-factor navigational risk assessment method for ship route optimization in costal offshore wind farm waters. *Ocean. Coast. Manag.* **2023**, *232*, 106428. [CrossRef]
47. Zhao, L.; Song, Y.; Xiong, J.; Xu, J.; Li, D.; Liu, F.; Shen, T. A time-delay neural network for ship-radiated noise recognition based on residual block and attention mechanism. *Digit. Signal Process.* **2024**, *149*, 104504. [CrossRef]
48. Hummel, H.I.; van der Mei, R.; Bhulai, S. A survey on machine learning in ship radiated noise. *Ocean Eng.* **2024**, *298*, 117252. [CrossRef]
49. Acharya, D.; Farazi, M.; Rolland, V.; Petersson, L.; Rosebrock, U.; Smith, D.; Ford, J.; Wang, D.; Tuck, G.N.; Little, L.R.; et al. Towards automatic anomaly detection in fisheries using electronic monitoring and automatic identification system. *Fish. Res.* **2024**, *272*, 106939. [CrossRef]
50. Meng, X.; Li, H.; Zhang, W.; Zhou, X.-Y.; Yang, X. Analyzing risk influencing factors of ship collision accidents: A data-driven Bayesian network model integrating physical knowledge. *Ocean. Coast. Manag.* **2024**, *256*, 107311. [CrossRef]
51. Park, M.-H.; Lee, C.-H.; Hur, J.-J.; Lee, W.-J. Prediction of fuel consumption and shaft torque using machine learning and analysis of engine curve diagrams. *Meas. J. Int. Meas. Confed.* **2025**, *248*, 116984. [CrossRef]
52. Xian, J.; Ma, J.; Mei, X.; Wu, H.; Saeed, N.; Han, D.; Marino, M.D.; Li, K.-C. Robust Coarse-to-fine 3D-Target-Localization Algorithm for Underwater-IoT-Based Networks: Design and Performance Evaluation Under Uncertain Multi-Parameter. *IEEE Internet Things J.* **2025**, 104131. [CrossRef]
53. Umar, Z.; Junaid, K.; Eunkyu, L.; Sajjad, H.; Salim, B.A.; Yahya, A.R.; Kyungsup, K. An Efficient Long Short-Term Memory and Gated Recurrent Unit Based Smart Vessel Trajectory Prediction Using Automatic Identification System Data. *Comput. Mater. Contin.* **2024**, *81*, 1789–1808. [CrossRef]
54. Wang, W.; Liu, B.; Tian, Q.; Xu, X.; Peng, Y.; Peng, S. Predicting dust pollution from dry bulk ports in coastal cities: A hybrid approach based on data decomposition and deep learning. *Environ. Pollut.* **2024**, *350*, 124053. [CrossRef]
55. Lin, C.; Zhen, R.; Tong, Y.; Yang, S.; Chen, S. Regional ship collision risk prediction: An approach based on encoder-decoder LSTM neural network model. *Ocean Eng.* **2024**, *296*, 117019. [CrossRef]
56. Xian, J.; Wu, H.; Mei, X.; Zhang, Y.; Chen, X.; Zhang, Q.; Liang, L. Novel energy-efficient opportunistic routing protocol for marine wireless sensor networks based on compressed sensing and power contro. *J. Ocean. Univ. China* **2022**, *21*, 1504–1516. [CrossRef]
57. Watanabe, E.; Samanta, S.; Kowatari, K.; Shibasaki, R. On Vessel Schedule Delay and Its Recoveries in Liner Services Using AIS Data. In Proceedings of the IEEE International Conference on Big Data, Washington, DC, USA, 15–18 December 2024. [CrossRef]

Journal of
Marine Science and Engineering

MDPI

Article

Maritime Port Freight Flow Optimization with Underground Container Logistics Systems Under Demand Uncertainty

Miaomiao Sun [1], Chengji Liang [1], Yu Wang [1,*] and Salvatore Antonio Biancardo [2,*]

1 Institute of Logistics Science and Engineering, Shanghai Maritime University, Shanghai 201306, China; mmsun@stu.shmtu.edu.cn (M.S.); liangcj@shmtu.edu.cn (C.L.)
2 Department of Civil, Construction and Environmental Engineering (DICEA), University of Naples Federico II, 80125 Naples, Italy
* Correspondence: wangyu@shmtu.edu.cn (Y.W.); salvatoreantonio.biancardo@unina.it (S.A.B.)

Abstract: As global trade and container transportation continue to grow, port collection and distribution systems face increasing challenges, including congestion, inefficiency, and environmental impact. Traditional ground-based transportation methods often exacerbate these issues, especially under uncertain demand conditions. This study aims to optimize freight flow allocation in port collection and distribution networks by integrating traditional and innovative transportation modes, including underground container logistics systems, under demand uncertainty. A stochastic optimization model is developed, incorporating transportation, environmental, carbon tax and subsidy, and congestion costs while satisfying various constraints, such as capacity limits, time constraints, and low-carbon transport requirements. The model is solved using a hybrid algorithm combining an improved Genetic Algorithm and Simulated Annealing (GA-SA) with Deep Q-Learning (DQN). Numerical experiments and case studies, particularly focusing on A Port, demonstrate that the proposed approach significantly reduces total operational costs, congestion, and environmental impacts while enhancing system robustness under uncertain demand conditions. The findings highlight the potential of underground logistics systems to improve port logistics efficiency, providing valuable insights for future port management strategies and the integration of sustainable transportation modes.

Keywords: underground container logistics system; port collection and distribution system; freight flow allocation; GA-SA-DQN; demand uncertainty; freight flow allocation

1. Introduction

In the context of deepening global economic integration, ports play a vital role in international trade, handling approximately 90% of global freight volume. As essential hubs that connect maritime and land transportation, ports are integral to global supply chains. According to the "Review of Maritime Transport 2024" by the United Nations Conference on Trade and Development (UNCTAD), global maritime trade increased by 2.4% to 12.3 billion tons in 2023, with projections of 2% growth in 2024 and an average annual growth rate of 2.4% through 2029 [1]. As a critical component of global trade, container trade grew by only 0.3% in 2023 but is expected to rebound by 3.5% in 2024, contingent upon supply chain stability [1]. Furthermore, according to UNCTAD's latest global trade update, global trade reached a record USD 33 trillion in 2024, representing a 3.7% increase (adding USD 1.2 trillion) year-on-year [2]. Despite the growth in container trade, logistical challenges persist, especially with the increasing complexity of port collection and distribution systems.

Port collection and distribution systems, serving as the link between ports and hinterlands, directly impact the stability and reliability of international supply chains. Panahi et al. (2022) demonstrate that efficient port collection and distribution systems are key factors in enhancing port competitiveness, directly affecting international trade logistics costs and supply chain efficiency [3]. Subsequently, Zhao et al. (2024) further propose a novel collaborative container logistics system integrating dry ports and water ports, which uses double-stack container trains for water-rail transport and constructs a multi-objective optimization model to balance costs, efficiency, and emissions [4]. Acciaro et al. (2023) point out that optimizing port collection and distribution systems not only relates to the competitiveness of ports themselves but also serves as a strategic pillar for promoting regional economic sustainable development [5]. Consequently, exploring innovative port collection and distribution models to improve system operational efficiency and service quality has become a focal research topic in both academic and practical domains. Despite these advancements, the optimization of freight flow allocation in port collection and distribution networks under uncertainty remains a critical challenge. Specifically, determining how to allocate freight flow between various transportation modes while considering demand uncertainty and ensuring operational resilience is a key problem in the field.

Facing the multiple challenges of traditional port collection and distribution systems, underground container logistics systems (UCLSs) have gradually gained attention from academia and industry as an innovative solution. A UCLS is physically defined as a set of underground depots and interconnected tunnels or pipelines that support 24 h all-weather goods movement and automated logistics operations.

In network flow optimization problems, two-stage robust optimization approaches considering demand uncertainty have been proven effective in addressing such challenges, providing more flexible solutions than single-stage optimization by deferring some flow decisions until after the realization of uncertain demand.

To address this challenge, this study proposes a stochastic chance-constrained optimization model that integrates transportation, environmental, and congestion costs with quadratic congestion functions and low-carbon-ratio constraints based on Shanghai's port layout. Additionally, we develop a hybrid GA-SA-DQN algorithm, combining heuristic global search capabilities with reinforcement learning for solving large-scale problems efficiently. This approach is evaluated through a port case study that demonstrates significant improvements in economic and environmental performance, including a 12.6% reduction in total costs, a 34.9% decrease in carbon emissions, and a 54.9% reduction in congestion.

This paper has three major contributions: a stochastic chance-constrained optimization model that integrates transportation, environmental, and congestion costs with quadratic congestion functions and low-carbon-ratio constraints based on Shanghai's port layout; a hybrid GA-SA-DQN optimization algorithm combining heuristic global search capabilities with reinforcement learning's adaptive optimization for solving complex large-scale problems; and an A Port case study demonstrating that UCLSs significantly improve economic and environmental performance even under demand uncertainty, reducing total costs by 12.6%, carbon emissions by 34.9%, and congestion index by 54.9%. The research results of this paper have important theoretical and practical value for promoting the green transformation and intelligent development of port collection and distribution systems. Theoretically, this paper extends the application of stochastic programming and chance constraint theory in port collection and distribution system optimization; practically, it provides scientific decision support tools for port managers, transportation planners, and policy makers, contributing to the formulation of more reasonable port collection and distribution development strategies.

The structure of this paper is arranged as follows: Section 2 elaborates on the research problem and basic assumptions; Section 3 establishes the mathematical optimization model; Section 4 introduces the design and implementation of the hybrid optimization algorithm; Section 5 verifies the effectiveness of the model and algorithm through numerical experiments and conducts sensitivity analysis; finally, Section 6 summarizes the research findings and suggests directions for future research.

2. Literature Review

2.1. Research on the Optimization of Maritime Port Freight Flow

The port collection and distribution system serves as a crucial bridge connecting ports with inland regions, and its optimization exerts a direct influence on port operational efficiency and service quality. In the backdrop of the continuous expansion of global trade and the steadily rising containerization levels, the optimization of the port collection and distribution system has emerged as a significant and pressing research topic. The existing optimization research can be systematically categorized into three hierarchical levels: strategic, tactical, and operational.

Traditional port collection and distribution systems face multiple challenges in addressing the growing demand for container transportation. First, ground transportation bottlenecks are increasingly prominent. First, ground transportation bottlenecks are increasingly prominent. Liu et al. (2024) reveal that the China-Europe liner shipping network demonstrates high resilience to individual port and service disruptions, but the structural differences among shipping alliances (e.g., Ocean Alliance) significantly influence vulnerability levels [6]. According to Liang et al. (2024), decarbonizing maritime transport through retrofitting vessels with green fuels presents a promising strategy for addressing these concerns [7]. According to Wang et al. (2022), relying predominantly on road haulage for port collection and distribution imposes significant constraints on infrastructure capacity and exacerbates environmental pressures [8]. Additionally, poor visibility due to fog or heavy rain can further degrade port-area monitoring systems, leading to delays and safety risks. Recent advances in ensemble-GAN-based image enhancement have shown promising results in restoring ship imagery under low-visibility conditions, thereby enabling more reliable situational awareness in port operations [9]. Furthermore, the container storage and handling issues are critical, as highlighted by Hu et al. (2021), who presented a solution for terminal operations by considering yard sharing [10]. Martinez-Moya et al.'s (2021) research indicates that port collection and distribution systems are a major source of port carbon emissions, with road transportation having significantly higher emission intensity than rail and waterway transportation [11].

Second, inadequate intermodal transport connections lead to low overall system coordination. In their 2023 study, Karam, Jensen, and Hussein emphasize that overcoming significant barriers—ranging from inadequate infrastructure and regulatory hurdles to operational inefficiencies—is crucial for optimizing port collection and distribution systems, and that targeted mitigation strategies are needed to enhance coordination and cut costs [12]. Additionally, the Intermodal Freight Transportation Market Size Report (n.d.) underscores the need for enhanced intermodal facilities and streamlined operations to improve the efficiency and capacity of multimodal transport networks, with the market expected to grow substantially in the coming years as these challenges are addressed [13]. Mamatok and Jin (2022) further point out that transfer time and costs between different transportation modes are key factors constraining the development of multimodal transportation [14]. To improve real-time traffic management around port areas, advanced sensing data preprocessing and ANN-based prediction schemes have been developed. For instance, Chen et al. (2020) employ denoising algorithms combined with neural networks to achieve high-accuracy short-term traffic flow forecasts, which can be directly applied to anticipate gate congestion and optimize drayage scheduling [15]. By means of a cost–benefit analysis of road–underground co-modality strategies, Liu et al. (2025) demonstrated that these approaches hold significant promise for improving the sustainability and efficiency of urban logistics [16].

Third, uncertainty factors increasingly affect port collection and distribution systems. Zhu et al. (2023) indicate that port container demand exhibits significant random fluctuations due to global economic volatility, trade policy changes, seasonal factors, and sudden events (such as the COVID-19 pandemic and geopolitical conflicts) [17]. Twiller et al. (2025) highlight that demand uncertainty in container shipping often leads to challenges in planning and operational efficiency and that traditional methods are inadequate in addressing these fluctuations [18]. Toygar et al.'s (2022) case analysis reveals that traditional deterministic optimization methods often fail when facing demand fluctuations, resulting in resource misallocation and increased operational costs [19]. Particularly in the post-pandemic era, port logistics activities such as empty container allocation face stronger uncertainty challenges, imposing higher requirements on the resilience and adaptability of port collection and distribution systems. Therefore, determining how to optimize port collection and distribution systems under uncertain demand conditions to improve their adaptability and resilience has become a key problem requiring urgent solutions.

2.2. Research on Underground Container Logistics

An underground container logistics system (UCLS) effectively alleviates ground traffic congestion, reduces environmental pollution, and enhances port collection and distribution efficiency by constructing specialized underground logistics channels and utilizing automation and intelligent technologies to achieve rapid, safe, and efficient container transportation [20].

Liu et al.'s (2023) research on urban underground logistics systems indicates that urban underground logistics systems (ULSs) are an important means of solving urban traffic problems with unique advantages [21]. China's freight transportation requires a new transportation mode; thus, ULSs have garnered increasing attention. Hu et al. (2023) systematically reviewed the development history of underground logistics systems and their latest development prospects in China, emphasizing the potential of underground logistics systems in solving urban logistics problems and promoting sustainable development [22]. Similarly, Liang et al. (2022) explored the integration of underground logistics systems in optimizing shipment equipment for enhanced efficiency [23]. Li and Yuen (2025) provided a systematic review on underground logistics systems, focusing on their design, impacts, and future directions [24]. They highlighted the significant role of ULSs in alleviating traffic congestion and reducing environmental pollution in urban areas, especially in port cities facing rapidly increasing container transportation demands. An underground logistics system, as a relatively new concept for container transportation, is designed to reduce congestion and pollution on the road caused by the sharply growing number of collections and distributions of containers in port cities.

Against the backdrop of explosive growth in freight demand, the negative impacts of the traditional road-system-based port-city logistics mode on urban transportation, environment, and safety have become prominent livelihood issues, as well as significant economic and social problems. Hou et al. (2024) proposed a metro-based underground logistics system (M-ULS) solution, noting that the global outbreak of COVID-19 has further exposed deficiencies in city logistics based on human and ground roads, such as poor emergency response capacity and a high risk of infection during transportation [25]. Sun and Hu et al. (2023) proposed an entropy-based fuzzy TOPSIS evaluation model for optimizing metro-based underground logistics system network planning, providing methodological support for implementing large-scale underground logistics systems [26].

From a sustainable development perspective, UCLSs offer significant environmental benefits. Xia et al.'s (2021) research indicates that introducing UCLSs can significantly reduce port collection and distribution carbon emissions, outperforming traditional road transportation [27]. Chen et al. (2023) explored decision-making mechanisms for the metro-based underground logistics system network expansion in Beijing, emphasizing the role of UCLSs in promoting sustainable city logistics and optimizing land use in urban areas [28]. An et al. (2024) applied Wasserstein distributionally robust optimization to train operation and freight assignment in metro-based underground logistics systems, providing an innovative method for addressing uncertainties in UCLS operations [29]. Wandel et al. (2023) conducted a feasibility study of an underground capsule pipeline logistics system, assessing its potential for urban applications and confirming its significant benefits in reducing congestion and environmental impact [30].

Particularly noteworthy is that UCLSs possess unique advantages in addressing demand uncertainty. Hou et al. (2025) demonstrate that compared to traditional ground transportation, underground logistics systems operate in a relatively enclosed and stable environment, are less affected by external factors, and can provide more stable and reliable transportation services [31]. An M-ULS, as an emerging solution, can effectively address the challenges faced by urban logistics under epidemic conditions. Gong et al. (2023) conducted a thematic literature review on the sustainable design and operations management of metro-based underground logistics systems, providing a theoretical foundation for further research on the coordinated optimization of underground logistics systems and traditional transportation modes [32].

3. Problem Statement

The port collection and distribution network freight flow allocation problem is essentially a decision-making problem that determines how to allocate container transportation demand among various transportation modes to achieve an optimization objective. Under uncertain demand conditions, this problem becomes more complex and challenging. As global trade demand continues to grow, port collection and distribution systems face a series of issues such as low transportation efficiency, severe congestion, and environmental pollution, particularly due to their reliance on traditional ground-based transportation methods. These problems are becoming more prominent as trade volumes increase.

The port collection and distribution network studied in this paper consists of logistics park warehouses (LPWs), coastal ports (CPs), and various transportation modes that connect them. These transportation modes include traditional road transport, rail transport, water transport, and the new shallow UCLS and deep UCLS. In the underground container logistics system, the topological structure of the port collection and distribution network is shown in Figure 1. It is assumed that the container transportation demand between LPWs and CPs follows a normal distribution as a random variable with a certain degree of uncertainty, making the freight flow allocation problem more complex, especially when demand fluctuates significantly.

The uniqueness and challenges of this problem are reflected in the following aspects:

(1) Multiple cost trade-offs: Decision-makers need to consider transportation costs, environmental costs, and congestion costs simultaneously. These objectives are interrelated and require comprehensive balancing. For example, reducing traffic congestion might increase transportation costs, while choosing low-carbon transportation methods might increase environmental costs.

(2) Demand uncertainty: Port container transportation demand is influenced by various factors and exhibits high uncertainty, requiring the freight flow allocation scheme to be sufficiently robust to adapt to different demand conditions and minimize the impact on the system.
(3) Multiple constraints: These include capacity constraints, time constraints, flow balance constraints, and low-carbon transportation ratio requirements, which add complexity to the problem. Finding the optimal solution while satisfying these various constraints is a significant challenge.
(4) Integration of new transportation modes: The underground container logistics system (UCLS) as a new transportation mode, presents a novel research topic regarding its coordinated optimization with traditional transportation modes. Determining how to effectively integrate the underground logistics system with traditional modes and allocate freight flow between them is an important focus of this study.

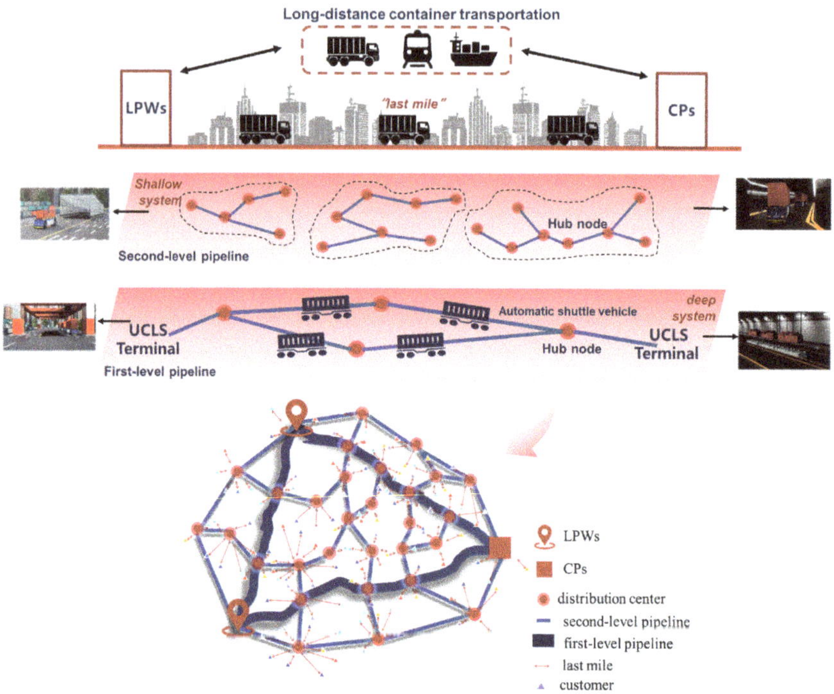

Figure 1. Schematic diagram of port collection and distribution network considering underground container logistics.

In consideration of the practical operations of the port transportation and distribution system, the following basic assumptions are made:

Assumption 1. *Goods are transported from logistics park warehouses (LPWs) to ports (CPs) using various transportation modes, including road, rail, and water transport and a UCLS.*

Assumption 2. *The freight demand for the port transportation system is a random variable that follows a normal distribution.*

$$d_j \sim N\left(\mu_j, \sigma_j^2\right) \tag{1}$$

where d_j represents the container cargo transportation demand of port j,μ_j represents the predicted mean, and σ_j is the standard deviation, reflecting the variability of demand.

Assumption 3. *The unit transportation costs, carbon emission coefficients, and capacity limits for each mode of transport are known and fixed parameters.*

Assumption 4. *Congestion costs are modeled as a quadratic function of the excess freight flow over the basic capacity of the road segment. The calculation formula is as follows:*

$$C_{congestion} = \beta_m \sum_{i,j,w} \max(0, x_{ijm} - Cap_{ijm})^2 \tag{2}$$

where $C_{congestion}$ is the congestion cost coefficient, representing the marginal cost increase due to the overloading of the transportation system, β_m represents emission reduction cost for transportation mode m, and Cap_{ijm} is the maximum available capacity for transportation from LPW i to port j using transportation mode m with a unit of TEU/day.

Assumption 5. *The distances and transport time limits between each LPW and the port are known parameters.*

Assumption 6. *The effects of handling costs and inventory costs are neglected, with the focus solely on transportation costs.*

Assumption 7. *The transfer costs between different transportation modes are not considered in this model.*

Table 1 lists the sets, constant parameters, and decision variables in the model of this paper.

Table 1. Sets, parameters, and decision variables.

Notations	Description
(1) Sets	
I	set of logistics park warehouses (LPWs), indexed by $i \in I$
J	set of coastal ports (CPs), indexed by $j \in J$
M	set of transportation modes, indexed by $m \in M$, where $m = 1$ represents road transportation, $m = 2$ represents rail transportation, $m = 3$ represents water transportation, $m = 4$ represents the shallow underground container logistics system, $m = 5$ represents the deep underground container logistics system
(2) Parameters	
d_j (TEU)	represents the container cargo transportation demand of port j, randomly varying
μ_j (TEU)	average cargo transportation demand at port j
$\widehat{d}_j$ (TEU)	maximum deviation of cargo demand at port j
Γ (TEU)	uncertainty level of transportation demand prediction
α	demand satisfaction level, i.e., the minimum confidence level for guaranteeing meeting transportation demand at the port under uncertain conditions
l_{ijm} (km)	distance from LPW i to port j using transportation mode m
a_i (TEU)	capacity of LPW i
Cap_{ijm} (TEU/day)	maximum available capacity for transportation from LPW i to port j using transportation mode m
v_m (km/h)	speed of transportation mode m
T_j (h)	maximum allowable time for all containers to reach port j
c_m (CNY/TEU·km)	unit transportation cost of mode m
e_m (kg/TEU·km)	carbon emission coefficient of transportation mode m
e_{uls}	carbon emission coefficient of the underground logistics system

Table 1. *Cont.*

Notations	Description
e_0	base emissions coefficient, typically using road transportation's emission coefficient as a benchmark
$E_{\max}$	government-set maximum emissions limit
E_{red}	government-set minimum emissions reduction
β_m	emission reduction cost for transportation mode m
s_c (CNY/kg)	government-specified carbon subsidy rate
t_c (CNY/kg)	government-specified carbon tax rate
B_g (CNY)	government-set emissions reduction target
A	a very large number
η	a non-negative number, used for ensuring the continuity of emission reduction compliance
(3) Decision variables	
y_{ijm}	binary variable, if LPW i to port j chooses transportation mode m, then $y_{ijm} = 1$; otherwise, $y_{ijm} = 0$
x_{ijm}	freight flow from LPW i to port j using transportation mode m, unit is TEU
z_s	binary variable, if cargo from LPW i to port j is selected for carbon taxation, then $z_s = 1$; otherwise, $z_s = 0$
z_t	binary variable, if cargo from LPW i to port j is selected for carbon subsidy, then $z_t = 1$; otherwise, $z_t = 0$

4. Model Development

4.1. Objective Function

Under uncertain demand conditions, when considering the containerized cargo flow of coastal ports (CPs) and the distribution of transportation networks, cost and efficiency are two important contributing factors. The total cost consists of the transportation cost, environmental cost, congestion cost, and carbon tax and subsidy cost. Efficiency, on the other hand, is mainly influenced by factors such as route congestion, which in turn affects the distribution of cargo flows and leads to longer transportation times and resource waste. This paper constructs the objective function from the following four perspectives.

(1) Transportation cost

Transportation costs are the direct economic expenditures in the port containerized transport system, including fuel costs, labor costs, and equipment costs. The transportation cost Z_1 is defined as the cost incurred from transporting the cargo flow from LPWs to the port, and it is calculated as

$$Z_1 = \sum_{i\in I}\sum_{j\in J}\sum_{m\in M} c_m \cdot l_{ijm} \cdot x_{ijm} \quad (3)$$

where c_m is the unit transportation cost of mode m, and l_{ijm} is the distance from LPWs i to port j using transportation mode m.

(2) Environmental cost

Environmental costs refer to the carbon emissions and other environmental pollution —specifically carbon dioxide (CO_2), nitrogen oxides (NO_x), sulfur oxides (SO_x), and particulate matter (PM)—resulting from transportation activities. This cost is calculated by estimating the external damage caused by emissions. The environmental cost Z_2 is calculated as

$$Z_2 = \sum_{i\in I}\sum_{j\in J}\sum_{m\in M} e_m \cdot l_{ijm} \cdot x_{ijm} \cdot p_e \quad (4)$$

where e_m is the carbon emission coefficient for transportation mode m, and p_e is the environmental cost per unit of carbon emissions.

(3) Congestion costs

The congestion cost Z_3 is calculated as

$$Z_3 = \sum_{i \in I} \sum_{j \in J} \sum_{m \in M} \beta_m \cdot \max(0, x_{ijm} - Cap_{ijm})^2 \tag{5}$$

(4) Carbon Tax and Carbon Subsidy Costs

The carbon tax and subsidy cost Z_4 is calculated as

$$Z_4 = \sum_{j \in J} z_t \cdot t_c \cdot \max(0, E_j - E_{\max}) - z_s \cdot s_c \cdot \max(0, E_{red} - E_j) \tag{6}$$

The total expected cost for optimizing the port container transportation system includes the transportation cost, environmental cost, congestion cost, and carbon tax/subsidy costs, as shown below:

$$\min Z = E[Z_1 + Z_2 + Z_3 + Z_4] \tag{7}$$

4.2. Constraint Condition

(1) Capacity constraints

The capacity constraint ensures that the cargo flow from LPW i to CP j using transportation mode m does not exceed the maximum capacity of that transportation mode, as shown below:

$$x_{ijm} \leq Cap_{ijm} \cdot y_{ijm}, \forall i \in I, j \in J, m \in M \tag{8}$$

This constraint ensures that cargo flow is only transported when the corresponding transportation mode is selected ($y_{ijm} = 1$).

(2) Flow balance constraints

The flow balance constraint ensures that the total cargo flow into CP j meets the demand d_j at CP j, as shown below:

$$d_j - \xi \leq \sum_{i \in I} \sum_{m \in M} x_{ijm} \leq d_j + \xi, \forall j \in J \tag{9}$$

where ξ is a positive small quantity, representing the tolerance range between the freight flow volume and the demand. The magnitude of this range can be adjusted according to the actual situation.

(3) Demand satisfaction probability constraints

This constraint expresses the probability that the total cargo flow from LPW i to CP j meets the demand d_j at CP j, ensuring that the system satisfies the demand with at least a probability of α. This is a stochastic chance constraint, reflecting uncertainty in demand:

$$P\left(\sum_{i \in I} \sum_{m \in M} x_{ijm} \geq d_j\right) \geq \alpha, \forall j \in J \tag{10}$$

(4) Time constraints

$$\frac{l_{ijm}}{v_m} \leq T_j \cdot y_{ijm}, \forall i \in I, j \in J, m \in M \tag{11}$$

This constraint ensures that the transportation time from LPW i to CP j does not exceed the maximum arrival time T_j permitted by the port.

(5) Low-carbon-transport-ratio constraints

This constraint requires that the proportion of freight flow borne by the underground logistics system (i.e., $m \in \{4,5\}$) is not lower than the preset low-carbon-proportion requirement η, promoting the green transformation of the transportation structure.

$$\frac{\sum\limits_{i\in I}\sum\limits_{j\in J}\sum\limits_{m\in\{4,5\}} x_{ijm}}{\sum\limits_{i\in I}\sum\limits_{j\in J}\sum\limits_{m\in M} x_{ijm}} \geq \eta \tag{12}$$

(6) Carbon-emission-related constraint

This constraint ensures that the port cannot receive carbon subsidies and pay carbon taxes simultaneously, that is, z_t and z_s cannot both be 1 at the same time.

$$z_t + z_s \leq 1, \forall j \in J \tag{13}$$

(7) Carbon emission calculation constraint

$$E_j = \sum_{i\in I}\sum_{m\in M} e_m \cdot l_{ijm} \cdot x_{ijm}, \forall j \in J \tag{14}$$

This constraint defines the total carbon emissions E_j at CP j, which is the sum of emissions generated by all transportation processes.

(8) Carbon subsidy constraints

This constraint represents the current carbon emissions E_j at CP j, where the total carbon emissions E_j cannot exceed the emissions reduction standard E_{red}, in order to activate carbon subsidies:

$$E_j \leq E_{red} + A(1 - z_s), \forall j \in J \tag{15}$$

This constraint indicates that only when the carbon emissions at CP j do not exceed the emissions reduction standard will the CP be eligible for carbon subsidies.

(9) Carbon tax constraints

This constraint represents the current carbon emissions E_j at CP j, where the total carbon emissions E_j cannot exceed the maximum emissions standard $E_{\max}$, in order to activate carbon taxes:

$$E_j \geq E_{\max} - A(1 - z_t), \forall j \in J \tag{16}$$

(10) LPW capacity constraints

This constraint ensures that the total cargo flow from LPW i does not exceed its capacity a_i:

$$\sum_{j\in J}\sum_{m\in M} x_{ijm} \leq a_i, \forall i \in I \tag{17}$$

This constraint ensures that the total amount of goods departing from LPW i does not exceed the maximum capacity a_i of this LPW.

(11) Decision variable non-negativity and binary constraints

This constraint ensures that the total cargo flow leaving LPW i does not exceed the maximum capacity a_i of the LPW:

$$x_{ijm} \geq 0, \forall i \in I, j \in J, m \in M \tag{18}$$

$$y_{ijm} \in \{0,1\}, \forall i \in I, j \in J, m \in M \tag{19}$$

$$y_{ijm} \in \{0,1\}, \forall i \in I, j \in J, m \in M \tag{20}$$

These constraints define the range of the decision variables, where x_{ijm} represents the non-negative cargo flow and y_{ijm} is a binary variable.

(12) Demand randomness constraints

This constraint defines the randomness of CP demand d_j, where the demand follows a normal distribution with mean μ_j and variance σ_j:

$$d_j \sim N\left(\mu_j, \sigma_j^2\right), \forall j \in J \tag{21}$$

Since the model contains stochastic chance constraints (Equation (10)), it is a Mixed-Integer Nonlinear Programming (MINLP) problem. To achieve an efficient solution, this section will transform the stochastic chance constraints into deterministic constraints. The specific steps are as follows:

$$P\left(\sum_{i\in I}\sum_{m\in M} x_{ijm} \geq d_j\right) \geq \alpha, \forall j \in J \tag{22}$$

Step 1: According to the theory of stochastic programming, perform an equivalent transformation on the stochastic chance constraints in Equation (10) to convert them into deterministic constraints. Define the variables and the objective. For the original constraint (22), proceed to the following step:

Step 2: Standardize the random variables. Since $d_j \sim N\left(\mu_j, \sigma_j^2\right)$, d_j can be standardized to $Z = \frac{d_j - \mu_j}{\sigma_j}$, where $Z \sim N(0,1)$. The original constraint (22) can be rewritten as

$$P\left(\sum_{i\in I}\sum_{m\in M} x_{ijm} \geq \mu_j + \sigma_j \cdot Z\right) \geq \alpha \tag{23}$$

This is equivalent to

$$P\left(Z \leq \frac{\sum\limits_{i\in I}\sum\limits_{m\in M} x_{ijm} - \mu_j}{\sigma_j}\right) \geq \alpha \tag{24}$$

Step 3: Quantile transformation. According to the definition of the standard normal distribution, the above equation holds if and only if

$$\frac{\sum\limits_{i\in I}\sum\limits_{m\in M} x_{ijm} - \mu_j}{\sigma_j} \geq \Phi^{-1}(\alpha) \tag{25}$$

where $\Phi^{-1}(\alpha)$ is the α quantile of the standard normal distribution.

Step 4: Rearrange the inequality. Multiply both sides by σ_j and move the terms:

$$\sum_{i\in I}\sum_{m\in M} x_{ijm} \geq \mu_j + \sigma_j \cdot \Phi^{-1}(\alpha), \forall j \in J \tag{26}$$

In summary, the stochastic chance constraint $P\left(\sum\limits_{i\in I}\sum\limits_{m\in M} x_{ijm} \geq d_j\right) \geq \alpha, \forall j \in J$ can be transformed into the deterministic constraint (26).

5. Methodology

In the previous section, we presented the research problem and the mathematical model. In this section, we will sequentially introduce the design concepts of the two core modules of the hybrid optimization algorithm—GA-SA and DQN—and ultimately demonstrate how they work together to form an efficient GA-SA-DQN framework. Due to the inclusion of stochastic chance constraints, nonlinear congestion costs, and various other constraints, the port container transportation network flow distribution model established in this paper under demand uncertainty is a complex Mixed-Integer Nonlinear Programming (MINLP) problem. The complexity of this problem primarily lies in the following aspects: first, the decision variables simultaneously include continuous variables (cargo flow distribution) and discrete variables (transportation mode selection); second, the congestion cost function in the objective function is nonlinear, which increases the difficulty of solving the problem; third, the opportunity constraints introduced by demand uncertainty make the solution space more complex. Traditional exact algorithms, such as branch-and-bound methods and outer approximation methods, often face challenges such as low computational efficiency and large memory requirements when dealing with large-scale MINLP problems, making it difficult to obtain satisfactory solutions within a reasonable time.

Therefore, this section designs a hybrid optimization method combining an improved Genetic Algorithm and Simulated Annealing Algorithm with Deep Q-Learning (GA-SA-DQN) to effectively solve this problem. This hybrid method makes full use of the global search capability of heuristic algorithms and the adaptive optimization capability of reinforcement learning, providing an effective approach to solving such complex optimization problems.

5.1. The Design of GA-SA

To address the complexity of the port container transportation network flow allocation problem under demand uncertainty, we propose an improved Genetic Algorithm combined with Simulated Annealing (GA-SA). This hybrid approach leverages the GA's global search capability while incorporating SA's ability to escape local optima.

5.1.1. Coding and Initialization

The coding method of solutions directly affects the search efficiency and solution quality of the algorithm. In the GA-SA algorithm, we adopt a two-part coding scheme to represent solutions. This coding can effectively express two types of decision variables: transportation mode selection and freight flow allocation.

The transportation mode selection matrix is represented as an $|I| \times |J| \times |M|$ binary matrix, corresponding to the decision variable y_{ijm}. Here, $y_{ijm} = 1$ indicates that transportation mode m from LPW i to CP j is selected, and $y_{ijm} = 0$ indicates that this transportation mode m is not selected.

The freight flow allocation matrix $|I| \times |J| \times |M|$ is represented as a real-valued matrix, corresponding to the decision variable x_{ijm}, which represents the quantity of goods transported from LPW i to CP j via transportation mode m.

5.1.2. Fitness Function

In this study, the fitness function directly uses the objective function of the model, that is, to minimize the total expected cost. To handle the constraint conditions, this study adopts the penalty function method, adding the constraint violation degree into the fitness function:

$$F(X,Y) = f(X,Y) + \sum_{k} w_k \cdot \max\left(0, g_k(X,Y)^2\right) \tag{27}$$

where $g_k(X,Y)$ represents the violation degree of the k-th constraint and w_k is the corresponding penalty weight. The calculation of the specific constraint violation degree is as follows:

(1) The violation degree of the flow balance constraint is $g_1(X,Y) = \left|\sum_{i\in I}\sum_{m\in M} x_{ijm} - d_j\right|$.

(2) The violation degree of the transportation mode selection and freight flow matching constraint is $g_2(X,Y) = \sum_{i\in I}\sum_{j\in J}\sum_{m\in M} \max(0, x_{ijm} - A\cdot y_{ijm})$, where A is a sufficiently large positive number.

(3) The violation degree of the chance constraint is $g_3(X,Y) = \max(0, \alpha - P\cdot x_{ijm} \leq Cap_{ijm})$.

The choice of penalty weight w_k has a significant impact on the algorithm's performance. This study adopts an adaptive penalty weight strategy. Initially, a relatively small penalty weight is set, and as the number of iterations increases, the penalty weight gradually increases to balance the feasibility and quality of the solution. The penalty weight update formula is $w_k^{(t+1)} = w_k^{(t)}\cdot(1+\delta\cdot\mathbf{1}(g_k(X^*,Y^*)>0))$, where δ is the growth coefficient, and $\mathbf{1}(\cdot)$ is the indicator function, and its value is either 0 or 1. When the current optimal solution (X^*,Y^*) violates the k-th constraint, the corresponding penalty weight will increase. This design of the adaptive penalty function can effectively guide the algorithm to converge towards the feasible region while maintaining the exploration of high-quality solutions.

5.1.3. Crossover Operation

The crossover operation in GA simulates the process of biological chromosome exchange and is the main means of generating new individuals. Considering the two-part characteristic of the encoding for this problem, different crossover operations are designed for the transportation mode selection matrix and the freight flow allocation matrix.

For the transportation mode selection matrix, a modified two-point crossover operation is adopted. Two crossover points are randomly selected, and the gene segments between these two points of two parent individuals are exchanged. Individuals that may violate the constraint "at least one transportation mode is selected for each OD pair" after crossover are repaired.

The specific repair operation is to check each OD pair (i,j). If no transportation mode is selected (that is, all y_{ijm} are 0), then a transportation mode m' is randomly selected and $y_{ijm}{}'$ is set to 1.

For the freight flow allocation matrix, arithmetic crossover is adopted.

$$x_{ijm}^{child1} = \alpha\cdot x_{ijm}^{parent1} + (1-\alpha)\cdot x_{ijm}^{parent2} \tag{28}$$

$$x_{ijm}^{child2} = (1-\alpha)\cdot x_{ijm}^{parent1} + \alpha\cdot x_{ijm}^{parent2} \tag{29}$$

where α is a randomly generated crossover coefficient, $\alpha\in[0,1]$. To increase the diversity of solutions, α can be a numerical value or a matrix, generating different crossover coefficients for each (i,j,m) combination.

After crossover, the freight flow allocation matrix needs to be normalized to ensure that the flow balance constraint is satisfied. The normalization method is

$$x_{ijm}^{norm} = x_{ijm}\cdot\frac{d_j}{\sum_{i\in I}\sum_{m\in M} x_{ijm}} \tag{30}$$

In addition, to improve the efficiency of the crossover operation, this study introduces a heuristic crossover strategy. For the freight flow allocation matrix, the crossover coefficient is adjusted according to the fitness values of the parent individuals.

$$\alpha = \frac{F\left(\left(X^{parent2}, Y^{parent2}\right)\right)}{F\left(\left(X^{parent1}, Y^{parent1}\right)\right) + F\left(\left(X^{parent2}, Y^{parent2}\right)\right)} \quad (31)$$

In this way, parent individuals with better fitness values contribute more to the offspring, which helps to retain superior genes.

5.1.4. Simulated Annealing

To prevent the algorithm from becoming trapped in local optima, this study introduces a Simulated Annealing mechanism, which accepts inferior solutions with a certain probability to enhance the global search ability of the algorithm. In each iteration, for a newly generated individual, if its fitness is superior to that of the current individual, it is directly accepted. If the fitness is inferior, it is accepted according to the following probability:

$$p = \begin{cases} 1, if\ F(X_{new}, Y_{new}) < F(X_{old}, Y_{old}) \\ \exp\left(-\frac{F(X_{new},Y_{new})-F(X_{old},Y_{old})}{T}\right), \text{otherwise} \end{cases} \quad (32)$$

where T is the current temperature, which controls the probability of accepting inferior solutions. The higher the temperature, the greater the probability of accepting inferior solutions, which is beneficial for global search. The lower the temperature, the more the algorithm tends to accept better solutions, which is beneficial for local optimization.

The annealing temperature decreases exponentially:

$$T_{k+1} = \lambda \cdot T_k \quad (33)$$

where T_k represents the temperature at the k-th iteration and $\lambda \in (0,1)$ is the cooling coefficient. The choice of the cooling coefficient λ is a trade-off. A larger λ (close to 1) makes the temperature decrease slowly, enhancing the global search ability but possibly leading to a slow convergence speed. A smaller λ makes the temperature drop rapidly, accelerating convergence but possibly becoming trapped in local optima.

This study adopts a segmented cooling strategy, and the cooling coefficient is adjusted according to the solution stage:

$$\lambda = \begin{cases} \lambda_1,\ if\ \mathrm{t} < \mathrm{T}_1 \\ \lambda_2 <\ if\ \mathrm{T}_1 < \mathrm{t} < \mathrm{T}_2 \\ \lambda_3, if\ \mathrm{t} > \mathrm{T}_2 \end{cases} \quad (34)$$

where $\lambda_1 > \lambda_2 > \lambda_3$, and T_1 and T_2 are preset iteration thresholds. This segmented cooling strategy can maintain a high global search ability in the initial stage of the algorithm, accelerate convergence in the middle stage, and perform fine-grained local optimization in the later stage.

In addition, to prevent the algorithm from becoming trapped in local optima prematurely, this study also introduces a temperature restart mechanism. When the optimal solution has not improved for multiple consecutive iterations (such as 10 times), the temperature is increased to a certain proportion (such as 50%) of the initial temperature to help the algorithm escape from local optima traps.

5.1.5. Local Search

To accelerate convergence and improve the quality of solutions, this study conducts local searches on the optimal individual after each iteration. The core idea of local search is to conduct small-scale exploration within the neighborhood of the current optimal solution, attempting to further improve the solution quality through local adjustments.

The specific steps of local search are as follows: Randomly select an OD pair (i, j). Adjust the freight flow allocation among different transportation modes for this OD pair. The specific method is as follows: Select two transportation modes m_1 and m_2, where $y_{ijm_1} = y_{ijm_2} = 1$. Calculate the adjustable freight flow volume $\Delta x = \min(x_{ijm_1}, Cap_{ijm_2} - x_{ijm_2})$. Transfer the freight flow from transportation mode m_1 to transportation mode m_2. Evaluate the adjusted solution. If it is better, accept the adjustment; otherwise, keep the original solution.

This local search strategy takes advantage of the characteristics of the problem. By reallocating freight flows among different transportation modes, it can quickly find local optimal solutions. To improve the efficiency of local search, heuristic selection can be carried out based on information such as transportation costs and congestion conditions, giving priority to adjusting those OD pairs and transportation modes that are likely to bring greater improvements. In addition, to prevent local searches from falling into a loop, a taboo list can be maintained to record the recently adjusted OD pair and transportation mode combinations, and these combinations will not be considered within a certain period (such as 10 local searches). Experience shows that local search can significantly improve the solution quality and the algorithm's convergence speed, especially when dealing with large-scale optimization problems with complex local structures. In conclusion, the GA-SA algorithm framework is shown in Figure 2. After completing the description of the optimization strategy for the GA-SA module, the next step is to focus on the specific construction of the DQN module, aiming to further enhance the learning ability of the solution.

5.2. The Design of DQN

To complement the GA-SA approach, we develop a Deep Q-Network (DQN) reinforcement learning algorithm that can learn optimal flow allocation strategies through interaction with the environment. The DQN approach offers advantages in handling the complex state space and uncertainty in the problem [33–39].

5.2.1. State Space Design

The design of the state space is key to the application of DQN, as it needs to include all relevant information about the problem while maintaining an appropriate dimensionality for effective learning by the neural network. In this study, the state space s captures the current status of the port transportation network and is defined as a multi-dimensional vector:

$$s = \{F, C, E, T, U\} \tag{35}$$

where $F = \left\{f_{ij}^m\right\}$ represents the current flow allocation matrix; f_{ij}^m represents the normalized flow from logistics park warehouse (LPW) i to coastal port (CP) j using mode m; $C = \{c_j\}$ represents the current congestion level at CP j, calculated as the ratio of incoming flow to the port's capacity; $E = \{e_j\}$ represents the current carbon emission level at CP j, normalized against the emission target; $T = \left\{t_{ij}^m\right\}$ represents the remaining available capacity for each transportation link (i, j, m); and $U = u_j$ represents the uncertainty factor for each port's demand, calculated as the ratio of standard deviation to mean demand $(\sigma_j / \overline{D_j})$.

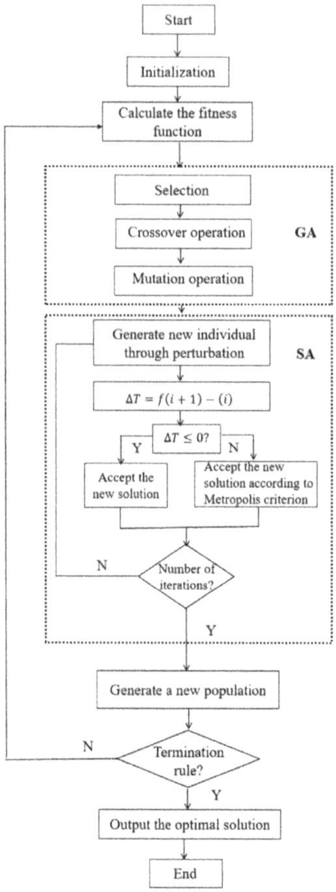

Figure 2. Flowchart of improved GA-SA.

5.2.2. Action Space Design

The action space A consists of discrete decisions to adjust the flow allocation. Each action $a \in A$ is defined as

$$a = \{i, j, m, \delta\} \tag{36}$$

where $i \in I$ represents the source LPWs, $j \in J$ represents the destination CPs, $m \in M$ represents the transportation mode, and $\delta \in \{\Delta_1, \Delta_2, \ldots, \Delta_k\}$ is a set of discrete adjustment values for the flow. This formulation enables the DQN agent to learn which transportation links should be adjusted and the extent of adjustment to optimize the objective function.

5.2.3. Reward Function Design

The reward function is designed to align with the optimization objective while providing immediate feedback for learning. The reward R for transitioning from state s to s' after taking action a is defined as

$$R(s, a, s') = -\Delta TC - \beta_1 \cdot \Delta EC - \beta_2 \cdot \Delta CC - \beta_3 \cdot \Delta CTSC + \beta_4 \cdot P_{feas} \tag{37}$$

where ΔTC is the change in transportation cost, ΔEC is the change in environmental cost, ΔCC is the change in congestion cost, and $\Delta CTSC$ is the change in carbon tax and subsidy cost. P_{feas} is the feasibility premium that offers a positive reward when the solution becomes more feasible or keeps its feasibility. $\beta_1, \beta_2, \beta_3, \beta_4$ are weighting coefficients used to balance different cost components.

5.2.4. Network Architecture and Training

The DQN utilizes a deep neural network to approximate the Q function. Its architecture is as follows: There is an input layer whose dimensionality corresponds to that of the state space. Then, there are three hidden layers, containing 256, 128, and 64 neurons, respectively, and these layers utilize ReLU activation functions. Finally, there is an output layer with a size equal to the action space, which represents the Q-values for each possible action. This architecture enables the DQN to process the state information from the environment, learn through the hidden layers, and output the estimated Q-values for different actions, enabling the agent to make optimal decisions to maximize cumulative rewards.

The network is trained using experience replay and a target network to stabilize learning. The experience replays buffer stores transitions (s, a, r, s') and randomly samples batches for training to break the correlation between consecutive samples. The target network parameters are updated periodically from the online network to reduce overestimation bias.

The loss function for training is the mean squared error between the predicted Q-values and the target Q-values:

$$L = E\left[\left(r + \gamma \max_{a'} Q(s', a'; \overline{\theta}) - Q(s, a; \theta)\right)^2\right] \quad (38)$$

where θ denotes the parameters of the online network, $\overline{\theta}$ represents the parameters of the target network, and γ is the discount factor for future rewards.

Check if the number of steps meets predefined requirements. If it meets them (Y), continue the cycle of interacting with the environment, storing experience, and sampling training the network. If not (N), the training may stop or move to the next state.

In summary, the basic framework of DQN is illustrated in Figure 3.

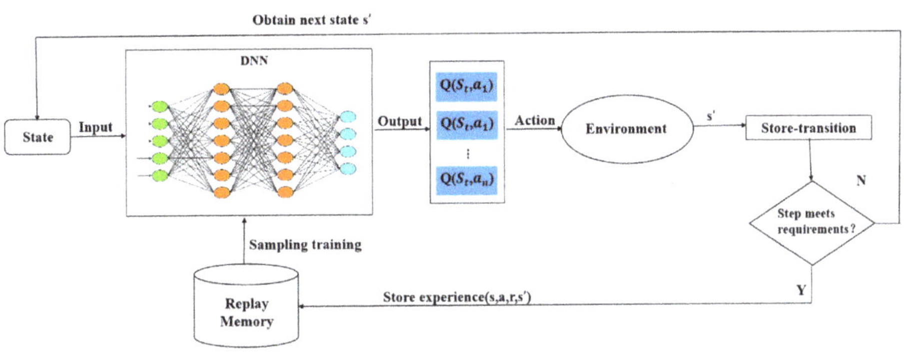

Figure 3. The basic framework of DQN.

5.3. The Framework of GA-SA-DQN

Based on the definitions of the state, action, and reward functions of DQN in the previous section, we will now organically integrate these two components to construct the hybrid GA-SA-DQN framework, thereby leveraging the advantages of both global search and deep learning. In the first stage, the improved GA-SA algorithm rapidly searches for potential high-quality solution regions from a large-scale solution space by simulating the biological evolution process. The crossover and mutation operations of GA provide extensive exploration capabilities in the solution space, while the probabilistic acceptance mechanism of SA helps the algorithm escape from local optimal traps. The main objective of this stage is to find solutions close to the global optimum, laying the foundation for subsequent fine-tuning optimization.

In the second stage, DQN, as a deep reinforcement learning method, can learn the optimal decision-making strategy through continuous interaction with the environment. In this stage, DQN takes the solutions obtained in the first stage as the initial states and conducts adaptive adjustment and optimization for demand uncertainties through a trial-and-error learning approach. The value function of DQN can evaluate the long-term benefits of different freight flow allocation strategies, thereby finding more robust solutions. The complete algorithm framework is shown in Figure 4.

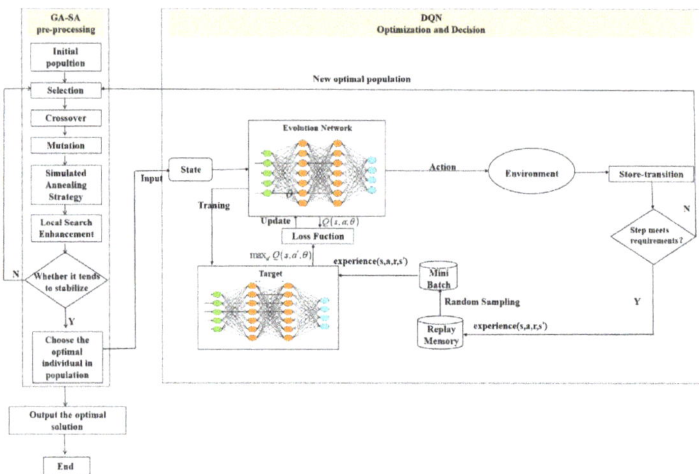

Figure 4. The hybrid algorithm framework based on GA-SA-DQN.

In summary, this section analyzes the design details of the GA-SA module and the DQN module step by step and finally presents their combined framework. The next section, based on the methods described here, will carry out a systematic evaluation of the algorithm's performance and compare it with existing approaches.

6. Numerical Experiments

Building on the hybrid algorithm discussed in Section 5, numerical experiments were conducted to evaluate the performance and effectiveness of the proposed approach. The experiments were performed using Python 3.10 on an Intel Quad Core 2.10 GHz processor with 12 GB of memory. These tests aimed to assess the algorithm's capability in optimizing freight flow distribution in port collection and distribution networks under uncertain demand conditions. The results from the previous theoretical model were applied to real-world data, as detailed in the case study, to validate the robustness and practical utility of the solution.

6.1. Algorithm Performance Testing

To comprehensively validate the effectiveness and robustness of the proposed GA-SA-DQN hybrid algorithm, this study designed a series of test instances with different scales and characteristics and compared them with several traditional algorithms.

Table 2 presents the instance configurations used for algorithm performance testing, including instances of different network scales.

Table 2. Configuration of test instances.

Instance Number	Number of LPWs	Number of CPs	Number of Transportation Modes
S1	2	2	4
S2	3	2	4
S3	3	3	4
S4	4	2	4
M1	4	3	4
M2	5	3	4
M3	5	4	4
M4	6	4	4
L1	8	4	4
L2	10	4	4
L3	10	5	4
L4	12	6	4

The demand data, distance matrix, cost parameters, etc., for each instance were randomly generated based on the characteristics of the actual port collection and distribution system. Meanwhile, sufficient diversity in the parameter distribution among different instances was ensured to test the algorithm's adaptability in various scenarios. To ensure the reliability of the results, each instance was solved independently 10 times, and the average results were taken for comparative analysis.

The main algorithm parameter settings used in the experiment are shown in Table 3.

Table 3. The main algorithm parameter settings.

Parameter	Value
population size	100
maximum number of iterations	1000
crossover probability	0.85
mutation probability	0.15
initial temperature	100
cooling rate	0.95
local search probability	0.2
neural network structure	[Input Layer, 256, 128, 64, Output Layer]
learning rate	0.001
discount factor	0.95
experience replay buffer size	10,000
target network update frequency	10
batch size	64
initial value of ε—greedy exploration	1.0
minimum value of ε—greedy exploration	0.01
decay rate of ε—greedy	0.995

To comprehensively evaluate the algorithm's performance, we adopt the following three key indicators: Solution quality is measured by the objective function value (total expected cost), which is compared with the exact solution obtained by CPLEX, and the percentage of relative deviation is calculated. Computational efficiency is measured by the solution time (in seconds), reflecting the running speed of the algorithm. Solution stability is measured by the standard deviation of the objective function values from 10 independent runs, reflecting the stability and reliability of the algorithm's results.

Table 4 shows the comparison results of the algorithm performance for all test instances.

Table 4. Comparison of algorithm performance for all test instances.

Instance	Algorithm	Objective Function Value (CNY 10,000)	Relative Deviation (%)	Computation Time (Seconds)	Standard Deviation (CNY 10,000)
S1	CPLEX	325.42	-	0.58	0.00
S1	GA	327.85	0.75	3.24	3.15
S1	SA	328.67	1.00	2.87	4.23
S1	GA-SA	326.18	0.23	2.35	1.84
S1	GA-SA-DQN	325.65	0.07	1.98	0.95
S2	CPLEX	428.56	-	0.92	0.00
S2	GA	432.13	0.11	4.56	4.26
S2	SA	433.74	0.73	4.12	5.34
S2	GA-SA	429.87	0.07	3.75	2.13
S2	GA-SA-DQN	429.15	0.13	3.05	1.24
S3	CPLEX	542.67	-	1.25	0.00
S3	GA	548.81	1.13	6.53	5.37
S3	SA	550.42	1.43	5.78	6.85
S3	GA-SA	544.32	0.30	4.67	2.45
S3	GA-SA-DQN	543.28	0.11	3.85	1.28
S4	CPLEX	653.89	-	2.34	0.00
S4	GA	662.57	0.41	8.46	7.24
S4	SA	665.21	0.65	7.65	8.56
S4	GA-SA	656.35	0.08	6.23	3.65
S4	GA-SA-DQN	654.72	0.58	5.12	1.87
M1	CPLEX	762.34	-	12.56	0.00
M1	GA	776.52	1.27	24.78	8.45
M1	SA	780.15	1.05	22.35	10.27
M1	GA-SA	768.53	0.04	16.42	5.32
M1	GA-SA-DQN	764.68	0.55	14.28	3.15
M2	CPLEX	865.27	-	26.34	0.00
M2	GA	883.75	2.14	35.67	11.28
M2	SA	886.92	2.50	32.45	13.56
M2	GA-SA	872.43	0.25	22.37	6.75
M2	GA-SA-DQN	868.56	0.38	18.64	4.23
M3	CPLEX	967.32	-	38.76	0.00
M3	GA	989.85	3.03	42.34	14.56
M3	SA	994.27	0.66	39.65	16.32
M3	GA-SA	976.24	0.66	25.48	8.83
M3	GA-SA-DQN	971.53	0.17	21.74	5.26
M4	CPLEX	1124.56	-	65.23	0.00
M4	GA	1153.28	1.48	68.52	18.74
M4	SA	1159.75	2.05	64.37	21.45
M4	GA-SA	1136.48	0.004	42.36	11.24
L1	CPLEX	1856.43	-	428.75	0.00
L1	GA	1932.56	3.17	165.34	32.56
L1	SA	1945.23	1.97	153.27	38.42
L1	GA-SA	1885.27	0.52	123.45	21.35
L1	GA-SA-DQN	1868.34	0.37	98.76	12.45
L2	CPLEX	2124.68	-	765.42	0.00
L2	GA	2228.54	2.07	214.56	45.78
L2	SA	2246.82	1.16	198.35	52.34
L2	GA-SA	2162.37	0.41	156.43	28.53
L2	GA-SA-DQN	2138.75	0.21	124.85	15.23
L3	CPLEX	2356.85	-	1245.67	0.00
L3	GA	2485.46	3.16	256.78	54.85
L3	SA	2508.32	1.73	238.45	63.62

Table 4. *Cont.*

Instance	Algorithm	Objective Function Value (CNY 10,000)	Relative Deviation (%)	Computation Time (Seconds)	Standard Deviation (CNY 10,000)
L3	GA-SA	2402.56	0.32	184.32	32.34
L3	GA-SA-DQN	2372.43	0.24	145.63	18.76
L4	CPLEX	2845.32	-	2856.45	0.00
L4	GA	3126.78	9.9	342.56	74.35
L4	SA	3658.24	28.6	318.67	85.24
L4	GA-SA	2608.67	8.3	242.85	42.56
L4	GA-SA-DQN	2467.85	13.3	187.42	24.35

As can be seen from Table 4, the performance differences among the algorithms gradually increase as the instance size increases. For smaller instance sizes (e.g., S1–S4), the differences among the algorithms are relatively small, but GA-SA-DQN always maintains the minimum solution bias and high computational efficiency. When the instance size increases (e.g., M1–M4 and L1–L4), the advantages of GA-SA-DQN become more obvious, especially in terms of solution quality and computation time.

For the largest L4 instance (12 LPWs, 6 CPs, and 4 modes of transportation), CPLEX takes about 47 min to obtain the exact solution, while GA-SA-DQN takes only about 3 min to obtain a solution that is less than 1% different from the exact solution, which is a reduction of computation time by about 93%. Also, the solution stability of GA-SA-DQN (USD 243,500 standard deviation) is significantly better than the other heuristics (USD 743,500 for GA and USD 852,400 for SA).

Figure 5 presents a comparison of the iterative processes of the various algorithms for the L4 problem instance. As observed, the GA-SA-DQN consistently outperforms the other methods by rapidly converging to a stable solution with minimal fluctuation in the objective function value, reinforcing its computational efficiency and reliability.

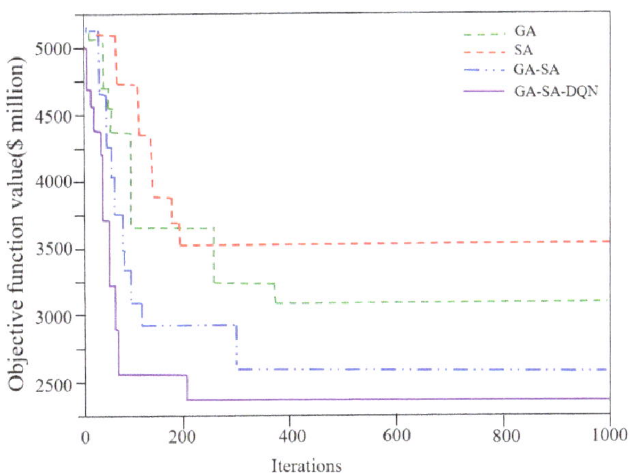

Figure 5. Comparison of algorithmic iterative processes for L4 problem instances.

6.2. Case Study

This paper selects A Port as a case study to examine the application of the proposed model and algorithm for a port transportation system. A Port is one of the largest container CPs in the world, with a container throughput of 47.57 million TEU in 2023 and a record-breaking 51.51 million TEU in 2024, marking an 8.3% increase. Its transportation system includes multiple port areas and inland logistics nodes, with a large throughput and significant transportation demand uncertainty, making it an ideal case for model validation. The case study focuses on A Port's main port areas and surrounding LPWs, including four port areas and five major LPWs. The transportation modes include road, rail, water, and the planned underground logistics system.

The main research parameters are sourced as follows:

(1) Transportation demand data: Estimated based on A Port's daily container throughput of 141,000 TEU in 2024.
(2) Distance data: Measured through electronic maps for the actual transportation distances between LPWs and CPs areas, considering the impact of newly opened highways.
(3) Transportation costs: Referencing the 2024 Shanghai transport price guidelines and the impact of fuel price fluctuations.
(4) Carbon emission factors: Using recommended values from the Transportation Industry Carbon Emission Calculation Methodology.
(5) Underground logistics system parameters: Based on data from the 2023–2024 Shanghai Urban Underground Space Research Institute.

The main parameters and the distance data for the major OD pairs in A Port will be presented next.

To comprehensively evaluate the effectiveness of the proposed model and algorithm, three scenarios were designed for comparison:

Scenario A (traditional scenario): Only considers road, rail, and water transportation modes, excluding an underground logistics system, and aims to minimize transportation costs.

Scenario B (environmentally prioritized scenario): Incorporates underground logistics system but excludes demand uncertainty, aiming to minimize total costs (including transportation, environmental, carbon tax and subsidy, and congestion costs).

Scenario C (proposed scenario): Includes an underground logistics system and accounts for demand uncertainty, aiming to minimize expected total costs.

6.2.1. Total Cost Analysis

The total cost comparison of the three scenarios is shown in Table 5.

Table 5. Total cost comparison of each scenario (unit: 10,000 CNY/day).

Cost Category	Scenario A	Scenario B	Scenario C
Transportation Cost	4562.8	4285.3	4318.6
Environmental Cost	783.5	498.2	512.7
Congestion Cost	625.4	267.8	293.2
Carbon Tax/Subsidy Cost	143.2	79.5	85.3
Total Cost	6114.9	5130.8	5209.8
Reduction Compared to Scenario A	-	16.1%	14.8%

In terms of cost structure, transportation costs accounted for the largest proportion of total costs in each scenario, followed by environmental costs and congestion costs. The total costs of Scenario B and Scenario C are reduced by 16.1% and 14.8%, respectively, compared to Scenario A, indicating that the introduction of an underground logistics system significantly reduces the total costs of the port consolidation and transportation system.

It is particularly noteworthy that the improvements in environmental and congestion costs are particularly significant for Scenarios B and C, with reductions of about 36.4% and 57.2% (Scenario B) and 34.6% and 53.1% (Scenario C), respectively. This suggests that underground logistics systems play an important role in reducing environmental costs (i.e., minimizing CO_2, NO_x, SO_x, and particulate matter emissions) and alleviating congestion. These improvements arise because the underground logistics system shifts a substantial share of short-haul, high-frequency container movements off crowded roadways and into a dedicated, electric-driven pipeline: by replacing diesel-powered trucks with continuous-flow vehicles, local air pollutants and greenhouse gas emissions fall sharply, and the need for curbside loading and unloading is all but eliminated. Consequently, road capacity during peak hours is freed up, average travel times become more predictable, and overall system resilience is enhanced. Moreover, the consolidation of cargo flows in the underground network reduces wear and tear on surface infrastructure and lowers accident risks, yielding downstream savings in maintenance and liability.

Meanwhile, Scenario C's total cost is only about 1.5% higher than Scenario B's because, under demand uncertainty, the model deliberately retains a slightly larger share of flexible road transport capacity to buffer against throughput fluctuations. This modest increase reflects an adaptive strategy that, although it incurs marginally higher costs, ensures robust service levels during demand spikes and thus guarantees more reliable port operations in practice. This finding confirms that the model, under conditions of demand uncertainty, can simultaneously reduce overall operational costs, alleviate congestion, and uphold environmental benefits, thereby fulfilling the study's original objectives.

6.2.2. Cargo Flow Distribution Results

Table 6 shows the distribution ratios of freight flows for different transportation modes under three scenarios.

Table 6. Distribution ratios of freight flows for different transportation modes under various scenarios.

Transportation Mode	Scenario A	Scenario B	Scenario C
Road	68.5%	37.2%	40.4%
Railway	15.8%	12.5%	13.2%
Waterway	15.7%	10.3%	11.4%
Underground Logistics System	0.0%	40.0%	35.0%
Total	100.0%	100.0%	100.0%

As can be seen from the results of cargo flow distribution, road transportation dominates in Scenario A, with a high proportion of 68.5%, while railroads and waterways account for 15.8% and 15.7%, respectively. After the introduction of the underground logistics system, the proportion of road transportation in Scenario B and Scenario C drops to 37.2% and 40.4%, respectively, while the UCLS accounts for 40.0% and 35.0%. This eases congestion around the port. Queues shrink. Speeds rise. Fewer breakdowns cut accident and repair costs. Rail and water shares drop slightly. They lose short-haul loads to the UCLS, not due to lower efficiency. It is worth noting that after considering demand uncertainty (Scenario C), the proportion of road transportation is slightly higher than that without considering demand uncertainty (Scenario B), while the proportion of the underground logistics system is slightly lower. This suggests that in the face of demand uncertainty, road transport is preferred due to its flexibility, while the underground logistics system is more suited to stable and predictable transportation demand.

Overall, the case studies show that the introduction of an underground logistics system can significantly improve the economic and environmental performance of port consolidation systems, even when demand uncertainty is taken into account. By optimizing the distribution of cargo flows between different modes of transport, the underground logistics system can effectively reduce total costs, carbon emissions, and traffic congestion and improve the overall efficiency and sustainability of the system.

This shift in modal share demonstrates that optimizing multimodal transport can markedly ease port-area congestion and boost operational efficiency. For port operators, it is therefore advisable to roll out the ULCS in staged phases—prioritizing deployment when costs remain manageable to free up surface capacity—while retaining a measured level of road transport reserve to handle demand surges and ensure service continuity and leveraging carbon-pricing incentives under supportive policies to drive more cargo onto the low-carbon ULCS.

6.3. Sensitivity Analysis

In order to deeply analyze the impact of each key parameter on the model results, this section conducts a sensitivity analysis, which mainly includes the impact of transportation cost changes, carbon emission factors, demand uncertainty, and low-carbon-ratio requirements.

6.3.1. Impact of Transportation Cost Changes

Transportation cost is one of the key factors affecting cargo flow distribution. This section analyzes the impact of changes in the unit transportation cost of various transportation modes on the cargo flow distribution. Figure 6 demonstrates the impact of changes in underground logistics system costs on cargo flow distribution.

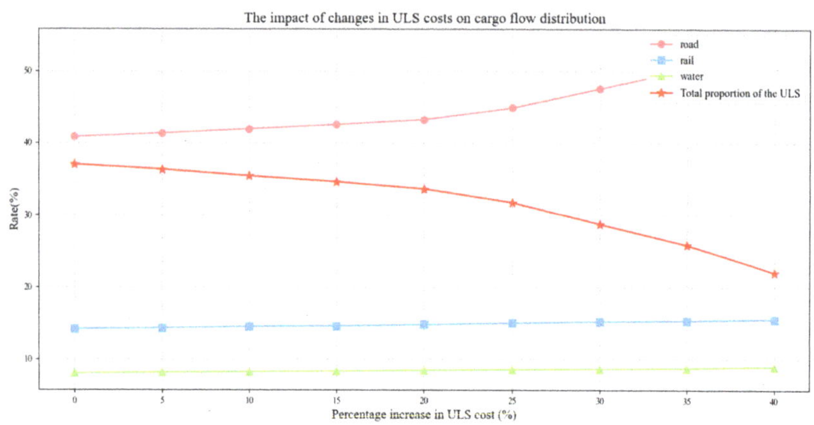

Figure 6. Influence analysis of the cost change of the underground logistics system.

From Figure 6, it is clear that as the cost of the underground logistics system increases, the proportion of cargo flow allocated to it decreases. With regard to road transportation, as the percentage of cost increase of the underground logistics system climbs gradually from 0 to 40%, the proportion of its transportation shows a slow increase, starting at about 40% and approaching 55% when the cost increases to 40%, which means that the cost of the underground logistics system rises, and more goods will tend to be transported by road transportation, and the status of road transportation in the distribution of cargo flow is also gradually improved. The proportion of rail transportation is almost stable in the whole process and always stays between 13 and 15%, which indicates that the change in the cost of the underground logistics system has little influence on the distribution of rail transportation, and the proportion of cargoes choosing rail transportation is basically not disturbed by it. The proportion of water transportation also remains stable, always at a low level of 8–9%, without significant fluctuations due to the increase in the cost of the underground logistics system, indicating that water transportation is minimally affected by changes in its cost.

When the percentage of cost increase starts to rise from 0, the total proportion of the underground logistics system's cargo flow distribution decreases significantly, starting at about 37%, and then decreases to about 22% when the cost increases to 40%, and when the cost increases to about 25%, the decreasing trend is more prominent. This shows that an increase in the cost of the underground logistics system will significantly reduce its share in the distribution of cargo flow, and as the cost rises, its use for cargo transportation will gradually reduce.

Overall, the impact of changes in the cost of underground logistics systems on the distribution of cargo flows is quite significant, with rising costs driving the shift of cargo transportation from underground logistics systems to road transportation, while rail and water transportation are less affected. This reflects the fact that the choice of transportation modes will be dynamically adjusted, driven by cost factors, with road transportation being more resilient to cost changes and the underground logistics system being less competitive when costs rise. In practice, if the construction cost of the underground logistics systems rises significantly, operators are advised to implement it in stages and use large-scale bidding to reduce unit costs; when costs are under control, they should accelerate rollout to maximize operational returns.

6.3.2. Impact of Carbon Emission Weightings

The carbon emission coefficient directly affects environmental costs and, in turn, influences cargo flow distribution decisions. Figure 7 shows the changes in cargo flow distribution across transportation modes under different carbon emission weightings.

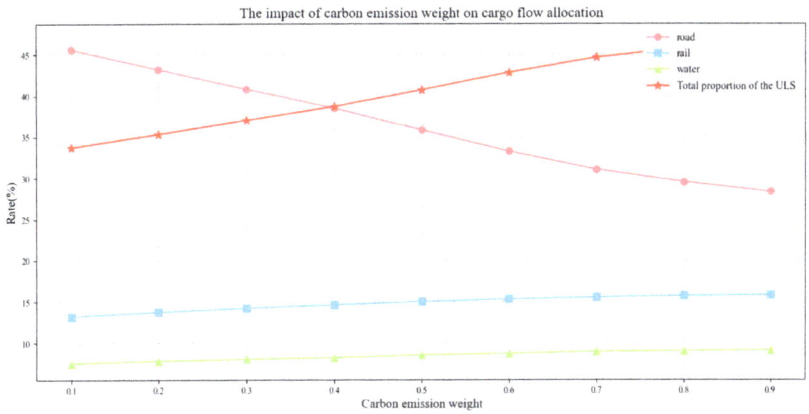

Figure 7. Influence analysis of carbon emission weighting.

From Figure 7, it can be seen that as the carbon emission weight increases (i.e., the share of environmental costs in total costs rises), the proportion of cargo flow allocated to the underground logistics system and rail/water transport gradually increases, while the proportion allocated to road transport decreases significantly. In particular, when the carbon emission weight increases from 0.1 to 0.5, the proportion of cargo flow allocated to the underground logistics system increases from 33.7% to 40.7%, a rise of 7 percentage points.

Overall, increasing the carbon emission weight positively impacts the promotion of the underground logistics system. In practice, increasing the carbon emission cost weight (e.g., through carbon taxes or carbon trading mechanisms) can effectively promote the development of green, low-carbon transportation modes. The analysis suggests that when the carbon emission weight reaches above 0.3, the underground logistics system will become the dominant transportation mode, which is crucial for developing a green port transportation system. Planners can incorporate a carbon-pricing mechanism into regional low-carbon development plans—linking it to port capacity allocation—so that policy levers drive the adoption of green transportation modes.

6.3.3. Impact of Demand Uncertainty

The uncertainty of port container demand is an important factor influencing cargo flow distribution strategies. Figure 8 shows the total cost changes under different demand volatility levels.

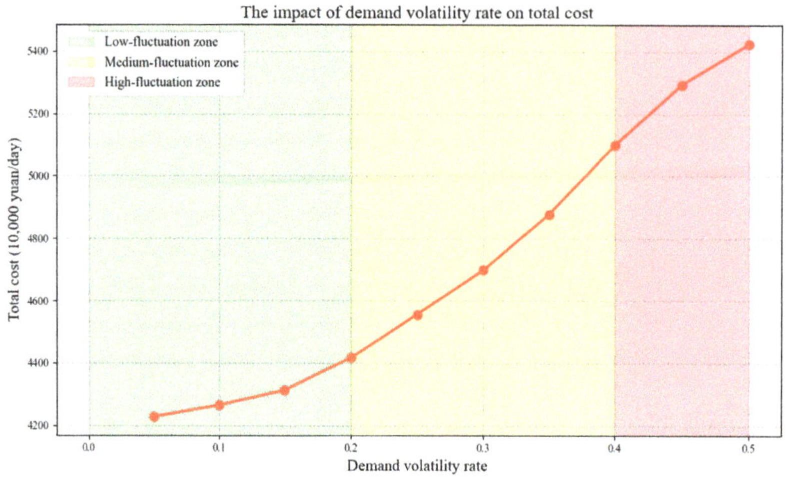

Figure 8. Influence analysis of demand uncertainty.

From Figure 8, it can be observed that as demand volatility increases, total cost increases nonlinearly. When volatility is below 0.2, the cost increases slowly; between 0.2 and 0.4, the cost growth accelerates; and when volatility exceeds 0.4, the cost growth slows down again. This nonlinear relationship indicates that for medium levels of demand uncertainty, the system requires more adjustments to maintain performance, leading to a more significant cost increase.

Figure 9 further demonstrates the changes in cargo flow distribution across different transportation modes under varying demand volatility levels.

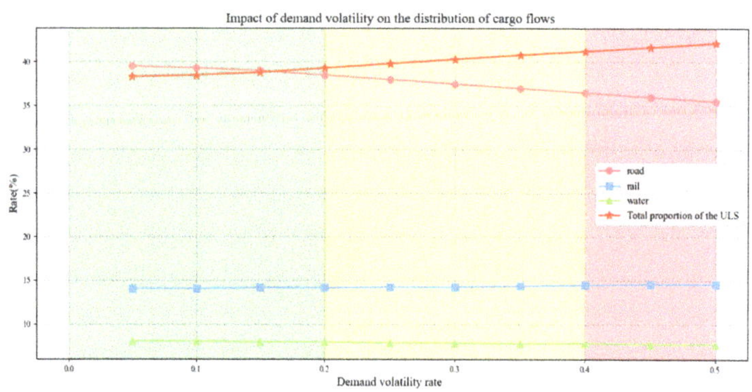

Figure 9. Influence analysis of the impact of different demand volatility rates on the freight flow.

From Figure 9, it can be seen that with the increase in demand volatility, the proportion of freight transported by the underground logistics system rises steadily from 38.3% to 42.2%, which shows a good adaptability to volatility. In contrast, the share of road transportation decreases gradually, indicating that it faces certain bottlenecks when demand uncertainty increases. This trend suggests that under high demand volatility, the underground logistics system, with its closed operation, high automation, and dispatch stability, is a more reliable transportation option, especially suitable for securing critical materials and core routes. Port schedulers should establish multiple capacity contingency plans and dynamically adjust the balance between the underground logistics system and road transport, ensuring stable service even when demand fluctuates.

6.3.4. Impact of Low-Carbon-Proportion Requirements

The low-carbon-proportion constraint is an important parameter in the model, directly influencing the cargo flow distribution across transportation modes. Figure 10 shows the cost and cargo flow distribution changes under different low-carbon-proportion requirements.

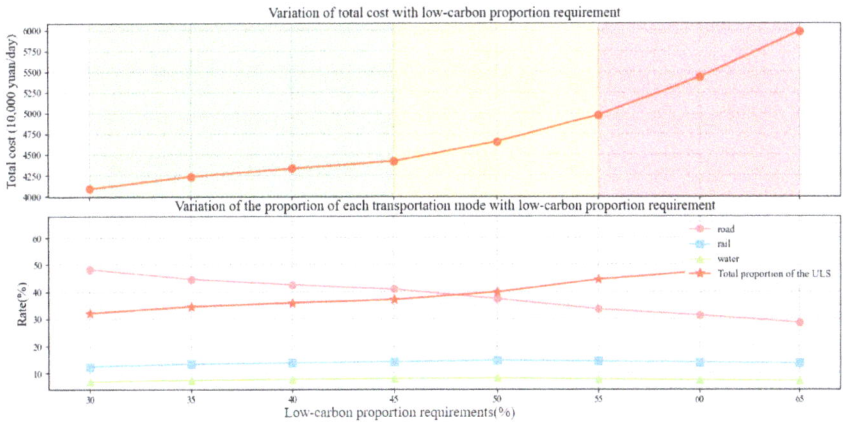

Figure 10. Influence analysis of the impact of changes in costs and freight flow allocation under different low-carbon-proportion requirements.

From Figure 10, it can be seen that as the low-carbon-proportion requirement increases, total costs rise in a stepwise manner. When the requirement increases from 30% to 45%, the cost increases relatively slowly, by about 7.4%. However, when the requirement exceeds 50%, the cost increases significantly, indicating that the system faces greater challenges in adapting to high low-carbon-proportion requirements.

In terms of cargo flow distribution, as the low-carbon-proportion requirement increases, the cargo flow proportion for the underground logistics system increases linearly, from 32.2% to 50.7%, while the proportion for road transport significantly decreases, from 48.3% to 28.5%. Notably, when the low-carbon-proportion requirement reaches 55% or higher, the proportion for rail and water transport also starts to decrease, indicating that under very high low-carbon requirements, the underground logistics system will replace some traditional low-carbon transportation modes.

Based on the sensitivity analysis, the following key conclusions can be drawn:

(1) Underground logistics system costs: The cost of an underground logistics system has a significant impact on its adoption. Controlling cost increases to within 25% is crucial for maintaining high usage rates.

(2) Carbon emission weight: increasing the carbon emission cost weight is an effective way to promote the development of underground logistics systems. When the weight exceeds 0.3, the underground logistics system will become the dominant transportation mode.

(3) Demand uncertainty: As demand uncertainty increases, the proportion of road transport rises, and the proportion of the underground logistics system slightly decreases. A shallow underground logistics system shows better adaptability.

(4) Low-carbon-proportion requirements: When the low-carbon-proportion requirement is below 45%, the system can adapt with a relatively small increase in cost. However, when the requirement exceeds 50%, costs rise significantly.

These conclusions provide important insights for the planning and operation of port transportation systems, particularly in terms of determining investment strategies for underground logistics systems, formulating carbon reduction policies, and addressing demand uncertainty. When setting low-carbon targets, a phased, incremental approach should be adopted; initial targets ought to lie within a range that allows a smooth transition, supported by subsidies and incentive policies to create sustainable backing for higher levels of green transformation.

7. Conclusions

This study makes three key contributions to the optimization of port cargo flow under uncertainty. First, by embedding an underground container-pipeline logistics system (UCLS) into a stochastic chance-constrained model, we provide a novel framework that simultaneously balances transportation costs, environmental impacts, congestion effects, and carbon tax/subsidy policies—extending beyond the largely deterministic, surface-only approaches in previous research. Second, our hybrid GA-SA-DQN algorithm not only accelerates convergence on large-scale instances but also delivers robust, high-quality solutions under demand volatility, outperforming both CPLEX and classic metaheuristics in runtime and stability. Third, through the empirical case of Port A, we quantify the economic–environmental co-benefits of UCLS: a 16% reduction in total cost, an over 34% cut in environmental expense, and a 53% decline in congestion cost, while maintaining operational resilience even when demand fluctuates.

Despite the contributions outlined above, our model makes several simplifying assumptions that should be addressed in future research. For example, the model omits handling costs, such as the costs associated with loading and unloading goods at transfer points, which can significantly impact the total cost in real-world port operations. Furthermore, transfer costs between different transportation modes (such as truck-to-rail or rail-to-ship transfers) are not considered in our analysis. These costs could influence the optimal cargo flow distribution, especially in multimodal systems where coordination between modes is essential.

Future research should explore the inclusion of handling and transfer costs, as these factors play a crucial role in the efficiency of port operations. Incorporating these costs would provide a more accurate and comprehensive model, better reflecting the complexities of multimodal logistics systems. Additionally, multi-objective optimization frameworks that account for these costs, along with environmental and economic factors, could further enhance the practical applicability and robustness of the model in real-world scenarios.

Author Contributions: Conceptualization, M.S., C.L. and S.A.B.; methodology, M.S.; software, M.S.; validation, M.S., C.L. and Y.W.; investigation, M.S.; resources, C.L.; data curation, M.S.; writing—original draft preparation, M.S.; writing—review and editing, C.L. and S.A.B.; visualization, Y.W.; supervision, C.L.; funding acquisition, C.L. and Y.W. All authors have read and agreed to the published version of the manuscript.

Funding: This research was funded by the National Industry and Technology Major Project of China (Grant No. QT1451-0415043), the National Natural Science Foundation of China (Grant No. 72271125), the Shanghai Sailing Program (Grant No. 21YF1416400), and the Shanghai Rising-Star Program (Grant No. 21QB1404800).

Data Availability Statement: The data used during this study are available from the first author.

Conflicts of Interest: The authors declare no conflicts of interest.

References

1. UNCTAD. *Review of Maritime Transport 2024*; United Nations: Geneva, Switzerland, 2024; Available online: https://unctad.org/publication/review-maritime-transport-2024 (accessed on 10 May 2025).
2. UNCTAD. Global Trade Hits Record $33 Trillion in 2024, Driven by Services and Developing Economies. 2025. Available online: https://unctad.org/news/global-trade-hits-record-33-trillion-2024-driven-services-and-developing-economies (accessed on 10 May 2025).
3. Panahi, R.; Ng, A.K.Y.; Afenyo, M.; Lau, Y.Y. Implications of a pandemic outbreak risk: A discussion on China's emergency logistics in the era of Coronavirus Disease 2019 (COVID-19). *J. Int. Logist. Trade* **2022**, *20*, 127–135. [CrossRef]
4. Zhao, Z.; Shen, M.; Chen, J.; Wang, X.; Wan, Z.; Hu, X.; Liu, W. Design and optimization of the collaborative container logistics system between a dry port and a water port. *Comput. Ind. Eng.* **2024**, *198*, 110654. [CrossRef]
5. Acciaro, M.; Arduino, G.; Borruso, G. Port sustainability in the transition to a low-carbon energy system: A review of digital technologies and decarbonization strategies. *Transp. Res. Part D Transp. Environ.* **2023**, *117*, 103615. [CrossRef]
6. Liu, Q.; Yang, Y.; Li, K.X.; Luo, M.; Xie, C. Liner shipping network vulnerability to component disruptions: A China-Europe container flow analysis. *Transp. Res. Part D Transp. Environ.* **2024**, *131*, 104232. [CrossRef]
7. Liang, C.; Sun, W.; Shi, J.; Wang, K.; Zhang, Y.; Lim, G. Decarbonizing Maritime Transport through Green Fuel-Powered Vessel Retrofitting: A Game-Theoretic Approach. *J. Mar. Sci. Eng.* **2024**, *12*, 1174. [CrossRef]
8. Wang, Y.; Liu, Y.; Chen, K.; Zhou, Q. Port hinterland intermodal container transport efficiency and its influencing factors analysis. *Transp. Policy* **2022**, *124*, 143–157. [CrossRef]
9. Chen, X.; Wei, C.; Xin, Z.; Zhao, J.; Xian, J. Ship Detection under Low-Visibility Weather Interference via an Ensemble Generative Adversarial Network. *J. Mar. Sci. Eng.* **2023**, *11*, 2065. [CrossRef]
10. Hu, X.; Liang, C.; Chang, D.; Zhang, Y. Container storage space assignment problem in two terminals with the consideration of yard sharing. *Adv. Eng. Inform.* **2021**, *47*, 101224. [CrossRef]
11. Martinez-Moya, J.; Vazquez-Paja, B.; Gimenez-Maldonado, J.A. Energy consumption minimization of a container terminal by the optimization of eco-efficient operational strategies. *Energy* **2021**, *222*, 119925. [CrossRef]
12. Karam, A.; Jensen, A.J.K.; Hussein, M. Analysis of the barriers to multimodal freight transport and their mitigation strategies. *Eur. Transp. Res. Rev.* **2023**, *15*, 43. [CrossRef]
13. Intermodal Freight Transportation Market Size Report. 2030. Available online: https://www.grandviewresearch.com/industry-analysis/intermodal-freight-transportation-market-report (accessed on 14 May 2025).
14. Mamatok, Y.; Jin, M. A transportation service network design problem with mixed temporal constraints. *Transp. Res. Part E Logist. Transp. Rev.* **2022**, *160*, 102689. [CrossRef]
15. Chen, X.; Wu, S.; Shi, C.; Huang, Y.; Yang, Y.; Ke, R.; Zhao, J. Sensing Data Supported Traffic Flow Prediction via Denoising Schemes and ANN: A Comparison. *IEEE Sens. J.* **2020**, *20*, 14317–14328. [CrossRef]
16. Liu, Q.; Hu, W.; Dong, J.; Yang, K.; Ren, R.; Chen, Z. Cost-benefit analysis of road-underground co-modality strategies for sustainable city logistics. *Transp. Res. Part D Transp. Environ.* **2025**, *139*, 104585. [CrossRef]
17. Zhu, L.; Li, W.; Sun, Z.B.; Chen, Y.; Chen, C.; Zhang, L.Y. Evaluating resilience in maritime port and shipping operations under various disruptions: A literature review. *Marit. Econ. Logist.* **2023**, *25*, 181–208. [CrossRef]
18. van Twiller, J.; Adulyasak, Y.; Delage, E.; Grbic, D.; Jensen, R.M. Navigating Demand Uncertainty in Container Shipping: Deep Reinforcement Learning for Enabling Adaptive and Feasible Master Stowage Planning. *arXiv* **2025**, arXiv:2502.12756. [CrossRef]
19. Toygar, A.; Yildirim, U.; Inegöl, G.M. Investigation of empty container shortage based on SWARA-ARAS methods in the COVID-19 era. *Eur. Transp. Res. Rev.* **2022**, *14*, 8. [CrossRef] [PubMed]
20. Liu, J.; Dong, C.; Li, S. Container shipping network optimization considering underground container transportation. *Marit. Policy Manag.* **2022**, *49*, 389–409. [CrossRef]
21. Liu, Y.; Feng, B.; Zhu, Y. Underground logistics network design: A review and bibliometric analysis. *Transp. Res. Part E Logist. Transp. Rev.* **2023**, *171*, 102978.
22. Hu, W.; Dong, J.; Ren, R.; Chen, Z. Underground logistics systems: Development overview and new prospects in China. *Front. Eng.* **2023**, *10*, 354–359. [CrossRef]
23. Liang, C.; Hu, X.; Shi, L.; Fu, H.; Xu, D. Joint dispatch of shipment equipment considering underground container logistics. *Comput. Ind. Eng.* **2022**, *165*, 16. [CrossRef]
24. Li, F.; Yuen, K.F. A systematic review on underground logistics system: Designs, impacts, and future directions. *Tunn. Undergr. Space Technol.* **2025**, *159*, 106483. [CrossRef]
25. Hou, L.; Hu, W.; Chen, Y.; Dong, J.; Ren, R.; Chen, Z. Measuring Effectiveness of Metro-Based Underground Logistics System in Sustaining City Logistics Performance During Public Health Emergencies: A Case Study of Shanghai. *Transp. Res. Rec.* **2024**, *2678*, 724–748. [CrossRef]
26. Sun, X.; Hu, W. Multi-objective optimization model for planning metro-based underground logistics system network: Nanjing case study. *J. Ind. Manag. Optim.* **2023**, *19*, 2607–2635. [CrossRef]

27. Xia, N.; Jing, Y.; Wang, W.; Shi, L. Assessment of the carbon emissions and energy efficiency of container terminal: An empirical analysis from China. *Sci. Rep.* **2021**, *11*, 1728. [CrossRef]
28. Chen, Z.; Hu, W.; Xu, Y.; Dong, J.; Yang, K.; Ren, R. Exploring decision-making mechanisms for the metro-based underground logistics system network expansion: An example of Beijing. *Tunn. Undergr. Space Technol.* **2023**, *139*, 105240. [CrossRef]
29. An, N.; Yang, K.; Chen, Y.; Yang, L. Wasserstein distributionally robust optimization for train operation and freight assignment in a metro-based underground logistics system. *Comput. Ind. Eng.* **2024**, *192*, 110228. [CrossRef]
30. Wandel, S.; Anderberg, S.; Larsson, F.E.; Johnsson, A.; Wells, A. Underground capsule pipeline logistic system Feasibility study of an urban application. *Transp. Res. Procedia* **2023**, *72*, 3126–3133. [CrossRef]
31. Hou, L.; Xu, Y.; Ren, R.; Yang, J.; Su, L. Optimization of three-dimensional urban underground logistics system alignment: A deep reinforcement learning approach. *Comput. Ind. Eng.* **2025**, *205*, 111185. [CrossRef]
32. Gong, D.; Tian, J.; Hu, W.; Dong, J.; Chen, Y.; Ren, R.; Chen, Z. Sustainable Design and Operations Management of Metro-Based Underground Logistics Systems: A Thematic Literature Review. *Buildings* **2023**, *13*, 1888. [CrossRef]
33. Vallidevi, K.; Jothi, S.; Karuppiah, S.V. HO-DQLN: Hybrid optimization-based deep Q-learning network for optimizing QoS requirements in service oriented model. *Expert Syst. Appl.* **2023**, *227*, 120188. [CrossRef]
34. Kim, S.; Jang, M.-G.; Kim, J.-K. Process design and optimization of single mixed-refrigerant processes with the application of deep reinforcement learning. *Appl. Therm. Eng.* **2023**, *223*, 120038. [CrossRef]
35. Xu, N.; Shi, Z.; Yin, S.; Xiang, Z. A hyper-heuristic with deep Q-network for the multi-objective unmanned surface vehicles scheduling problem. *Neurocomputing* **2024**, *596*, 127943. [CrossRef]
36. Xue, F.; Hai, Q.; Dong, T.; Cui, Z.; Gong, Y. A deep reinforcement learning based hybrid algorithm for efficient resource scheduling in edge computing environment. *Inf. Sci.* **2022**, *608*, 362–374. [CrossRef]
37. Zhang, J.; Cheng, L.; Liu, C.; Zhao, Z.; Mao, Y. Cost-aware scheduling systems for real-time workflows in cloud: An approach based on Genetic Algorithm and Deep Reinforcement Learning. *Expert Syst. Appl.* **2023**, *234*, 120972. [CrossRef]
38. Zhao, F.; Liu, Y.; Zhu, N.; Xu, T.; Jonrinaldi. A selection hyper-heuristic algorithm with Q-learning mechanism. *Appl. Soft Comput.* **2023**, *147*, 110815. [CrossRef]
39. Liu, X.; Qiu, L.; Fang, Y.; Wang, K.; Li, Y.; Rodríguez, J. Event-Driven Based Reinforcement Learning Predictive Controller Design for Three-Phase NPC Converters Using Online Approximators. *IEEE Trans. Power Electron.* **2024**, *40*, 4914–4926. [CrossRef]

MDPI AG
Grosspeteranlage 5
4052 Basel
Switzerland
Tel.: +41 61 683 77 34

Journal of Marine Science and Engineering Editorial Office
E-mail: jmse@mdpi.com
www.mdpi.com/journal/jmse

www.ingramcontent.com/pod-product-compliance
Lightning Source LLC
Chambersburg PA
CBHW061505220926
51850CB00004B/176
9783725857234